Probability and Statistics in Engineering

Probability and Statistics in Engineering

Fourth Edition

William W. Hines
Professor Emeritus
School of Industrial and Systems Engineering
Georgia Institute of Technology

Douglas C. Montgomery
Professor of Engineering and Statistics
Department of Industrial Engineering
Arizona State University

David M. Goldsman
Professor
School of Industrial and Systems Engineering
Georgia Institute of Technology

Connie M. Borror
Senior Lecturer
Department of Industrial Engineering
Arizona State University

John Wiley & Sons, Inc.

Acquisitions Editor *Wayne Anderson*
Associate Editor *Jenny Welter*
Marketing Manager *Katherine Hepburn*
Senior Production Editor *Valerie A. Vargas*
Senior Designer *Dawn Stanley*
Cover Image *Alfredo Pasieka/Photo Researchers*
Production Management Services *Argosy Publishing*

This book was set in 10/12 Times Roman by Argosy Publishing and printed and bound by Hamilton Printing. The cover was printed by Phoenix Color.

This book is printed on acid-free paper. 

To order books or for customer service please call 1(800)-CALL WILEY (225-5945).

***Library of Congress Cataloging in Publication Data*:**
Probability and statistics in engineering / William W. Hines … [et al.]. -- 4th ed.

p. cm.
Includes bibliographical references.
1. Engineering--Statistical methods. I. Hines, William W.

TA340 .H55 2002 519.2--dc21 2002026703
ISBN 0-471-24087-7 (cloth: acid-free paper)

Printed in the United States of America

10 9 8 7 6 5 4 3 2 1

PREFACE to the 4th Edition

This book is written for a first course in applied probability and statistics for undergraduate students in engineering, physical sciences, and management science curricula. We have found that the text can be used effectively as a two-semester sophomore- or junior-level course sequence, as well as a one-semester refresher course in probability and statistics for first-year graduate students.

The text has undergone a major overhaul for the fourth edition, especially with regard to many of the statistics chapters. The idea has been to make the book more accessible to a wide audience by including more motivational examples, real-world applications, and useful computer exercises. With the aim of making the course material easier to learn and easier to teach, we have also provided a convenient set of course notes, available on the Web site www.wiley.com/college/hines. For instructors adopting the text, the complete solutions are also available on a password-protected portion of this Web site.

Structurally speaking, we start the book off with probability theory (Chapter 1) and progress through random variables (Chapter 2), functions of random variables (Chapter 3), joint random variables (Chapter 4), discrete and continuous distributions (Chapters 5 and 6), and normal distribution (Chapter 7). Then we introduce statistics and data description techniques (Chapter 8). The statistics chapters follow the same rough outline as in the previous edition, namely, sampling distributions (Chapter 9), parameter estimation (Chapter 10), hypothesis testing (Chapter 11), single- and multifactor design of experiments (Chapters 12 and 13), and simple and multiple regression (Chapters 14 and 15). Subsequent special-topics chapters include nonparametric statistics (Chapter 16), quality control and reliability engineering (Chapter 17), and stochastic processes and queueing theory (Chapter 18). Finally, there is an entirely new chapter, on statistical techniques for computer simulation (Chapter 19)—perhaps the first of its kind in this type of statistics text.

The chapters that have seen the most substantial evolution are Chapters 8–14. The discussion in Chapter 8 on descriptive data analysis is greatly enhanced over that of the previous edition's. We also expanded the discussion on different types of interval estimation in Chapter 10. In addition, an emphasis has been placed on real-life computer data analysis examples. Throughout the book, we incorporated other structural changes. In all chapters, we included new examples and exercises, including numerous computer-based exercises.

A few words on Chapters 18 and 19. Stochastic processes and queueing theory arise naturally out of probability, and we feel that Chapter 18 serves as a good introduction to the subject—normally taught in operations research, management science, and certain engineering disciplines. Queueing theory has garnered a great deal of use in such diverse fields as telecommunications, manufacturing, and production planning. Computer simulation, the topic of Chapter 19, is perhaps the most widely used tool in operations research and management science, as well as in a number of physical sciences. Simulation marries all the tools of probability and statistics and is used in everything from financial analysis to factory control and planning. Our text provides what amounts to a simulation minicourse, covering the areas of Monte Carlo experimentation, random number and variate generation, and simulation output data analysis.

We are grateful to the following individuals for their help during the process of completing the current revision of the text. Christos Alexopoulos (Georgia Institute of Technology), Michael Caramanis (Boston University), David R. Clark (Kettering University), J. N. Hool (Auburn University), John S. Ramberg (University of Arizona), and Edward J. Williams (University of Michigan – Dearborn), served as reviewers and provided a great deal of valuable feedback. Beatriz Valdés (Argosy Publishing) did a wonderful job supervising the typesetting and page proofing of the text, and Jennifer Welter at Wiley provided great leadership at every turn. Everyone was certainly a pleasure to work with. Of course, we thank our families for their infinite patience and support throughout the endeavor.

Hines, Montgomery, Goldsman, and Borror

Contents

16. Nonparametric Statistics 491

17. Statistical Quality Control and Reliability Engineering 507

18. Stochastic Processes and Queueing 555

19. Computer Simulation 576

Appendix 597

Chapter 1

An Introduction to Probability

1-1 INTRODUCTION

Since professionals working in engineering and applied science are often engaged in both the analysis and the design of systems where system component characteristics are nondeterministic, the understanding and utilization of probability is essential to the description, design, and analysis of such systems. Examples reflecting probabilistic behavior are abundant, and in fact, true deterministic behavior is rare. To illustrate, consider the description of a variety of product quality or performance measurements: the operational lifespan of mechanical and/or electronic systems; the pattern of equipment failures; the occurrence of natural phenomena such as sun spots or tornados; particle counts from a radioactive source; travel times in delivery operations; vehicle accident counts during a given day on a section of freeway; or customer waiting times in line at a branch bank.

The term *probability* has come to be widely used in everyday life to quantify the degree of belief in an event of interest. There are abundant examples, such as the statements that "there is a 0.2 probability of rain showers" and "the probability that brand X personal computer will survive 10,000 hours of operation without repair is 0.75." In this chapter we introduce the basic structure, elementary concepts, and methods to support precise and unambiguous statements like those above.

The formal study of probability theory apparently originated in the seventeenth and eighteenth centuries in France and was motivated by the study of games of chance. With little formal mathematical understructure, people viewed the field with some skepticism; however, this view began to change in the nineteenth century, when a probabilistic model (description) was developed for the behavior of molecules in a liquid. This became known as Brownian motion, since it was Robert Brown, an English botanist, who first observed the phenomenon in 1827. In 1905, Albert Einstein explained Brownian motion under the hypothesis that particles are subject to the continual bombardment of molecules of the surrounding medium. These results greatly stimulated interest in probability, as did the emergence of the telephone system in the latter part of the nineteenth and early twentieth centuries. Since a physical connecting system was necessary to allow for the interconnection of individual telephones, with call lengths and interdemand intervals displaying large variation, a strong motivation emerged for developing probabilistic models to describe this system's behavior.

Although applications like these were rapidly expanding in the early twentieth century, it is generally thought that it was not until the 1930s that a rigorous mathematical structure for probability emerged. This chapter presents basic concepts leading to and including a definition of probability as well as some results and methods useful for problem solution. The emphasis throughout Chapters 1–7 is to encourage an understanding and appreciation of the subject, with applications to a variety of problems in engineering and science. The reader should recognize that there is a large, rich field of mathematics related to probability that is beyond the scope of this book.

Indeed, our objectives in presenting the basic probability topics considered in the current chapter are threefold. First, these concepts enhance and enrich our basic understanding of the world in which we live. Second, many of the examples and exercises deal with the use of probability concepts to model the behavior of real-world systems. Finally, the probability topics developed in Chapters 1–7 provide a foundation for the statistical methods presented in Chapters 8–16 and beyond. These statistical methods deal with the analysis and interpretation of data, drawing inference about populations based on a sample of units selected from them, and with the design and analysis of experiments and experimental data. A sound understanding of such methods will greatly enhance the professional capability of individuals working in the data-intensive areas commonly encountered in this twenty-first century.

1-2 A REVIEW OF SETS

To present the basic concepts of probability theory, we will use some ideas from the theory of sets. A *set* is an aggregate or collection of objects. Sets are usually designated by capital letters, A, B, C, and so on. The members of the set A are called the elements of A. In general, when x is an element of A we write $x \in A$, and if x is not an element of A we write $x \notin A$. In specifying membership we may resort either to *enumeration* or to a *defining property*. These ideas are illustrated in the following examples. Braces are used to denote a set, and the colon within the braces is shorthand for the term "such that."

Example 1-1

The set whose elements are the integers 5, 6, 7, 8 is a finite set with four elements. We could denote this by

$$A = \{5, 6, 7, 8\}.$$

Note that $5 \in A$ and $9 \notin A$ are both true.

Example 1-2

If we write $V = \{a, e, i, o, u\}$ we have defined the set of vowels in the English alphabet. We may use a defining property and write this using a symbol as

$$V = \{*: * \text{ is a vowel in the English alphabet}\}.$$

Example 1-3

If we say that A is the set of all real numbers between 0 and 1 inclusive, we might also denote A by a defining property as

$$A = \{x: x \in R, 0 \leq x \leq 1\},$$

where R is the set of all real numbers.

Example 1-4

The set $B = \{-3, +3\}$ is thc same set as

$$B = \{x: x \in R, x^2 = 9\},$$

where R is again the set of real numbers.

Example 1-5

In the real plane we can consider points (x, y) that lie on a given line A. Thus, the condition for inclusion for A requires (x, y) to satisfy $ax + by = c$, so that

$$A = \{(x, y): x \in R, y \in R, ax + by = c\},$$

where R is the set of real numbers.

The *universal set* is the set of all objects under consideration, and it is generally denoted by U. Another special set is the *null set* or *empty set*, usually denoted by $\varnothing$. To illustrate this concept, consider a set

$$A = \{x: x \in R, x^2 = -1\}.$$

The universal set here is R, the set of real numbers. Obviously, set A is empty, since there are no real numbers having the defining property $x^2 = -1$. We should point out that the set $\{0\} \neq \varnothing$.

If two sets are considered, say A and B, we call A a *subset* of B, denoted $A \subset B$, if each element in A is also an element of B. The sets A and B are said to be *equal* ($A = B$) if and only if $A \subset B$ and $B \subset A$. As direct consequences of this we may show the following:

1. For any set A, $\varnothing \subset A$.
2. For a given U, A considered in the context of U satisfies the relation $A \subset U$.
3. For a given set A, $A \subset A$ (a reflexive relation).
4. If $A \subset B$ and $B \subset C$, then $A \subset C$ (a transitive relation).

An interesting consequence of set equality is that the order of element listing is immaterial. To illustrate, let $A = \{a, b, c\}$ and $B = \{c, a, b\}$. Obviously $A = B$ by our definition. Furthermore, when defining properties are used, the sets may be equal although the defining properties are outwardly different. As an example of the second consequence, we let $A = \{x: x \in R$, where x is an even, prime number$\}$ and $B = \{x: x + 3 = 5\}$. Since the integer 2 is the only even prime, $A = B$.

We now consider some operations on sets. Let A and B be any subsets of the universal set U. Then the following hold:

1. The *complement* of A (with respect to U) is the set made up of the elements of U that do not belong to A. We denote this complementary set as $\overline{A}$. That is,

$$\overline{A} = \{x: x \in U, x \notin A\}.$$

2. The *intersection* of A and B is the set of elements that belong to both A *and* B. We denote the intersection as $A \cap B$. In other words,

$$A \cap B = \{x: x \in A \text{ and } x \in B\}.$$

We should also note that $A \cap B$ is a *set*, and we could give this set some designator, such as C.

3. The *union* of A and B is the set of elements that belong to *at least one* of the sets A and B. If D represents the union, then

$$D = A \cup B = \{x: x \in A \text{ or } x \in B \text{ (or both)}\}.$$

These operations are illustrated in the following examples.

Example 1-6

Let U be the set of letters in the alphabet, that is, $U = \{*: * \text{ is a letter of the English alphabet}\}$; and let $A = \{*: * \text{ is a vowel}\}$ and $B = \{*: * \text{ is one of the letters } a, b, c\}$. As a consequence of the definitions,

$$\begin{aligned} \overline{A} &= \text{the set of consonants}, \\ \overline{B} &= \{d, e, f, g, \ldots, x, y, z\}, \\ A \cup B &= \{a, b, c, e, i, o, u\}, \\ A \cap B &= \{a\}. \end{aligned}$$

Example 1-7

If the universal set is defined as $U = \{1, 2, 3, 4, 5, 6, 7\}$, and three subsets, $A = \{1, 2, 3\}$, $B = \{2, 4, 6\}$, $C = \{1, 3, 5, 7\}$, are defined, then we see immediately from the definitions that

$$\overline{A} = \{4, 5, 6, 7\}, \quad \overline{B} = \{1, 3, 5, 7\} = C, \quad \overline{C} = \{2, 4, 6\} = B,$$
$$A \cup B = \{1, 2, 3, 4, 6\}, \quad A \cup C = \{1, 2, 3, 5, 7\}, \quad B \cup C = U,$$
$$A \cap B = \{2\}, \quad A \cap C = \{1, 3\}, \quad B \cap C = \varnothing.$$

The *Venn diagram* can be used to illustrate certain set operations. A rectangle is drawn to represent the universal set U. A subset A of U is represented by the region within a circle drawn inside the rectangle. Then $\overline{A}$ will be represented by the area of the rectangle outside of the circle, as illustrated in Fig. 1-1. Using this notation, the intersection and union are illustrated in Fig. 1-2.

The operations of intersection and union may be extended in a straightforward manner to accommodate any finite number of sets. In the case of three sets, say A, B, and C, $A \cup B \cup C$ has the property that $A \cup (B \cup C) = (A \cup B) \cup C$, which obviously holds since both sides have identical members. Similarly, we see that $A \cap B \cap C = (A \cap B) \cap C = A \cap (B \cap C)$. Some important laws obeyed by sets relative to the operations previously defined are listed below.

Identity laws: $A \cup \varnothing = A$, $\quad A \cap U = A$,
$A \cup U = U$, $\quad A \cap \varnothing = \varnothing$.

De Morgan's law: $\overline{A \cup B} = \overline{A} \cap \overline{B}$, $\quad \overline{A \cap B} = \overline{A} \cup \overline{B}$.

Associative laws: $A \cup (B \cup C) = (A \cup B) \cup C$,
$A \cap (B \cap C) = (A \cap B) \cap C$.

Distributive laws: $A \cup (B \cap C) = (A \cup B) \cap (A \cup C)$,
$A \cap (B \cup C) = (A \cap B) \cup (A \cap C)$.

The reader is asked in Exercise 1-2 to illustrate some of these statements with Venn diagrams. Formal proofs are usually more lengthy.

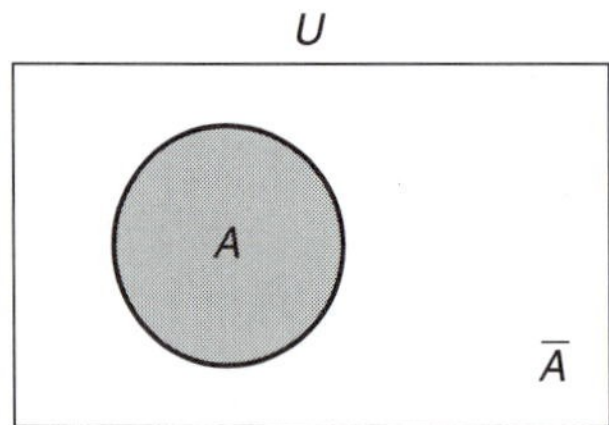

Figure 1-1 A set in a Venn diagram.

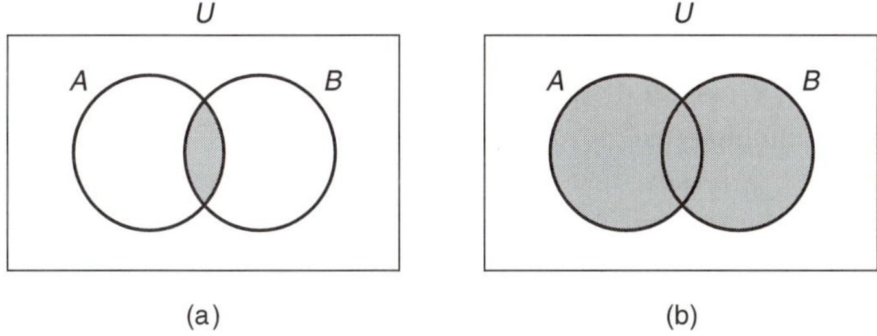

Figure 1-2 The intersection and union of two sets in a Venn diagram. (*a*) The intersection shaded. (*b*) The union shaded.

In the case of more than three sets, we use a subscript to generalize. Thus, if n is a positive integer, and $B_1, B_2, \ldots, B_n$ are given sets, then $B_1 \cap B_2 \cap \cdots \cap B_n$ is the set of elements belonging to *all* of the sets, and $B_1 \cup B_2 \cup \cdots \cup B_n$ is the set of elements that belong to *at least one* of the given sets.

If A and B are sets, then the set of all ordered pairs (a, b) such that $a \in A$ and $b \in B$ is called the *Cartesian product set* of A and B. The usual notation is $A \times B$. We thus have

$$A \times B = \{(a, b)\colon a \in A \text{ and } b \in B\}.$$

Let r be a positive integer greater than 1, and let $A_1, \ldots, A_r$ represent sets. Then the Cartesian product set is given by

$$A_1 \times A_2 \times \cdots \times A_r = \{(a_1, a_2, \ldots, a_r)\colon a_j \in A_j \text{ for } j = 1, 2, \ldots, r\}.$$

Frequently, the *number* of elements in a set is of some importance, and we denote by $n(A)$ the number of elements in set A. If the number is *finite*, we say we have a *finite set*. Should the set be infinite, such that the elements can be put into a one-to-one correspondence with the natural numbers, then the set is called a *denumerably infinite set*. The *nondenumerable set* contains an infinite number of elements that cannot be enumerated. For example, if $a < b$, then the set $A = \{x \in R, a \le x \le b\}$ is a nondenumerable set.

A set of particular interest is called the *power set*. The elements of this set are the subsets of a set A, and a common notation is $\{0, 1\}^A$. For example if $A = \{1, 2, 3\}$, then

$$\{0, 1\}^A = \{\varnothing, \{1\}, \{2\}, \{3\}, \{1, 2\}, \{1, 3\}, \{2, 3\}, \{1, 2, 3\}\}.$$

1-3 EXPERIMENTS AND SAMPLE SPACES

Probability theory has been motivated by real-life situations where an experiment is performed and the experimenter observes an outcome. Furthermore, the outcome may not be predicted with certainty. Such experiments are called *random experiments*. The concept of a random experiment is considered mathematically to be a primitive notion and is thus not otherwise defined; however, we can note that random experiments have some common characteristics. First, while we cannot predict a particular outcome with certainty, we can describe the *set of possible outcomes*. Second, from a conceptual point of view, the experiment is one that could be repeated under conditions that remain unchanged, with the outcomes appearing in a haphazard manner; however, as the number of repetitions increases, certain patterns in the frequency of outcome occurrence emerge.

We will often consider idealized experiments. For example, we may rule out the outcome of a coin toss when the coin lands on edge. This is more for convenience than out of necessity. The set of possible outcomes is called the *sample space*, and these outcomes define the particular idealized experiment. The symbols $\mathscr{E}$ and $\mathscr{S}$ are used to represent the random experiment and the associated sample space.

Following the terminology employed in the review of sets and set operations, we will classify sample spaces (and thus random experiments). A *discrete sample space* is one in which there is a finite number of outcomes or a countably (denumerably) infinite number of outcomes. Likewise, a *continuous sample space* has nondenumerable (uncountable) outcomes. These might be real numbers on an interval or real pairs contained in the product of intervals, where measurements are made on two variables following an experiment.

To illustrate random experiments with an associated sample space, we consider several examples.

Example 1-8

$\mathscr{E}_1$: Toss a true coin and observe the "up" face.

$\mathscr{S}_1$: $\{H, T\}$.

Note that this set is finite.

Example 1-9

$\mathscr{E}_2$: Toss a true coin three times and observe the sequence of heads and tails.

$\mathscr{S}_2$: $\{HHH, HHT, HTH, HTT, THH, THT, TTH, TTT\}$.

Example 1-10

$\mathscr{E}_3$: Toss a true coin three times and observe the total number of heads.

$\mathscr{S}_3$: $\{0, 1, 2, 3\}$.

Example 1-11

$\mathscr{E}_4$: Toss a pair of dice and observe the up faces.

$\mathscr{S}_4$: $\{(1, 1), (1, 2), (1, 3), (1, 4), (1, 5), (1, 6),$
$(2, 1), (2, 2), (2, 3), (2, 4), (2, 5), (2, 6),$
$(3, 1), (3, 2), (3, 3), (3, 4), (3, 5), (3, 6),$
$(4, 1), (4, 2), (4, 3), (4, 4), (4, 5), (4, 6),$
$(5, 1), (5, 2), (5, 3), (5, 4), (5, 5), (5, 6),$
$(6, 1), (6, 2), (6, 3), (6, 4), (6, 5), (6, 6)\}$.

Example 1-12

$\mathscr{E}_5$: An automobile door is assembled with a large number of spot welds. After assembly, each weld is inspected, and the total number of defectives is counted.

$\mathscr{S}_5$: $\{0, 1, 2, \ldots, K\}$, where K = the total number of welds in the door.

Example 1-13

$\mathscr{E}_6$: A cathode ray tube is manufactured, put on life test, and aged to failure. The elapsed time (in hours) at failure is recorded.

$\mathscr{S}_6$: $\{t: t \in R, t \geq 0\}$.

This set is uncountable.

Example 1-14

$\mathscr{E}_7$: A monitor records the emission count from a radioactive source in one minute.

$\mathscr{S}_7$: $\{0, 1, 2, \ldots\}$.

This set is countably infinite.

Example 1-15

$\mathscr{E}_8$: Two key solder joints on a printed circuit board are inspected with a probe as well as visually, and each joint is classified as good, G, or defective, D, requiring rework or scrap.

$\mathscr{S}_8$: $\{GG, GD, DG, DD\}$.

Example 1-16

$\mathscr{E}_9$: In a particular chemical plant the volume produced per day for a particular product ranges between a minimum value, b, and a maximum value, c, which corresponds to capacity. A day is randomly selected and the amount produced is observed.

$\mathscr{S}_9$: $\{x: x \in R, b \leq x \leq c\}$.

Example 1-17

$\mathscr{E}_{10}$: An extrusion plant is engaged in making up an order for pieces 20 feet long. Inasmuch as the trim operation creates scrap at both ends, the extruded bar must exceed 20 feet. Because of costs involved, the amount of scrap is critical. A bar is extruded, trimmed, and finished, and the total length of scrap is measured.

$\mathscr{S}_{10}$: $\{x: x \in R, x > 0\}$.

Example 1-18

$\mathscr{E}_{11}$: In a missile launch, the three components of velocity are mentioned from the ground as a function of time. At 1 minute after launch these are printed for a control unit.

$\mathscr{S}_{11}$: $\{(v_x, v_y, v_z): v_x, v_y, v_z$ are real numbers$\}$.

Example 1-19

$\mathscr{E}_{12}$: In the preceding example, the velocity components are continuously recorded for 5 minutes.

$\mathscr{S}_{12}$: The space is complicated here, as we have all possible realizations of the functions $v_x(t)$, $v_y(t)$, and $v_z(t)$ for $0 \leq t \leq 5$ minutes to consider.

All these examples have the characteristics required of random experiments. With the exception of Example 1-19, the description of the sample space is straightforward, and although repetition is not considered, ideally we could repeat the experiments. To illustrate the phenomena of random occurrence, consider Example 1-8. Obviously, if $\mathscr{E}_1$ is repeated indefinitely, we obtain a sequence of *heads* and *tails.* A pattern emerges as we continue the experiment. Notice that since the coin is true, we should obtain heads approximately one-half of the time. In recognizing the *idealization* in the model, we simply agree on a theoretical possible set of outcomes. In $\mathscr{E}_1$, we ruled out the possibility of having the coin land on edge, and in $\mathscr{E}_6$, where we recorded the elapsed time to failure, the idealized sample space consisted of all nonnegative real numbers.

1-4 EVENTS

An event, say A, is associated with the sample space of the experiment. The sample space is considered to be the universal set so that event A is simply a subset of $\mathscr{S}$. Note that both $\varnothing$ and $\mathscr{S}$ are subsets of $\mathscr{S}$. As a general rule, a capital letter will be used to denote an event. For a finite sample space, we note that the set of all subsets is the *power set*, and more generally, we require that if $A \subset \mathscr{S}$, then $\overline{A} \subset \mathscr{S}$, and if $A_1, A_2, \ldots$ is a sequence of mutually exclusive events in $\mathscr{S}$, as defined below, then $\bigcup_{i=1}^{\infty} A_i \subset \mathscr{S}$. The following events relate to experiments $\mathscr{E}_1, \mathscr{E}_2, \ldots, \mathscr{E}_{10}$, described in the preceding section. These are provided for illustration only; many other events could have been described for each case.

$\mathscr{E}_1$. A:	The coin toss yields a head $\{H\}$.
$\mathscr{E}_2$. A:	All the coin tosses give the same face $\{HHH, TTT\}$.
$\mathscr{E}_3$. A:	The total number of heads is two $\{2\}$.
$\mathscr{E}_4$. A:	The sum of the "up" faces is seven $\{(1, 6), (2, 5), (3, 4), (4, 3), (5, 2), (6, 1)\}$.
$\mathscr{E}_5$. A:	The number of defective welds does not exceed 5 $\{0, 1, 2, 3, 4, 5\}$.
$\mathscr{E}_6$. A:	The time to failure is greater than 1000 hours $\{t: t > 1000\}$.
$\mathscr{E}_7$. A:	The count is exactly two $\{2\}$.
$\mathscr{E}_8$. A:	Neither weld is bad $\{GG\}$.
$\mathscr{E}_9$. A:	The volume produced is between $a > b$ and c $\{x: x \in R, b < a < x < c\}$.
$\mathscr{E}_{10}$. A:	The scrap does not exceed one foot $\{x: x \in R, 0 < x \leq 1\}$.

Since an event is a set, the set operations defined for events and the laws and properties of Section 1-2 hold. If the intersections for all combinations of two or more events among k events considered are empty, then the k events are said to be *mutually exclusive* (or *disjoint*). If there are two events, A and B, then they are mutually exclusive if $A \cap B = \varnothing$. With $k = 3$, we would require $A_1 \cap A_2 = \varnothing$, $A_1 \cap A_3 = \varnothing$, $A_2 \cap A_3 = \varnothing$, and $A_1 \cap A_2 \cap A_3 = \varnothing$, and this case is illustrated in Fig.1-3. We emphasize that these multiple events are associated with one experiment.

1-5 PROBABILITY DEFINITION AND ASSIGNMENT

An axiomatic approach is taken to define probability as a *set function* where the elements of the domain are sets and the elements of the range are real numbers between 0 and 1. If event A is an element in the domain of this function, we use customary functional notation, $P(A)$, to designate the corresponding element in the range.

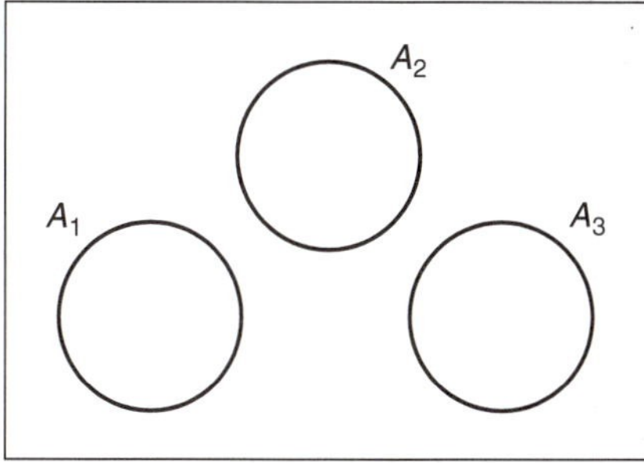

Figure 1-3 Three mutually exclusive events.

Definition

If an experiment $\mathscr{E}$ has sample space $\mathscr{S}$ and an event A is defined on $\mathscr{S}$, then $P(A)$ is a real number called the probability of event A or the probability of A, and the function $P(\cdot)$ has the following properties:

1. $0 \le P(A) \le 1$ for each event A of $\mathscr{S}$.
2. $P(\mathscr{S}) = 1$.
3. For any finite number k of mutually exclusive events defined on $\mathscr{S}$,

$$P\left(\bigcup_{i=1}^{k} A_i\right) = \sum_{i=1}^{k} P(A_i).$$

4. If $A_1, A_2, A_3, \ldots$ is a denumerable sequence of mutually exclusive events defined on $\mathscr{S}$, then

$$P\left(\bigcup_{i=1}^{\infty} A_i\right) = \sum_{i=1}^{\infty} P(A_i).$$

Note that the properties of the definition given do not tell the experimenter *how* to assign probabilities; however, they do restrict the way in which the assignment may be accomplished. In practice, probability is assigned on the basis of (1) estimates obtained from previous experience or prior observations, (2) an analytical consideration of experimental conditions, or (3) assumption.

To illustrate the assignment of probability based on experience, we consider the repetition of the experiment and the relative frequency of the occurrence of the event of interest.

This notion of relative frequency has intuitive appeal, and it involves the conceptual repetition of an experiment and a counting of both the number of repetitions and the number of times the event in question occurs. More precisely, $\mathscr{E}$ is repeated m times and two events are denoted A and B. We let m_A and m_B be the number of times A and B occur in the m repetitions.

Definition

The value $f_A = m_A/m$ is called the *relative frequency* of event A. It has the following properties:

1. $0 \le f_A \le 1$.
2. $f_A = 0$ if and only if A never occurs, and $f_A = 1$ if and only if A occurs on every repetition.
3. If A and B are mutually exclusive events, then $f_{A \cup B} = f_A + f_B$.

As m becomes large, f_A tends to stabilize. That is, as the number of repetitions of the experiment increases, the relative frequency of event A will vary less and less (from repetition to repetition). The concept of relative frequency and the tendency toward stability lead to one method for assigning probability. If an experiment $\mathscr{E}$ has sample space $\mathscr{S}$ and an event A is defined, and if the relative frequency f_A approaches some number p_A as the number of repetitions increases, then the number p_A is ascribed to A as its probability, that is, as $m \to \infty$,

$$P(A) = \frac{m_A}{m} = p_A. \tag{1-1}$$

In practice, something less than infinite replication must obviously be accepted. As an example, consider a simple coin-tossing experiment $\mathscr{E}$ in which we sequentially toss a fair coin and observe the outcome—either heads (H) or tails (T)—arising from each trial. If the observational process is considered as a random experiment such that for a particular repetition the sample space is $S = \{H, T\}$, we define the event $A = \{H\}$, where this event is defined before the observations are made. Suppose that after $m = 100$ repetitions of $\mathscr{E}$, we observe $m_A = 43$, resulting in a relative frequency of $f_A = 0.43$, a seemingly low value. Now suppose that we instead conduct $m = 10{,}000$ repetitions of $\mathscr{E}$ and this time we observe $m_A = 4{,}924$, so that the relative frequency is in this case $f_A = 0.4924$. Since we now have at our disposal 10,000 observations instead of 100, everyone should be more comfortable in assigning the updated relative frequency of 0.4924 to $P(A)$. The stability of f_A as m gets large is only an intuitive notion at this point; we will be able to be more precise later.

A method for computing the probability of an event A is as follows. Suppose the sample space has a finite number, n, of elements, e_i, and the probability assigned to an outcome is $p_i = P(E_i)$, where $E_i = \{e_i\}$ and

$$p_i \geq 0, \qquad i = 1, 2, \ldots, n,$$

while

$$p_1 + p_2 + \cdots + p_n = 1,$$

$$P(A) = \sum_{i:e_i \in A} p_i. \tag{1-2}$$

This is a statement that the probability of event A is the sum of the probabilities associated with the outcomes making up event A, and this result is simply a consequence of the definition of probability. The practitioner is still faced with assigning probabilities to the outcomes, e_i. It is noted that the sample space will not be finite if, for example, the elements e_i of $\mathscr{S}$ are countably infinite in number. In this case, we note that

$$p_i \geq 0, \qquad i = 1, 2, \ldots,$$

$$\sum_{i=1}^{\infty} p_i = 1.$$

However, equation 1-2 may be used without modification.

If the sample space is finite and has n *equally likely* outcomes so that $p_1 = p_2 = \cdots = p_n = 1/n$, then

$$P(A) = \frac{n(A)}{n} \tag{1-3}$$

and $n(A)$ outcomes are contained in A. Counting methods useful in determining n and $n(A)$ will be presented in Section 1-6.

Example 1-20

Suppose the coin in Example 1-9 is biased so that the outcomes of the sample space $\mathcal{S} = \{HHH, HHT, HTH, HTT, THH, THT, TTH, TTT\}$ have probabilities $p_1 = \frac{1}{27}, p_2 = \frac{2}{27}, p_3 = \frac{2}{27}, p_4 = \frac{4}{27}, p_5 = \frac{2}{27}, p_6 = \frac{4}{27}, p_7 = \frac{4}{27}, p_8 = \frac{8}{27}$, where $e_1 = HHH$, $e_2 = HHT$, etc. If we let event A be the event that all tosses yield the same face, then $P(A) = \frac{1}{27} + \frac{8}{27} = \frac{1}{3}$.

Example 1-21

Suppose that in Example 1-14 we have prior knowledge that

$$\begin{aligned} p_i &= \frac{e^{-2} \cdot 2^{i-1}}{(i-1)!} \qquad && i = 1, 2, \dots, \\ &= 0, && \text{otherwise.} \end{aligned}$$

where p_i is the probability that the monitor will record a count outcome of i–1 during a 1-minute interval. If we consider event A as the event containing the outcomes 0 and 1, then $A = \{0, 1\}$, and $P(A) = p_1 + p_2 = e^{-2} + 2e^{-2} = 3e^{-2} \cong 0.406$.

Example 1-22

Consider Example 1-9, where a true coin is tossed three times, and consider event A, where all coins show the same face. By equation 1-3,

$$P(A) = \frac{n(A)}{n} = \frac{2}{8},$$

since there are eight total outcomes and two are favorable to event A. The coin was assumed to be true, so all eight possible outcomes are equally likely.

Example 1-23

Assume that the dice in Example 1-11 are true, and consider an event A where the sum of the up faces is 7. Using the results of equation 1-3 we note that there are 36 outcomes, of which six are favorable to the event in question, so that $P(A) = \frac{1}{6}$.

Note that Examples 1-22 and 1-23 are extremely simple in two respects: the sample space is of a highly restricted type, and the counting process is easy. Combinatorial methods frequently become necessary as the counting becomes more involved. Basic counting methods are reviewed in Section 1-6. Some important theorems regarding probability follow.

Theorem 1-1

If $\varnothing$ is the empty set, then $P(\varnothing) = 0$.

Proof Note that $\mathcal{S} = \mathcal{S} \cup \varnothing$ and $\mathcal{S}$ and $\varnothing$ are mutually exclusive. Then $P(\mathcal{S}) = P(\mathcal{S}) + P(\varnothing)$ from property 4; therefore $P(\varnothing) = 0$.

Theorem 1-2

$P(\overline{A}) = 1 - P(A)$.

Proof Note that $\mathcal{S} = A \cup \overline{A}$ and A and $\overline{A}$ are mutually exclusive. Then $P(\mathcal{S}) = P(A) + P(\overline{A})$ from property 4, but from property 2, $P(\mathcal{S}) = 1$; therefore $P(\overline{A}) = 1 - P(A)$.

Theorem 1-3

$P(A \cup B) = P(A) + P(B) - P(A \cap B)$.

Proof Since $A \cup B = A \cup (B \cap \overline{A})$, where A and $(B \cap \overline{A})$ are mutually exclusive, and $B = (A \cap B) \cup (B \cap \overline{A})$, where $(A \cap B)$ and $(B \cap \overline{A})$ are mutually exclusive, then $P(A \cup B) = P(A) + P(B \cap \overline{A})$, and $P(B) = P(A \cap B) + P(B \cap \overline{A})$. Subtracting, $P(A \cup B) - P(B) = P(A) - P(A \cap B)$, and thus $P(A \cup B) = P(A) + P(B) - P(A \cap B)$.

The Venn diagram shown in Fig. 1-4 is helpful in following the argument of the proof for Theorem 1-3. We see that the "double counting" of the hatched region in the expression $P(A) + P(B)$ is corrected by subtraction of $P(A \cap B)$.

Theorem 1-4

$P(A \cup B \cup C) = P(A) + P(B) + P(C) - P(A \cap B) - P(A \cap C) - P(B \cap C) + P(A \cap B \cap C)$.

Proof We may write $A \cup B \cup C = (A \cup B) \cup C$ and use Theorem 1-3 since $A \cup B$ is an event. The reader is asked to provide the details in Exercise 1-32.

Theorem 1-5

$$P(A_1 \cup A_2 \cup \cdots \cup A_k) = \sum_{i=1}^{k} P(A_i) - \sum_{i<j=2}^{k} P(A_i \cap A_j) + \sum_{i<j<r=3}^{k} P(A_i \cap A_j \cap A_r) + \cdots + (-1)^{k-1} P(A_1 \cap A_2 \cap \cdots \cap A_k).$$

Proof Refer to Exercise 1-33.

Theorem 1-6

If $A \subset B$, then $P(A) \le P(B)$.

Proof
If $A \subset B$, then $B = A \cup (\overline{A} \cap B)$ and $P(B) = P(A) + P(\overline{A} \cap B) \ge P(A)$, since $P(\overline{A} \cap B) \ge 0$.

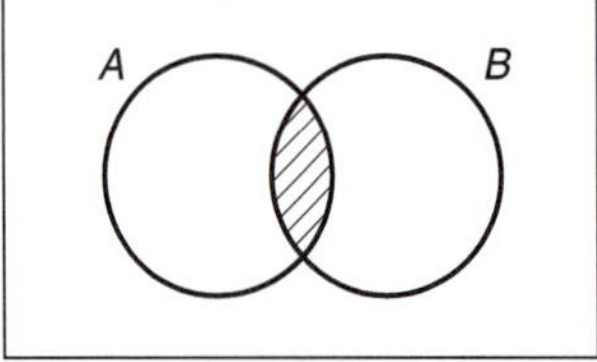

Figure 1-4 Venn diagram for two events.

Example 1-24

If A and B are mutually exclusive events, and if it is known that $P(A) = 0.20$ while $P(B) = 0.30$, we can evaluate several probabilities:

1. $P(\overline{A}) = 1 - P(A) = 0.80$.
2. $P(\overline{B}) = 1 - P(B) = 0.70$.
3. $P(A \cup B) = P(A) + P(B) = 0.2 + 0.3 = 0.5$.
4. $P(A \cap B) = 0$.
5. $P(\overline{A} \cap \overline{B}) = P(\overline{A \cup B})$, by De Morgan's law
$$= 1 - P(A \cup B)$$
$$= 1 - [P(A) + P(B)] = 0.5.$$

Example 1-25

Suppose events A and B are not mutually exclusive and we know that $P(A) = 0.20$, $P(B) = 0.30$, and $P(A \cap B) = 0.10$. Then evaluating the same probabilities as before, we obtain

1. $P(\overline{A}) = 1 - P(A) = 0.80$.
2. $P(\overline{B}) = 1 - P(B) = 0.70$.
3. $P(A \cup B) = P(A) + P(B) - P(A \cap B) = 0.2 + 0.3 - 0.1 = 0.4$.
4. $P(A \cap B) = 0.1$.
5. $P(\overline{A} \cap \overline{B}) = P(\overline{A \cup B}) = 1 - [P(A) + P(B) - P(A \cap B)] = 0.6$.

Example 1-26

Suppose that in a certain city 75% of the residents jog (J), 20% like ice cream (I), and 40% enjoy music (M). Further, suppose that 15% jog *and* like ice cream, 30% jog *and* enjoy music, 10% like ice cream *and* music, and 5% do all three types of activities. We can consolidate all of this information in the simple Venn diagram in Fig. 1-5 by starting from the last piece of data, $P(J \cap I \cap M) = 0.05$, and "working our way out" of the center.

1. Find the probability that a random resident will engage in at least one of the three activities. By Theorem 1-4,

$$P(J \cup I \cup M) = P(J) + P(I) + P(M) - P(J \cap I) - P(J \cap M) - P(I \cap M) + P(J \cap I \cap M)$$
$$= 0.75 + 0.20 + 0.40 - 0.15 - 0.30 - 0.10 + 0.05 = 0.85.$$

This answer is also immediate by adding up the components of the Venn diagram.

2. Find the probability that a resident engages in precisely one type of activity. By the Venn diagram, we see that the desired probability is

$$P(J \cap \overline{I} \cap \overline{M}) + P(\overline{J} \cap I \cap \overline{M}) + P(\overline{J} \cap \overline{I} \cap M) = 0.35 + 0 + 0.05 = 0.40.$$

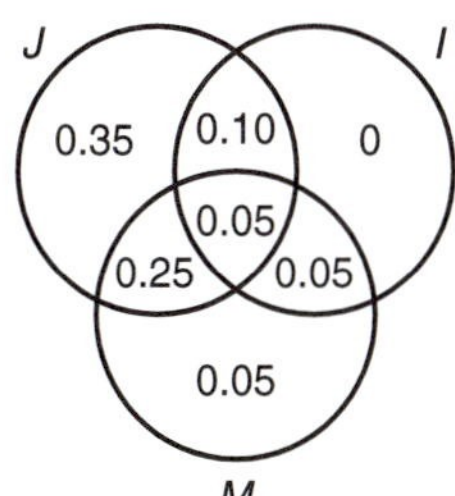

Figure 1-5 Venn diagram for Example 1-26.

1-6 FINITE SAMPLE SPACES AND ENUMERATION

Experiments that give rise to a finite sample space have already been discussed, and the methods for assigning probabilities to events associated with such experiments have been presented. We can use equations 1-1, 1-2, and 1-3 and deal either with "equally likely" or with "not equally likely" outcomes. In some situations we will have to resort to the relative frequency concept and successive trials (experimentation) to *estimate* probabilities, as indicated in equation 1-1, with some finite m. In this section, however, we deal with equally likely outcomes and equation 1-3. Note that this equation represents a special case of equation 1-2, where $p_1 = p_2 = \cdots = p_n = 1/n$.

In order to assign probabilities, $P(A) = n(A)/n$, we must be able to determine both n, the number of *outcomes*, and $n(A)$, the *number of outcomes favorable to event A*. If there are n outcomes in $\mathcal{S}$, then there are 2^n possible subsets that are the elements of the power set, $\{0, 1\}^A$.

The requirement for the n outcomes to be equally likely is an important one, and there will be numerous applications where the experiment will specify that one (or more) item(s) is (are) selected at *random* from a population group of N items without replacement.

If n represents the sample size ($n \leq N$) and the selection is random, then each possible selection (sample) is equally likely. It will soon be seen that there are $N!/[n!(N - n)!]$ such samples, so the probability of getting a particular sample must be $n!(N - n)!/N!$. It should be carefully noted that one sample differs from another if one (or more) item appears in one sample and not the other. The population items must thus be identifiable. In order to illustrate, suppose a population has four chips ($N = 4$) labeled a, b, c, and d. The sample size is to be two ($n = 2$). The *possible* results of the selection, disregarding order, are elements of $\mathcal{S} = \{ab, ac, ad, bc, bd, cd\}$. If the sampling process is random, the probability of obtaining each possible sample is $\frac{1}{6}$. The mechanics of selecting random samples vary a great deal, and devices such as pseudorandom number generators, *random number tables*, and *icosahedron dice* are frequently used, as will be discussed at a later point.

It becomes obvious that we need *enumeration* methods for evaluating n and $n(A)$ for experiments yielding equally likely outcomes; the following sections, 1-6.1 through 1-6.5, review basic enumeration techniques and results useful for this purpose.

1-6.1 Tree Diagram

In simple experiments, a tree diagram may be useful in the enumeration of the sample space. Consider Example 1-9, where a true coin is tossed three times. The set of possible outcomes could be found by taking all the paths in the tree diagram shown in Fig. 1-6. It should be noted that there are 2 outcomes to each trial, 3 trials, and $2^3 = 8$ outcomes $\{HHH, HHT, HTH, HTT, THH, THT, TTH, TTT\}$.

1-6.2 Multiplication Principle

If sets $A_1, A_2, \ldots, A_k$ have, respectively, $n_1, n_2, \ldots, n_k$ elements, then there are $n_1 \cdot n_2 \cdot \cdots \cdot n_k$ ways to select an element first from A_1, then from A_2, ..., and finally from A_k.

In the special case where $n_1 = n_2 = \cdots = n_k = n$, there are n^k possible selections. This was the situation encountered in the coin-tossing experiment of Example 1-9.

Suppose we consider some compound experiment $\mathcal{E}$ consisting of k experiments, $\mathcal{E}_1, \mathcal{E}_2, \ldots, \mathcal{E}_k$. If the sample spaces $\mathcal{S}_1, \mathcal{S}_2, \ldots, \mathcal{S}_k$ contain $n_1, n_2, \ldots, n_k$ outcomes, respectively, then there are $n_1 \cdot n_2 \cdot \cdots \cdot n_k$ outcomes to $\mathcal{E}$. In addition, if the n_j outcomes of $\mathcal{S}_j$ are equally likely for $j = 1, 2, \ldots, k$, then the $n_1 \cdot n_2 \cdot \cdots \cdot n_k$ outcomes of $\mathcal{E}$ are equally likely.

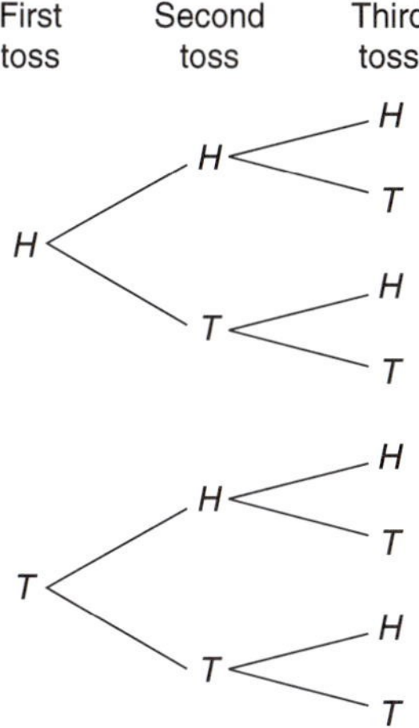

Figure 1-6 A tree diagram for tossing a true coin three times.

Example 1-27

Suppose we toss a true coin and cast a true die. Since the coin and the die are true, the two outcomes to $\mathscr{E}_1$, $\mathscr{S}_1 = \{H, T\}$, are equally likely and the six outcomes to $\mathscr{E}_2$, $\mathscr{S}_2 = \{1,2,3,4,5,6\}$, are equally likely. Since $n_1 = 2$ and $n_2 = 6$, there are 12 outcomes to the total experiment and all the outcomes are equally likely. Because of the simplicity of the experiment in this case, a tree diagram permits an easy and complete enumeration. Refer to Fig. 1-7.

Example 1-28

A manufacturing process is operated with very little "in-process inspection." When items are completed they are transported to an inspection area, and four characteristics are inspected, each by a different inspector. The first inspector rates a characteristic according to one of four ratings. The second inspector uses three ratings, and the third and fourth inspectors use two ratings each. Each inspector marks the rating on the item identification tag. There would be a total of $4 \cdot 3 \cdot 2 \cdot 2 = 48$ ways in which the item may be marked.

1-6.3 Permutations

A permutation is an ordered arrangement of distinct objects. One permutation differs from another if the order of arrangement differs or if the content differs. To illustrate, suppose we

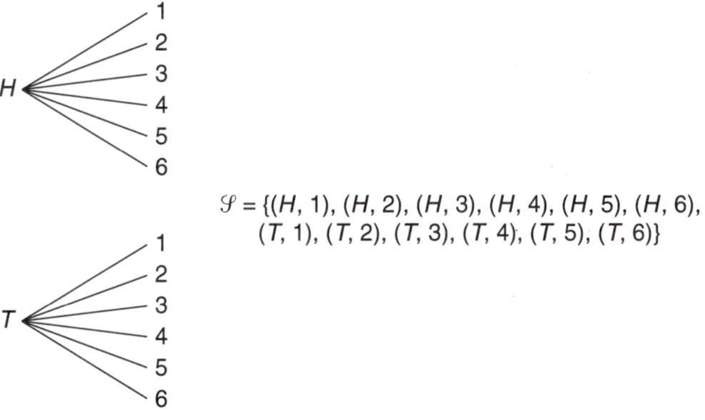

Figure 1-7 The tree diagram for Example 1-27.

again consider four distinct chips labeled a, b, c, and d. Suppose we wish to consider all permutations of these chips taken one at a time. These would be

$$\begin{array}{c} a \\ b \\ c \\ d \end{array}$$

If we are to consider all permutations taken two at a time, these would be

$$\begin{array}{cc} ab & bc \\ ba & cb \\ ac & bd \\ ca & db \\ ad & cd \\ da & dc \end{array}$$

Note that permutations ab and ba differ because of a difference in order of the objects, while permutations ac and ab differ because of content differences. In order to generalize, we consider the case where there are n distinct objects from which we plan to select permutations of r objects ($r \le n$). The number of such permutations, P_r^n, is given by

$$\begin{aligned} P_r^n &= n(n-1)(n-2)(n-3)\cdot\dots\cdot(n-r+1) \\ &= \frac{n!}{(n-r)!}. \end{aligned}$$

This is a result of the fact that there are n ways to select the first object, $(n-1)$ ways to select the second, ..., $[n-(r-1)]$ ways to select the rth and the application of the multiplication principle. Note that $P_n^n = n!$ and $0! = 1$.

Example 1-29

A major league baseball team typically has 25 players. A line-up consists of nine of these players in a particular order. Thus, there are $P_9^{25} \cong 7.41 \times 10^{11}$ possible lineups.

1-6.4 Combinations

A combination is an arrangement of distinct objects where one combination differs from another only if the content of the arrangement differs. Here order does not matter. In the case of the four lettered chips a, b, c, d, the combinations of the chips, taken two at a time, are

$$\begin{array}{c} ab \\ ac \\ ad \\ bc \\ bd \\ cd \end{array}$$

We are interested in determining the number of combinations when there are n distinct objects to be selected r at a time. Since the number of permutations was the number of ways to select r objects from the n *and* then permute the r objects, we note that

$$P_r^n = r! \cdot \binom{n}{r}, \tag{1-4}$$

where $\binom{n}{r}$ represents the number of *combinations*. It follows that

$$\binom{n}{r} = P_r^n / r! = \frac{n!}{r!(n-r)!}. \tag{1-5}$$

In the illustration with the four chips where $r = 2$, the reader may readily verify that $P_2^4 = 12$ and $\binom{4}{2} = 6$, as we found by complete enumeration.

For present purposes, $\binom{n}{r}$ is defined where n and r are integers such that $0 \le r \le n$; however, the terms $\binom{n}{r}$ may be generally defined for real n and any nonnegative integer r. In this case we write

$$\binom{n}{r} = \frac{n(n-1)(n-2)\cdot\cdots\cdot(n-r+1)}{r!}.$$

The reader will recall the *binomial theorem*:

$$(a+b)^n = \sum_{r=0}^{n} \binom{n}{r} a^r b^{n-r}. \tag{1-6}$$

The numbers $\binom{n}{r}$ are thus called binomial coefficients.

Returning briefly to the definition of random sampling from a finite population without replacement, there were N objects with n to be selected. There are thus $\binom{N}{n}$ different samples. If the sampling process is random, each possible outcome has probability $1/\binom{N}{n}$ of being the one selected.

Two identities that are often helpful in problem solutions are

$$\binom{n}{r} = \binom{n}{n-r} \tag{1-7}$$

and

$$\binom{n}{r} = \binom{n-1}{r-1} + \binom{n-1}{r}. \tag{1-8}$$

To develop the result shown in equation 1-7, we note that

$$\binom{n}{r} = \frac{n!}{r!(n-r)!} = \frac{n!}{(n-r)!r!} = \binom{n}{n-r},$$

and to develop the result shown in equation 1-8, we expand the right-hand side and collect terms.

To verify that a finite collection of n elements has 2^n subsets, as indicated earlier, we see that

$$2^n = (1+1)^n = \sum_{r=0}^{n} \binom{n}{r} = \binom{n}{0} + \binom{n}{1} + \cdots + \binom{n}{n}$$

from equation 1-6. The right side of this relationship gives the total number of subsets since $\binom{n}{0}$ is the number of subsets with 0 elements. $\binom{n}{1}$ is the number with one element, ..., and $\binom{n}{n}$ is the number with n elements.

Example 1-30

A production lot of size 100 is known to be 5% defective. A random sample of 10 items is selected without replacement. In order to determine the probability that there will be no defectives in the sample, we resort to counting both the number of possible samples and the number of samples favorable to event A, where event A is taken to mean that there are no defectives. The number of possible samples is $\binom{100}{10} = \frac{100!}{10!90!}$. The number "favorable to A" is $\binom{5}{0} \cdot \binom{95}{10}$, so that

$$P(A) = \frac{\binom{5}{0}\binom{95}{10}}{\binom{100}{10}} = \frac{\frac{5!}{0!5!}\frac{95!}{10!85!}}{\frac{100!}{10!90!}} = 0.58375.$$

To generalize the preceding example, we consider the case where the population has N items of which D belong to some class of interest (such as defective). A random sample of size n is selected without replacement. If A denotes the event of obtaining exactly r items from the class of interest in the sample, then

$$P(A) = \frac{\binom{D}{r}\binom{N-D}{N-r}}{\binom{N}{n}}, \qquad r = 0, 1, 2, \ldots, \min(n, D). \tag{1-9}$$

Problems of this type are often referred to as hypergeometric sampling problems.

Example 1-31

An NBA basketball team typically has 12 players. A starting team consists of five of these players in no particular order. Thus, there are $\binom{12}{5} = 792$ possible starting teams.

Example 1-32

One obvious application of counting methods lies in calculating probabilities for poker hands. Before proceeding, we remind the reader of some standard terminology. The *rank* of a particular card drawn from a standard 52-card deck can be 2, 3, …, *Q, K, A,* while the possible *suits* are ♣, ♢, ♡, ♠. In poker, we draw five cards at random from a deck. The number of possible hands is $\binom{52}{5} = 2{,}598{,}960$.

1. We first calculate the probability of obtaining two pairs, for example A♡, A♣, 3♡, 3♢, 10♠. We proceed as follows.

   (a) Select two ranks (e.g., A, 3). We can do this $\binom{13}{2}$ ways.
   (b) Select two suits for first pair (e.g., ♡, ♣). There are $\binom{4}{2}$ ways.
   (c) Select two suits for second pair (e.g., ♡, ♢). There are $\binom{4}{2}$ ways.
   (d) Select remaining card to complete the hand. There are 44 ways.

   Thus, the number of ways to select two pairs is

   $$n(2 \text{ pairs}) = \binom{13}{2}\binom{4}{2}\binom{4}{2}44 = 123{,}552,$$

   and so

   $$P(2 \text{ pairs}) = \frac{123{,}552}{2{,}598{,}960} \approx 0.0475.$$

2. Here we calculate the probability of obtaining a full house (one pair, one three-of-a-kind), for example $A\heartsuit$, $A\clubsuit$, $3\heartsuit$, $3\diamondsuit$, $3\spadesuit$.

(a) Select two *ordered* ranks (e.g., A, 3). There are $P_{13,2}$ ways. Indeed, the ranks must be *ordered* since "three A's, two 3's" differs from "two A's, three 3's."
(b) Select two suits for the pair (e.g., $\heartsuit$, $\clubsuit$). There are $\binom{4}{2}$ ways.
(c) Select three suits for the three-of-a-kind (e.g., $\heartsuit$, $\diamondsuit$, $\spadesuit$). $\binom{4}{3}$ ways.

Thus, the number of ways to select a full house is

$$n(\text{full house}) = 13\cdot 12\binom{4}{2}\binom{4}{3} = 3744,$$

and so

$$P(\text{full house}) = \frac{3744}{2{,}598{,}960} \approx 0.00144.$$

3. Finally, we calculate the probability of a flush (all five cards from same suit). How many ways can we obtain this event?

(a) Select a suit. There are $\binom{4}{1}$ ways.
(b) Select five cards from that suit. There are $\binom{13}{5}$ ways.

Then we have

$$P(\text{flush}) = \frac{5148}{2{,}598{,}960} \approx 0.00198.$$

1-6.5 Permutations of Like Objects

In the event that there are k distinct classes of objects, and the objects within the classes are not distinct, the following result is obtained, where n_1 is the number in the first class, n_2 is the number in the second class, ..., n_k is the number in the kth class, and $n_1 + n_2 + \cdots + n_k = n$:

$$P^{n}_{n_1,n_2,\ldots,n_k} = \frac{n!}{n_1!\cdot n_2!\cdot\cdots\cdot n_k!}. \tag{1-10}$$

Example 1-33

Consider the word "TENNESSEE." The number of ways to arrange the letters in this word is

$$\frac{n!}{n_T!n_E!n_N!n_S!} = \frac{9!}{1!4!2!2!} = 3780,$$

where we use the obvious notation. Problems of this type are often referred to as multinomial sampling problems.

The counting methods presented in this section are primarily to support probability assignment where there is a finite number of equally likely outcomes. It is important to remember that this is a special case of the more general types of problems encountered in probability applications.

1-7 CONDITIONAL PROBABILITY

As noted in Section 1-4, an event is associated with a sample space, and the event is represented by a subset of $\mathcal{S}$. The probabilities discussed in Section 1-5 all relate to the entire sample space. We have used the symbol $P(A)$ to denote the probability of these events; however, we could have used the symbol $P(A|\mathcal{S})$, read as "the probability of A, given sample space $\mathcal{S}$." In this section we consider the probability of events where the event is *conditioned* on some subset of the sample space.

Some illustrations of this idea should be helpful. Consider a group of 100 persons of whom 40 are college graduates, 20 are self-employed, and 10 are both college graduates and self-employed. Let B represent the set of college graduates and A represent the set of self-employed, so that $A \cap B$ is the set of college graduates who are self-employed. From the group of 100, one person is to be randomly selected. (Each person is given a number from 1 to 100, and 100 chips with the same numbers are agitated, with one being selected by a blindfolded outsider.) Then, $P(A) = 0.2$, $P(B) = 0.4$, and $P(A \cap B) = 0.1$ if the entire sample space is considered. As noted, it may be more instructive to write $P(A|\mathcal{S})$, $P(B|\mathcal{S})$, and $P(A \cap B|\mathcal{S})$ in such a case. Now suppose the following event is considered: self-employed *given* that the person is a college graduate $(A|B)$. Obviously the sample space is reduced in that only college graduates are considered (Fig. 1-8). The probability $P(A|B)$ is thus given by

$$\frac{n(A\cap B)}{n(B)} = \frac{n(A\cap B)/n}{n(B)/n} = P(A|B) = \frac{P(A\cap B)}{P(B)} = \frac{0.1}{0.4} = 0.25.$$

The reduced sample space consists of the set of all subsets of $\mathcal{S}$ that belong to B. Of course, $A \cap B$ satisfies the condition.

As a second illustration, consider the case where a sample of size 2 is randomly selected from a lot of size 10. It is known that the lot has seven good and three bad items. Let A be the event that the first item selected is good, and B be the event that the second item selected is good. If the items are selected *without replacement*, that is, the first item is not replaced before the second item is selected, then

$$P(A) = \frac{7}{10}$$

and

$$P(B|A) = \frac{6}{9}.$$

If the first item is replaced before the second item is selected, the conditional probability $P(B|A) = P(B) = \frac{7}{10}$, and the events A and B resulting from the two selection experiments comprising $\mathcal{E}$ are said to be *independent*. A formal definition of conditional probability

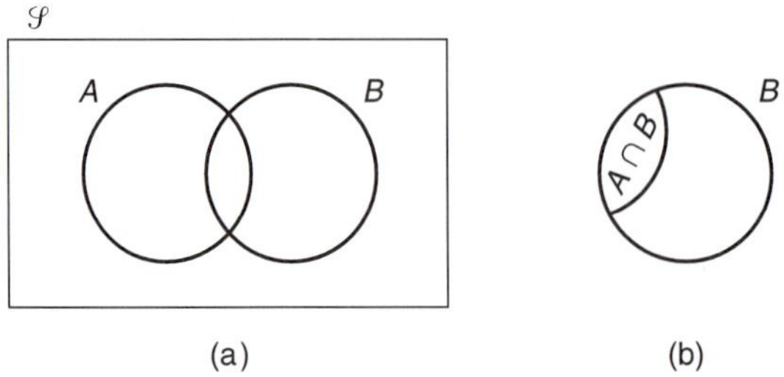

Figure 1-8 Conditional probability. *(a)* Initial sample space. *(b)* Reduced sample space.

$P(A|B)$ will be given later, and independence will be discussed in detail. The following examples will help to develop some intuitive feeling for conditional probability.

Example 1-34

Recall Example 1-11 where two dice are tossed, and assume that each die is true. The 36 possible outcomes were enumerated. If we consider two events,

$$A = \{(d_1, d_2): d_1 + d_2 = 4\},$$

$$B = \{(d_1, d_2): d_2 \geq d_1\},$$

where d_1 is the value of the up face of the first die and d_2 is the value of the up face of the second die, then $P(A) = \frac{3}{36}$, $P(B) = \frac{21}{36}$, $P(A \cap B) = \frac{2}{36}$, $P(B|A) = \frac{2}{3}$, and $P(A|B) = \frac{2}{21}$. The probabilities were obtained from a direct consideration of the sample space and the counting of outcomes. Note that

$$P(A|B) = \frac{P(A \cap B)}{P(B)} \quad \text{and} \quad P(B|A) = \frac{P(A \cap B)}{P(A)}.$$

Example 1-35

In World War II an early operations research effort in England was directed at establishing search patterns for U-boats by patrol flights or *sorties*. For a time, there was a tendency to concentrate the flights on in-shore areas, as it had been believed that more sightings took place in-shore. The research group studied 1000 sortie records with the following result (the data are fictitious):

	In-shore	Off-shore	Total
Sighting	80	20	100
No sighting	820	80	900
Total sorties	900	100	1000

Let S_1: There was a sighting.
S_2: There was no sighting.
B_1: In-shore sortie.
B_2: Off-shore sortie.

We see immediately that

$$P(S_1|B_1) = \frac{80}{900} = 0.0889,$$

$$P(S_1|B_2) = \frac{20}{100} = 0.20,$$

which indicates a search strategy counter to prior practice.

Definition

We may define the conditional probability of event A given event B as

$$P(A|B) = \frac{P(A \cap B)}{P(B)} = \quad \text{if} \quad P(B) > 0. \tag{1-11}$$

This definition results from the intuitive notion presented in the preceding discussion. The conditional probability $P(\cdot|\cdot)$ satisfies the properties required of probabilities. That is,

1. $0 \le P(A|B) \le 1$.
2. $P(\mathcal{S}|B) = 1$.
3. $P\left(\bigcup_{i=1}^{k} A_i|B\right) = \sum_{i=1}^{k} P(A_i|B)$ if $A_i \cap A_j = \varnothing$ for $i \neq j$.
4. $P\left(\bigcup_{i=1}^{\infty} A_i|B\right) = \sum_{i=1}^{\infty} P(A_i|B)$

   for $A_1, A_2, A_3, \ldots$ a denumerable sequence of disjoint events.

In practice we may solve problems by using equation 1-11 and calculating $P(A \cap B)$ and $P(B)$ with respect to the original sample space (as was illustrated in Example 1-35) or by considering the probability of A with respect to the reduced sample space B (as was illustrated in Example 1-34).

A restatement of equation 1-11 leads to what is often called the *multiplication rule*, that is

$$P(A \cap B) = P(B) \cdot P(A|B), \qquad P(B) > 0,$$

and

$$P(A \cap B) = P(A) \cdot P(B|A), \qquad P(A) > 0. \tag{1-12}$$

The second statement is an obvious consequence of equation 1-11 with the conditioning on event A rather than event B.

It should be noted that if A and B are *mutually exclusive* as indicated in Fig. 1-9, then $A \cap B = \varnothing$ so that $P(A|B) = 0$ and $P(B|A) = 0$.

In the other extreme, if $B \subset A$, as shown in Fig. 1-10, then $P(A|B) = 1$. In the first case, A and B cannot occur simultaneously, so knowledge of the occurrence of B tells us that A does not occur. In the second case, if B occurs, A must occur. On the other hand, there are many cases where the events are totally unrelated, and knowledge of the occurrence of one has no bearing on and yields no information about the other. Consider, for example, the experiment where a true coin is tossed twice. Event A is the event that the first toss results in a "heads," and event B is the event that the second toss results in a "heads." Note that $P(A) = \frac{1}{2}$, since the coin is true, and $P(B|A) = \frac{1}{2}$, since the coin is true and it has no memory. The occurrence of event A did not in any way affect the occurrence of B, and if we wanted to find the probability of A and B occurring, that is, $P(A \cap B)$, we find that

$$P(A \cap B) = P(A) \cdot P(B|A) = \frac{1}{2} \cdot \frac{1}{2} = \frac{1}{4}.$$

We may observe that if we had no knowledge about the occurrence or nonoccurrence of A, we have $P(B) = P(B|A)$, as in this example.

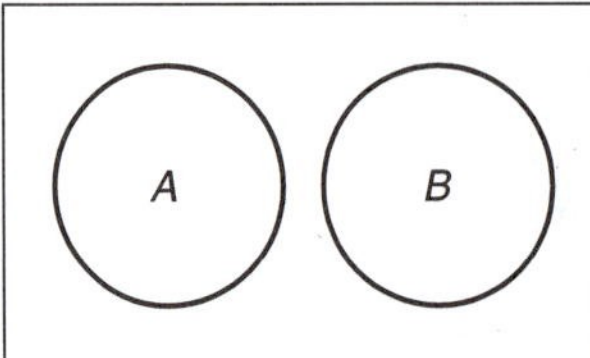

Figure 1-9 Mutually exclusive events.

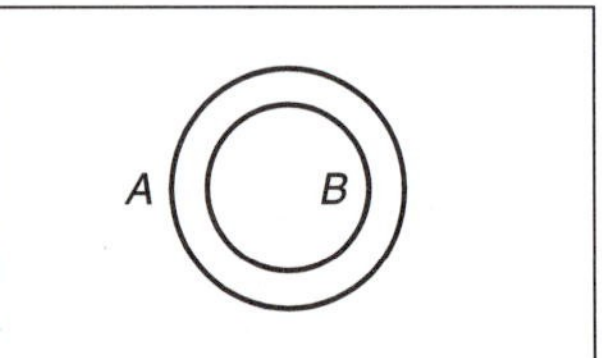

Figure 1-10 Event B as a subset of A.

Informally speaking, two events are considered to be *independent* if the probability of the occurrence of one is not affected by the occurrence or nonoccurrence of the other. This leads to the following definition.

Definition

A and B are independent if and only if

$$P(A \cap B) = P(A) \cdot P(B). \tag{1-13}$$

An immediate consequence of this definition is that if A and B are independent events, then from equation 1-12,

$$P(A|B) = P(A) \quad \text{and} \quad P(B|A) = P(B). \tag{1-14}$$

The following theorem is sometimes useful. The proof is given here only for the first part.

Theorem 1-7

If A and B are independent events, then the following holds:

1. A and $\overline{B}$ are independent events.
2. $\overline{A}$ and $\overline{B}$ are independent events.
3. $\overline{A}$ and B are independent events.

Proof (Part 1)

$$\begin{aligned} P(A \cap \overline{B}) &= P(A) \cdot P(\overline{B}|A) \\ &= P(A) \cdot [1 - P(B|A)] \\ &= P(A) \cdot [1 - P(B)] \\ &= P(A) \cdot P(\overline{B}). \end{aligned}$$

In practice, there are many situations where it may not be easy to determine whether two events are independent; however, there are numerous other cases where the requirements may be either justified or approximated from a physical consideration of the experiment. A sampling experiment will serve to illustrate.

Example 1-36

Suppose a random sample of size 2 is to be selected from a lot of size 100, and it is known that 98 of the 100 items are good. The sample is taken in such a manner that the first item is observed and replaced before the second item is selected. If we let

A: First item observed is good,

B: Second item observed is good,

and if we want to determine the probability that both items are good, then

$$P(A \cap B) = P(A) \cdot P(B) = \frac{98}{100} \cdot \frac{98}{100} = 0.9604.$$

If the sample is taken "without replacement" so that the first item is not replaced before the second item is selected, then

$$P(A \cap B) = P(A) \cdot P(B|A) = \frac{98}{100} \cdot \frac{97}{99} = 0.9602.$$

The results are obviously very close, and one common practice is to assume the events independent when the *sampling fraction* (sample size/population size) is small, say less than 0.1.

Example 1-37

The field of *reliability engineering* has developed rapidly since the early 1960s. One type of problem encountered is that of estimating system reliability given subsystem reliabilities. Reliability is defined here as the probability of proper functioning for a stated period of time. Consider the structure of a simple serial system, shown in Fig. 1-11. The system functions if and only if both subsystems function. If the subsystems survive independently, then.

$$\text{System reliability} = R_s = R_1 \cdot R_2,$$

where R_1 and R_2 are the reliabilities for subsystems 1 and 2, respectively. For example, if $R_1 = 0.90$ and $R_2 = 0.80$, then $R_s = 0.72$.

Example 1-37 illustrates the need to generalize the concept of independence to more than two events. Suppose the system consisted of three subsystems or perhaps 20 subsystems. What conditions would be required in order to allow the analyst to obtain an estimate of system reliability by obtaining the product of the subsystem reliabilities?

Definition

The k events $A_1, A_2, \ldots, A_k$ are mutually independent if and only if the probability of the intersection of any $2, 3, \ldots, k$ of these sets is the product of their respective probabilities.

Stated more precisely, we require that for $r = 2, 3, \ldots, k$,

$$P(A_{i_1} \cap A_{i_2} \cap \cdots \cap A_{i_r}) = P(A_{i_1}) \cdot P(A_{i_2}) \cdot \cdots \cdot P(A_{i_r})$$
$$= \prod_{j=1}^{r} P\left(A_{i_j}\right).$$

In the case of serial system reliability calculations where mutual independence may reasonably be assumed, the system reliability is a product of subsystem reliabilities

$$R_s = R_1 R_2 \cdots R_k \tag{1-15}$$

In the foregoing definition there are $2^k - k - 1$ conditions to be satisfied. Consider three events, A, B, and C. These are independent if and only if $P(A \cap B) = P(A) \cdot P(B)$, $P(A \cap C) = P(A) \cdot P(C)$, $P(B \cap C) = P(B) \cdot P(C)$, and $P(A \cap B \cap C) = P(A) \cdot P(B) \cdot P(C)$. The following example illustrates a case where events are pairwise independent but not mutually independent.

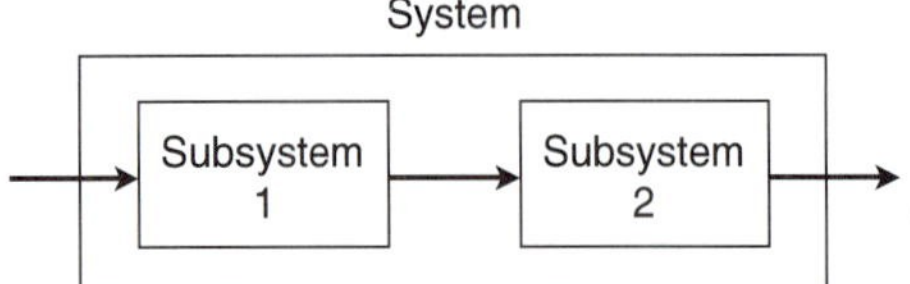

Figure 1-11 A simple serial system.

Example 1-38

Suppose the sample space, with equally likely outcomes, for a particular experiment is as follows:

$$\mathcal{S} = \{(0, 0, 0), (0, 1, 1), (1, 0, 1), (1, 1, 0)\}.$$

Let A_0: First digit is zero.
A_1: First digit is one.
B_0: Second digit is zero.
B_1: Second digit is one.
C_0: Third digit is zero.
C_1: Third digit is one.

It follows that

$$P(A_0) = P(A_1) = P(B_0) = P(B_1) = P(C_0) = P(C_1) = \frac{1}{2},$$

and it is easily seen that

$$P(A_i \cap B_j) = \frac{1}{4} = P(A_i) \cdot P(B_j), \qquad i = 0,1, \; j = 0,1,$$

$$P(A_i \cap C_j) = \frac{1}{4} = P(A_i) \cdot P(C_j), \qquad i = 0,1, \; j = 0,1,$$

$$P(B_i \cap C_j) = \frac{1}{4} = P(B_i) \cdot P(C_j), \qquad i = 0,1, \; j = 0,1.$$

However, we note that

$$P(A_0 \cap B_0 \cap C_0) = \frac{1}{4} \neq P(A_0) \cdot P(B_0) \cdot P(C_0),$$

$$P(A_0 \cap B_0 \cap C_1) = 0 \neq P(A_0) \cdot P(B_0) \cdot P(C_1),$$

and there are other triplets to which this violation could be extended.

The concept of *independent experiments* is introduced to complete this section. If we consider two experiments denoted $\mathscr{E}_1$ and $\mathscr{E}_2$ and let A_1 and A_2 be arbitrary events defined on the respective sample spaces $\mathcal{S}_1$ and $\mathcal{S}_2$ of the two experiments, then the following definition can be given.

Definition

If $P(A_1 \cap A_2) = P(A_1) \cdot P(A_2)$, then $\mathscr{E}_1$ and $\mathscr{E}_2$ are said to be independent experiments.

1-8 PARTITIONS, TOTAL PROBABILITY, AND BAYES' THEOREM

A *partition* of the sample space may be defined as follows.

Definition

If $B_1, B_2 \ldots, B_k$ are disjoint subsets of $\mathcal{S}$ (mutually exclusive events), and if $B_1 \cup B_2 \cup \cdots \cup B_k = \mathcal{S}$, then these subsets are said to form a partition of $\mathcal{S}$.

When the experiment is performed, one and only one of the events, B_i, occurs if we have a partition of $\mathcal{S}$.

Example 1-39

A particular binary "word" consists of five "bits," b_1, b_2, b_3, b_4, b_5, where $b_i = 0, 1, i = 1, 2, 3, 4, 5$. An experiment consists of transmitting a "word," and it follows that there are 32 possible words. If the events are

$$B_1 = \{(0, 0, 0, 0, 0), (0, 0, 0, 0, 1)\},$$
$$B_2 = \{(0, 0, 0, 1, 0), (0, 0, 0, 1, 1), (0, 0, 1, 0, 0), (0, 0, 1, 0, 1), (0, 0, 1, 1, 0), (0, 0, 1, 1, 1)\},$$
$$B_3 = \{(0, 1, 0, 0, 0), (0, 1, 0, 0, 1), (0, 1, 0, 1, 0), (0, 1, 0, 1, 1), (0, 1, 1, 0, 0), (0, 1, 1, 0, 1), (0, 1, 1, 1, 0), (0, 1, 1, 1, 1)\},$$
$$B_4 = \{(1, 0, 0, 0, 0), (1, 0, 0, 0, 1), (1, 0, 0, 1, 0), (1, 0, 0, 1, 1), (1, 0, 1, 0, 0), (1, 0, 1, 0, 1), (1, 0, 1, 1, 0), (1, 0, 1, 1, 1)\},$$
$$B_5 = \{(1, 1, 0, 0, 0), (1, 1, 0, 0, 1), (1, 1, 0, 1, 0), (1, 1, 0, 1, 1), (1, 1, 1, 0, 0), (1, 1, 1, 0, 1), (1, 1, 1, 1, 0)\},$$
$$B_6 = \{(1, 1, 1, 1, 1)\},$$

then $\mathcal{S}$ is partitioned by the events B_1, B_2, B_3, B_4, B_5, and B_6.

In general, if k events B_i $(i = 1, 2, \ldots, k)$ form a partition and A is an arbitrary event with respect to $\mathcal{S}$, then we may write

$$A = (A \cap B_1) \cup (A \cap B_2) \cup \cdots \cup (A \cap B_k)$$

so that

$$P(A) = P(A \cap B_1) + P(A \cap B_2) + \cdots + P(A \cap B_k),$$

since the events $(A \cap B_i)$ are pairwise mutually exclusive. (See Fig. 1-12 for $k = 4$.) It does not matter that $A \cap B_i = \varnothing$ for some or all of the i, since $P(\varnothing) = 0$.

Using the results of equation 1-12 we can state the following theorem.

Theorem 1-8

If $B_1, B_2, \ldots, B_k$ represents a partition of $\mathcal{S}$ and A is an arbitrary event on $\mathcal{S}$, then the *total probability* of A is given by

$$P(A) = P(B_1) \cdot P(A \mid B_1) + P(A \mid B_2) + \cdots + P(B_k) \cdot P(A \mid B_k) = \sum_{i=1}^{k} P(B_i)P(A \mid B_i).$$

The result of Theorem 1-8, also known as the law of total probability, is very useful, as there are numerous practical situations in which $P(A)$ cannot be computed directly. However,

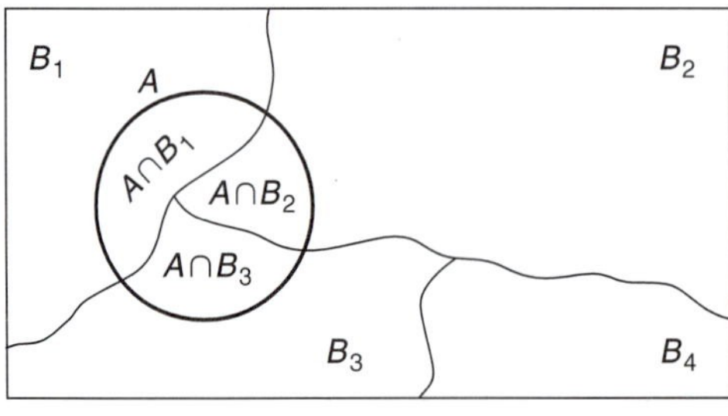

Figure 1-12 Partition of $\mathcal{S}$.

with the information that B_i has occurred, it is possible to evaluate $P(A|B_i)$ and thus determine $P(A)$ when the values $P(B_i)$ are obtained.

Another important result of the total probability law is known as *Bayes' theorem.*

Theorem 1-9

If $B_1, B_2, \ldots, B_k$ constitute a partition of the sample space $\mathscr{S}$ and A is an arbitrary event on $\mathscr{S}$, then for $r = 1, 2, \ldots, k$,

$$P(B_r|A) = \frac{P(B_r) \cdot P(A|B_r)}{\sum_{i=1}^{k} P(B_i) \cdot P(A|B_i)}. \tag{1-16}$$

Proof

$$P(B_r|A) = \frac{P(B_r \cap A)}{P(A)}$$

$$= \frac{P(B_r) \cdot P(A|B_r)}{\sum_{i=1}^{k} P(B_i) \cdot P(A|B_i)}.$$

The numerator is a result of equation 1-12 and the denominator is a result of Theorem 1.8.

Example 1-40

Three facilities supply microprocessors to a manufacturer of telemetry equipment. All are supposedly made to the same specifications. However, the manufacturer has for several years tested the microprocessors, and records indicate the following information:

Supplying Facility	Fraction Defective	Fraction Supplied
1	0.02	0.15
2	0.01	0.80
3	0.03	0.05

The manufacturer has stopped testing because of the costs involved, and it may be reasonably assumed that the fractions that are defective and the inventory mix are the same as during the period of record keeping. The director of manufacturing randomly selects a microprocessor, takes it to the test department, and finds that it is defective. If we let A be the event that an item is defective, and B_i be the event that the item came from facility i ($i = 1, 2, 3$), then we can evaluate $P(B_i|A)$. Suppose, for instance, that we are interested in determining $P(B_3|A)$. Then

$$P(B_3|A) = \frac{P(B_3) \cdot P(A|B_3)}{P(B_1) \cdot P(A|B_1) + P(B_2) \cdot P(A|B_2) + P(B_3) \cdot P(A|B_3)}$$

$$= \frac{(0.05)(0.03)}{(0.15)(0.02) + (0.80)(0.01) + (0.05)(0.03)} = \frac{3}{25}.$$

1-9 SUMMARY

This chapter has introduced the concept of random experiments, the sample space and events, and has presented a formal definition of probability. This was followed by methods to assign probability to events. Theorems 1-1 to 1-6 provide results for dealing with the probability of special events. Finite sample spaces with their special properties were discussed, and enumeration methods were reviewed for use in assigning probability to events in the case of equally likely experimental outcomes. Conditional probability was defined and illustrated, along with the concept of independent events. In addition, we considered partitions of the sample space, total probability, and Bayes' theorem. The concepts presented in this chapter form an important background for the rest of the book.

1-10 EXERCISES

1-1. Television sets are given a final inspection following assembly. Three types of defects are identified, critical, major, and minor defects, and are coded A, B, and C, respectively, by a mail-order house. Data are analyzed with the following results.

Sets having only critical defects	2%
Sets having only major defects	5%
Sets having only minor defects	7%
Sets having only critical *and* major defects	3%
Sets having only critical *and* minor defects	4%
Sets having only major *and* minor defects	3%
Sets having all three types of defects	1%

(a) What fraction of the sets has no defects?

(b) Sets with either critical defects or major defects (or both) get a complete rework. What fraction falls in this category?

1-2. Illustrate the following properties by shadings or colors on Venn diagrams.

(a) *Associative laws*: $A \cup (B \cup C) = (A \cup B) \cup C$, $A \cap (B \cap C) = (A \cap B) \cap C$.

(b) *Distributive laws*: $A \cup (B \cap C) = (A \cup B) \cap (A \cup C)$, $A \cap (B \cup C) = (A \cap B) \cup (A \cap C)$.

(c) If $A \subset B$, then $A \cap B = A$.

(d) If $A \subset B$, then $A \cup B = B$.

(e) If $A \cap B = \varnothing$, then $A \subset \overline{B}$.

(f) If $A \subset B$ and $B \subset C$, then $A \subset C$.

1-3. Consider a universal set consisting of the integers 1 through 10, or $U = \{1, 2, 3, 4, 5, 6, 7, 8, 9, 10\}$. Let $A = \{2, 3, 4\}$, $B = \{3, 4, 5\}$, and $C = \{5, 6, 7\}$. By enumeration, list the membership of the following sets.

(a) $\overline{\overline{A}} \cap B$.

(b) $\overline{A} \cup B$.

(c) $\overline{\overline{A} \cap \overline{B}}$.

(d) $\overline{A \cap (B \cap C)}$.

(e) $\overline{A \cap (B \cup C)}$.

1-4. A flexible circuit is selected at random from a production run of 1000 circuits. Manufacturing defects are classified into three different types, labeled A, B, and C. Type A defects occur 2% of the time, type B defects occur 1% of the time, and type C occur 1.5% of the time. Furthermore, it is known that 0.5% have both type A and B defects, 0.6% have both A and C defects, and 0.4% have B and C defects, while 0.2% have all three defects. What is the probability that the flexible circuit selected has at least one of the three types of defects?

1-5. In a human-factors laboratory, the reaction times of human subjects are measured as the elapsed time from the instant a position number is displayed on a digital display until the subject presses a button located at the position indicated. Two subjects are involved, and times are measured in seconds for each subject (t_1, t_2). What is the sample space for this experiment? Present the following events as subsets and mark them on a diagram: $(t_1 + t_2)/2 \le 0.15$, $\max(t_1, t_2) \le 0.15$, $|t_1 - t_2| \le 0.06$.

1-6. During a 24-hour period, a computer is to be accessed at time X, used for some processing, and exited at time $Y \ge X$. Take X and Y to be measured in hours on the time line with the beginning of the 24-hour period as the origin. The experiment is to observe X and Y.

(a) Describe the sample space $\mathscr{S}$.

(b) Sketch the following events in the X, Y plane.

(i) The time of use is 1 hour or less.

(ii) The access is before t_1 and the exit is after t_2, where $0 \leq t_1 < t_2 \leq 24$.

(iii) The time of use is less than 20% of the period.

1-7. Diodes from a batch are tested one at a time and marked either defective or nondefective. This is continued until either two defective items are found or five items have been tested. Describe the sample space for this experiment.

1-8. A set has four elements, $A = \{a, b, c, d\}$. Describe the power set $\{0, 1\}^A$.

1-9. Describe the sample space for each of the following experiments.

(a) A lot of 120 battery lids for pacemaker cells is known to contain a number of defectives because of a problem with the barrier material applied to the glassed-in feed-through. Three lids are randomly selected (without replacement) and are carefully inspected following a cut down.

(b) A pallet of 10 castings is known to contain one defective and nine good units. Four castings are randomly selected (without replacement) and inspected.

1-10. The production manager of a certain company is interested in testing a finished product, which is available in lots of size 50. She would like to rework a lot if she can be reasonably sure that 10% of the items are defective. She decides to select a random sample of 10 items without replacement and rework the lot if it contains one or more defective items. Does this procedure seem reasonable?

1-11. A trucking firm has a contract to ship a load of goods from city *W* to city *Z*. There are no direct routes connecting *W* to *Z*, but there are six roads from *W* to *X* and five roads from *X* to *Z*. How many total routes are there to be considered?

1-12. A state has one million registered vehicles and is considering using license plates with six symbols where the first three are letters and the last three are digits. Is this scheme feasible?

1-13. The manager of a small plant wishes to determine the number of ways he can assign workers to the first shift. He has 15 workers who can serve as operators of the production equipment, eight who can serve as maintenance personnel, and four who can be supervisors. If the shift requires six operators, two maintenance personnel, and one supervisor, how many ways can the first shift be manned?

1-14. A production lot has 100 units of which 20 are known to be defective. A random sample of four units is selected without replacement. What is the probability that the sample will contain no more than two defective units?

1-15. In inspecting incoming lots of merchandise, the following inspection rule is used where the lots contain 300 units. A random sample of 10 items is selected. If there is no more than one defective item in the sample, the lot is accepted. Otherwise it is returned to the vendor. If the fraction defective in the original lot is p', determine the probability of accepting the lot as a function of p'.

1-16. In a plastics plant 12 pipes empty different chemicals into a mixing vat. Each pipe has a five-position gauge that measures the rate of flow into the vat. One day, while experimenting with various mixtures, a solution is obtained that emits a poisonous gas. The settings on the gauges were not recorded. What is the probability of obtaining this same solution when randomly experimenting again?

1-17. Eight equally skilled men and women are applying for two jobs. Because the two new employees must work closely together, their personalities should be compatible. To achieve this, the personnel manager has administered a test and must compare the scores for each possibility. How many comparisons must the manager make?

1-18. By accident, a chemist combined two laboratory substances that yielded a desirable product. Unfortunately, her assistant did not record the names of the ingredients. There are forty substances available in the lab. If the two in question must be located by successive trial-and-error experiments, what is the maximum number of tests that might be made?

1-19. Suppose, in the previous problem, a known catalyst was used in the first accidental reaction. Because of this, the order in which the ingredients are mixed is important. What is the maximum number of tests that might be made?

1-20. A company plans to build five additional warehouses at new locations. Ten locations are under consideration. How many total possible choices are there for the set of five locations?

1-21. Washing machines can have five kinds of major and five kinds of minor defects. In how many ways can one major and one minor defect occur? In how many ways can two major and two minor defects occur?

1-22. Consider the diagram at the top of the next page of an electronic system, which shows the probabilities of the system components operating properly. The entire system operates if assembly III and at least one of the components in each of assemblies I and II

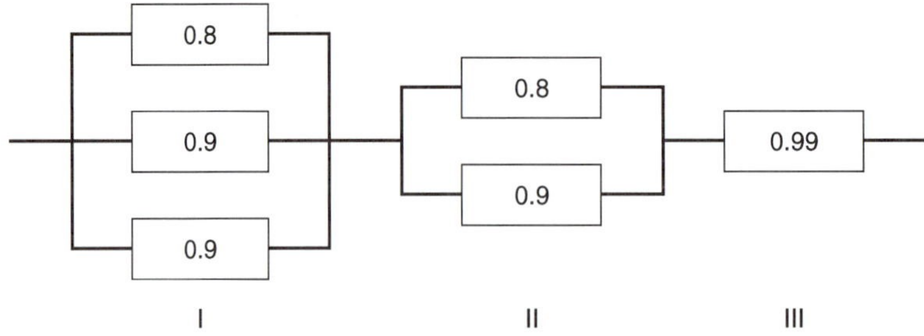

operates. Assume that the components of each assembly operate independently and that the assemblies operate independently. What is the probability that the entire system operates?

1-23. How is the probability of system operation affected if, in the foregoing problem, the probability of successful operation for the component in assembly III changes from 0.99 to 0.9?

1-24. Consider the series-parallel assembly shown below. The values R_i $(i = 1, 2, 3, 4, 5)$ are the reliabilities for the five components shown, that is, R_i = probability that unit i will function properly. The components operate (and fail) in a mutually independent manner and the assembly fails only when the path from A to B is broken. Express the assembly reliability as a function of R_1, R_2, R_3, R_4, and R_5.

1-25. A political prisoner is to be exiled to either Siberia or the Urals. The probabilities of being sent to these places are 0.6 and 0.4, respectively. It is also known that if a resident of Siberia is selected at random the probability is 0.5 that he will be wearing a fur coat, whereas the probability is 0.7 that a resident of the Urals will be wearing one. Upon arriving in exile, the first person the prisoner sees is not wearing a fur coat. What is the probability he is in Siberia?

1-26. A braking device designed to prevent automobile skids may be broken down into three series subsystems that operate independently: an electronics system, a hydraulic system, and a mechanical activator. On a particular braking, the reliabilities of these units are approximately 0.995, 0.993, and 0.994, respectively. Estimate the system reliability.

1-27. Two balls are drawn from an urn containing m balls numbered from 1 to m. The first ball is kept if it is numbered 1 and returned to the urn otherwise. What is the probability that the second ball drawn is numbered 2?

1-28. Two digits are selected at random from the digits 1 through 9 and the selection is without replacement (the same digit cannot be picked on both selections). If the sum of the two digits is even, find thc probability that both digits are odd.

1-29. At a certain university, 20% of the men and 1% of the women are over 6 feet tall. Furthermore, 40% of the students are women. If a student is randomly picked and is observed to be over 6 feet tall, what is the probability that the student is a woman?

1-30. At a machine center there are four automatic screw machines. An analysis of past inspection records yields the following data.

Machine	Percent Production	Percent Defectives Produced
1	15	4
2	30	3
3	20	5
4	35	2

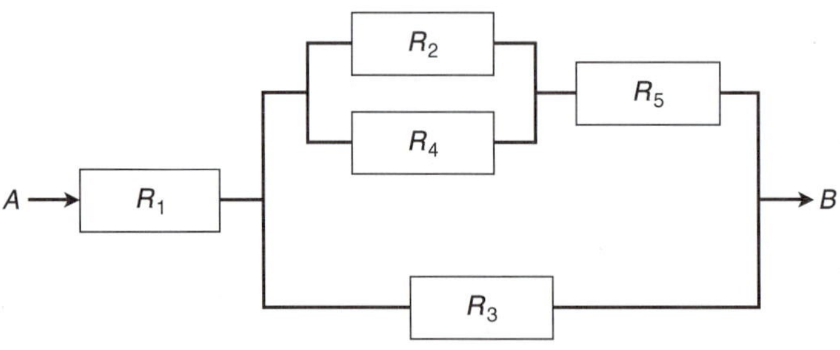

Machines 2 and 4 are newer and more production has been assigned to them than to machines 1 and 3. Assume that the current inventory mix reflects the production percentages indicated.

(a) If a screw is randomly picked from inventory, what is the probability that it will be defective?

(b) If a screw is picked and found to be defective, what is the probability that it was produced by machine 3?

1-31. A point is selected at random inside a circle. What is the probability that the point is closer to the center than to the circumference?

1-32. Complete the details of the proof for Theorem 1-4 in the text.

1-33. Prove Theorem 1-5.

1-34. Prove the second and third parts of Theorem 1-7.

1-35. Suppose there are n people in a room. If a list is made of all their birthdays (the specific month and day of the month), what is the probability that two or more persons have the same birthday? Assume there are 365 days in the year and that each day is equally likely to occur for any person's birthday. Let B be the event that two or more persons have the same birthday. Find $P(B)$ and $P(\overline{B})$ for $n = 10, 20, 21, 22, 23, 24, 25, 30, 40, 50$, and 60.

1-36. In a certain dice game, players continue to throw two dice until they either win or lose. The player wins on the first throw if the sum of the two upturned faces is either 7 or 11 and loses if the sum is 2, 3, or 12. Otherwise, the sum of the faces becomes the player's "point." The player continues to throw until the first succeeding throw on which he makes his point (in which case he wins), or until he throws a 7 (in which case he loses). What is the probability that the player with the dice will eventually win the game?

1-37. The industrial engineering department of the XYZ Company is performing a work sampling study on eight technicians. The engineer wishes to randomize the order in which he visits the technicians' work areas. In how many ways may he arrange these visits?

1-38. A hiker leaves point A shown in the figure below, choosing at random one path from AB, AC, AD, and AE. At each subsequent junction she chooses another path at random. What is the probability that she arrives at point X?

1-39. Three printers do work for the publications office of Georgia Tech. The publications office does not negotiate a contract penalty for late work, and the data below reflect a large amount of experience with these printers.

Printer i	Fraction of Contracts Held by Printer i	Fraction of Deliveries More than One Month Late
1	0.2	0.2
2	0.3	0.5
3	0.5	0.3

A department observes that its recruiting booklet is more than a month late. What is the probability that the contract is held by printer 3?

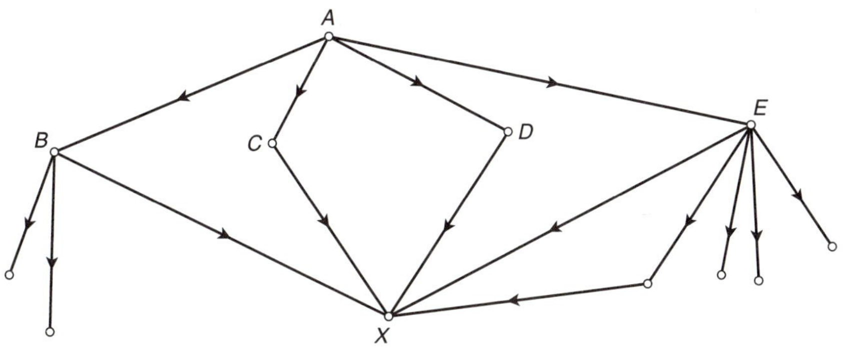

1-40. Following aircraft accidents, a detailed investigation is conducted. The probability that an accident due to structural failure is correctly identified is 0.9 and the probability that an accident that is not due to structural failure being identified incorrectly as due to structural failure is 0.2. If 25% of all aircraft accidents are due to structural failures, find the probability that an aircraft accident is due to structural failure *given* that it has been diagnosed as due to structural failure.

Chapter 2

One-Dimensional Random Variables

2-1 INTRODUCTION

The objectives of this chapter are to introduce the concept of random variables, to define and illustrate probability distributions and cumulative distribution functions and to present useful characterizations for random variables.

When describing the sample space of a random experiment, it is not necessary to specify that an individual outcome be a number. In several examples we observe this, such as in Example 1-9, where a true coin is tossed three times and the sample space is $\mathscr{S} = \{HHH, HHT, HTH, HTT, THH, THT, TTH, TTT\}$, or Example 1-15, where probes of solder joints yield a sample space $\mathscr{S} = \{GG, GD, DG, DD\}$.

In most experimental situations, however, we are interested in numerical outcomes. For example, in the illustration involving coin tossing we might assign some real number x to every element of the sample space. In general, we want to assign a real number x to every outcome e of the sample space $\mathscr{S}$. A functional notation will be used initially, so that $x = X(e)$, where X is the function. The domain of X is $\mathscr{S}$, and the numbers in the range are real numbers. The function X is called a random variable. Figure 2-1 illustrates the nature of this function.

Definition

If $\mathscr{E}$ is an experiment having sample space $\mathscr{S}$, and X is a *function* that assigns a real number $X(e)$ to every outcome $e \in \mathscr{S}$, then $X(e)$ is called a *random variable.*

Example 2-1

Consider the coin-tossing experiment discussed in the preceding paragraphs. If X is the number of heads showing, then $X(HHH) = 3$, $X(HHT) = 2$, $X(HTH) = 2$, $X(HTT) = 1$, $X(THH) = 2$, $X(THT) = 1$, $X(TTH) = 1$, and $X(TTT) = 0$. The range space $R_X = \{x: x = 0, 1, 2, 3\}$ in this example (see Fig. 2-2).

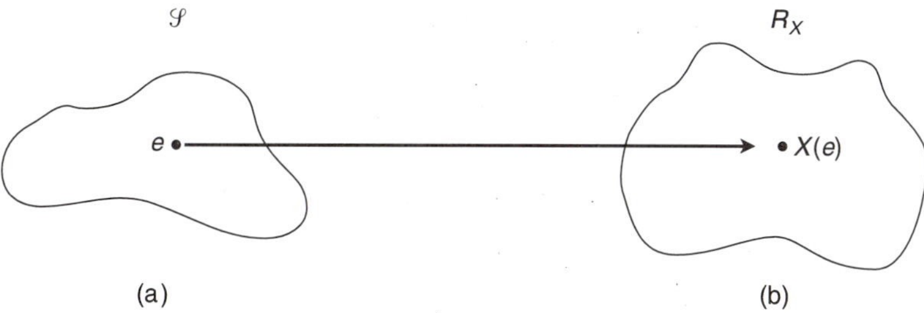

Figure 2-1 The concept of a random variable. (*a*) $\mathscr{S}$: The sample space of $\mathscr{E}$. (*b*) R_X: The range space of X.

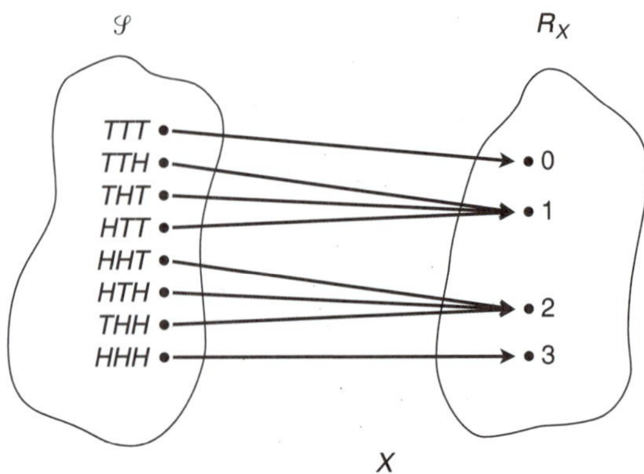

Figure 2-2 The number of heads in three coin tosses.

The reader should recall that for all functions and for every element in the domain, there is exactly one value in the range. In the case of the random variable, for every outcome $e \in \mathcal{S}$ there corresponds exactly one value $X(e)$. It should be noted that different values of e may lead to the same x, as was the case where $X(TTH) = 1$, $X(THT) = 1$, and $X(HTT) = 1$ in the preceding example.

Where the outcome in $\mathcal{S}$ is already the numerical characteristic desired, then $X(e) = e$, the identity function. Example 1-13, in which a cathode ray tube was aged to failure, is a good example. Recall that $\mathcal{S} = \{t: t \geq 0\}$. If X is the time to failure, then $X(t) = t$. Some authors call this type of sample space a *numerical-valued phenomenon.*

The range space, R_X, is made up of all possible values of X, and in subsequent work it will not be necessary to indicate the functional nature of X. Here we are concerned with events that are associated with R_X, and the random variable X will induce probabilities onto these events. If we return again to the coin-tossing experiment for illustration and assume the coin to be true, there are eight equally likely outcomes: *HHH, HHT, HTH, HTT, THH, THT, TTH,* and *TTT*, each having probability $\frac{1}{8}$. Now suppose A is the event "exactly two heads" and, as previously, we let X represent the number of heads (see Fig. 2-2). The event that $(X = 2)$ relates to R_X, not $\mathcal{S}$; however, $P_X(X = 2) = P(A) = \frac{3}{8}$, since $A = \{HHT, HTH, THH\}$ is the equivalent event in $\mathcal{S}$, and probability was defined on events in the sample space. The random variable X induced the probability of $\frac{3}{8}$ to the event $(X = 2)$. Note that parentheses will be used to denote an event in the range of the random variable, and in general we will write $P_X(X = x)$.

In order to generalize this notion, consider the following definition.

Definition

If $\mathcal{S}$ is the sample space of an experiment $\mathcal{E}$ and a random variable X with range space R_X is defined on $\mathcal{S}$, and furthermore if event A is an event in $\mathcal{S}$ while event B is an event in R_X, then A and B are equivalent events if

$$A = \{e \in \mathcal{S}: X(e) \in B\}.$$

Figure 2-3 illustrates this concept.

More simply, if event A in $\mathcal{S}$ consists of all outcomes in $\mathcal{S}$ for which $X(e) \in B$, then A and B are equivalent events. Whenever A occurs, B occurs; and whenever B occurs, A occurs. Note that A and B are associated with different spaces.

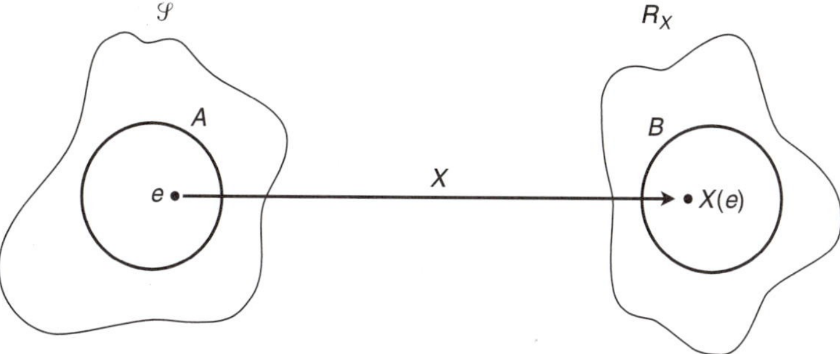

Figure 2-3 Equivalent events.

Definition

If A is an event in the sample space and B is an event in the range space R_X of the random variable X, then we define the probability of B as

$$P_X(B) = P(A), \qquad \text{where} \qquad A = \{e \in \mathcal{S}: X(e) \in B\}.$$

With this definition, we may assign probabilities to events in R_X in terms of probabilities defined on events in $\mathcal{S}$, and we will *suppress* the function X, so that $P_X(X = 2) = \frac{3}{8}$ in the familiar coin-tossing example means that there is an event $A = \{HHT, HTH, THH\} = \{e: X(e) = 2\}$ in the sample space with probability $\frac{3}{8}$. In subsequent work, we will not deal with the nature of the function X, since we are interested in the values of the range space and their associated probabilities. While outcomes in the sample space may not be real numbers, it is noted again that all elements of the range of X are real numbers.

An alternative but similar approach makes use of the inverse of the function X. We would simply define $X^{-1}(B)$ as

$$X^{-1}(B) = \{e \in \mathcal{S}: X(e) \in B\}$$

so that

$$P_X(B) = P(X^{-1}(B)) = P(A).$$

Table 2-1 Equivalent Events

Some Events in R_Y	Equivalent Events in $\mathcal{S}$	Probability
$Y = 2$	{(1, 1)}	$\frac{1}{36}$
$Y = 3$	{(1, 2), (2, 1)}	$\frac{2}{36}$
$Y = 4$	{(1, 3), (2, 2), (3, 1)}	$\frac{3}{36}$
$Y = 5$	{(1, 4), (2, 3), (3, 2), (4, 1)}	$\frac{4}{36}$
$Y = 6$	{(1, 5), (2, 4), (3, 3), (4, 2), (5, 1)}	$\frac{5}{36}$
$Y = 7$	{(1, 6), (2, 5), (3, 4), (4, 3), (5, 2), (6, 1)}	$\frac{6}{36}$
$Y = 8$	{(2, 6), (3, 5), (4, 4), (5, 3), (6, 2)}	$\frac{5}{36}$
$Y = 9$	{(3, 6), (4, 5), (5, 4), (6, 3)}	$\frac{4}{36}$
$Y = 10$	{(4, 6), (5, 5), (6, 4)}	$\frac{3}{36}$
$Y = 11$	{(5, 6), (6, 5)}	$\frac{2}{36}$
$Y = 12$	{(6, 6)}	$\frac{1}{36}$

The following examples illustrate the sample space–range space relationship, and the concern with the range space rather than the sample space is evident, since numerical results are of interest.

Example 2-2

Consider the tossing of two true dice as described in Example 1-11. (The sample space was described in Chapter 1.) Suppose we define a random variable Y as the sum of the "up" faces. Then $R_Y = \{2, 3, 4, 5, 6, 7, 8, 9, 10, 11, 12\}$, and the probabilities are $\left(\frac{1}{36}, \frac{2}{36}, \frac{3}{36}, \frac{4}{36}, \frac{5}{36}, \frac{6}{36}, \frac{5}{36}, \frac{4}{36}, \frac{3}{36}, \frac{2}{36}, \frac{1}{36}\right)$, respectively. Table 2-1 shows equivalent events. The reader will recall that there are 36 outcomes, which, since the dice are true, are equally likely.

Example 2-3

One hundred cardiac pacemakers were placed on life test in a saline solution held as close to body temperature as possible. The test is functional, with pacemaker output monitored by a system providing for output signal conversion to digital form for comparison against a design standard. The test was initiated on July 1, 1997. When a pacer output varies from the standard by as much as 10%, this is considered a failure and the computer records the date and the time of day (d, t). If X is the random variable "time to failure," then $\mathcal{S} = \{(d, t): d = \text{date}, t = \text{time}\}$ and $R_X = \{x: x \geq 0\}$. The random variable X is the total number of elapsed time units since the module went on test. We will deal directly with X and its probability law. This concept will be discussed in the following sections.

2-2 THE DISTRIBUTION FUNCTION

As a convention we will use a lowercase of the same letter to denote a particular value of a random variable. Thus $(X = x)$, $(X < x)$, $(X \leq x)$ are events in the range space of the random variable X, where x is a real number. The probability of the event $(X \leq x)$ may be expressed as a function of x as

$$F_X(x) = P_X(X \leq x). \tag{2-1}$$

This function F_X is called the *distribution function* or *cumulative function* or *cumulative distribution* function (CDF) of the random variable X.

Example 2-4

In the case of the coin-tossing experiment, the random variable X assumed four values, 0, 1, 2, 3, with probabilities $\frac{1}{8}, \frac{3}{8}, \frac{3}{8}, \frac{1}{8}$. We can state $F_X(x)$ as follows:

$$\begin{aligned} F_X(x) &= 0, & x &< 0, \\ &= \frac{1}{8}, & 0 &\leq x < 1, \\ &= \frac{4}{8}, & 1 &\leq x < 2, \\ &= \frac{7}{8}, & 2 &\leq x < 3, \\ &= 1, & x &\geq 3. \end{aligned}$$

A graphical representation is as shown in Fig. 2-4.

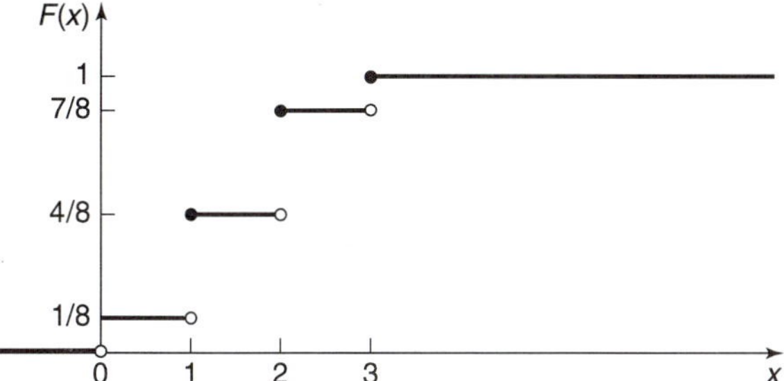

Figure 2-4 A distribution function for the number of heads in tossing three true coins.

Example 2-5

Again recall Example 1-13, when a cathode ray tube is aged to failure. Now $\mathcal{S} = \{t: t \geq 0\}$, and if we let X represent the elapsed time in hours to failure, then the event $(X \leq x)$ is in the range space of X. A mathematical model that assigns probability to $(X \leq x)$ is

$$\begin{aligned} F_X(x) &= 0, & x &\leq 0, \\ &= 1 - e^{-\lambda x}, & x &> 0, \end{aligned}$$

where λ is a positive number called the failure rate (failures/hour). The use of this "exponential" model in practice depends on certain assumptions about the failure process. These assumptions will be presented in more detail later. A graphical representation of the cumulative distribution for the time to failure for the CRT is shown in Fig. 2-5.

Example 2-6

A customer enters a bank where there is a common waiting line for all tellers, with the individual at the head of the line going to the first teller that becomes available. Thus, as the customer enters, the waiting time prior to moving to the teller is assigned to the random variable X. If there is no one in line at the time of arrival and a teller is free, the waiting time is zero, but if others are waiting or all tellers are busy, then the waiting time will assume some positive value. Although the mathematical form of F_X depends on assumptions about this service system, a general graphical representation is as illustrated in Fig. 2-6.

Cumulative distribution functions have the following properties, which follow directly from the definition:

1. $0 \leq F_X(x) \leq 1, \qquad -\infty < x < \infty.$

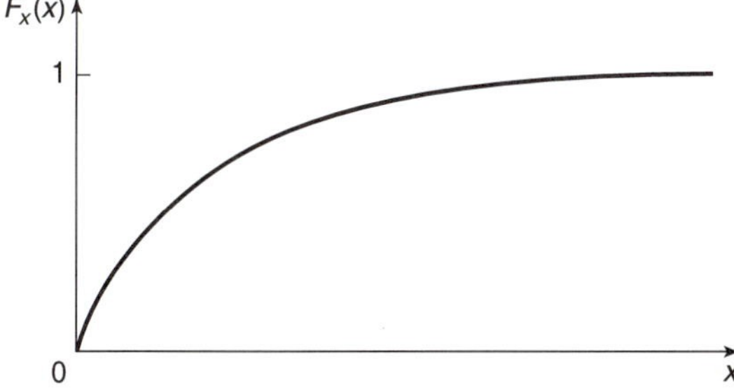

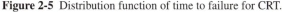

Figure 2-5 Distribution function of time to failure for CRT.

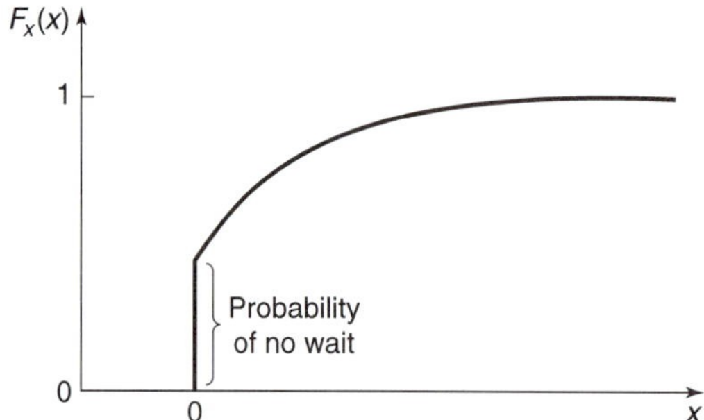

Figure 2-6 Waiting-time distribution function.

2. $\lim_{x\to\infty} F_X(x) = 1$,
   $\lim_{x\to-\infty} F_X(x) = 0$.
3. The function is nondecreasing. That is, if $x_1 \le x_2$, then $F_X(x_1) \le F_X(x_2)$.
4. The function is continuous from the right. That is, for all x and $\delta > 0$,

$$\lim_{\delta\to 0}\left[F_X(x+\delta) - F_X(x)\right] = 0.$$

In reviewing the last three examples, we note that in Example 2-4, the values of x for which there is an increase in $F_X(x)$ were integers, and where x is not an integer, then $F_X(x)$ has the value that it had at the nearest integer x to the left. In this case, $F_X(x)$ has a *saltus* or jump at the values 0, 1, 2, and 3 and proceeds from 0 to 1 in a series of such jumps. Example 2-5 illustrates a different situation, where $F_X(x)$ proceeds smoothly from 0 to 1 and is continuous everywhere but not differentiable at $x = 0$. Finally Example 2-6 illustrates a situation where there is a saltus at $x = 0$, and for $x > 0$, $F_X(x)$ is continuous.

Using a simplified form of results from the *Lebesgue decomposition theorem*, it is noted that we can represent $F_X(x)$ as the sum of two component functions, say $G_X(x)$ and $H_X(x)$, or

$$F_X(x) = G_X(x) + H_X(x), \tag{2-2}$$

where $G_X(x)$ is continuous and $H_X(x)$ is a right-hand continuous step function with jumps coinciding with those of $F_X(x)$, and $H_X(-\infty) = 0$. If $G_X(x) \equiv 0$ for all x, then X is called a *discrete random variable,* and if $H_X(x) \equiv 0$, then X is called a *continuous random variable.* Where neither situation holds, X is called a *mixed type random variable*, and although this was illustrated in Example 2-6, this text will concentrate on purely discrete and continuous random variables, since most of the engineering and management applications of statistics and probability in this book relate either to counting or to simple measurement.

2-3 DISCRETE RANDOM VARIABLES

Although discrete random variables may result from a variety of experimental situations, in engineering and applied science they are often associated with counting. If X is a discrete random variable, then $F_X(x)$ will have at most a countably infinite number of jumps and $R_X = \{x_1, x_2, \ldots, x_k, \ldots\}$.

Example 2-7

Suppose that the number of working days in a particular year is 250 and that the records of employees are marked for each day they are absent from work. An experiment consists of randomly selecting a record to observe the days marked absent. The random variable X is defined as the number of days absent, so that $R_X = \{0, 1, 2, \ldots, 250\}$. This is an example of a discrete random variable with a finite number of possible values.

Example 2-8

A Geiger counter is connected to a gas tube in such a way that it will record the background radiation count for a selected time interval $[0, t]$. The random variable of interest is the count. If X denotes the random variable, then $R_X = \{0, 1, 2, \ldots, k, \ldots\}$, and we have, at least conceptually, a countably infinite range space (outcomes can be placed in a one-to-one correspondence with the natural numbers) so that the random variable is discrete.

Definition

If X is a discrete random variable, we associate a number $p_X(x_i) = P_X(X = x_i)$ with each outcome x_i in R_X for $i = 1, 2, \ldots, n, \ldots$, where the numbers $p_X(x_i)$ satisfy the following:

1. $p_X(x_i) \geq 0$ for all i.
2. $\sum_{i=1}^{\infty} p_X(x_i) = 1$.

We note immediately that

$$p_X(x_i) = F_X(x_i) - F_X(x_{i-1}) \tag{2-3}$$

and

$$F_X(x_i) = P_X(X \leq x_i) = \sum_{x \leq x_i} p_X(x). \tag{2-4}$$

The function p_X is called the *probability function* or *probability mass function* or *probability law* of the random variable, and the collection of pairs $[(x_i, p_X(x_i)), i = 1, 2, \ldots]$ is called the *probability distribution* of X. The function p_X is usually presented in either *tabular, graphical,* or *mathematical* form, as illustrated in the following examples.

Example 2-9

For the coin-tossing experiment of Example 1-9, where X = the number of heads, the probability distribution is given in both tabular and graphical form in Fig. 2-7. It will be recalled that $R_X = \{0, 1, 2, 3\}$.

Tabular Presentation

x	$p(x)$
0	1/8
1	3/8
2	3/8
3	1/8

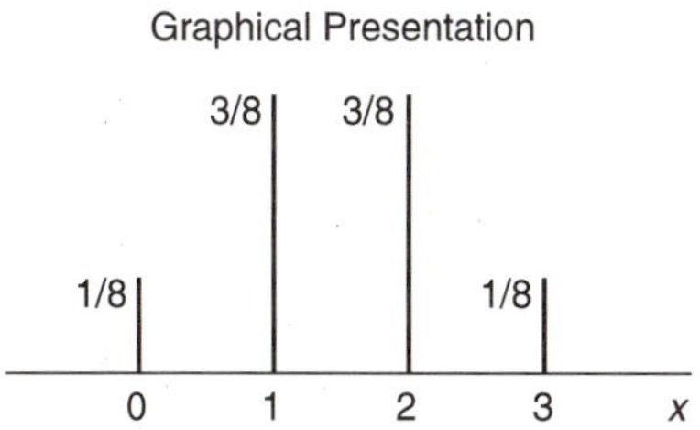

Graphical Presentation

Figure 2-7 Probability distribution for coin-tossing experiment.

Example 2-10

Suppose we have a random variable X with a probability distribution given by the relationship

$$p_X(x) = \binom{n}{x} p^x (1-p)^{n-x}, \qquad x = 0, 1, \ldots, n,$$
$$= 0, \qquad \text{otherwise,} \tag{2-5}$$

where n is a positive integer and $0 < p < 1$. This relationship is known as the *binomial distribution*, and it will be studied in more detail later. Although it would be possible to display this model in graphical or tabular form for particular n and p by evaluating $p_X(x)$ for $x = 0, 1, 2, \ldots, n$, this is seldom done in practice.

Example 2-11

Recall the earlier discussion of random sampling from a finite population without replacement. Suppose there are N objects of which D are defective. A random sample of size n is selected without replacement, and if we let X represent the number of defectives in the sample, then

$$p_X(x) = \frac{\binom{D}{x}\binom{N-D}{n-x}}{\binom{N}{n}}, \qquad x = 0, 1, 2, \ldots, \min(n, D),$$
$$= 0, \qquad \text{otherwise,} \tag{2-6}$$

This distribution is known as the *hypergeometric distribution.* In a particular case, suppose $N = 100$ items, $D = 5$ items, and $n = 4$; then

$$p_X(x) = \frac{\binom{5}{x}\binom{95}{4-x}}{\binom{100}{4}}, \qquad x = 0, 1, 2, 3, 4,$$
$$= 0, \qquad \text{otherwise,}$$

In the event that either tabular or graphical presentation is desired, this would be as shown in Fig. 2-8; however, unless there is some special reason to use these forms, we will use the mathematical relationship.

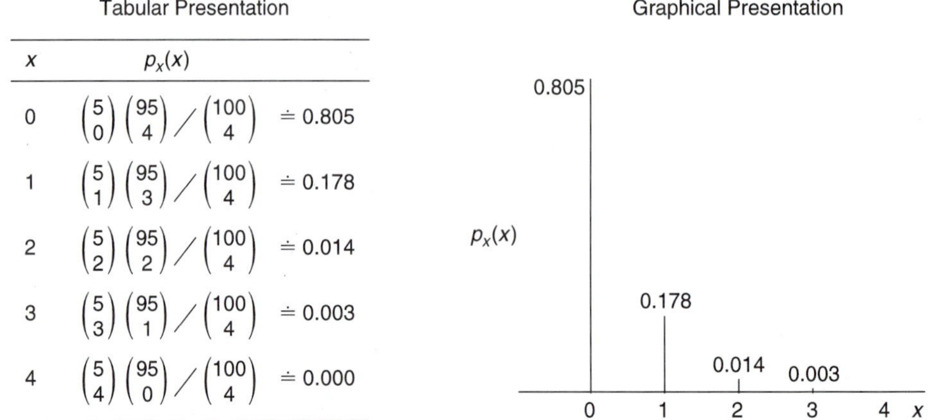

Tabular Presentation

x	$p_X(x)$	
0	$\binom{5}{0}\binom{95}{4}\Big/\binom{100}{4}$	$\doteq 0.805$
1	$\binom{5}{1}\binom{95}{3}\Big/\binom{100}{4}$	$\doteq 0.178$
2	$\binom{5}{2}\binom{95}{2}\Big/\binom{100}{4}$	$\doteq 0.014$
3	$\binom{5}{3}\binom{95}{1}\Big/\binom{100}{4}$	$\doteq 0.003$
4	$\binom{5}{4}\binom{95}{0}\Big/\binom{100}{4}$	$\doteq 0.000$

Figure 2-8 Some hypergeometric probabilities $N = 100$, $D = 5$, $n = 4$.

Example 2-12

In Example 2-8, where the Geiger counter was prepared for detecting the background radiation count, we might use the following relationship, which has been experimentally shown to be appropriate:

$$\begin{aligned} p_X(x) &= e^{-\lambda t}(\lambda t)^x/x!, \qquad x = 0, 1, 2, \ldots, \qquad \lambda > 0, \\ &= 0 \qquad \text{otherwise.} \end{aligned} \tag{2-7}$$

This is called the *Poisson distribution*, and at a later point it will be derived analytically. The parameter λ is the mean rate in "hits" per unit time, and x is the number of these "hits."

These examples have illustrated some discrete probability distributions and alternate means of presenting the pairs $[(x_i, p_X(x_i)), i = 1, 2, \ldots]$. In later sections a number of probability distributions will be developed, each from a set of postulates motivated from considerations of *real-world phenomena.*

A general graphical presentation of the discrete distribution from Example 2-12 is given in Fig. 2-9. This geometric interpretation is often useful in developing an intuitive feeling for discrete distributions. There is a close analogy to mechanics if we consider the probability distribution as a mass of one unit distributed over the real line in amounts $p_X(x_i)$ at points x_i, $i = 1, 2, \ldots, n$. Also, utilizing equation 2-3, we note the following useful result where $b \geq a$:

$$P_X(a < X \leq b) = F_X(b) - F_X(a). \tag{2-8}$$

2-4 CONTINUOUS RANDOM VARIABLES

Recall from Section 2-2 that where $H_X(x) \equiv 0$, X is called *continuous.* Then $F_X(x)$ is continuous, $F_X(x)$ has derivative $f_X(x) = (d/dx)\, F_X(x)$ for all x (with the exception of possibly a countable number of values), and $f_X(x)$ is piecewise continuous. Under these conditions, the range space R_X will consist of one or more intervals.

An interesting difference from the case of discrete random variables is that for $\delta > 0$,

$$P_X(X = x) = \lim_{\delta \to 0}\left[F_X(x+\delta) - F_X(x)\right] = 0. \tag{2-9}$$

We define the *probability density function* $f_X(x)$ as

$$f_X(x) = \frac{d}{dx} F_X(x), \tag{2-10}$$

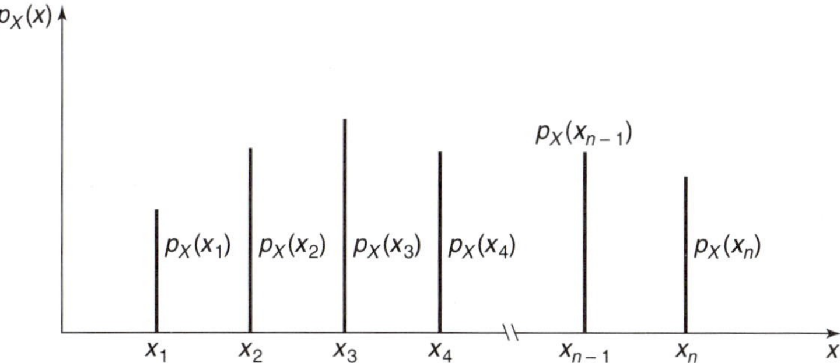

Figure 2-9 Geometric interpretation of a probability distribution.

and it follows that

$$F_X(x) = \int_{-\infty}^{x} f_X(t)dt. \tag{2-11}$$

We also note the close correspondence in this form with equation 2-4 with an integral replacing a summation symbol, and the following properties of $f_X(x)$:

1. $f_X(x) \geq 0$ for all $x \in R_X$.
2. $\int_{R_X} f_X(x)dx = 1$.
3. $f_X(x)$ is piecewise continuous.
4. $f_X(x) = 0$ if x is not in the range R_X.

These concepts are illustrated in Fig. 2-10. This definition of a density function stipulates a function f_X defined on R_X such that

$$P\{e \in \mathscr{S}: a \leq X(e) \leq b\} = P_x(a \leq X \leq b) = \int_a^b f_X(x)dx, \tag{2-12}$$

where e is an outcome in the sample space. We are concerned only with R_X and f_X. It is important to realize that $f_X(x)$ does not represent the probability of anything, and that only when the function is integrated between two points does it yield a probability.

Some comments about equation 2-9 may be useful, as this result may be counterintuitive. If we allow X to assume all values in some interval, then $P_X(X = x_0) = 0$ is not equivalent to saying the event $(X = x_0)$ in R_X is impossible. Recall that if $A = \varnothing$, then $P_X(A) = 0$; however, although $P_X(X = x_0) = 0$, the fact that the set $A = \{x: x = x_0\}$ is not empty clearly indicates that the converse is not true.

An immediate result of this is that $P_X(a \leq X \leq b) = P_X(a < X \leq b) = P_X(a < X < b) = P_X(a \leq X < b)$, where X is continuous, all given by $F_X(b) - F_X(a)$.

Example 2-13

The time to failure of the cathode ray tube described in Example 1-13 has the following probability density function:

$$\begin{aligned} f_T(t) &= \lambda e^{-\lambda t}, && t \geq 0, \\ &= 0 && \text{otherwise,} \end{aligned}$$

where $\lambda > 0$ is a constant known as the failure rate. This probability density function is called the *exponential density*, and experimental evidence has indicated that it is appropriate to describe the time

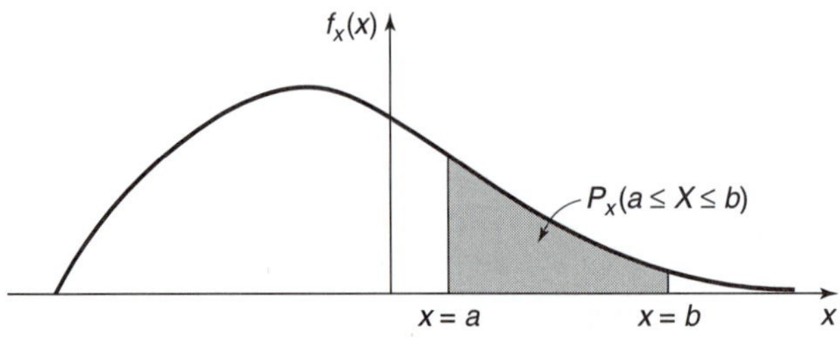

Figure 2-10 Hypothetical probability density function.

to failure (a real-world occurrence) for some types of components. In this example suppose we want to find $P_T(T \geq 100 \text{ hours})$. This is equivalent to stating $P_T(100 \leq T < \infty)$, and

$$\begin{aligned} P_T(T \geq 100) &= \int_{100}^{\infty} \lambda e^{-\lambda t}\, dt \\ &= e^{-100\lambda}. \end{aligned}$$

We might again employ the concept of conditional probability and determine $P_T(T \geq 100 | T > 99)$, the probability the tube lives at least 100 hours given that it has lived beyond 99 hours. From our earlier work,

$$P_T(T \geq 100 | T > 99) = \frac{P_T(T \geq 100 \text{ and } T > 99)}{P_T(T > 99)}$$

$$= \frac{\int_{100}^{\infty} \lambda e^{-\lambda t}\, dt}{\int_{99}^{\infty} \lambda e^{-\lambda t}\, dt} = \frac{e^{-100\lambda}}{e^{-99\lambda}} = e^{-\lambda}.$$

Example 2-14

A random variable X has the *triangular* probability density function given below and shown graphically in Fig. 2-11:

$$\begin{aligned} f_X(x) &= x, && 0 \leq x < 1, \\ &= 2 - x, && 1 \leq x < 2, \\ &= 0, && \text{otherwise.} \end{aligned}$$

The following are calculated for illustration:

1. $P_X\left(-1 < X < \frac{1}{2}\right) = \int_{-1}^{0} 0\, dx + \int_{0}^{1/2} x dx = \frac{1}{8}.$
2. $P_X\left(X \leq \frac{3}{2}\right) = \int_{-\infty}^{0} 0\, dx + \int_{0}^{1} x dx + \int_{1}^{3/2} (2 - x) dx$

   $= 0 + \frac{1}{2} + \left(2x - \frac{x^2}{2}\right)\Big|_1^{3/2} = \frac{7}{8}.$
3. $P_X(X \leq 3) = 1.$
4. $P_X(X \geq 2.5) = 0.$
5. $P_X\left(\frac{1}{4} < X < \frac{3}{2}\right) = \int_{1/4}^{1} x dx + \int_{1}^{3/2} (2 - x) dx$

   $= \frac{15}{32} + \frac{3}{8} = \frac{27}{32}.$

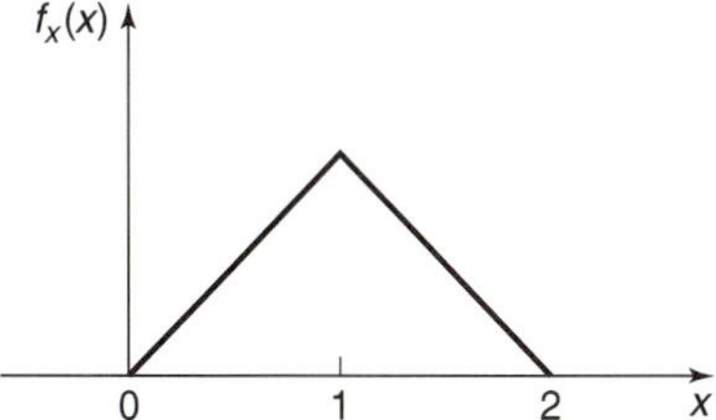

Figure 2-11 Triangular density function.

In describing probability density functions, a mathematical model is usually employed. A graphical or geometric presentation may also be useful. The area under the density function corresponds to probability, and the total area is one. Again the student familiar with mechanics might consider the probability of one to be distributed over the real line according to f_X. In Fig. 2-12, the intervals (a, b) and (b, c) are of the same length; however, the probability associated with (a, b) is greater.

2-5 SOME CHARACTERISTICS OF DISTRIBUTIONS

While a discrete distribution is completely specified by the pairs $[(x_i, p_X(x_i)); i = 1, 2, \ldots, n, \ldots]$, and a probability density function is likewise specified by $[(x, f_X(x)); x \in R_X]$, it is often convenient to work with some descriptive characteristics of the random variable. In this section we introduce two widely used descriptive measures, as well as a general expression for other similar measures. The first of these is the *first moment* about the origin. This is called the *mean* of the random variable and is denoted by the Greek letter μ, where

$$\begin{aligned}\mu &= \sum_i x_i p_X(x_i) && \text{for discrete } X,\\ &= \int_{-\infty}^{\infty} x f_X(x)\,dx && \text{for continuous } X. \qquad (2\text{-}13)\end{aligned}$$

This measure provides an indication of *central tendency* in the random variable.

Example 2-15

Returning to the coin-tossing experiment where X represents the number of heads and the probability distribution is as shown in Fig. 2-7, the calculation of μ yields

$$\mu = \sum_{i=1}^{4} x_i p_X(x_i) = 0\cdot\left(\frac{1}{8}\right) + 1\cdot\left(\frac{3}{8}\right) + 2\cdot\left(\frac{3}{8}\right) + 3\cdot\left(\frac{1}{8}\right) = \frac{3}{2},$$

as indicated in Fig. 2-13. In this particular example, because of symmetry, the value μ could have been easily determined from inspection. Note that the mean value in this example cannot be obtained as the output from a single trial.

Example 2-16

In Example 2-14, a density f_X was defined as

$$\begin{aligned} f_X(x) &= x, && 0 \le x \le 1,\\ &= 2 - x, && 1 \le x \le 2,\\ &= 0, && \text{otherwise.}\end{aligned}$$

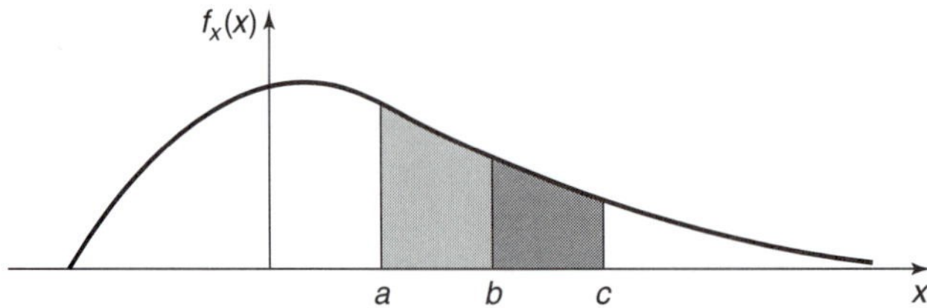

Figure 2-12 A density function.

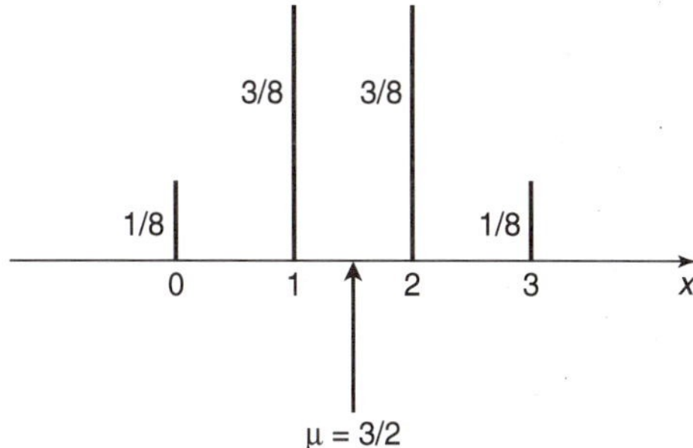

Figure 2-13 Calculation of the mean.

The mean is determined by

$$\mu = \int_0^1 x \cdot x dx + \int_1^2 x \cdot (2 - x) dx$$
$$+ \int_{-\infty}^0 x \cdot 0 dx + \int_2^{\infty} x \cdot 0 dx = 1,$$

another result that we could have determined via symmetry.

Another measure describes the spread or dispersion of the probability associated with elements in R_X. This measure is called the *variance*, denoted by σ^2, and is defined as follows:

$$\sigma^2 = \sum_i (x_i - \mu)^2 p_X(x_i) \qquad \text{for discrete } X,$$
$$= \int_{-\infty}^{\infty} (x - \mu)^2 f_X(x) dx \qquad \text{for continuous } X. \tag{2-14}$$

This is the second moment about the mean, and it corresponds to the moment of inertia in mechanics. Consider Fig. 2-14, where two hypothetical discrete distributions are shown in graphical form. Note that the mean is one in both cases. The variance for the discrete random variable shown in Fig. 2-14*a* is

$$\sigma^2 = (0-1)^2 \cdot \left(\frac{1}{4}\right) + (1-1)^2 \cdot \left(\frac{1}{2}\right) + (2-1)^2 \cdot \left(\frac{1}{4}\right) = \frac{1}{2},$$

and the variance of the discrete random variable shown in Fig. 2-14*b* is

$$\sigma^2 = (-1-1)^2 \cdot \frac{1}{5} + (0-1)^2 \cdot \frac{1}{5} + (1-1)^2 \cdot$$
$$+ (2-1)^2 \cdot \frac{1}{5} + (3-1)^2 \cdot \frac{1}{5} = 2,$$

which is four times as great as the variance of the random variable shown in Fig. 2-14*a*.

If the units on the random variable are, say feet, then the units of the mean are the same, but the units on the variance would be feet squared. Another measure of dispersion, called the *standard deviation*, is defined as the positive square root of the variance and denoted σ, where

$$\sigma = \sqrt{\sigma^2}. \tag{2-15}$$

It is noted that the units of σ are the same as those of the random variable, and a small value for σ indicates little dispersion whereas a large value indicates greater dispersion.

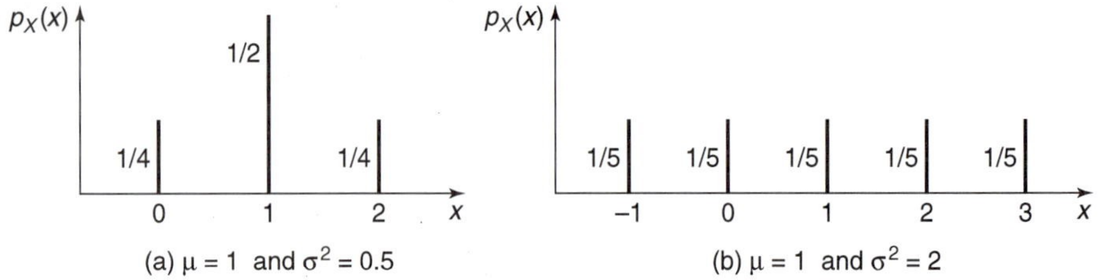

Figure 2-14 Some hypothetical distributions.

An alternate form of equation 2-14 is obtained by algebraic manipulation as

$$\begin{aligned} \sigma^2 &= \sum_i x_i^2 p_X(x_i) - \mu^2 && \text{for discrete } X, \\ &= \int_{-\infty}^{\infty} x^2 f_X(x)dx - \mu^2 && \text{for continuous } X. \end{aligned} \tag{2-16}$$

This simply indicates that the second moment about the mean is equal to the second moment about the origin less the square of the mean. The reader familiar with engineering mechanics will recognize that the development leading to equation 2-16 is of the same nature as that leading to the *theorem of moments* in mechanics.

Example 2-17

1. *Coin tossing—Example 2-9.* Recall that $\mu = \frac{3}{2}$ from Example 2-15 and

$$\sigma^2 = \left(0 - \frac{3}{2}\right)^2 \cdot \frac{1}{8} + \left(1 - \frac{3}{2}\right)^2 \cdot \frac{3}{8} + \left(2 - \frac{3}{2}\right)^2 \cdot \frac{3}{8} + \left(3 - \frac{3}{2}\right)^2 \cdot \frac{1}{8} = \frac{3}{4}.$$

Using the alternate form,

$$\sigma^2 = \left[0^2 \cdot \frac{1}{8} + 1^2 \cdot \frac{3}{8} + 2^2 \cdot \frac{3}{8} + 3^2 \cdot \frac{1}{8}\right] - \left(\frac{3}{2}\right)^2 = \frac{3}{4},$$

which is only slightly easier.

2. *Binomial distribution—Example 2-10.* From equation 2-13 we may show that $\mu = np$, and

$$\sigma^2 = \sum_{x=0}^{n} (x - np)^2 \binom{n}{x} p^x (1-p)^{n-x}$$

or

$$\sigma^2 = \left[\sum_{x=0}^{n} x^2 \cdot \binom{n}{x} p^x (1-p)^{n-x}\right] - (np)^2,$$

which (after some algebra) simplifies to

$$\sigma^2 = np\,(1-p).$$

3. *Exponential distribution—Example 2-13.* Consider the density function $f_X(x)$, where

$$\begin{aligned} f_X(x) &= 2e^{-2x}, && x \ge 0, \\ &= 0 && \text{otherwise.} \end{aligned}$$

Then, using integration by parts,

$$\mu = \int_0^\infty x \cdot 2e^{-2x} dx = \frac{1}{2}$$

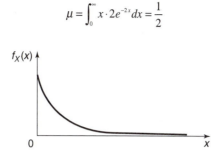

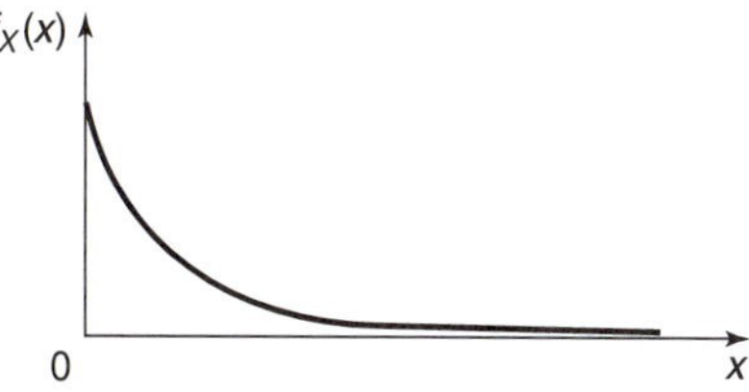

and

$$\sigma^2 = \int_0^\infty x^2 \cdot 2e^{-2x} dx - \left(\frac{1}{2}\right)^2 = \frac{1}{2} - \frac{1}{4} = \frac{1}{4}.$$

4. Another density is $g_X(x)$, where

$$\begin{aligned} g_X(x) &= 16xe^{-4x}, \qquad & x \geq 0, \\ &= 0, & \text{otherwise.} \end{aligned}$$

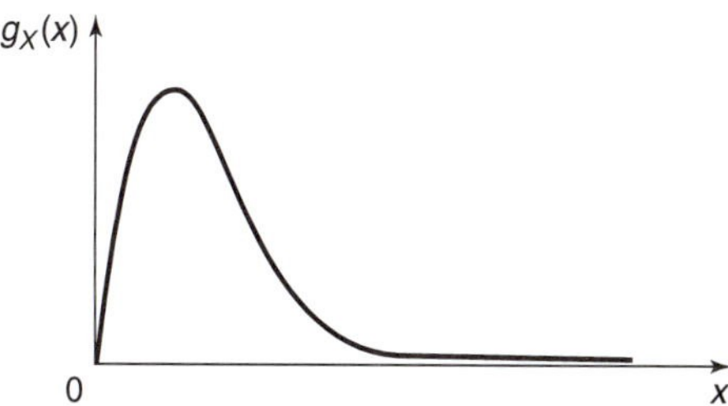

Then

$$\mu = \int_0^\infty x \cdot 16xe^{-4x} dx = \frac{1}{2}$$

and

$$\sigma^2 = \int_0^\infty x^2 \cdot 16xe^{-4x} dx - \left(\frac{1}{2}\right)^2 = \frac{1}{8}.$$

Note that the mean is the same for the densities in parts 3 and 4, with part 3 having a variance twice that of part 4.

In the development of the mean and variance, we used the terminology "mean of the random variable" and "variance of the random variable." Some authors use the terminology "mean of the distribution" and "variance of the distribution." Either terminology is acceptable. Also, where several random variables are being considered, it is often convenient to use a subscript on μ and σ, for example, μ_X and σ_X.

In addition to the mean and variance, other moments are also frequently used to describe distributions. That is, the moments of a distribution describe that distribution, measure its properties, and, in certain circumstances, specify it. Moments about the origin are called *origin moments* and are denoted μ_k' for the kth origin moment, where

$$\mu_k' = \sum_i x_i^k p_X(x_i) \qquad \text{for discrete } X,$$

$$= \int_{-\infty}^{\infty} x^k f_X(x)dx \qquad \text{for continuous } X,$$

$$k = 0, 1, 2, \ldots. \tag{2-17}$$

Moments about the mean are called *central moments* and are denoted μ_k, where

$$\mu_k = \sum_i (x_i - \mu)^k p_X(x_i) \qquad \text{for discrete } X,$$

$$= \int_{-\infty}^{\infty} (x - \mu)^k f_X(x)\,dx \qquad \text{for continuous } X,$$

$$k = 0, 1, 2, \ldots. \tag{2-18}$$

Note that the mean $\mu = \mu_1'$ and the variance is $\sigma^2 = \mu_2$. Central moments may be expressed in terms of origin moments by the relationship

$$\mu_k = \sum_{j=0}^{k} (-1)^j \binom{k}{j} \mu^j \mu_{k-j}', \qquad k = 0,\ 1,\ 2,\ \ldots. \tag{2-19}$$

2-6 CHEBYSHEV'S INEQUALITY

In earlier sections of this chapter, it was pointed out that a small variance, σ^2, indicates that large deviations from the mean, μ, are improbable. *Chebyshev's inequality* gives us a way of understanding how the variance measures the probability of deviations about μ.

Theorem 2-1

Let X be a random variable (discrete or continuous), and let k be some positive number. Then

$$P_X(|X - \mu| \geq k\sigma) \leq \frac{1}{k^2}. \tag{2-20}$$

Proof For continuous X and a constant $K > 0$, consider

$$\sigma^2 = \int_{-\infty}^{\infty} (x - \mu)^2 \cdot f_X(x)\,dx = \int_{-\infty}^{\mu - \sqrt{K}} (x - \mu)^2 \cdot f_X(x)\,dx$$

$$+ \int_{\mu - \sqrt{K}}^{\mu + \sqrt{K}} (x - \mu)^2 \cdot f_X(x)\,dx + \int_{\mu + \sqrt{K}}^{\infty} (x - \mu)^2 \cdot f_X(x)\,dx.$$

Since

$$\int_{\mu - \sqrt{K}}^{\mu + \sqrt{K}} (x - \mu)^2 \cdot f_X(x)\,dx \geq 0,$$

it follows that

$$\sigma^2 \geq \int_{-\infty}^{\mu - \sqrt{K}} (x - \mu)^2 \cdot f_X(x)dx + \int_{\mu + \sqrt{K}}^{\infty} (x - \mu)^2 \cdot f_X(x)\,dx.$$

Now, $(x - \mu)^2 \geq K$ if and only if $|x - \mu| \geq \sqrt{K}$; therefore,

$$\sigma^2 \geq \int_{-\infty}^{\mu - \sqrt{K}} K f_X(x)dx + \int_{\mu + \sqrt{K}}^{\infty} K f_X(x)dx$$

$$= K\left[P_X\left(X \leq \mu - \sqrt{K}\right) + P_X\left(X \geq \mu + \sqrt{K}\right)\right]$$

and

$$P_X\left(|X-\mu| \geq \sqrt{K}\right) \leq \frac{\sigma^2}{K},$$

so that if $k = \sqrt{K}/\sigma$, then

$$P_X\left(|X-\mu| \geq k\sigma\right) \leq \frac{1}{k^2}.$$

The proof for discrete X is quite similar.

An alternate form of this inequality,

$$P_X\left(|X-\mu| < k\sigma\right) \geq 1-\frac{1}{k^2} \tag{2-21}$$

or

$$P_X\left(\mu - k\sigma < X < \mu + k\sigma\right) \geq 1-\frac{1}{k^2},$$

is often useful.

The usefulness of Chebyshev's inequality stems from the fact that so little knowledge about the distribution of X is required. Only μ and σ^2 must be known. However, Chebyshev's inequality is a weak statement, and this detracts from its usefulness. For example, $P_X(|X - \mu| \geq \sigma) \leq 1$, which we knew before we started! If the precise form of $f_X(x)$ or $p_X(x)$ is known, then a more powerful statement can be made.

Example 2-18

From an analysis of company records, a materials control manager estimates that the mean and standard deviation of the "lead time" required in ordering a small valve are 8 days and 1.5 days, respectively. He does not know the distribution of lead time, but he is willing to assume the estimates of the mean and standard deviation to be absolutely correct. The manager would like to determine a time interval such that the probability is at least $\frac{8}{9}$ that the order will be received during that time. That is,

$$1-\frac{1}{k^2}=\frac{8}{9}$$

so that $k = 3$ and $\mu \pm k\sigma$ gives $8 \pm 3(1.5)$ or [3.5 days to 12.5 days]. It is noted that this interval may very well be too large to be of any value to the manager, in which case he may elect to learn more about the distribution of lead times.

2-7 SUMMARY

This chapter has introduced the idea of random variables. In most engineering and management applications, these are either discrete or continuous; however, Section 2-2 illustrates a more general case. A vast majority of the discrete variables to be considered in this book result from counting processes, whereas the continuous variables are employed to model a variety of measurements. The mean and variance as measures of central tendency and dispersion, and as characterizations of random variables, were presented along with more general moments of higher order. The Chebyshev inequality is presented as a bounding probability that a random variable lies between $\mu - k\sigma$ and $\mu + k\sigma$.

2-8 EXERCISES

2-1. A five-card poker hand may contain from zero to four aces. If X is the random variable denoting the number of aces, enumerate the range space of X. What are the probabilities associated with each possible value of X?

2-2. A car rental agency has either 0, 1, 2, 3, 4, or 5 cars returned each day, with probabilities $\frac{1}{6}, \frac{1}{6}, \frac{1}{3}, \frac{1}{12}, \frac{1}{6}$, and $\frac{1}{12}$, respectively. Find the mean and the variance of the number of cars returned.

2-3. A random variable X has the probability density function ce^{-x}. Find the proper value of c, assuming $0 \le X < \infty$. Find the mean and the variance of X.

2-4. The cumulative distribution function that a television tube will fail in t hours is $1 - e^{-ct}$, where c is a parameter dependent on the manufacturer and $t \ge 0$. Find the probability density function of T, the life of the tube.

2-5. Consider the three functions given below. Determine which functions are distribution functions (CDFs).

(a) $F_X(x) = 1 - e^{-x}, \quad 0 < x < \infty.$

(b) $G_X(x) = e^{-x}, \quad 0 \le x < \infty,$
$\quad = 0, \quad x < 0.$

(c) $H_X(x) = e^{x}, \quad -\infty < x \le 0,$
$\quad = 1, \quad x > 0.$

2-6. Refer to Exercise 2-5. Find the probability density function corresponding to the functions given, if they are distribution functions.

2-7. Which of the following functions are discrete probability distributions?

(a)
$$\begin{aligned} p_X(x) &= \frac{1}{3}, & x &= 0,\\ &= \frac{2}{3}, & x &= 1,\\ &= 0, & &\text{otherwise.}\end{aligned}$$

(b)
$$\begin{aligned} p_X(x) &= \binom{5}{x}\left(\frac{2}{3}\right)^x\left(\frac{1}{3}\right)^{5-x}, & x &= 0,1,2,3,4,5,\\ &= 0, & &\text{otherwise.}\end{aligned}$$

2-8. The demand for a product is $-1, 0, +1, +2$ per day with probabilities $\frac{1}{5}, \frac{1}{10}, \frac{2}{5}, \frac{3}{10}$, respectively. A demand of -1 implies a unit is returned. Find the expected demand and the variance. Sketch the distribution function (CDF).

2-9. The manager of a men's clothing store is concerned over the inventory of suits, which is currently 30 (all sizes). The number of suits sold from now to the end of the season is distributed as

$$\begin{aligned} p_X(x) &= \frac{e^{-20}20^x}{x!}, & x &= 0,1,2,\ldots,\\ &= 0, & &\text{otherwise.}\end{aligned}$$

Find the probability that he will have suits left over at the season's end.

2-10. A random variable X has a CDF of the form

$$\begin{aligned} F_X(x) &= 1 - \left(\frac{1}{2}\right)^{x+1}, & x &= 0,1,2,\ldots,\\ &= 0, & x &< 0.\end{aligned}$$

(a) Find the probability function for X.

(b) Find $P_X(0 < X \le 8)$.

2-11. Consider the following probability density function:

$$\begin{aligned} f_X(x) &= kx, & 0 &\le x < 2,\\ &= k(4-x), & 2 &\le x \le 4,\\ &= 0, & &\text{otherwise.}\end{aligned}$$

(a) Find the value of k for which f is a probability density function.

(b) Find the mean and variance of X.

(c) Find the cumulative distribution function.

2-12. Rework the above problem, except let the probability density function be defined as

$$\begin{aligned} f_X(x) &= kx, & 0 &\le x < a,\\ &= k(2a-x), & a &\le x \le 2a,\\ &= 0, & &\text{otherwise.}\end{aligned}$$

2-13. The manager of a job shop does not know the probability distribution of the time required to complete an order. However, from past performance she has been able to estimate the mean and variance as 14 days and 2 $(\text{days})^2$, respectively. Find an interval such that the probability is at least 0.75 that an order is finished during that time.

2-14. The continuous random variable T has the probability density function $f(t) = kt^2$ for $-1 \le t \le 0$. Find the following:

(a) The appropriate value of k.

(b) The mean and variance of T.

(c) The cumulative distribution function.

2-15. A discrete random variable X has probability function $p_X(x)$, where

$$\begin{aligned} p_X(x) &= k(1/2)^x, & x &= 1, 2, 3,\\ &= 0, & &\text{otherwise.}\end{aligned}$$

(a) Find k.

(b) Find the mean and variance of X.

(c) Find the cumulative distribution function $F_X(x)$.

2-16. The discrete random variable N ($N = 0, 1, \ldots$) has probabilities of occurrence of kr^n ($0 < r < 1$). Find the appropriate value of k.

2-17. The postal service requires, on the average, 2 days to deliver a letter across town. The variance is estimated to be 0.4 (day)^2. If a business executive wants at least 99% of his letters delivered on time, how early should he mail them?

2-18. Two different real estate developers, A and B, own parcels of land being offered for sale. The probability distributions of selling prices per parcel are shown in the following table.

Price

	\$1000	\$1050	\$1100	\$1150	\$1200	\$1350
A	0.2	0.3	0.1	0.3	0.05	0.05
B	0.1	0.1	0.3	0.3	0.1	0.1

Assuming that A and B are operating independently, compute the following:

(a) The expected selling price of A and of B.

(b) The expected selling price of A given that the B selling price is \$1150.

(c) The probability that A and B both have the same selling price.

2-19. Show that the probability function for the sum of values obtained in tossing two dice may be written as

$$
\begin{aligned}
p_X(x) &= \frac{x-1}{36}, \qquad x = 2, 3, \ldots, 6, \\
&= \frac{13-x}{36}, \qquad x = 7, 8, \ldots, 12.
\end{aligned}
$$

2-20. Find the mean and variance of the random variable whose probability function is defined in the previous problem.

2-21. A continuous random variable X has a density function

$$
\begin{aligned}
f_X(x) &= \frac{2x}{9}, \qquad 0 < x < 3, \\
&= 0 \qquad \text{otherwise.}
\end{aligned}
$$

(a) Develop the CDF for X.

(b) Find the mean of X and the variance of X.

(c) Find μ_3'.

(d) Find a value m such that $P_X(X \geq m) = P(X \leq m)$. This is called the *median* of X.

2-22. Suppose X takes on the values 5 and -5 with probabilities $\frac{1}{2}$. Plot the quantity $P[|X - \mu| \geq k\sigma]$ as a function of k (for $k > 0$). On the same set of axes, plot the same probability determined by Chebyshev's inequality.

2-23. Find the cumulative distribution function associated with

$$
\begin{aligned}
f_X(x) &= \frac{x}{t^2}\exp\left(-\frac{x^2}{2t^2}\right), \qquad t > 0, x \geq 0, \\
&= 0, \qquad \text{otherwise.}
\end{aligned}
$$

2-24. Find the cumulative distribution function associated with

$$
f_X(x) = \frac{1}{\sigma\pi}\frac{1}{\left\{1 + \left[(x-\mu)^2/\sigma^2\right]\right\}}, \quad -\infty < x < \infty.
$$

2-25. Consider the probability density function $f_Y(y) = k \sin y$, $0 \leq y \leq \pi/2$. What is the appropriate value of k? Find the mean of the distribution.

2-26. Show that central moments can be expressed in terms of origin moments by equation 2-19. *Hint*: See Chapter 3 of Kendall and Stuart (1963).

Chapter 3

Functions of One Random Variable and Expectation

3-1 INTRODUCTION

Engineers and management scientists are frequently interested in the behavior of some function, say H, of a random variable X. For example, suppose the circular cross-sectional area of a copper wire is of interest. The relationship $Y = \pi X^2/4$, where X is the diameter, gives the cross-sectional area. Since X is a random variable, Y also is a random variable, and we would expect to be able to determine the probability distribution of $Y = H(X)$ if the distribution of X is known. The first portion of this chapter will be concerned with problems of this type. This is followed by the concept of expectation, a notion employed extensively throughout the remaining chapters of this book. Approximations are developed for the mean and variance of functions of random variables, and the moment-generating function, a mathematical device for producing moments and describing distributions, is presented with some example illustrations.

3-2 EQUIVALENT EVENTS

Before presenting some specific methods used in determining the probability distribution of a function of a random variable, the concepts involved should be more precisely formulated.

Consider an experiment $\mathscr{E}$ with sample space $\mathscr{S}$. The random variable X is defined on $\mathscr{S}$, assigning values to the outcomes e in $\mathscr{S}$, $X(e) = x$, where the values x are in the range space R_X of X. Now if $Y = H(X)$ is defined so that the values $y = H(x)$ in R_Y, the range space of Y, are real, then Y is a random variable, since for every outcome $e \in \mathscr{S}$, a value y of the random variable Y is determined; that is, $y = H[X(e)]$. This notion is illustrated in Fig. 3-1.

If C is an event associated with R_Y and B is an event in R_X, then B and C are *equivalent events* if they occur together, that is, if $B = \{x \in R_X: H(x) \in C\}$. In addition, if A is an event associated with $\mathscr{S}$ and, furthermore, A and B are equivalent, then A and C are equivalent events.

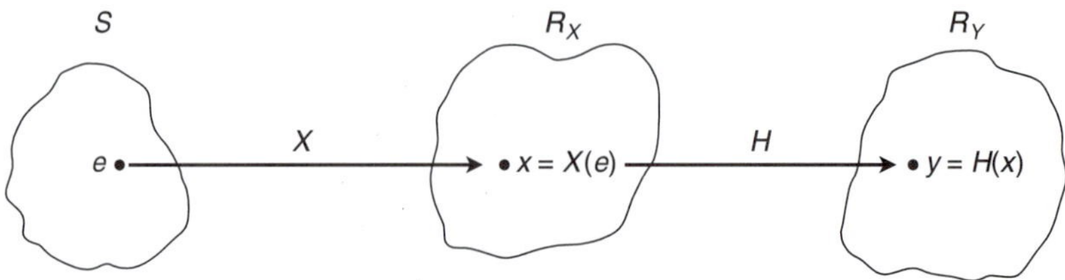

Figure 3-1 A function of a random variable.

Definition

If X is a random variable (defined on $\mathscr{S}$) having range space R_X, and if H is a real-valued function, so that $Y = H(X)$ is a random variable with range space R_Y, then for any event $C \subset R_Y$, we define

$$P_Y(C) = P_X(\{x \in R_X: H(x) \in C\}). \tag{3-1}$$

It is noted that these probabilities relate to probabilities in the sample space. We could write

$$P_Y(C) = P(\{e \in \mathscr{S}: H[X(e)] \in C\}).$$

However, equation 3-1 indicates the *method* to be used in problem solution. We find the event B in R_X that is equivalent to event C in R_Y; then we find the probability of event B.

Example 3-1

In the case of the cross-sectional area Y of a wire, suppose we know that the diameter of a wire has density function

$$\begin{aligned} f_X(x) &= 200, & 1.000 \le x \le 1.005, \\ &= 0, & \text{otherwise.} \end{aligned}$$

Let $Y = (\pi/4)X^2$ be the cross-sectional area of the wire, and suppose we want to find $P_Y(Y \le (1.01)\pi/4)$. The equivalent event is determined. $P_Y(Y \le (1.01)\pi/4) = P_X[(\pi/4)X^2 \le (1.01)\pi/4] = P_X(|X| \le \sqrt{1.01})$. The event $\{x \in R_X : |x| \le \sqrt{1.01}\}$ is in the range space R_X, and since $f_X(x) = 0$, for all $x < 1.0$, we calculate

$$\begin{aligned} P_X\left(|X| \le \sqrt{1.01}\right) &= P_X\left(1.0 \le X \le \sqrt{1.01}\right) = \int_{1.000}^{\sqrt{1.01}} 200\,dx \\ &= 0.9975. \end{aligned}$$

Example 3-2

In the case of the Geiger counter experiment of Example 2-12, we used the distribution given in equation 2-7:

$$\begin{aligned} p_X(x) &= e^{-\lambda t}(\lambda t)^x/x!, & x = 0, 1, 2, \ldots, \\ &= 0, & \text{otherwise.} \end{aligned}$$

Recall that λ, where $\lambda > 0$, represents the mean "hit" rate and t is the time interval for which the counter is operated. Now suppose we wish to find $P_Y(Y \le 5)$, where

$$Y = 2X + 2.$$

Proceeding as in the previous example,

$$\begin{aligned} P_Y(Y \le 5) &= P_X(2X + 2 \le 5) = P_X\left(X \le \frac{3}{2}\right) \\ &= \left[p_X(0) + p_X(1)\right] = \left[e^{-\lambda t}(\lambda t)^0/0!\right] + \left[e^{-\lambda t}(\lambda t)^1/1!\right] \\ &= e^{-\lambda t}[1 + \lambda t]. \end{aligned}$$

The event $\{x \in R_X: x \le \frac{3}{2}\}$ is in the range space of X, and we have the function p_X to work with in that space.

3-3 FUNCTIONS OF A DISCRETE RANDOM VARIABLE

Suppose that both X and Y are discrete random variables, and let $x_{i_1}, x_{i_2}, \ldots, x_{i_k}, \ldots$, represent the values of X such that $H(x_{i_j}) = y_i$ for some set of index values, $\Omega_i = \{j: j = 1, 2, \ldots, s_i\}$.

The probability distribution for Y is denoted by $p_Y(y_i)$ and is given by

$$p_Y(y_i) = P_Y(Y = y_i) = \sum_{j \in \Omega_i} p_X(x_{i_j}). \tag{3-2}$$

For example, in Fig. 3-2, where $s_i = 4$, the probability of y_i is $p_Y(y_i) = p_X(x_{i_1}) + p_X(x_{i_2}) + p_X(x_{i_3}) + p_X(x_{i_4})$.

In the special case where H is such that for each y there is exactly one x, then $p_Y(y_i) = p_X(x_i)$, where $y_i = H(x_i)$. To illustrate these concepts, consider the following examples.

Example 3-3

In the coin-tossing experiment where X represented the number of heads, recall that X assumed four values, 0, 1, 2, 3, with probabilities $\frac{1}{8}, \frac{3}{8}, \frac{3}{8}, \frac{1}{8}$. If $Y = 2X - 1$, then the possible values of Y are -1, 1, 3, 5, and $p_Y(-1) = \frac{1}{8}$, $p_Y(1) = \frac{3}{8}$, $p_Y(3) = \frac{3}{8}$, $p_Y(5) = \frac{1}{8}$. In this case, H is such that for each y there is exactly one x.

Example 3-4

X is as in the previous example; however, suppose now that $Y = |X - 2|$, so that the possible values of Y are 0, 1, 2, as indicated in Fig. 3-3. In this case

$$p_Y(0) = p_X(2) = \frac{3}{8},$$

$$p_Y(1) = p_X(1) + p_X(3) = \frac{4}{8},$$

$$p_Y(2) = p_X(0) = \frac{1}{8}.$$

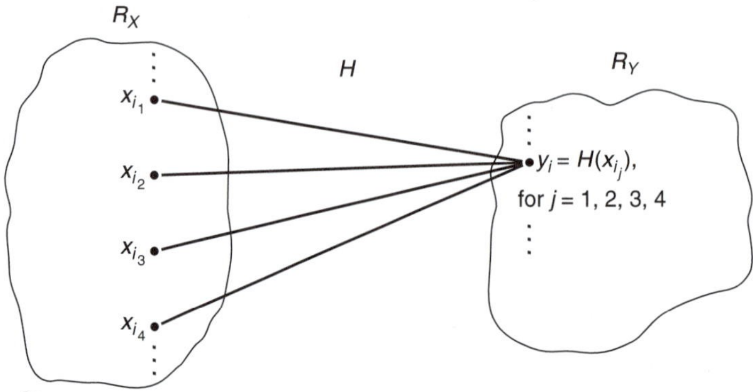

Figure 3-2 Probabilities in R_Y.

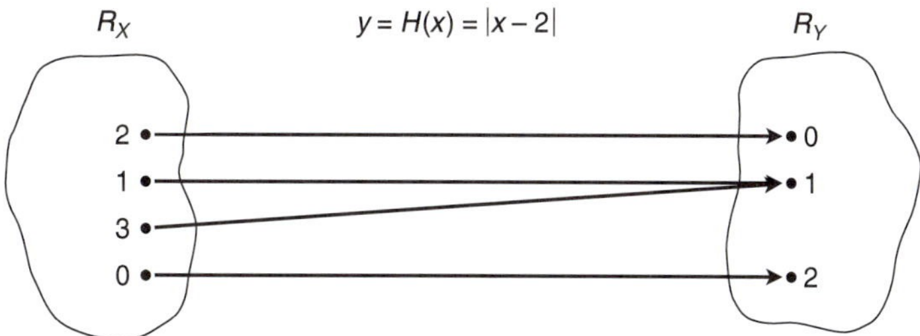

Figure 3-3 An example function *H*.

In the event that X is continuous but Y is discrete, the formulation for $p_Y(y_i)$ is

$$p_Y(y_i) = \int_B f_X(x)dx, \tag{3-3}$$

where the event B is the event in R_X that is equivalent to the event $(Y = y_i)$ in R_Y.

Example 3-5

Suppose X has the exponential probability density function given by

$$\begin{aligned} f_X(x) &= \lambda e^{-\lambda x}, && x \geq 0, \\ &= 0 && \text{otherwise.} \end{aligned}$$

Furthermore, if

$$\begin{aligned} Y &= 0 && \text{for } X \leq 1/\lambda, \\ &= 1 && \text{for } X > 1/\lambda, \end{aligned}$$

then

$$p_Y(0) = \int_0^{1/\lambda} \lambda e^{-\lambda x}\, dx = -e^{-\lambda x}\Big|_0^{1/\lambda} = 1 - e^{-1} \simeq 0.6321$$

and

$$p_Y(1) = 1 - P_Y(0) \simeq 0.3679.$$

3-4 CONTINUOUS FUNCTIONS OF A CONTINUOUS RANDOM VARIABLE

If X is a continuous random variable with probability density function f_X, and H is also continuous, then $Y = H(X)$ is a continuous random variable. The probability density function for the random variable Y will be denoted f_Y and it may be found by performing three steps.

1. Obtain the CDF of Y, $F_Y(y) = P_Y(Y \leq y)$, by finding the event B in R_X, which is equivalent to the event $(Y \leq y)$ in R_Y.
2. Differentiate $F_Y(y)$ with respect to y to obtain the probability density function $f_Y(y)$.
3. Find the range space of the new random variable.

Example 3-6

Suppose that the random variable X has the following density function:

$$f_X(x) = x/8, \quad 0 \le x \le 4,$$
$$= 0, \quad \text{otherwise.}$$

If $Y = H(X)$ is the random variable for which the density f_Y is desired, and $H(x) = 2x + 8$, as shown in Fig. 3-4, then we proceed according to the steps given above.

1. $F_Y(y) = P_Y(Y \le y) = P_X(2X + 8 \le y)$
$$= P_X(X \le (y-8)/2)$$
$$= \int_0^{(y-8)/2} (x/8)\,dx = \frac{x^2}{16}\bigg|_0^{(y-8)/2} = \frac{1}{64}\left(y^2 - 16y + 64\right).$$
2. $f_Y(y) = F_Y'(y) = \dfrac{y}{32} - \dfrac{1}{4}.$
3. If $x = 0$, $y = 8$, and if $x = 4$, $y = 16$, so then we have

$$f_Y(y) = \frac{y}{32} - \frac{1}{4}, \quad 8 \le y \le 16,$$
$$= 0, \quad \text{otherwise.}$$

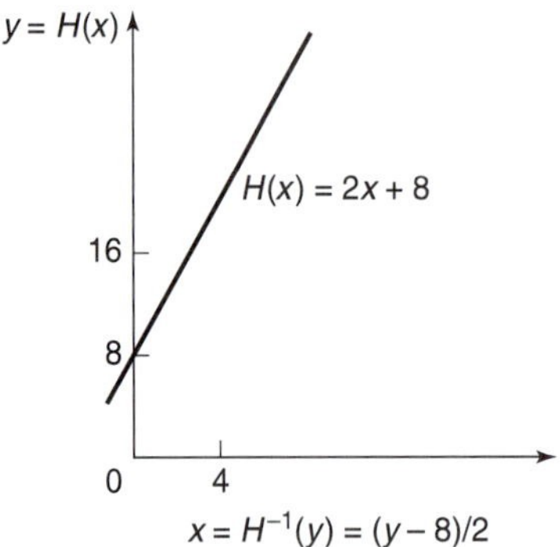

Figure 3-4 The function $H(x) = 2x + 8$.

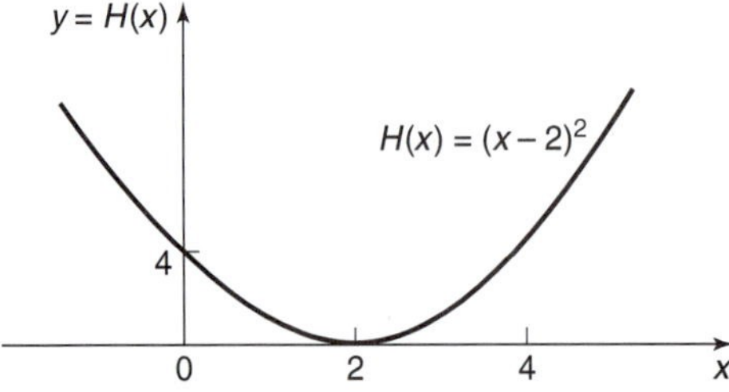

Figure 3-5 The function $H(x) = (x - 2)^2$.

Example 3-7

Consider the random variable X defined in Example 3-6, and suppose $Y = H(X) = (X - 2)^2$, as shown in Fig. 3-5. Proceeding as in Example 3-6, we find the following:

1. $$F_Y(y) = P_Y(Y \le y) = P_X((X-2)^2 \le y) = P_X\left(-\sqrt{y} \le X - 2 \le \sqrt{y}\right)$$
$$= P_X\left(2-\sqrt{y} \le X \le 2+\sqrt{y}\right)$$
$$= \int_{2-\sqrt{y}}^{2+\sqrt{y}} \frac{x}{8}\,dx = \frac{x^2}{16}\bigg|_{2-\sqrt{y}}^{2+\sqrt{y}}$$
$$= \frac{1}{16}\left[\left(4+4\sqrt{y}+y\right)-\left(4-4\sqrt{y}+y\right)\right]$$
$$= \frac{1}{2}\sqrt{y}.$$

2. $$f_Y(y) = F_Y'(y) = \frac{1}{4\sqrt{y}}.$$

3. If $x = 2$, $y = 0$, and if $x = 0$ or $x = 4$, then $y = 4$. However, f_Y is not defined for $y = 0$; therefore,

$$\begin{aligned} f_Y(y) &= \frac{1}{4\sqrt{y}}, & 0 < y \le 4, \\ &= 0 & \text{otherwise.} \end{aligned}$$

In Example 3-6, the event in R_X equivalent to $(Y \le y)$ in R_Y was $[X \le (y-8)/2]$; and in Example 3-7, the event in R_X equivalent to $(Y \le y)$ in R_Y was $(2 - \sqrt{y} \le X \le 2 + \sqrt{y})$. In the first example, the function H is a strictly increasing function of x, while in the second example this is not the case.

Theorem 3-1

If X is a continuous random variable with probability density function f_X that satisfies $f_X(x) > 0$ for $a < x < b$, and if $y = H(x)$ is a continuous strictly increasing or strictly decreasing function of x, then the random variable $Y = H(X)$ has density function

$$f_Y(y) = f_X(x)\cdot\left|\frac{dx}{dy}\right|, \tag{3-4}$$

with $x = H^{-1}(y)$ expressed in terms of y. If H is increasing, then $f_Y(y) > 0$ if $H(a) < y < H(b)$; and if H is decreasing, then $f_Y(y) > 0$ if $H(b) < y < H(a)$.

Proof (Given only for H increasing. A similar argument holds for H decreasing.)

$$\begin{aligned} F_Y(y) &= P_Y(Y \le y) = P_X(H(X) \le y) \\ &= P_X[X \le H^{-1}(y)] \\ &= F_X[H^{-1}(y)]. \end{aligned}$$

$$f_Y(y) = F_Y'(y) = \frac{dF_X(x)}{dx}\cdot\frac{dx}{dy}, \qquad \text{by the chain rule,}$$

$$= f_X(x)\cdot\frac{dx}{dy}, \qquad \text{where } x = H^{-1}(y).$$

Example 3-8

In Example 3-6, we had

$$\begin{aligned} f_X(x) &= x/8, & 0 \le x \le 4, \\ &= 0, & \text{otherwise,} \end{aligned}$$

and $H(x) = 2x + 8$, which is a strictly increasing function. Using equation 3-4,

$$f_Y(y) = f_X(x) \cdot \left|\frac{dx}{dy}\right| = \frac{y-8}{16} \cdot \frac{1}{2},$$

since $x = (y - 8)/2$. $H(0) = 8$ and $H(4) = 16$; therefore,

$$\begin{aligned} f_Y(y) &= \frac{y}{32} - \frac{1}{4}, & 8 \le y \le 16, \\ &= 0, & \text{otherwise.} \end{aligned}$$

3-5 EXPECTATION

If X is a random variable and $Y = H(X)$ is a function of X, then the *expected value* of $H(X)$ is defined as follows:

$$E[H(X)] = \sum_i H(x_i) \cdot p_X(x_i) \quad \text{for } X \text{ discrete,} \tag{3-5}$$

$$E[H(X)] = \int_{-\infty}^{\infty} H(x) \cdot f_X(x)\,dx \quad \text{for } X \text{ continuous.} \tag{3-6}$$

In the case where X is continuous, we restrict H so that $Y = H(X)$ is a continuous random variable. In reality, we ought to regard equations 3-5 and 3-6 as *theorems* (not definitions), and these results have come to be known as *the law of the unconscious statistician.*

The mean and variance, presented earlier, are special applications of equations 3-5 and 3-6. If $H(X) = X$, we see that

$$E[H(X)] = E(X) = \mu. \tag{3-7}$$

Therefore, the expected value of the random variable X is just the mean, μ.

If $H(X) = (X - \mu)^2$, then

$$E[H(X)] = E((X - \mu)^2) = \sigma^2. \tag{3-8}$$

Thus, the variance of the random variable X may be defined in terms of expectation. Since the variance is utilized extensively, it is customary to introduce a variance *operator* V that is defined in terms of the expected value operator E:

$$V[H(X)] = E([H(X) - E(H(X))]^2). \tag{3-9}$$

Again, in the case where $H(X) = X$,

$$\begin{aligned} V(X) &= E[(X - E(X))^2] \\ &= E(X^2) - [E(X)]^2, \end{aligned} \tag{3-10}$$

which is the *variance* of X, denoted σ^2.

The origin moments and central moments discussed in the previous chapter may also be expressed using the expected value operator as

$$\mu_k' = E(X^k) \tag{3-11}$$

and

$$\mu_k = E[(X - E(X))^k].$$

There is a special linear function H that should be considered at this point. Suppose that $H(X) = aX + b$, where a and b are constants. Then for discrete X, we have

$$\begin{aligned} E(aX+b) &= \sum_i (ax_i + b)p_X(x_i) \\ &= a\sum_i x_i p_X(x_i) + b\sum_i p_X(x_i) \\ &= aE(x) + b, \end{aligned} \tag{3-12}$$

and the same result is obtained for continuous X; namely,

$$\begin{aligned} E(aX+b) &= \int_{-\infty}^{\infty} (ax+b)f_X(x)\,dx \\ &= a\int_{-\infty}^{\infty} x f_X(x)\,dx + b\int_{-\infty}^{\infty} f_X(x)\,dx \\ &= aE(X) + b. \end{aligned} \tag{3-13}$$

Using equation 3-9,

$$\begin{aligned} V(aX+b) &= E[(aX+b-E(aX+b))^2] \\ &= E[(aX+b-aE(X)-b)^2] \\ &= aE[(X-E(X))^2] \\ &= a^2V(X). \end{aligned} \tag{3-14}$$

In the further special case where $H(X) = b$, a constant, the reader may readily verify that

$$E(b) = b \tag{3-15}$$

and

$$V(b) = 0. \tag{3-16}$$

These results show that a linear shift of size b only affects the expected value, not the variance.

Example 3-9

Suppose X is a random variable such that $E(X) = 3$ and $V(X) = 5$. In addition, let $H(X) = 2X - 7$. Then

$$E[H(X)] = E(2X-7) = 2E(X) - 7 = -1$$

and

$$V[H(X)] = V(2X-7) = 4V(X) = 20.$$

The next examples give additional illustrations of calculations involving expectation and variance.

Example 3-10

Suppose a contractor is about to bid on a job requiring X days to complete, where X is a random variable denoting the number of days for job completion. Her profit, P, depends on X; that is, $P = H(X)$. The probability distribution of X, $(x, p_X(x))$, is as follows:

x	$p_X(x)$
3	$\frac{1}{8}$
4	$\frac{5}{8}$
5	$\frac{2}{8}$
otherwise,	0.

Using the notion of expected value, we calculate the mean and variance of X as

$$E(X) = 3 \cdot \frac{1}{8} + 4 \cdot \frac{5}{8} + 5 \cdot \frac{2}{8} = \frac{33}{8}$$

and

$$V(X) = \left[3^2 \cdot \frac{1}{8} + 4^2 \cdot \frac{5}{8} + 5^2 \cdot \frac{2}{8}\right] - \left(\frac{33}{8}\right)^2 = \frac{23}{64}.$$

If the function $H(X)$ is given as

x	$H(x)$
3	\$10,000
4	2,500
5	−7,000

then the expected value of $H(X)$ is

$$E[H(X)] = 10{,}000 \cdot \left(\frac{1}{8}\right) + 2500 \cdot \left(\frac{5}{8}\right) - 7000 \cdot \left(\frac{2}{8}\right) = \$1062.50$$

and the contractor would view this as the average profit that she would obtain if she bid this job many, many times (actually an infinite number of times), where H remained the same and the random variable X behaved according to the probability function p_X. The variance of $P = H(X)$ can readily be calculated as

$$\begin{aligned} V[H(X)] &= \left[(10{,}000)^2 \cdot \frac{1}{8} + (2500)^2 \cdot \frac{5}{8} + (-7000)^2 \cdot \frac{2}{8}\right] - (1062.5)^2 \\ &\simeq \$27.53 \cdot 10^6. \end{aligned}$$

Example 3-11

A well-known simple inventory problem is "the newsboy problem," described as follows. A newsboy buys papers for 15 cents each and sells them for 25 cents each, and he cannot return unsold papers. Daily demand has the following distribution and each day's demand is independent of the previous day's demand.

Number of customers x	23	24	25	26	27	28	29	30
Probability, $p_X(x)$	0.01	0.04	0.10	0.10	0.25	0.25	0.15	0.10

If the newsboy stocks too many papers, he suffers a loss attributable to the excess supply. If he stocks too few papers, he loses profit because of the excess demand. It seems reasonable for the newsboy to stock some number of papers so as to minimize the *expected* loss. If we let s represent the number of papers stocked, X the daily demand, and $L(X, s)$ the newsboy's loss for a particular stock level s, then the loss is simply

$$\begin{aligned} L(X, s) &= 0.10(X - s) \qquad \text{if } X > s, \\ &= 0.15(s - X) \qquad \text{if } X \leq s, \end{aligned}$$

and for a given stock level, s, the expected loss is

$$E[L(X,s)] = \sum_{x=23}^{s} 0.15(s-x) \cdot p_X(x) + \sum_{x=s+1}^{30} 0.10(x-s) \cdot p_X(x)$$

and the $E[L(X, s)]$ is evaluated for some different values of s.

For $s = 26$,

$$\begin{aligned} E[L(X, 26)] &= 0.15[(26-23)(0.01) + (26-24)(0.04) + (26-25)(0.10) \\ &\quad + (26-26)(0.10)] + 0.10[(27-26)(0.25) + (28-26)(0.25) \\ &\quad + (29-26)(0.15) + (30-26)(0.10)] \\ &= \$0.1915. \end{aligned}$$

For $s = 27$,

$$\begin{aligned} E[L(X, 27)] &= 0.15[(27-23)(0.01) + (27-24)(0.04) + (27-25)(0.10) \\ &\quad + (27-26)(0.10) + (27-27)(0.25)] + 0.10[(28-27)(0.25) \\ &\quad + (29-27)(0.15) + (30-27)(0.10)] \\ &= \$0.1540. \end{aligned}$$

For $s = 28$,

$$\begin{aligned} E[L(X, 28)] &= 0.15[(28-23)(0.01) + (28-24)(0.04) + (28-25)(0.10) \\ &\quad + (28-26)(0.10) + (28-27)(0.25) + (28-28)(0.25)] \\ &\quad + 0.10[(29-28)(0.15) + (30-28)(0.10)] \\ &= \$0.1790 \end{aligned}$$

Thus, the newsboy's policy should be to stock 27 papers if he desires to minimize his expected loss.

Example 3-12

Consider the redundant system shown in the diagram below.

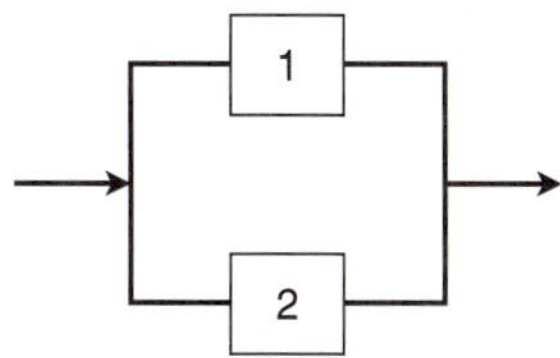

At least one of the units must function, the redundancy is *standby* (meaning that the second unit does not operate until the first fails), switching is perfect, and the system is nonmaintained. It can be shown that under certain conditions, when the time to failure for each of the units of this system has an exponential distribution, then the time to failure for the system has the following probability density function.

$$f_X(x) = \lambda^2 x e^{-\lambda x}, \qquad x > 0, \lambda > 0,$$
$$= 0, \qquad \text{otherwise,}$$

where λ is the "failure rate" parameter of the component exponential models. The mean time to failure (MTTF) for this system is

$$E(X) = \int_0^\infty x \cdot \lambda^2 x e^{-\lambda x} dx = \frac{2}{\lambda}.$$

Hence, the redundancy doubles the expected life. The terms "mean time to failure" and "expected life" are synonymous.

3-6 APPROXIMATIONS TO *E*[*H*(*X*)] AND *V*[*H*(*X*)]

In cases where $H(X)$ is very complicated, the evaluation of the expectation and variance may be difficult. Often, approximations to $E[H(X)]$ and $V[H(X)]$ may be obtained by utilizing a Taylor series expansion. This technique is sometimes called the delta method. To estimate the mean, we expand the function H to three terms, where the expansion is about $x = \mu$. If $Y = H(X)$, then

$$Y = H(\mu) + (X - \mu)H'(\mu) + \frac{(X-\mu)^2}{2} \cdot H''(\mu) + R,$$

where R is the remainder. We use equations 3-12 through 3-16 to perform

$$\begin{aligned} E(Y) &= E[H(\mu)] + E[H'(\mu)(X-\mu)] \\ &\quad + E\left[\frac{1}{2}H''(\mu)(X-\mu)^2\right] + E(R) \\ &= H(\mu) + \frac{1}{2}H''(\mu)V(X) + E(R) \\ &\cong H(\mu) + \frac{1}{2}H''(\mu)\sigma^2. \end{aligned} \tag{3-17}$$

Using only the first two terms and grouping the third into the remainder so that

$$Y = H(\mu) + (X - \mu) \cdot H'(\mu) + R_1,$$

where

$$R_1 = R + \frac{(X-\mu)^2}{2} \cdot H''(\mu),$$

then an approximation for the variance of Y is determined as

$$\begin{aligned} V(Y) &\simeq V[H(\mu)] + V[(X-\mu) \cdot H'(\mu)] + V(R_1) \\ &\simeq 0 + V(X) \cdot [H'(\mu)]^2 \\ &= [H'(\mu)]^2 \cdot \sigma^2. \end{aligned} \tag{3-18}$$

If the variance of X, σ^2, is large and the mean, μ, is small, there may be a rather large error in this approximation.

Example 3-13

The surface tension of a liquid is represented by T (dyne/centimeter), and under certain conditions, $T \simeq 2(1 - 0.005X)^{1.2}$, where X is the liquid temperature in degrees centigrade. If X has probability density function f_X, where

$$f_X(x) = 3000x^{-4}, \qquad x \geq 10,$$
$$= 0 \qquad \text{otherwise,}$$

then

$$E(T) = \int_{10}^{\infty} 2(1-0.005x)^{1.2} \cdot 3000x^{-4}\, dx$$

and

$$V(T) = \int_{10}^{\infty} 4(1-0.005x)^{2.4} \cdot 3000x^{-4} dx - \left(E(T)\right)^2.$$

In order to determine these values, it is necessary to evaluate

$$\int_{10}^{\infty} \frac{(1-0.005x)^{1.2}}{x^4} dx \qquad \text{and} \qquad \int_{10}^{\infty} \frac{(1-0.005x)^{2.4}}{x^4} dx.$$

Since the evaluation is difficult, we use the approximations given by equations 3-17 and 3-18. Note that

$$\mu = E(X) = \int_{10}^{\infty} x \cdot 3000x^{-4} dx = -1500x^{-2}\Big|_{10}^{\infty} = 15°C$$

and

$$\sigma^2 = V(X) = E\left(X^2\right) - [E(X)]^2 = \int_{10}^{\infty} x^2 \cdot 3000x^{-4} dx - 15^2 = 75(°C)^2.$$

Since

$$H(X) = 2(1 - 0.005X)^{1.2},$$

then

$$H'(X) = -\,0.012(1 - 0.005X)^{0.2}$$

and

$$H''(X) = 0.000012(1 - 0.005X)^{-0.8}.$$

Thus,

$$H(15) = 2[1 - 0.005(15)]^{1.2} = 1.82,$$

$$H'(15) \simeq -\,0.012,$$

and

$$H''(15) \simeq 0.$$

Using equations 3-17 and 3-18

$$E(T) \simeq H(15) + \frac{1}{2} H''(15) \cdot \sigma^2 = 1.82$$

and

$$V(T) \simeq [H'(15)]^2 \cdot \sigma^2 = [-\,0.012]^2 \cdot 75 \simeq 0.0108.$$

An alternative approach to this sort of approximation utilizes digital simulation and statistical methods to estimate $E(Y)$ and $V(Y)$. Simulation will be discussed in general in Chapter 19, but the essence of this approach is as follows.

1. Produce n independent realizations of the random variable X, where X has probability distribution p_X or f_X. Call these realizations $x_1, x_2, \ldots, x_n$. (The notion of independence will be discussed further in Chapter 4.)
2. Use x_i to compute independent realizations of Y; namely, $y_1 = H(x_1)$, $y_2 = H(x_2)$, …, $y_n = H(x_n)$.
3. Estimate $E(Y)$ and $V(Y)$ from the $y_1, y_2, \ldots, y_n$ values. For example, the natural estimator for $E(Y)$ is the sample mean $\bar{y} \equiv \sum_{i=1}^{n} y_i / n$. (Such statistical estimation problems will be treated in detail in Chapters 9 and 10.)

As a preview of this approach, we give a taste of some of the details. First, how do we generate the n realizations of X that are subsequently used to obtain $y_1, y_2, \ldots, y_n$? The most important technique, called the *inverse transform method*, relies on a remarkable result.

Theorem 3-2

Suppose that X is a random variable with CDF $F_X(x)$. Then the random variable $F_X(X)$ has a uniform distribution on [0, 1]; that is, it has probability density function

$$\begin{aligned} f_U(u) &= 1, \qquad 0 \le u \le 1, \\ &= 0 \qquad \text{otherwise.} \end{aligned} \tag{3-19}$$

Proof Suppose that X is a continuous random variable. (The discrete case is similar.) Since X is continuous, its CDF $F_X(x)$ has a unique inverse. Thus, the CDF of the random variable $F_X(X)$ is

$$P(F_X(X) \le u) = P(X \le F_X^{-1}(u)) = F_X(F_X^{-1}(u)) = u.$$

Taking the derivative with respect to u, we obtain the probability density function of $F_X(X)$,

$$\frac{d}{du} P\left(F_X(X) \le u\right) = 1,$$

which matches equation 3-19, and completes the proof.

We are now in a position to describe the inverse transform method for generating random variables. According to Theorem 3-2, the random variable $F_X(X)$ has a uniform distribution on [0,1]. Suppose that we can somehow generate such a uniform [0,1] random variate U. The inverse transform method proceeds by setting $F_X(X) = U$ and solving for X via the equation

$$X = F_X^{-1}(U). \tag{3-20}$$

We can then generate independent realizations of Y by setting $Y_i = H(X_i) = H(F_X^{-1}(U_i))$, $i = 1, 2, \ldots, n$, where $U_1, U_2, \ldots, U_n$ are independent realizations of the uniform [0,1] distribution. We defer the question of generating U until Chapter 19, when we discuss computer simulation techniques. Suffice it for now to say that there are a variety of methods available for this task, the most widely used actually being a *pseudo*-uniform generation method—one that generates numbers that *appear* to be independent and uniform on [0,1] but that are actually calculated from a deterministic algorithm.

Example 3-14

Suppose that U is uniform on [0,1]. We will show how to use the inverse transform method to generate an exponential random variate with parameter λ, that is, the continuous distribution having probability density function $f_X(x) = \lambda e^{-\lambda x}$, for $x \geq 0$. The CDF of the exponential random variable is

$$F_X(x) = \int_0^x f_X(t)\,dt = 1 - e^{-\lambda x}.$$

Therefore, by the inverse transform theorem, we can set

$$F_X(X) = 1 - e^{-\lambda X} = U$$

and solve for X,

$$X = F_X^{-1}(U) = -\frac{1}{\lambda}\ln(1-U),$$

where we obtain the inverse after a bit of algebra. In other words, $-(1/\lambda)\ln(1-U)$ yields an exponential random variable with parameter λ.

We remark that it is not always possible to find F_X^{-1} in closed form, so the usefulness of equation 3-20 is sometimes limited. Luckily, many other random variate generation schemes are available, and taken together, they span all of the commonly used probability distributions.

3-7 THE MOMENT-GENERATING FUNCTION

It is often convenient to utilize a special function in finding the moments of a probability distribution. This special function, called the *moment-generating function*, is defined as follows.

Definition

Given a random variable X, the moment-generating function $M_X(t)$ of its probability distribution is the expected value of e^{tX}. Expressed mathematically,

$$M_X(t) = E(e^{tX}) \tag{3-21}$$

$$= \sum_i e^{tx_i} p_X(x_i) \quad \text{discrete } X, \tag{3-22}$$

$$= \int_{-\infty}^{\infty} e^{tx} f_X(x)\,dx \quad \text{continuous } X. \tag{3-23}$$

For certain probability distributions, the moment-generating function may not exist for all real values of t. However, for the probability distributions treated in this book, the moment-generating function always exists.

Expanding e^{tX} as a power series in t we obtain

$$e^{tX} = 1 + tX + \frac{t^2X^2}{2!} + \cdots + \frac{t^rX^r}{r!} + \cdots.$$

On taking expectations we see that

$$M_X(t) = E\left[e^{tX}\right] = 1 + E(X)\cdot t + E\left(X^2\right)\cdot\frac{t^2}{2} + \cdots + E\left(X^r\right)\cdot\frac{t^r}{r!} + \cdots$$

so that

$$M_X(t) = 1 + \mu_1'\cdot t + \mu_2'\cdot\frac{t^2}{2!} + \cdots + \mu_r'\cdot\frac{t^r}{r!} + \cdots. \tag{3-24}$$

Thus, we see that *when $M_X(t)$ is written as a power series in t*, the coefficient of $t^r/r!$ in the expansion is the rth moment about the origin. One procedure, then, for using the moment-generating function would be the following:

1. Find $M_X(t)$ analytically for the particular distribution.
2. Expand $M_X(t)$ as a power series in t and obtain the coefficient of $t^r/r!$ as the rth origin moment.

The main difficulty in using this procedure is the expansion of $M_X(t)$ as a power series in t.

If we are only interested in the first few moments of the distribution, then the process of determining these moments is usually made easier by noting that the rth derivative of $M_X(t)$, with respect to t, evaluated at $t = 0$, is just

$$\frac{d^r}{dt^r} M_X(t)\Big|_{t=0} = E\left[X^r e^{tX}\right]_{t=0} = \mu_r', \tag{3-25}$$

assuming we can interchange the operations of differentiation and expectation. So, a second procedure for using the moment-generating function is the following:

1. Determine $M_X(t)$ analytically for the particular distribution.
2. Find $\mu_r' = \frac{d^r}{dt^r} M_X(t)\Big|_{t=0}$.

Moment-generating functions have many interesting and useful properties. Perhaps the most important of these properties is that the moment-generating function is unique when it exists, so that if we know the moment-generating function, we may be able to determine the form of the distribution.

In cases where the moment-generating function does not exist, we may utilize the characteristic function, $C_X(t)$, which is defined to be the expectation of e^{itX}, where $i = \sqrt{-1}$. There are several advantages to using the characteristic function rather than the moment-generating function, but the principal one is that $C_X(t)$ always exists for all t. However, for simplicity, we will use only the moment-generating function.

Example 3-15

Suppose that X has a *binomial distribution*, that is,

$$\begin{aligned} p_X(x) &= \binom{n}{x} p^x (1-p)^{n-x}, \qquad && x = 0,1,2,\ldots,n, \\ &= 0, && \text{otherwise,} \end{aligned}$$

where $0 < p < 1$ and n is a positive integer. The moment-generating function $M_X(t)$ is

$$\begin{aligned} M_X(t) &= \sum_{x=0}^{n} e^{tx} \binom{n}{x} p^x (1-p)^{n-x} \\ &= \sum_{x=0}^{n} \binom{n}{x} \left(pe^t\right)^x (1-p)^{n-x}. \end{aligned}$$

This last summation is recognized as the binomial expansion of $[pe^t + (1-p)]^n$, so that

$$M_X(t) = [pe^t + (1-p)]^n.$$

Taking the derivatives, we obtain

$$M_X'(t) = npe^t[1 + p(e^t - 1)]^{n-1}$$

and

$$M''_X(t) = npe^t(1 - p + npe^t)[1 + p(e^t - 1)]^{n-2}.$$

Thus

$$\mu'_1 = \mu = M'_X(t)\big|_{t=0} = np$$

and

$$\mu'_2 = M''_X(t)\big|_{t=0} = np(1 - p + np).$$

The second *central moment* may be obtained using $\sigma^2 = \mu'_2 - \mu^2 = np(1 - p)$.

Example 3-16

Assume X to have the following *gamma distribution*:

$$\begin{aligned} f_X(x) &= \frac{a^b}{\Gamma(b)} x^{b-1} e^{-ax}, \quad 0 \le x < \infty, a > 0, b > 0, \\ &= 0, \qquad\qquad \text{otherwise,} \end{aligned}$$

where $\Gamma(b) \equiv \int_0^\infty e^{-y} y^{b-1}\, dy$ is the *gamma function.*

The moment-generating function is

$$M_X(t) = \int_0^\infty \frac{a^b}{\Gamma(b)} e^{x(t-a)} x^{b-1}\, dx,$$

which, if we let $y = x(a - t)$, becomes

$$M_X(t) = \frac{a^b}{\Gamma(b)(a-t)^b} \int_0^\infty e^{-y} y^{b-1}\, dy.$$

Since the integral on the right is just $\Gamma(b)$, we obtain

$$M_X(t) = \frac{a^b}{(a-t)^b} = \left(1 - \frac{t}{a}\right)^{-b} \qquad \text{for } t < a.$$

Now using the power series expansion for

$$\left(1 - \frac{t}{a}\right)^{-b},$$

we find

$$M_X(t) = 1 + b\frac{t}{a} + \frac{b(b+1)}{2!}\left(\frac{t}{a}\right)^2 + \cdots,$$

which gives the moments

$$\mu'_1 = \frac{b}{a} \qquad \text{and} \qquad \mu'_2 = \frac{b(b+1)}{a^2}.$$

3-8 SUMMARY

This chapter first introduced methods for determining the probability distribution of a random variable that arises as a function of another random variable with known distribution. That is, where $Y = H(X)$, and either X is discrete with known distribution $p_X(x)$ or X is continuous with known density $f_X(x)$, methods were presented for obtaining the probability distribution of Y.

The expected value operator was introduced in general terms for $E[H(X)]$, and it was shown that $E(X) = \mu$, the mean, and $E[(X-\mu)^2] = \sigma^2$, the variance. The variance operator V was given as $V(X) = E(X^2) - [E(X)]^2$. Approximations were developed for $E[H(X)]$ and $V[H(X)]$ that are useful when exact methods prove difficult. We also showed how to use the inverse transform method to generate realizations of random variables and to estimate their means.

The moment-generating function was presented and illustrated for the moments μ_r' of a probability distribution. It was noted that $E(X^r) = \mu_r'$.

3-9 EXERCISES

3-1. A robot positions 10 units in a chuck for machining as the chuck is indexed. If the robot positions the unit improperly, the unit falls away and the chuck position remains open, thus resulting in a cycle that produces fewer than 10 units. A study of the robot's past performance indicates that if X = number of open positions,

$$
\begin{aligned}
p_X(x) &= 0.6, && x = 0,\\
&= 0.3, && x = 1,\\
&= 0.1, && x = 2,\\
&= 0.0, && \text{otherwise.}
\end{aligned}
$$

If the loss due to empty positions is given by $Y = 20X^2$, find the following:

(a) $p_Y(y)$.

(b) $E(Y)$ and $V(Y)$.

3-2. The content of magnesium in an alloy is a random variable given by the following probability density function:

$$
\begin{aligned}
f_X(x) &= \frac{x}{18}, && 0 \le x \le 6,\\
&= 0, && \text{otherwise.}
\end{aligned}
$$

The profit obtained from this alloy is $P = 10 + 2X$.

(a) Find the probability distribution of P.

(b) What is the expected profit?

3-3. A manufacturer of color television sets offers a 1-year warranty of free replacement if the picture tube fails. He estimates the time to failure, T, to be a random variable with the following probability distribution (in units of years):

$$
\begin{aligned}
f_T(t) &= \frac{1}{4}e^{-t/4} && t > 0,\\
&= 0, && \text{otherwise.}
\end{aligned}
$$

(a) What percentage of the sets will he have to service?

(b) If the profit per sale is \$200 and the replacement of a picture tube costs \$200, find the expected profit of the business.

3-4. A contractor is going to bid a project, and the number of days, X, required for completion follows the probability distribution given as

$$
\begin{aligned}
p_X(x) &= 0.1, && x = 10,\\
&= 0.3, && x = 11,\\
&= 0.4, && x = 12,\\
&= 0.1, && x = 13,\\
&= 0.1, && x = 14,\\
&= 0, && \text{otherwise.}
\end{aligned}
$$

The contractor's profit is $Y = 2000(12 - X)$.

(a) Find the probability distribution of Y.

(b) Find $E(X)$, $V(X)$, $E(Y)$, and $V(Y)$.

3-5. Assume that a continuous random variable X has probability density function

$$
\begin{aligned}
f_X(x) &= 2xe^{-x^2}, && x \ge 0,\\
&= 0, && \text{otherwise.}
\end{aligned}
$$

Find the probability distribution of $Z = X^2$.

3-6. In developing a random digit generator, an important property sought is that each digit D_i follows the following *discrete uniform distribution*,

$$
\begin{aligned}
p_{D_i}(d) &= \frac{1}{10}, && d = 0,1,2,3,\ldots,9,\\
&= 0, && \text{otherwise.}
\end{aligned}
$$

(a) Find $E(D_i)$ and $V(D_i)$.

(b) If $y = \lfloor D_i - 4.5 \rfloor$, where $\lfloor\ \rfloor$ is the greatest integer ("round down") function, find $p_Y(y)$, $E(Y)$, $V(Y)$.

3-7. The percentage of a certain additive in gasoline determines the wholesale price. If A is a random variable representing the percentage, then $0 \le A \le 1$. If the percentage of A is less than 0.70 the gasoline is low-test and sells for 92 cents per gallon. If the percentage of A is greater than or equal to 0.70 the gasoline is high-test and sells for 98 cents per gallon. Find the expected revenue per gallon where $f_A(a) = 1, 0 \le a \le 1$; otherwise, $f_A(a) = 0$.

3-8. The probability function of the random variable X,

$$f_X(x) = \frac{1}{\theta} e^{-(1/\theta)(x-\beta)}, \quad x \ge \beta, \theta > 0,$$
$$= 0, \quad \text{otherwise}$$

is known as the *two-parameter exponential distribution.* Find the moment-generating function of X. Evaluate $E(X)$ and $V(X)$ using the moment-generating function.

3-9. A random variable X has the following probability density function:

$$f_X(x) = e^{-x}, \quad x > 0,$$
$$= 0, \quad \text{otherwise.}$$

(a) Develop the density function for $Y = 2X^2$.

(b) Develop the density for $V = X^{1/2}$.

(c) Develop the density for $U = \ln X$.

3-10. A two-sided rotating antenna receives signals. The rotational position (angle) of the antenna is denoted X, and it may be assumed that this position at the time a signal is received is a random variable with the density below. Actually, the randomness lies in the signal.

$$f_X(x) = \frac{1}{2\pi}, \quad 0 \le x \le 2\pi,$$
$$= 0, \quad \text{otherwise.}$$

The signal can be received if $Y > y_0$, where $Y = \tan X$ For instance, $y_0 = 1$ corresponds to $\frac{\pi}{4} < X < \frac{\pi}{2}$ and $\frac{5\pi}{4} < X < \frac{3\pi}{2}$. Find the density function for Y.

3-11. The demand for antifreeze in a season is considered to be a uniform random variable X, with density

$$f_X(x) = 10^{-6}, \quad 10^6 \le x \le 2 \times 10^6,$$
$$= 0, \quad \text{otherwise,}$$

where X is measured in liters. If the manufacturer makes a 50 cent profit on each liter she sells in the fall of the year, and if she must carry any excess over to the next year at a cost of 25 cents per liter, find the "optimum" stock level for a particular fall season.

3-12. The acidity of a certain product, measured on an arbitrary scale, is given by the relationship

$$A = (3 + 0.05G)^2,$$

where G is the amount of one of the constituents having probability distribution

$$f_G(g) = \frac{g}{8}, \quad 0 \le g \le 4,$$
$$= 0, \quad \text{otherwise.}$$

Evaluate $E(A)$ and $V(A)$ by using the approximations derived in this chapter.

3-13. Suppose that X has the uniform probability density function

$$f_X(x) = 1, \quad 1 \le x \le 2,$$
$$= 0, \quad \text{otherwise.}$$

(a) Find the probability density function of $Y = H(X)$ where $H(x) = 4 - x^2$.

(b) Find the probability density function of $Y = H(X)$ where $H(x) = e^x$.

3-14. Suppose that X has the exponential probability density function

$$f_X(x) = e^{-x}, \quad x \ge 0,$$
$$= 0, \quad \text{otherwise.}$$

Find the probability density function of $Y = H(X)$ where

$$H(x) = \frac{3}{(1+x)^2}.$$

3-15. A used-car salesman finds that he sells either 1, 2, 3, 4, 5, or 6 cars per week with equal probability.

(a) Find the moment-generating function of X.

(b) Using the moment-generating function, find $E(X)$ and $V(X)$.

3-16. Let X be a random variable with probability density function

$$f_X(x) = ax^2 e^{-bx^2}, \quad x > 0,$$
$$= 0, \quad \text{otherwise.}$$

(a) Evaluate the constant a.

(b) Suppose a new function $Y = 18X^2$ is of interest. Find an approximate value for $E(Y)$ and for $V(Y)$.

3-17. Assume that Y has the exponential probability density function

$$f_Y(y) = e^{-y}, \quad y > 0,$$
$$= 0, \quad \text{otherwise.}$$

Find the approximate values of $E(X)$ and $V(X)$ where

$$X = \sqrt{Y^2 + 36}.$$

3-18. The concentration of reactant in a chemical process is a random variable having probability distribution

$$f_R(r) = 6r(1 - r), \quad 0 \le r \le 1,$$
$$= 0, \quad \text{otherwise.}$$

The profit associated with the final product is $P = \$1.00 + \$3.00R$. Find the expected value of P. What is the probability distribution of P?

3-19. The repair time (in hours) for a certain electronically controlled milling machine follows the density function

$$f_X(x) = 4xe^{-2x}, \quad x > 0,$$
$$= 0, \quad \text{otherwise.}$$

Determine the moment-generating function for X and use this function to evaluate $E(X)$ and $V(X)$.

3-20. The cross-sectional diameter of a piece of bar stock is circular with diameter X. It is known that $E(X) = 2$ cm and $V(X) = 25 \times 10^{-6}$ cm^2. A cutoff tool cuts wafers that are exactly 1 cm thick, and this is constant. Find the expected volume of a wafer.

3-21. If a random variable X has moment-generating function $M_X(t)$, prove that the random variable $Y = aX + b$ has moment-generating function $e^{tb} M_X(at)$.

3-22. Consider the *beta distribution* probability density function

$$f_X(x) = k(1-x)^{a-1}x^{b-1}, \quad 0 \le x \le 1,\ a > 0,\ b > 0,$$
$$= 0, \quad \text{otherwise.}$$

(a) Evaluate the constant k.

(b) Find the mean.

(c) Find the variance.

3-23. The probability distribution of a random variable X is given by

$$p_X(x) = 1/2, \quad x = 0,$$
$$= 1/4, \quad x = 1,$$
$$= 1/8, \quad x = 2,$$
$$= 1/8, \quad x = 3,$$
$$= 0, \quad \text{otherwise.}$$

(a) Determine the mean and variance of X from the moment-generating function.

(b) If $Y = (X - 2)^2$, find the CDF for Y.

3-24. The third moment about the mean is related to the asymmetry, or skewness, of the distribution and is defined as

$$\mu_3 = E(X - \mu_1')^3.$$

Show that $\mu_3 = \mu_3' - 3\mu_2'\mu_1' + 2(\mu_1')^3$. Show that for a symmetric distribution, $\mu_3 = 0$.

3-25. Let f be a probability density function for which the rth order moment μ_r' exists. Prove that all moments of order less than r also exist.

3-26. A set of constants k_r, called *cumulants*, may be used instead of moments to characterize a probability distribution. If $M_X(t)$ is the moment-generating function of a random variable X, then the cumulants are defined by the generating function

$$\psi_X(t) = \log M_X(t).$$

Thus, the rth cumulant is given by

$$k_r = \left.\frac{d^r \psi_X(t)}{dt^r}\right|_{t=0}.$$

Find the cumulants of the *normal distribution* whose density function is

$$f(x) = \frac{1}{\sigma\sqrt{2\pi}} \exp\left\{-\frac{1}{2}\left(\frac{x-\mu}{\sigma}\right)^2\right\}, -\infty < x < \infty.$$

3-27. Using the inverse transform method, produce 20 realizations of the variable X described by $p_X(x)$ in Exercise 3-23.

3-28. Using the inverse transform method, produce 10 realizations of the random variable T in Exercise 3-3.

Chapter 4

Joint Probability Distributions

4-1 INTRODUCTION

In many situations we must deal with two or more random variables simultaneously. For example, we might select fabricated sheet steel specimens and measure shear strength and weld diameter of spot welds. Thus, both weld shear strength and weld diameter are the random variables of interest. Or we may select people from a certain population and measure their height and weight.

The objective of this chapter is to formulate *joint probability distributions* for two or more random variables and to present methods for obtaining both *marginal* and *conditional* distributions. *Conditional expectation* is defined as well as the *regression of the mean.* We also present a definition of *independence* for random variables, and *covariance* and *correlation* are defined. Functions of two or more random variables are presented, and a special case of *linear combinations* is presented with its corresponding moment-generating function. Finally the *law of large numbers* is discussed.

Definition

If $\mathscr{S}$ is the sample space associated with an experiment $\mathscr{E}$, and $X_1, X_2, \ldots, X_k$ are functions, each assigning a real number $X_1(e), X_2(e) \ldots, X_k(e)$ to every outcome e, we call $[X_1, X_2, \ldots, X_k]$ a *k-dimensional random vector* (see Fig. 4-1).

The range space of the random vector $[X_1, X_2, \ldots, X_k]$ is the set of all possible values of the random vector. This may be represented as $R_{X_1 \times X_2 \times \ldots \times X_k}$, where

$$R_{X_1 \times X_2 \times \ldots \times X_k} = \{[x_1, x_2, \ldots, x_k]: x_1 \in R_{X_1}, x_2 \in R_{X_2}, \ldots, x_k \in R_{X_k}\}$$

This is the Cartesian product of the range space sets for the components. In the case where $k = 2$, that is, where we have a two-dimensional random vector, as in the earlier illustrations, $R_{X_1 \times X_2}$ is a subset of the Euclidean plane.

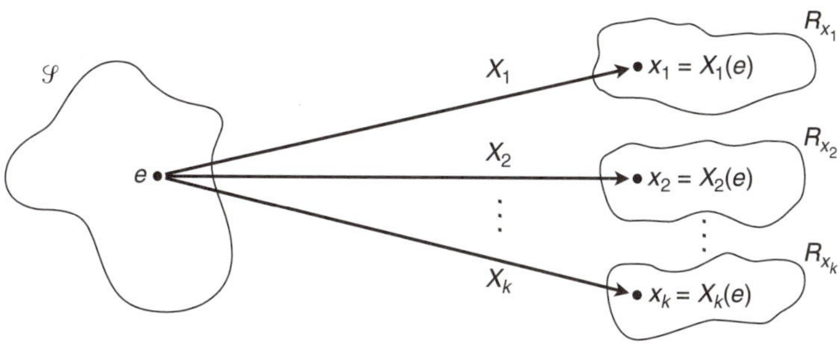

Figure 4-1 A *k*-dimensional random vector.

4-2 JOINT DISTRIBUTION FOR TWO-DIMENSIONAL RANDOM VARIABLES

In most of our considerations here, we will be concerned with two-dimensional random vectors. Sometimes the equivalent term *two-dimensional random variables* will be used.

If the possible values of $[X_1, X_2]$ are either finite or countably infinite in number, then $[X_1, X_2]$ will be a *two-dimensional discrete random vector.* The possible values of $[X_1, X_2]$ are $[x_{1i}, x_{2j}]$, $i = 1, 2, \ldots, j = 1, 2, \ldots$.

If the possible values of $[X_1, X_2]$ are some uncountable set in the Euclidean plane, then $[X_1, X_2]$ will be a *two-dimensional continuous random vector.* For example, if $a \le x_1 \le b$ and $c \le x_2 \le d$, we would have $R_{X_1 \times X_2} = \{[x_1, x_2]: a \le x_1 \le b,\ c \le x_2 \le d\}$.

It is also possible for one component to be discrete and the other continuous; however, here we consider only the case where both are discrete or both are continuous.

Example 4-1

Consider the case where weld shear strength and weld diameter are measured. If we let X_1 represent diameter in inches and X_2 represent strength in pounds, and if we know $0 \le x_1 < 0.25$ inch while $0 \le x_2 \le 2000$ pounds, then the range space for $[X_1, X_2]$ is the set $\{[x_1, x_2]: 0 \le x_1 < 0.25,\ 0 \le x_2 \le 2000\}$. This space is shown graphically in Fig. 4-2.

Example 4-2

A small pump is inspected for four quality-control characteristics. Each characteristic is classified as good, minor defect (not affecting operation) or major defect (affecting operation). A pump is to be selected and the defects counted. If X_1 = the number of minor defects and X_2 = the number of major defects, we know that $x_1 = 0, 1, 2, 3, 4$ and $x_2 = 0, 1, \ldots, 4 - x_1$ because only four characteristics are inspected. The range space for $[X_1, X_2]$ is thus $\{[0, 0], [0, 1], [0, 2], [0, 3], [0, 4], [1, 0], [1, 1], [1, 2], [1, 3], [2, 0], [2, 1], [2, 2], [3, 0], [3, 1], [4, 0]\}$. These possible outcomes are shown in Fig. 4-3.

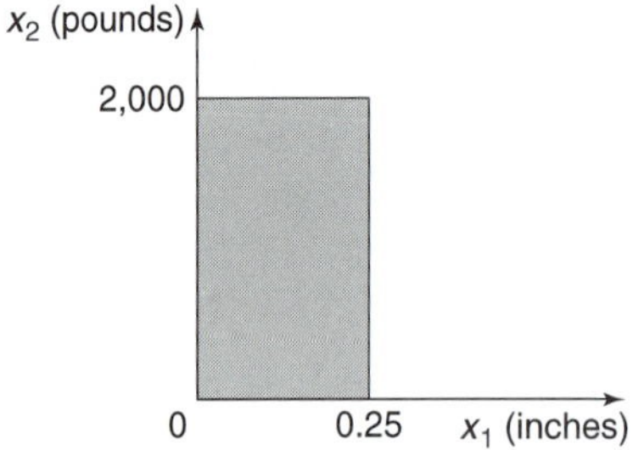

Figure 4-2 The range space of $[X_1, X_2]$, where X_1 is weld diameter and X_2 is shear strength.

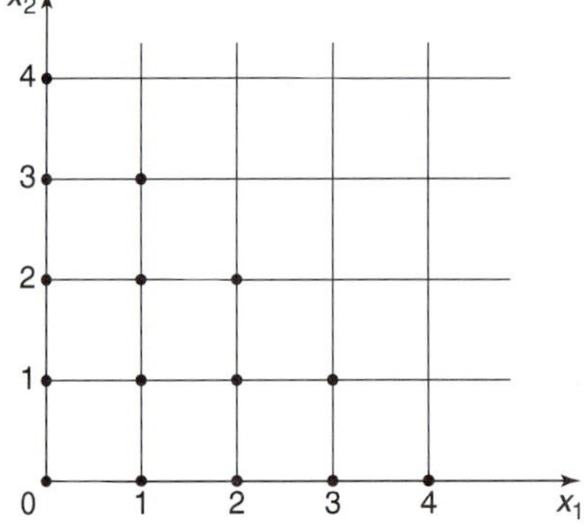

Figure 4-3 The range space of $[X_1, X_2]$, where X_1 is the number of minor defects and X_2 is the number of major defects. The range space is indicated by heavy dots.

In presenting the joint distributions in the definition that follows and throughout the remaining sections of this chapter, where no ambiguity is introduced, we shall simplify the notation by omitting the subscript on the symbols used to specify these joint distributions. Thus, if $\mathbf{X} = [X_1, X_2]$, $p_{\mathbf{X}}(x_1, x_2) = p(x_1, x_2)$ and $f_{\mathbf{X}}(x_1, x_2) = f(x_1, x_2)$.

Definition

Bivariate probability functions are as follows.

1. *Discrete case.* To each outcome $[x_1, x_2]$ of $[X_1, X_2]$, we associate a number,

$$p(x_1, x_2) = P(X_1 = x_1 \text{ and } X_2 = x_2),$$

where

$$p(x_1, x_2) \geq 0, \qquad \text{for all } x_1, x_2,$$

and

$$\sum_{x_1} \sum_{x_2} p(x_1, x_2) = 1. \tag{4-1}$$

The values $([x_1, x_2], p(x_1, x_2))$ for all i, j make up the *probability distribution* of $[X_1, X_2]$.

2. *Continuous case.* If $[X_1, X_2]$ is a continuous random vector with range space R in the Euclidean plane, then f, the *joint density function*, has the following properties:

$$f(x_1, x_2) \geq 0, \qquad \text{for all } (x_1, x_2) \in R$$

and

$$\iint_R f(x_1, x_2)\, dx_1 dx_2 = 1.$$

A probability statement is then of the form

$$P(a_1 \leq X_1 \leq b_1, a_2 \leq X_2 \leq b_2) = \int_{a_2}^{b_2} \int_{a_1}^{b_1} f(x_1, x_2)\, dx_1\, dx_2$$

(see Fig. 4-4).

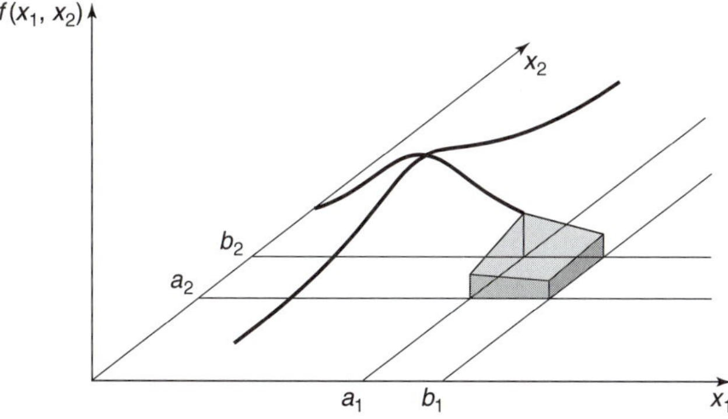

Figure 4-4 A bivariate density function, where $P(a_1 \leq X_1 \leq b_1, a_2 \leq X_2 \leq b_2)$ is given by the shaded volume.

It should again be noted that $f(x_1, x_2)$ does not represent the probability of anything, and the convention that $f(x_1, x_2) = 0$ for $(x_1, x_2) \notin R$ will be employed so that the second property may be written

$$\int_{-\infty}^{\infty}\int_{-\infty}^{\infty} f(x_1, x_2)\, dx_1\, dx_2 = 1.$$

In the case where $[X_1, X_2]$ is discrete, we might present the probability distribution of $[X_1, X_2]$ in tabular, graphical, or mathematical form. In the case where $[X_1, X_2]$ is continuous, we usually employ a mathematical relationship to present the probability distribution; however, a graphical presentation may occasionally be helpful.

Example 4-3

A hypothetical probability distribution is shown in both tabular and graphical form in Fig. 4-5 for the random variables defined in Example 4-2.

Example 4-4

In the case of the weld diameters represented by X_1 and tensile strength represented by X_2, we might have a uniform distribution as by

x_2 \ x_1	0	1	2	3	4
0	1/30	1/30	2/30	3/30	1/30
1	1/30	1/30	3/30	4/30	
2	1/30	2/30	3/30		
3	1/30	3/30			
4	3/30				

(a)

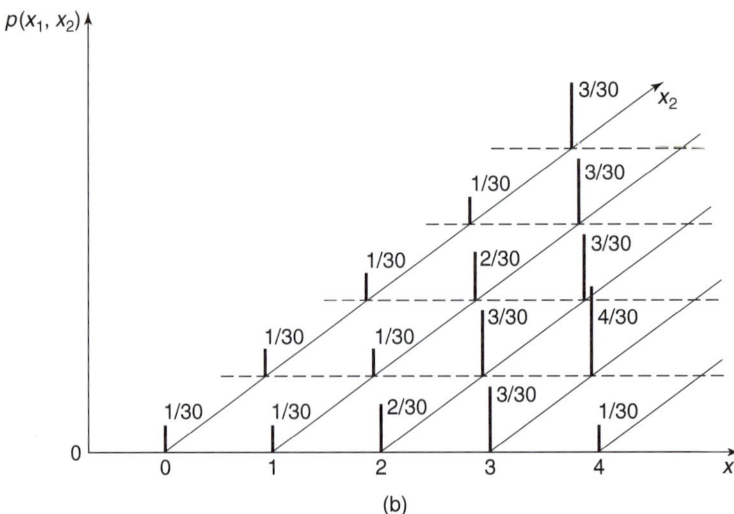

(b)

Figure 4-5 Tabular and graphical presentation of a bivariate probability distribution. (*a*) Tabulated values are $p(x_1, x_2)$. (*b*) Graphical presentation of discrete bivariate distribution.

$$f(x_1,x_2)=\frac{1}{500}, \qquad 0\le x_1<0.25, 0\le x_2\le 2000,$$
$$=0, \qquad \text{otherwise.}$$

The range space was shown in Fig. 4-2, and if we add another dimension to display graphically $y=f(x_1, x_2)$, then the distribution would appear as in Fig. 4-6. In the univariate case, area corresponded to probability; in the bivariate case, volume under the surface represents the probability.

For example, suppose we wish to find $P(0.1 \le X_1 \le 0.2, 100 \le X_2 \le 200)$. This probability would be found by integrating $f(x_1, x_2)$ over the region $0.1 \le x_1 \le 0.2$, $100 \le x_2 \le 200$. That is,

$$\int_{100}^{200}\int_{0.1}^{0.2}\frac{1}{500}dx_1\,dx_2=\frac{1}{50}.$$

4-3 MARGINAL DISTRIBUTIONS

Having defined the bivariate probability distribution, sometimes called the *joint probability distribution* (or in the continuous case the *joint density*), a natural question arises as to the distribution of X_1 or X_2 alone. These distributions are called *marginal distributions.* In the discrete case, the marginal distribution of X_1 is

$$p_1(x_1)=\sum_{x_2}p(x_1,x_2) \quad \text{for all } x_1 \tag{4-2}$$

and the marginal distribution of X_2 is

$$p_2(x_2)=\sum_{x_1}p(x_1,x_2) \quad \text{for all } x_2. \tag{4-3}$$

Example 4-5

In Example 4-2 we considered the joint discrete distribution shown in Fig. 4-5. The marginal distributions are shown in Fig. 4-7. We see that, $[x_1, p_1(x_1)]$ is a univariate distribution and it is the distribution of X_1 (the number of minor defects) alone. Likewise $[x_2, p_2(x_2)]$ is a univariate distribution and it is the distribution of X_2 (the number of major defects) alone.

If $[X_1, X_2]$ is a continuous random vector, the marginal distribution of X_1 is

$$f_1(x_1)=\int_{-\infty}^{\infty}f(x_1,x_2)\,dx_2 \tag{4-4}$$

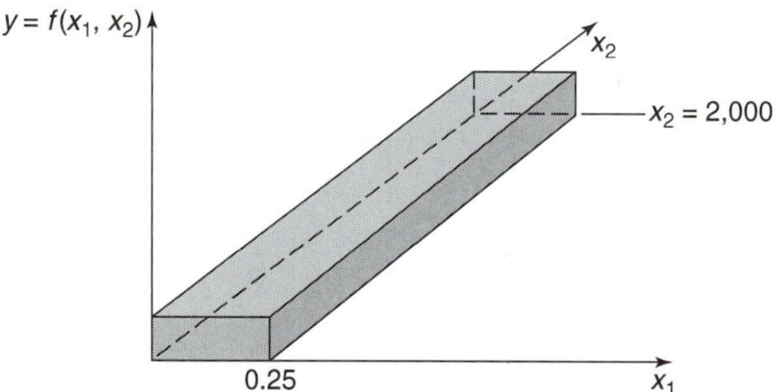

Figure 4-6 A bivariate uniform density.

x_2 \ x_1	0	1	2	3	4	$p_2(x_2)$
0	1/30	1/30	2/30	3/30	1/30	8/30
1	1/30	1/30	3/30	4/30		9/30
2	1/30	2/30	3/30			6/30
3	1/30	3/30				4/30
4	3/30					3/30
$p_1(x_1)$	7/30	7/30	8/30	7/30	1/30	$\sum_x p(x) = 1$

(a)

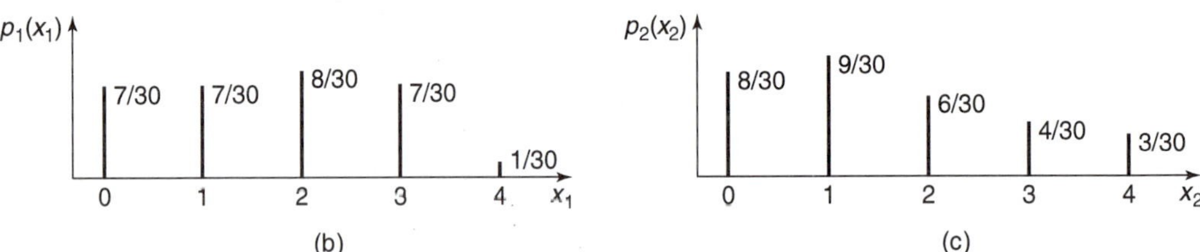

(b) (c)

Figure 4-7 Marginal distributions for discrete $[X_1, X_2]$. (*a*) Marginal distributions—tabular form. (*b*) Marginal distribution $(x_1, p_1(x_1))$. (*c*) Marginal distribution $(x_2, p_2(x_2))$.

and the marginal distribution of X_2 is

$$f_2(x_2) = \int_{-\infty}^{\infty} f(x_1, x_2)\, dx_1. \tag{4-5}$$

The function f_1 is the probability density function for X_1 alone, and the function f_2 is the density function for X_2 alone.

Example 4-6

In Example 4-4, the joint density of $[X_1, X_2]$ was given by

$$\begin{aligned} f(x_1, x_2) &= \frac{1}{500}, && 0 \le x_1 < 0.25,\ 0 \le x_2 \le 2000, \\ &= 0, && \text{otherwise.} \end{aligned}$$

The marginal distributions of X_1 and X_2 are

$$\begin{aligned} f_1(x_1) &= \int_0^{2000} \frac{1}{500}\, dx_2 = 4, && 0 \le x_1 < 0.25, \\ &= 0, && \text{otherwise.} \end{aligned}$$

and

$$\begin{aligned} f_2(x_2) &= \int_0^{.25} \frac{1}{500}\, dx_1 = \frac{1}{2000}, && 0 \le x_2 \le 2000, \\ &= 0, && \text{otherwise.} \end{aligned}$$

These are shown graphically in Fig. 4-8.

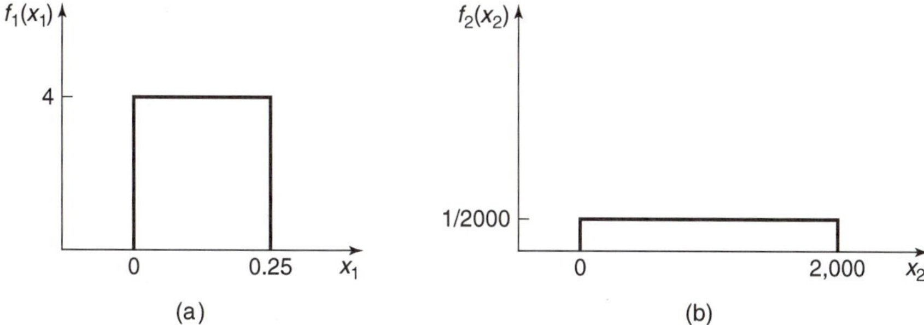

Figure 4-8 Marginal distributions for bivariate uniform vector $[X_1, X_2]$. (*a*) Marginal distribution of X_1. (*b*) Marginal distribution of X_2.

Sometimes the marginals do not come out so nicely. This is the case, for example, when the range space of $[X_1, X_2]$ is not rectangular.

Example 4-7

Suppose that the joint density of $[X_1, X_2]$ is given by

$$\begin{aligned} f(x_1, x_2) &= 6x_1, \qquad 0 < x_1 < x_2 < 1, \\ &= 0, \qquad \text{otherwise.} \end{aligned}$$

Then the marginal of X_1 is

$$\begin{aligned} f_1(x_1) &= \int_{-\infty}^{\infty} f(x_1, x_2)\, dx_2 \\ &= \int_{x_1}^{1} 6x_1\, dx_2 \\ &= 6x_1(1 - x_1) \qquad \text{for } 0 < x_1 < 1, \end{aligned}$$

and the marginal of X_2 is

$$\begin{aligned} f_2(x_2) &= \int_{-\infty}^{\infty} f(x_1, x_2)\, dx_1 \\ &= \int_{0}^{x_2} 6x_1\, dx_1 \\ &= 3x_2^2 \qquad \text{for } 0 < x_2 < 1. \end{aligned}$$

The expected values and variances of X_1 and X_2 are determined from the marginal distributions exactly as in the univariate case. Where $[X_1, X_2]$ is *discrete*, we have

$$E(X_1) = \mu_1 = \sum_{x_1} x_1 p_1(x_1) = \sum_{x_1}\sum_{x_2} x_1 p(x_1, x_2), \tag{4-6}$$

$$\begin{aligned} V(X_1) = \sigma_1^2 &= \sum_{x_1} (x_1 - \mu_1)^2 p_1(x_1) \\ &= \sum_{x_1} x_1^2 p_1(x_1) - \mu_1^2 \\ &= \sum_{x_1}\sum_{x_2} x_1^2 p(x_1, x_2) - \mu_1^2, \end{aligned} \tag{4-7}$$

and, similarly,

$$E(X_2) = \mu_2 = \sum_{x_2} x_2 p_2(x_2) = \sum_{x_1}\sum_{x_2} x_2 p(x_1, x_2), \tag{4-8}$$

$$\begin{aligned} V(X_2) = \sigma_2^2 &= \sum_{x_2} (x_2 - \mu_2)^2 p_2(x_2) \\ &= \sum_{x_2} x_2^2 p_2(x_2) - \mu_2^2 \\ &= \sum_{x_1}\sum_{x_2} x_2^2 p(x_1, x_2) - \mu_2^2. \end{aligned} \tag{4-9}$$

Example 4-8

In Example 4-5 and Fig. 4-7 marginal distributions for X_1 and X_2 were given. Working with the marginal distribution of X_1 shown in Fig. 4-7*b*, we may calculate:

$$E(X_1) = \mu_1 = 0 \cdot \frac{7}{30} + 1 \cdot \frac{7}{30} + 2 \cdot \frac{8}{30} + 3 \cdot \frac{7}{30} + 4 \cdot \frac{1}{30} = \frac{8}{5}$$

and

$$\begin{aligned} V(X_1) = \sigma_1^2 &= \left[0^2 \cdot \frac{7}{30} + 1^2 \cdot \frac{7}{30} + 2^2 \cdot \frac{8}{30} + 3^2 \cdot \frac{7}{30} + 4^2 \cdot \frac{1}{30}\right] - \left[\frac{8}{5}\right]^2 \\ &= \frac{103}{75} \end{aligned}$$

The mean and variance of X_2 could also be determined using the marginal distribution of X_2.

Equations 4-6 through 4-9 show that the mean and variance of X_1 and X_2, respectively, may be determined from the marginal distributions or directly from the joint distribution. In practice, if the marginal distribution has already been determined, it is usually easier to make use of it.

In the case where $[X_1, X_2]$ is *continuous*, then

$$E(X_1) = \mu_1 = \int_{-\infty}^{\infty} x_1 f_1(x_1)dx_1 = \int_{-\infty}^{\infty}\int_{-\infty}^{\infty} x_1 f(x_1, x_2)\,dx_2 dx_1, \tag{4-10}$$

$$\begin{aligned} V(X_1) = \sigma_1^2 &= \int_{-\infty}^{\infty} (x_1 - \mu_1)^2 f_1(x_1)dx_1 \\ &= \int_{-\infty}^{\infty} x_1^2 f_1(x_1)dx_1 - \mu_1^2 \\ &= \int_{-\infty}^{\infty}\int_{-\infty}^{\infty} x_1^2 f(x_1, x_2)dx_2 dx_1 - \mu_1^2, \end{aligned} \tag{4-11}$$

and

$$E(X_2) = \mu_2 = \int_{-\infty}^{\infty} x_2 f_2(x_2)dx_2 = \int_{-\infty}^{\infty}\int_{-\infty}^{\infty} x_2 f(x_1, x_2)\,dx_1 dx_2, \tag{4-12}$$

$$\begin{aligned} V(X_2) = \sigma_2^2 &= \int_{-\infty}^{\infty} (x_2 - \mu_2)^2 f_2(x_2)dx_2 \\ &= \int_{-\infty}^{\infty} x_2^2 f_2(x_2)dx_2 - \mu_2^2 \\ &= \int_{-\infty}^{\infty}\int_{-\infty}^{\infty} x_2^2 f(x_1, x_2)dx_1 dx_2 - \mu_2^2. \end{aligned} \tag{4-13}$$

Again, in equations 4-10 through 4-13, observe that we may use either the marginal densities or the joint density in the calculations.

Example 4-9

In Example 4-4, the joint density of weld diameters, X_1, and shear strength, X_2, was given as

$$f(x_1, x_2) = \frac{1}{500}, \qquad 0 \le x_1 < 0.25, 0 \le x_2 \le 2000,$$
$$= 0, \qquad \text{otherwise.}$$

and the marginal densities for X_1 and X_2 were given in Example 4-6 as

$$f_1(x_1) = 4, \qquad 0 \le x_1 < 0.25,$$
$$= 0, \qquad \text{otherwise,}$$

and

$$f_2(x_2) = \frac{1}{2000}, \qquad 0 \le x_2 < 2000,$$
$$= 0, \qquad \text{otherwise.}$$

Working with the marginal densities, the mean and variance of X_1 are thus

$$E(X_1) = \mu_1 = \int_0^{0.25} x_1 \cdot 4\,dx_1 = 2(0.25)^2 = 0.125$$

and

$$V(X_1) = \sigma_1^2 = \int_0^{0.25} x_1^2 \cdot 4\,dx_1 - (0.125)^2 = \frac{4}{3}(0.25)^3 - (0.125)^2 \simeq 5.21 \times 10^{-3}.$$

4-4 CONDITIONAL DISTRIBUTIONS

When dealing with two jointly distributed random variables it may be of interest to find the distribution of one of these variables, given a particular value of the other. That is, we may wish to find the distribution of X_1 given that $X_2 = x_2$. For example, what is the distribution of a person's weight given that he is a particular height? This probability distribution would be called the *conditional* distribution of X_1 given that $X_2 = x_2$.

Suppose that the random vector $[X_1, X_2]$ is discrete. From the definition of conditional probability, it is easily seen that the conditional probability distributions are

$$p_{X_2|x_1}(x_2) = \frac{p(x_1, x_2)}{p_1(x_1)} \qquad \text{for all } x_1, x_2 \tag{4-14}$$

and

$$p_{X_1|x_2}(x_1) = \frac{p(x_1, x_2)}{p_2(x_2)} \qquad \text{for all } x_1, x_2, \tag{4-15}$$

where $p_1(x_1) > 0$ and $p_2(x_2) > 0$.

It should be noted that there are as many conditional distributions of X_2 for given X_1 as there are values x_1 with $p_1(x_1) > 0$, and there are as many conditional distributions of X_1 for given X_2 as there are values x_2 with $p_2(x_2) > 0$.

Example 4-10

Consider the counting of minor and major defects of the small pumps in Example 4-2 and Fig. 4-7. There will be five conditional distributions of X_2, one for each value of X_1. They are shown in Fig. 4-9. The distribution $p_{X_2|0}(x_2)$, for $X_1 = 0$, is shown in Fig. 4-9*a*. Figure 4-9*b* shows the distribution $p_{X_2|1}(x_2)$. Other conditional distributions could likewise be determined for $X_1 = 2$, 3, and 4, respectively. The distribution of X_1 given that $X_2 = 3$ is

$$p_{X_1|3}(0) = \frac{1/30}{4/30} = \frac{1}{4},$$
$$p_{X_1|3}(1) = \frac{3/30}{4/30} = \frac{3}{4},$$
$$p_{X_1|3}(x_1) = 0, \qquad \text{otherwise.}$$

If $[X_1, X_2]$ is a continuous random vector, the conditional densities are

$$f_{X_2|x_1}(x_2) = \frac{f(x_1, x_2)}{f_1(x_1)} \tag{4-16}$$

and

$$f_{X_1|x_2}(x_1) = \frac{f(x_1, x_2)}{f_2(x_2)}, \tag{4-17}$$

where $f_1(x_1) > 0$ and $f_2(x_2) > 0$.

Example 4-11

Suppose the joint density of $[X_1, X_2]$ is the function f presented here and shown in Fig. 4-10:

$$f(x_1, x_2) = x_1^2 + \frac{x_1 x_2}{3}, \qquad 0 < x_1 \le 1, 0 \le x_2 \le 2,$$
$$= 0, \qquad \text{otherwise.}$$

x_2	0	1	2	3	4
$p_{x_2\|0}(x_2) = p(0, x_2)/p_1(0)$	$\frac{p(0,0)}{p_1(0)}$	$\frac{p(0,1)}{p_1(0)}$	$\frac{p(0,2)}{p_1(0)}$	$\frac{p(0,3)}{p_1(0)}$	$\frac{p(0,4)}{p_1(0)}$
Quotient	$\frac{1/30}{7/30} = \frac{1}{7}$	$\frac{1/30}{7/30} = \frac{1}{7}$	$\frac{1/30}{7/30} = \frac{1}{7}$	$\frac{1/30}{7/30} = \frac{1}{7}$	$\frac{3/30}{7/30} = \frac{3}{7}$

(a)

x_2	0	1	2	3	4
$p_{x_2\|1}(x_2) = p(1, x_2)/p_1(1)$	$\frac{p(1,0)}{p_1(1)}$	$\frac{p(1,1)}{p_1(1)}$	$\frac{p(1,2)}{p_1(1)}$	$\frac{p(1,3)}{p_1(1)}$	$\frac{p(1,4)}{p_1(1)}$
Quotient	$\frac{1/30}{7/30} = \frac{1}{7}$	$\frac{1/30}{7/30} = \frac{1}{7}$	$\frac{2/30}{7/30} = \frac{2}{7}$	$\frac{3/30}{7/30} = \frac{3}{7}$	$\frac{0}{7/30} = 0$

(b)

Figure 4-9 Some examples of conditional distributions.

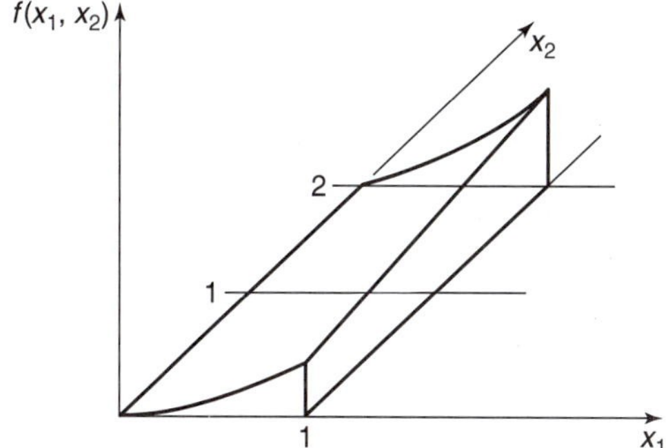

Figure 4-10 A bivariate density function.

The marginal densities are $f_1(x_1)$ and $f_2(x_2)$. These are determined as

$$f_1(x_1) = \int_0^2 \left(x_1^2 + \frac{x_1 x_2}{3} \right) dx_2 = 2x_1^2 + \frac{2}{3}x_1, \quad 0 < x_1 \le 1,$$
$$= 0, \qquad \text{otherwise,}$$

and

$$f_2(x_2) = \int_0^1 \left(x_1^2 + \frac{x_1 x_2}{3} \right) dx_1 = \frac{1}{3} + \frac{x_2}{6}, \quad 0 < x_2 \le 2,$$
$$= 0, \qquad \text{otherwise.}$$

The marginal densities are shown in Fig. 4-11.

The conditional densities may be determined using equations 4-16 and 4-17 as

$$f_{X_2|x_1}(x_2) = \frac{x_1^2 + \dfrac{x_1 x_2}{3}}{2x_1^2 + \dfrac{2}{3}x_1} = \frac{1}{2} \cdot \frac{3x_1 + x_2}{3x_1 + 1}, \quad 0 < x_1 \le 1, 0 \le x_2 \le 2,$$
$$= 0, \qquad \text{otherwise,}$$

and

$$f_{X_1|x_2}(x_1) = \frac{x_1^2 + \dfrac{x_1 x_2}{3}}{\dfrac{1}{3} + \dfrac{x_2}{6}} = \frac{x_1(3x_1 + x_2)}{1 + (x_2/2)}, \quad 0 < x_1 \le 1, 0 \le x_2 \le 2,$$
$$= 0, \qquad \text{otherwise.}$$

Note that for $f_{X_2|x_1}(x_2)$, there are an infinite number of these conditional densities, one for each value $0 < x_1 < 1$. Two of these, $f_{X_2|(1/2)}(x_2)$ and $f_{X_2|1}(x_2)$, are shown in Fig. 4-12. Also, for $f_{X_1|x_2}(x_1)$, there are an

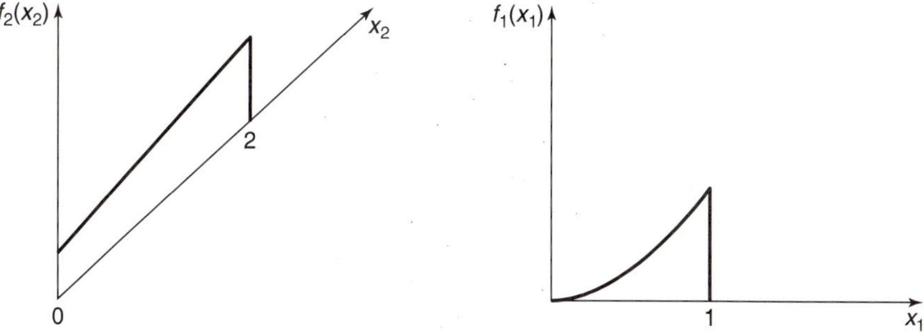

Figure 4-11 The marginal densities for Example 4-11.

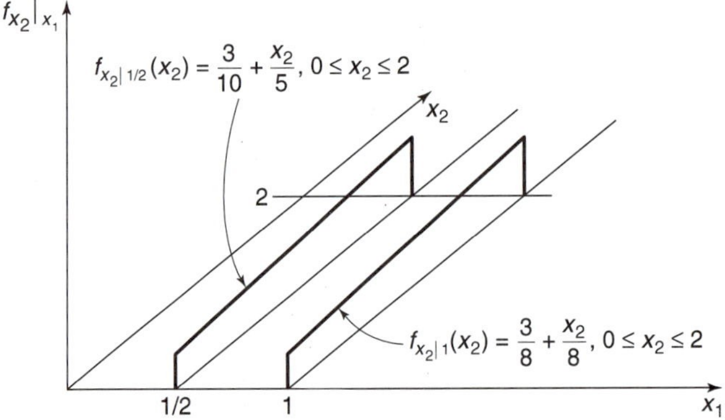

Figure 4-12 Two conditional densities $f_{X_2|1/2}$ and $f_{X_2|1}$, from Example 4-11.

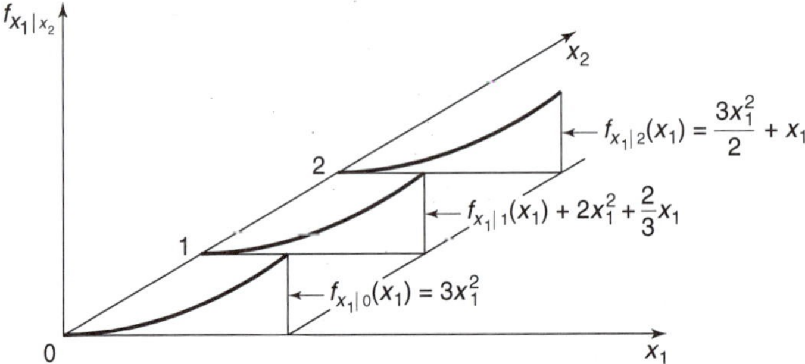

Figure 4-13 Three conditional densities $f_{X_1|0}$, $f_{X_1|1}$, and $f_{X_1|2}$, from Example 4-11.

infinite number of these conditional densities, one for each value $0 \le x_2 \le 2$. Three of these are shown in Fig. 4-13.

4-5 CONDITIONAL EXPECTATION

In this section we are interested in such questions as determining a person's expected weight given that he is a particular height. More generally we want to find the expected value of X_1 given information about X_2. If $[X_1, X_2]$ is a discrete random vector, the *conditional expectations* are

$$E(X_1|x_2) = \sum_{x_1} x_1 p_{X_1|x_2}(x_1) \tag{4-18}$$

and

$$E(X_2|x_1) = \sum_{x_2} x_2 p_{X_2|x_1}(x_2). \tag{4-19}$$

Note that there will be an $E(X_1|x_2)$ for each value of x_2. The value of each $E(X_1|x_2)$ will depend on the value x_2, which is in turn governed by the probability function. Similarly, there will be as many values of $E(X_2|x_1)$ as there are values x_1, and the value of $E(X_2|x_1)$ will depend on the value x_1 determined by the probability function.

Example 4-12

Consider the probability distribution of the discrete random vector $[X_1, X_2]$, where X_1 represents the number of orders for a large turbine in July and X_2 represents the number of orders in August. The joint distribution as well as the marginal distributions are given in Fig. 4-14. We consider the three conditional distributions $p_{X_2|0}$, $p_{X_2|1}$, and $p_{X_2|2}$ and the conditional expected values of each:

$$\begin{aligned} p_{X_2|0}(x_2) &= \frac{1}{6}, & x_2 &= 0, \\ &= \frac{2}{6}, & x_2 &= 1, \\ &= \frac{2}{6}, & x_2 &= 2, \\ &= \frac{1}{6}, & x_2 &= 3, \\ &= 0, & &\text{otherwise,} \end{aligned}$$

$$E(X_2|0) = 0 \cdot \frac{1}{6} + 1 \cdot \frac{2}{6} + 2 \cdot \frac{2}{6} + 3 \cdot \frac{1}{6} = 1.5$$

$$\begin{aligned} p_{X_2|1}(x_2) &= \frac{1}{10}, & x_2 &= 0, \\ &= \frac{5}{10}, & x_2 &= 1, \\ &= \frac{3}{10}, & x_2 &= 2, \\ &= \frac{1}{10}, & x_2 &= 3, \\ &= 0, & &\text{otherwise.} \end{aligned}$$

$$E(X_2|1) = 0 \cdot \frac{1}{10} + 1 \cdot \frac{5}{10} + 2 \cdot \frac{3}{10} + 3 \cdot \frac{1}{10} = 1.4$$

$$\begin{aligned} p_{X_2|2}(x_2) &= \frac{1}{2}, & x_2 &= 0, \\ &= \frac{1}{4}, & x_2 &= 1, \\ &= \frac{1}{4}, & x_2 &= 2, \\ &= 0, & x_2 &= 3, \\ &= 0, & &\text{otherwise} \end{aligned}$$

$$E(X_2|2) = 0 \cdot \frac{1}{2} + 1 \cdot \frac{1}{4} + 2 \cdot \frac{1}{4} + 3 \cdot 0 = .75$$

If $[X_1, X_2]$ is a continuous random vector, the *conditional expectations* are

$$E(X_1|x_2) = \int_{-\infty}^{\infty} x_1 \cdot f_{X_1|x_2}(x_1)\,dx_1 \tag{4-20}$$

and

$$E(X_2|x_1) = \int_{-\infty}^{\infty} x_2 \cdot f_{X_2|x_1}(x_2)\,dx_2, \tag{4-21}$$

and in each case there will be an infinite number of values that the expected value may take. In equation 4-20, there will be one value of $E(X_1|x_2)$ for each value x_2 and in equation 4-21, there will be one value of $E(X_2|x_1)$ for each value x_1.

Example 4-13

In Example 4-11, we considered a joint density f, where

$$\begin{aligned} f(x_1, x_2) &= x_1^2 + \frac{x_1 x_2}{3}, & &0 < x_1 \le 1, 0 \le x_2 \le 2, \\ &= 0, & &\text{otherwise.} \end{aligned}$$

$x_2 \backslash x_1$	0	1	2	$p_2(x_2)$
0	0.05	0.05	0.10	0.2
1	0.10	0.25	0.05	0.4
2	0.10	0.15	0.05	0.3
3	0.05	0.05	0.00	0.1
$p_1(x_1)$	0.3	0.5	0.2	

Figure 4-14 Joint and marginal distributions of $[X_1, X_2]$. Values in body of table are $p(x_1, x_2)$.

The conditional densities were

$$f_{X_2/x_1}(x_2) = \frac{1}{2} \cdot \frac{3x_1 + x_2}{3x_1 + 1} \qquad 0 < x_1 \le 1, 0 \le x_2 \le 2,$$

and

$$f_{X_2/x_1}(x_1) = \frac{x_1(3x_1 + x_2)}{1 + (x_2/2)}, \qquad 0 < x_1 \le 1, 0 \le x_2 \le 2.$$

Then, using equation 4-21, the $E(X_2|x_1)$ is determined as

$$\begin{aligned} E(X_2|x_1) &= \int_0^2 x_2 \cdot \frac{1}{2} \cdot \frac{3x_1 + x_2}{3x_1 + 1}\, dx_2 \\ &= \frac{9x_1 + 4}{9x_1 + 3}. \end{aligned}$$

It should be noted that this is a function of x_1. For the two conditional densities shown in Fig. 4-12, where $x_1 = \frac{1}{2}$ and $x_1 = 1$, the corresponding expected values are $E(X_2|\frac{1}{2}) = \frac{17}{15}$ and $E(X_2|1) = \frac{13}{12}$.

Since $E(X_2|x_1)$ is a function of x_1, and x_1 is a realization of the random variable X_1, $E(X_2|X_1)$ is a random variable, and we may consider the expected value of $E(X_2|X_1)$, that is, $E[E(X_2|X_1)]$. The inner operator is the expectation of X_2 given $X_1 = x_1$, and the outer expectation is with respect to the marginal density of X_1. This suggests the following "double expectation" result.

Theorem 4-1

$$E[E(X_2|X_1)] = E(X_2) = \mu_2 \tag{4-22}$$

and

$$E[E(X_1|X_2)] = E(X_1) = \mu_1. \tag{4-23}$$

Proof Suppose that X_1 and X_2 are continuous random variables. (The discrete case is similar.) Since the random variable $E(X_2|X_1)$ is a function of X_1, the law of the unconscious statistician (see Section 3-5) says that

$$\begin{aligned} E[E(X_2|X_1)] &= \int_{-\infty}^{\infty} E(X_2|x_1) f_1(x_1)\, dx_1 \\ &= \int_{-\infty}^{\infty} \left(\int_{-\infty}^{\infty} x_2 f_{X_2|x_1}(x_2)\, dx_2 \right) f_1(x_1)\, dx_1 \\ &= \int_{-\infty}^{\infty} \int_{-\infty}^{\infty} x_2 f_{X_2|x_1}(x_2) f_1(x_1)\, dx_1 dx_2 \\ &= \int_{-\infty}^{\infty} \int_{-\infty}^{\infty} x_2 \frac{f(x_1, x_2)}{f_1(x_1)} f_1(x_1)\, dx_1 dx_2 \\ &= \int_{-\infty}^{\infty} x_2 \int_{-\infty}^{\infty} f(x_1, x_2)\, dx_1 dx_2 \\ &= \int_{-\infty}^{\infty} x_2 f_2(x_2)\, dx_2 \\ &= E(X_2), \end{aligned}$$

which is equation 4-22. Equation 4-23 is derived similarly, and the proof is complete.

Example 4-14

Consider yet again the joint probability density function from Example 4-11.

$$f(x_1, x_2) = x_1^2 + \frac{x_1 x_2}{3}, \qquad 0 < x_1 \le 1, 0 \le x_2 \le 2,$$
$$= 0, \qquad \text{otherwise.}$$

In that example, we derived expressions for the marginal densities $f_1(x_1)$ and $f_2(x_2)$. We also derived $E(X_2|x_1) = (9x_1 + 4)/(9x_1 + 3)$ in Example 4-13. Thus,

$$\begin{aligned} E\left[E(X_2|X_1)\right] &= \int_{-\infty}^{\infty} E(X_2|x_1) f_1(x_1) dx_1 \\ &= \int_0^1 \left(\frac{9x_1 + 4}{9x_1 + 3}\right) \cdot \left(2x_1^2 + \frac{2}{3}x_1\right) dx_1 \\ &= \frac{10}{9}. \end{aligned}$$

Note that this is also $E(X_2)$, since

$$E(X_2) = \int_{-\infty}^{\infty} x_2 f_2(x_2) dx_2 = \int_0^2 x_2 \cdot \left(\frac{1}{3} + \frac{x_2}{6}\right) dx_2 = \frac{10}{9}.$$

The *variance operator* may be applied to conditional distributions exactly as in the univariate case.

4-6 REGRESSION OF THE MEAN

It has been observed previously that $E(X_2|x_1)$ is a value of the random variable $E(X_2|X_1)$ for a particular $X_1 = x_1$, and it is a function of x_1. The graph of this function is called the *regression* of X_2 on X_1. Alternatively, the function $E(X_1|x_2)$ would be called the *regression* of X_1 on X_2. This is demonstrated in Fig. 4-15.

Example 4-15

In Example 4-13 we found $E(X_2|x_1)$ for the bivariate density of Example 4-11, that is,

$$f(x_1, x_2) = x_1^2 + \frac{x_1 x_2}{3}, \qquad 0 < x_1 \le 1, 0 \le x_2 \le 2,$$
$$= 0 \qquad \text{otherwise.}$$

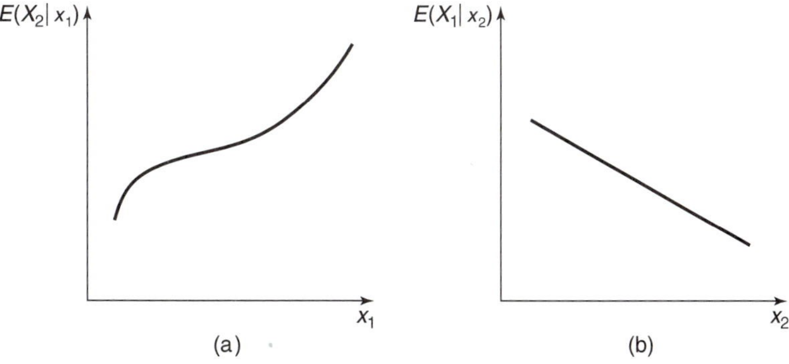

Figure 4-15 Some regression curves. (*a*) Regression of X_2 on X_1. (*b*) Regression of X_1 on X_2.

The result was

$$E(X_2|x_1) = \frac{9x_1+4}{9x_1+3}.$$

In a like manner, we may find

$$E(X_1|x_2) = \int_0^1 x_1 \cdot \left[\frac{3x_1^2+x_1x_2}{1+(x_2/2)}\right] dx_1$$
$$= \frac{4x_2+9}{6x_2+12}.$$

Regression will be discussed further in Chapters 14 and 15.

4-7 INDEPENDENCE OF RANDOM VARIABLES

The notions of independence and independent random variables are very useful and important statistical concepts. In Chapter 1 the idea of independent events was introduced and a formal definition of this concept was presented. We are now concerned with defining *independent random variables.* When the outcome of one variable, say X_1, does not influence the outcome of X_2, and vice versa, we say the random variables X_1 and X_2 are independent.

Definition

1. If $[X_1, X_2]$ is a discrete random vector, then we say that X_1 and X_2 are independent if and only if

$$p(x_1, x_2) = p_1(x_1) \cdot p_2(x_2) \tag{4-24}$$

for all x_1 and x_2.

2. If $[X_1, X_2]$ is a continuous random vector, then we say that X_1 and X_2 are independent if and only if

$$f(x_1, x_2) = f_1(x_1) \cdot f_2(x_2) \tag{4-25}$$

for all x_1 and x_2.

Utilizing this definition and the properties of conditional probability distributions we may extend the concept of independence to a theorem.

Theorem 4-2

1. Let $[X_1, X_2]$ be a discrete random vector. Then

$$p_{X_2|x_1}(x_2) = p_2(x_2)$$

and

$$p_{X_1|x_2}(x_1) = p_1(x_1)$$

for all x_1 and x_2 if and only if X_1 and X_2 are independent.

2. Let $[X_1, X_2]$ be a continuous random vector. Then

$$f_{X_2|x_1}(x_2) = f_2(x_2)$$

and

$$f_{X_1|x_2}(x_1) = f_1(x_1)$$

for all x_1 and x_2 if and only if X_1 and X_2 are independent.

Proof We consider here only the continuous case. We see that

$$f_{X_2|x_1}(x_2) = f_2(x_2) \text{ for all } x_1 \text{ and } x_2$$

if and only if

$$\frac{f(x_1, x_2)}{f_1(x_1)} = f_2(x_2) \qquad \text{for all } x_1 \text{ and } x_2$$

if and only if

$$f(x_1, x_2) = f_1(x_1)f_2(x_2) \text{ for all } x_1 \text{ and } x_2$$

if and only if X_1 and X_2 are independent, and we are done.

Note that the requirement for the joint distribution to be factorable into the respective marginal distributions is somewhat similar to the requirement that, for independent events, the probability of the intersection of events equals the product of the event probabilities.

Example 4-16

A city transit service receives calls from broken-down buses and a wrecker crew must haul the buses in for service. The joint distribution of the number of calls received on Mondays and Tuesdays is given in Fig. 4-16, along with the marginal distributions. The variable X_1 represents the number of calls on Mondays and X_2 represents the number of calls on Tuesdays. A quick inspection will show that X_1 and X_2 are independent, since the joint probabilities are the product of the appropriate marginal probabilities.

4-8 COVARIANCE AND CORRELATION

We have noted that $E(X_1) = \mu_1$ and $V(X_1) = \sigma_1^2$ are the mean and variance of X_1. They may be determined from the marginal distribution of X_1. In a similar manner, μ_2 and σ_2^2 are the mean and variance of X_2. Two measures used in describing the *degree of association* between X_1 and X_2 are the *covariance* of $[X_1, X_2]$ and the *correlation coefficient.*

$x_2 \backslash x_1$	0	1	2	3	4	$p_2(x_2)$
0	0.02	0.04	0.06	0.04	0.04	0.2
1	0.02	0.04	0.06	0.04	0.04	0.2
2	0.01	0.02	0.03	0.02	0.02	0.1
3	0.04	0.08	0.12	0.08	0.08	0.4
4	0.01	0.02	0.03	0.02	0.02	0.1
$p_1(x_1)$	0.1	0.2	0.3	0.2	0.2	

Figure 4-16 Joint probabilities for wrecker calls.

Definition

If $[X_1, X_2]$ is a two-dimensional random variable, the *covariance*, denoted σ_{12}, is

$$\text{Cov}(X_1, X_2) = \sigma_{12} = E[(X_1 - E(X_1))(X_2 - E(X_2))] \tag{4-26}$$

and the *correlation coefficient*, denoted ρ, is

$$\rho = \frac{\text{Cov}(X_1, X_2)}{\sqrt{V(X_1)} \cdot \sqrt{V(X_2)}} = \frac{\sigma_{12}}{\sigma_1 \cdot \sigma_2}. \tag{4-27}$$

The covariance is measured in the units of X_1 times the units of X_2. The correlation coefficient is a dimensionless quantity that measures the linear association between two random variables. By performing the multiplication operations in equation 4-26 before distributing the outside expected value operator across the resulting quantities, we obtain an alternate form for the covariance as follows:

$$\text{Cov}(X_1, X_2) = E(X_1 \cdot X_2) - E(X_1) \cdot E(X_2). \tag{4-28}$$

Theorem 4-3

If X_1 and X_2 are independent, then $\rho = 0$.

Proof We again prove the result for the continuous case. If X_1 and X_2 are independent,

$$\begin{aligned} E(X_1 \cdot X_2) &= \int_{-\infty}^{\infty}\int_{-\infty}^{\infty} x_1 x_2 \cdot f(x_1, x_2)dx_1dx_2 \\ &= \int_{-\infty}^{\infty}\int_{-\infty}^{\infty} x_1 f_1(x_1)\cdot x_2 f_2(x_2)dx_1dx_2 \\ &= \left[\int_{-\infty}^{\infty} x_1 f_1(x_1)dx_1\right]\cdot\left[\int_{-\infty}^{\infty} x_2 f_2(x_2)dx_2\right] \\ &= E(X_1)\cdot E(X_2). \end{aligned}$$

Thus $\text{Cov}(X_1, X_2) = 0$ from equation 4-28, and $\rho = 0$ from equation 4-27. A similar argument would be used for $[X_1, X_2]$ discrete.

The converse of the theorem is not necessarily true, and we may have $\rho = 0$ without the variables being *independent.* If $\rho = 0$, the random variables are said to be *uncorrelated.*

Theorem 4-4

The value of ρ will be on the interval $[-1, +1]$, that is,

$$-1 \le \rho \le +1.$$

Proof Consider the function Q defined below and illustrated in Fig. 4-17.

$$\begin{aligned} Q(t) &= E[(X_1 - E(X_1)) + t(X_2 - E(X_2))]^2 \\ &= E[X_1 - E(X_1)]^2 + 2tE[(X_1 - E(X_1))(X_2 - E(X_2))] \\ &\qquad + t^2E[X_2 - E(X_2)]^2. \end{aligned}$$

Since $Q(t) \ge 0$, the discriminant of $Q(t)$ must be ≤ 0, so that

$$\{2E[(X_1 - E(X_1))(X_2 - E(X_2))]\}^2 - 4E[X_2 - E(X_2)]^2E[X_1 - E(X_1)]^2 \le 0.$$

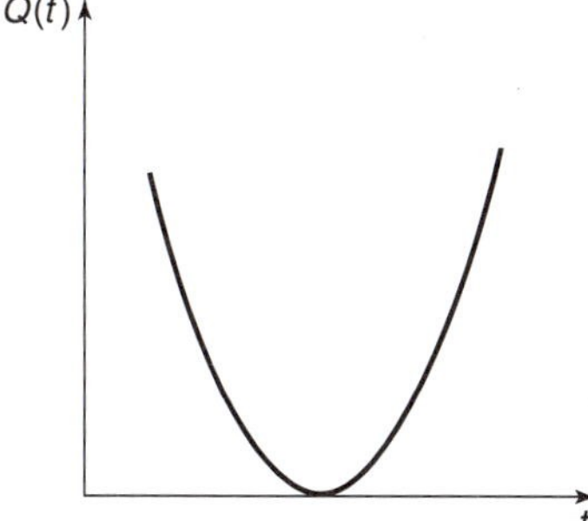

Figure 4-17 The quadratic $Q(t)$.

It follows that

$$4[\mathrm{Cov}(X_1, X_2)]^2 - 4V(X_2) \cdot V(X_1) \le 0,$$

so

$$\frac{[\mathrm{Cov}(X_1, X_2)]^2}{V(X_1)V(X_2)} \le 1$$

and

$$-1 \le \rho \le +1. \tag{4-29}$$

A correlation of $\rho \approx 1$ indicates a "high positive correlation," such as one might find between the stock prices of Ford and General Motors (two related companies). On the other hand, a "high negative correlation," $\rho \approx -1$, might exist between snowfall and temperature.

Example 4-17

Recall Example 4-7, which examined a continuous random vector $[X_1, X_2]$ with the joint probability density function

$$\begin{aligned} f(x_1, x_2) &= 6x_1, && 0 < x_1 < x_2 < 1, \\ &= 0, && \text{otherwise.} \end{aligned}$$

The marginal of X_1 is

$$\begin{aligned} f_1(x_1) &= 6x_1(1 - x_1), && \text{for } 0 < x_1 < 1, \\ &= 0, && \text{otherwise,} \end{aligned}$$

and the marginal of X_2 is

$$\begin{aligned} f_2(x_2) &= 3x_2^2, && \text{for } 0 < x_2 < 1, \\ &= 0, && \text{otherwise.} \end{aligned}$$

These facts yield the following results:

$$E(X_1) = \int_0^1 6x_1^2(1 - x_1)dx_1 = 1/2,$$

$$E(X_1^2) = \int_0^1 6x_1^3(1 - x_1)dx_1 = 3/10,$$

$$V(X_1) = E(X_1^2) - [E(X_1)]^2 = 1/20,$$

$$E(X_2) = \int_0^1 3x_2^3 dx_2 = 3/4,$$

$$E(X_2^2) = \int_0^1 3x_2^4 dx_2 = 3/5,$$

$$V(X_2) = E(X_2^2) - [E(X_2)]^2 = 0.39.$$

Further, we have

$$E(X_1X_2) = \int_{-\infty}^{\infty}\int_{-\infty}^{\infty} x_1x_2 f(x_1, x_2)\, dx_1 dx_2$$

$$= \int_0^1\int_0^{x_2} 6x_1^2 x_2\, dx_1\, dx_2 = 2/5.$$

This implies that

$$\text{Cov}(X_1, X_2) = E(X_1X_2) - E(X_1)E(X_2) = 1/40,$$

and then

$$\rho = \frac{\text{Cov}(X_1, X_2)}{\sqrt{V(X_1)\cdot V(X_2)}} = 0.179.$$

Example 4-18

A continuous random vector $[X_1, X_2]$ has density function f as given below:

$$\begin{aligned} f(x_1, x_2) &= 1, && -x_2 < x_1 < +x_2,\ 0 < x_2 < 1, \\ &= 0, && \text{otherwise.} \end{aligned}$$

This function is shown in Fig. 4-18.

The marginal densities are

$$\begin{aligned} f_1(x_1) &= 1 - x_1 && \text{for } 0 < x_1 < 1, \\ &= 1 + x_1 && \text{for } -1 < x_1 < 0, \\ &= 0, && \text{otherwise,} \end{aligned}$$

and

$$\begin{aligned} f_2(x_2) &= 2x_2, && 0 < x_2 < 1, \\ &= 0, && \text{otherwise.} \end{aligned}$$

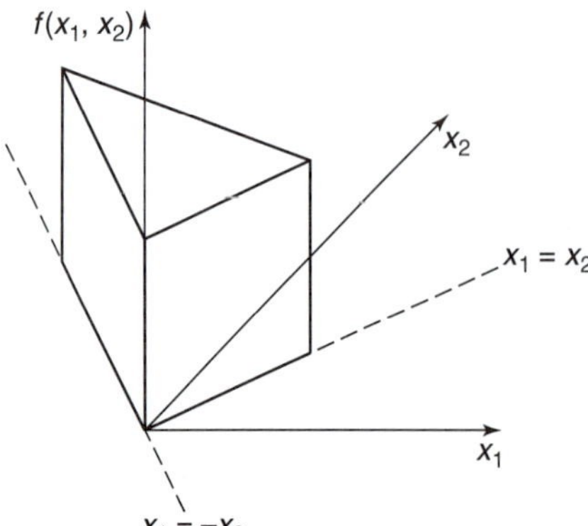

Figure 4-18 A joint density from Example 4-18.

Since $f(x_1, x_2) \neq f_1(x_1) \cdot f_2(x_2)$, the variables are *not independent.* If we calculate the covariance, we obtain

$$\operatorname{Cov}(X_1, X_2) = \int_0^1 \int_{-x_2}^{x_2} x_1 \cdot x_2 \cdot 1 \, dx_1 \, dx_2 - 0 = 0,$$

and thus $\rho = 0$, so the variables are *uncorrelated* although they are not independent.

Finally, it is noted that if X_2 is related to X_1 linearly, that is, $X_2 = A + BX_1$, then $\rho^2 = 1$. If $B > 0$, then $\rho = +1$; and if $B < 0$, $\rho = -1$. Thus, as we observed earlier, the correlation coefficient is a measure of linear association between two random variables.

4-9 THE DISTRIBUTION FUNCTION FOR TWO-DIMENSIONAL RANDOM VARIABLES

The distribution function of the random vector $[X_1, X_2]$ is F, where

$$F(x_1, x_2) = P(X_1 \leq x_1, X_2 \leq x_2). \tag{4-30}$$

This is the probability over the shaded region in Fig. 4-19.

If $[X_1, X_2]$ is discrete, then

$$F(x_1, x_2) = \sum_{t_1 \leq x_1} \sum_{t_2 \leq x_2} p(t_1, t_2), \tag{4-31}$$

and if $[X_1, X_2]$ is continuous, then

$$F(x_1, x_2) = \int_{-\infty}^{x_2} \int_{-\infty}^{x_1} f(t_1, t_2) \, dt_1 dt_2. \tag{4-32}$$

Example 4-19

Suppose X_1 and X_2 have the following density:

$$\begin{aligned} f(x_1, x_2) &= 24x_1x_2, && x_1 > 0, x_2 > 0, x_1 + x_2 < 1, \\ &= 0, && \text{otherwise.} \end{aligned}$$

Looking at the Euclidean plane shown in Fig. 4-20 we see several cases that must be considered.

1. $x_1 \leq 0$, $F(x_1, x_2) = 0$.
2. $x_2 \leq 0$, $F(x_1, x_2) = 0$.

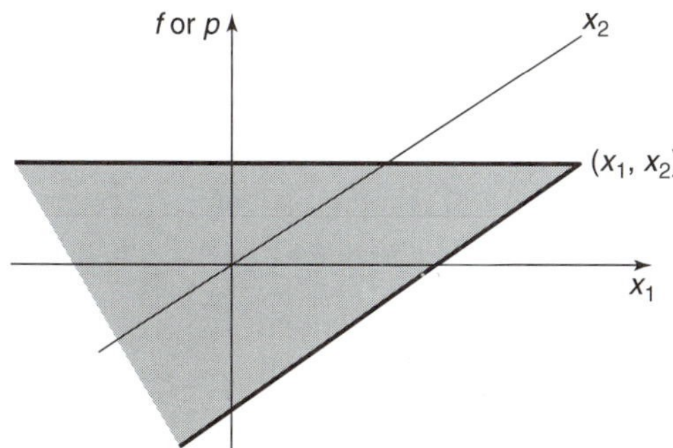

Figure 4-19 Domain of integration or summation for $F(x_1, x_2)$.

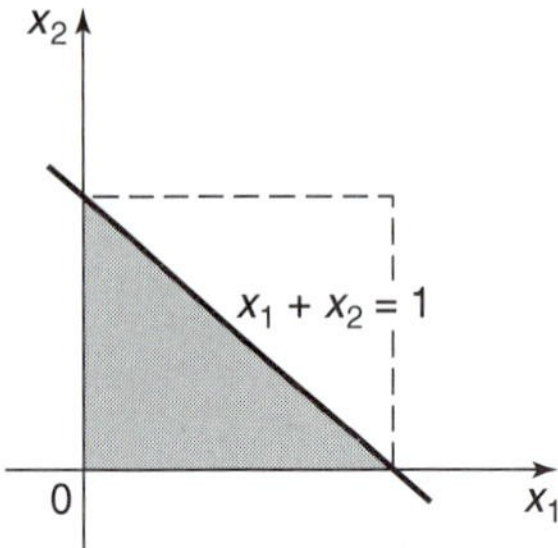

Figure 4-20 The domain of F, Example 4-19.

3. $0 < x_1 < 1$ and $x_1 + x_2 < 1$,

$$F(x_1,x_2)=\int_0^{x_2}\int_0^{x_1} 24t_1t_2\,dt_1\,dt_2$$
$$=6x_1^2\cdot x_2^2.$$

4. $0 < x_1 < 1$ and $1 - x_1 \le x_2 \le 1$,

$$F(x_1,x_2)=\int_0^{1-x_1}\int_0^{x_1} 24t_1t_2\,dt_1\,dt_2+\int_{1-x_1}^{x_2}\int_0^{1-t} 24t_1t_2\,dt_1\,dt_2$$
$$=3x_1^4-8x_1^3+6x_1^2+3x_2^4-8x_2^3+6x_2^2-1.$$

5. $0 < x_1 < 1$ and $x_2 > 1$,

$$F(x_1,x_2)=\int_0^{x_1}\int_0^{1-t_1} 24t_1t_2\,dt_2\,dt_1$$
$$=6x_1^2-8x_1^3+3x_1^4.$$

6. $0 \le x_2 \le 1$ and $x_1 \ge 1$,

$$F(x_1, x_2) = 6x_2^2 - 8x_2^3 + 3x_2^4.$$

7. $x_1 \ge 1$ and $x_2 \ge 1$,

$$F(x_1, x_2) = 1.$$

The function F has properties analogous to those discussed in the one-dimensional case. We note that when X_1 and X_2 are continuous,

$$\frac{\partial^2 F(x_1,x_2)}{\partial x_1\,\partial x_2}=f(x_1,x_2)$$

if the derivatives exist.

4-10 FUNCTIONS OF TWO RANDOM VARIABLES

Often we will be interested in functions of several random variables; however, at present, this section will concentrate on functions of two random variables, say $Y = H(X_1, X_2)$. Since $X_1 = X_1(e)$ and $X_2 = X_2(e)$, we see that $Y = H[X_1(e), X_2(e)]$ clearly depends on the outcome of the original experiment and, thus, Y is a random variable with range space R_Y.

The problem of finding the distribution of Y is somewhat more involved than in the case of functions of one variable; however, if $[X_1, X_2]$ is discrete, the procedure is straightforward if X_1 and X_2 take on a relatively small number of values.

Example 4-20

If X_1 represents the number of defective units produced by machine No. 1 in 1 hour and X_2 represents the number of defective units produced by machine No. 2 in the same hour, then the joint distribution might be presented as in Fig. 4-21. Furthermore, suppose the random variable $Y = H(X_1, X_2)$, where $H(x_1, x_2) = 2x_1 + x_2$. It follows that $R_Y = \{0, 1, 2, 3, 4, 5, 6, 7, 8, 9\}$. In order to determine, say, $P(Y = 0) = p_Y(0)$, we note that $Y = 0$ if and only if $X_1 = 0$ and $X_2 = 0$; therefore, $p_Y(0) = 0.02$.

We note that $Y = 1$ if and only if $X_1 = 0$ and $X_2 = 1$; therefore $p_Y(1) = 0.06$. We also note that $Y = 2$ if and only if either $X_1 = 0, X_2 = 2$ or $X_1 = 1, X_2 = 0$; so $p_Y(2) = 0.10 + 0.03 = 0.13$. Using similar logic, we obtain the rest of the distribution, as follows:

y_i	$p_Y(y_i)$
0	0.02
1	0.06
2	0.13
3	0.11
4	0.19
5	0.15
6	0.21
7	0.07
8	0.05
9	0.01
otherwise	0

In the case where the random vector is continuous with joint density function $f(x_1, x_2)$ and $H(x_1, x_2)$ is continuous, then $Y = H(X_1, X_2)$ is a continuous, one-dimensional random variable. The general procedure for the determination of the density function of Y is outlined below.

1. We are given $Y = H_1(X_1, X_2)$.
2. Introduce a second random variable $Z = H_2(X_1, X_2)$. The function H_2 is selected for convenience, but we want to be able to solve $y = H_1(x_1, x_2)$ and $z = H_2(x_1, x_2)$ for x_1 and x_2 in terms of y and z.
3. Find $x_1 = G_1(y, z)$, and $x_2 = G_2(y, z)$.
4. Find the following partial derivatives (we assume they exist and are continuous):

$$\frac{\partial x_1}{\partial y} \quad \frac{\partial x_1}{\partial z} \quad \frac{\partial x_2}{\partial y} \quad \frac{\partial x_2}{\partial z}.$$

$x_2 \backslash x_1$	0	1	2	3	$p_2(x_2)$
0	0.02	0.03	0.04	0.01	0.1
1	0.06	0.09	0.12	0.03	0.3
2	0.10	0.15	0.20	0.05	0.5
3	0.02	0.03	0.04	0.01	0.1
$p_1(x_1)$	0.2	0.3	0.4	0.1	

Figure 4-21 Joint distribution of defectives produced on two machines $p(x_1, x_2)$.

5. The joint density of $[Y, Z]$, denoted $\ell(y, z)$, is found as follows:

$$\ell(y, z) = f[G_1(y, z), G_2(y, z)] \cdot |J(y, z)|, \tag{4-33}$$

where $J(y, z)$, called the *Jacobian* of the transformation, is given by the determinant

$$J(y,z) = \begin{vmatrix} \partial x_1/\partial y & \partial x_1/\partial z \\ \partial x_2/\partial y & \partial x_2/\partial z \end{vmatrix}. \tag{4-34}$$

6. The density of Y, say g_Y, is then found as

$$g_Y(y) = \int_{-\infty}^{\infty} \ell(y,z)dz. \tag{4-35}$$

Example 4-21

Consider the continuous random vector $[X_1, X_2]$ with the following density:

$$\begin{aligned} f(x_1, x_2) &= 4e^{-2(x_1 + x_2)}, & x_1 > 0, x_2 > 0, \\ &= 0, & \text{otherwise.} \end{aligned}$$

Suppose we are interested in the distribution of $Y = X_1/X_2$. We will let $y = x_1/x_2$ and choose $z = x_1 + x_2$ so that $x_1 = yz/(1 + y)$ and $x_2 = z/(1 + y)$. It follows that

$$\partial x_1/\partial y = \frac{z}{(1+y)^2} \quad \text{and} \quad \partial x_1/\partial z = \frac{y}{1+y},$$

$$\partial x_2/\partial y = \frac{-z}{(1+y)^2} \quad \text{and} \quad \partial x_2/\partial z = \frac{1}{1+y}.$$

Therefore,

$$J(y,z) = \begin{vmatrix} \dfrac{z}{(1+y)^2} & \dfrac{y}{1+y} \\ \dfrac{-z}{(1+y)^2} & \dfrac{1}{(1+y)} \end{vmatrix} = \frac{z}{(1+y)^3} + \frac{zy}{(1+y)^3} = \frac{z}{(1+y)^2}$$

and

$$\begin{aligned} f[G_1(y, z), G_2(y, z)] &= 4e^{\{-2[yz/(1 + y) + z/(1 + y)]\}} \\ &= 4e^{-2z}. \end{aligned}$$

Thus,

$$\ell(y,z) = 4e^{-2z} \cdot \frac{z}{(1+y^2)}$$

and

$$\begin{aligned} g_Y(y) &= \int_0^{\infty} 4e^{-2z}\left[z/(1+y)^2\right]dz \\ &= \frac{1}{(1+y)^2}, \qquad y > 0, \\ &= 0, \qquad \text{otherwise.} \end{aligned}$$

4-11 JOINT DISTRIBUTIONS OF DIMENSION $n > 2$

Should we have three or more random variables, the random vector will be denoted $[X_1, X_2, \ldots, X_n]$, and extensions will follow from the two-dimensional case. We will assume that the variables are continuous; however, the results may readily be extended to the discrete case by substituting the appropriate summation operations for integrals. We assume the existence of a joint density f such that

$$f(x_1, x_2, \ldots, x_n) \geq 0 \tag{4-36}$$

and

$$\int_{-\infty}^{\infty}\int_{-\infty}^{\infty} \cdots \int_{-\infty}^{\infty} f(x_1, x_2, \ldots, x_n)\, dx_n \cdots dx_2 dx_1 = 1.$$

Thus,

$$P(a_1 \leq X_1 \leq b_1, a_2 \leq X_2 \leq b_2, \ldots, a_n \leq X_n \leq b_n)$$
$$= \int_{a_1}^{b_1}\int_{a_2}^{b_2} \cdots \int_{a_n}^{b_n} f(x_1, x_2, \ldots, x_n)\, dx_n \cdots dx_2 dx_1. \tag{4-37}$$

The marginal densities are determined as follows:

$$f_1(x_1) = \int_{-\infty}^{\infty}\int_{-\infty}^{\infty} \cdots \int_{-\infty}^{\infty} f(x_1, x_2, \ldots, x_n)\, dx_n \cdots dx_2,$$

$$f_2(x_2) = \int_{-\infty}^{\infty}\int_{-\infty}^{\infty} \cdots \int_{-\infty}^{\infty} f(x_1, x_2, \ldots, x_n)\, dx_n \cdots dx_3 dx_1,$$

$$f_n(x_n) = \int_{-\infty}^{\infty}\int_{-\infty}^{\infty} \cdots \int_{-\infty}^{\infty} f(x_1, x_2, \ldots, x_n)\, dx_{n-1} \cdots dx_2 dx_1.$$

The integration is over all variables having a subscript different from the one for which the marginal density is required.

Definition

The variables $[X_1, X_2, \ldots, X_n]$ are *independent* random variables if and only if for all $[x_1, x_2, \ldots, x_n]$

$$f(x_1, x_2, \ldots, x_n) = f_1(x_1) \cdot f_2(x_2) \cdot \cdots \cdot f_n(x_n). \tag{4-38}$$

The expected value of, say, X_1 is

$$\mu_1 = E(X_1) = \int_{-\infty}^{\infty}\int_{-\infty}^{\infty} \cdots \int_{-\infty}^{\infty} x_1 \cdot f(x_1, x_2, \cdots, x_n)\, dx_1 dx_2 \cdots dx_n \tag{4-39}$$

and the variance is

$$V(X_1) = \int_{-\infty}^{\infty}\int_{-\infty}^{\infty} \cdots \int_{-\infty}^{\infty} (x_1 - \mu_1)^2 \cdot f(x_1, x_2, \cdots, x_n)\, dx_1 dx_2 \cdots dx_n. \tag{4-40}$$

We recognize these as the mean and variance, respectively, of the marginal distribution of X_1.

In the two-dimensional case considered earlier, geometric interpretations were instructive; however, in dealing with n-dimensional random vectors, the range space is the Euclidean n-space, and graphical presentations are thus not possible. The marginal distributions are, however, in one dimension and the conditional distribution for one variable given values for the other variables is in one dimension. The conditional distribution of X_1 given values $(x_2, x_3, \ldots, x_n)$ is denoted

$$f_{X_1|x_2,\ldots,x_n}(x_1) = \frac{f(x_1, x_2, \ldots, x_n)}{\int_{-\infty}^{\infty} f(x_1, x_2, \ldots, x_n) dx_1}, \tag{4-41}$$

and the expected value of X_1 for given $(x_2, \ldots, x_n)$ is

$$E(X_1 | x_2, x_3, \ldots, x_n) = \int_{-\infty}^{\infty} x_1 \cdot f_{X_1|x_2,\ldots,x_n}(x_1) dx_1. \tag{4-42}$$

The hypothetical graph of $E(X_1|x_2, x_3, \ldots, x_n)$ as a function of the vector $[x_2, x_3, \ldots, x_n]$ is called the *regression of* X_1 *on* $(X_2, X_3, \ldots, X_n)$.

4-12 LINEAR COMBINATIONS

The consideration of general functions of random variables, say $X_1, X_2, \ldots, X_n$, is beyond the scope of this text. However, there is one particular function of the form $Y = H(X_1, \ldots, X_n)$, where

$$H(X_1, X_2, \ldots, X_n) = a_0 + a_1X_1 + \cdots + a_nX_n, \tag{4-43}$$

that is of interest. The a_i are real constants for $i = 0, 1, 2, \ldots, n$. This is called a *linear combination* of the variables $X_1, X_2, \ldots, X_n$. A special situation occurs when $a_0 = 0$ and $a_1 = a_2 = \cdots = a_n = 1$, in which case we have a *sum* $Y = X_1 + X_2 + \cdots + X_n$.

Example 4-22

Four resistors are connected in series as shown in Fig. 4-22. Each resistor has a resistance that is a random variable. The resistance of the assembly may be denoted Y, where $Y = X_1 + X_2 + X_3 + X_4$.

Example 4-23

Two parts are to be assembled as shown in Fig. 4-23. The clearance can be expressed as $Y = X_1 - X_2$ or $Y = (1)X_1 + (-1)X_2$. Of course, a negative clearance would mean interference. This is a linear combination with $a_0 = 0$, $a_1 = 1$, and $a_2 = -1$.

Example 4-24

A sample of 10 items is randomly selected from the output of a process that manufactures a small shaft used in electric fan motors, and the diameters are to be measured with a value called the sample mean, calculated as

$$\overline{X} = \frac{1}{10}(X_1 + X_2 + \cdots + X_{10}).$$

The value $\overline{X} = \frac{1}{10}X_1 + \frac{1}{10}X_2 + \cdots + \frac{1}{10}X_{10}$ is a linear combination with $a_0 = 0$ and $a_1 = a_2 = \cdots = a_{10} = \frac{1}{10}$.

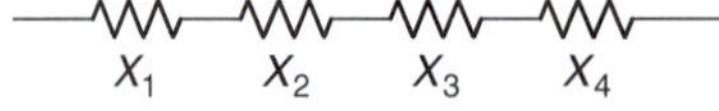

Figure 4-22 Resistors in series.

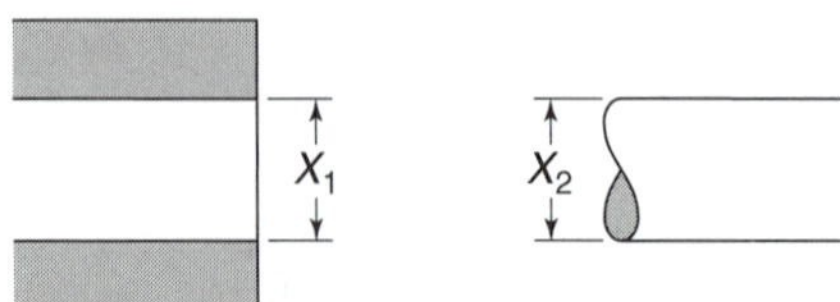

Figure 4-23 A simple assembly.

Let us next consider how to determine the mean and variance of linear combinations. Consider the sum of two random variables,

$$Y = X_1 + X_2. \tag{4-44}$$

The mean of Y or $\mu_Y = E(Y)$ is given as

$$E(Y) = E(X_1) + E(X_2). \tag{4-45}$$

However, the variance calculation is not so obvious.

$$\begin{aligned}
V(Y) &= E[Y - E(Y)]^2 = E(Y^2) - [E(Y)]^2 \\
&= E[(X_1 + X_2)^2] - [E(X_1 + X_2)]^2 \\
&= E[X_1^2 + 2X_1X_2 + X_2^2] - [E(X_1) + E(X_2)]^2 \\
&= E(X_1^2) + 2E(X_1X_2) + E(X_2^2) - [E(X_1)]^2 - 2E(X_1) \cdot E(X_2) - [E(X_2)]^2 \\
&= \{E(X_1^2) - [E(X_1)]^2\} + \{E(X_2^2) - [E(X_2)]^2\} + 2[E(X_1X_2) - E(X_1) \cdot E(X_2)] \\
&= V(X_1) + V(X_2) + 2\text{Cov}(X_1, X_2),
\end{aligned}$$

or

$$\sigma_Y^2 = \sigma_1^2 + \sigma_2^2 + 2\sigma_{12}. \tag{4-46}$$

These results generalize to any linear combination

$$Y = a_0 + a_1X_1 + a_2X_2 + \cdots + a_nX_n \tag{4-47}$$

as follows:

$$\begin{aligned}
E(Y) &= a_0 + \sum_{i=1}^{n} a_i E(X_1) \\
&= a_0 + \sum_{i=1}^{n} a_i \mu_i,
\end{aligned} \tag{4-48}$$

where $E(X_i) = \mu_i$, and

$$V(Y) = \sum_{i=1}^{n} a_i^2 V(X_i) + \sum_{\substack{i=1 \\ i \neq j}}^{n} \sum_{j=1}^{n} a_i a_j \sigma_{ij} \text{ Cov}(X_i, X_j) \tag{4-49}$$

or

$$\sigma_y^2 = \sum_{i=1}^{n} a_i^2 \sigma_i^2 + \sum_{\substack{i=1 \\ i \neq j}}^{n} \sum_{j=1}^{n} a_i a_j \sigma_{ij}.$$

If the *variables are independent*, the expression for the variance of Y is greatly simplified, as all the covariance terms are zero. In this situation the variance of Y is simply

$$V(Y) = \sum_{i=1}^{n} a_i^2 \cdot V(X_i) \tag{4-50}$$

or

$$\sigma_Y^2 = \sum_{i=1}^{n} a_i^2 \sigma_i^2.$$

Example 4-25

In Example 4-22, four resistors were connected in series so that $Y = X_1 + X_2 + X_3 + X_4$ was the resistance of the assembly, where X_1 was the resistance of the first resistor, and so on. The mean and variance of Y in terms of the means and variances of the components may be easily calculated. If the resistors are selected randomly for the assembly, it is reasonable to assume that X_1, X_2, X_3, and X_4 are independent,

$$\mu_Y = \mu_1 + \mu_2 + \mu_3 + \mu_4$$

and

$$\sigma_Y^2 = \sigma_1^2 + \sigma_2^2 + \sigma_3^2 + \sigma_4^2.$$

We have said nothing yet about the distribution of Y; however, given the mean and variance of X_1, X_2, X_3, and X_4, we may readily calculate the mean and the variance of Y since the variables are independent.

Example 4-26

In Example 4-23, where two components were to be assembled, suppose the joint distribution of $[X_1, X_2]$ is

$$\begin{aligned} f(x_1, x_2) &= 8e^{-(2x_1 + 4x_2)}, & x_1 \geq 0, x_2 \geq 0, \\ &= 0, & \text{otherwise.} \end{aligned}$$

Since $f(x_1, x_2)$ can be easily factored as

$$\begin{aligned} f(x_1, x_2) &= [2e^{-2x_1}] \cdot [4e^{-4x_2}] \\ &= f_1(x_1) \cdot f_2(x_2), \end{aligned}$$

X_1 and X_2 are independent. Furthermore, $E(X_1) = \mu_1 = \frac{1}{2}$, and $E(X_2) = \mu_2 = \frac{1}{4}$. We may calculate the variances

$$V(X_1) = \sigma_1^2 = \int_0^\infty x_1^2 \cdot 2e^{-2x_1} dx_1 - \left(\frac{1}{2}\right)^2 = \frac{1}{4}$$

and

$$V(X_2) = \sigma_2^2 = \int_0^\infty x_2^2 \cdot 4e^{-4x_2} dx_2 - \left(\frac{1}{4}\right)^2 = \frac{1}{16}.$$

We denoted the clearance $Y = X_1 - X_2$ or in the example, so that $E(Y) = \mu_1 - \mu_2 = \frac{1}{2} - \frac{1}{4} = \frac{1}{4}$ and $V(Y) = (1)^2 \cdot \sigma_1^2 + (-1)^2 \cdot \sigma_2^2 = \frac{1}{4} + \frac{1}{16} = \frac{5}{16}$.

Example 4-27

In Example 4-24, we might expect the random variables $X_1, X_2, \ldots, X_{10}$ to be independent because of the random sampling process. Furthermore, the distribution for each variable X_i is identical. This is shown in Fig. 4-24. In the earlier example, the linear combination of interest was the *sample mean*

$$\overline{X} = \frac{1}{10} X_1 + \cdots + \frac{1}{10} X_{10}.$$

It follows that

$$\begin{aligned} E(\overline{X}) = \mu_{\overline{X}} &= \frac{1}{10} \cdot E(X_1) + \frac{1}{10} \cdot E(X_2) + \cdots + \frac{1}{10} \cdot E(X_{10}) \\ &= \frac{1}{10}\mu + \frac{1}{10}\mu + \cdots + \frac{1}{10}\mu \\ &= \mu. \end{aligned}$$

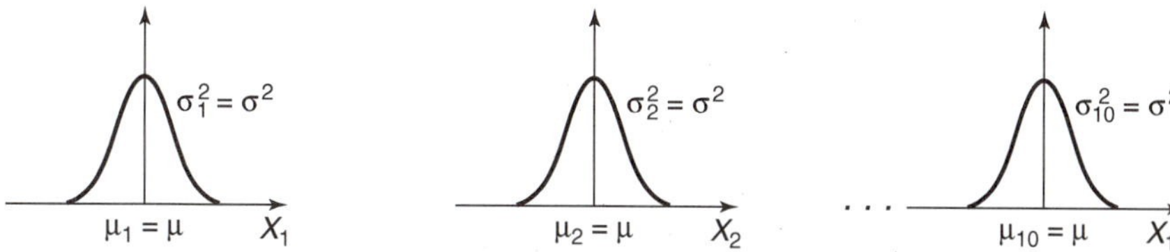

Figure 4-24 Some identical distributions.

Furthermore,

$$
\begin{aligned}
V(\overline{X}) = \sigma_{\overline{X}}^2 &= \left(\frac{1}{10}\right)^2 \cdot V(X_1) + \left(\frac{1}{10}\right)^2 \cdot V(X_2) + \cdots + \left(\frac{1}{10}\right)^2 \cdot V(X_{10}) \\
&= \left(\frac{1}{10}\right)^2 \cdot \sigma^2 + \left(\frac{1}{10}\right)^2 \cdot \sigma^2 + \cdots + \left(\frac{1}{10}\right)^2 \cdot \sigma^2 \\
&= \frac{\sigma^2}{10}.
\end{aligned}
$$

4-13 MOMENT-GENERATING FUNCTIONS AND LINEAR COMBINATIONS

In the case where $Y = aX$, it is easy to show that

$$M_Y(t) = M_X(at). \tag{4-51}$$

For sums of *independent random variables* $Y = X_1 + X_2 + \cdots + X_n$,

$$M_Y(t) = M_{X_1}(t) \cdot M_{X_2}(t) \cdot \cdots \cdot M_{X_n}(t). \tag{4-52}$$

This property has considerable use in statistics. If the linear combination is of the general form $Y = a_0 + a_1X_1 + \cdots + a_nX_n$ and the variables $X_1, \ldots, X_n$ are independent, then

$$M_Y(t) = e^{a_0 t}[M_{X_1}(a_1 t) \cdot M_{X_2}(a_2 t) \cdot \cdots \cdot M_{X_n}(a_n t)].$$

Linear combinations are to be of particular significance in later chapters, and we will discuss them again at greater length.

4-14 THE LAW OF LARGE NUMBERS

A special case arises in dealing with sums of independent random variables where each variable may take only two values, 0 and 1. Consider the following formulation. An experiment $\mathscr{E}$ consists of n independent experiments (trials) $\mathscr{E}_j$, $j = 1, 2, \ldots, n$. There are only two outcomes, success, $\{S\}$, and failure, $\{F\}$, to each trial, so that the sample space $\mathscr{S}_j = \{S, F\}$. The probabilities

$$P\{S\} = p$$

and

$$P\{F\} = 1 - p = q$$

remain constant for $j = 1, 2, \ldots, n$. We let

$$X_j = \begin{cases} 0 & \text{if the } j\text{th trial results in failure,} \\ 1 & \text{if the } j\text{th trial results in success,} \end{cases}$$

and

$$Y = X_1 + X_2 + \cdots + X_n.$$

Thus Y represents the number of successes in n trials, and Y/n is an approximation (or estimator) for the unknown probability p. For convenience, we will let $\hat{p} = Y/n$. Note that this value corresponds to the term f_A used in the relative frequency definition of Chapter 1.

The *law of large numbers* states that

$$P\left[\left|\hat{p} - p\right| < \epsilon\right] \geq 1 - \frac{p(1-p)}{n\epsilon^2}, \tag{4-53}$$

or equivalently

$$P\left[\left|\hat{p} - p\right| \geq \epsilon\right] \leq \frac{p(1-p)}{n\epsilon^2}. \tag{4-54}$$

To indicate the proof, we note that

$$E(Y) = n \cdot E(X_j) = n[(0 \cdot q) + (1 \cdot p)] = np$$

and

$$V(Y) = nV(X_j) = n[(0^2 \cdot q) + (1^2 \cdot p) - (p)^2] = np(1 - p).$$

Since $\hat{p} = Y/n$, we have

$$E(\hat{p}) = \frac{1}{n} \cdot E(Y) = p \tag{4-55}$$

and

$$V(\hat{p}) = \left[\frac{1}{n}\right]^2 \cdot V(Y) = \frac{p(1-p)}{n}.$$

Using Chebyshev's inequality,

$$P\left[\left|\hat{p} - p\right| < k\sqrt{\frac{p(1-p)}{n}}\right] \geq 1 - \frac{1}{k^2}, \tag{4-56}$$

so if

$$\epsilon = k\sqrt{\frac{p(1-p)}{n}}$$

then we obtain equation 4-53.

Thus for arbitrary $\epsilon > 0$, as $n \to \infty$,

$$P\left[\left|\hat{p} - p\right| < \epsilon\right] \to 1.$$

Equation 4-53 may be rewritten, with an obvious notation, as

$$P\left[\left|\hat{p} - p\right| < \epsilon\right] \geq 1 - \alpha. \tag{4-57}$$

We may now fix both ϵ and α in equation 4-57 and determine the value of n required to satisfy the probability statement as

$$n \geq \frac{p(1-p)}{\epsilon^2 \alpha}. \tag{4-58}$$

Example 4-28

A manufacturing process operates so that there is a probability p that each item produced is defective, and p is unknown. A random sample of n items is to be selected to *estimate* p. The estimator to be used is $\hat{p} = Y/n$, where

$$X_j = \begin{cases} 0, & \text{if the } j\text{th item is good,} \\ 1, & \text{if the } j\text{th item is defective,} \end{cases}$$

and

$$Y = X_1 + X_2 + \cdots + X_n.$$

It is desired that the probability be at least 0.95 that the error, $|\hat{p} - p|$, not exceed 0.01. In order to determine the required value of n, we note that $\epsilon = 0.01$, and $\alpha = 0.05$; however, p is unknown. Equation 4-58 indicates that

$$n \geq \frac{p(1-p)}{(0.01)^2 \cdot (0.05)}.$$

Since p is unknown, the worst possible case must be assumed [note that $p(1 - p)$ is maximum when $p = \frac{1}{2}$. This yields

$$n \geq \frac{(0.5)(0.5)}{(0.01)^2(0.05)} = 50,000,$$

a very large number indeed.

Example 4-28 demonstrates why the law of large numbers sometimes requires large sample sizes. The requirements of $\epsilon = 0.01$ and $\alpha = 0.05$ to give a probability of 0.95 of the departure $|\hat{p} - p|$ being less than 0.01 seem reasonable; however, the resulting sample size is very large. In order to resolve problems of this nature we must know the distribution of the random variables involved ($\hat{p}$ in this case). The next three chapters will consider in detail a number of the more frequently encountered distributions.

4-15 SUMMARY

This chapter has presented a number of topics related to jointly distributed random variables and functions of jointly distributed variables. The examples presented illustrated these topics, and the exercises that follow will allow the student to reinforce these concepts.

A great many situations encountered in engineering, science, and management involve situations where several related random variables simultaneously bear on the response being observed. The approach presented in this chapter provides the structure for dealing with several aspects of such problems.

4-16 EXERCISES

4-1. A refrigerator manufacturer subjects his finished products to a final inspection. Of interest are two categories of defects: scratches or flaws in the porcelain finish, and mechanical defects. The number of each type of defect is a random variable. The results of inspecting 50 refrigerators are shown in the following table, where X represents the occurrence of finish defects and Y represents the occurrence of mechanical defects.

(a) Find the marginal distributions of X and Y.

(b) Find the probability distribution of mechanical defects given that there are no finish defects.

Y \ X	0	1	2	3	4	5
0	11/50	4/50	2/50	1/50	1/50	1/50
1	8/50	3/50	2/50	1/50	1/50	
2	4/50	3/50	2/50	1/50		
3	3/50	1/50				
4	1/50					

(c) Find the probability distribution of finish defects given that there are no mechanical defects.

4-2. An inventory manager has accumulated records of demand for her company's product over the last 100 days. The random variable X represents the number of orders received per day and the random variable Y represents the number of units per order. Her data are shown in the table at the bottom of this page.

(a) Find the marginal distributions of X and Y.

(b) Find all conditional distributions for Y given X.

4-3. Let X_1 and X_2 be the scores on a general intelligence test and an occupational preference test, respectively. The probability density function of the random variables $[X_1, X_2]$ is given by

$$f(x_1,x_2)=\frac{k}{1000}, \qquad 0\le x_1\le 100, 0\le x_2\le 10,$$
$$=0, \qquad \text{otherwise.}$$

(a) Find the appropriate value of k.

(b) Find the marginal densities of X_1 and X_2.

(c) Find an expression for the cumulative distribution function $F(x_1, x_2)$.

4-4. Consider a situation in which the surface tension and acidity of a chemical product are measured. These variables are coded such that surface tension is measured on a scale $0 \le X_1 \le 2$, and acidity is measured on a scale $2 \le X_2 \le 4$. The probability density function of $[X_1, X_2]$ is

$$f(x_1, x_2) = k(6 - x_1 - x_2), \qquad 0\le x_1\le 2, 2\le x_2\le 4,$$
$$=0, \qquad \text{otherwise.}$$

(a) Find the appropriate value of k.

(b) Calculate the probability that $X_1 < 1, X_2 < 3$.

(c) Calculate the probability that $X_1 + X_2 \le 4$.

(d) Find the probability that $X_1 < 1.5$.

(e) Find the marginal densities of both X_1 and X_2.

4-5. Consider the density function

$$f(w, x, y, z) = 16wxyz, \qquad 0 \le w, x, y, z \le 1$$
$$=0, \qquad \text{otherwise.}$$

(a) Compute the probability that $W \le \frac{2}{3}$ and $Y \le \frac{1}{2}$.

(b) Compute the probability that $X \le \frac{1}{2}$ and $Z \le \frac{1}{4}$.

(c) Find the marginal density of W.

4-6. Suppose the joint density of $[X, Y]$ is

$$f(x,y)=\frac{1}{8}(6-x-y), \qquad 0\le x\le 2, 2\le y\le 4,$$
$$=0, \qquad \text{otherwise.}$$

Find the conditional densities $f_{X|y}(x)$ and $f_{Y|x}(y)$.

4-7. For the data in Exercise 4-2 find the expected number of units per order given that there are three orders per day.

4-8. Consider the probability distribution of the discrete random vector $[X_1, X_2]$, where X_1 represents the number of orders for aspirin in August at the neighborhood drugstore and X_2 represents the number of orders in September. The joint distribution is shown in the table on the next page.

(a) Find the marginal distributions.

(b) Find the expected sales in September given that sales in August were either 51, 52, 53, 54, or 55.

4-9. Assume that X_1 and X_2 are coded scores on two intelligence tests, and the probability density function of $[X_1, X_2]$ is given by

$$f(x_1, x_2) = 6x_1^2x_2, \qquad 0\le x_1\le 1, 0\le x_2\le 1,$$
$$=0, \qquad \text{otherwise.}$$

Find the expected value of the score on test No. 2 given the score on test No. 1. Also, find the expected

y \ x	1	2	3	4	5	6	7	8	9
1	10/100	6/100	3/100	2/100	1/100	1/100	1/100	1/100	1/100
2	8/100	5/100	3/100	2/100	1/100	1/100	1/100		
3	8/100	5/100	2/100	1/100	1/100				
4	7/100	4/100	2/100	1/100	1/100				
5	6/100	3/100	1/100	1/100					
6	5/100	3/100	1/100	1/100					

X_2 \ X_1	51	52	53	54	55
51	0.06	0.05	0.05	0.01	0.01
52	0.07	0.05	0.01	0.01	0.01
53	0.05	0.10	0.10	0.05	0.05
54	0.05	0.02	0.01	0.01	0.03
55	0.05	0.06	0.05	0.01	0.03

value of the score on test No. 1 given the score on test No. 2.

4-10. Let

$$f(x_1, x_2) = 4x_1x_2e^{-(x_1^2+x_2^2)}, \quad x_1, x_2 > 0$$
$$= 0, \quad \text{otherwise.}$$

(a) Find the marginal distributions of X_1 and X_2.

(b) Find the conditional probability distributions of X_1 and X_2.

(c) Find expressions for the conditional expectations of X_1 and X_2.

4-11. Assume that $[X, Y]$ is a continuous random vector and that X and Y are independent such that $f(x, y) = g(x)h(y)$. Define a new random variable $Z = XY$. Show that the probability density function of Z, $\ell_Z(z)$, is given by

$$\ell_Z(z) = \int_{-\infty}^{\infty} g(t)h\left(\frac{z}{t}\right)\left|\frac{1}{t}\right|dt.$$

Hint: Let $Z = XY$ and $T = X$ and find the Jacobian for the transformation to the joint probability density function of Z and T, say $r(z, t)$. Then integrate $r(z, t)$ with respect to t.

4-12. Use the result of the previous problem to find the probability density function of the area of a rectangle $A = S_1S_2$, where the sides are of random length. Specifically, the sides are independent random variables such that

$$g_{S_1}(s_1) = 2s_1, \quad 0 \le s_1 \le 1,$$
$$= 0 \quad \text{otherwise,}$$

and

$$h_{S_2}(s_2) = \frac{1}{8}s_2, \quad 0 \le s_2 \le 4,$$
$$= 0, \quad \text{otherwise.}$$

Some care must be taken in determining the limits of integration because the variable of integration cannot assume negative values.

4-13. Assume that $[X, Y]$ is a continuous random vector and that X and Y are independent such that $f(x, y) = g(x)h(y)$. Define a new random variable $Z = X/Y$. Show that the probability density function of Z, $\ell_Z(z)$, is given by

$$\ell_Z(z) = \int_{-\infty}^{\infty} g(uz)h(u)|u|du.$$

Hint: Let $Z = X/Y$ and $U = Y$, and find the Jacobian for the transformation to the joint probability density function of Z and U, say $r(z, u)$. Then integrate $r(z, u)$ with respect to u.

4-14. Suppose we have a simple electrical circuit in which Ohm's law $V = IR$ holds. We wish to find the probability distribution of resistance given that the probability distributions of voltage (V) and current (I) are known to be

$$g_V(v) = e^{-v}, \quad v \ge 0,$$
$$= 0, \quad \text{otherwise;}$$
$$h_I(i) = 3e^{-3i}, \quad i \ge 0,$$
$$= 0, \quad \text{otherwise.}$$

Use the results of the previous problem, and assume that V and I are independent random variables.

4-15. Demand for a certain product is a random variable having a mean of 20 units per day and a variance of 9. We define the lead time to be the time that elapses between the placement of an order and its arrival. The lead time for the product is fixed at 4 days. Find the expected value and the variance of *lead time demand*, assuming demands to be independently distributed.

4-16. Prove the discrete case of Theorem 4-2.

4-17. Let X_1 and X_2 be random variables such that $X_2 = A + BX_1$. Show that $\rho^2 = 1$ and that $\rho = -1$ if $B < 0$ while $\rho = +1$ if $B > 0$.

4-18. Let X_1 and X_2 be random variables such that $X_2 = A + BX_1$. Show that the moment-generating function for X_2 is

$$M_{X_2}(t) = e^{At}M_{X_1}(Bt).$$

4-19. Let X_1 and X_2 be distributed according to

$$f(x_1, x_2) = 2, \quad 0 \le x_1 \le x_2 \le 1,$$
$$= 0, \quad \text{otherwise.}$$

Find the correlation coefficient between X_1 and X_2.

4-20. Let X_1 and X_2 be random variables with correlation coefficient ρ_{X_1,X_2}. Suppose we define two new random variables $U = A + BX_1$ and $V = C + DX_2$, where A, B, C, and D are constants. Show that $\rho_{UV} = (BD/|BD|)\rho_{X_1,X_2}$.

4-21. Consider the data shown in Exercise 4-1. Are X and Y independent? Calculate the correlation coefficient.

4-22. A couple wishes to sell their house. The minimum price that they are willing to accept is a random variable, say X, where $s_1 \le X \le s_2$. A population of buyers is interested in the house. Let Y, where $p_1 \le Y \le p_2$, denote the maximum price they are willing to pay. Y is also a random variable. Assume that the joint distribution of $[X, Y]$ is $f(x, y)$.

(a) Under what circumstances will a sale take place?

(b) Write an expression for the probability of a sale taking place.

(c) Write an expression for the expected price of the transaction.

4-23. Let $[X, Y]$ be uniformly distributed over the semicircle in the following diagram. Thus $f(x, y) = 2/\pi$ if $[x, y]$ is in the semicircle.

(a) Find the marginal distributions of X and Y.

(b) Find the conditional probability distributions.

(c) Find the conditional expectations.

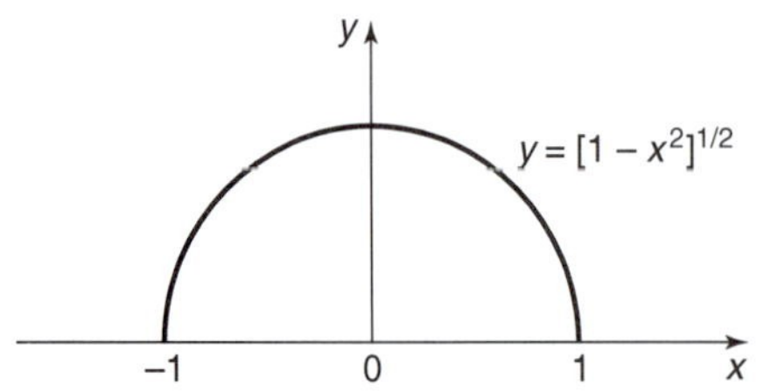

4-24. Let X and Y be independent random variables. Prove that $E(X|y) = E(X)$ and that $E(Y|x) = E(Y)$.

4-25. Show that, in the discrete case,

$$E[E(X|Y)] = E(X),$$
$$E[E(Y|X)] = E(Y).$$

4-26. Consider the two independent random variables S and D, whose probability densities are

$$f_S(s) = \frac{1}{30}, \quad 10 \le s \le 40,$$
$$= 0 \quad \text{otherwise;}$$

$$g_D(d) = \frac{1}{20}, \quad 10 \le d \le 30,$$
$$= 0. \quad \text{otherwise.}$$

Find the probability distribution of the new random variable

$$W = S + D.$$

4-27. If

$$f(x, y) = x + y, \quad 0 < x < 1, 0 < y < 1,$$
$$= 0, \quad \text{otherwise,}$$

find the following:

(a) $E[X|y]$.

(b) $E[X]$.

(c) $E[Y]$.

4-28. For the bivariate distribution,

$$f(x,y) = \frac{k(1+x+y)}{(1+x)^4(1+y)^4}, \quad 0 \le x < \infty, 0 \le y < \infty,$$
$$= 0, \quad \text{otherwise.}$$

(a) Evaluate the constant k.

(b) Find the marginal distribution of X.

4-29. For the bivariate distribution,

$$f(x,y) = \frac{k}{(1+x+y)^n}, \quad x \ge 0, y \ge 0, n > 2,$$
$$= 0, \quad \text{otherwise.}$$

(a) Evaluate the constant k.

(b) Find $F(x, y)$.

4-30. The manager of a small bank wishes to determine the proportion of the time a particular teller is busy. He decides to observe the teller at n randomly spaced intervals. The estimator of the degree of gainful employment is to be Y/n, where

$$X_i = \begin{cases} 0, & \text{if on the } i\text{th observation, the teller is idle,} \\ 1, & \text{if on the } i\text{th observation, the teller is busy,} \end{cases}$$

and $Y = \sum_{i=1}^{n} X_i$. It is desired to estimate $p = P(X_i = 1)$ so that the error of the estimate does not exceed 0.05 with probability 0.95. Determine the necessary value of n.

4-31. Given the following joint distributions, determine whether X and Y are independent.

(a) $g(x, y) = 4xye^{-(x^2+y^2)}, \quad x \ge 0, y \ge 0.$

(b) $f(x, y) = 3x^2y^{-3}, \quad 0 \le x \le y \le 1.$

(c) $f(x, y) = 6(1 + x + y)^{-4}, \quad x \ge 0, y \ge 0.$

4-32. Let $f(x, y, z) = h(x)h(y)h(z)$, $x \ge 0$, $y \ge 0$, $z \ge 0$. Determine the probability that a point drawn at random will have a coordinate (x, y, z) that does not satisfy either $x > y > z$ or $x < y < z$.

4-33. Suppose that X and Y are random variables denoting the fraction of a day that a request for merchandise occurs and the receipt of a shipment occurs, respectively. The joint probability density function is

$$f(x, y) = 1, \quad 0 \le x \le 1, 0 \le y \le 1,$$
$$= 0, \quad \text{otherwise.}$$

(a) What is the probability that both the request for merchandise and the receipt of an order occur during the first half of the day?

(b) What is the probability that a request for merchandise occurs after its receipt? Before its receipt?

4-34. Suppose that in Problem 4-33 the merchandise is highly perishable and must be requested during the

$\frac{1}{4}$-day interval after it arrives. What is the probability that merchandise will not spoil?

4-35. Let X be a continuous random variable with probability density function $f(x)$. Find a general expression for the new random variable Z, where

(a) $Z = a + bX$.

(b) $Z = 1/X$.

(c) $Z = \ln X$.

(d) $Z = e^X$.

Chapter 5

Some Important Discrete Distributions

5-1 INTRODUCTION

In this chapter we present several discrete probability distributions, developing their analytical form from certain basic assumptions about real-world phenomena. We also present some examples of their application. The distributions presented have found extensive application in engineering, operations research, and management science. Four of the distributions, the *binomial*, the *geometric*, the *Pascal*, and the *negative binomial*, stem from a *random process* made up of sequential *Bernoulli trials*. The *hypergeometric distribution*, the *multinomial distribution*, and the *Poisson distribution* will also be presented in this chapter.

When we are dealing with one random variable and no ambiguity is introduced, the symbol for the random variable will once again be omitted in the specification of the probability distributions and cumulative distribution function; thus, $p_X(x) = p(x)$ and $F_X(x) = F(x)$. This practice will be continued throughout the text.

5-2 BERNOULLI TRIALS AND THE BERNOULLI DISTRIBUTION

There are many problems in which the experiment consists of n trials or subexperiments. Here we are concerned with an individual trial that has as its two possible outcomes *success*, S, or *failure*, F. For each trial we thus have the following:

$\mathscr{E}_j$: Perform an experiment (the jth) and observe the outcome.
$\mathscr{S}_j$: $\{S, F\}$.

For convenience, we will define a random variable $X_j = 1$ if $\mathscr{E}_j$ results in $\{S\}$ and $X_j = 0$ if $\mathscr{E}_j$ results in F (see Fig. 5-1).

The n Bernoulli trials $\mathscr{E}_1, \mathscr{E}_2, \ldots, \mathscr{E}_n$ are called a Bernoulli process if the trials are independent, each trial has only two possible outcomes, say S or F, and the probability of success remains constant from trial to trial. That is,

$$p(x_1, x_2, \ldots, x_n) = p_1(x_1) \cdot p_2(x_2) \cdot \cdots \cdot p_n(x_n)$$

and

$$p_j(x_j) = p(x_j) = \begin{cases} p & x_j = 1,\ j = 1, 2, \ldots, n, \\ 1 - p = q, & x_j = 0,\ j = 1, 2, \ldots, n, \\ 0 & \text{otherwise.} \end{cases} \tag{5-1}$$

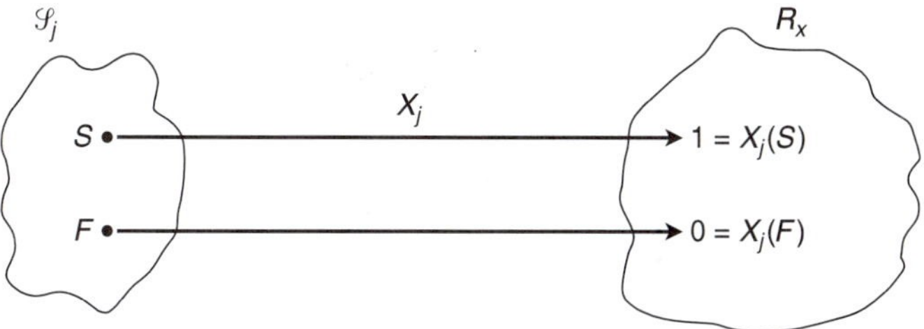

Figure 5-1 A Bernoulli trial.

For one trial, the distribution given in equation 5-1 and Fig. 5-2 is called the *Bernoulli distribution.*

The mean and variance are

$$E(X_j) = (0 \cdot q) + (1 \cdot p) = p$$

and

$$V(X_j) = [(0^2 \cdot q) + (1^2 \cdot p)] - p^2 = p(1-p) = pq. \tag{5-2}$$

The moment-generating function may be shown to be

$$M_{X_j}(t) = q + pe^t. \tag{5-3}$$

Example 5-1

Suppose we consider a manufacturing process in which a small steel part is produced by an automatic machine. Furthermore, each part in a production run of 1000 parts may be classified as defective or good when inspected. We can think of the production of a part as a single trial that results in success (say a defective) or failure (a good item). If we have reason to believe that the machine is just as likely to produce a defective on one run as on another, and if the production of a defective on one run is neither more nor less likely because of the results on the previous runs, then it would be quite reasonable to assume that the production run is a Bernoulli process with 1000 trials. The probability, p, of a defective being produced on one trial is called the *process average fraction defective.*

Note that in the preceding example the assumption of a Bernoulli process is a *mathematical idealization* of the actual real-world situation. Effects of tool wear, machine adjustment, and instrumentation difficulties were ignored. The real world was approximated by a model that did not consider all factors, but nevertheless, the approximation is good enough for useful results to be obtained.

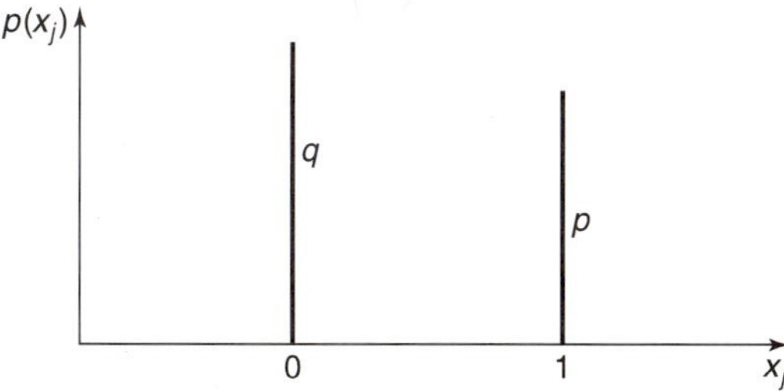

Figure 5-2 The Bernoulli distribution.

We are going to be primarily concerned with a series of Bernoulli trials. In this case the experiment $\mathscr{E}$ is denoted $\{(\mathscr{E}_1, \mathscr{E}_2, \ldots, \mathscr{E}_n): \mathscr{E}_j$ are independent Bernoulli trials, $j = 1, 2, \ldots, n\}$. The sample space is

$$\mathscr{S} = \{(x_1, \ldots, x_n): x_i = S \text{ or } F,\ i = 1, \ldots, n\}.$$

Example 5-2

Suppose an experiment consists of three Bernoulli trials and the probability of success is p on each trial (see Fig. 5-3). The random variable X is given by $X = \sum_{j=1}^{3} X_j$. The distribution of X can be determined as follows:

x	$p(x)$
0	$P\{FFF\} = q \cdot q \cdot q = q^3$
1	$P\{FFS\} + P\{FSF\} + P\{SFF\} = 3pq^2$
2	$P\{FSS\} + P\{SFS\} + P\{SSF\} = 3p^2q$
3	$P\{SSS\} = p^3$

5-3 THE BINOMIAL DISTRIBUTION

The random variable X that denotes *the number of successes in* n *Bernoulli trials has a binomial distribution* given by $p(x)$, where

$$\begin{aligned} p(x) &= \binom{n}{x} p^x (1-p)^{n-x}, & x &= 0, 1, 2, \ldots, n, \\ &= 0, & &\text{otherwise.} \end{aligned} \tag{5-4}$$

Example 5-2 illustrates a binomial distribution with $n = 3$. The parameters of the binomial distribution are n and p, where n is a positive integer and $0 \le p \le 1$. A simple derivation is outlined below. Let

$$p(x) = P\{\text{“}x \text{ successes in } n \text{ trials”}\}.$$

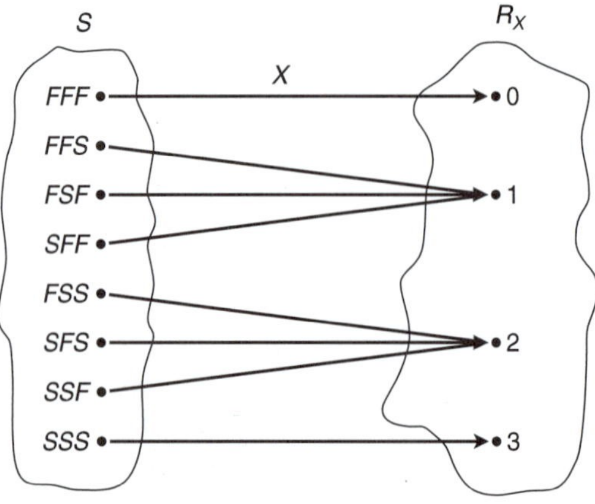

Figure 5-3 Three Bernoulli trials.

The probability of the *particular outcome* in $\mathscr{S}$ with Ss for the first x trials and Fs for the last $n-x$ trials is

$$P\left(\overbrace{SSS\ldots SS}^{x}\;\overbrace{FF\ldots FF}^{n-x}\right)=p^x q^{n-x}$$

(where $q = 1 - p$), due to the independence of the trials. There are $\binom{n}{x}=\frac{n!}{x!(n-x)!}$ outcomes having exactly x Ss and $(n-x)$ Fs; therefore,

$$\begin{aligned} p(x) &= \binom{n}{x} p^x q^{n-x}, \qquad x = 0, 1, 2, \ldots, n, \\ &= 0, \qquad \text{otherwise.} \end{aligned}$$

Since $q = 1 - p$, this last expression is the binomial distribution.

5-3.1 Mean and Variance of the Binomial Distribution

The mean of the binomial distribution may be determined as

$$\begin{aligned} E(X) &= \sum_{x=0}^{n} x \cdot \frac{n!}{x!(n-x)!} p^x q^{n-x} \\ &= np \sum_{x=1}^{n} x \cdot \frac{(n-1)!}{(x-1)!(n-x)!} p^{x-1} q^{n-x}, \end{aligned}$$

and letting $y = x - 1$,

$$E(X) = np \sum_{y=0}^{n-1} \frac{(n-1)!}{y!(n-1-y)!} p^y q^{n-1-y},$$

so that

$$E(X) = np. \tag{5-5}$$

Using a similar approach, we can find

$$\begin{aligned} E\big(X(X-1)\big) &= \sum_{x=0}^{n} \frac{x(x-1)n!}{x!(n-x)!} p^x q^{n-x} \\ &= n(n-1)p^2 \sum_{x=2}^{n} \frac{(n-2)!}{(x-2)!(n-x)!} p^{x-2} q^{n-x} \\ &= n(n-1)p^2 \sum_{y=0}^{n-2} \frac{(n-2)!}{y!(n-y-2)!} p^y q^{n-y-2} \\ &= n(n-1)p^2, \end{aligned}$$

so that

$$\begin{aligned} V(X) &= E(X^2) - (E(X))^2 \\ &= E(X(X-1)) + E(X) - (E(X))^2 \\ &= n(n-1)p^2 + np - (np)^2 \\ &= npq. \end{aligned} \tag{5-6}$$

An easier approach to find the mean and variance is to consider X as a sum of n independent Bernoulli random variables, each with mean p and variance pq, so that $X = X_1 + X_2 + \cdots + X_n$. Then

$$E(X) = p + p + \cdots + p = np$$

and

$$V(X) = pq + pq + \cdots + pq = npq.$$

The moment-generating function for the binomial distribution is

$$M_X(t) = (pe^t + q)^n. \tag{5-7}$$

Example 5-3

A production process represented schematically by Fig. 5-4 produces thousands of parts per day. On the average, 1% of the parts are defective and this average does not vary with time. Every hour, a random sample of 100 parts is selected from a conveyor and several characteristics are observed and measured on each part; however, the inspector classifies the part as either good or defective. If we consider the sampling as $n = 100$ Bernoulli trials with $p = 0.01$, the total number of defectives in the sample, X, would have a binomial distribution

$$\begin{aligned} p(x) &= \binom{100}{x}(0.01)^x(0.99)^{100-x}, \qquad && x = 0, 1, 2, \ldots, 100, \\ &= 0 && \text{otherwise.} \end{aligned}$$

Suppose the inspector has instructions to stop the process if the sample has more than two defectives. Then, $P(X > 2) = 1 - P(X \le 2)$, and we may calculate

$$\begin{aligned} P(X \le 2) &= \sum_{x=0}^{2} \binom{100}{x}(0.01)^x(0.99)^{100-x} \\ &= (0.99)^{100} + 100(0.01)^1(0.99)^{99} + 4950(0.01)^2(0.99)^{98} \\ &\simeq 0.92. \end{aligned}$$

Thus, the probability of the inspector stopping the process is approximately $1 - 0.92 = 0.08$. The mean number of defectives that would be found is $E(X) = np = 100(0.01) = 1$, and the variance is $V(X) = npq = 0.99$.

5-3.2 The Cumulative Binomial Distribution

The cumulative binomial distribution or the distribution function, F, is

$$F(x) = \sum_{k=0}^{x} \binom{n}{k} p^k (1-p)^{n-k}. \tag{5-8}$$

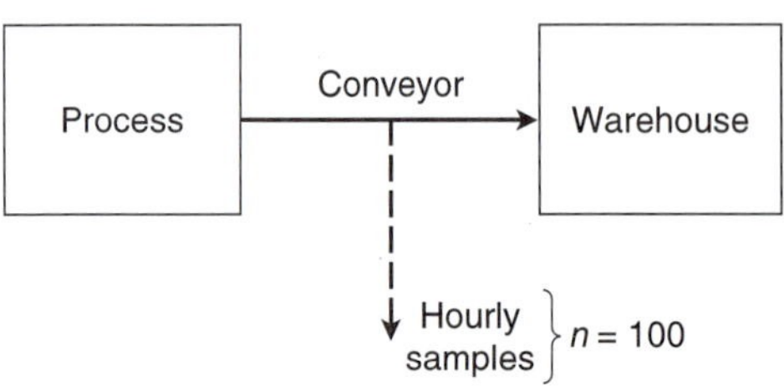

Figure 5-4 A sampling situation with attribute measurement.

The function is readily calculated by such packages as Excel and Minitab. For example, suppose that $n = 10, p = 0.6$, and we are interested in calculating $F(6) = \Pr(X \leq 6)$. Then the Excel function call BINOMDIST(6,10,0.5,TRUE) gives the result $F(6) = 0.6177$. In Minitab, all we need do is go to Calc/Probability Distributions/Binomial, and click on "Cumulative probability" to obtain the same result.

5-3.3 An Application of the Binomial Distribution

Another random variable, first noted in the law of large numbers, is frequently of interest. It is the proportion of successes and is denoted by

$$\hat{p} = X/n, \tag{5-9}$$

where X has a binomial distribution with parameters n and p. The mean, variance, and moment-generating function are

$$E(\hat{p}) = \frac{1}{n} \cdot E(X) = \frac{1}{n} np = p, \tag{5-10}$$

$$V(\hat{p}) = \left(\frac{1}{n}\right)^2 \cdot V(X) = \left(\frac{1}{n}\right)^2 npq = \frac{pq}{n}, \tag{5-11}$$

$$M(t) = M_X\left(\frac{t}{n}\right) = \left(pe^{t/n} + q\right)^n. \tag{5-12}$$

In order to evaluate, say, $P(\hat{p} \leq p_0)$, where p_0 is some number between 0 and 1, we note that

$$P(\hat{p} \leq p_0) = P\left(\frac{X}{n} \leq p_0\right) = P(X \leq np_0).$$

Since np_0 is possibly not an integer,

$$P(\hat{p} \leq p_0) = P(X \leq np_0) = \sum_{x=0}^{\lfloor np_0 \rfloor} \binom{n}{x} p^x q^{n-x}, \tag{5-13}$$

where $\lfloor\ \rfloor$ indicates the "greatest integer contained in" function.

Example 5-4

From a flow of product on a conveyor belt between production operations J and $J + 1$, a random sample of 200 units is taken every 2 hours (see Fig. 5-5). Past experience has indicated that if the unit is not properly degreased, the painting operation will not be successful, and, furthermore, on the average 5% of the units are not properly degreased. The manufacturing manager has grown accustomed to accepting the 5%, but he strongly feels that 6% is bad performance and 7% is totally unacceptable. He decides to plot the fraction defective in the samples, that is, $\hat{p}$. If the process average stays at 5%, he would know that $E(\hat{p}) = 0.05$. Knowing enough about probability to understand that $\hat{p}$ will vary, he asks the quality-control department to determine the $P(\hat{p} > 0.07 \mid p = 0.05)$. This is done as follows:

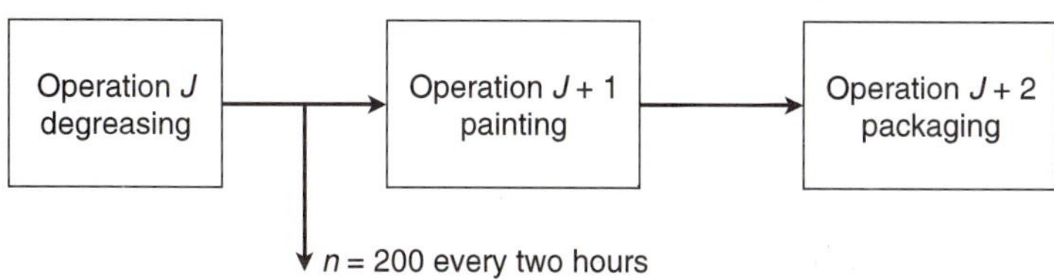

Figure 5-5 Sequential production operations.

$$\begin{aligned} P(\hat{p} > 0.07 | p = 0.05) &= 1 - P(\hat{p} \le 0.07 | p = 0.05) \\ &= 1 - P(X \le 200(0.07) | p = 0.05) \\ &= 1 - \sum_{k=0}^{14} \binom{200}{k} (0.05)^k (0.95)^{200-k} \\ &= 1 - 0.922 = 0.078. \end{aligned}$$

Example 5-5

An industrial engineer is concerned about the excessive "avoidable delay" time that one machine operator seems to have. The engineer considers two activities as "avoidable delay time" and "not avoidable delay time." She identifies a time-dependent variable as follows:

$$\begin{aligned} X(t) &= 1, \qquad \text{avoidable delay,} \\ &= 0, \qquad \text{otherwise.} \end{aligned}$$

A particular realization of $X(t)$ for 2 days (960 minutes) is shown in Fig. 5-6.

Rather than have a time study technician continuously analyze this operation, the engineer elects to use "work sampling," randomly selects n points on the 960-minute span, and estimates the fraction of time the "avoidable delay" category exists. She lets $X_i = 1$ if $X(t) = 1$ at the time of the ith observation and $X_i = 0$ if $X(t) = 0$ at the time of the ith observation. The statistic

$$\hat{P} = \frac{\sum_{i=1}^{n} X_i}{n}$$

is to be evaluated. Of course, $\hat{P}$ is a random variable having a mean equal to p, variance equal to pq/n, and a standard deviation equal to $\sqrt{pq/n}$. The procedure outlined is not necessarily the best way to go about such a study, but it does illustrate one utilization of the random variable $\hat{P}$.

In summary, analysts must be sure that the phenomenon they are studying may be reasonably considered to be a series of Bernoulli trials in order to use the binomial distribution to describe X, the number of successes in n trials. It is often useful to visualize the graphical presentation of the binomial distribution, as shown in Fig. 5-7. The values $p(x)$ increase to a point and then decrease. More precisely, $p(x) > p(x - 1)$ for $x < (n + 1)p$, and $p(x) < p(x - 1)$ for $x > (n + 1)p$. If $(n + 1)p$ is an integer, say m, then $p(m) = p(m - 1)$.

5-4 THE GEOMETRIC DISTRIBUTION

The geometric distribution is also related to a sequence of Bernoulli trials except that the number of trials is not fixed, and, in fact, the random variable of interest, denoted X, is defined to be the number of trials required to achieve the first success. The sample space

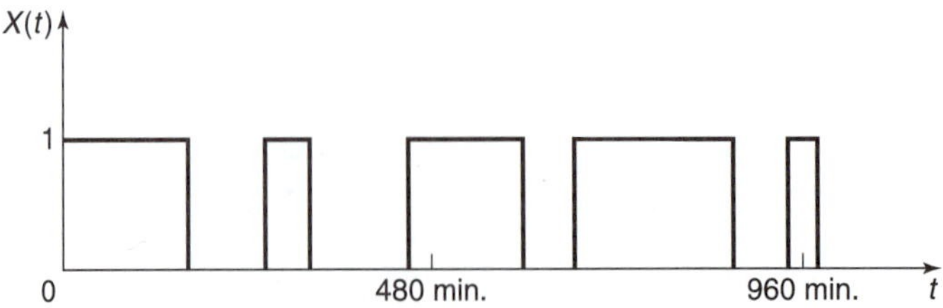

Figure 5-6 A realization of $X(t)$, Example 5-5.

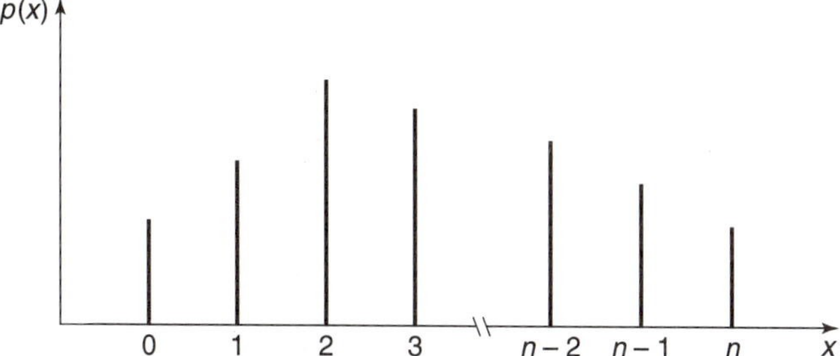

Figure 5-7 The binomial distribution.

and range space for X are illustrated in Fig. 5-8. The range space for X is $R_X = \{1, 2, 3, \ldots\}$, and the distribution of X is given by

$$\begin{aligned} p(x) &= q^{x-1}p, \qquad && x = 1, 2, \ldots, \\ &= 0 && \text{otherwise.} \end{aligned} \tag{5-14}$$

It is easy to verify that this is a probability distribution since

$$\sum_{x=1}^{\infty} pq^{x-1} = p\sum_{k=0}^{\infty} q^k = p \cdot \left[\frac{1}{1-q}\right] = 1$$

and

$$p(x) \geq 0 \qquad \text{for all } x.$$

5-4.1 Mean and Variance of the Geometric Distribution

The mean and variance of the geometric distribution are easily found as follows:

$$\mu = E(X) = \sum_{x=1}^{\infty} x \cdot p \cdot q^{x-1} = p \cdot \frac{d}{dq}\sum_{x=1}^{\infty} q^x,$$

or

$$\mu = p\frac{d}{dq}\left[\frac{q}{1-q}\right] = \frac{1}{p}, \tag{5-15}$$

$$\sigma^2 = V(X) = \sum_{x=1}^{\infty} x^2 \cdot pq^{x-1} - \left(\frac{1}{p}\right)^2 = p\sum_{x=1}^{\infty} x^2 q^{x-1} - \frac{1}{p^2},$$

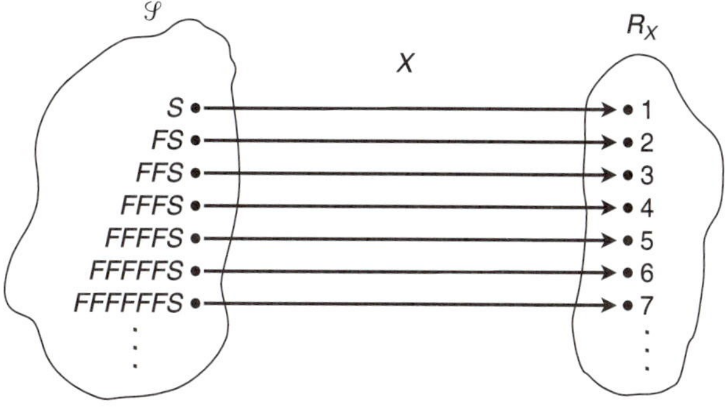

Figure 5-8 Sample space and range space for X.

or, after some algebra,

$$\sigma^2 = q/p^2. \tag{5-16}$$

The moment-generating function is

$$M_X(t) = \frac{pe^t}{1 - qe^t}. \tag{5-17}$$

Example 5-6

A certain experiment is to be performed until a successful result is obtained. The trials are independent and the cost of performing the experiment is \$25,000; however, if a failure results, it costs \$5000 to "set up" for the next trial. The experimenter would like to determine the expected cost of the project. If X is the number of trials required to obtain a successful experiment, then the cost function would be

$$\begin{aligned} C(X) &= \$25{,}000X + \$5000(X - 1) \\ &= 30{,}000X - 5000. \end{aligned}$$

Then

$$\begin{aligned} E[C(X)] &= \$30{,}000 \cdot E(X) - E(\$5000) \\ &= \left[30{,}000 \cdot \frac{1}{p}\right] - 5000. \end{aligned}$$

If the probability of success on a single trial is, say, 0.25, then the $E[C(X)] = \$30{,}000/0.25 - \$5000 = \$115{,}000$. This may or may not be acceptable to the experimenter. It should also be recognized that it is possible to continue indefinitely without having a successful experiment. Suppose that the experimenter has a maximum of \$500,000. He may wish to find the probability that the experimental work would cost more than this amount, that is,

$$\begin{aligned} P(C(X) > \$500{,}000) &= P(\$30{,}000X - \$5000 > \$500{,}000) \\ &= P\left(X > \frac{505{,}000}{30{,}000}\right) \\ &= P(X > 16.833) \\ &= 1 - P(X \le 16) \\ &= 1 - \sum_{x=1}^{16} 0.25(0.75)^{x-1} \\ &\simeq 0.01. \end{aligned}$$

The experimenter may not be at all willing to run the risk (probability 0.01) of spending the available \$500,000 without getting a successful run.

The geometric distribution decreases, that is, $p(x) < p(x - 1)$ for $x = 2, 3, \ldots$. This is shown graphically in Fig. 5-9.

An interesting and useful property of the geometric distribution is that it has no memory, that is,

$$P(X > x + s \mid X > s) = P(X > x). \tag{5-18}$$

The geometric distribution is the only discrete distribution having this *memoryless property.*

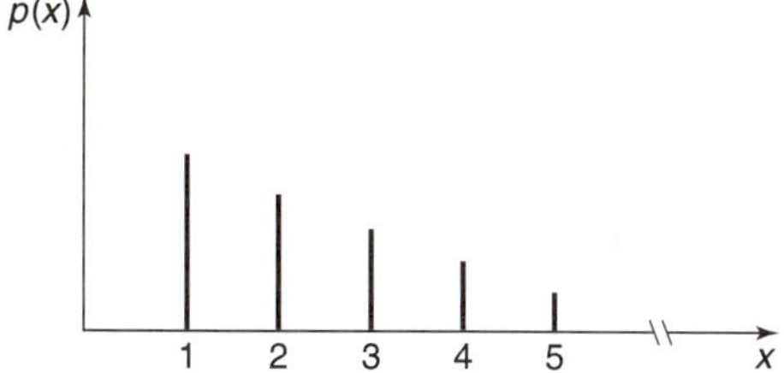

Figure 5-9 The geometric distribution.

Example 5-7

Let X denote the number of tosses of a fair die until we observe a 6. Suppose we have already tossed the die five times without seeing a 6. The probability that more than two additional tosses will be required is

$$\begin{aligned} P(X>7 \mid X>5) &= P(X>2) \\ &= 1-P(X\le 2) \\ &= 1-\sum_{x=1}^{2} p(x) \\ &= 1-\sum_{x=1}^{2} q^{x-1}p \\ &= 1-\frac{1}{6}\left(1+\frac{5}{6}\right) \\ &= \frac{25}{36}. \end{aligned}$$

5-5 THE PASCAL DISTRIBUTION

The *Pascal distribution* also has its basis in Bernoulli trials. It is a logical extension of the geometric distribution. In this case, the random variable X denotes the trial on which the rth success occurs, where r is an integer. The probability mass function of X is

$$\begin{aligned} p(x) &= \binom{x-1}{r-1} p^r q^{x-r}, \qquad x = r, r+1, r+2, \ldots, \\ &= 0, \qquad \text{otherwise.} \end{aligned} \tag{5-19}$$

The term $p^r q^{x-r}$ arises from the probability associated with exactly one outcome in $\mathscr{S}$ that has $(x-r)$ Fs (failures) and r Ss (successes). In order for this outcome to occur, there must be $r-1$ successes in the $x-1$ repetitions before the last outcome, which is always success. There are thus $\binom{x-1}{r-1}$ arrangements satisfying this condition, and therefore the distribution is as shown in equation 5-19.

The development thus far has been for integer values of r. If we have arbitrary $r>0$ and $0<p<1$, the distribution of equation 5-19 is known as the *negative binomial distribution.*

5-5.1 Mean and Variance of the Pascal Distribution

If X has a Pascal distribution, as illustrated in Fig. 5-10, the mean, variance, and moment-generating function are:

$$\mu = r/p, \tag{5-20}$$

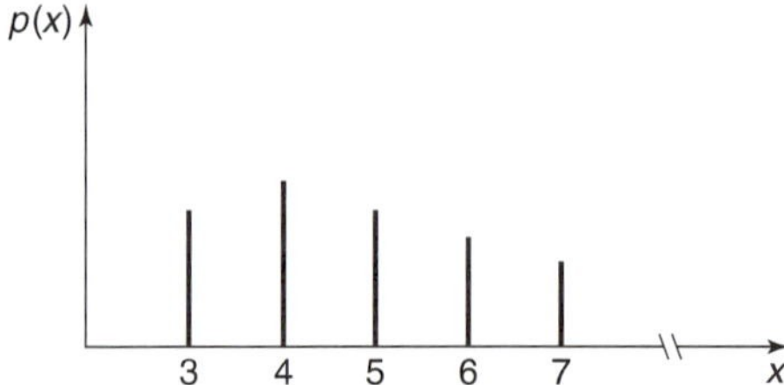

Figure 5-10 An example of the Pascal distribution.

$$\sigma^2 = rq/p^2, \tag{5-21}$$

and

$$M_X(t) = \left(\frac{pe^t}{1 - qe^t} \right)^r. \tag{5-22}$$

Example 5-8

The president of a large corporation makes decisions by throwing darts at a board. The center section is marked "yes" and represents a success. The probability of his hitting a "yes" is 0.6, and this probability remains constant from throw to throw. The president continues to throw until he has three "hits." We denote X as the number of the trial on which he experiences the third hit. The mean is 3/0.6 = 5, meaning that on the average it will take five throws. The president's decision rule is simple. If he gets three hits on or before the fifth throw he decides in favor of the question. The probability that he will decide in favor is therefore

$$\begin{aligned} P(X \le 5) &= p(3) + p(4) + p(5) \\ &= \binom{2}{2}(0.6)^3(0.4)^0 + \binom{3}{2}(0.6)^3(0.4)^1 + \binom{4}{2}(0.6)^3(0.4)^2 \\ &= 0.6826. \end{aligned}$$

5-6 THE MULTINOMIAL DISTRIBUTION

An important and useful higher dimensional random variable has a distribution known as the *multinomial distribution*. Assume an experiment $\mathscr{E}$ with sample space $\mathscr{S}$ is partitioned into k mutually exclusive events, say $B_1, B_2, \ldots, B_k$. We consider n independent repetitions of $\mathscr{E}$ and let $p_i = P(B_i)$ be constant from trial to trial, for $i = 1, 2, \ldots, k$. If $k = 2$, we have Bernoulli trials, as described earlier. The random vector $[X_1, X_2, \ldots, X_k]$ has the following distribution, where X_i is the number of times B_i occurs in the n repetitions of $\mathscr{E}$, $i = 1, 2, \ldots, k$.

$$p(x_1, x_2, \ldots, x_k) = \left[\frac{n!}{x_1! x_2! \cdots x_k!} \right] p_1^{x_1} p_2^{x_2} \cdots p_k^{x_k} \tag{5-23}$$

for $x_i = 0, 1, 2, \ldots, n$, $i = 1, 2, \ldots, k$, and where $\sum_{i=1}^{k} x = n$.

It should be noted that $X_1, X_2, \ldots, X_k$ are not independent random variables, since $\sum_{i=1}^{k} x = n$ for any n repetitions.

It turns out that the mean and variance of X_i, a particular component, are

$$E(X_i) = np_i \tag{5-24}$$

and

$$V(X_i) = np_i(1 - p_i). \tag{5-25}$$

Example 5-9

Mechanical pencils are manufactured by a process involving a large amount of labor in the assembly operations. This is highly repetitive work and incentive pay is involved. Final inspection has revealed that 85% of the product is good, 10% is defective but may be reworked, and 5% is defective and must be scrapped. These percentages remain constant over time. A random sample of 20 items is selected, and if we let

X_1 = number of good items,
X_2 = number of defective but reworkable items,
X_3 = number of items to be scrapped,

then

$$p(x_1, x_2, x_3) = \frac{(20)!}{x_1!x_2!x_3!}(0.85)^{x_1}(0.10)^{x_2}(0.05)^{x_3}.$$

Suppose we want to evaluate this probability function for $x_1 = 18$, $x_2 = 2$, and $x_3 = 0$ (we must have $x_1 + x_2 + x_3 = 20$); then

$$\begin{aligned} p(18,2,0) &= \frac{(20)!}{(18)!2!0!}(0.85)^{18}(0.10)^2(0.05)^0 \\ &= 0.102. \end{aligned}$$

5-7 THE HYPERGEOMETRIC DISTRIBUTION

In an earlier section an example presented the hypergeometric distribution. We will now formally develop this distribution and further illustrate its application. Suppose there is some finite population with N items. Some number D ($D \le N$) of the items fall into a class of interest. The particular class will, of course, depend on the situation under consideration. It might be defectives (vs. nondefectives) in the case of a production lot, or persons with blue eyes (vs. not blue eyed) in a classroom with N students. A random sample of size n is selected *without replacement*, and the random variable of interest, X, is the number of items in the sample that belong to the class of interest. The distribution of X is

$$\begin{aligned} p(x) &= \frac{\binom{D}{x}\binom{N-D}{n-x}}{\binom{N}{n}}, && x = 0, 1, 2, \ldots, \min(n, D), \\ &= 0 && \text{otherwise.} \end{aligned} \tag{5-26}$$

The hypergeometric's probability mass function is available in many popular software packages. For instance, suppose that $N = 20$, $D = 8$, $n = 4$, and $x = 1$. Then the Excel function

$$\text{HYPGEOMDIST}(x, n, D, N) = \text{HYPGEOMDIST}(1, 4, 8, 20) = \frac{\binom{8}{1}\binom{12}{3}}{\binom{20}{8}} = 0.3633.$$

5-7.1 Mean and Variance of the Hypergeometric Distribution

The mean and variance of the hypergeometric distribution are

$$E(X) = n \cdot \left[\frac{D}{N}\right] \tag{5-27}$$

and

$$V(X) = n \cdot \left[\frac{D}{N}\right] \cdot \left[1 - \frac{D}{N}\right] \cdot \left[\frac{N-n}{N-1}\right]. \tag{5-28}$$

Example 5-10

In a receiving inspection department, lots of a pump shaft are periodically received. The lots contain 100 units and the following *acceptance sampling plan* is used. A random sample of 10 units is selected without replacement. The lot is accepted if the sample has no more than one defective. Suppose a lot is received that is $p'(100)$ percent defective. What is the probability that it will be accepted?

$$P(\text{accept lot}) = P(X \le 1) = \frac{\sum_{x=0}^{1} \binom{100p'}{x}\binom{100[1-p']}{10-x}}{\binom{100}{10}}$$

$$= \frac{\binom{100p'}{0}\binom{100[1-p']}{10} + \binom{100p'}{1}\binom{100[1-p']}{9}}{\binom{100}{10}}.$$

Obviously the probability of accepting the lot is a function of the lot quality, p'.
If $p' = 0.05$, then

$$P(\text{accept lot}) = \frac{\binom{5}{0}\binom{95}{10} + \binom{5}{1}\binom{95}{9}}{\binom{100}{10}} = 0.923.$$

5-8 THE POISSON DISTRIBUTION

One of the most useful discrete distributions is the *Poisson distribution.* The Poisson distribution may be developed in two ways, and both are instructive insofar as they indicate the circumstances where this random variable may be expected to apply in practice. The first development involves the definition of a *Poisson process.* The second development shows the Poisson distribution to be *a limiting form of the binomial distribution.*

5-8.1 Development from a Poisson Process

In defining the Poisson process, we initially consider a collection of arbitrary, time-oriented occurrences, often called "arrivals" or "births" (see Fig. 5-11). The random variable of interest, say X_t, is the number of arrivals that occur on the interval $[0, t]$. The range space $R_{X_t} = \{0, 1, 2, \ldots\}$. In developing the distribution of X_t it is necessary to make some assumptions, the plausibility of which is supported by considerable empirical evidence.

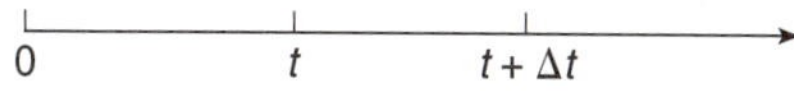

Figure 5-11 The time axis.

The first assumption is that the number of arrivals during *nonoverlapping* time intervals are *independent* random variables. Second, we make the assumption that there exists a positive quantity λ such that for any small time interval, Δt, the following *postulates* are satisfied.

1. *The probability that exactly one arrival will occur in an interval of width* Δt *is approximately* $\lambda \cdot \Delta t$. The approximation is in the sense that the probability is $(\lambda \cdot \Delta t) + o_1(\Delta t)$ where the function $[o_1(\Delta t)/\Delta t] \to 0$ as $\Delta t \to 0$.
2. *The probability that exactly zero arrivals will occur in the interval is approximately* $1 - (\lambda \cdot \Delta t)$. Again this is in the sense that it is equal to $1 - (\lambda \cdot \Delta t) + o_2(\Delta t)$ and $[o_2(\Delta t)/\Delta t] \to 0$ as $\Delta t \to 0$.
3. *The probability that two or more arrivals occur in the interval is equal to a quantity* $o_3(\Delta t)$, *where* $[o_3(\Delta t)/\Delta t] \to 0$ *as* $\Delta t \to 0$.

The parameter λ is sometimes called the mean arrival rate or mean occurrence rate. In the development to follow, we let

$$p(x) = P(X_t = x) = p_x(t), \qquad x = 0, 1, 2, \ldots. \tag{5-29}$$

We fix time at t and obtain

$$p_0(t + \Delta t) \simeq [1 - \lambda \cdot \Delta t] \cdot p_0(t),$$

so that

$$\frac{p_0(t + \Delta t) - p_0(t)}{\Delta t} \simeq -\lambda\, p_0(t)$$

and

$$\lim_{\Delta t \to 0}\left[\frac{p_0(t + \Delta t) - p_0(t)}{\Delta t}\right] = p_0'(t) = -\lambda\, p_0(t). \tag{5-30}$$

For $x > 0$,

$$p_x(t + \Delta t) \simeq \lambda \cdot \Delta t\, p_{x-1}(t) + [1 - \lambda \cdot \Delta t] \cdot p_x(t),$$

so that

$$\frac{p_x(t + \Delta t) - p_x(t)}{\Delta t} = \lambda \cdot p_{x-1}(t) - \lambda \cdot p_x(t)$$

and

$$\lim_{\Delta t \to 0}\left[\frac{p_x(t + \Delta t) - p_x(t)}{\Delta t}\right] = p_x'(t) = \lambda \cdot p_{x-1}(t) - \lambda \cdot p_x(t). \tag{5-31}$$

Summarizing, we have a system of differential equations:

$$p_0'(t) = -\lambda p_0(t) \tag{5-32a}$$

and

$$p_x'(t) = \lambda p_{x-1}(t) - \lambda p_x(t), \qquad x = 1, 2, \ldots. \tag{5-32b}$$

The solution to these equations is

$$p_x(t) = (\lambda t)^x e^{-\lambda t}/x!, \qquad x = 0, 1, 2, \ldots . \tag{5-33}$$

Thus, for fixed t, we let $c = \lambda t$ and obtain the Poisson distribution as

$$\begin{aligned} p(x) &= \frac{c^x e^{-c}}{x!}, \quad x = 0, 1, 2, \ldots, \\ &= 0, \qquad \text{otherwise.} \end{aligned} \tag{5-34}$$

Note that this distribution was developed as a *consequence* of certain assumptions; thus, when the assumptions hold or approximately hold, the Poisson distribution is an appropriate model. There are many real-world phenomena for which the Poisson model is appropriate.

5-8.2 Development of the Poisson Distribution from the Binomial

To show how the Poisson distribution may also be developed as a limiting form of the binomial distribution with $c = np$, we return to the binomial distribution

$$p(x) = \frac{n!}{x!(n-x)!} p^x (1-p)^{n-x}, \qquad x = 0, 1, 2, \ldots, n.$$

If we let $np = c$, so that $p = c/n$ and $1 - p = 1 - c/n = (n - c)/n$, and if we then replace terms involving p with the corresponding terms involving c, we obtain

$$\begin{aligned} p(x) &= \frac{n(n-1)(n-2)\cdots(n-x+1)}{x!}\left[\frac{c}{n}\right]^x \left[\frac{n-c}{n}\right]^{n-x} \\ &= \frac{c^x}{x!}\left[(1)\left(1-\frac{1}{n}\right)\left(1-\frac{2}{n}\right)\cdots\left(1-\frac{x-1}{n}\right)\right]\left(1-\frac{c}{n}\right)^n\left(1-\frac{c}{n}\right)^{-x}. \end{aligned} \tag{5-35}$$

In letting $n \to \infty$ and $p \to 0$ in such a way that $np = c$ remains fixed, the terms $(1 - \frac{1}{n})$, $(1 - \frac{2}{n})$, $\ldots$, $(1 - \frac{x-1}{n})$ all approach 1, as does $(1 - \frac{c}{n})^{-x}$. Now we know that $(1 - \frac{c}{n})^n \to e^c$ as $n \to \infty$. Thus, the limiting form of equation 5-35 is $p(x) = (c^x/x!) \cdot e^{-c}$, which is the Poisson distribution.

5-8.3 Mean and Variance of the Poisson Distribution

The *mean* of the Poisson distribution is c and the variance is also c, as seen below.

$$\begin{aligned} E(X) &= \sum_{x=0}^{\infty} \frac{xe^{-c}c^x}{x!} = \sum_{x=1}^{\infty} \frac{e^{-c}c^x}{(x-1)!} \\ &= ce^{-c}\left[1 + \frac{c}{1!} + \frac{c^2}{2!} + \cdots\right] \\ &= ce^{-c} \cdot e^c \\ &= c. \end{aligned} \tag{5-36}$$

Similarly,

$$E(X^2) = \sum_{x=0}^{\infty} \frac{x^2 \cdot e^{-c}c^x}{x!} = c^2 + c,$$

so that

$$\begin{aligned} V(X) &= E(X^2) - [E(X)]^2 \\ &= c, \end{aligned} \tag{5-37}$$

The moment-generating function is

$$M_X(t) = e^{c(e^t-1)}. \tag{5-38}$$

The utility of this generating function is illustrated in the proof of the following theorem.

Theorem 5-1

If $X_1, X_2, \ldots, X_k$ are independently distributed random variables, each having a Poisson distribution with parameter c_i, $i = 1, 2, \ldots, k$, and $Y = X_1 + X_2 + \cdots + X_k$, then Y has a Poisson distribution with parameter

$$c = c_1 + c_2 + \cdots + c_k.$$

Proof The moment-generating function of X_i is

$$M_{X_i}(t) = e^{c_i(e^t-1)},$$

and since $M_Y(t) = M_{X1}(t) \cdot M_{X_2}(t) \cdot \cdots \cdot M_{X_k}(t)$, then

$$M_Y(t) = e^{(c_1+c_2+\cdots+c_k)(e^t-1)},$$

which is recognized as the moment-generating function of a Poisson random variable with parameter $c = c_1 + c_2 + \cdots + c_k$.

This reproductive property of the Poisson distribution is highly useful. Simply, it states that sums of independent Poisson random variables are distributed according to the Poisson distribution.

A brief tabulation for the Poisson distribution is given in Table I of the Appendix. Most statistical software packages automatically calculate Poisson probabilities.

Example 5-11

Suppose a retailer determines that the number of orders for a certain home appliance in a particular period has a Poisson distribution with parameter c. She would like to determine the stock level K for the beginning of the period so that there will be a probability of at least 0.95 of supplying all customers who order the appliance during the period. She does not wish to back-order merchandise or resupply the warehouse during the period. If X represents the number of orders, the dealer wishes to determine K such that

$$P(X \le K) \ge 0.95$$

or

$$P(X > K) \le 0.05,$$

so that

$$\sum_{x=K+1}^{\infty} e^{-c}c^x/x! \le 0.05.$$

The solution may be determined directly from tables of the Poisson distribution and is obviously a function of c.

Example 5-12

The attainable sensitivity for electronic amplifiers and apparatus is limited by noise or spontaneous current fluctuations. In vacuum tubes, one noise source is shot noise due to the random emission of electrons from the heated cathode. Assume that the potential difference between the anode and cathode is large enough to ensure that all electrons emitted by the cathode have high velocity—high enough to preclude spare charge (accumulation of electrons between the cathode and anode). Under these conditions, and defining an arrival to be an emission of an electrode from the cathode, Davenport and Root (1958) showed that the number of electrons, X, emitted from the cathode in time t has a Poisson distribution given by

$$\begin{aligned} p(x) &= (\lambda t)^x e^{-\lambda t}/x!, & x &= 0, 1, 2, \ldots, \\ &= 0, & & \text{otherwise.} \end{aligned}$$

The parameter λ is the mean rate of emission of electrons from the cathode.

5-9 SOME APPROXIMATIONS

It is often useful to approximate one distribution using another, particularly when the approximation is easier to manipulate. The two approximations considered in this section are as follows:

1. The binomial approximation to the hypergeometric distribution.
2. The Poisson approximation to the binomial distribution.

For the hypergeometric distribution, if the *sampling fraction* n/N is small, say less than 0.1, then the binomial distribution with parameters $p = D/N$ and n provides a good approximation. The smaller the ratio n/N, the better the approximation.

Example 5-13

A production lot of 200 units has eight defectives. A random sample of 10 units is selected, and we want to find the probability that the sample will contain exactly one defective. The true probability is

$$P(X=1) = \frac{\binom{8}{1}\binom{192}{9}}{\binom{200}{10}} = 0.288.$$

Since $n/N = \frac{10}{200} = 0.05$ is small, we let $p = \frac{8}{200} = 0.04$ and use the binomial approximation

$$p(1) = \binom{10}{1}(0.04)^1(0.96)^9 = 0.277.$$

In the case of the Poisson approximation to the binomial, we indicated earlier that for large n and small p, the approximation is satisfactory. In utilizing this approximation we let $c = np$. In general, p should be less than 0.1 in order to apply the approximation. The smaller p and the larger n, the better the approximation.

Example 5-14

The probability that a particular rivet in the wing surface of a new aircraft is defective is 0.001. There are 4000 rivets in the wing. What is the probability that not more than six defective rivets will be installed?

$$P(X \le 6) = \sum_{x=0}^{6} \binom{4000}{x}(0.001)^x(0.999)^{4000-x}.$$

Using the Poisson approximation,

$$c = 4000(0.001) = 4$$

and

$$P(X \le 6) = \sum_{x=0}^{6} e^{-4} 4^x / x! = 0.889.$$

5-10 GENERATION OF REALIZATIONS

Schemes exist for using random numbers, as is described in Section 3-6, to generate realizations of most common random variables.

With Bernoulli trials, we might first generate a value u_i as the ith realization of a uniform [0,1] random variable U, where

$$\begin{aligned} f(u) &= 1, \qquad 0 \le u \le 1, \\ &= 0, \qquad \text{otherwise}, \end{aligned}$$

and independence among the sequence U_i is maintained. Then if $u_i \le p$, we let $X_i = 1$, and if $u_i > p$, $X_i = 0$. Thus if $Y = \sum_{i=1}^{n} X_i$, Y will follow a binomial distribution with parameters n and p, and this entire process might be repeated to produce a series of values of Y, that is, realizations from the binomial distribution with parameters n and p.

Similarly, we could produce geometric variates by sequentially generating values u_i and counting the number of trials until $u_i \le p$. At the point this condition is met, the trial number is assigned to the random variable X, and the entire process is repeated to produce a series of realizations of a geometric random variable.

Also, a similar scheme may be used for Pascal random variable realizations, where we proceed testing $u_i \le p$ until this condition has been satisfied r times, at which point the trial number is assigned to X, and once again, the entire process is repeated to obtain subsequent realizations.

Realizations from a Poisson distribution with parameter $\lambda t = c$ may be obtained by employing a technique based on the so-called acceptance-rejection method. The approach is to sequentially generate values u_i as described above until the product $u_1 \cdot u_2 \cdots u_{k+1} < e^{-c}$ is obtained, at which point we assign $X \leftarrow k$, and once again, this process is repeated to obtain a sequence of realizations.

See Chapter 19 for more details on the generation of discrete random variables.

5-11 SUMMARY

The distributions presented in this chapter have wide use in engineering, scientific, and management applications. The selection of a specific discrete distribution will depend on how well the assumptions underlying the distribution are met by the phenomenon to be modeled. The distributions presented here were selected because of their wide applicability.

A summary of these distributions is presented in Table 5-1.

5-12 EXERCISES

5-1. An experiment consists of four independent Bernoulli trials with probability of success p on each trial. The random variable X is the number of successes. Enumerate the probability distribution of X.

5-2. Six independent space missions to the moon are planned. The estimated probability of success on each mission is 0.95. What is the probability that at least five of the planned missions will be successful?

Table 5-1 Summary of Discrete Distributions

Distribution	Parameters	Probability Function $p(x)$	Mean	Variance	Moment-Generating Function
Bernoulli	$0 < p < 1$	$p(x) = p^x \cdot q^{1-x}, \quad x = 0, 1$ $= 0, \quad$ otherwise	p	pq	$pe^t + q$
Binomial	$n = 1, 2, \ldots$ $0 < p < 1$	$p(x) = \binom{n}{x} p^x q^{n-x}, \quad x = 0, 1, 2, \ldots, n$ $= 0. \quad$ otherwise	np	npq	$(pe^t + q)^n$
Geometric	$0 < p < 1$	$p(x) = pq^{x-1}, \quad x - 0, 1, 2, \ldots$ $= 0, \quad$ otherwise	$1/p$	q/p^2	$pe^t / (1 - qe^t)$
Pascal (Neg. binomial)	$0 < p < 1$ $r = 1, 2, \ldots \; (r > 0)$	$p(x) = \binom{x-1}{r-1} p^r q^{x-r}, \quad x = r, \; r+1, r+2, \ldots$ $= 0, \quad$ otherwise	r/p	rq/p^2	$\left[\dfrac{pe^t}{1-qe^t}\right]^r$
Hypergeometric	$N = 1, 2, \ldots$ $n = 1, 2, \ldots, N$ $D = 1, 2, \ldots, N$	$p(x) = \dfrac{\binom{D}{x}\binom{N-D}{n-x}}{\binom{N}{n}}, \quad x = 0, 1, 2, \ldots, \min(n, D)$ $= 0, \quad$ otherwise	$n\left[\dfrac{D}{N}\right]$	$n\left[\dfrac{D}{N}\right]\left[1 - \dfrac{D}{N}\right]\left[\dfrac{N-n}{N-1}\right]$	See Kendall and Stuart (1963)
Poisson	$c > 0$	$p(x) = e^{-c}c^x / x!, \quad x = 0, 1, 2, \ldots$ $= 0, \quad$ otherwise	c	c	$e^{c(e^t-1)}$

5-3. The *XYZ* Company has planned sales presentations to a dozen important customers. The probability of receiving an order as a result of such a presentation is estimated to be 0.5. What is the probability of receiving four or more orders as the result of the meetings?

5-4. A stockbroker calls her 20 most important customers every morning. If the probability is one in three of making a transaction as the result of such a call, what are the chances of her handling 10 or more transactions?

5-5. A production process that manufactures transistors operates, on the average, at 2% fraction defective. Every 2 hours a random sample of size 50 is taken from the process. If the sample contains more than two defectives the process must be stopped. Determine the probability that the process will be stopped by the sampling scheme.

5-6. Find the mean and variance of the binomial distribution using the moment-generating function (see equation 5-7).

5-7. A production process manufacturing turn-indicator dash lights is known to produce lights that are 1% defective. Assume this value remains unchanged and assume a sample of 100 such lights is randomly selected. Find $P(\hat{p} \leq 0.03)$, where $\hat{p}$ is the sample fraction defective.

5-8. Suppose a random sample of size 200 is taken from a process that is 0.07 fraction defective. What is the probability that $\hat{p}$ will exceed the true fraction defective by one standard deviation? By two standard deviations? By three standard deviations?

5-9. Five cruise missiles have been built by an aerospace company. The probability of a successful firing is, on any one test, 0.95. Assuming independent firings, what is the probability that the first failure occurs on the fifth firing?

5-10. A real estate agent estimates his probability of selling a house to be 0.10. He has to see four clients today. If he is successful on the first three calls, what is the probability that his fourth call is unsuccessful?

5-11. Suppose five independent identical laboratory experiments are to be undertaken. Each experiment is extremely sensitive to environmental conditions, and there is only a probability p that it will be completed successfully. Plot, as a function of p, the probability that the fifth experiment is the first failure. Find mathematically the value of p that maximizes the probability of the fifth trial being the first unsuccessful experiment.

5-12. The *XYZ* Company plans to visit potential customers until a substantial sale is made. Each sales presentation costs \$1000. It costs \$4000 to travel to the next customer and set up a new presentation.

(a) What is the expected cost of making a sale if the probability of making a sale after any presentation is 0.10?

(b) If the expected profit from each sale is \$15,000, should the trips be undertaken?

(c) If the budget for advertising is only \$100,000, what is the probability that this sum will be spent without getting an order?

5-13. Find the mean and variance of the geometric distribution using the moment-generating function.

5-14. A submarine's probability of sinking an enemy ship with any one firing of its torpedoes is 0.8. If the firings are independent, determine the probability of a sinking within the first two firings. Within the first three.

5-15. In Atlanta the probability that a thunderstorm will occur on any day during the spring is 0.05. Assuming independence, what is the probability that the first thunderstorm occurs on April 25? Assume spring begins on March 21.

5-16. A potential customer enters an automobile dealership every hour. The probability of a salesperson concluding a transaction is 0.10. She is determined to keep working until she has sold three cars. What is the probability that she will have to work exactly 8 hours? More than 8 hours?

5-17. A personnel manager is interviewing potential employees in order to fill two jobs. The probability of an interviewee having the necessary qualifications and accepting an offer is 0.8. What is the probability that exactly four people must be interviewed? What is the probability that fewer than four people must be interviewed?

5-18. Show that the moment-generating function of the Pascal random variable is as given by equation 5-22. Use it to determine the mean and variance of the Pascal distribution.

5-19. The probability that an experiment has a successful outcome is 0.80. The experiment is to be repeated until five successful outcomes have occurred. What is the expected number of repetitions required? What is the variance?

5-20. A military commander wishes to destroy an enemy bridge. Each flight of planes he sends out has a probability of 0.8 of scoring a direct hit on the bridge. It takes four direct hits to completely destroy the bridge. If he can mount seven assaults before the bridge becomes tactically unimportant, what is the probability that the bridge will be destroyed?

5-21. Three companies, X, Y, and Z, have probabilities of obtaining an order for a particular type of merchandise of 0.4, 0.3, and 0.3, respectively. Three orders are to be awarded independently. What is the probability that one company receives all the orders?

5-22. Four companies are interviewing five college students for positions after graduation. Assuming all five receive offers from each company and assuming the probabilities of the companies hiring a new employee are equal, what is the probability that one company gets all of the new employees? None of them?

5-23. We are interested in the weight of bags of feed. Specifically, we need to know if any of the four events below has occurred:

$$T_1 = (X \le 10), \qquad p(T_1) = 0.2,$$
$$T_2 = (10 < X \le 11), \qquad p(T_2) = 0.2,$$
$$T_3 = (11 < X \le 11.5), \qquad p(T_3) = 0.2,$$
$$T_4 = (11.5 < X), \qquad p(T_4) = 0.4.$$

If 10 bags are selected at random, what is the probability of four being less than or equal to 10 pounds, one being greater than 10 but less than or equal to 11 pounds, and two being greater than 11.5 pounds?

5-24. In Problem 5-23 what is the probability that all 10 bags weigh more than 11.5 pounds? What is the probability that five bags weigh more than 11.5 pounds and the remaining five weigh less than 10 pounds?

5-25. A lot of 25 color television tubes is subjected to an acceptance testing procedure. The procedure consists of drawing five tubes at random, without replacement, and testing them. If two or fewer tubes fail, the remaining ones are accepted. Otherwise the lot is rejected. Assume the lot contains four defective tubes.

(a) What is the exact probability of lot acceptance?

(b) What is the probability of lot acceptance computed from the binomial distribution with $p = \frac{4}{25}$?

5-26. Suppose that in Exercise 5-25 the lot size had been 100. Would the binomial approximation be satisfactory in this case?

5-27. A purchaser receives small lots ($N = 25$) of a high-precision device. She wishes to reject the lot 95% of the time if it contains as many as seven defectives. Suppose she decides that the presence of one defective in the sample is sufficient to cause rejection. How large should her sample size be?

5-28. Show that the moment-generating function of the Poisson random variable is as given by equation 5-38.

5-29. The number of automobiles passing through a particular intersection per hour is estimated to be 25. Find the probability that fewer than 10 vehicles pass through during any 1-hour interval. Assume that the number of vehicles follows a Poisson distribution.

5-30. Calls arrive at a telephone switchboard such that the number of calls per hour follows a Poisson distribution with a mean of 10. The current equipment can handle up to 20 calls without becoming overloaded. What is the probability of such an overload occurring?

5-31. The number of red blood cells per square unit visible under a microscope follows a Poisson distribution with a mean of 4. Find the probability that more than five such blood cells are visible to the observer.

5-32. Let X_t be the number of vehicles passing through an intersection during a length of time t. The random variable X_t is Poisson distributed with a parameter λt. Suppose an automatic counter has been installed to count the number of passing vehicles. However, this counter is not functioning properly, and each passing vehicle has a probability p of not being counted. Let Y_t be the number of vehicles counted during t. Find the probability distribution of Y_t.

5-33. A large insurance company has discovered that 0.2% of the U.S. population is injured as a result of a particular type of accident. This company has 15,000 policyholders carrying coverage against such an accident. What is the probability that three or fewer claims will be filed against those policies next year? Five or more claims?

5-34. Maintenance crews arrive at a tool crib requesting a particular spare part according to a Poisson distribution with parameter $\lambda = 2$. Three of these spare parts are normally kept on hand. If more than three orders occur, the crews must journey a considerable distance to central stores.

(a) On a given day, what is the probability that such a journey must be made?

(b) What is the expected demand per day for spare parts?

(c) How many spare parts must be carried if the tool crib is to service all incoming crews 90% of the time?

(d) What is the expected number of crews serviced daily at the tool crib?

(e) What is the expected number of crews making the journey to central stores?

5-35. A loom experiences one yarn breakage approximately every 10 hours. A particular style of cloth is being produced that will take 25 hours on this loom. If three or more breaks are required to render the product unsatisfactory, find the probability that this style of cloth is finished with acceptable quality.

5-36. The number of people boarding a bus at each stop follows a Poisson distribution with parameter λ. The bus company is surveying its usages for scheduling purposes and has installed an automatic counter on each bus. However, if more than 10 people board at any one stop, the counter cannot record the excess and merely registers 10. If X is the number of riders recorded, find the probability distribution of X.

5-37. A mathematics textbook has 200 pages on which typographical errors in the equations could occur. If there are in fact five errors randomly dispersed among these 200 pages, what is the probability that a random sample of 50 pages will contain at least one error? How large must the random sample be to assure that at least three errors will be found with 90% probability?

5-38. The probability of a vehicle having an accident at a particular intersection is 0.0001. Suppose that 10,000 vehicles per day travel through this intersection. What is the probability of no accidents occurring? What is the probability of two or more accidents?

5-39. If the probability of being involved in an auto accident is 0.01 during any year, what is the probability of having two or more accidents during any 10-year driving period?

5-40. Suppose that the number of accidents to employees working on high-explosive shells over a period of time (say 5 weeks) is taken to follow a Poisson distribution with parameter $\lambda = 2$.

(a) Find the probabilities of 1, 2, 3, 4, or 5 accidents.

(b) The Poisson distribution has been freely applied in the area of industrial accidents. However, it frequently provides a poor "fit" to actual historical data. Why might this be true? *Hint:* See Kendall and Stuart (1963), pp. 128–30.

5-41. Use either your favorite computer language or the random integers in Table XV in the Appendix (and scale them by multiplying by 10^{-5} to get uniform [0,1] random number realizations) to do the following:

(a) Produce five realizations of a binomial random variable with $n = 8$, $p = 0.5$.

(b) Produce ten realizations of a geometric distribution with $p = 0.4$.

(c) Produce five realizations of a Poisson random variable with $c = 0.15$.

5-42. If $Y = X^{1.3}$ and X follows a geometric distribution with a mean of 6, use uniform [0,1] random number realizations, and produce five realizations of Y.

5-43. With Exercise 5-42 above, use your computer to do the following:

(a) Produce 500 realizations of Y.

(b) Calculate $\bar{y} = \frac{1}{500}(y_1 + y_2 + \cdots + y_{500})$, the mean from this sample.

5-44. Prove the memoryless property of the geometric distribution.

Chapter 6

Some Important Continuous Distributions

6-1 INTRODUCTION

We will now study several important continuous probability distributions. They are the uniform, exponential, gamma, and Weibull distributions. In Chapter 7 the normal distribution, and several other probability distributions closely related to it, will be presented. The normal distribution is perhaps the most important of all continuous distributions. The reason for postponing its study is that the normal distribution is important enough to warrant a separate chapter.

It has been noted that the range space for a continuous random variable X consists of an interval or a set of intervals. This was illustrated in an earlier chapter, and it was observed that an idealization is involved. For example, if we are measuring the time to failure for an electronic component or the time to process an order through an information system, the measurement devices used are such that there are only a finite number of possible outcomes; however, we will idealize and assume that time may take *any* value on some interval. Once again we will simplify the notation where no ambiguity is introduced, and we let $f(x) = f_X(x)$ and $F(x) = F_X(x)$.

6-2 THE UNIFORM DISTRIBUTION

The uniform density function is defined as

$$
\begin{aligned}
f(x) &= \frac{1}{\beta - \alpha}, && \alpha \le x \le \beta, \\
&= 0, && \text{otherwise,}
\end{aligned} \qquad (6\text{-}1)
$$

where α and β are real constants with $\alpha < \beta$. The density function is shown in Fig. 6-1. Since a uniformly distributed random variable has a probability density function that is

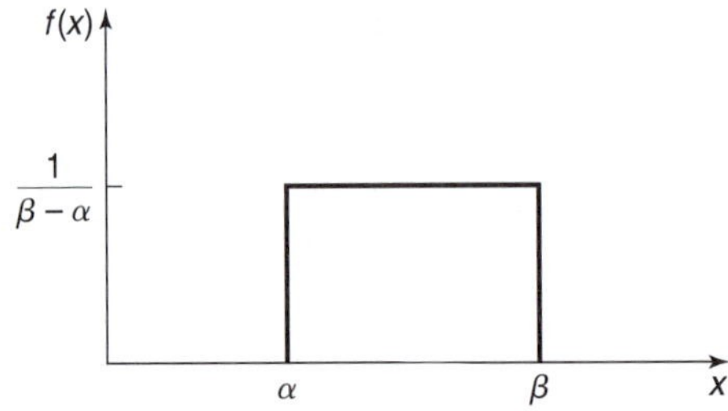

Figure 6-1 A uniform density.

constant over some interval of definition, the constant must be the reciprocal of the length of the interval in order to satisfy the requirement that

$$\int_{-\infty}^{\infty} f(x)\,dx = 1.$$

A uniformly distributed random variable represents the continuous analog to equally likely outcomes in the sense that for any subinterval $[a, b]$, where $\alpha \le a < b \le \beta$, the $P(a \le X \le b)$ depends only on the length $b - a$.

$$P(a \le X \le b) = \int_a^b \frac{dx}{\beta - \alpha} = \frac{b-a}{\beta - \alpha}.$$

The statement that we *choose a point at random on* $[\alpha, \beta]$ simply means that the value chosen, say Y, is uniformly distributed on $[\alpha, \beta]$.

6-2.1 Mean and Variance of the Uniform Distribution

The *mean* and *variance* of the uniform distribution are

$$E(X) = \int_\alpha^\beta \frac{x\,dx}{\beta - \alpha} = \frac{\beta + \alpha}{2} \tag{6-2}$$

(which is obvious by symmetry) and

$$V(X) = \int_\alpha^\beta \frac{x^2 dx}{\beta - \alpha} - \left[\frac{\beta + \alpha}{2}\right]^2$$

$$= \frac{(\beta - \alpha)^2}{12}. \tag{6-3}$$

The moment-generating function $M_X(t)$ is found as follows:

$$M_X(t) = E\left(e^{tX}\right) = \int_\alpha^\beta e^{tx} \cdot \frac{1}{\beta - \alpha} dx = \frac{1}{t(\beta - \alpha)} e^{tx}\Big|_\alpha^\beta$$

$$= \frac{e^{t\beta} - e^{t\alpha}}{t(\beta - \alpha)}. \tag{6-4}$$

For a uniformly distributed random variable, the distribution function $F(x) = P(X \le x)$ is given by equation 6-5, and its graph is shown in Fig. 6-2.

$$\begin{aligned} F(x) &= 0, && x < \alpha, \\ &= \int_\alpha^x \frac{dx}{\beta - \alpha} = \frac{x - \alpha}{\beta - \alpha}, && \alpha \le x < \beta, \\ &= 1, && x \ge \beta. \end{aligned} \tag{6-5}$$

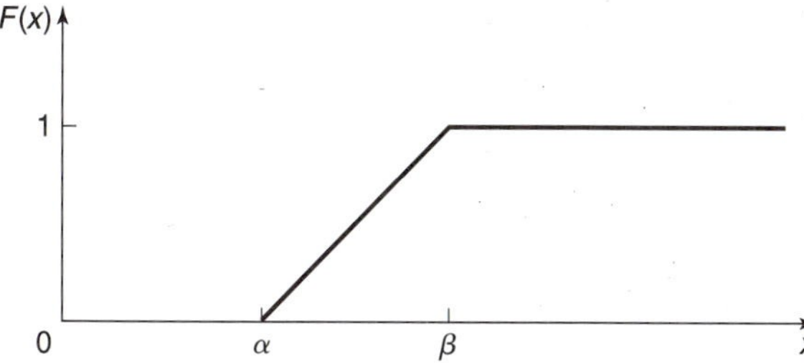

Figure 6-2 Distribution function for the uniform random variable.

Example 6-1

A point is chosen at random on the interval [0, 10]. Suppose we wish to find the probability that the point lies between $\frac{3}{2}$ and $\frac{7}{2}$. The density of the random variable X is $f(x) = \frac{1}{10}$, $0 \le x \le 10$, and $f(x) = 0$ otherwise. Hence, $P(\frac{3}{2} \le X \le \frac{7}{2}) = \frac{2}{10}$.

Example 6-2

Numbers of the form *NN.N* are "rounded off" to the nearest integer. The round-off procedure is such that if the decimal part is less than 0.5, the round is "down" by simply dropping the decimal part; however, if the decimal part is greater than 0.5, the round is up, that is, the new number is $\lfloor NN.N \rfloor + 1$ where $\lfloor\ \rfloor$ is the "greatest integer contained in" function. If the decimal part is exactly 0.5, a coin is tossed to determine which way to round. The round-off error, X, is defined as the difference between the number before rounding and the number after rounding. These errors are commonly distributed according to the uniform distribution on the interval [–0.5, + 0.5]. That is,

$$
\begin{aligned}
f(x) &= 1, \qquad -0.5 \le x \le +0.5, \\
&= 0, \qquad \text{otherwise.}
\end{aligned}
$$

Example 6-3

One of the special features of many simulation languages is a simple automatic procedure for using the uniform distribution. The user declares a mean and modifier (e.g., 500, 100). The compiler immediately creates a routine to produce realizations of a random variable X uniformly distributed on [400, 600].

In the special case where $\alpha = 0$, $\beta = 1$, the uniform variable is said to be uniform on [0, 1] and a symbol U is often used to describe this special variable. Using the results from equations 6-2 and 6-3, we note that $E(U) = \frac{1}{2}$ and $V(U) = \frac{1}{12}$. If $U_1, U_2, \ldots, U_k$ is a sequence of such variables, where the variables are mutually independent, the values $U_1, U_2, \ldots, U_k$ are called *random numbers* and a realization $u_1, u_2, \ldots, u_k$ is properly called a *random number realization*; however, in common usage, the term "random numbers" is often given to the realizations.

6-3 THE EXPONENTIAL DISTRIBUTION

The exponential distribution has density function

$$
\begin{aligned}
f(x) &= \lambda e^{-\lambda x}, \qquad x \ge 0, \\
&= 0, \qquad \text{otherwise,}
\end{aligned}
\tag{6-6}
$$

where the parameter λ is a real, positive constant. A graph of the exponential density is shown in Fig. 6-3.

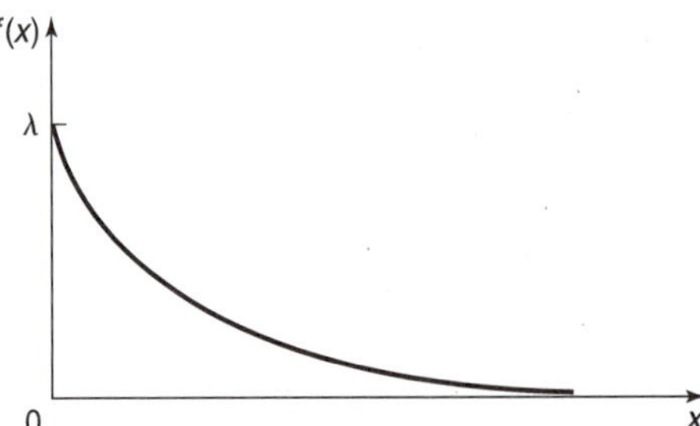

Figure 6-3 The exponential density function.

6-3.1 The Relationship of the Exponential Distribution to the Poisson Distribution

The exponential distribution is closely related to the Poisson distribution, and an explanation of this relationship should help the reader develop an understanding of the kinds of situations for which the exponential density is appropriate.

In developing the Poisson distribution from the Poisson postulates and the Poisson process, we fixed time at some value t, and we developed the distribution of the *number of occurrences in the interval* $[0, t]$. We denoted this random variable X, and the distribution was

$$\begin{aligned} p(x) &= e^{-\lambda t}(\lambda t)^x/x!, && x = 0, 1, 2, \ldots, \\ &= 0, && \text{otherwise.} \end{aligned} \tag{6-7}$$

Now consider $p(0)$, which is the probability of no occurrences on $[0, t]$. This is given by

$$p(0) = e^{-\lambda t}. \tag{6-8}$$

Recall that we originally fixed time at t. Another interpretation of $p(0) = e^{-\lambda t}$ is that this is the probability that the time to the first occurrence is greater than t. Considering this time as a random variable T, we note that

$$p(0) = P(T > t) = e^{-\lambda t}, \qquad t \geq 0. \tag{6-9}$$

If we now let time vary and consider the random variable T as the time to occurrence, then

$$F(t) = P(T \leq t) = 1 - e^{-\lambda t}, \qquad t \geq 0. \tag{6-10}$$

And since $f(t) = F'(t)$, we see that the density is

$$\begin{aligned} f(t) &= \lambda e^{-\lambda t}, && t \geq 0, \\ &= 0, && \text{otherwise.} \end{aligned} \tag{6-11}$$

This is the exponential density of equation 6-6. Thus, the relationship between the exponential and Poisson distributions may be stated as follows: if the number of occurrences has a Poisson distribution as shown in equation 6-7, then the time between successive occurrences has an exponential distribution as shown in equation 6-11. For example, if the number of orders for a certain item received per week has a Poisson distribution, then the time between orders would have an exponential distribution. One variable is discrete (the count) and the other (time) is continuous.

In order to verify that f is a density function, we note that $f(x) \geq 0$ for all x and

$$\int_0^\infty \lambda e^{-\lambda x}\,dx = -e^{-\lambda x}\Big|_0^\infty = 1.$$

6-3.2 Mean and Variance of the Exponential Distribution

The *mean* and *variance* of the exponential distribution are

$$E(X) = \int_0^\infty x\lambda e^{-\lambda x}\,dx = -xe^{-\lambda x}\Big|_0^\infty + \int_0^\infty e^{-\lambda x}\,dx = 1/\lambda \tag{6-12}$$

and

$$\begin{aligned} V(X) &= \int_0^\infty x^2 \lambda e^{-\lambda x}\,dx - (1/\lambda)^2 \\ &= \left[-x^2 e^{-\lambda x}\Big|_0^\infty + 2\int_0^\infty xe^{-\lambda x}\,dx\right] - (1/\lambda)^2 = 1/\lambda^2. \end{aligned} \tag{6-13}$$

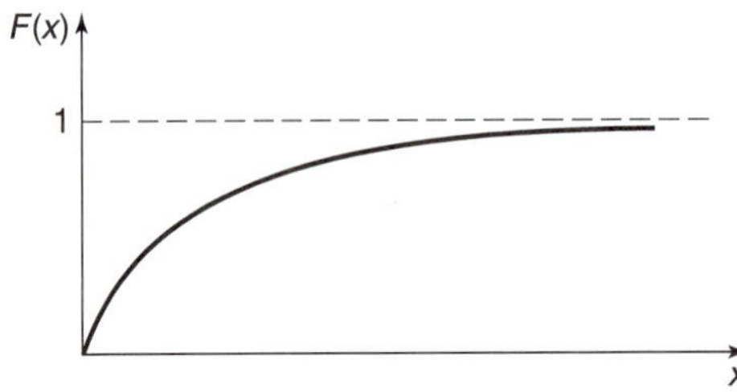

Figure 6-4 Distribution function for the exponential.

The standard deviation is $1/\lambda$, and thus the mean and the standard deviation are equal.

The moment-generating function is

$$M_X(t) = \left(1 - \frac{t}{\lambda}\right)^{-1} \tag{6-14}$$

provided $t < \lambda$.

The cumulative distribution function F can be obtained by integrating equation 6-6 as follows:

$$\begin{aligned} F(x) &= 0, & x < 0, \\ &= \int_0^x \lambda e^{-\lambda t}\,dt = 1 - e^{-\lambda x}, & x \geq 0. \end{aligned} \tag{6-15}$$

Figure 6-4 depicts the distribution function of equation 6-15.

Example 6-4

An electronic component is known to have a useful life represented by an exponential density with failure rate of 10^{-5} failures per hour (i.e., $\lambda = 10^{-5}$). The mean time to failure, $E(X)$, is thus 10^5 hours. Suppose we want to determine the fraction of such components that would fail before the mean life or expected life:

$$\begin{aligned} P\left(T \leq \frac{1}{\lambda}\right) &= \int_0^{1/\lambda} \lambda e^{-\lambda x}\,dx = -e^{-\lambda x}\Big|_0^{1/\lambda} = 1 - e^{-1} \\ &= 0.63212. \end{aligned}$$

This result holds for any value of λ greater than zero. In our example, 63.212% of the items would fail before 10^5 hours (see Fig. 6-5).

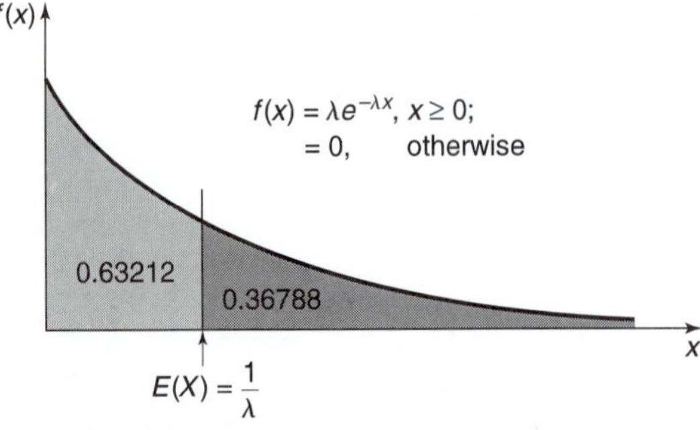

Figure 6-5 The mean of an exponential distribution.

Example 6-5

Suppose a designer is to make a decision between two manufacturing processes for the manufacture of a certain component. Process A costs C dollars per unit to manufacture a component. Process B costs $k \cdot C$ dollars per unit to manufacture a component, where $k > 1$. Components have an exponential time to failure density with a failure rate of 200^{-1} failures per hour for process A, while components from process B have a failure rate of 300^{-1} failures per hour. The mean lives are thus 200 hours and 300 hours, respectively, for the two processes. Because of a warranty clause, if a component lasts for fewer than 400 hours, the manufacturer must pay a penalty of K dollars. Let X be the time to failure of each component. Thus, the component costs are

$$\begin{aligned} C_A &= C && \text{if } X \geq 400, \\ &= C + K && \text{if } X < 400, \end{aligned}$$

and

$$\begin{aligned} C_B &= kC && \text{if } X \geq 400, \\ &= kC + K && \text{if } X < 400. \end{aligned}$$

The expected costs are

$$\begin{aligned} E(C_A) &= (C+K)\int_0^{400} 200^{-1}e^{-x/200}dx + C\int_{400}^{\infty} 200^{-1}e^{-x/200}dx \\ &= (C+K)\left[-e^{-x/200}\Big|_0^{400}\right] + C\left[-e^{-x/200}\Big|_{400}^{\infty}\right] \\ &= (C+K)\left[1-e^{-2}\right] + C\left[e^{-2}\right] \\ &= C + K\left(1-e^{-2}\right) \end{aligned}$$

and

$$\begin{aligned} E(C_B) &= (kC+K)\int_0^{400} 300^{-1}e^{-x/300}dx + kC\int_{400}^{\infty} 300^{-1}e^{-x/300}dx \\ &= (kC+K)\left[1-e^{-4/3}\right] + kC\left[e^{-4/3}\right] \\ &= kC + K\left(1-e^{-4/3}\right). \end{aligned}$$

Therefore, if $k < 1 - K/C(e^{-2} - e^{-4/3})$, then $E(C_A) > E(C_B)$, and it is likely that the designer would select process B.

6-3.3 Memoryless Property of the Exponential Distribution

The exponential distribution has an interesting and unique memoryless property for continuous variables; that is,

$$\begin{aligned} P(X > x+s \mid X > x) &= \frac{P(X > x+s)}{P(X > x)} \\ &= \frac{e^{-\lambda(x+s)}}{e^{-\lambda x}} = e^{-\lambda s}, \end{aligned}$$

so that

$$P(X > x+s \mid X > x) = P(X > s). \tag{6-16}$$

For example, if a cathode ray tube has an exponential time to failure distribution and at time x it is observed to be still functioning, then the *remaining* life has the same exponential failure distribution as the tube had at time zero.

6-4 THE GAMMA DISTRIBUTION

6-4.1 The Gamma Function

A function used in the definition of a gamma distribution is the gamma function defined by

$$\Gamma(n) = \int_0^\infty x^{n-1} e^{-x} dx \qquad \text{for } n > 0. \tag{6-17}$$

An important recursive relationship that may easily be shown on integrating equation 6-17 by parts is

$$\Gamma(n) = (n-1)\Gamma(n-1). \tag{6-18}$$

If n is a *positive integer*, then

$$\Gamma(n) = (n-1)!, \tag{6-19}$$

since $\Gamma(1) = \int_0^\infty e^{-x} dx = 1$. Thus, the gamma function is a generalization of the factorial. The reader is asked in Exercise 6-17 to verify that

$$\Gamma\left(\frac{1}{2}\right) = \int_0^\infty x^{-1/2} e^{-x} dx = \sqrt{\pi}. \tag{6-20}$$

6-4.2 Definition of the Gamma Distribution

With the use of the gamma function, we are now able to introduce the gamma probability density function as

$$\begin{aligned} f(x) &= \frac{\lambda}{\Gamma(r)} (\lambda x)^{r-1} e^{-\lambda x}, && x > 0, \\ &= 0, && \text{otherwise.} \end{aligned} \tag{6-21}$$

The parameters are $r > 0$ and $\lambda > 0$. The parameter r is usually called the *shape parameter*, and λ is called the *scale parameter.* Figure 6-6 shows several gamma distributions, for $\lambda = 1$ and various r. It should be noted that $f(x) \geq 0$ for all x, and

$$\begin{aligned} \int_{-\infty}^{\infty} f(x)dx &= \int_0^\infty \frac{\lambda}{\Gamma(r)} (\lambda x)^{r-1} e^{-\lambda x} dx \\ &= \frac{1}{\Gamma(r)} \int_0^\infty y^{r-1} e^{-y} dy = \frac{1}{\Gamma(r)} \cdot \Gamma(r) = 1. \end{aligned}$$

The cumulative distribution function (CDF) of the gamma distribution is analytically intractable but is readily obtained from software packages such as Excel and Minitab. In particular, the Excel function GAMMADIST(x, r, $1/\lambda$, TRUE) gives the CDF $F(x)$. For exam-

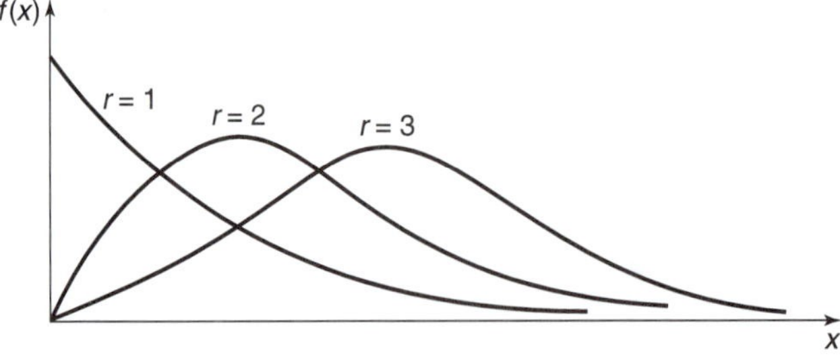

Figure 6-6 Gamma distribution for $\lambda = 1$.

ple, GAMMADIST(15, 5.5, 4.2, TRUE) returns a value of $F(15) = 0.2126$ for the case $r = 5.5$, $\lambda = 1/4.2 = 0.238$. The Excel function GAMMAINV gives the inverse of the CDF.

6-4.3 Relationship Between the Gamma Distribution and the Exponential Distribution

There is a close relationship between the exponential distribution and the gamma distribution. Namely, if $r = 1$ the gamma distribution reduces to the exponential distribution. This follows from the general definition that *if the random variable X is the sum of r independent, exponentially distributed random variables, each with parameter λ, then X has a gamma density with parameters r and λ.* That is to say, if

$$X = X_1 + X_2 + \cdots + X_r, \tag{6-22}$$

where X_j has probability density function

$$\begin{aligned} g(x) &= \lambda e^{-\lambda x}, \qquad && x \geq 0, \\ &= 0, && \text{otherwise,} \end{aligned}$$

and where the X_j are mutually independent, then X has the density given in equation 6-21. In many applications of the gamma distribution that we will consider, r will be a positive integer, and we may use this knowledge to good advantage in developing the distribution function. Some authors refer to the special case in which r is a positive integer as the *Erlang distribution.*

6-4.4 Mean and Variance of the Gamma Distribution

We may show that the *mean* and *variance* of the gamma distribution are

$$E(X) = r/\lambda \tag{6-23}$$

and

$$V(X) = r/\lambda^2. \tag{6-24}$$

Equations 6-23 and 6-24 represent the mean and variance regardless of whether or not r is an integer; however, when r is an integer and the interpretation given in equation 6-22 is made, it is obvious that

$$E(X) = \sum_{j=1}^{r} E(X_j) = r \cdot 1/\lambda = r/\lambda$$

and

$$V(X) = \sum_{j=1}^{r} V(X_j) = r \cdot 1/\lambda^2 = r/\lambda^2$$

from a direct application of the expected value and variance operators to the sum of independent random variables.

The moment-generating function for the gamma distribution is

$$M_X(t) = \left(1 - \frac{t}{\lambda}\right)^{-r}. \tag{6-25}$$

Recalling that the moment-generating function for the exponential distribution was $[1 - (t/\lambda)]^{-1}$, this result is expected, since

$$M_{(X_1+X_2+\cdots+X_r)}(t) = \prod_{j=1}^{r} M_{X_j}(t) = \left[\left(1-\frac{t}{\lambda}\right)^{-1}\right]^r. \tag{6-26}$$

The distribution function, F, is

$$\begin{aligned} F(x) &= 1 - \int_x^\infty \frac{\lambda}{\Gamma(r)}(\lambda t)^{r-1} e^{-\lambda t}\,dt, && x > 0, \\ &= 0, && x \le 0. \end{aligned} \tag{6-27}$$

If r is a positive integer, then equation 6-27 may be integrated by parts, giving

$$F(x) = 1 - \sum_{k=0}^{r-1} e^{-\lambda x}(\lambda x)^k / k!, \qquad x > 0, \tag{6-28}$$

which is the sum of Poisson terms with mean λx. Thus, tables of the cumulative Poisson may be used to evaluate the distribution function of the gamma.

Example 6-6

A redundant system operates as shown in Fig. 6-7. Initially unit 1 is on line, while unit 2 and unit 3 are on standby. When unit 1 fails, the decision switch (DS) switches unit 2 on until it fails and then unit 3 is switched on. The decision switch is assumed to be perfect, so that the system life X may be represented as the sum of the subsystem lives $X = X_1 + X_2 + X_3$. If the subsystem lives are independent of one another, and if the subsystems each have a life X_j, $j = 1, 2, 3$, having density $g(x) = (1/100)e^{-x/100}$, $x \ge 0$, then X will have a gamma density with $r = 3$ and $\lambda = 0.01$. That is,

$$\begin{aligned} f(x) &= \frac{0.01}{2!}(0.01x)^2 e^{-0.01x}, && x > 0, \\ &= 0, && \text{otherwise.} \end{aligned}$$

The probability that the system will operate at least x hours is denoted $R(x)$ and is called the *reliability function.* Here,

$$\begin{aligned} R(x) = 1 - F(x) &= \sum_{k=0}^{2} e^{-0.01x}(0.01x)^k / k! \\ &= e^{-0.01x}\left[1 + (0.01x) + (0.01x)^2/2\right]. \end{aligned}$$

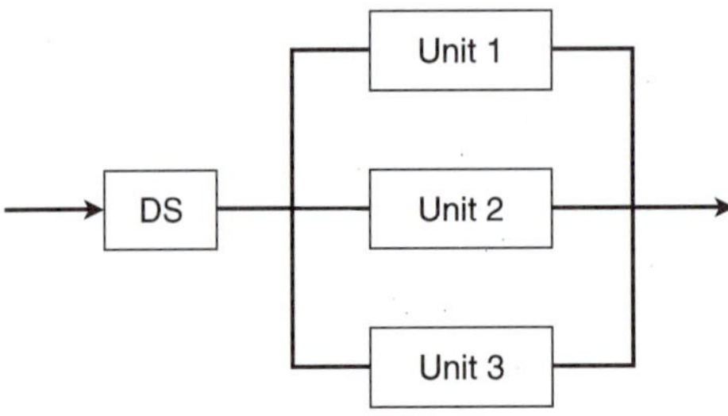

Figure 6-7 A standby redundant system.

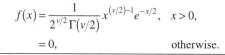

Example 6-7

For a gamma distribution with $\lambda = \frac{1}{2}$ and $r = v/2$, where v is a positive integer, the *chi-square distribution with v degrees of freedom* results:

$$f(x) = \frac{1}{2^{v/2}\Gamma(v/2)} x^{(v/2)-1} e^{-x/2}, \quad x > 0,$$
$$= 0, \qquad \text{otherwise.}$$

This distribution will be discussed further in Chapter 8.

6-5 THE WEIBULL DISTRIBUTION

The Weibull distribution has been widely applied to many random phenomena. The principal utility of the Weibull distribution is that it affords an excellent approximation to the probability law of many random variables. One important area of application has been as a model for time to failure in electrical and mechanical components and systems. This is discussed in Chapter 17. The density function is

$$f(x) = \frac{\beta}{\delta}\left(\frac{x-\gamma}{\delta}\right)^{\beta-1} \exp\left[-\left(\frac{x-\gamma}{\delta}\right)^{\beta}\right], \quad x \geq \gamma,$$
$$= 0, \qquad \text{otherwise.} \tag{6-29}$$

Its parameters are $\gamma(-\infty < \gamma < \infty)$, the location parameter; $\delta > 0$, the scale parameter; and $\beta > 0$, the shape parameter. By appropriate selection of these parameters, this density function will closely approximate many observational phenomena.

Figure 6-8 shows some Weibull densities for $\gamma = 0$, $\delta = 1$, and $\beta = 1, 2, 3, 4$. Note that when $\gamma = 0$ and $\beta = 1$, the Weibull distribution reduces to an exponential density with $\lambda = 1/\delta$. Although the exponential distribution is a special case of both the gamma and Weibull distributions, the gamma and Weibull in general are noninterchangeable.

6-5.1 Mean and Variance of the Weibull Distribution

The *mean* and *variance* of the Weibull distribution can be shown to be

$$E(X) = \gamma + \delta\Gamma\left(1 + \frac{1}{\beta}\right) \tag{6-30}$$

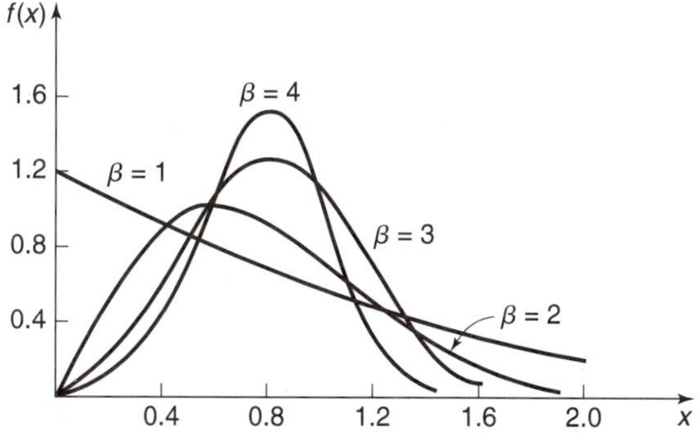

Figure 6-8 Weibull densities for $\gamma = 0$, $\delta = 1$, and $\beta = 1, 2, 3, 4$.

and

$$V(X)=\delta^2\left\{\Gamma\left(1+\frac{2}{\beta}\right)-\left[\Gamma\left(1+\frac{1}{\beta}\right)\right]^2\right\}. \tag{6-31}$$

The distribution function has the relatively simple form

$$F(x)=1-\exp\left[-\left(\frac{x-\gamma}{\delta}\right)^{\beta}\right] \qquad x\geq\gamma. \tag{6-32}$$

The Weibull CDF $F(x)$ is conveniently provided by software packages such as Excel and Minitab. For the case $\lambda = 0$, the Excel function WEIBULL (x, β, γ, TRUE) returns $F(x)$.

Example 6-8

The time-to-failure distribution for electronic subassemblies is known to have a Weibull density with $\gamma = 0$, $\beta = \frac{1}{2}$, and $\delta = 100$. The fraction expected to survive to, say, 400 hours is thus

$$1-F(400)=e^{-\sqrt{400/100}}=0.1353.$$

The same result could have been obtained in Excel via the function call 1 – WEIBULL (400, 1/2, 100, TRUE).

The mean time to failure is

$$E(X) = 0 + 100(2) = 200 \text{ hours}.$$

Example 6-9

Berrettoni (1964) presented a number of applications of the Weibull distribution. The following are examples of natural processes having a probability law closely approximated by the Weibull distribution. The random variable is denoted X in the examples.

1. Corrosion resistance of magnesium alloy plates.
   X: Corrosion weight loss of 10^2 mg/(cm^2)(day) when magnesium alloy plates are immersed in an inhibited aqueous 20% solution of $MgBr_2$.
2. Return goods classified according to number of weeks after shipment.
   X: Length of period (10^{-1} weeks) until a customer returns the defective product after shipment.
3. Number of downtimes per shift.
   X: Number of downtimes per shift (times 10^{-1}) occurring in a continuous automatic and complicated assembly line.
4. Leakage failure in dry-cell batteries.
   X: Age (years) when leakage starts.
5. Reliability of capacitors.
   X: Life (hours) of 3.3-μF, 50-V, solid tantalum capacitors operating at an ambient temperature of 125°C, where the rated catalogue voltage is 33 V.

6-6 GENERATION OF REALIZATIONS

Suppose for now that $U_1, U_2, \ldots,$ are independent uniform [0, 1] random variables. We will show how to use these uniforms to generate other random variables.

If we desire to produce realizations of a uniform random variable on $[\alpha, \beta]$, this is simply accomplished using

$$x_i = \alpha + u_i(\beta - \alpha), \quad i = 1, 2, \ldots . \tag{6-33}$$

If we seek realizations of an exponential random variable with parameter λ, the *inverse transform method* yields

$$x_i = \frac{-1}{\lambda}\ln(u_i), \qquad i = 1, 2, \ldots . \tag{6-34}$$

Similarly, using the same method, realizations of a Weibull random variable with parameters γ, β, δ are obtained using

$$x_i = \gamma + \delta(-\ln u_i)^{1/\beta}, \; i = 1, 2, \ldots . \tag{6-35}$$

The generation of gamma variable realizations usually employs a technique known as the *acceptance–rejection method*, and a variety of these methods have been used. If we wish to produce realizations from a gamma variable with parameters $r > 1$ and $\lambda > 0$, one approach, suggested by Cheng (1977), is as follows:

Step 1. Let $a = (2r - 1)^{1/2}$ and $b = 2r - \ln 4 + 1/a$.

Step 2. Generate u_1, u_2 as uniform [0, 1] random number realizations.

Step 3. Let $y = r[u_1/(1 - u_1)]^a$.

Step 4a. If $y > b - \ln(u_1^2 u_2)$, reject y and return to Step 2.

Step 4b. If $y \le b - \ln(u_1^2 u_2)$ assign $x \leftarrow (y/\lambda)$.

For more details on these and other random-variate generation techniques, see Chapter 19.

6-7 SUMMARY

This chapter has presented four widely used density functions for continuous random variables. The *uniform, exponential, gamma*, and *Weibull* distributions were presented along with underlying assumptions and example applications. Table 6-1 presents a summary of these distributions.

6-8 EXERCISES

6-1. A point is chosen at random on the line segment [0, 4]. What is the probability that it lies between $\frac{1}{2}$ and $1\frac{3}{4}$? Between $2\frac{1}{4}$ and $3\frac{3}{8}$?

6-2. The opening price of a particular stock is uniformly distributed on the interval $[35\frac{3}{4}, 44\frac{1}{4}]$. What is the probability that, on any given day, the opening price is less than 40? Between 40 and 42?

6-3. The random variable X is uniformly distributed on the interval [0, 2]. Find the distribution of the random variable $Y = 5 + 2X$.

6-4. A real estate broker charges a fixed fee of \$50 plus a 6% commission on the landowner's profit. If this profit is uniformly distributed between \$0 and \$2000, find the probability distribution of the broker's total fees.

6-5. Use the moment-generating function for the uniform density (as given by equation 6-4) to generate the mean and variance.

6-6. Let X be uniformly distributed and symmetric about zero with variance 1. Find the appropriate values for α and β.

6-7. Show how the uniform density function can be used to generate variates from the empirical probability distribution described below:

Table 6-1 Summary of Continuous Divisions

Density	Parameters	Density Function $f(x)$	Mean	Variance	Moment-Generating Function
Uniform	α, β $\beta > \alpha$	$f(x) = \dfrac{1}{\beta - \alpha}, \quad \alpha \le x \le \beta$ $= 0, \quad$ otherwise.	$(\alpha + \beta) / 2$	$(\beta - \alpha)^2 / 12$	$\dfrac{e^{t\beta} - e^{t\alpha}}{t(\beta - \alpha)}$
Exponential	$\lambda > 0$	$f(x) = \lambda e^{-\lambda x}, \quad x > 0$ $= 0, \quad$ otherwise.	$1 / \lambda$	$1 / \lambda^2$	$(1 - t / \lambda)^{-1}$
Gamma	$r > 0$ $\lambda > 0$	$f(x) = \dfrac{\lambda}{\Gamma(r)} (\lambda x)^{r-1} e^{-\lambda x}, \quad x > 0$ $= 0, \quad$ otherwise.	r / λ	r / λ^2	$(1 - t / \lambda)^{-r}$
Weibull	$-\infty < \gamma < \infty$ $\delta > 0$ $\beta > 0$	$f(x) = \dfrac{\beta}{\delta} \left(\dfrac{x - \gamma}{\delta} \right)^{\beta - 1} \exp\left[-\left(\dfrac{x - \gamma}{\delta} \right)^{\beta} \right], \quad x \ge \gamma$ $= 0, \quad$ otherwise.	$\gamma + \delta \cdot \Gamma\left(\dfrac{1}{\beta} + 1 \right)$	$\delta^2 \left\{ \Gamma\left(\dfrac{2}{\beta} + 1 \right) - \left[\Gamma\left(\dfrac{1}{\beta} + 1 \right) \right]^2 \right\}$	

y	$p(y)$
1	0.3
2	0.2
3	0.4
4	0.1

Hint: Apply the inverse transform method.

6-8. The random variable X is uniformly distributed over the interval [0, 4]. What is the probability that the roots of $y^2 + 4Xy + X + 1 = 0$ are real?

6-9. Verify that the moment-generating function of the exponential distribution is as given by equation 6-14. Use it to generate the mean and variance.

6-10. The engine and drive train of a new car is guaranteed for 1 year. The mean life of an engine and drive train is estimated to be 3 years, and the time to failure has an exponential density. The realized profit on a new car is \$1000. Including costs of parts and labor the dealer must pay \$250 to repair each failure. What is the expected profit per car?

6-11. For the data in Exercise 6-10, what percentage of cars will experience failure in the engine and drive train during the first 6 months of use?

6-12. Let the length of time a machine will operate be an exponentially distributed random variable with probability density function $f(t) = \theta e^{-\theta t}$, $t \geq 0$. Suppose an operator for this machine must be hired for a predetermined and fixed length of time, say Y. She is paid d dollars per time period during this interval. The net profit from operating this machine, exclusive of labor costs, is r dollars per time period that it is operating. Find the value of Y that maximizes the expected total profit obtained.

6-13. The time to failure of a television tube is estimated to be exponentially distributed with a mean of 3 years. A company offers insurance on these tubes for the first year of usage. On what percentage of policies will they have to pay a claim?

6-14. Is there an exponential density that satisfies the following condition?

$$P\{X \leq 2\} = \tfrac{2}{3}P\{X \leq 3\}$$

If so, find the value of λ.

6-15. Two manufacturing processes are under consideration. The per-unit cost for process I is C, while for process II it is $3C$. Products from both processes have exponential time-to-failure densities with mean rates of 25^{-1} failures per hour and 35^{-1} failures per hour from I and II, respectively. If a product fails before 15 hours it must be replaced at a cost of Z dollars. Which process would you recommend?

6-16. A transistor has an exponential time-to-failure distribution with a mean time to failure of 20,000 hours. The transistor has already lasted 20,000 hours in a particular application. What is the probability that the transistor fails by 30,000 hours?

6-17. Show that $\Gamma(\frac{1}{2}) = \sqrt{\pi}$.

6-18. Prove the gamma function properties given by equations 6-18 and 6-19.

6-19. A ferry boat will take its customers across a river when 10 cars are aboard. Experience shows that cars arrive at the ferry boat independently and at a mean rate of seven per hour. Find the probability that the time between consecutive trips will be at least 1 hour.

6-20. A box of candy contains 24 bars. The time between demands for these candy bars is exponentially distributed with a mean of 10 minutes. What is the probability that a box of candy bars opened at 8:00 A.M. will be empty by noon?

6-21. Use the moment-generating function of the gamma distribution (as given by equation 6-25) to find the mean and variance.

6-22. The life of an electronic system is $Y = X_1 + X_2 + X_3 + X_4$, the sum of the subsystem component lives. The subsystems are independent, each having exponential failure densities with a mean time between failures of 4 hours. What is the probability that the system will operate for at least 24 hours?

6-23. The replenishment time for a certain product is known to be gamma distributed with a mean of 40 and a variance of 400. Find the probability that an order is received within the first 20 days after it is ordered. Within the first 60 days.

6-24. Suppose a gamma distributed random variable is defined over the interval $u \leq x < \infty$ with density function

$$f(x) = \frac{\lambda^r}{\Gamma(r)}(x-u)^{r-1}e^{-\lambda(x-u)}, \quad x \geq u, \lambda \geq 0, r > 0,$$
$$= 0, \quad \text{otherwise.}$$

Find the mean of this *three-parameter* gamma distribution.

6-25. The beta probability distribution is defined by

$$f(x) = \frac{\Gamma(\lambda + r)}{\Gamma(\lambda)\Gamma(r)}x^{\lambda-1}(1-x)^{r-1}, \quad 0 \leq x \leq 1, \lambda > 0, r > 0,$$
$$= 0, \quad \text{otherwise.}$$

(a) Graph the distribution for $\lambda > 1$, $r > 1$.
(b) Graph the distribution for $\lambda < 1$, $r < 1$.
(c) Graph the distribution for $\lambda < 1$, $r \geq 1$.
(d) Graph the distribution for $\lambda \geq 1$, $r < 1$.
(e) Graph the distribution for $\lambda = r$.

6-26. Show that when $\lambda = r = 1$ the beta distribution reduces to the uniform distribution.

6-27. Show that when $\lambda = 2$, $r = 1$ or $\lambda = 1$, $r = 2$ the beta distribution reduces to a triangular probability distribution. Graph the density function.

6-28. Show that if $\lambda = r = 2$ the beta distribution reduces to a parabolic probability distribution. Graph the density function.

6-29. Find the mean and variance of the beta distribution.

6-30. Find the mean and variance of the Weibull distribution.

6-31. The diameter of steel shafts is Weibull distributed with parameters $\gamma = 1.0$ inches, $\beta = 2$, and $\delta = 0.5$. Find the probability that a randomly selected shaft will not exceed 1.5 inches in diameter.

6-32. The time to failure of a certain transistor is known to be Weibull distributed with parameters $\gamma = 0$, $\beta = \frac{1}{3}$, and $\delta = 400$. Find the fraction expected to survive 600 hours.

6-33. The time to leakage failure in a certain type of dry-cell battery is expected to have a Weibull distribution with parameters $\gamma = 0$, $\beta = \frac{1}{2}$, and $\delta = 400$. What is the probability that a battery will survive beyond 800 hours of use?

6-34. Graph the Weibull distribution with $\gamma = 0$, $\delta = 1$, and $\beta = 1, 2, 3$, and 4.

6-35. The time to failure density for a small computer system has a Weibull density with $\gamma = 0$, $\beta = \frac{1}{4}$, and $\delta = 200$.

(a) What fraction of these units will survive to 1000 hours?

(b) What is the mean time to failure?

6-36. A manufacturer of a commercial television monitor guarantees the picture tube for 1 year (8760 hours). The monitors are used in airport terminals for flight schedules, and they are in continuous use with power on. The mean life of the tubes is 20,000 hours, and they follow an exponential time-to-failure density. It costs the manufacturer \$300 to make, sell, and deliver a monitor that will be sold for \$400. It costs \$150 to replace a failed tube, including materials and labor. The manufacturer has no replacement obligation beyond the first replacement. What is the manufacturer's expected profit?

6-37. The lead time for orders of diodes from a certain manufacturer is known to have a gamma distribution with a mean of 20 days and a standard deviation of 10 days. Determine the probability of receiving an order within 15 days of the placement date.

6-38. Use random numbers generated from your favorite computer language or from scaling the random integers in Table XV of the Appendix by multiplying by 10^{-5}, and do the following:

(a) Produce 10 realizations of a variable that is uniform on [10, 20].

(b) Produce five realizations of an exponential random variable with a parameter of $\lambda = 2 \times 10^{-5}$.

(c) Produce five realizations of a gamma variable with $r = 2$ and $\lambda = 4$.

(d) Produce 10 realizations of a Weibull variable with $\gamma = 0$, $\beta = 1/2$, $\delta = 100$.

6-39. Use the random number generation schemes suggested in Exercise 6-38, and do the following:

(a) Produce 10 realizations of $Y = 2X^{0.3}$, where X follows an exponential distribution with a mean of 10.

(b) Produce 10 realizations of $Y = \sqrt{X_1}/\sqrt{X_2}$, where X_1 is gamma with $r = 2$, $\lambda = 4$ and X_2 is uniform on [0, 1].

Chapter 7

The Normal Distribution

7-1 INTRODUCTION

In this chapter we consider the normal distribution. This distribution is very important in both the theory and application of statistics. We also discuss the lognormal and bivariate normal distributions.

The normal distribution was first studied in the eighteenth century, when the patterns in errors of measurement were observed to follow a symmetrical, bell-shaped distribution. It was first presented in mathematical form in 1733 by DeMoivre, who derived it as a limiting form of the binomial distribution. The distribution was also known to Laplace no later than 1775. Through historical error, it has been attributed to Gauss, whose first published reference to it appeared in 1809, and the term *Gaussian distribution* is frequently employed. Various attempts were made during the eighteenth and nineteenth centuries to establish this distribution as the underlying probability law for all continuous random variables; thus, the name *normal* came to be applied.

7-2 THE NORMAL DISTRIBUTION

The normal distribution is in many respects the cornerstone of statistics. A random variable X is said to have a normal distribution with mean μ $(-\infty < \mu < \infty)$ and variance $\sigma^2 > 0$ if it has the density function

$$f(x) = \frac{1}{\sigma\sqrt{2\pi}} e^{-(1/2)[(x-\mu)/\sigma]^2}, \qquad -\infty < x < \infty. \tag{7-1}$$

The distribution is illustrated graphically in Fig. 7-1. The normal distribution is used so extensively that the shorthand notation $X \sim N(\mu, \sigma^2)$ is often employed to indicate that the random variable X is normally distributed with mean μ and variance σ^2.

7-2.1 Properties of the Normal Distribution

The normal distribution has several important properties.

$$\left.\begin{array}{ll} \textbf{1.}\ \int_{-\infty}^{\infty} f(x)\,dx = 1 & \\ \textbf{2.}\ f(x) \ge 0 & \text{for all } x \end{array}\right\} \text{required of all density functions.} \tag{7-2}$$

3. $\lim_{x\to\infty} f(x) = 0$ and $\lim_{x\to-\infty} f(x) = 0$.

4. $f(\mu + x) = f(\mu - x)$. The density is symmetric about μ.

5. The maximum value of f occurs at $x = \mu$.

6. The points of inflection of f are at $x = \mu \pm \sigma$.

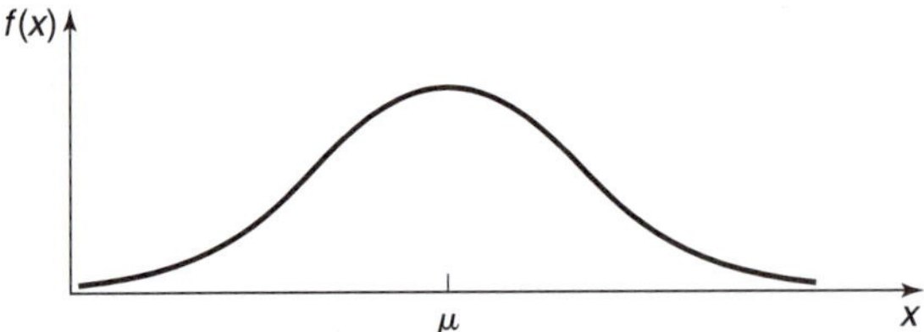

Figure 7-1 The normal distribution.

Property 1 may be demonstrated as follows. Let $y = (x - \mu)/\sigma$ in equation 7-1 and denote the integral I. That is,

$$I = \frac{1}{\sqrt{2\pi}} \int_{-\infty}^{\infty} e^{-(1/2)y^2}\, dy.$$

Our proof that $\int_{-\infty}^{\infty} f(x)\,dx = 1$ will consist of showing that $I^2 = 1$ and then inferring that $I = 1$, since f must be everywhere positive. Defining a second normally distributed variable, Z, we have

$$\begin{aligned} I^2 &= \frac{1}{\sqrt{2\pi}} \int_{-\infty}^{\infty} e^{-(1/2)y^2}\, dy \frac{1}{\sqrt{2\pi}} \int_{-\infty}^{\infty} e^{-(1/2)z^2}\, dz \\ &= \frac{1}{2\pi} \int_{-\infty}^{\infty} \int_{-\infty}^{\infty} e^{-(1/2)\left(y^2 + z^2\right)}\, dy\, dz. \end{aligned}$$

On changing to polar coordinates with the transformation of variables $y = r \sin\theta$ and $z = r \cos\theta$, the integral becomes

$$\begin{aligned} I^2 &= \frac{1}{2\pi} \int_0^{\infty} \int_0^{2\pi} r e^{-(1/2)r^2}\, d\theta dr \\ &= \int_0^{\infty} r e^{-(1/2)r^2}\, dr = 1, \end{aligned}$$

completing the proof.

7-2.2 Mean and Variance of the Normal Distribution

The mean of the normal distribution may be determined easily. Since

$$E(X) = \int_{-\infty}^{\infty} \frac{x}{\sigma\sqrt{2\pi}} e^{-(1/2)[(x-\mu)/\sigma]^2}\, dx,$$

and if we let $z = (x - \mu)/\sigma$, we obtain

$$\begin{aligned} E(X) &= \int_{-\infty}^{\infty} \frac{1}{\sqrt{2\pi}} (\mu + \sigma z) e^{-z^2/2} dz \\ &= \mu \int_{-\infty}^{\infty} \frac{1}{\sqrt{2\pi}} e^{-z^2/2} dz + \sigma \int_{-\infty}^{\infty} \frac{1}{\sqrt{2\pi}} z e^{-z^2/2} dz. \end{aligned}$$

Since the integrand of the first integral is that of a normal density with $\mu = 0$ and $\sigma^2 = 1$, the value of the first integral is one. The second integral has value zero, that is,

$$\int_{-\infty}^{\infty} \frac{1}{\sqrt{2\pi}} z e^{-z^2/2} dz = -\frac{1}{\sqrt{2\pi}} e^{-z^2/2} \Big|_{-\infty}^{\infty} = 0,$$

and thus

$$E(X) = \mu[1] + \sigma[0].$$
$$= \mu \tag{7-3}$$

In retrospect, this result makes sense via a symmetry argument.

To find the variance we must evaluate

$$V(X) = E\left[(X-\mu)^2\right] = \int_{-\infty}^{\infty} (x-\mu)^2 \frac{1}{\sigma\sqrt{2\pi}} e^{-(1/2)[(x-\mu)/\sigma]^2} dx,$$

and letting $z = (x - \mu)/\sigma$, we obtain

$$\begin{aligned} V(X) &= \int_{-\infty}^{\infty} \sigma^2 z^2 \frac{1}{\sqrt{2\pi}} e^{-z^2/2} dz = \sigma^2 \int_{-\infty}^{\infty} \frac{z^2}{\sqrt{2\pi}} e^{-z^2/2} dz \\ &= \sigma^2 \left[\frac{-ze^{-z^2/2}}{\sqrt{2\pi}} \bigg|_{-\infty}^{\infty} + \int_{-\infty}^{\infty} \frac{1}{\sqrt{2\pi}} e^{-z^2/2} dz \right] \\ &= \sigma^2[0+1], \end{aligned}$$

so that

$$V(X) = \sigma^2. \tag{7-4}$$

In summary the mean and variance of the normal density given in equation 7-1 are μ and σ^2, respectively.

The *moment-generating function* for the normal distribution can be shown to be

$$M_X(t) = \exp\left[t\mu + \tfrac{\sigma^2 t^2}{2}\right]. \tag{7-5}$$

For the development of equation 7-5, see Exercise 7-10.

7-2.3 The Normal Cumulative Distribution Function

The distribution function F is

$$F(x) = P(X \le x) = \int_{-\infty}^{x} \frac{1}{\sigma\sqrt{2\pi}} e^{-(1/2)[(u-\mu)/\sigma]^2} du. \tag{7-6}$$

It is impossible to evaluate this integral without resorting to numerical methods, and even then the evaluation would have to be accomplished for each pair (μ, σ^2). However, a simple transformation of variables, $z = (x - \mu)/\sigma$, allows the evaluation to be independent of μ and σ. That is,

$$\begin{aligned} F(x) = P(X \le x) = P\left(Z \le \frac{X-\mu}{\sigma}\right) &= \int_{-\infty}^{(x-\mu)/\sigma} \frac{1}{\sqrt{2\pi}} e^{-z^2/2} dz \\ &= \int_{-\infty}^{(x-\mu)/\sigma} \varphi(z) dz = \Phi\left(\frac{x-\mu}{\sigma}\right). \end{aligned} \tag{7-7}$$

7-2.4 The Standard Normal Distribution

The probability density function in equation 7-7 above,

$$\varphi(z) = \frac{1}{\sqrt{2\pi}} e^{-z^2/2}, \qquad -\infty < z < \infty,$$

is that of a normal distribution with mean 0 and variance 1; that is, $Z \sim N(0, 1)$, and we say that Z has a *standard normal distribution.* A graph of the probability density function is shown in Fig. 7-2. The corresponding distribution function is Φ, where

$$\Phi(z) = \int_{-\infty}^{z} \frac{1}{\sqrt{2\pi}} e^{-u^2/2} du, \tag{7-8}$$

and this function has been well tabulated. A table of the integral in equation 7-8 has been provided in Table II of the Appendix. In fact, many software packages such as Excel and Minitab provide functions to evaluate $\Phi(z)$. For instance, the Excel call NORMSDIST(z) does just this task. As an example, we find that NORMSDIST(1.96) = 0.9750. The Excel function NORMSINV returns the inverse CDF. For example, NORMSINV(0.975) = 1.960. The functions NORMDIST(x, μ, σ, TRUE) and NORMINV give the CDF and inverse CDF of the $N(\mu, \sigma^2)$ distribution.

7-2.5 Problem-Solving Procedure

The procedure for solving practical problems involving the evaluation of cumulative normal probabilities is actually very simple. For example, suppose that $X \sim N(100, 4)$ and we wish to find the probability that X is less than or equal to 104; that is, $P(X \leq 104) = F(104)$. Since the standard normal random variable is

$$Z = \frac{X - \mu}{\sigma},$$

we can *standardize* the point of interest $x = 104$ to obtain

$$z = \frac{x - \mu}{\sigma} = \frac{104 - 100}{2} = 2.$$

Now the probability that the *standard* normal random variable Z is less than or equal to 2 is equal to the probability that the *original* normal random variable X is less than or equal to 104. Expressed mathematically,

$$F(x) = \Phi\left(\frac{x - \mu}{\sigma}\right) = \Phi(z)$$

or

$$F(104) = \Phi(2).$$

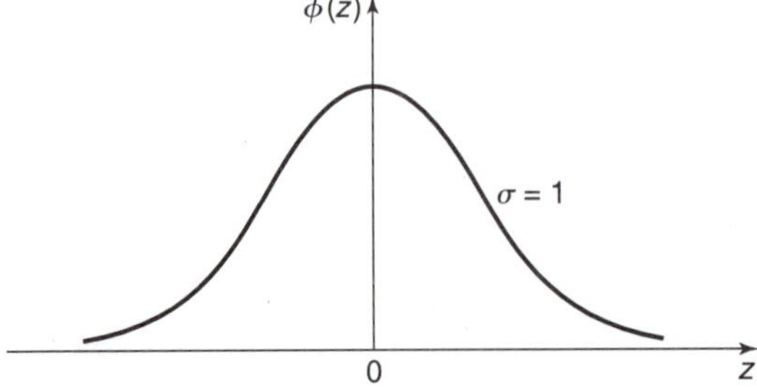

Figure 7-2 The standard normal distribution.

Appendix Table II contains cumulative standard normal probabilities for various values of z. From this table, we can read

$$\Phi(2) = 0.9772.$$

Note that in the relationship $z = (x - \mu)/\sigma$, the variable z measures the departure of x from the mean μ in standard deviation (σ) units. For instance, in the case just considered, $F(104) = \Phi(2)$, which indicates that 104 is *two* standard deviations ($\sigma = 2$) above the mean. In general, $x = \mu + \sigma z$. In solving problems, we sometimes need to use the symmetry property of φ in addition to the tables. It is helpful to make a sketch if there is any confusion in determining exactly which probabilities are required, since the area under the curve and over the interval of interest is the probability that the random variable will lie on the interval.

Example 7-1

The breaking strength (in newtons) of a synthetic fabric is denoted X, and it is distributed as $N(800, 144)$. The purchaser of the fabric requires the fabric to have a strength of at least 772 nt. A fabric sample is randomly selected and tested. To find $P(X \geq 772)$, we first calculate

$$\begin{aligned} P(X < 772) &= P\left(\frac{X-\mu}{\sigma} < \frac{772-800}{12}\right) \\ &= P(Z < -2.33) \\ &= \Phi(-2.33) = 0.01. \end{aligned}$$

Hence the desired probability, $P(X \geq 772)$, equals 0.99. Figure 7-3 shows the calculated probability relative to both X and Z. We have chosen to work with the random variable Z because its distribution function is tabulated.

Example 7-2

The time required to repair an automatic loading machine in a complex food-packaging operation of a production process is X minutes. Studies have shown that the approximation $X \sim N(120, 16)$ is quite good. A sketch is shown in Fig. 7-4. If the process is down for more than 125 minutes, all equipment

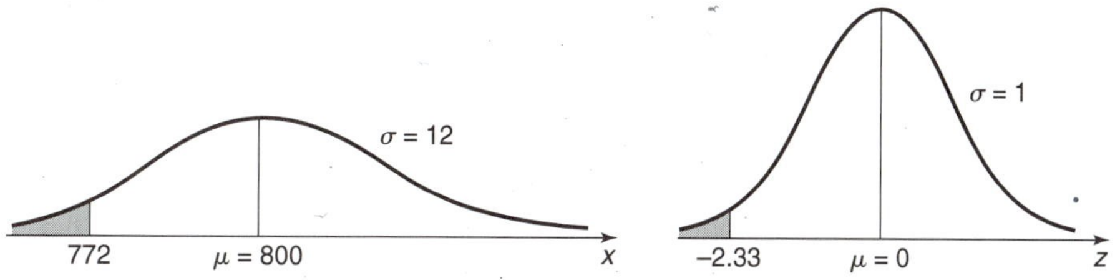

Figure 7-3 $P(X < 772)$, where $X \sim N(800, 144)$.

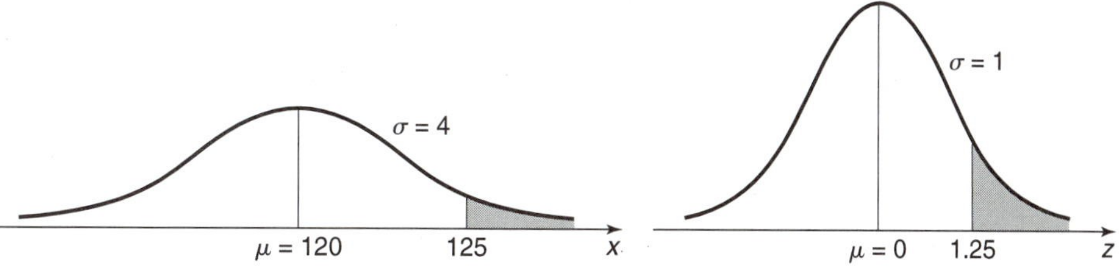

Figure 7-4 $P(X > 125)$, where $X \sim N(120, 16)$.

must be cleaned, with the loss of all product in process. The total cost of product loss and cleaning associated with the long downtime is $10,000. In order to determine the probability of this occurring, we proceed as follows:

$$\begin{aligned} P(X > 125) &= P\left(Z > \frac{125-120}{4}\right) = P(Z > 1.25) \\ &= 1 - \Phi(1.25) \\ &= 1 - 0.8944 \\ &= 0.1056. \end{aligned}$$

Thus, given a breakdown of the packaging machine, the expected cost is $E(C) = 0.1056(10{,}000 + C_{R_1}) + 0.8944(C_{R_1})$, where C is the total cost and C_{R_1} is the repair cost. Simplified, $E(C) = C_{R_1} + 1056$. Suppose the management can reduce the mean of the service time distribution to 115 minutes by adding more maintenance personnel. The new cost for repair will be $C_{R_2} > C_{R_1}$; however,

$$\begin{aligned} P(X > 125) &= P\left(Z > \frac{125-115}{4}\right) = P(Z > 2.5) \\ &= 1 - \Phi(2.5) \\ &= 1 - 0.9938 \\ &= 0.0062, \end{aligned}$$

so that the new expected cost would be $C_{R_2} + 62$, and one would logically make the decision to add to the maintenance crew if

$$C_{R_2} + 62 < C_{R_1} + 1056$$

or

$$C_{R_2} - C_{R_1} < \$994.$$

It is assumed that the frequency of breakdowns remains unchanged.

Example 7-3

The pitch diameter of the thread on a fitting is normally distributed with a mean of 0.4008 cm and a standard deviation of 0.0004 cm. The design specifications are 0.4000 ± 0.0010 cm. This is illustrated in Fig. 7-5. Notice that the process is operating with the mean not equal to the nominal specifications. We desire to determine what fraction of product is within tolerance. Using the approach employed previously,

$$\begin{aligned} P(0.399 \le X \le 0.401) &= P\left(\frac{0.3990-0.4008}{0.0004} \le Z \le \frac{0.4010-0.4008}{0.0004}\right) \\ &= P(-4.5 \le Z \le 0.5) \\ &= \Phi(0.5) - \Phi(-4.5) \\ &= 0.6915 - 0.0000 \\ &= 0.6915. \end{aligned}$$

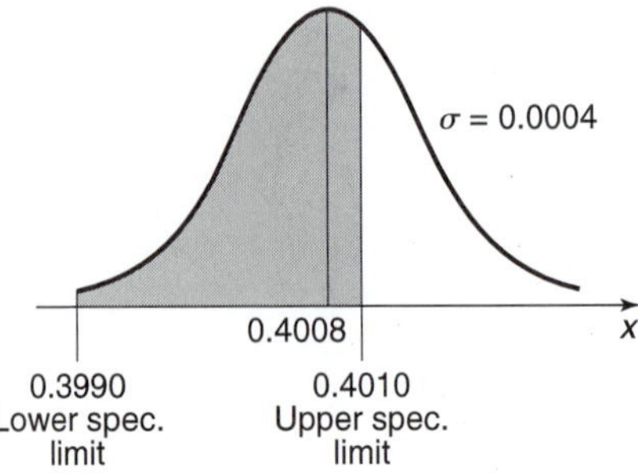

Figure 7-5 Distribution of thread pitch diameters.

As process engineers study the results of such calculations, they decide to replace a worn cutting tool and adjust the machine producing the fittings so that the new mean falls directly at the nominal value of 0.4000. Then,

$$\begin{aligned} P(0.3990 \le X \le 0.4010) &= P\left(\frac{0.3990-0.4}{0.0004} \le Z \le \frac{0.4010-0.4}{0.0004}\right) \\ &= P(-2.5 \le Z \le +2.5) \\ &= \Phi(2.5) - \Phi(-2.5) \\ &= 0.9938 - 0.0062 \\ &= 0.9876. \end{aligned}$$

We see that with the adjustments, 98.76% of the fittings will be within tolerance. The distribution of adjusted machine pitch diameters is shown in Fig. 7-6.

The previous example illustrates a concept important in quality engineering. Operating a process at the nominal level is generally superior to operating the process at some other level, if there are two-sided specification limits.

Example 7-4

Another type of problem involving the use of tables of the normal distribution sometimes arises. Suppose, for example, that $X \sim N(50, 4)$. Furthermore, suppose we want to determine a value of X, say x, such that $P(X > x) = 0.025$. Then,

$$P(X > x) = P\left(Z > \frac{x-50}{2}\right) = 0.025$$

or

$$P\left(Z \le \frac{x-50}{2}\right) = 0.975,$$

so that, reading the normal table "backward," we obtain

$$\frac{x-50}{2} = 1.96 = \Phi^{-1}(0.975)$$

and thus

$$x = 50 + 2(1.96) = 53.92.$$

There are several symmetric intervals that arise frequently. Their probabilities are

$$\begin{aligned} P(\mu - 1.00\sigma \le X \le \mu + 1.00\sigma) &= 0.6826, \\ P(\mu - 1.645\sigma \le X \le \mu + 1.645\sigma) &= 0.90, \\ P(\mu - 1.96\sigma \le X \le \mu + 1.96\sigma) &= 0.95, \\ P(\mu - 2.57\sigma \le X \le \mu + 2.57\sigma) &= 0.99, \\ P(\mu - 3.00\sigma \le X \le \mu + 3.00\sigma) &= 0.9978. \end{aligned} \tag{7-9}$$

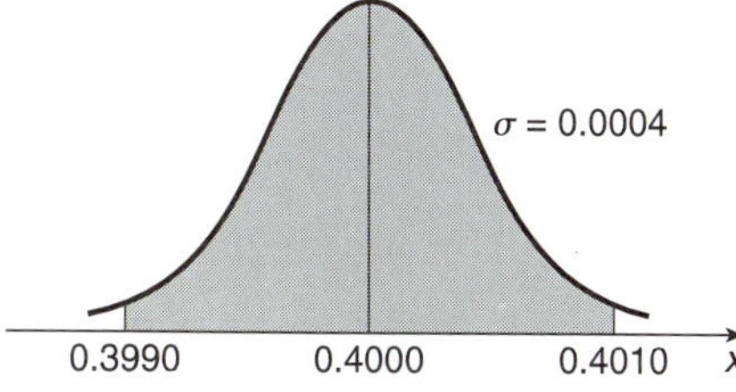

Figure 7-6 Distribution of adjusted machine pitch diameters.

7-3 THE REPRODUCTIVE PROPERTY OF THE NORMAL DISTRIBUTION

Suppose we have n independent, normal random variables $X_1, X_2, \ldots, X_n$, where $X_i \sim N(\mu_i, \sigma_i^2)$, for $i = 1, 2, \ldots, n$. It was shown earlier that if

$$Y = X_1 + X_2 + \cdots + X_n, \tag{7-10}$$

then

$$E(Y) = \mu_Y = \sum_{i=1}^{n} \mu_i \tag{7-11}$$

and

$$V(Y) = \sigma_Y^2 = \sum_{i=1}^{n} \sigma_i^2.$$

Using moment-generating functions, we see that

$$\begin{aligned} M_Y(t) &= M_{X_1}(t) \cdot M_{X_2}(t) \cdot \cdots \cdot M_{X_n}(t) \\ &= \left[e^{\mu_1 t + \sigma_1^2 t^2/2}\right] \cdot \left[e^{\mu_2 t + \sigma_2^2 t^2/2}\right] \cdot \cdots \cdot \left[e^{\mu_n t + \sigma_n^2 t^2/2}\right]. \end{aligned} \tag{7-12}$$

Therefore,

$$M_Y(t) = e^{\left[(\mu_1 + \mu_2 + \cdots + \mu_n)t + \left(\sigma_1^2 + \sigma_2^2 + \cdots + \sigma_n^2\right)t^2/2\right]}, \tag{7-13}$$

which is the moment-generating function of a normally distributed random variable with mean $\mu_1 + \mu_2 + \cdots + \mu_n$ and variance $\sigma_1^2 + \sigma_2^2 + \cdots + \sigma_n^2$. Therefore, by the uniqueness property of the moment-generating function, we see that Y is normal with mean μ_Y and variance σ_Y^2.

Example 7-5

An assembly consists of three linkage components, as shown in Fig. 7-7. The properties X_1, X_2, and X_3 are given below, with means in centimeters and variance in square centimeters.

$$X_1 \sim N(12, 0.02)$$
$$X_2 \sim N(24, 0.03)$$
$$X_3 \sim N(18, 0.04)$$

Links are produced by different machines and operators, so we have reason to assume that X_1, X_2, and X_3 are independent. Suppose we want to determine $P(53.8 \leq Y \leq 54.2)$. Since $Y = X_1 + X_2 + X_3$, Y is distributed normally with mean $\mu_Y = 12 + 24 + 18 = 54$ and variance $\sigma^2 = \sigma_1^2 + \sigma_2^2 + \sigma_3^2 = 0.02 + 0.03 + 0.04 = 0.09$. Thus,

$$\begin{aligned} P(53.8 \leq Y \leq 54.2) &= P\left(\frac{53.8 - 54}{0.3} \leq Z \leq \frac{54.2 - 54}{0.3}\right) \\ &= P\left(-\frac{2}{3} \leq Z \leq +\frac{2}{3}\right) \\ &= \Phi(0.667) - \Phi(-0.667) \\ &= 0.748 - 0.252 \\ &= 0.496. \end{aligned}$$

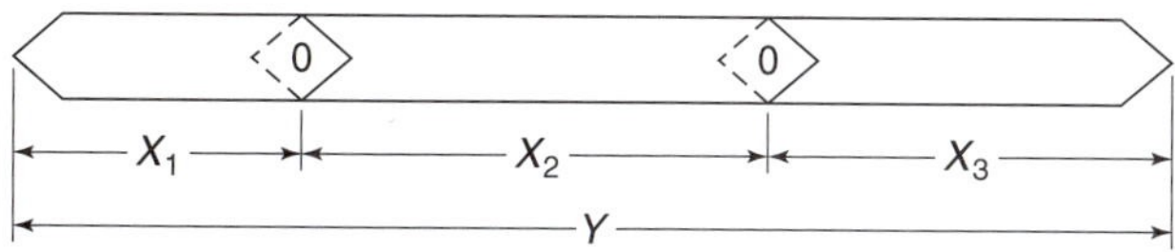

Figure 7-7 A linkage assembly.

These results can be generalized to linear combinations of independent normal variables. Linear combinations of the form

$$Y = a_0 + a_1X_1 + \cdots + a_nX_n \tag{7-14}$$

were presented earlier and we found that $\mu_Y = a_0 + \sum_{i=1}^{n} a_i\mu_i$. When the variables are independent, $\sigma_Y^2 = \sum_{i=1}^{n} a_i^2\sigma_i^2$. Again, if $X_1, X_2, \ldots, X_n$ are independent and normally distributed, then $Y \sim N(\mu_Y, \sigma_Y^2)$.

Example 7-6

A shaft is to be assembled into a bearing, as shown in Fig. 7-8. The clearance is $Y = X_1 - X_2$. Suppose

$$X_1 \sim N(1.500, 0.0016)$$

and

$$X_2 \sim N(1.480, 0.0009).$$

Then,

$$\begin{aligned}\mu_Y &= a_1\mu_1 + a_2\mu_2\\ &= (1)(1.500) + (-1)(1.480)\\ &= 0.02\end{aligned}$$

and

$$\begin{aligned}\sigma_Y^2 &= a_1^2\sigma_1^2 + a_2^2\sigma_2^2\\ &= (1)^2(0.0016) + (-1)^2(0.0009)\\ &= 0.0025,\end{aligned}$$

so that

$$\sigma_Y = 0.05.$$

When the parts are assembled, there will be interference if $Y < 0$, so

$$\begin{aligned}P(\text{interference}) = P(Y<0) &= P\left(Z < \frac{0-0.02}{0.05}\right)\\ &= \Phi(-0.4) = 0.3446.\end{aligned}$$

This indicates that 34.46% of all assemblies attempted would meet with failure. If the designer feels that the *nominal clearance* $\mu_Y = 0.02$ is as large as it can be made for the assembly, then the only way to reduce the 34.46% figure is to reduce the variance of the distributions. In many cases, this can be accomplished by the overhaul of production equipment, better training of production operators, and so on.

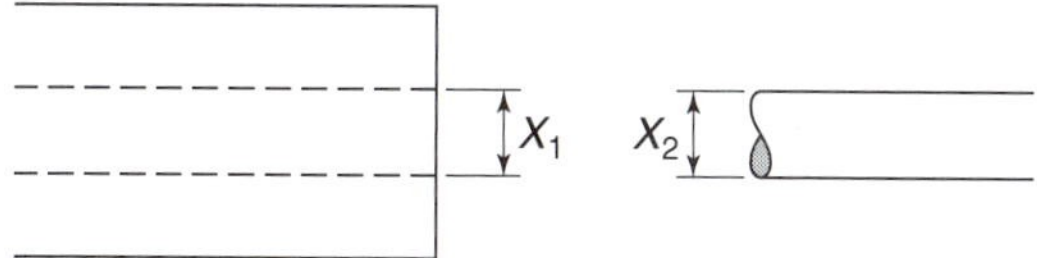

Figure 7-8 An assembly.

7-4 THE CENTRAL LIMIT THEOREM

If a random variable Y is the *sum of n independent random variables* that satisfy certain general conditions, then for sufficiently large n, Y is approximately normally distributed. We state this as a theorem—the most important theorem in all of probability and statistics.

Theorem 7-1 Central Limit Theorem

If $X_1, X_2, \ldots, X_n$ is a sequence of n independent random variables with $E(X_i) = \mu_i$ and $V(X_i) = \sigma_i^2$ (both finite) and $Y = X_1 + X_2 + \cdots + X_n$, then under some general conditions

$$Z_n = \frac{Y - \sum_{i=1}^{n} \mu_i}{\sqrt{\sum_{i=1}^{n} \sigma_i^2}} \tag{7-15}$$

has an approximate $N(0,1)$ distribution as n approaches infinity. If F_n is the distribution function of Z_n, then

$$\lim_{n\to\infty} \frac{F_n(z)}{\Phi(z)} = 1, \quad \text{for all } z. \tag{7-16}$$

The "general conditions" mentioned in the theorem are informally summarized as follows. The terms X_i, taken individually, contribute a negligible amount to the variance of the sum, and it is not likely that a single term makes a large contribution to the sum.

The proof of this theorem, as well as a rigorous discussion of the necessary assumptions, is beyond the scope of this presentation. There are, however, several observations that should be made. The fact that Y is approximately normally distributed when the X_i terms may have *essentially any distribution* is the basic underlying reason for the importance of the normal distribution. In numerous applications, the random variable being considered may be represented as the sum of n independent random variables, some of which may be measurement error, some due to physical considerations, and so on, and thus the normal distribution provides a good approximation.

A special case of the central limit theorem arises when each of the components has the same distribution.

Theorem 7-2

If $X_1, X_2, \ldots, X_n$ is a sequence of n independent, identically distributed random variables with $E(X_i) = \mu$ and $V(X_i) = \sigma^2$, and $Y = X_1 + X_2 + \cdots + X_n$, then

$$Z_n = \frac{Y - n\mu}{\sigma\sqrt{n}} \tag{7-17}$$

has an approximate $N(0,1)$ distribution in the same sense as equation 7-16.

Under the restriction that $M_X(t)$ exists for real t, a straightforward proof may be presented for this form of the central limit theorem. Many mathematical statistics texts present such a proof.

The question immediately encountered in practice is the following: How large must n be to get reasonable results using the normal distribution to approximate the distribution of Y? This is not an easy question to answer, since the answer depends on the characteristics of the distribution of the X_i terms as well as the meaning of "reasonable results." From a practical standpoint, some very crude rules of thumb can be given where the distribution of the X_i terms falls into one of three arbitrarily selected groups, as follows:

1. Well behaved—The distribution of X_i does not radically depart from the normal distribution. There is a bell-shaped density that is nearly symmetric. For this case, practitioners in quality control and other areas of application have found that n should be at least 4. That is, $n \geq 4$.
2. Reasonably behaved—The distribution of X_i has no prominent mode, and it appears much as a uniform density. In this case, $n \geq 12$ is a commonly used rule.
3. Ill behaved—The distribution has most of its measure in the tails, as in Fig. 7-9. In this case it is most difficult to say; however, in many practical applications, $n \geq 100$ should be satisfactory.

Example 7-7

Small parts are packaged 250 to the crate. Part weights are independent random variables with a mean of 0.5 pound and a standard deviation of 0.10 pound. Twenty crates are loaded to a pallet. Suppose we wish to find the probability that the parts on a pallet will exceed 2510 pounds in weight. (Neglect both pallet and crate weight.) Let

$$Y = X_1 + X_2 + \cdots + X_{5000}$$

represent the total weight of the parts, so that

$$\mu_Y = 5000(0.5) = 2500,$$
$$\sigma_Y^2 = 5000(0.01) = 50,$$

and

$$\sigma_Y = \sqrt{50} = 7.071.$$

Then

$$P(Y > 2510) = P\left(Z > \frac{2510 - 2500}{7.071}\right),$$
$$= 1 - \Phi(1.41) = 0.08.$$

Note that we did not know the distribution of the individual part weights.

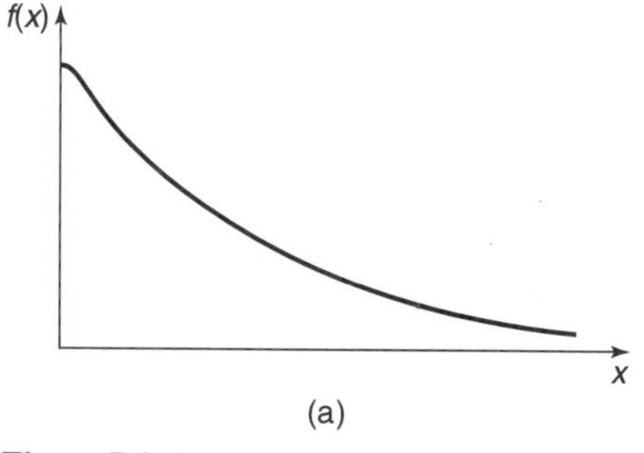

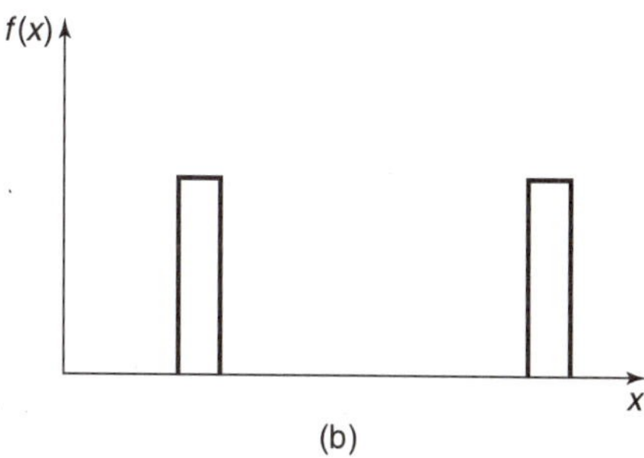

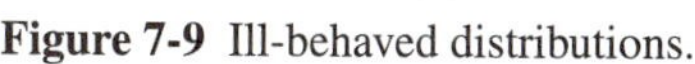
Figure 7-9 Ill-behaved distributions.

Table 7-1 Activity Mean Times and Variances (in Weeks and Weeks²)

Activity	Mean	Variance	Activity	Mean	Variance
1	2.7	1.0	9	3.1	1.2
2	3.2	1.3	10	4.2	0.8
3	4.6	1.0	11	3.6	1.6
4	2.1	1.2	12	0.5	0.2
5	3.6	0.8	13	2.1	0.6
6	5.2	2.1	14	1.5	0.7
7	7.1	1.9	15	1.2	0.4
8	1.5	0.5	16	2.8	0.7

Example 7-8

In a construction project, a network of major activities has been constructed to serve as the basis for planning and scheduling. On a *critical path* there are 16 activities. The means and variances are given in Table 7-1.

The activity times may be considered independent and the project time is the sum of the activity times on the critical path, that is, $Y = X_1 + X_2 + \cdots + X_{16}$, where Y is the project time and X_i is the time for the ith activity. Although the distributions of the X_i are unknown, the distributions are fairly well behaved. The contractor would like to know (a) the expected completion time, and (b) a project time corresponding to a probability of 0.90 of having the project completed. Calculating μ_Y and σ_Y^2, we obtain

$$\mu_Y = 49 \text{ weeks},$$
$$\sigma_Y^2 = 16 \text{ weeks}^2.$$

The expected completion time for the project is thus 49 weeks. In determining the time y_0 such that the probability is 0.9 of having the project completed by that time, Fig. 7-10 may be helpful.

We may calculate

$$P(Y \leq y_0) = 0.90$$

or

$$P\left(Z \leq \frac{y_0 - 49}{4}\right) = 0.90,$$

so that

$$\frac{y_0 - 49}{4} = 1.282 = \Phi^{-1}(0.90)$$

and

$$y_0 = 49 + 1.282(4)$$
$$= 54.128 \text{ weeks}.$$

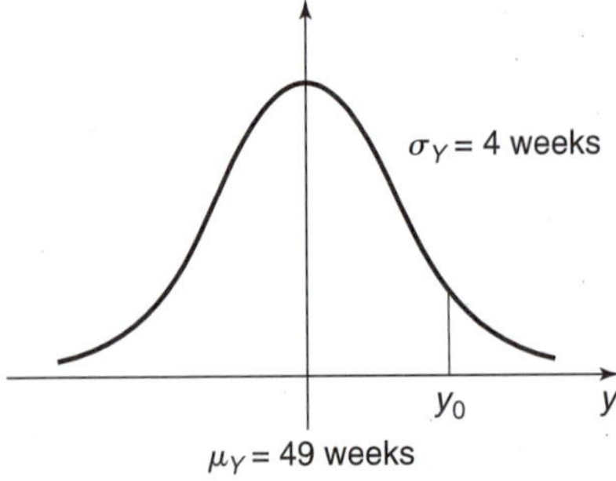

Figure 7-10 Distribution of project times.

7-5 THE NORMAL APPROXIMATION TO THE BINOMIAL DISTRIBUTION

In Chapter 5, the binomial approximation to the hypergeometric distribution was presented, as was the Poisson approximation to the binomial distribution. In this section we consider the normal approximation to the binomial distribution. Since the binomial is a discrete probability distribution, this may seem to go against intuition; however, a limiting process is involved, keeping p of the binomial distribution fixed and letting $n \to \infty$. The approximation is known as the DeMoivre-Laplace approximation.

We recall the binomial distribution as

$$\begin{aligned} p(x) &= \frac{n!}{x!(n-x)!} p^x q^{n-x}, && x = 0, 1, 2, \ldots, n, \\ &= 0, && \text{otherwise.} \end{aligned}$$

Stirling's approximation to $n!$ is

$$n! \simeq (2\pi)^{1/2} e^{-n} n^{n+(1/2)}. \tag{7-18}$$

The error

$$\frac{n! - (2\pi)^{1/2} e^{-n} n^{n+(1/2)}}{n!} \to 0 \tag{7-19}$$

as $n \to \infty$. Using Stirling's formula to approximate the terms involving $n!$ in the binomial model, we eventually find that, for large n,

$$P(X = x) \simeq \frac{1}{\sqrt{np(1-p)}\sqrt{2\pi}} e^{(-1/2)(x-np)^2/(npq)} \tag{7-20}$$

so that

$$P(X \le x) \simeq \Phi\left(\frac{x - np}{\sqrt{npq}}\right) = \int_{-\infty}^{(x-np)/\sqrt{npq}} \frac{1}{\sqrt{2\pi}} e^{-z^2/2} dz. \tag{7-21}$$

This result makes sense in light of the central limit theorem and the fact that X is the sum of independent Bernoulli trials (so that $E(X) = np$ and $V(X) = npq$). Thus, the quantity $(X - np)/\sqrt{npq}$ *approximately* has a $N(0,1)$ distribution. If p is close to $\frac{1}{2}$ and $n > 10$, the approximation is fairly good; however, for other values of p, the value of n must be larger. In general, experience indicates that the approximation is fairly good as long as $np > 5$ for $p \le \frac{1}{2}$ or when $nq > 5$ when $p > \frac{1}{2}$.

Example 7-9

In sampling from a production process that produces items of which 20% are defective, a random sample of 100 items is selected each hour of each production shift. The number of defectives in a sample is denoted X. To find, say, $P(X \le 15)$ we may use the normal approximation as follows:

$$\begin{aligned} P(X \le 15) &= P\left(Z \le \frac{15 - 100 \cdot 0.2}{\sqrt{100(0.2)(0.8)}}\right) \\ &= P(Z \le -1.25) = \Phi(-1.25) = 0.1056. \end{aligned}$$

Since the binomial distribution is discrete and the normal distribution is continuous, it is common practice to use a *half-interval correction* or *continuity correction.* In fact, this is a necessity in calculating $P(X = x)$. The usual procedure is to go a half-unit on either side of the integer x, depending on the interval of interest. Several cases are shown in Table 7-2.

Table 7-2 Continuity Corrections

Quantity Desired from Binomial Distribution	With Continuity Correction	In Terms of the Distribution Function Φ
$P(X = x)$	$P\left(x - \frac{1}{2} \le X \le x + \frac{1}{2}\right)$	$\Phi\left(\frac{x + \frac{1}{2} - np}{\sqrt{npq}}\right) - \Phi\left(\frac{x - \frac{1}{2} - np}{\sqrt{npq}}\right)$
$P(X \le x)$	$P\left(X \le x + \frac{1}{2}\right)$	$\Phi\left(\frac{x + \frac{1}{2} - np}{\sqrt{npq}}\right)$
$P(X < x) = P(X \le x - 1)$	$P\left(X \le x - 1 + \frac{1}{2}\right)$	$\Phi\left(\frac{x - \frac{1}{2} - np}{\sqrt{npq}}\right)$
$P(X \ge x)$	$P\left(X \ge x - \frac{1}{2}\right)$	$1 - \Phi\left(\frac{x - \frac{1}{2} - np}{\sqrt{npq}}\right)$
$P(X > x) = P(X \ge x + 1)$	$P\left(X \ge x + 1 - \frac{1}{2}\right)$	$1 - \Phi\left(\frac{x + \frac{1}{2} - np}{\sqrt{npq}}\right)$
$P(a \le X \le b)$	$P\left(a - \frac{1}{2} \le X \le b + \frac{1}{2}\right)$	$\Phi\left(\frac{b + \frac{1}{2} - np}{\sqrt{npq}}\right) - \Phi\left(\frac{a - \frac{1}{2} - np}{\sqrt{npq}}\right)$

Example 7-10

Using the data from Example 7-9, where we had $n = 100$ and $p = 0.2$, we evaluate $P(X = 15)$, $P(X \le 15)$, $P(X < 18)$, $P(X \ge 22)$, and $P(18 < X < 21)$.

1. $P(X = 15) \simeq P(14.5 \le X \le 15.5) = \Phi\left(\frac{15.5 - 20}{4}\right) - \Phi\left(\frac{14.5 - 20}{4}\right)$
$= \Phi(-1.125) - \Phi(-1.375) \simeq 0.046.$

2. $P(X \le 15) \simeq \Phi\left(\frac{15.5 - 20}{4}\right) \simeq 0.130.$

3. $P(X < 18) = P(X \le 17) \simeq \Phi\left(\frac{17.5 - 20}{4}\right) \simeq 0.266.$

4. $P(X \ge 22) \simeq 1 - \Phi\left(\frac{21.5 - 20}{4}\right) \simeq 0.354.$

5. $P(18 < X < 21) = P(19 \le X \le 20)$
$\simeq \Phi\left(\frac{20.5 - 20}{4}\right) - \Phi\left(\frac{18.5 - 20}{4}\right)$
$\simeq 0.550 - 0.354 = 0.196.$

As discussed in Chapter 5, the random variable $\hat{p} = X/n$ where X has a binomial distribution with parameters p and n, is often of interest. Interest in this quantity stems primarily from sampling applications, where a random sample of n observations is made, with each observation classified success or failure, and where X is the number of successes in the sample. The quantity $\hat{p}$ is simply the sample fraction of successes. Recall that we showed that

$$E(\hat{p}) = p \tag{7-22}$$

and

$$V(\hat{p}) = \frac{pq}{n}.$$

In addition to the DeMoivre-Laplace approximation, note that the quantity

$$Z = \frac{\hat{p} - p}{\sqrt{pq/n}} \tag{7-23}$$

has an approximate $N(0,1)$ distribution. This result has proved useful in many applications, including those in the areas of quality control, work measurement, reliability engineering, and economics. The results are much more useful than those from the law of large numbers.

Example 7-11

Instead of timing the activity of a maintenance mechanic over the period of a week to determine the fraction of his time spent in an activity classification called "secondary but necessary," a technician elects to use a *work sampling* study, randomly picking 400 time points over the week, taking a flash observation at each, and classifying the activity of the maintenance mechanic. The value X will represent the number of times the mechanic was involved in a "secondary but necessary" activity and $\hat{p} = X/400$. If the true fraction of time that he is involved in this activity is 0.2, we determine the probability that $\hat{p}$, the estimated fraction, falls between 0.15 and 0.25. That is,

$$\begin{aligned} P(0.1 \le \hat{p} \le 0.3) &\simeq \Phi\left(\frac{0.25 - 0.2}{\sqrt{0.16/400}}\right) - \Phi\left(\frac{0.15 - 0.2}{\sqrt{0.16/400}}\right) \\ &= \Phi(2.5) - \Phi(-2.5) \\ &\simeq 0.9876. \end{aligned}$$

7-6 THE LOGNORMAL DISTRIBUTION

The lognormal distribution is the distribution of a random variable whose logarithm follows the normal distribution. Some practitioners hold that the lognormal distribution is as fundamental as the normal distribution. It arises from the combination of random terms by a multiplicative process.

The lognormal distribution has been applied in a wide variety of fields, including the physical sciences, life sciences, social sciences, and engineering. In engineering applications, the lognormal distribution has been used to describe "time to failure" in reliability engineering and "time to repair" in maintainability engineering.

7-6.1 Density Function

We consider a random variable X with range space $R_X = \{x: 0 < x < \infty\}$, where $Y = \ln X$ is normally distributed with mean μ_Y and variance σ_Y^2, that is,

$$E(Y) = \mu_Y \qquad \text{and} \qquad V(Y) = \sigma_Y^2.$$

The density function of X is

$$\begin{aligned} f(x) &= \frac{1}{x\sigma_Y\sqrt{2\pi}} e^{(-1/2)[(\ln x - \mu_Y)/\sigma_Y]^2}, && x > 0, \\ &= 0 && \text{otherwise.} \end{aligned} \tag{7-24}$$

The lognormal distribution is shown in Fig. 7-11. Notice that, in general, the distribution is skewed, with a long tail to the right. The Excel functions LOGNORMDIST and LOGINV provide the CDF and inverse CDF, respectively, of the lognormal distribution.

7-6.2 Mean and Variance of the Lognormal Distribution

The mean and variance of the lognormal distribution are

$$E(X) = \mu_X = e^{\mu_Y + (1/2)\sigma_Y^2} \tag{7-25}$$

and

$$V(X) = \sigma_X^2 = e^{2\mu_Y + \sigma_Y^2}\left(e^{\sigma_Y^2} - 1\right) = \mu_X^2\left(e^{\sigma_Y^2} - 1\right). \tag{7-26}$$

In some applications of the lognormal distribution, it is important to know the values of the median and the mode. The median, which is the value $\tilde{x}$ such that $P(X \le \tilde{x}) = 0.5$, is

$$\tilde{x} = e^{\mu_Y}. \tag{7-27}$$

The mode is the value of x for which $f(x)$ is maximum, and for the lognormal distribution, the mode is

$$\text{MO} = e^{\mu_Y - \sigma_Y^2}. \tag{7-28}$$

Figure 7-11 shows the relative location of the mean, median, and mode for the lognormal distribution. Since the distribution has a right skew, generally we will find that mode < median < mean.

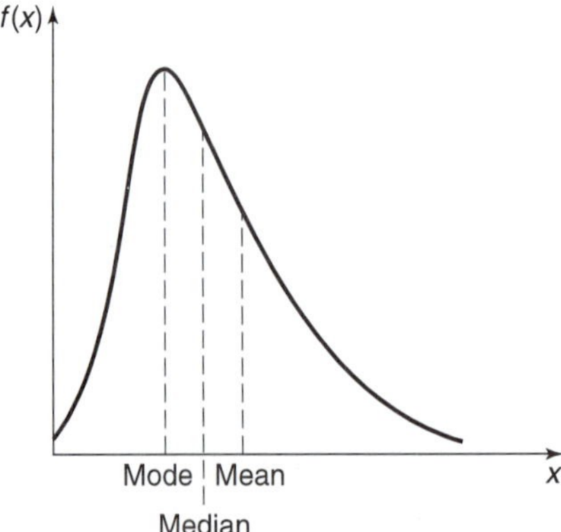

Figure 7-11 The lognormal distribution.

7-6.3 Properties of the Lognormal Distribution

While the normal distribution has additive reproductive properties, the lognormal distribution has multiplicative reproductive properties. Some of the more important properties are as follows:

1. If X has a lognormal distribution with parameters μ_Y and σ_Y^2 and if a, b, and d are constants such that $b = e^d$, then $W = bX^a$ has a lognormal distribution with parameters $(d + a\mu_Y)$ and $(a\sigma_Y)^2$.
2. If X_1 and X_2 are independent lognormal variables, with parameters $(\mu_{Y_1} + \mu_{Y_1}^2)$ and $(\mu_{Y_2} + \mu_{Y_2}^2)$ respectively, then $W = X_1 \cdot X_2$ has a lognormal distribution with parameters $[(\mu_{Y_1} + \mu_{Y_2}), (\sigma_{Y_1}^2 + \sigma_{Y_2}^2)]$.
3. If $X_1, X_2, \ldots, X_n$ is a sequence of n independent lognormal variates, with parameters $(\mu_{Y_j}, \sigma_{Y_j}^2), j = 1, 2, \ldots, n$, respectively, and $\{a_j\}$ is a sequence of constants while $b = e^d$ is a single constant, then the product

$$W = b\prod_{j=1}^{n} X_j^{a_j}$$

has a lognormal distribution with parameters

$$\left(d + \sum_{j=1}^{n} a_j \mu_{Y_j}\right) \quad \text{and} \quad \left(\sum_{j=1}^{n} a_j^2 \sigma_{Y_j}^2\right).$$

4. If $X_1, X_2, \ldots, X_n$ are independent lognormal variates, each with the same parameters (μ_Y, σ_Y^2), then the geometric mean

$$\left(\prod_{j=1}^{n} X_j\right)^{1/n}$$

has a lognormal distribution with parameters μ_Y and σ_Y^2/n.

Example 7-12

The random variable $Y = \ln X$ has a $N(10, 4)$ distribution, so X has a lognormal distribution with a mean and variance of

$$E(X) = e^{10 + (1/2)4} = e^{12} \simeq 162{,}754$$

and

$$V(X) = e^{[2(10) + 4]}(e^4 - 1),$$
$$= e^{24}(e^4 - 1) \simeq 53.598e^{24},$$

respectively. The mode and median are

$$\text{mode} = e^6 \simeq 403.43$$

and

$$\text{median} = e^{10} \simeq 22{,}026.$$

In order to determine a specific probability, say $P(X \le 1000)$, we use the transform $P(\ln X \le \ln 1000) = P(Y \le \ln 1000)$:

$$P(Y \le \ln 1000) = P\left(Z \le \frac{\ln 1000 - 10}{2}\right)$$
$$= \Phi(-1.55) = 0.0611.$$

Example 7-13

Suppose

$$\begin{aligned}
Y_1 &= \ln X_1 \sim N(4, 1),\\
Y_2 &= \ln X_2 \sim N(3, 0.5),\\
Y_3 &= \ln X_3 \sim N(2, 0.4),\\
Y_4 &= \ln X_4 \sim N(1, 0.01),
\end{aligned}$$

and furthermore suppose X_1, X_2, X_3, and X_4 are independent random variables. The random variable W defined as follows represents a critical performance variable on a telemetry system:

$$W = e^{1.5}[X_1^{2.5} X_2^{0.2} X_3^{0.7} X_4^{3.1}].$$

By reproductive property 3, W will have a lognormal distribution with parameters

$$1.5 + (2.5 \cdot 4 + 0.2 \cdot 3 + 0.7 \cdot 2 + 3.1 \cdot 1) = 16.6$$

and

$$(2.5)^2 \cdot 1 + (0.2)^2 \cdot 0.5 + (0.7)^2 \cdot 0.4 + (3.1)^2 \cdot (0.01) = 6.562,$$

respectively. That is to say, $\ln W \sim N(16.6, 6.562)$. If the specifications on W are, say, 20,000–600,000, we could determine the probability that W would fall within specifications as follows:

$$\begin{aligned}
&P(20{,}000 \le W \le 600 \cdot 10^3)\\
&\quad = P\left[\ln(20{,}000) \le \ln W < \ln(600 \cdot 10^3)\right]\\
&\quad = \Phi\left(\frac{\ln 600 \cdot 10^3 - 16.6}{\sqrt{6.526}}\right) - \Phi\left(\frac{\ln 20{,}000 - 16.6}{\sqrt{6.526}}\right)\\
&\quad \simeq \Phi(-1.290) - \Phi(-2.621) = 0.0985 - 0.0044\\
&\quad = 0.0941.
\end{aligned}$$

7-7 THE BIVARIATE NORMAL DISTRIBUTION

Up to this point, all of the continuous random variables have been of one dimension. A very important two-dimensional probability law that is a generalization of the one-dimensional normal probability law is called the *bivariate normal distribution.* If $[X_1, X_2]$ is a bivariate normal random vector, then the joint density function of $[X_1, X_2]$ is

$$f(x_1, x_2) = \frac{1}{2\pi\sigma_1\sigma_2\sqrt{1-\rho^2}} \exp\left\{-\frac{1}{2(1-\rho^2)}\left[\left(\frac{x_1-\mu_1}{\sigma_1}\right)^2 - 2\rho\left(\frac{x_1-\mu_1}{\sigma_1}\right)\left(\frac{x_2-\mu_2}{\sigma_2}\right) + \left(\frac{x_2-\mu_2}{\sigma_2}\right)^2\right]\right\} \tag{7-29}$$

for $-\infty < x_1 < \infty$ and $-\infty < x_2 < \infty$.

The joint probability $P(a_1 \le X_1 \le b_1, a_2 \le X_2 \le b_2)$ is defined as

$$\int_{a_2}^{b_2}\int_{a_1}^{b_1} f(x_1, x_2)\, dx_1 dx_2 \tag{7-30}$$

and is represented by the volume under the surface and over the region $\{(x_1, x_2): a_1 \le x_1 \le b_1, a_2 \le x_2 \le b_2\}$, as shown in Fig. 7-12. Owen (1962) has provided a table of probabilities. The bivariate normal density has five parameters. These are μ_1, μ_2, σ_1, σ_2, and ρ, the cor-

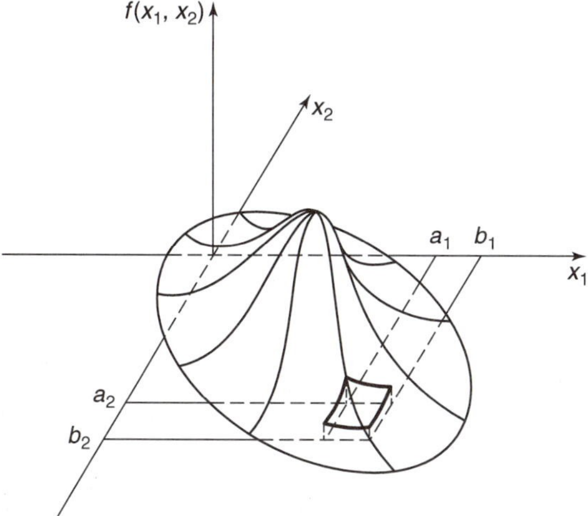

Figure 7-12 The bivariate normal density.

relation coefficient between X_1 and X_2, such that $-\infty < \mu_1 < \infty$, $-\infty < \mu_2 < \infty$, $\sigma_1 > 0$, $\sigma_2 > 0$, and $-1 < \rho < 1$.

The marginal densities f_1 and f_2 are given, respectively, as

$$f_1(x_1) = \int_{-\infty}^{\infty} f(x_1, x_2)\, dx_2 = \frac{1}{\sigma_1\sqrt{2\pi}} e^{-(1/2)[(x_1-\mu_1)/\sigma_1]^2} \tag{7-31}$$

for $-\infty < x_1 < \infty$ and

$$f_2(x_2) = \int_{-\infty}^{\infty} f(x_1, x_2)\, dx_1 = \frac{1}{\sigma_2\sqrt{2\pi}} e^{-(1/2)[(x_2-\mu_2)/\sigma_2]^2} \tag{7-32}$$

for $-\infty < x_2 < \infty$.

We note that these marginal densities are normal; that is,

$$X_1 \sim N(\mu_1, \sigma_1^2) \tag{7-33}$$

and

$$X_2 \sim N(\mu_2, \sigma_2^2),$$

so that

$$\begin{aligned} E(X_1) &= \mu_1, \\ E(X_2) &= \mu_2, \\ V(X_1) &= \sigma_1^2, \\ V(X_2) &= \sigma_2^2. \end{aligned} \tag{7-34}$$

The correlation coefficient ρ is the ratio of the covariance to $[\sigma_1 \cdot \sigma_2]$. The covariance is

$$\sigma_{12} = \int_{-\infty}^{\infty}\int_{-\infty}^{\infty} (x_1 - \mu_1)(x_2 - \mu_2) f(x_1, x_2)\, dx_1\, dx_2.$$

Thus

$$\rho = \frac{\sigma_{12}}{\sigma_1 \cdot \sigma_2} \tag{7-35}$$

The conditional distributions $f_{X_2|x_1}(x_2)$ and $f_{X_1|x_2}(x_1)$ are also important. These conditional densities are normal, as shown here:

$$f_{X_2|x_1}(x_2) = \frac{f(x_1, x_2)}{f_1(x_1)}$$

$$= \frac{1}{\sigma_2 \sqrt{2\pi}\sqrt{1-\rho^2}} \exp\left[\frac{-1}{2\sigma_2^2(1-\rho^2)}\left\{x_2 - \left[\mu_2 + \rho(\sigma_2 \;\; \sigma_1)(x_1 - \mu_1)\right]\right\}^2\right] \quad (7\text{-}36)$$

for $-\infty < x_2 < \infty$ and

$$f_{X_1|x_2}(x_1) = \frac{f(x_1, x_2)}{f_2(x_2)}$$

$$= \frac{1}{\sigma_1 \sqrt{2\pi}\sqrt{1-\rho^2}} \exp\left[\frac{-1}{2\sigma_1^2(1-\rho^2)}\left\{x_1 - \left[\mu_1 + \rho(\sigma_1 \;\; \sigma_2)(x_2 - \mu_2)\right]\right\}^2\right] \quad (7\text{-}37)$$

for $-\infty < x_1 < \infty$. Figure 7-13 illustrates some of these conditional densities.

We first consider the distribution $f_{X_2|x_1}$. The mean and variance are

$$E(X_2|x_1) = \mu_2 + \rho(\sigma_2/\sigma_1)(x_1 - \mu_1) \quad (7\text{-}38)$$

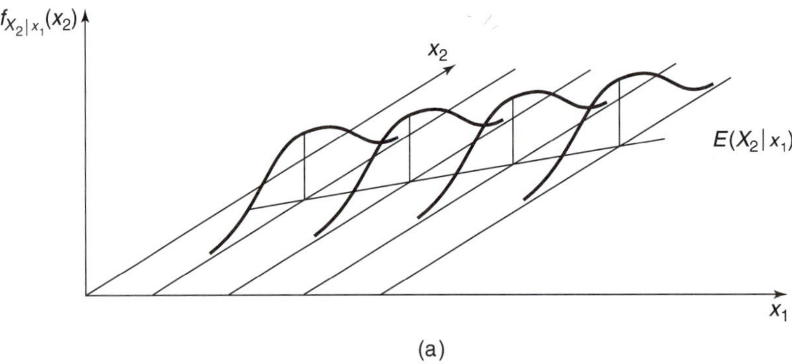

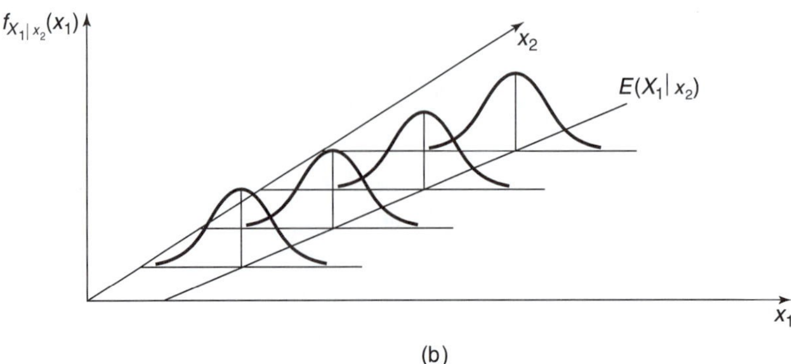

Figure 7-13 Some typical conditional distributions. (*a*) Some example conditional distributions of X_2 for a few values of x_1. (*b*) Some example conditional distributions of X_1 for a few values of x_2.

and

$$V(X_2|x_1) = \sigma_2^2(1 - \rho^2) \tag{7-39}$$

Furthermore, $f_{X_2|x_1}$ is normal; that is,

$$X_2|x_1 \sim N[\mu_2 + \rho(\sigma_2/\sigma_1)(x_1 - \mu_1), \sigma_2^2(1 - \rho^2)]. \tag{7-40}$$

The locus of expected values of X_2 for given x_1, as shown in equation 7-38, is called the *regression of* X_2 *on* X_1, and it is linear. Also, the *variance in the conditional distributions is constant* for all x_1.

In the case of the distribution $f_{X_1|x_2}$, the results are similar. That is,

$$E(X_1|x_2) = \mu_1 + \rho(\sigma_1/\sigma_2)(x_2 - \mu_2), \tag{7-41}$$

$$V(X_1|x_2) = \sigma_1^2(1 - \rho^2), \tag{7-42}$$

and

$$X_1|x_2 \sim N[\mu_1 + \rho(\sigma_1/\sigma_2)(x_2 - \mu_2), \sigma_1^2(1 - \rho^2)]. \tag{7-43}$$

In the bivariate normal distribution we observe that if $\rho = 0$, the joint density may be factored into the product of the marginal densities and so X_1 and X_2 are independent. Thus, for a bivariate normal density, zero correlation and independence are equivalent. If planes parallel to the x_1, x_2 plane are passed through the surface shown in Fig. 7-12, the contours cut from the bivariate normal surface are ellipses. The student may wish to show this property.

Example 7-14

In an attempt to substitute a nondestructive testing procedure for a destructive test, an extensive study was made of shear strength, X_2, and weld diameter, X_1, of spot welds, with the following findings.

1. $[X_1, X_2]$ ~ bivariate normal.
2. $\mu_1 = 0.20$ inch, $\mu_2 = 1100$ pounds, $\sigma_1^2 = 0.02$ inch^2, $\sigma_2^2 = 525$ pounds^2, and $\rho = 0.9$.

The regression of X_2 on X_1 is thus

$$\begin{aligned} E(X_2|x_1) &= \mu_2 + \rho(\sigma_2/\sigma_1)(x_1 - \mu_1) \\ &= 1100 + 0.9\left(\frac{\sqrt{525}}{\sqrt{0.02}}\right)(x_1 - 0.2) \\ &= 145.8x_1 + 1070.84, \end{aligned}$$

and the variance is

$$\begin{aligned} V(X_2|x_1) &= \sigma_2^2(1 - \rho^2) \\ &= 525(0.19) = 99.75. \end{aligned}$$

In studying these results, the manager of manufacturing notes that since $\rho = 0.9$, that is, close to 1, weld diameter is highly correlated with shear strength. The specification on shear strength calls for a value greater than 1080. If a weld has a diameter of 0.18, he asks: "What is the probability that the strength specification will be met?" The process engineer notes that $E(X_2|0.18) = 1097.05$; therefore,

$$\begin{aligned} P(X_2 \geq 1080) &= P\left(Z \geq \frac{1080 - 1097.05}{\sqrt{99.75}}\right) \\ &= 1 - \Phi(-1.71) = 0.9564, \end{aligned}$$

and he recommends a policy such that if the weld diameter is not less than 0.18, the weld will be classified as satisfactory.

Example 7-15

In developing an admissions policy for a large university, the office of student testing and evaluation has noted that X_1, the combined score on the college board examinations, and X_2, the student grade point average at the end of the freshman year, have a bivariate normal distribution. A grade point of 4.0 corresponds to *A*. A study indicates that

$$\begin{aligned} \mu_1 &= 1300, \\ \mu_2 &= 2.3, \\ \sigma_1^2 &= 6400, \\ \sigma_2^2 &= 0.25, \\ \rho &= 0.6. \end{aligned}$$

Any student with a grade point average less than 1.5 is automatically dropped at the end of the freshman year; however, an average of 2.0 is considered to be satisfactory.

An applicant takes the college board exams, receives a combined score of 900, and is not accepted. An irate parent argues that the student will do satisfactory work and, specifically, will have better than a 2.0 grade point average at the end of the freshman year. Considering only the probabilistic aspects of the problem, the director of admissions wants to determine $P(X_2 \geq 2.0|x_1 = 900)$. Noting that

$$\begin{aligned} E\left(X_2|900\right) &= 2.3+(0.6)\left(\frac{0.5}{80}\right)(900-1300) \\ &= 0.8 \end{aligned}$$

and

$$V(X_2|900) = 0.16,$$

the director calculates

$$1-\Phi\left(\frac{2.0-0.8}{0.4}\right) = 0.0013,$$

which predicts only a very slim chance of the parent's claim being valid.

7-8 GENERATION OF NORMAL REALIZATIONS

We will consider both direct and approximate methods for generating realizations of a standard normal variable *Z*, where $Z \sim N(0, 1)$. Recall that $X = \mu + \sigma Z$, so realizations of $X \sim N(\mu, \sigma^2)$ are easily obtained as $x = \mu + \sigma z$.

The direct method calls for generating uniform [0, 1] random number realizations in pairs: u_1 and u_2. Then, using the methods of Chapter 4, it turns out that

$$\begin{aligned} z_1 &= (-2 \ln u_1)^{1/2}\cos(2\pi u_2), \\ z_2 &= (-2 \ln u_1)^{1/2}\sin(2\pi u_2) \end{aligned} \tag{7-44}$$

are realizations of independent, $N(0, 1)$ variables. The values $x = \mu + \sigma z$ follow directly, and the process is repeated until the desired number of realizations of *X* are obtained.

An approximate method that makes use of the Central Limit Theorem is as follows:

$$z = \sum_{i=1}^{12} u_i - 6. \tag{7-45}$$

With this procedure, we would begin by generating 12 uniform [0, 1] random number realizations, adding them and subtracting 6. This entire process is repeated until the desired number of realizations is obtained.

Although the direct method is exact and usually preferable, approximate values are sometimes acceptable.

7-9 SUMMARY

This chapter has presented the normal distribution with a number of example applications. The *normal* distribution, the related *standard normal*, and the *lognormal* distributions are univariate, while the *bivariate normal* gives the joint density of two related normal random variables.

The normal distribution forms the basis on which a great deal of the work in statistical inference rests. The wide application of the normal distribution makes it particularly important.

7-10 EXERCISES

7-1. Let Z be a standard normal random variable and calculate the following probabilities, using sketches where appropriate:

(a) $P(0 \le Z \le 2)$.

(b) $P(-1 \le Z \le +1)$.

(c) $P(Z \le 1.65)$.

(d) $P(Z \ge -1.96)$.

(e) $P(|Z| > 1.5)$.

(f) $P(-1.9 \le Z \le 2)$.

(g) $P(Z \le 1.37)$.

(h) $P(|Z| \le 2.57)$.

7-2. Let $X \sim N(10, 9)$. Find $P(X \le 8)$, $P(X \ge 12)$, $P(2 \le X \le 10)$.

7-3. In each part below, find the value of c that makes the probability statement true.

(a) $\Phi(c) = 0.94062$.

(b) $P(|Z| \le c) = 0.95$.

(c) $P(|Z| \le c) = 0.99$.

(d) $P(Z \le c) = 0.05$.

7-4. If $P(Z \ge z_\alpha) = \alpha$, determine z_α for $\alpha = 0.025$, $\alpha = 0.005$, $\alpha = 0.05$, and $\alpha = 0.0014$.

7-5. If $X \sim N(80, 10^2)$, compute the following:

(a) $P(X \le 100)$.

(b) $P(X \le 80)$.

(c) $P(75 \le X \le 100)$.

(d) $P(75 \le X)$.

(e) $P(|X - 80| \le 19.6)$.

7-6. The life of a particular type of dry-cell battery is normally distributed with a mean of 600 days and a standard deviation of 60 days. What fraction of these batteries would be expected to survive beyond 680 days? What fraction would be expected to fail before 560 days?

7-7. The personnel manager of a large company requires job applicants to take a certain test and achieve a score of 500. If the test scores are normally distributed with a mean of 485 and a standard deviation of 30, what percentage of the applicants pass the test?

7-8. Experience indicates that the development time for a photographic printing paper is distributed as $X \sim N$(30 seconds, 1.21 seconds^2). Find the following: (a) The probability that X is at least 28.5 seconds. (b) The probability that X is at most 31 seconds. (c) The probability that X differs from its expected value by more than 2 seconds.

7-9. A certain type of light bulb has an output known to be normally distributed with mean of 2500 end footcandles and a standard deviation of 75 end footcandles. Determine a lower specification limit such that only 5% of the manufactured bulbs will be defective.

7-10. Show that the moment-generating function for the normal distribution is as given by equation 7-5. Use it to generate the mean and variance.

7-11. If $X \sim N(\mu, \sigma^2)$, show that $Y = aX + b$, where a and b are real constants, is also normally distributed. Use the methods outlined in Chapter 3.

7-12. The inside diameter of a piston ring is normally distributed with a mean of 12 cm and a standard deviation of 0.02 cm.

(a) What fraction of the piston rings will have diameters exceeding 12.05 cm?

(b) What inside diameter value c has a probability of 0.90 being exceeded?

(c) What is the probability that the inside diameter will fall between 11.95 and 12.05?

7-13. A plant manager orders a process shutdown and setting readjustment whenever the pH of the final product falls above 7.20 or below 6.80. The sample pH is normally distributed with unknown μ and a standard deviation of $\sigma = 0.10$. Determine the following probabilities:

(a) Of readjusting when the process is operating as intended with $\mu = 7.0$.

(b) Of readjusting when the process is slightly off target with the mean pH 7.05.

(c) Of failing to readjust when the process is too alkaline and the mean pH is $\mu = 7.25$.

(d) Of failing to readjust when the process is too acidic and the mean pH is $\mu = 6.75$.

7-14. The price being asked for a certain security is distributed normally with a mean of \$50.00 and a standard deviation of \$5.00. Buyers are willing to pay an amount that is also normally distributed with a mean of \$45.00 and a standard deviation of \$2.50. What is the probability that a transaction will take place?

7-15. The specifications for a capacitor are that its life must be between 1000 and 5000 hours. The life is known to be normally distributed with a mean of 3000 hours. The revenue realized from each capacitor is \$9.00; however, a failed unit must be replaced at a cost of \$3.00 to the company. Two manufacturing processes can produce capacitors having satisfactory mean lives. The standard deviation for process A is 1000 hours and for process B it is 500 hours. However, process A manufacturing costs are only half those for B. What value of process manufacturing cost is critical, so far as dictating the use of process A or B?

7-16. The diameter of a ball bearing is a normally distributed random variable with mean μ and a standard deviation of 1. Specifications for the diameter are $6 \leq X \leq 8$, and a ball bearing within these limits yields a profit of C dollars. However, if $X < 6$, then the profit is $-R_1$ dollars, or if $X > 8$, the profit is $-R_2$ dollars. Find the value of μ that maximizes the expected profit.

7-17. In the preceding exercise, find the optimum value of μ if $R_1 = R_2 = R$.

7-18. Use the results of Exercise 7-16 with $C = \$8.00$, $R_1 = \$2.00$, and $R_2 = \$4.00$. What is the value of μ that maximizes the expected profit?

7-19. The Rockwell hardness of a particular alloy is normally distributed with a mean of 70 and a standard deviation of 4.

(a) If a specimen is acceptable only if its hardness is between 62 and 72, what is the probability that a randomly chosen specimen has an acceptable hardness?

(b) If the acceptable range of hardness was $(70 - c, 70 + c)$, for what value of c would 95% of all specimens have acceptable hardness?

(c) Where the acceptable range is as in (a) and the hardness of each of nine randomly selected specimens is independently determined, what is the expected number of acceptable specimens among the nine specimens?

7-20. Prove that $E(Z_n) = 0$ and $V(Z_n) = 1$, where Z_n is as defined in Theorem 7-2.

7-21. Let $X_i (i = 1, 2, \ldots, n)$ be independent and identically distributed random variables with mean μ and variance σ^2. Consider the sample mean,

$$\bar{X} = \frac{1}{n}(X_1 + X_2 + \cdots + X_n) = \frac{1}{n}\sum_{i=1}^{n} X_i.$$

Show that $E(\bar{X}) = \mu$ and $V(\bar{X}) = \sigma^2/n$.

7-22. A shaft with an outside diameter (O.D.) $\sim N(1.20, 0.0016)$ is inserted into a sleeve bearing having an inside diameter (I.D.) that is $N(1.25, 0.0009)$. Determine the probability of interference.

7-23. An assembly consists of three components placed side by side. The length of each component is normally distributed with a mean of 2 inches and a standard deviation of 0.2 inch. Specifications require that all assemblies be between 5.7 and 6.3 inches long. How many assemblies will pass these requirements?

7-24. Find the mean and variance of the linear combination

$$Y = X_1 + 2X_2 + X_3 + X_4,$$

where $X_1 \sim N(4, 3)$, $X_2 \sim N(4, 4)$, $X_3 \sim N(2, 4)$, and $X_4 \sim N(3, 2)$. What is the probability that $15 \leq Y \leq 20$?

7-25. Round-off error has a uniform distribution on $[-0.5, +0.5]$ and round-off errors are independent. A sum of 50 numbers is calculated where each is rounded before adding. What is the probability that the total round-off error exceeds 5?

7-26. One hundred small bolts are packed in a box. Each bolt weights 1 ounce, with a standard deviation of 0.01 ounce. Find the probability that a box weighs more than 102 ounces.

7-27. An automatic machine is used to fill boxes with soap powder. Specifications require that the boxes weigh between 11.8 and 12.2 ounces. The only data available about machine performance concern the average content of groups of nine boxes. It is known that the average content is 11.9 ounces with a standard deviation of 0.05 ounce. What fraction of the boxes produced is defective? Where should the mean be located in order to minimize this fraction defective? Assume the weight is normally distributed.

7-28. A bus travels between two cities, but visits six intermediate cities on the route. The means and standard deviations of the travel times are as follows:

City Pairs	Mean Time (hours)	Standard Deviation (hours)
1 – 2	3	0.4
2 – 3	4	0.6
3 – 4	3	0.3
4 – 5	5	1.2
5 – 6	7	0.9
6 – 7	5	0.4
7 – 8	3	0.4

What is the probability that the bus completes its journey within 32 hours?

7-29. A production process produces items, of which 8% are defective. A random sample of 200 items is selected every day and the number of defective items, say X, is counted. Using the normal approximation to the binomial find the following:

(a) $P(X \le 16)$.

(b) $P(X = 15)$.

(c) $P(12 \le X \le 20)$.

(d) $P(X = 14)$.

7-30. In a work-sampling study it is often desired to find the necessary number of observations. Given that $p = 0.1$, find the necessary n such that $P(0.05 \le \hat{p} \le 0.15) = 0.95$.

7-31. Use random numbers generated from your favorite computer package or from scaling the random integers in Table XV of the Appendix by multiplying by 10^{-5} to generate six realizations of a $N(100, 4)$ variable, using the following:

(a) The direct method.

(b) The approximate method.

7-32. Consider a linear combination $Y = 3X_1 - 2X_2$, where X_1 is $N(10, 3)$ and X_2 is uniformly distributed on [0, 20]. Generate six realizations of the random variable Y, where X_1 and X_2 are independent.

7-33. If $Z \sim N(0, 1)$, generate five realizations of Z^2.

7-34. If $Y = \ln X$ and $Y \sim N(\mu_Y, \sigma_Y^2)$, develop a procedure for generating realizations of X.

7-35. If $Y = X_1^{1/2}/X_2^2$, where $X_1 \sim N(\mu_1, \sigma_1^2)$ and $X_2 \sim N(\mu_2, \sigma_2^2)$ and where X_1 and X_2 are independent, develop a generator for producing realizations of Y.

7-36. The brightness of light bulbs is normally distributed with a mean of 2500 footcandles and a standard deviation of 50 footcandles. The bulbs are tested and all those brighter than 2600 footcandles are placed in a special high-quality lot. What is the probability distribution of the remaining bulbs? What is their expected brightness?

7-37. The random variable $Y = \ln X$ has a $N(50, 25)$ distribution. Find the mean, variance, mode, and median of X.

7-38. Suppose independent random variables Y_1, Y_2, Y_3 are such that

$$Y_1 = \ln X_1 \sim N(4, 1),$$
$$Y_2 = \ln X_2 \sim N(3, 1),$$
$$Y_3 = \ln X_3 \sim N(2, 0.5).$$

Find the mean and variance of $W = e^2 X_1^2 X_2^{1.5} X_3^{1.28}$. Determine a set of specifications L and R such that

$$P(L \le W \le R) = 0.90.$$

7-39. Show that the density function for a lognormally distributed random variable X is given by equation 7-24.

7-40. Consider the bivariate normal density

$$f(x_1, x_2) = \Delta \exp\left\{-\frac{1}{2(1-\rho^2)}\left[\frac{x_1^2}{\sigma_1^2} - \frac{2\rho x_1 x_2}{\sigma_1\sigma_2} + \frac{x_2^2}{\sigma_2^2}\right]\right\}$$

$$-\infty < x_1 < \infty, \; -\infty < x_2 < \infty,$$

where Δ is chosen so that f is a probability distribution. Are the random variables X_1 and X_2 independent? Define two new random variables:

$$Y_1 = \frac{1}{(1-\rho^2)^{1/2}}\left(\frac{X_1}{\sigma_1} - \frac{\rho X_2}{\sigma_2}\right),$$

$$Y_2 = \frac{X_2}{\sigma_2}.$$

Show that the two new random variables are independent.

7-41. The life of a tube (X_1) and the filament diameter (X_2) are distributed as a bivariate normal random variable with the parameters $\mu_1 = 2000$ hours, $\mu_2 = 0.10$ inch, $\sigma_1^2 = 2500$ hours^2, $\sigma_2^2 = 0.01$ inch^2, and $\rho = 0.87$. The quality-control manager wishes to determine the life of each tube by measuring the filament diameter. If a filament diameter is 0.098, what is the probability that the tube will last 1950 hours?

7-42. A college professor has noticed that grades on each of two quizzes have a bivariate normal distribution with the parameters $\mu_1 = 75$, $\mu_2 = 83$, $\sigma_1^2 = 25$, $\sigma_2^2 = 16$, and $\rho = 0.8$. If a student receives a grade of 80 on the first quiz, what is the probability that she will do better on the second one? How is the answer affected by making $\rho = -0.8$?

7-43. Consider the surface $y = f(x_1, x_2)$, where f is the bivariate normal density function.

(a) Prove that y = constant cuts the surface in an ellipse.

(b) Prove that y = constant with $\rho = 0$ and $\sigma_1^2 = \sigma_2^2$ cuts the surface as a circle.

7-44. Let X_1 and X_2 be independent random variables, each following a normal density with a mean of zero and variance σ^2. Find the distribution of

$$R = \sqrt{X_1^2 + X_2^2}$$

The resulting distribution is known as the *Rayleigh* distribution and is frequently used to model the distribution of radial error in a plane. *Hint*: Let $X_1 = R\cos\theta$ and $X_2 = R\sin\theta$. Obtain the joint probability distribution of R and θ, then integrate out θ.

7-45. Using a method similar to that in Exercise 7-44, obtain the distribution of

$$R = \sqrt{X_1^2 + X_2^2 + \cdots + X_n^2}$$

where $X_i \sim N(0, \sigma^2)$ and is independent.

7-46. Let the independent random variables $X_i \sim N(0, \sigma^2)$ for $i = 1, 2$. Find the probability distribution of

$$C = \frac{X_1}{X_2}.$$

We say that C follows the *Cauchy* distribution. Try to compute $E(C)$.

7-47. Let $X \sim N(0, 1)$. Find the probability distribution of $Y = X^2$. Y is said to follow the *chi-square* distribution with one degree of freedom. It is an important distribution in statistical methodology.

7-48. Let the independent random variables $X_i \sim N(0, 1)$ for $i = 1, 2, \ldots, n$. Show that the probability distribution of $Y = \sum_{i=1}^{n} X_i^2$ follows a chi-square distribution with n degrees of freedom.

7-49. Let $X \sim N(0, 1)$. Define a new random variable $Y = |X|$. Then, find the probability distribution of Y. This is often called the *half-normal* distribution.

Chapter 8

Introduction to Statistics and Data Description

8-1 THE FIELD OF STATISTICS

Statistics deals with the collection, presentation, analysis, and use of data to solve problems, make decisions, develop estimates, and design and develop both products and procedures. An understanding of basic statistics and statistical methods would be useful to anyone in this information age; however, since engineers, scientists, and those working in management science are routinely engaged with data, a knowledge of statistics and basic statistical methods is particularly vital. In this intensely competitive, high-tech world economy of the first decade of the twenty-first century, the expectations of consumers regarding product quality, performance, and reliability have increased significantly over expectations of the recent past. Furthermore, we have come to expect high levels of performance from logistical systems at all levels, and the operation as well as the refinement of these systems is largely dependent on the collection and use of data. While those who are employed in various "service industries" deal with somewhat different problems, at a basic level they, too, are collecting data to be used in solving problems and improving the "service" so as to become more competitive in attracting market share.

Statistical methods are used to present, describe, and understand *variability*. In observing a variable value or several variable values repeatedly, where these values are assigned to units by a process, we note that these repeated observations tend to yield different results. While this chapter will deal with data presentation and description issues, the following chapters will utilize the probability concepts developed in prior chapters to model and develop an understanding of variability and to utilize this understanding in presenting inferential statistics topics and methods.

Virtually all real-world processes exhibit variability. For example, consider situations where we select several castings from a manufacturing process and measure a critical dimension (such as a vane opening) on each part. If the measuring instrument has sufficient resolution, the vane openings will be different (there will be variability in the dimension). Alternatively, if we count the number of defects on printed circuit boards, we will find variability in the counts, as some boards will have few defects and others will have many defects. This variability extends to *all* environments. There is variability in the thickness of oxide coatings on silicon wafers, the hourly yield of a chemical process, the number of errors on purchase orders, the flow time required to assemble an aircraft engine, and the therms of natural gas billed to the residential customers of a distributing utility in a given month.

Almost all experimental activity reflects similar variability in the data observed, and in Chapter 12 we will employ statistical methods not only to analyze experimental data but also to construct effective experimental designs for the study of processes.

Why does variability occur? Generally, variability is the result of changes in the conditions under which observations are made. In a manufacturing context, these changes may be differences in specimens of material, differences in the way people do the work, differences in process variables, such as temperature, pressure, or holding time, and differences in *environmental factors*, such as relative humidity. Variability also occurs because of the measurement system. For example, the measurement obtained from a scale may depend on where the test item is placed on the pan. The process of selecting units for observation may also cause variability. For example, suppose that a lot of 1000 integrated circuit chips has exactly 100 defective chips. If we inspected all 1000 chips, and if our inspection process was perfect (no inspection or measurement error), we would find all 100 defective chips. However, suppose that we select 50 chips. Now some of the chips will likely be defective; and we would expect the sample to be about 10% defective, but it could be 0% or 2% or 12% defective, depending on the specific chips selected.

The field of statistics consists of methods for describing and modeling variability, and for making decisions when variability is present. In *inferential statistics,* we usually want to make a decision about some *population*. The term *population* refers to the collection of measurements on all elements of a *universe* about which we wish to draw conclusions or make decisions. In this text, we make a distinction between the universe and the populations, in that the universe is composed of the set of *elementary units* or simply *units*, while a population is the set of numerical or categorical variable values for one variable associated with each of the universe units. Obviously, there may be several populations associated with a given universe. An example is the universe consisting of the residential class customers of an electric power company whose accounts are active during part of the month of August 2003. Example populations might be the set of energy consumption (kilowatt-hour) values billed to these customers in the August 2003 bill, the set of customer demands (kilowatts) at the instant the company experiences the August peak demand, and the set made up of dwelling category, such as single-family unattached, apartments, mobile home, etc.

Another example is a universe which consists of all the power supplies for a personal computer manufactured by an electronics company during a given period. Suppose that the manufacturer is interested specifically in the output voltage of each power supply. We may think of the output voltage levels in the power supplies as such a population. In this case, each population value is a numerical measurement, such as 5.10 or 5.24. The data in this case would be referred to as *measurement data*. On the other hand, the manufacturer may be interested in whether or not each power supply produced an output voltage that conforms to the requirements. We may then visualize the population as consisting of *attribute data*, in which each power supply is assigned a value of one if the unit is nonconforming and a value of zero if it conforms to requirements. Both measurement data and attribute data are called *numerical data*. Furthermore, it is convenient to consider measurement data as being either *continuous data* or *discrete data* depending on the nature of the process assigning values to unit variables. Yet another type of data is called *categorical data*. Examples are gender, day of the week when the observation is taken, make of automobile, etc. Finally, we have *unit identifying data,* which are alphanumeric and used to identify the universe and sample units. These data might neither exist nor have statistical interpretation; however, in some situations they are essential identifiers for universe and sample units. Examples would be social security numbers, account numbers in a bank, VIN numbers for automobiles, and serial numbers of cardiac pacemakers. In this book we will present techniques for dealing with both measurement and attribute data; however, categorical data are also considered.

In most applications of statistics, the available data result from a *sample* of units selected from a universe of interest, and these data reflect measurement or classification of one or more variables associated with the sampled units. The sample is thus a subset of the units, and the measurement or classification values of these units are subsets of the respective universe populations.

Figure 8-1 presents an overview of the data acquisition activity. It is convenient to think of a *process* that produces *units* and assigns values to the variables associated with the units. An example would be the manufacturing process for the power supplies. The power supplies in this case are the units, and the output voltage values (perhaps along with other variable values) may be thought of as being assigned by the process. Oftentimes, a probabilistic model or a model with some probabilistic components is *assumed* to represent the value assignment process. As indicated earlier, the set of units is referred to as the *universe*, and this set may have a finite or an infinite membership of units. Furthermore, in some cases this set exists only in concept; however, we can describe the elements of the set without enumerating them, as was the case with the sample space, $\mathscr{S}$, associated with a random experiment presented in Chapter 1. The set of values assigned or that may be assigned to a specific variable becomes the *population* for that variable.

Now, we illustrate with a few additional examples that reflect several universe structures and different aspects of observing processes and population data. First, continuing with the power supplies, consider the production from one specific shift that consists of 300 units, each having some voltage output. To begin, consider a sample of 10 units selected from these 300 units and tested with voltage measurements, as previously described for sample units. The universe here is finite, and the set of voltage values for these universe units (the population) is thus finite. If our interest is simply to describe the sample results, we have done that by enumeration of the results, and both graphical and quantitative methods for further description are given in the following sections of this chapter. On the other

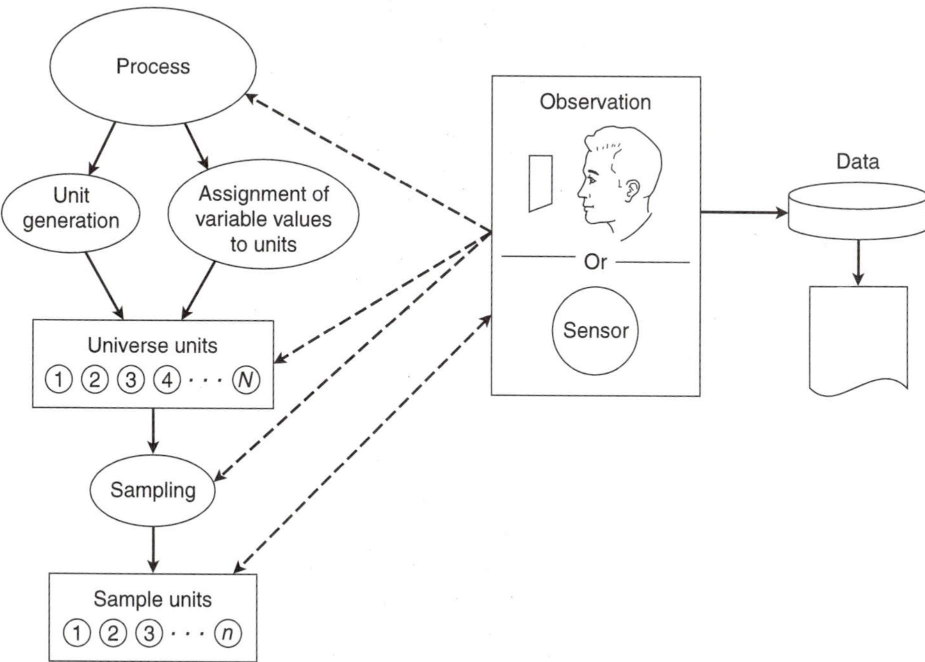

Figure 8-1 From process to data.

hand, if we wish to employ the sample results to draw statistical inference about the population consisting of 300 voltage values, this is called an *enumerative study*, and careful attention must be given to the method of *sample selection* if valid inference is to be made about the population. The key to much of what may be accomplished in such a study involves either simple or more complex forms of *random sampling* or at least *probability sampling*. While these concepts will be developed further in the next chapter, it is noted at this point that the application of simple random sampling from a finite universe results in an equal inclusion probability for each unit of the universe. In the case of a finite universe, where the sampling is done *without replacement* and the sample size, usually denoted n, is the same as the universe size, N, then this sample is called a *census*.

Suppose our interest lies not in the voltage values for units produced in this specific shift but rather in the process or process variable assignment model. Not only must great care be given to sampling methods, we must also make assumptions about the stability of the process and the structure of the process model during the period of sampling. In essence, the universe unit variable values may be thought of as a realization of the process, and a random sample from such a universe, measuring voltage values, is equivalent to a random sample on the process or process model voltage variable or population. With our example, we might assume that unit voltage, E, is distributed as $N(\mu, \sigma^2)$ during sample selection. This is often called an *analytic study*, and our objective might be, for example, to estimate μ and/or σ^2. Once again, random sampling is to be rigorously defined in the following chapter.

Even though a universe sometimes exists only in concept, defined by a verbal description of the membership, it is generally useful to think carefully about the entire process of observation activity and the conceptual universe. Consider an example where the ingredients of concrete specimens have been specified to achieve high strength with early cure time. A batch is formulated, five specimens are poured into cylindrical molds, and these are cured according to test specifications. Following the cure, these test cylinders are subjected to longitudinal load until rupture, at which point the rupture strength is recorded. These test units, with the resulting data, are considered sample data from the strength measurements associated with the universe units (each with an assigned population value) that might have been, but were never actually, produced. Again, as in the prior example, inference often relates to the process or process variable assignment model, and model assumptions may become crucial to attaining meaningful inference in this analytic study.

Finally, consider measurements taken on a fluid in order to measure some variable or characteristic. Specimens are drawn from well-mixed (we hope) fluid, with each specimen placed in a specimen container. Some difficulty arises in universe identification. A convention here is to again view the universe as made up of all possible specimens that might have been selected. Similarly, an alternate view may be taken that the sample values represent a realization of the process value assignment model. Getting meaningful results from such an analytic study once again requires close attention to sampling methods and to process or process model assumptions.

Descriptive statistics is the branch of statistics that deals with organization, summarization, and presentation of data. Many of the techniques of descriptive statistics have been in use for over 200 years, with origins in surveys and census activity. Modern computer technology, particularly computer graphics, has greatly expanded the field of descriptive statistics in recent years. *The techniques of descriptive statistics can be applied either to entire finite populations or to samples*, and these methods and techniques are illustrated in the following sections of this chapter. A wide selection of software is available, ranging in focus, sophistication, and generality from simple spreadsheet functions such as those found in Microsoft Excel®, to the more-comprehensive but "user friendly" Minitab®, to large, comprehensive, flexible systems such as SAS. Many other options are also available.

In enumerative and analytic studies, the objective is to make a conclusion or draw inference about a finite population or about a process or variable assignment model. This activity is called *inferential statistics*, and most of the techniques and methods employed have been developed within the past 90 years. Subsequent chapters will focus on these topics.

8-2 DATA

Data are collected and stored electronically as well as by human observation using traditional records or files. Formats differ depending on the observer or observational process and reflect individual preference and ease of recording. In large-scale studies, where unit identification exists, data must often be obtained by stripping the required information from several files and merging it into a file format suitable for statistical analysis.

A table or spreadsheet-type format is often convenient, and it is also compatible with most software analysis systems. Rows are typically assigned to units observed, and columns present numerical or categorical data on one or more variables. Furthermore, a column may be assigned for a sequence or order index as well as for other unit identifying data. In the case of the power supply voltage measurements, no order or sequence was intended so that any permutation of the 10 data elements is the same. In other contexts, for example where the index relates to the time of observation, the position of the element in the sequence may be quite important. A common practice in both situations described is to employ an index, say $i = 1, 2, \ldots, n$, to serve as a unit identifier where n units are observed. Also, an alphabetical character is usually employed to represent a given variable value; thus if e_i equals the value of the ith voltage measurement in the example, $e_1 = 5.10$, $e_2 = 5.24$, $e_3 = 5.14$, ..., $e_{10} = 5.11$. In a table format for these data there would be 10 rows (11 if a headings row is employed) and one or two columns, depending on whether the index is included and presented in a column.

Where several variables and/or categories are to be associated with each unit, each of these may be assigned a column, as shown in Table 8-1 (from Montgomery and Runger, 2003) which presents measurements of pull strength, wire length, and die height made on each of 25 sampled units in a semiconductor manufacturing facility. In situations where categorical classification is involved, a column is provided for each categorical variable and a category code or identifier is recorded for the unit. An example is shift of manufacture with identifiers D, N, G for day, night, and graveyard shifts, respectively.

8-3 GRAPHICAL PRESENTATION OF DATA

In this section we will present a few of the many graphical and tabular methods for summarizing and displaying data. In recent years, the availability of computer graphics has resulted in a rapid expansion of visual displays for observational data.

8-3.1 Numerical Data: Dot Plots and Scatter Plots

When we are concerned with one of the variables associated with the observed units, the data are often called univariate data, and *dot plots* provide a simple, attractive display that reflects spread, extremes, centering, and voids or gaps in the data. A horizontal line is scaled so that the range of data values is accommodated. Each observation is then plotted as a dot directly above this scaled line, and where multiple observations have the same value, the dots are simply stacked vertically at that scale point. Where the number of units is relatively small, say $n < 30$, or where there are relatively few distinct values represented

Table 8-1 Wire Bond Pull Strength Data

Observation Number	Pull Strength (y)	Wire Length (x_1)	Die Height (x_2)
1	9.95	2	50
2	24.45	8	110
3	31.75	11	120
4	35.00	10	550
5	25.02	8	295
6	16.86	4	200
7	14.38	2	375
8	9.60	2	52
9	24.35	9	100
10	27.50	8	300
11	17.08	4	412
12	37.00	11	400
13	41.95	12	500
14	11.66	2	360
15	21.65	4	205
16	17.89	4	400
17	69.00	20	600
18	10.30	1	585
19	34.93	10	540
20	46.59	15	250
21	44.88	15	290
22	54.12	16	510
23	56.63	17	590
24	22.13	6	100
25	21.15	5	400

in the data set, dot plots are effective displays. Figure 8-2 shows the univariate dot plots or marginal dot plots for each of the three variables with data presented in Table 8-1. These plots were produced using Minitab®.

Where we wish to jointly display results for two variables, the bivariate equivalent of the dot plot is called a *scatter plot.* We construct a simple rectangular coordinate graph, assigning the horizontal axis to one of the variables and the vertical axis to the other. Each observation is then plotted as a point in this plane. Figure 8-3 presents scatter plots for pull strength vs. wire length and for pull strength vs. die height for the data in Table 8-1. In order to accommodate data pairs that are identical, and thus that fall at the same point on the plane, one convention is to employ alphabet characters as plot symbols; so A is the displayed plot point where one data point falls at a specific point on the plane, B is the displayed plot point if two fall at a specific point of the plane, etc. Another approach for this, which is useful where values are close but not identical, is to assign a randomly generated, small, positive or negative quantity sometimes called *jitter* to one or both variables in order to make the plots better displays. While scatter plots show the region of the plane where the data points fall, as well as the data density associated with this region, they also suggest possible association between the variables. Finally, we note that the usefulness of these plots is not limited to small data sets.

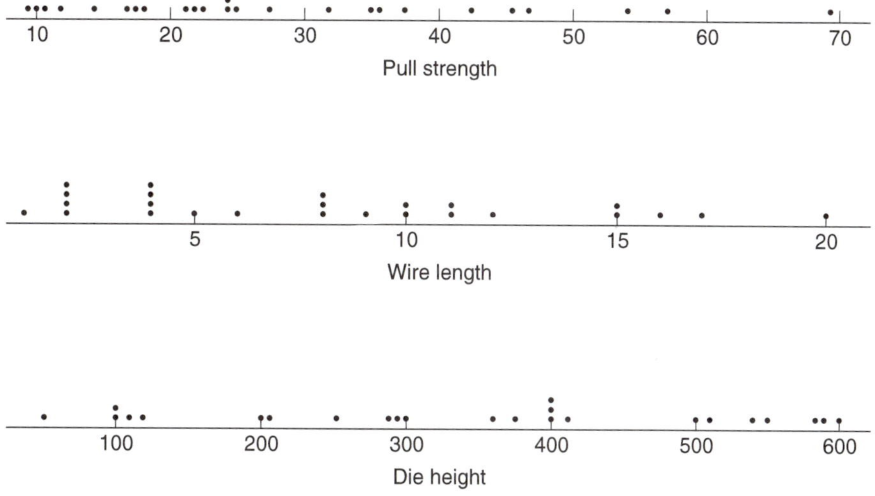

Figure 8-2 Dot plots for pull strength, wire length, and die height.

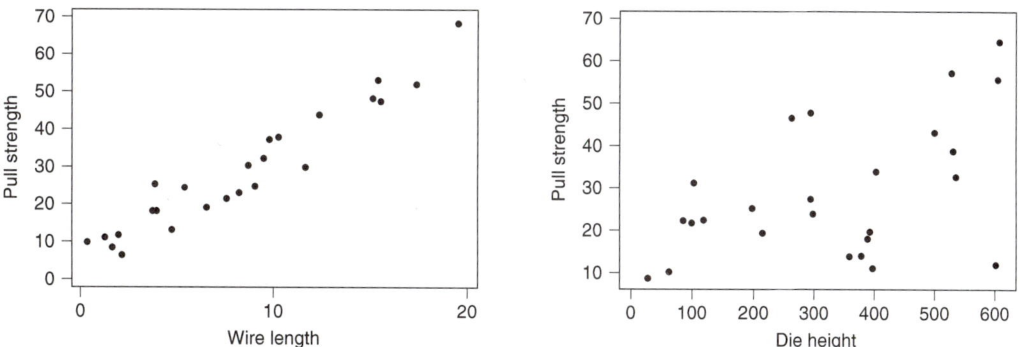

Figure 8-3 Scatter plots for pull strength vs. wire length and for pull strength vs. die height (from Minitab®).

To extend the dimensionality and graphically display the joint data pattern for three variables, a *three-dimensional scatter plot* may be employed, as illustrated in Figure 8-4 for the data in Table 8-1. Another option for a display, not illustrated here, is the *bubble plot,* which is presented in two dimensions, with the third variable reflected in the dot (now called bubble) diameter that is assigned to be proportional to the magnitude of the third variable. As was the case with scatter plots, these plots also suggest possible associations between the variables involved.

8-3.2 Numerical Data: The Frequency Distribution and Histogram

Consider the data in Table 8-2. These data are the strengths in pounds per square inch (psi) of 100 glass, nonreturnable 1-liter soft drink bottles. These observations were obtained by testing each bottle until failure occurred. The data were recorded in the order in which the bottles were tested, and in this format they do not convey very much information about bursting strength of the bottles. Questions such as "what is the average bursting strength?"

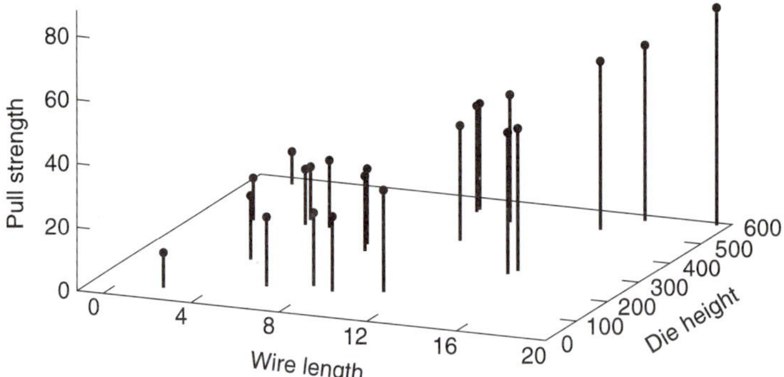

Figure 8-4 Three-dimensional plots for pull strength, wire length, and die height.

Table 8-2 Bursting Strength in Pounds per Square Inch for 100 Glass, 1-Liter, Nonreturnable Soft Drink Bottles

265	197	346	280	265	200	221	265	261	278
205	286	317	242	254	235	176	262	248	250
263	274	242	260	281	246	248	271	260	265
307	243	258	321	294	328	263	245	274	270
220	231	276	228	223	296	231	301	337	298
268	267	300	250	260	276	334	280	250	257
260	281	208	299	308	264	280	274	278	210
234	265	187	258	235	269	265	253	254	280
299	214	264	267	283	235	272	287	274	269
215	318	271	293	277	290	283	258	275	251

or "what percentage of the bottles burst below 230 psi?" are not easy to answer when the data are presented in this form.

A frequency distribution is a more useful summary of data than the simple enumeration given in Table 8-2. To construct a frequency distribution, we must divide the range of the data into intervals, which are usually called *class intervals.* If possible, the class intervals should be of equal width, to enhance the visual information in the frequency distribution. Some judgment must be used in selecting the number of class intervals in order to give a reasonable display. The number of class intervals used depends on the number of observations and the amount of scatter or dispersion in the data. A frequency distribution that uses either too few or too many class intervals will not be very informative. We generally find that between 5 and 20 intervals is satisfactory in most cases, and that the number of class intervals should increase with n. Choosing a number of class intervals approximately equal to the square root of the number of observations often works well in practice.

A frequency distribution for the bursting strength data in Table 8-2 is shown in Table 8-3. Since the data set contains 100 observations, we suspect that about $\sqrt{100} = 10$ class intervals will give a satisfactory frequency distribution. The largest and smallest data values are 346 and 176, respectively, so the class intervals must cover at least $346 - 176 = 170$ psi units on the scale. If we want the lower limit for the first interval to begin slightly below the smallest data value and the upper limit for the last cell to be slightly above the largest data value, then we might start the frequency distribution at 170 and end it at 350. This is an interval of 180 psi units. Nine class intervals, each of width 20 psi, gives a reasonable frequency distribution, and the frequency distribution in Table 8-3 is thus based on nine class intervals.

Table 8-3 Frequency Distribution for the Bursting Strength Data in Table 8-2

Class Interval (psi)	Tally	Frequency	Relative Frequency	Cumulative Relative Frequency
$170 \le x < 190$	II	2	0.02	0.02
$190 \le x < 210$	IIII	4	0.04	0.06
$210 \le x < 230$	~~IIII~~ II	7	0.07	0.13
$230 \le x < 250$	~~IIII~~ ~~IIII~~ III	13	0.13	0.26
$250 \le x < 270$	~~IIII~~ ~~IIII~~ ~~IIII~~ ~~IIII~~ ~~IIII~~ ~~IIII~~ II	32	0.32	0.58
$270 \le x < 290$	~~IIII~~ ~~IIII~~ ~~IIII~~ ~~IIII~~ IIII	24	0.24	0.82
$290 \le x < 310$	~~IIII~~ ~~IIII~~ I	11	0.11	0.93
$310 \le x < 330$	IIII	4	0.04	0.97
$330 \le x < 350$	III	3	0.03	1.00
		100	1.00	

The fourth column in Table 8-3 contains the *relative frequency distribution.* The relative frequencies are found by dividing the observed frequency in each class interval by the total number of observations. The last column in Table 8-3 expresses the relative frequencies on a cumulative basis. Frequency distributions are often easier to interpret than tables of data. For example, from Table 8-3 it is very easy to see that most of the bottles burst between 230 and 290 psi, and that 13% of the bottles burst below 230 psi.

It is also helpful to present the frequency distribution in graphical form, as shown in Fig. 8-5. Such a display is called a *histogram.* To draw a histogram, use the horizontal axis to represent the measurement scale and draw the boundaries of the class intervals. The vertical axis represents the frequency (or relative frequency) scale. If the class intervals are of equal width, then the *heights* of the rectangles drawn on the histogram are proportional to the frequencies. If the class intervals are of unequal width, then it is customary to draw rectangles whose *areas* are proportional to the frequencies. In this case, the result is called a

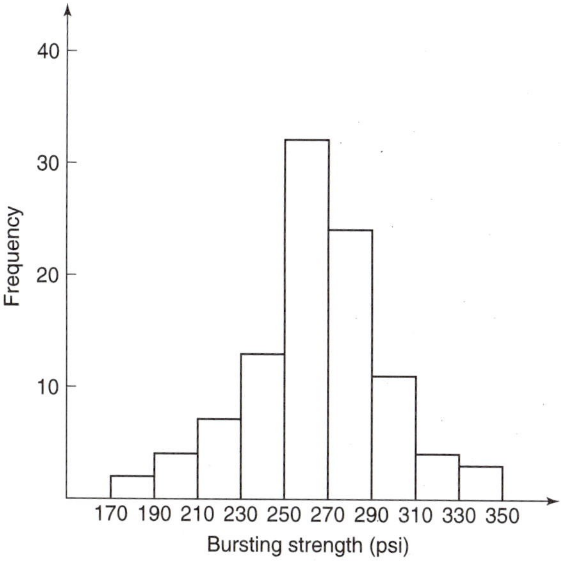

Figure 8-5 Histogram of bursting strength for 100 glass, 1-liter, nonreturnable soft drink bottles.

density histogram. For a histogram displaying relative frequency on the vertical axis, the rectangle heights are calculated as

$$\text{rectangle height} = \frac{\text{class relative frequency}}{\text{class width}}.$$

When we find there are multiple empty class intervals after grouping data into equal-width intervals, one option is to merge the empty intervals with contiguous intervals, thus creating some wider intervals. The density histogram resulting from this may produce a more attractive display. However, histograms are easier to interpret when the class intervals are of equal width. The histogram provides a visual impression of the shape of the distribution of the measurements, as well as information about centering and the scatter or dispersion of the data.

In passing from the original data to either a frequency distribution or a histogram, a certain amount of information has been lost in that we no longer have the individual observations. On the other hand, this information loss is small compared to the ease of interpretation gained in using the frequency distribution and histogram. In cases where the data assume only a few distinct values, a dot plot is perhaps a better graphical display.

Where observed data are of a discrete nature, such as is found in counting processes, then two choices are available for constructing a histogram. One option is to center the rectangles on the integers reflected in the count data, and the other is to collapse the rectangle into a vertical line placed directly over these integers. In both cases, the height of the rectangle or the length of the line is either the frequency or relative frequency of the occurrence of the value in question.

In summary, the histogram is a very useful graphic display. A histogram can give the decision maker a good understanding of the data and is very useful in displaying the *shape, location*, and *variability* of the data. However, the histogram does not allow individual data points to be identified, because all observations falling in a cell are indistinguishable.

8-3.3 The Stem-and-Leaf Plot

Suppose that the data are represented by $x_1, x_2, \ldots, x_n$, and that each number x_i consists of at least two digits. To construct a stem-and-leaf plot, we divide each number x_i into two parts: a stem, consisting of one or more of the leading digits, and a leaf, consisting of the remaining digits. For example, if the data consist of the percentage of defective information between 0 and 100 on lots of semiconductor wafers, then we could divide the value 76 into the stem 7 and the leaf 6. In general, we should choose relatively few stems in comparison with the number of observations. It is usually best to choose between 5 and 20 stems. Once a set of stems has been chosen, they are listed along the left-hand margin of the display, and beside each stem all leaves corresponding to the observed data values are listed in the order in which they are encountered in the data set.

Example 8-1

To illustrate the construction of a stem-and-leaf plot, consider the bottle-bursting-strength data in Table 8-2. To construct a stem-and-leaf plot, we select as stem values the numbers 17, 18, 19, ..., 34. The resulting stem-and-leaf plot is presented in Fig. 8-6. Inspection of this display immediately reveals that most of the bursting strengths lie between 220 and 330 psi, and that the central value is somewhere between 260 and 270 psi. Furthermore, the bursting strengths are distributed approximately symmetrically about the central value. Therefore, the stem-and-leaf plot, like the histogram, allows us to determine quickly some important features of the data that were not immediately obvious in the original display, Table 8-2. Note that here the original numbers are not lost, as occurs in a

Stem	Leaf	Frequency
17	6	1
18	7	1
19	7	1
20	0,5,8	3
21	0,4,5	3
22	1,0,8,3	4
23	5,1,1,4,5,5	6
24	2,8,2,6,8,3,5	7
25	4,0,8,0,0,7,8,3,4,8,1	11
26	5,5,5,1,2,3,0,0,5,3,8,7,0,0,4,5,9,5,4,7,9	21
27	8,4,1,4,0,6,6,4,8,2,4,1,7,5	14
28	0,6,1,0,1,0,0,3,7,3	10
29	4,6,8,9,9,3,0	7
30	7,1,0,8	4
31	7,8	2
32	1,8	2
33	7,4	2
34	6	1
		100

Figure 8-6 Stem-and-leaf plot for the bottle-bursting-strength data in Table 8-2.

histogram. Sometimes, in order to assist in finding percentiles, we order the leaves by magnitude, producing an *ordered* stem-and-leaf plot, as in Fig. 8-7. For instance, since $n = 100$ is an even number, the *median*, or "middle" observation (see Section 8-4.1), is the average of the two observations with ranks 50 and 51, or

$$\tilde{x} = (265 + 265)/2 = 265.$$

The *tenth percentile* is the observation with rank $(0.1)(100) + 0.5 = 10.5$ (halfway between the 10th and 11th observations), or $(220 + 221)/2 = 220.5$. The *first quartile* is the observation with rank $(0.25)(100) + 0.5 = 25.5$ (halfway between the 25th and 26th observations), or $(248 + 248)/2 = 248$, and the third quartile is the observation with rank $(0.75)(100) + 0.5 = 75.5$ (halfway between the 75th and 76th observations), or $(280 + 280)/2 = 280$. The first and third quartiles are occasionally denoted by the symbols Q1 and Q3, respectively, and the *interquartile range* IQR = Q3 – Q1 may be used as a measure of variability. For the bottle-bursting-strength data, the interquartile range is IQR = Q3 – Q1 = 280 – 248 = 32. The stem-and-leaf displays in Figs. 8-6 and 8-7 are equivalent to a histogram with 18 class intervals. In some situations, it may be desirable to provide more classes or stems. One way to do this would be to modify the original stems as follows: divide stem 5 (say) into two new stems, 5* and 5• Stem 5* has leaves 0, 1, 2, 3, and 4, and stem 5• has leaves 5, 6, 7, 8, and 9. This will double the number of original stems. We could increase the number of original stems by five by defining five new stems: 5* with leaves 0 and 1, 5*t* (for twos and threes) with leaves 2 and 3, 5*f* (for fours and fives) with leaves 4 and 5, 5*s* (for sixes and sevens) with leaves 6 and 7, and 5• with leaves 8 and 9.

8-3.4 The Box Plot

A box plot displays the three quartiles, the minimum, and the maximum of the data on a rectangular box, aligned either horizontally or vertically. The box encloses the interquartile range with the left (or lower) line at the first quartile Q1 and the right (or upper) line at the third quartile Q3. A line is drawn through the box at the second quartile (which is the 50th

Stem	Leaf	Frequency
17	6	1
18	7	1
19	7	1
20	0,5,8	3
21	0,4,5	3
22	0,1,3,8	4
23	1,1,4,5,5,5	6
24	2,2,3,5,6,8,8	7
25	0,0,0,1,3,4,4,7,8,8,8	11
26	0,0,0,0,1,2,3,3,4,4,5,5,5,5,5,5,7,7,8,9,9	21
27	0,1,1,2,4,4,4,4,5,6,6,7,8,8	14
28	0,0,0,0,1,1,3,3,6,7	10
29	0,3,4,6,8,9,9	7
30	0,1,7,8	4
31	7,8	2
32	1,8	2
33	4,7	2
34	6	1
		100

Figure 8-7 Ordered stem-and-leaf plot for the bottle-bursting-strength data.

percentile or the median) Q2 = $\tilde{x}$. A line at either end extends to the extreme values. These lines, sometimes called whiskers, may extend only to the 10th and 90th percentiles or the 5th and 95th percentiles in large data sets. Some authors refer to the box plot as the box-and-whisker plot. Figure 8-8 presents the box plot for the bottle-bursting-strength data. This box plot indicates that the distribution of bursting strengths is fairly symmetric around the central value, because the left and right whiskers and the lengths of the left and right boxes around the median are about the same.

The box plot is useful in comparing two or more samples. To illustrate, consider the data in Table 8-4. The data, taken from Messina (1987), represent viscosity readings on three different mixtures of raw material used on a manufacturing line. One of the objectives of the study that Messina discusses is to compare the three mixtures. Figure 8-9 presents the box plots for the viscosity data. This display permits easy interpretation of the data. Mixture 1 has higher viscosity than mixture 2, and mixture 2 has higher viscosity than mixture 3. The distribution of viscosity is not symmetric, and the maximum viscosity reading from mixture 3 seems unusually large in comparison to the other readings. This observation may be an *outlier*, and it possibly warrants further examination and analysis.

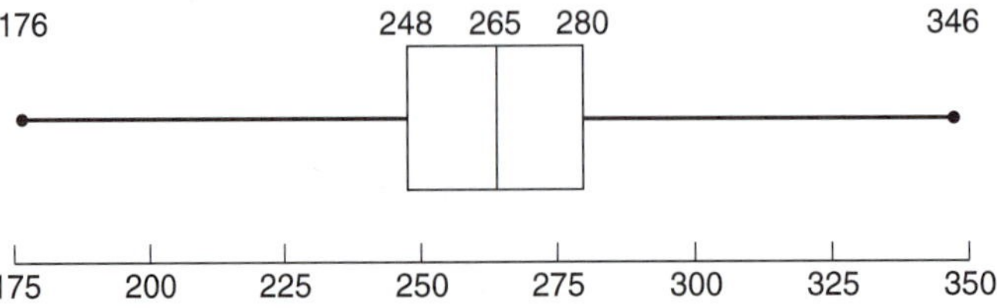

Figure 8-8 Box plot for the bottle-bursting-strength data.

Table 8-4 Viscosity Measurements for Three Mixtures

Mixture 1	Mixture 2	Mixture 3
22.02	21.49	20.33
23.83	22.67	21.67
26.67	24.62	24.67
25.38	24.18	22.45
25.49	22.78	22.28
23.50	22.56	21.95
25.90	24.46	20.49
24.98	23.79	21.81

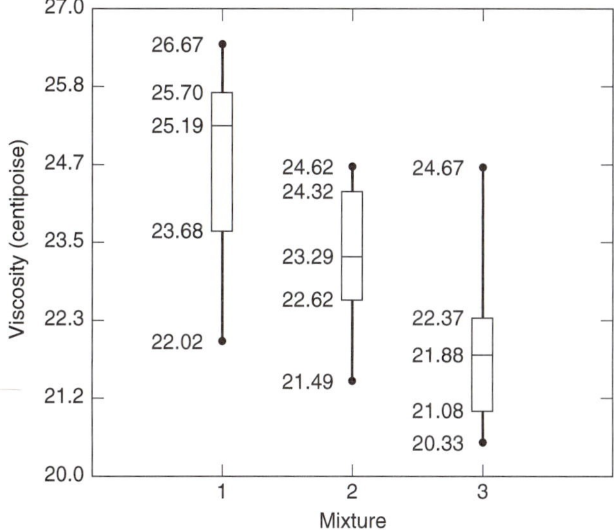

Figure 8-9 Box plots for the mixture-viscosity data in Table 8-4.

8-3.5 The Pareto Chart

A Pareto diagram is a bar graph for count data. It displays the frequency of each count on the vertical axis and the category of classification on the horizontal axis. We always arrange the categories in *descending order of frequency of occurrence*; that is, the most frequently occurring is on the left, followed by the next most frequently occurring type, and so on.

Figure 8-10 presents a Pareto diagram for the production of transport aircraft by the Boeing Commercial Airplane Company in the year 2000. Notice that the 737 was the most popular model, followed by the 777, the 757, the 767, the 717, the 747, the MD-11, and the MD-90. The line on the Pareto chart connects the cumulative percentages of the k most frequently produced models ($k = 1, 2, 3, 4, 5$). In this example, the two most frequently produced models account for approximately 69% of the total airplanes manufactured in 2000. One feature of these charts is that the horizontal scale is not necessarily numeric. Usually, categorical classifications are employed as in the airplane production example.

The Pareto chart is named for an Italian economist who theorized that in certain economies the majority of the wealth is held by a minority of the people. In count data, the "Pareto principle" frequently occurs, hence the name for the chart.

Pareto charts are very useful in the analysis of *defect data* in manufacturing systems. Figure 8-11 presents a Pareto chart showing the frequency with which various types of defects

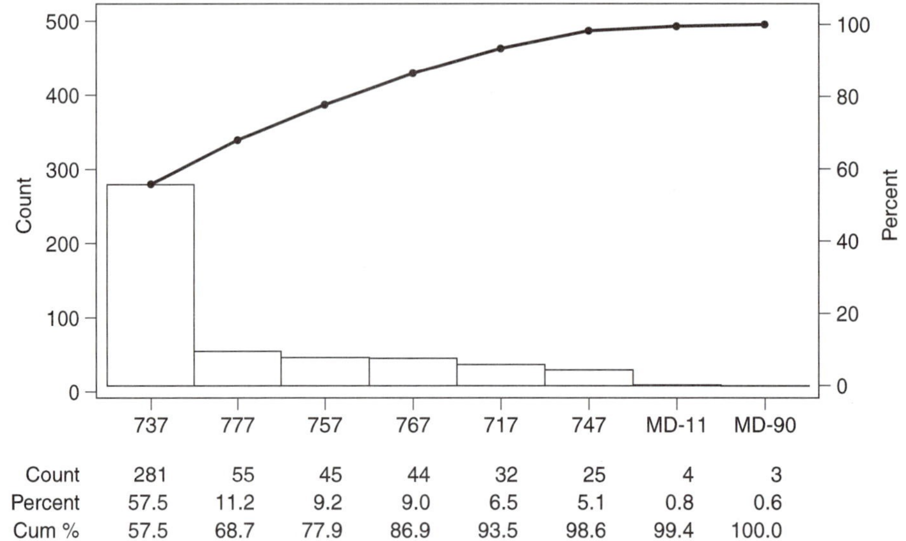

	737	777	757	767	717	747	MD-11	MD-90
Count	281	55	45	44	32	25	4	3
Percent	57.5	11.2	9.2	9.0	6.5	5.1	0.8	0.6
Cum %	57.5	68.7	77.9	86.9	93.5	98.6	99.4	100.0

Figure 8-10 Airplane production in 2000. (*Source*: Boeing Commercial Airplane Company.)

occur on metal parts used in a structural component of an automobile door frame. Notice how the Pareto chart highlights the relatively few types of defects that are responsible for most of the observed defects in the part. The Pareto chart is an important part of a quality-improvement program because it allows management and engineering to focus attention on the most critical defects in a product or process. Once these critical defects are identified, corrective actions to reduce or eliminate these defects must be developed and implemented. This is easier to do, however, when we are sure that we are attacking a legitimate problem: it is much easier to reduce or eliminate frequently occurring defects than rare ones.

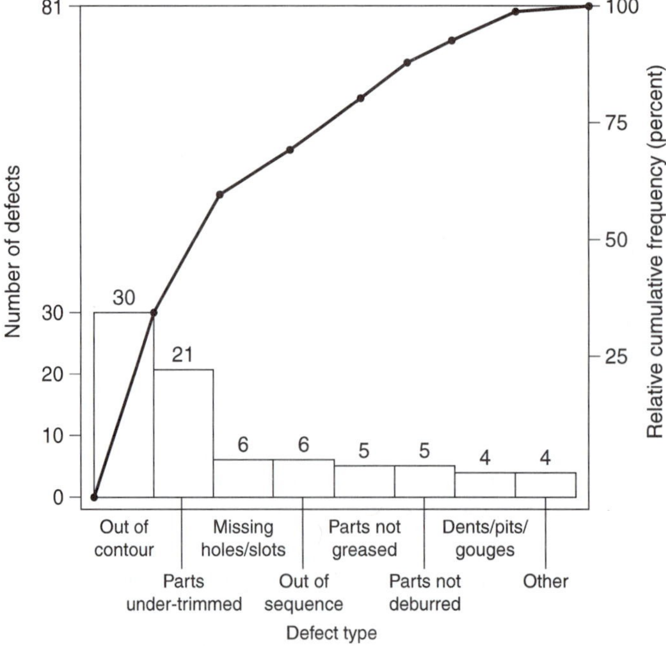

Figure 8-11 Pareto chart of defects in door structural elements.

8-3.6 Time Plots

Virtually everyone should be familiar with time plots, since we view them daily in media presentations. Examples are historical temperature profiles for a given city, the closing Dow Jones Industrials Index for each trading day, each month, each quarter, etc., and the plot of the yearly Consumer Price Index for all urban consumers published by the Bureau of Labor Statistics. Many other time-oriented data are routinely gathered to support inferential statistical activity. Consider the electric power demand, measured in kilowatts, for a given office building and presented as hourly data for each of the 24 hours of the day in which the supplying utility experiences a summer peak demand from all customers. Demand data such as these are gathered using a time of use or "load research" meter. In concept, a kilowatt is a continuous variable. The meter sampling interval is very short, however, and the hourly data that are obtained are truly averages over the meter sampling intervals contained within each hour.

Usually with time plots, time is represented on the horizontal axis, and the vertical scale is calibrated to accommodate the range of values represented in the observational results. The kilowatt hourly demand data are displayed in Fig. 8-12. Ordinarily, when series like these display averages over a time interval, the variation displayed in the data is a function of the length of the averaging interval, with shorter intervals producing more variability. For example, in the case of the kilowatt data, if we used a 15-minute interval, there would be 96 points to plot, and the variation in data would appear much greater.

8-4 NUMERICAL DESCRIPTION OF DATA

Just as graphs can improve the display of data, numerical descriptions are also of value. In this section, we present several important numerical measures for describing the characteristics of data.

8-4.1 Measures of Central Tendency

The most common measure of central tendency, or location, of the data is the ordinary arithmetic mean. Because we usually think of the data as being obtained from a *sample* of units,

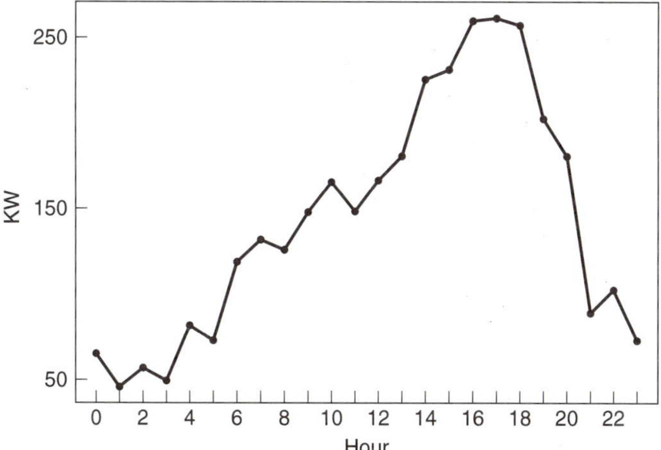

Figure 8-12 Summer peak day hourly kilowatt (KW) demand data for an office building.

we will refer to the arithmetic mean as the *sample mean.* If the observations in a sample of size n are $x_1, x_2, \ldots, x_n$, then the sample mean is

$$\bar{x} = \frac{x_1 + x_2 + \cdots + x_n}{n} = \frac{\sum_{i=1}^{n} x_i}{n}. \tag{8-1}$$

For the bottle-bursting-strength data in Table 8-2, the sample mean is

$$\bar{x} = \frac{\sum_{i=1}^{100} x_i}{100} = \frac{26,406}{100} = 264.06.$$

From examination of Fig. 8-5, it seems that the sample mean 264.06 psi is a "typical" value of bursting strength, since it occurs near the middle of the data, where the observations are concentrated. However, this impression can be misleading. Suppose that the histogram looked like Fig. 8-13. The mean of these data is still a measure of central tendency, but it does not necessarily imply that most of the observations are concentrated around it. In general, if we think of the observations as having unit mass, the sample mean is just the center of mass of the data. This implies that the histogram will just exactly balance if it is supported at the sample mean.

The sample mean $\bar{x}$ represents the average value of all the observations in the sample. We can also think of calculating the average value of all the observations in a finite *population.* This average is called the *population mean*, and as we here seen in previous chapters, it is denoted by the Greek letter μ. When there are a finite number of possible observations (say N) in the population, then the population mean is

$$\mu = \frac{\tau_x}{N}, \tag{8-2}$$

where $\tau_x = \sum_{i=1}^{N} x_i$ is the finite population total for the population.

In the following chapters dealing with statistical inference, we will present methods for making inferences about the population mean that are based on the sample mean. For example, we will use the sample mean as a *point estimate* of μ.

Another measure of central tendency is the *median*, or the point at which the sample is divided into two equal halves. Let $x_{(1)}, x_{(2)}, \ldots, x_{(n)}$ denote a sample arranged in increasing order of magnitude; that is, $x_{(1)}$ denotes the smallest observation, $x_{(2)}$ denotes the second

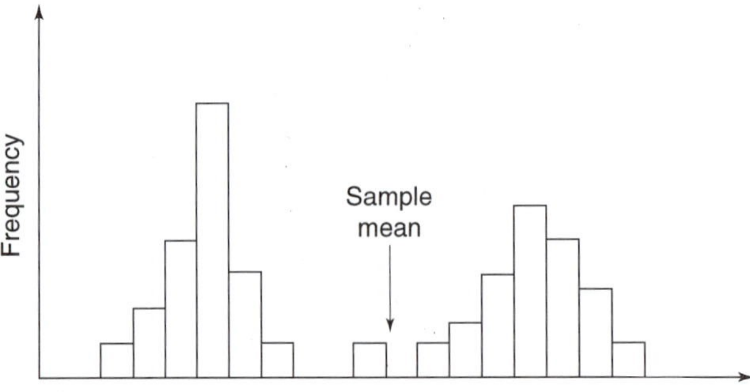

Figure 8-13 A histogram.

smallest observation, ..., and $x_{(n)}$ denotes the largest observation. Then the median is defined mathematically as

$$\tilde{x} = \begin{cases} x_{((n+1)/2)} & n \text{ odd}, \\ \dfrac{x_{(n/2)} + x_{((n/2)+1)}}{2} & n \text{ even}. \end{cases} \tag{8-3}$$

The median has the advantage that it is not influenced very much by extreme values. For example, suppose that the sample observations are

$$1, 3, 4, 2, 7, 6, \text{ and } 8.$$

The sample mean is 4.43, and the sample median is 4. Both quantities give a reasonable measure of the central tendency of the data. Now suppose that the next-to-last observation is changed, so that the data are

$$1, 3, 4, 2, 7, 2519, \text{ and } 8.$$

For these data, the sample mean is 363.43. Clearly, in this case the sample mean does not tell us very much about the central tendency of most of the data. The median, however, is still 4, and this is probably a much more meaningful measure of central tendency for the majority of the observations.

Just as $\tilde{x}$ is the middle value in a sample, there is a middle value in the population. We define $\tilde{\mu}$ as the median of the population; that is, $\tilde{\mu}$ is a value of the associated random variable such that half the population lies below $\tilde{\mu}$ and half lies above.

The *mode* is the observation that occurs most frequently in the sample. For example, the mode of the sample data

$$2, 4, 6, 2, 5, 6, 2, 9, 4, 5, 2, \text{ and } 1$$

is 2, since it occurs four times, and no other value occurs as often. There may be more than one mode.

If the data are symmetric, then the mean and median coincide. If, in addition, the data have only one mode (we say the data are unimodal), then the mean, median, and mode may all coincide. If the data are skewed (asymmetric, with a long tail to one side), then the mean, median, and mode will not coincide. Usually we find that mode < median < mean if the distribution is skewed to the right, while mode > median > mean if the distribution is skewed to the left (see to Fig. 8-14).

The distribution of the sample mean is well-known and relatively easy to work with. Furthermore, the sample mean is usually more stable than the sample median, in the sense that it does not vary as much from sample to sample. Consequently, many analytical statistical techniques use the sample mean. However, the median and mode may also be helpful descriptive measures.

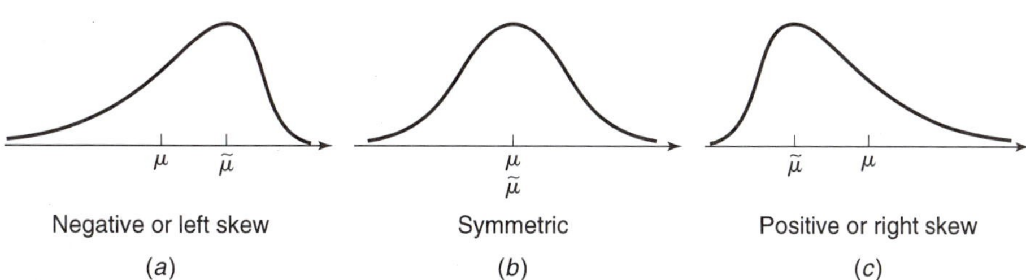

Figure 8-14 The mean and median for symmetric and skewed distributions.

8-4.2 Measures of Dispersion

Central tendency does not necessarily provide enough information to describe data adequately. For example, consider the bursting strengths obtained from two samples of six bottles each:

Sample 1:	230	250	245	258	265	240
Sample 2:	190	228	305	240	265	260

The mean of both samples is 248 psi. However, note that the scatter or dispersion of Sample 2 is much greater than that of Sample 1 (see Fig. 8-15). In this section, we define several widely used measures of dispersion.

The most important measure of dispersion is the *sample variance.* If $x_1, x_2, \ldots, x_n$ is a sample of n observations, then the sample variance is

$$s^2 = \frac{\sum_{i=1}^{n}(x_i - \bar{x})^2}{n-1} = \frac{S_{xx}}{n-1}. \tag{8-4}$$

Note that computation of s^2 requires calculation of $\bar{x}$, n subtractions, and n squaring and adding operations. The deviations $x_i - \bar{x}$ may be rather tedious to work with, and several decimals may have to be carried to ensure numerical accuracy. A more efficient (yet equivalent) computational formula for calculating S_{xx} is

$$S_{xx} = \sum_{i=1}^{n} x_i^2 - \frac{1}{n}\left(\sum_{i=1}^{n} x_i\right)^2 \tag{8-5}$$

The formula for S_{xx} presented in equation 8-5 requires only one computational pass; but care must again be taken to keep enough decimals to prevent round-off error.

To see how the sample variance measures dispersion or variability, refer to Fig. 8-16 which shows the deviations $x_i - \bar{x}$ for the second sample of six bottle-bursting strengths. The greater the amount of variability in the bursting-strength data, the larger in absolute magnitude *some* of the deviations $x_i - \bar{x}$. Since the deviations $x_i - \bar{x}$ will always sum to zero, we must use a measure of variability that changes the negative deviations to nonnegative quantities. Squaring the deviations is the approach used in the sample variance. Consequently, if s^2 is small, then there is relatively little variability in the data, but if s^2 is large, the variability is relatively large.

The units of measurements for the sample variance are the square of the original units of the variable. Thus, if x is measured in pounds per square inch (psi), the units for the sample variance are $(\text{psi})^2$.

Example 8-2

We will calculate the sample variance of the bottle-bursting strengths for the second sample in Fig. 8-15. The deviations $x_i - \bar{x}$ for this sample are shown in Fig. 8-16.

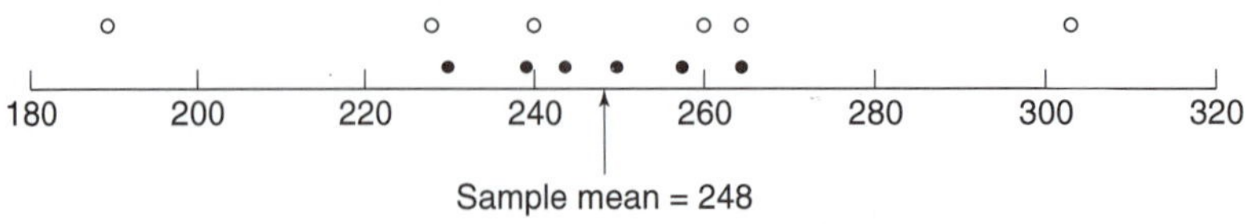

Figure 8-15 Bursting-strength data.

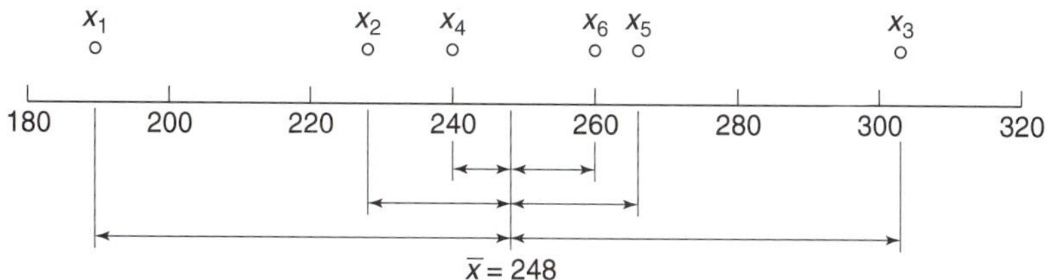

Figure 8-16 How the sample variance measures variability through the deviations $x_i - \bar{x}$.

Observations	$x_i - \bar{x}$	$(x_i - \bar{x})^2$
$x_1 = 190$	−58	3364
$x_2 = 228$	−20	400
$x_3 = 305$	57	3249
$x_4 = 240$	−8	64
$x_5 = 265$	17	289
$x_6 = 260$	12	144
$\bar{x} = 248$	Sum = 0	Sum = 7510

From equation 8-4,

$$s^2 = \frac{S_{xx}}{n-1} = \frac{\sum_{i=1}^{n}(x_i - \bar{x})^2}{n-1} = \frac{7510}{5} = 1502(\text{psi})^2.$$

We may also calculate S_{xx} from the formulation given in equation 8-5, so that

$$s^2 = \frac{S_{xx}}{n-1} = \frac{\sum_{i=1}^{n} x_i^2 - \frac{1}{n}\left(\sum_{i=1}^{n} x_i\right)^2}{n-1} = \frac{367{,}534 - (1488)^2/6}{5} = 1502(\text{psi})^2.$$

If we calculate the sample variance of the bursting strength for the Sample 1 values, we find that $s^2 = 158$ (psi)^2. This is considerably smaller than the sample variance of Sample 2, confirming our initial impression that Sample 1 has less variability than Sample 2.

Because s^2 is expressed in the square of the original units, it is not easy to interpret. Furthermore, variability is a more difficult and unfamiliar concept than location or central tendency. However, we can solve the "curse of dimensionality" by working with the (positive) square root of the variance, s, called the *sample standard deviation.* This gives a measure of dispersion expressed in the same units as the original variable.

Example 8-3

The sample standard deviation of the bottle-bursting strengths for the Sample 2 bottles in Example 8-2 and Fig. 8-15 is

$$s = \sqrt{s^2} = \sqrt{1502} = 38.76 \text{ psi}.$$

For the Sample 1 bottles, the standard deviation of bursting strength is

$$s = \sqrt{158} = 12.57 \text{ psi}.$$

Example 8-4

Compute the sample variance and sample standard deviation of the bottle-bursting-strength data in Table 8-2. Note that

$$\sum_{i=1}^{100} x_i^2 = 7{,}074{,}258.00 \qquad \text{and} \qquad \sum_{i=1}^{100} x_i = 26{,}406.$$

Consequently,

$$S_{xx} = 7{,}074{,}258.00 - (26{,}406)^2/100 = 101{,}489.85 \text{ and } s^2 = 101{,}489.85/99 = 1025.15 \text{ psi}^2,$$

so that the sample standard deviation is

$$s = \sqrt{1025.15} = 32.02 \text{ psi}.$$

When the population is finite and consists of N values we may define the population variance as

$$\sigma^2 = \frac{\sum_{i=1}^{N}(x_i - \mu)^2}{N}, \tag{8-6}$$

which is simply the mean of the average squared departures of the data values from the population mean. A closely related quantity, $\tilde{\sigma}^2$, is also sometimes called the population variance and is defined as

$$\tilde{\sigma}^2 = \frac{N}{N-1}\cdot\sigma^2. \tag{8-7}$$

Obviously, as N gets large, $\tilde{\sigma}^2 \to \sigma^2$, and oftentimes the use of $\tilde{\sigma}^2$ simplifies some of the algebraic formulation presented in Chapters 9 and 10. Where several populations are to be observed, a subscript may be employed to identify the population characteristics and descriptive measures, e.g., μ_x, σ_x^2, s_x^2, etc., if the x variable is being described.

We noted that the sample mean may be used to make inferences about the population mean. Similarly, the sample variance may be used to make inferences about the population variance. We observe that the divisor for the sample variance, s^2, is the sample size minus 1, $(n-1)$. If we actually knew the true value of the population mean μ, then we could define the sample variance as the average squared deviation of the sample observations about μ. In practice, the value of μ is almost never known, and so the sum of the squared deviations about the sample average $\bar{x}$ must be used instead. However, the observations x_i tend to be closer to their average $\bar{x}$ than to the population mean μ, so to compensate for this we use as a divisor $n-1$ rather than n.

Another way to think about this is to consider the sample variance s^2 as being based on $n-1$ *degrees of freedom.* The term *degrees of freedom* results from the fact that the n deviations $x_1 - \bar{x}, x_2 - \bar{x}, \ldots, x_n - \bar{x}$ always sum to zero, so specifying the values of any $n-1$ of these quantities automatically determines the remaining one. Thus, only $n-1$ of the n deviations $x_i - \bar{x}$ are independent.

Another useful measure of dispersion is the sample range

$$R = \max(x_i) - \min(x_i). \tag{8-8}$$

The sample range is very simple to compute, but it ignores all the information in the sample between the smallest and largest observations. For small sample sizes, say $n \le 10$, this information loss is not too serious in some situations. The range traditionally has had widespread application in statistical quality control, where sample sizes of 4 or 5 are common and computational simplicity is a major consideration; however, that advantage has been largely diminished by the widespread use of electronic measurement, and data storage and analysis systems, and as we will later see, the sample variance (or standard deviation) provides a "better" measure of variability. We will briefly discuss the use of the range in statistical quality-control problems in Chapter 17.

Example 8-5

Calculate the ranges of the two samples of bottle-bursting-strength data from Section 8-4, shown in Fig. 8-15. For the first sample, we find that

$$R_1 = 265 - 230 = 35,$$

whereas for the second sample

$$R_2 = 305 - 190 = 115.$$

Note that the range of the second sample is much larger than the range of the first, implying that the second sample has greater variability than the first.

Occasionally, it is desirable to express variation as a fraction of the mean. A measure of relative variation called the *sample coefficient of variation* is defined as

$$CV = \frac{s}{\bar{x}}. \tag{8-9}$$

The coefficient of variation is useful when comparing the variability of two or more data sets that differ considerably in the magnitude of the observations. For example, the coefficient of variation might be useful in comparing the variability of daily electricity usage within samples of single-family residences in Atlanta, Georgia, and Butte, Montana, during July.

8-4.3 Other Measures for One Variable

Two other measures, both dimensionless, are provided by spreadsheet or statistical software systems and are called *skewness* and *kurtosis* estimates. The notion of skewness was graphically illustrated in Fig. 8-14. These characteristics are population or population model characteristics and they are defined in terms of moments, μ_k, as described in Chapter 2, Section 2-5.

$$\text{skewness} \quad \beta_3 = \frac{\mu_3}{\sigma^3} \qquad \text{and} \qquad \text{kurtosis} \quad \beta_4 = \frac{\mu_4}{\sigma^4} - 3, \tag{8-10}$$

where, in the case of finite populations, the *kth central moment* is defined as

$$\mu_k = \frac{\sum_{i=1}^{N} (x_i - \mu)^k}{N}. \tag{8-11}$$

As discussed earlier, skewness reflects the degree of symmetry about the mean; and negative skew results from an asymmetric tail toward smaller values of the variable, while positive skew results from an asymmetric tail extending toward the larger values of the variable.

Symmetric variables, such as those described by the normal and uniform distributions, have skewness equal zero. The exponential distribution, for example, has skewness $\beta_3 = 2$.

Kurtosis describes the relative peakedness of a distribution as compared to a normal distribution, where a negative value is associated with a relatively flat distribution and a positive value is associated with relatively peaked distributions. For example, the kurtosis measure for a uniform distribution is –1.2, while for a normal variable, the kurtosis is zero.

If the data being analyzed represent measurement of a variable made on sample units, the sample estimates of β_3 and β_4 are as shown in equations 8-12 and 8-13. These values may be calculated from Excel® worksheet functions using the SKEW and KURT functions.

$$\text{skewness} \quad \beta_3 = \frac{n}{(n-1)(n-2)} \cdot \frac{\sum_{i=1}^{n}(x_i - \bar{x})^3}{s^3}; \quad n > 2. \tag{8-12}$$

$$\text{kurtosis} \quad \beta_4 = \frac{(n)(n+1)}{(n-1)(n-2)(n-3)} \cdot \frac{\sum_{i=1}^{n}(x_i - \bar{x})^4}{s^4} - 3; \quad n > 3. \tag{8-13}$$

If the data represent measurements on all the units of a finite population or a census, then equations 8-10 and 8-11 should be utilized directly to determine these measures.

For the pull strength data shown in Table 8-1, the worksheet functions return values of 0.865 and 0.161 for the skewness and kurtosis measures, respectively.

8-4.4 Measuring Association

One measure of association between two numerical variables in sample data is called the *Pearson* or *simple correlation coefficient*, and it is usually denoted r. Where data sets contain a number of variables and the correlation coefficient for only one pair of variables, designated x and y, is to be presented, a subscript notation such as r_{xy} may be used to designate the simple correlation between variable x and variable y. The correlation coefficient is

$$r_{xy} = \frac{S_{xy}}{\left(S_{xx} \cdot S_{xy}\right)^{1/2}}, \tag{8-14}$$

where S_{xx} is as shown in equation 8-5 and S_{yy} is similarly defined for the y variable, replacing x with y in equation 8-5, while

$$S_{xy} = \sum_{i=1}^{n}(x_i - \bar{x})(y_i - \bar{y}) = \left(\sum_{i=1}^{n} x_i y_i\right) - \frac{1}{n}\left(\sum_{i=1}^{n} x_i\right) \cdot \left(\sum_{i=1}^{n} y_i\right). \tag{8-15}$$

Now we will return to the wire bond pull strength data as presented in Table 8-1, with the scatter plots shown in Fig. 8-3 for both pull strength (variable y) vs. wire length (variable x_1) and for pull strength vs. die height (variable x_2). For the sake of distinguishing between the two correlation coefficients, we let r_1 be used for y vs. x_1, and r_2 for y vs. x_2. The correlation coefficient is a dimensionless measure which lies on the interval [–1, +1]. It is a measure of linear association between the two variables of the pair. As the strength of linear association increases, $|r| \to 1$. A positive association means that larger x_1 values have larger y values associated with them, and the same is true for x_2 and y. In other sets of data, where the larger x values have smaller y values associated with them, the correlation coefficient is negative. When the data reflect no linear association, the correlation is zero. In the case of the pull strength data, $r_1 = 0.982$, and $r_2 = 0.493$. The calculations were made using Minitab®, selecting Stat>Basic Statistics>Correlation. Both are obviously positive. It is important to note however, that the above result *does not imply causality.* We cannot claim that increasing x_1 or x_2 causes an increase in y. This important point is often missed—with the potential for major misinterpretation—as it may well be that a fourth, unobserved variable is the causative variable influencing all the observed variables.

In the case where the data represent measures on variables associated with an entire finite universe, equations 8-14 and 8-15 may be employed after replacing $\bar{x}$ by μ_x, the finite population mean for the x population and $\bar{y}$ by μ_y, the finite population mean for the y population, while replacing the sample size n in the formulation by N, the universe size. In this case, it is customary to use the symbol ρ_{xy} to represent this *finite population correlation coefficient* between variables x and y.

8-4.5 Grouped Data

If the data are in a frequency distribution, it is necessary to modify the computing formulas for the measures of central tendency and dispersion given in Sections 8-4.1 and 8-4.2. Suppose that for each of p distinct values of x, say $x_1, x_2, \ldots, x_p$, the observed frequency is f_j. Then the sample mean and sample variance may be computed as

$$\bar{x} = \frac{\sum_{j=1}^{p} f_j x_j}{\sum_{j=1}^{p} f_j} = \frac{\sum_{j=1}^{p} f_j x_j}{n} \tag{8-16}$$

and

$$s^2 = \frac{\sum_{j=1}^{p} f_j x_j^2 - \frac{1}{n}\left(\sum_{j=1}^{p} f_j x_j\right)^2}{n-1}, \tag{8-17}$$

respectively.

A similar situation arises where original data have been either lost or destroyed but the grouped data have been preserved. In such cases, we can approximate the important data moments by using a convention, which assigns each data value the value representing the midpoint of the class interval into which the observations were classified, so that all sample values falling in a particular interval are assigned the same value. This may be done easily, and the resulting *approximate* mean and variance values are as follows in equations 8-18 and 8-19. If m_j denotes the midpoint of the jth class interval and there are c class intervals, then the sample mean and sample variance are approximately

$$\bar{x} \simeq \frac{\sum_{j=1}^{c} f_j m_j}{\sum_{j=1}^{c} f_j} = \frac{\sum_{j=1}^{c} f_j m_j}{n} \tag{8-18}$$

and

$$s^2 \simeq \frac{\sum_{j=1}^{c} f_j m_j^2 - \frac{1}{n}\left(\sum_{j=1}^{c} f_j m_j\right)^2}{n-1}. \tag{8-19}$$

Example 8-6

To illustrate the use of equations 8-18 and 8-19 we compute the mean and variance of bursting strength for the data in the frequency distribution of Table 8-3. Note that there are $c = 9$ class intervals, and that $m_1 = 180, f_1 = 2, m_2 = 200, f_2 = 4, m_3 = 220, f_3 = 7, m_4 = 240, f_4 = 13, m_5 = 260, f_5 = 32, m_6 = 280, f_6 = 24, m_7 = 300, f_7 = 11, m_8 = 320, f_8 = 4, m_9 = 340$, and $f_9 = 3$. Thus,

$$\bar{x} \simeq \frac{\sum_{j=1}^{9} f_j m_j}{n} = \frac{26,460}{100} = 264.60 \text{ psi}$$

and

$$s^2 \simeq \frac{\sum_{j=1}^{9} f_j m_j^2 - \frac{1}{100}\left(\sum_{j=1}^{9} f_j m_j\right)^2}{99} = \frac{7,091,900 - (26,460)^2/100}{99} = 914.99\ (\text{psi})^2.$$

Notice that these are very close to the values obtained from the ungrouped data.

When the data are grouped in class intervals, it is also possible to approximate the median and mode. The median is approximately

$$\tilde{x} = L_M + \left(\frac{\frac{n+1}{2} - T}{f_M}\right)\Delta, \tag{8-20}$$

where L_M is the lower limit of the class interval containing the median (called the median class), f_M is the frequency in the median class, T is the total of all frequencies in the class intervals preceding the median class, and Δ is the width of the median class. The mode, say *MO,* is approximately

$$MO = L_{MO} + \left(\frac{a}{a+b}\right)\Delta, \tag{8-21}$$

where L_{MO} is the lower limit of the modal class (the class interval with the greatest frequency), a is the absolute value of the difference in frequency between the modal class and the preceding class, b is the absolute value of the difference in frequency between the modal class and the following class, and Δ is the width of the modal class.

8-5 SUMMARY

This chapter has provided an introduction to the field of statistics, including the notions of process, universe, population, sampling, and sample results called data. Furthermore, a variety of commonly used displays have been described and illustrated. These were the dot plot, frequency distribution, histogram, stem-and-leaf plot, Pareto chart, box plot, scatter plot, and time plot.

We have also introduced quantitative measures for summarizing data. The mean, median and mode describe *central tendency*, or *location*, while the variance, standard deviation, range, and interquartile range describe *dispersion* or *spread* in the data. Furthermore, measures of skew and kurtosis were presented to describe *asymmetry* and *peakedness,* respectively. We also presented the correlation coefficient to describe the strength of *linear association* between two variables. Subsequent chapters will focus on utilizing sample results to draw inferences about the process or process model or about the universe.

8-6 EXERCISES

8-1. The shelf life of a high-speed photographic film is being investigated by the manufacturer. The following data are available.

Life (days)	Life (days)	Life (days)	Life (days)
126	129	134	141
131	132	136	145
116	128	130	162
125	126	134	129
134	127	120	127
120	122	129	133
125	111	147	129
150	148	126	140
130	120	117	131
149	117	143	133

Construct a histogram and comment on the properties of the data.

8-2. The percentage of cotton in a material used to manufacture men's shirts is given below. Construct a histogram for the data. Comment on the properties of the data.

34.2	33.6	33.8	34.7	37.8	32.6	35.8	34.6
33.1	34.7	34.2	33.6	36.6	33.1	37.6	33.6
34.5	35.0	33.4	32.5	35.4	34.6	37.3	34.1
35.6	35.4	34.7	34.1	34.6	35.9	34.6	34.7
34.3	36.2	34.6	35.1	33.8	34.7	35.5	35.7
35.1	36.8	35.2	36.8	37.1	33.6	32.8	36.8
34.7	35.1	35.0	37.9	34.0	32.9	32.1	34.3
33.6	35.3	34.9	36.4	34.1	33.5	34.5	32.7

8-3. The following data represent the yield on 90 consecutive batches of ceramic substrate to which a metal coating has been applied by a vapor-deposition process. Construct a histogram for these data and comment on the properties of the data.

94.1	87.3	94.1	92.4	84.6	85.4
93.2	84.1	92.1	90.6	83.6	86.6
90.6	90.1	96.4	89.1	85.4	91.7
91.4	95.2	88.2	88.8	89.7	87.5
88.2	86.1	86.4	86.4	87.6	84.2
86.1	94.3	85.0	85.1	85.1	85.1
95.1	93.2	84.9	84.0	89.6	90.5
90.0	86.7	87.3	93.7	90.0	95.6
92.4	83.0	89.6	87.7	90.1	88.3
87.3	95.3	90.3	90.6	94.3	84.1
86.6	94.1	93.1	89.4	97.3	83.7
91.2	97.8	94.6	88.6	96.8	82.9
86.1	93.1	96.3	84.1	94.4	87.3
90.4	86.4	94.7	82.6	96.1	86.4
89.1	87.6	91.1	83.1	98.0	84.5

8-4. An electronics company manufactures power supplies for a personal computer. They produce several hundred power supplies each shift, and each unit is subjected to a 12-hour burn-in test. The number of units failing during this 12-hour test each shift is shown below.

(a) Construct a frequency distribution and histogram.

(b) Find the sample mean, sample variance, and sample standard deviation.

3	6	4	7	6	7
4	7	8	2	1	4
2	9	4	6	4	8
5	10	10	9	13	7
6	14	14	10	12	3
10	13	8	7	10	6
5	10	12	9	2	7
4	9	4	16	5	8
3	8	5	11	7	4
11	10	14	13	10	12
9	3	2	3	4	6
2	2	8	13	2	17
7	4	6	3	2	5
8	6	10	7	6	10
4	4	8	3	4	8
2	10	6	2	10	9
6	8	4	9	8	11
5	7	6	4	14	7
4	14	15	13	6	2
3	13	4	3	4	8
2	12	7	6	4	10
8	5	5	5	8	7
10	4	3	10	7	4
9	6	2	6	9	3
11	5	6	7	2	6

8-5. Consider the shelf life data in Exercise 8-1. Compute the sample mean, sample variance, and sample standard deviation.

8-6. Consider the cotton percentage data in Exercise 8-2. Find the sample mean, sample variance, sample standard deviation, sample median, and sample mode.

8-7. Consider the yield data in Exercise 8-3. Calculate the sample mean, sample variance, and sample standard deviation.

8-8. An article in *Computers and Industrial Engineering* (2001, p. 51) describes the time-to-failure data (in hours) for jet engines. Some of the data are reproduced below.

Engine #	Failure Time	Engine #	Failure Time
1	150	14	171
2	291	15	197
3	93	16	200
4	53	17	262
5	2	18	255
6	65	19	286
7	183	20	206
8	144	21	179
9	223	22	232
10	197	23	165
11	187	24	155
12	197	25	203
13	213		

(a) Construct a frequency distribution and histogram for these data.

(b) Calculate the sample mean, sample median, sample variance, and sample standard deviation.

8-9. For the time-to-failure data in Exercise 8-8, suppose the fifth observation (2 hours) is discarded. Construct a frequency distribution and a histogram for the remaining data, and calculate the sample mean, sample median, sample variance, and sample standard deviation. Compare the results with those obtained in Exercise 8-8. What impact has removal of this observation had on the summary statistics?

8-10. An article in *Technometrics* (Vol. 19, 1977, p. 425) presents the following data on motor fuel octane ratings of several blends of gasoline:

88.5, 87.7, 83.4, 86.7, 87.5, 91.5, 88.6, 100.3,
95.6, 93.3, 94.7, 91.1, 91.0, 94.2, 87.8, 89.9,
88.3, 87.6, 84.3, 86.7, 88.2, 90.8, 88.3, 98.8,
94.2, 92.7, 93.2, 91.0, 90.3, 93.4, 88.5, 90.1,
89.2, 88.3, 85.3, 87.9, 88.6, 90.9, 89.0, 96.1,
93.3, 91.8, 92.3, 90.4, 90.1, 93.0, 88.7, 89.9,
89.8, 89.6, 87.4, 88.4, 88.9, 91.2, 89.3, 94.4,
92.7, 91.8, 91.6, 90.4, 91.1, 92.6, 89.8, 90.6,
91.1, 90.4, 89.3, 89.7, 90.3, 91.6, 90.5, 93.7,
92.7, 92.2, 92.2, 91.2, 91.0, 92.2, 90.0, 90.7.

(a) Construct a stem-and-leaf plot.

(b) Construct a frequency distribution and histogram.

(c) Calculate the sample mean, sample variance, and sample standard deviation.

(d) Find the sample median and sample mode.

(e) Determine the skewness and kurtosis measures.

8-11. Consider the shelf-life data in Exercise 8-1. Construct a stem-and-leaf plot for these data. Construct an ordered stem-and-leaf plot. Use this plot to find the 65th and 95th percentiles.

8-12. Consider the cotton percentage data in Exercise 8-2.

(a) Construct a stem-and-leaf plot.

(b) Calculate the sample mean, sample variance, and sample standard deviation.

(c) Construct an ordered stem-and-leaf plot.

(d) Find the median and the first and third quartiles.

(e) Find the interquartile range.

(f) Determine the skewness and kurtosis measures.

8-13. Consider the yield data in Exercise 8-3.

(a) Construct an ordered stem-and-leaf plot.

(b) Find the median and the first and third quartiles.

(c) Calculate the interquartile range.

8-14. Construct a box plot for the shelf-life data in Exercise 8-1. Interpret the data using this plot.

8-15. Construct a box plot for the cotton percentage data in Exercise 8-2. Interpret the data using this plot.

8-16. Construct a box plot for the yield data in Exercise 8-3. Compare it to the histogram (Exercise 8-3) and the stem-and-leaf plot (Exercise 8-13). Interpret the data.

8-17. An article in the *Electrical Manufacturing & Coil Winding Conference Proceedings* (1995, p. 829) presents the results for the number of returned shipments for a record-of-the-month club. The company is interested in the reason for a returned shipment. The results are shown below. Construct a Pareto chart and interpret the data.

Reason	Number of Customers
Refused	195,000
Wrong selection	50,000
Wrong answer	68,000
Canceled	5,000
Other	15,000

8-18. The following table contains the frequency of occurrence of final letters in an article in the *Atlanta Journal.* Construct a histogram from these data. Do any of the numerical descriptors in this chapter have any meaning for these data?

a	12	n	19
b	11	o	13
c	11	p	1
d	20	q	0
e	25	r	15
f	13	s	18
g	12	t	20
h	12	u	0
i	8	v	0
j	0	w	41
k	2	x	0
l	11	y	15
m	12	z	0

8-19. Show the following:

(a) That $\sum_{i=1}^{n}(x_i - \bar{x}) = 0$.

(b) That $\sum_{i=1}^{n}(x_i - \bar{x})^2 = \sum_{i=1}^{n} x_i^2 - n\bar{x}^2$.

8-20. The weight of bearings produced by a forging process is being investigated. A sample of six bearings provided the weights 1.18, 1.21, 1.19, 1.17, 1.20, and 1.21 pounds. Find the sample mean, sample variance, sample standard deviation, and sample median.

8-21. The diameter of eight automotive piston rings is shown below. Calculate the sample mean, sample variance, and sample standard deviation.

74.001 mm	73.998 mm
74.005	74.000
74.003	74.006
74.001	74.002

8-22. The thickness of printed circuit boards is a very important characteristic. A sample of eight boards had the following thicknesses (in thousands of an inch): 63, 61, 65, 62, 61, 64, 60, and 66. Calculate the sample mean, sample variance, and sample standard deviation. What are the units of measurement for each statistic?

8-23. Coding the Data. Consider the printed circuit board thickness data in Exercise 8-22.

(a) Suppose that we subtract a constant 63 from each number. How are the sample mean, sample variance, and sample standard deviation affected?

(b) Suppose that we multiply each number by 100. How are the sample mean, sample variance, and sample standard deviation affected?

8-24. Coding the Data. Let $y_i = a + bx_i$, $i = 1, 2, \ldots, n$, where a and b are nonzero constants. Find the relationship between $\bar{x}$ and $\bar{y}$, and between s_x and s_y.

8-25. Consider the quantity $\sum_{i=1}^{n}(x_i - a)^2$ For what value of a is this quantity minimized?

8-26. The Trimmed Mean. Suppose that the data are arranged in increasing order, LN% of the observations removed from each end, and the sample mean of the remaining numbers calculated. The resulting quantity is called a *trimmed mean.* The trimmed mean generally lies between the sample mean $\bar{x}$ and the sample median $\tilde{x}$ (why?).

(a) Calculate the 10% trimmed mean for the yield data in Exercise 8-3.

(b) Calculate the 20% trimmed mean for the yield data in Exercise 8-3 and compare it with the quantity found in part (a).

8-27. The Trimmed Mean. Suppose that LN is not an integer. Develop a procedure for obtaining a trimmed mean.

8-28. Consider the shelf-life data in Exercise 8-1. Construct a frequency distribution and histogram using a class interval width of 2. Compute the approximate mean and standard deviation from the frequency distribution and compare it with the exact values found in Exercise 8-5.

8-29. Consider the following frequency distribution.

(a) Calculate the sample mean, variance, and standard deviation.

(b) Calculate the median and mode.

x_i	115	116	117	118	119	120	121	122	123	124
f_i	4	6	9	13	15	19	20	18	15	10

8-30. Consider the following frequency distribution.

(a) Calculate the sample mean, variance, and standard deviation.

(b) Calculate the median and mode.

x_i	−4	−3	−2	−1	0	1	2	3	4

8-31. For the two sets of data in Exercises 8-29 and 8-30, compute the sample coefficients of variation.

8-32. Compute the approximate sample mean, sample variance, sample median, and sample mode from the data in the following frequency distribution:

Class Interval	Frequency
$10 \le x < 20$	121
$20 \le x < 30$	165
$30 \le x < 40$	184
$40 \le x < 50$	173
$50 \le x < 60$	142
$60 \le x < 70$	120
$70 \le x < 80$	118
$80 \le x < 90$	110
$90 \le x < 100$	90

8-33. Compute the approximate sample mean, sample variance, median, and mode from the data in the following frequency distribution:

Class Interval	Frequency
$-10 \le x < 0$	3
$0 \le x < 10$	8
$10 \le x < 20$	12
$20 \le x < 30$	16
$30 \le x < 40$	9
$40 \le x < 50$	4
$50 \le x < 60$	2

8-34. Compute the approximate sample mean, sample standard deviation, sample variance, median, and mode for the data in the following frequency distribution:

Class Interval	Frequency
$600 \le x < 650$	41
$650 \le x < 700$	46
$700 \le x < 750$	50
$750 \le x < 800$	52
$800 \le x < 850$	60
$850 \le x < 900$	64
$900 \le x < 950$	65
$950 \le x < 1000$	70
$1000 \le x < 1050$	72

8-35. An article in the *International Journal of Industrial Ergonomics* (1999, p. 483) describes a study conducted to determine the relationship between exhaustion time and distance covered until exhaustion for several wheelchair exercises performed on a 400-m outdoor track. The time and distances for ten participants are as follows:

Exhaustion Time (in seconds)	Distance Covered until Exhaustion (in meters)
610	1373
310	698
720	1440
990	2228
1820	4550
475	713
890	2003
390	488
745	1118
885	1991

Determine the simple (Pearson) correlation between time and distance. Interpret your results.

8-36. An electric utility which serves 850,332 residential customers on May 1, 2002, selects a sample of 120 customers randomly and installs a time-of-use meter or "load research meter" at each selected residence. At the time of installation, the technician also records the residence size (sq. ft.). During the summer peak demand period, the company experienced peak demand at 5:31 P.M. on July 30, 2002. July bills for usage (kwh) were sent to all customers, including those in the sample group. Due to cyclical billing, the July bills do not reflect the same time-of-use period for all customers; however, each customer is assigned a usage for July billing. The time-of-use meters have memory and they record time specific demand (kw) by time interval; so for the sampled customers, average 15-minute demand is available for the time interval 5:00–5:45 P.M. on July 30, 2002. The data file Loaddata contains the kwh, kw, and sq. ft. data for each of the 120 sampled residences. This file is available at www.wiley.com/college/hines. Using Minitab® or other software, do the following:

(a) Construct scatter plots of

1. kw vs. sq. ft. and
2. kw vs. kwh,

and comment on the observed displays specifically in regard to the nature of the association in parts 1 and 2 above, and for each part, the general pattern of observed variation in kw across the range of sq. ft. and the range of kwh.

(b) Construct the three-dimensional plot of kw vs. kwh and sq. ft.

(c) Construct histograms for kwh, kw, and sq. ft. data, and comment on the patterns observed.

(d) Construct a stem-and-leaf plot for the kwh data and compare to the histogram for the kwh data.

(e) Determine the sample mean, median, and mode for kwh, kw, and sq. ft.

(f) Determine the sample standard deviation for kwh, kw, and sq. ft.

(g) Determine the first and third quartiles for kwh, kw, and sq. ft.

(h) Determine the skewness and kurtosis measures for kwh, kw, and sq. ft., and compare to the measures for a normal distribution.

(i) Determine the simple (Pearson) correlation between kw and sq.ft. and between kw and kwh. Interpret.

Chapter 9

Random Samples and Sampling Distributions

In this chapter, we begin our study of statistical inference. Recall that statistics is the science of drawing conclusions about a population based on an analysis of sample data from that population. There are many different ways to take a sample from a population. Furthermore, the conclusions that we can draw about the population often depend on how the sample is selected. Generally, we want the sample to be *representative* of the population. One important method of selecting a sample is *random sampling.* Most of the statistical techniques that we present in the book assume that the sample is a random sample. In this chapter we will define a *random sample* and introduce several probability distributions useful in analyzing the information in sample data.

9-1 RANDOM SAMPLES

To define a random sample, let X be a random variable with probability distribution $f(x)$. Then the set of n observations $X_1, X_2, \ldots, X_n$, taken on the random variable X, and having numerical outcomes $x_1, x_2, \ldots, x_n$, is called a *random sample* if the observations are obtained by observing X independently under unchanging conditions for n times. Note that the observations $X_1, X_2, \ldots, X_n$ in a random sample are independent random variables with the same probability distribution $f(x)$. That is, the marginal distributions of $X_1, X_2, \ldots, X_n$ are $f(x_1)$, $f(x_2), \ldots, f(x_n)$, respectively, and by independence the joint probability distribution of the random sample is

$$g(x_1, x_2, \ldots, x_n) = f(x_1) \cdot f(x_2) \cdot \cdots \cdot f(x_n). \tag{9-1}$$

Definition

$X_1, X_2, \ldots, X_n$ is a *random sample* of size n if (a) the Xs are independent random variables, and (b) every observation X_i has the same probability distribution.

To illustrate this definition, suppose that we are investigating the bursting strength of glass, 1-liter soft drink bottles, and that bursting strength in the population of bottles is normally distributed. Then we would expect each of the observations on bursting strength X_1, $X_2, \ldots, X_n$ in a random sample of n bottles to be independent random variables with exactly the same normal distribution.

It is not always easy to obtain a random sample. Sometimes we may use tables of uniform random numbers. At other times, the engineer or scientist cannot easily use formal procedures to help ensure randomness and must rely on other selection methods. A *judgment sample* is one chosen from the population by the objective judgment of an individual.

Since the accuracy and statistical behavior of judgment samples cannot be described, they should be avoided.

Example 9-1

Suppose we wish to take a random sample of five batches of raw material out of 25 available batches. We may number the batches with the integers 1 to 25. Now, using Table XV of the Appendix, arbitrarily choose a row and column as a starting point. Read down the chosen column, obtaining two digits at a time, until five acceptable numbers are found (an acceptable number lies between 1 and 25). To illustrate, suppose the above process gives us a sequence of numbers that reads 37, 48, 55, **02**, **17**, 61, 70, 43, **21**, 82, 73, **13**, 60, **25**. The bold numbers specify which batches of raw material are to be chosen as the random sample.

We first present some specifics on sampling from finite universes. Subsequent sections will describe various sampling distributions. Sections marked by an asterisk may be omitted without loss of continuity.

*9-1.1 Simple Random Sampling from a Finite Universe

When sampling n items *without replacement* from a universe of size N, there are $\binom{N}{n}$ possible samples, and if the selection probabilities are $\pi_k = 1/\binom{N}{n}$, for $k = 1, 2, \ldots, \binom{N}{n}$, then this is *simple random sampling*. Note that each universe unit appears in exactly $\binom{N-1}{n-1}$ of the possible samples, so each unit has inclusion probability of $\binom{N-1}{n-1}/\binom{N}{n} = \frac{n}{N}$.

As will be seen later, sampling without replacement is more "efficient" than sampling with replacement for estimating the finite population mean or total; however, we briefly discuss *simple random sampling with replacement* for a basis of comparison. In this case, there are N^n possible samples, and we select each with probability $\pi_k = 1/N^n$, for $k = 1, 2, \ldots, N^n$. In this situation, a universe unit may appear in no samples or in as many as n, so the notion of inclusion probability is less meaningful; however, if we consider the probability that a specific unit will be selected *at least once*, that is obviously $1 - (1 - \frac{1}{N})^n$, since for each unit the probability of selection on a given observation is $1/N$, a constant, and the n selections are independent, so that these observations may be considered Bernoulli trials.

Example 9-2

Consider a universe consisting of five units numbered 1,2,3,4,5. In sampling without replacement, we will employ a sample of size two and enumerate the possible samples as

(1,2), (1,3), (1,4), (1,5), (2,3), **(2,4)**, (2,5), (3,4), (3,5), (4,5).

Note that there are $\binom{5}{2} = 10$ possible samples. If we select one from these, where each has an equal selection probability of 0.1 assigned, that is simple random sampling. Consider these possible samples to be numbered 1, 2, …, 0, where 0 represents number 10. Now, go to Table XV (in the Appendix), showing random integers, and with eyes closed place a finger down. Read the first digit from the five-digit integer presented. Suppose we pick the integer which is in row 7 of column 4. The first digit is 6, so the sample consists of units 2 and 4. An alternate to using the table is to cast a single icosohedron die and pick the sample corresponding to the outcome. Notice also that each unit appears in exactly four of the possible samples, and thus the inclusion probability for each unit is 0.4, which is simply $n/N = 2/5$.

---

*May be omitted on first reading.

It is usually not feasible to enumerate the set of possible samples. For example, if $N = 100$ and $n = 25$, there would be more than 2.43×10^{23} possible samples, so other selection procedures which maintain the properties described in the definition must be employed. The most commonly used procedure is to first number the universe units from 1 to N, then use realizations from a random number process to sequentially pick numbers from 1 to N, discarding duplicates until n units have been picked if we are sampling without replacement, keeping duplicates if sampling is with replacement. Recall from Chapter 6 that the term *random numbers* is used to describe a sequence of mutually independent variables $U_1, U_2, \ldots$, which are identically distributed as uniform on [0,1]. We employ a realization $u_1, u_2, \ldots$, which in sequentially selecting units as members of the sample is roughly outlined as

$$(\text{Unit Number})_i = \lfloor N \cdot u_j \rfloor + 1, \quad j = 1, 2, \ldots, J, i = 1, 2, \ldots, n, \quad i < j,$$

where J is the trial number on which the nth, or final, unit is selected. In sampling without replacement, $J \geq n$, and $\lfloor\ \rfloor$ is the greatest integer contained function. When sampling with replacement, $J = n$.

*9-1.2 Stratified Random Sampling of a Finite Universe

In finite-universe sampling, sometimes an explanatory or auxiliary variable(s) is available that has known value(s) for each unit of the universe. These may be either numerical or categorical variables or both. If we are able to use the auxiliary variables to establish criteria for assigning each universe unit to exactly one of the resulting *strata*, before sampling begins, then simple random samples may be selected within each stratum, and the sampling is independent from stratum to stratum, which allows us to later combine, with appropriate weights, stratum "statistics" such as the means and variances obtained from various strata. In general, this is an "efficient" scheme in the sense of estimation of population means and totals if, following the classification, the variance in the variable we seek to measure is small within strata while the differences in the stratum mean values are large. If L strata are formed, then the stratum sizes are $N_1, N_2, \ldots, N_L$, and $N_1 + N_2 + \cdots + N_L = N$, while the sample sizes are $n_1, n_2, \ldots, n_L$, and $n_1 + n_2 + \cdots + n_L = n$. It is noted that the inclusion probabilities are constant *within* strata, as n_h/N_h for stratum h, but they may differ greatly across strata. Two commonly used methods for allocating the overall sample to strata are proportional allocation and Neyman optimal allocation, where proportional allocation is

$$n_h = n(N_h/N), h = 1,2, \ldots, L, \tag{9-2}$$

and Neyman allocation is

$$n_h = n\left[\frac{N_h \cdot \tilde{\sigma}_h}{\sum_{h=1}^{L} N_h \cdot \tilde{\sigma}_h}\right], \qquad h = 1,2,\ldots,L. \tag{9-3}$$

The values $\tilde{\sigma}_h$ are standard deviation values within strata, and when designing the sampling study, they are usually unknown for the variables to be observed in the study; however, they may often be calculated for an explanatory or auxiliary variable where at least one of these is numeric, and if there is a "reasonable" correlation, $|\rho| > 0.6$, between the variable to be measured and such an auxiliary variable, then these surrogate standard deviation values, denoted σ_h', will produce an allocation which will be reasonably close to optimal.

*May be omitted on first reading.

Example 9-3

In seeking to address growing consumer concerns regarding the quality of claim processing, a national managed health care/health insurance company has identified several characteristics to be monitored on a monthly basis. The most important of these to the company are, first, the size of the mean "financial error," which is the absolute value of the error (overpay or underpay) for a claim and, second, the fraction of correctly filed claims paid to providers within 14 days. Three processing centers are eastern, mid-America, and western. Together, these centers process about 450,000 claims per month, and it has been observed that they differ in accuracy and timeliness. Furthermore, the correlation between the total dollar amount of the claim and the financial error overall is historically about 0.65–0.80. If a sample of size 1000 claims is to be drawn monthly, strata might be formed using centers E, MA, and W and total claim sizes as \$0–\$200, \$201–\$900, \$901–\$max claim. Therefore there would be nine strata. And if for a given year there were 441,357 claims, these may be easily assigned to strata, since they are naturally grouped by center, and each center uses the same processing system to record data by claim amount, thus allowing ease of classification by claim size. At this point, the 1000-unit planned sample is allocated to the nine strata as shown in Table 9-1. The allocation in italics, shown first, employs proportional allocation while the second one uses "optimal" allocation. The values represent the standard deviation in the claim-amount metric within the stratum, since the standard deviation in "financial error" is unknown. After deciding on the allocation to use, nine independent, simple random samples, without replacement, would be selected, yielding 1000 claim forms to be inspected. Stratum identifying subscripts are not shown in Table 9-1.

It is noted that proportional allocation results in an equal inclusion probability for all units, not just within strata, while an optimal allocation draws larger samples from strata in which the product of the internal variability as measured by standard deviation (often in an auxiliary variable) and the number of units in the stratum is large. Thus, inclusion probabilities are the same for all units assigned to a stratum but may differ greatly across strata.

9-2 STATISTICS AND SAMPLING DISTRIBUTIONS

A *statistic* is any function of the observations in a random sample that does not depend on unknown parameters. The process of drawing conclusions about populations based on

Table 9-1 Data for Health Insurance Example 9-3

	Claim Amount			
Center	\$0 – \$200	\$201 – \$900	\$901 – \$Max Claim	All
E	$N = 132{,}365$ $\sigma' = \$42$ $n =$ *300* / 29	$N = 41{,}321$ $\sigma' = \$255$ $n =$ *94* / 54	$N = 10{,}635$ $\sigma' = \$6781$ $n =$ *24* / 371	184,321 *418* / 454
MA	$N = 96{,}422$ $\sigma' = \$31$ $n =$ *218* / 15	$N = 31{,}869$ $\sigma' = \$210$ $n =$ *72* / 35	$N = 6{,}163$ $\sigma' = \$5128$ $n =$ *14* / 163	134,454 *304* / 213
W	$N = 82{,}332$ $\sigma' = \$57$ $n =$ *187* / 24	$N = 33{,}793$ $\sigma' = \$310$ $n =$ *76* / 54	$N = 6{,}457$ $\sigma' = \$7674$ $n =$ *15* / 255	122,582 *278* / 333
All	$N = 311{,}119$ $n =$ *705* / 68	$N = 106{,}983$ $n =$ *242* / 143	$N = 23{,}255$ $n =$ *53* / 789	441,357 *1000* / 1000

sample data makes considerable use of statistics. The procedures require that we understand the probabilistic behavior of certain statistics. In general, we call the probability distribution of a statistic a *sampling distribution.* There are several important sampling distributions that will be used extensively in subsequent chapters. In this section, we define and briefly illustrate these sampling distributions. First, we give some relevant definitions and additional motivation.

A *statistic* is now defined as a value determined by a function of the values observed in a sample. For example, if $X_1, X_2, \ldots, X_n$ represent values to be observed in a probability sample of size n on a single random variable X, then $\overline{X}$ and S^2, as described in Chapter 8 in equations 8-1 and 8-4, are statistics. Furthermore, the same is true for the median, the mode, the sample range, the sample skewness measure, and the sample kurtosis. Note that capital letters are used here as this reference is to random variables, not specific numerical results, as was the case in Chapter 8.

9-2.1 Sampling Distributions

Definition

The *sampling distribution of a statistic* is the density function or probability function that describes the probabilistic behavior of the statistic in repeated sampling from the same universe or on the same process variable assignment model.

Examples have been presented earlier, in Chapters 5–7. Recall that random sampling with sample size n on a process variable X provides results $X_1, X_2, \ldots, X_n$, which are mutually independent random variables, all with a common distribution function. Thus, the sample mean, $\overline{X}$, is a linear combination of n independent variables. If $E(X) = \mu$ and $V(X) = \sigma^2$, then recall that $E(\overline{X}) = \mu$ and $V(\overline{X}) = \sigma^2/n$. And if X is a measurement variable, the density function of $\overline{X}$ is the sampling distribution of this statistic, $\overline{X}$. There are several important sampling distributions that will be used extensively in subsequent chapters. In this section we will describe and briefly illustrate these sampling distributions.

The form of a sampling distribution depends on the stability assumption as well as on the form of the process variable model. In Chapter 7, we observed that if $X \sim N(\mu, \sigma^2)$, then $\overline{X} \sim N(\mu, \sigma^2/n)$, and this is the sampling distribution of $\overline{X}$ for $n \geq 1$. Now, if the process variable assignment model takes some form other than the normal model illustrated here (e.g., the exponential model), and if all stationarity assumptions hold, then mathematical analysis may yield a closed form for the sampling distribution of $\overline{X}$. In this example case, if we employ an exponential process model for X, the resulting sampling distribution for X is a form of the gamma distribution. In Chapter 7, the important Central Limit Theorem was presented. Recall that if the moment-generating function $M_X(t)$ exists for all t, and if

$$Z_n = \frac{\overline{X} - \mu}{\sigma/\sqrt{n}}, \text{ then } \lim_{n \to \infty} \frac{F_n(z)}{\Phi(z)} = 1, \tag{9-4}$$

where $F_n(z)$ is the cumulative distribution function of Z_n, and $\Phi(z)$ is the CDF for the standard normal variable, Z. Simply stated, as $n \to \infty$, $Z_n \to N(0, 1)$ random variable, and this result has enormous utility in applied statistics. However, in applied work, a question arises regarding how large the sample size n must be to employ the $N(0,1)$ model as the sampling distribution of Z_n, or equivalently stated, to describe the sampling distribution of $\overline{X}$ as $N(\mu, \sigma^2/n)$. This is an important question, as the exact form of the process variable assignment model is usually unknown. Furthermore, any response must be conditioned even when it is based on simulation evidence, experience, or "accepted practice." Assuming process stability, a general suggestion is that if the skewess measure is close to zero (implying that the variable assignment model is symmetric or very nearly so), then $\overline{X}$ approaches normality quickly, say for $n \approx 10$, but this depends also on the standardized kurtosis of X. For example, if $\beta_3 \approx 0$ and $|\beta_4| < 0.75$, then a sample size of $n = 5$ may be quite adequate for many applications. However, it should be noted that the tail behavior of $\overline{X}$ may deviate somewhat from that predicted by a normal model.

Where there is considerable skew present, application of the Central Limit Theorem describing the sampling distribution must be interpreted with care. A "rule of thumb" that has been successfully used in survey sampling, where such variable behavior is common ($\beta_3 > 0$), is that

$$n > 25(\beta_3)^2. \tag{9-5}$$

For instance, returning to an exponential process variable assignment model, this rule suggests that a sample size of $n > 100$ is required for employing a normal distribution to describe the behavior of $\overline{X}$, since $\beta_3 = 2$.

Definition

The *standard error* of a statistic is the standard deviation of its sampling distribution. If the standard error involves unknown parameters whose values can be estimated, substitution of these estimates into the standard error results in an *estimated standard error.*

To illustrate this definition, suppose we are sampling from a normal distribution with mean μ and variance σ^2. Now the distribution of $\overline{X}$ is normal with mean μ and variance σ^2/n, and so the *standard error* of $\overline{X}$ is

$$\frac{\sigma}{\sqrt{n}}.$$

If we did not know σ but substituted the sample standard deviation s into the above, then the *estimated standard error* of $\overline{X}$ is

$$\frac{s}{\sqrt{n}}.$$

Example 9-4

Suppose we collect data on the tension bond strength of a modified portland cement mortar. The ten observations are

$$16.85, 16.40, 17.21, 16.35, 16.52, 17.04, 16.96, 17.15, 16.59, 16.57,$$

where tension bond strength is measured in units of kgf/cm^2. We assume that the tension bond strength is well-described by a normal distribution. The sample average is

$$\bar{x} = 16.76 \text{ kgf/cm}^2.$$

First, suppose we know (or are willing to assume) that the standard deviation of tension bond strength is $\sigma = 0.25$ kgf/cm^2. Then the *standard error* of the sample average is

$$\sigma/\sqrt{n} = 0.25/\sqrt{10} = 0.079 \text{ kgf/cm}^2.$$

If we are unwilling to assume that $\sigma = 0.25$ kgf/cm^2, we could use the *sample standard deviation* $s =$ 0.316 kgf/cm^2 to obtain the *estimated standard error* as follows:

$$s/\sqrt{n} = 0.316/\sqrt{10} = 0.0999 \text{ kgf/cm}^2.$$

*9-2.2 Finite Populations and Enumerative Studies

In the case where sampling may be conceptually repeated on the same finite universe of units, the sampling distribution of $\overline{X}$ is interpreted in a manner similar to that of sampling a process, except the issue of process stability is not a concern. Any inference to be drawn is to be about the specific collection of unit population values. Ordinarily, in such situations, sampling is without replacement, as this is more efficient. The general notion of *expectation* is different in such studies in that the expected value of a statistic $\hat{\theta}$ is defined as

$$E^0(\hat{\theta}) = \sum_{k=1}^{\binom{N}{n}} \pi_k \cdot \hat{\theta}_k, \tag{9-6}$$

where $\hat{\theta}_k$ is the value of the statistic if possible sample k is selected. In simple random sampling, recall that $\pi_k = 1/\binom{N}{n}$.

Now in the case of the sample mean statistic, $\overline{X}$, we have $E^0(\overline{X}) = \mu_x$, where μ_x is the finite population mean of the random variable X. Also, under simple random sampling without replacement,

$$V(\overline{X}) = E^0[\overline{X}]^2 - \mu_x^2 = \frac{\tilde{\sigma}_x^2}{n}\left(1 - \frac{n}{N}\right), \tag{9-7}$$

where $\tilde{\sigma}_x^2$ is as defined in equation 8-7. The ratio n/N is called the sampling fraction, and it represents the fraction of the population measures to be included in the sample. Concise proofs of the results shown in equations 9-6 and 9-7 are given by Cochran (1977, p. 22). Oftentimes, in studies of this sort, the objective is to estimate the population total (see equation 8-2) as well as the mean. The "mean per unit," or mpu, estimate of the total is simply $\tilde{\tau}_x = N \cdot \overline{X}$, and the variance of this statistic is obviously $V(\hat{\tau}_x) = N^2 \cdot V(\overline{X})$. While necessary and sufficient conditions for the distribution of $\overline{X}$ to approach normality have been developed, these are of little practical utility, and the rule given by equation 9-5 has been widely employed.

In sampling with replacement, the quanitities $E^0(\overline{X}) = \mu_x$, and $V^0(\overline{X}) = \tilde{\sigma}_x^2/n$.

Where stratification has been employed in a finite population enumeration study, there are two statistics of common interest. These are the aggregate sample mean and the estimate of population total. The mean is given by

$$\hat{\mu}_x = \frac{1}{N}\sum_{h=1}^{L} N_h \cdot \overline{X}_h = \sum_{h=1}^{L} W_h \cdot \overline{X}_h, \tag{9-8}$$

and the estimate of total is $\hat{\tau}_x$, where

$$\hat{\tau}_x = N \cdot \hat{\mu}_x. \tag{9-9}$$

In these formulations, $\overline{X}_h$ is the sample mean for stratum h, and W_h, given by (N_h/N), is called the stratum weight for stratum h. Note that both of these statistics are expressed as simple linear combinations of the independent, stratum statistics $\overline{X}_h$, and both are mpu estimators. The variance of these statistics is given by

$$V(\hat{\mu}_x) = \sum_{h=1}^{L} W_h^2 \cdot \left[\frac{\tilde{\sigma}_x^2}{n_h} \cdot \left(1 - \frac{n_h}{N_h}\right)\right],$$

$$V(\hat{\tau}_x) = N^2 \cdot V(\hat{\mu}_x). \tag{9-10}$$

*May be omitted on first reading.

The within-stratum variance terms, $\tilde{\sigma}^2_{x_h}$, may be estimated by the sample variance terms, $S^2_{x_h}$, for the respective strata. The sampling distributions of the aggregate mean and total estimate statistics for such stratified, enumeration studies are stated only for situations where sample sizes are large enough to employ the limiting normality indicated by the Central Limit Theorem. Note that these statistics are linear combinations across observations within strata and across strata. The result is that in such cases we take

$$\hat{\mu}_x \sim N(\mu_x, V(\hat{\mu}_x)) \tag{9-11}$$

and

$$\hat{\tau}_x \sim N(\tau_x, V(\hat{\tau}_x)).$$

Example 9-5

Suppose the sampling plan in Example 9-3 utilizes the "optimal" allocation as shown in Table 9-1. The stratum sample means and sample standard deviations on the financial error variable are calculated with the results shown in Table 9-2, where the units of measurement are error $/claim.

Then utilizing the results presented in equations 9-8 and 9-9, the aggregate statistics when evaluated are $\bar{x} = \$18.36$, and $\hat{\tau}_x = \$8,103,315$. Utilizing equations 9-10 and employing the within-stratum sample variance values $s^2_{x_h}$, to estimate the within-stratum variances $\tilde{\sigma}^2_{x_h}$ the estimates for the variance in the sampling distribution of the sample mean and the estimate of total are $V(\hat{\mu}_x) = 0.574$ and $V(\hat{\tau}_x) = 1.118 \times 10^{11}$, and the estimates of the respective standard errors are thus \$0.785 and \$334,408 for the mean and total estimator distributions.

9-3 THE CHI-SQUARE DISTRIBUTION

Many other useful sampling distributions can be defined in terms of normal random variables. The chi-square distribution is defined below.

Table 9-2 Sample Statistics for Health Insurance Example

	Claim Amount			
Center	\$0 – \$200	\$201 – \$900	\$901 – \$Max Claim	All
E	$N = 132,365$	$N = 41,321$	$N = 10,635$	184,321
	$n = 29$	$n = 54$	$n = 371$	454
	$\bar{x} = 6.25$	$\bar{x} = 34.10$	$\bar{x} = 91.65$	
	$s_x = 4.30$	$s_x = 28.61$	$s_x = 81.97$	
MA	$N = 96,422$	$N = 31,869$	$N = 6,163$	134,454
	$n = 15$	$n = 35$	$n = 163$	213
	$\bar{x} = 5.30$	$\bar{x} = 22.00$	$\bar{x} = 72.00$	
	$s_x = 4.82$	$s_x = 16.39$	$s_x = 56.67$	
W	$N = 82,332$	$N = 33,793$	$N = 6,457$	131,225
	$n = 24$	$n = 54$	$n = 255$	333
	$\bar{x} = 10.52$	$\bar{x} = 46.28$	$\bar{x} = 124.91$	
	$s_x = 9.86$	$s_x = 31.23$	$s_x = 109.42$	
All	$N = 311,119$	$N = 106,983$	$N = 31,898$	441,357
	$n = 67$	$n = 143$	$n = 789$	$n = 1,000$

Theorem 9-1

Let $Z_1, Z_2, \ldots, Z_k$ be normally and independently distributed random variables, with mean $\mu = 0$ and variance $\sigma^2 = 1$. Then the random variable

$$\chi^2 = Z_1^2 + Z_2^2 + \cdots + Z_k^2$$

has the probability density function

$$\begin{aligned} f(u) &= \frac{1}{2^{k/2}\Gamma\left(\frac{k}{2}\right)} u^{(k/2)-1} e^{-u/2}, \qquad && u > 0, \\ &= 0, && \text{otherwise} \end{aligned} \tag{9-12}$$

and is said to follow the chi-square distribution with k degrees of freedom, abbreviated χ_k^2.

For the proof of Theorem 9-1, see Exercises 7-47 and 7-48.

The mean and variance of the χ_k^2 distribution are

$$\mu = k \tag{9-13}$$

and

$$\sigma^2 = 2k. \tag{9-14}$$

Several chi-square distributions are shown in Fig. 9-1. Note that the chi-square random variable is nonnegative, and that the probability distribution is skewed to the right. However, as k increases, the distribution becomes more symmetric. As $k \to \infty$, the limiting form of the chi-square distribution is the normal distribution.

The percentage points of the χ_k^2 distribution are given in Table III of the Appendix. Define $\chi_{\alpha,k}^2$ as the percentage point or value of the chi-square random variable with k degrees of freedom such that the probability that χ_k^2 exceeds this value is α. That is,

$$P\{\chi_k^2 \ge \chi_{\alpha,k}^2\} = \int_{\chi_{\alpha,k}^2}^{\infty} f(u)\,du = \alpha.$$

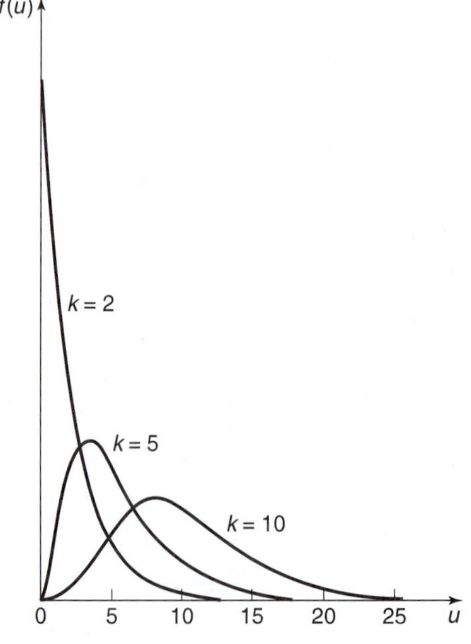

Figure 9-1 Several χ^2 distributions.

This probability is shown as the shaded area in Fig. 9-2. To illustrate the use of Table III, note that

$$P\{\chi^2_{10} \geq \chi^2_{0.05,10}\} = P\{\chi^2_{10} \geq 18.31\} = 0.05.$$

That is, the 5% point of the chi-square distribution with 10 degrees of freedom is $\chi^2_{0.05,10} = 18.31$.

Like the normal distribution, the chi-square distribution has an important reproductive property.

Theorem 9-2 Additivity Theorem of Chi-Square

Let $\chi^2_1, \chi^2_2, \ldots, \chi^2_p$ be independent chi-square random variables with $k_1, k_2, \ldots, k_p$ degrees of freedom, respectively. Then the quantity

$$Y = \chi^2_1 + \chi^2_2 + \cdots + \chi^2_p$$

follows the chi-square distribution with degrees of freedom equal to

$$k = \sum_{i=1}^{p} k_i.$$

Proof Note that each chi-square random variable χ^2_i can be written as the sum of the squares of k_i standard normal random variables, say

$$\chi^2_i = \sum_{j=1}^{k_i} Z^2_{ij} \qquad i = 1, 2, \ldots, p.$$

Therefore,

$$Y = \sum_{i=1}^{p} \chi^2_i = \sum_{i=1}^{p}\sum_{j=1}^{k_i} Z^2_{ij}$$

and since all the random variables Z_{ij} are independent because the χ^2_i are independent, Y is just the sum of the squares of $k = \sum_{i=1}^{p} k_i$ independent standard normal random variables. From Theorem 9-1, it follows that Y is a chi-square random variable with k degrees of freedom.

Example 9-6

As an example of a statistic that follows the chi-square distribution, suppose that $X_1, X_2, \ldots, X_n$ is a random sample from a normal population, with mean μ and variance σ^2. The function of the sample variance

$$\frac{(n-1)S^2}{\sigma^2}$$

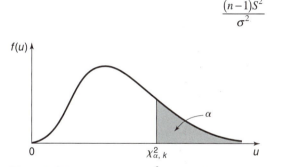

Figure 9-2 Percentage point $\chi^2_{\alpha,k}$ of the chi-square distribution.

is distributed as χ^2_{n-1}. We will use this random variable extensively in Chapters 10 and 11. We will see in those chapters that because the distribution of this random variable is chi square, we can construct confidence interval estimates and test statistical hypotheses about the variance of a normal population.

To illustrate heuristically *why* the distribution of the random variable in Example 9-6, $(n-1)S^2/\sigma^2$, is chi-square, note that

$$\frac{(n-1)S^2}{\sigma^2} = \frac{\sum_{i=1}^{n}(X_i - \overline{X})^2}{\sigma^2}. \tag{9-15}$$

If $\overline{X}$ in equation 9-15 were replaced by μ, then the distribution of

$$\frac{\sum_{i=1}^{n}(X_i - \mu)^2}{\sigma^2}$$

is χ^2_n, because each term $(X_i - \mu)/\sigma$ is an independent standard normal random variable. Now consider the following:

$$\begin{aligned}
\sum_{i=1}^{n}(X_i - \mu)^2 &= \sum_{i=1}^{n}\left[(X_i - \overline{X}) + (\overline{X} - \mu)\right]^2 \\
&= \sum_{i=1}^{n}(X_i - \overline{X})^2 + \sum_{i=1}^{n}(\overline{X} - \mu)^2 + 2(\overline{X} - \mu)\sum_{i=1}^{n}(X_i - \overline{X}) \\
&= \sum_{i=1}^{n}(X_i - \overline{X})^2 + n(\overline{X} - \mu)^2.
\end{aligned}$$

Therefore,

$$\frac{\sum_{i=1}^{n}(X_i - \mu)^2}{\sigma^2} = \frac{\sum_{i=1}^{n}(X_i - \overline{X})^2}{\sigma^2} + \frac{(\overline{X} - \mu)^2}{\sigma^2/n}$$

or

$$\frac{\sum_{i=1}^{n}(X_i - \mu)^2}{\sigma^2} = \frac{(n-1)S^2}{\sigma^2} + \frac{(\overline{X} - \mu)^2}{\sigma^2/n}. \tag{9-16}$$

Since $\overline{X}$ is normally distributed with mean μ and variance σ^2/n, the quantity $(\overline{X} - \mu)^2/(\sigma^2/n)$ is distributed as χ^2_1. Furthermore, it can be shown that the random variables $\overline{X}$ and S^2 are independent. Therefore, since $\sum_{i=1}^{n}(X_i - \mu)^2/\sigma^2$ is distributed as χ^2_n, it seems logical to use the additivity property of the chi-square distribution (Theorem 9-2) and conclude that the distribution of $(n-1)S^2/\sigma^2$ is χ^2_{n-1}.

9-4 THE *t* DISTRIBUTION

Another important sampling distribution is the *t* distribution, sometimes called the student *t* distribution.

Theorem 9-3

Let $Z \sim N(0, 1)$ and V be a chi-square random variable with k degrees of freedom. If Z and V are independent, then the random variable

$$T = \frac{Z}{\sqrt{V/k}}$$

has the probability density function

$$f(t)=\frac{\Gamma[(k+1)/2]}{\sqrt{\pi k}\,\Gamma(k/2)}\cdot\frac{1}{\left[\left(t^2/k\right)+1\right]^{(k+1)/2}},\qquad -\infty<t<\infty, \tag{9-17}$$

and is said to follow the t distribution with k degrees of freedom, abbreviated t_k.

Proof Since Z and V are independent, their joint density function is

$$f(z,v)=\frac{v^{(k/2)-1}}{\sqrt{2\pi}\,2^{k/2}\Gamma\left(\frac{k}{2}\right)}e^{-\left(z^2+v\right)/2},\qquad -\infty<z<\infty, 0<v<\infty.$$

Using the method of Section 4-10 we define a new random variable $U = V$. Thus, the inverse solutions of

$$t=\frac{z}{\sqrt{v/k}}$$

and

$$u=v$$

are

$$z=t\sqrt{\frac{u}{k}}$$

and

$$v=u.$$

The Jacobian is

$$J=\begin{vmatrix}\sqrt{\dfrac{u}{k}} & \dfrac{t}{2\sqrt{uk}}\\ 0 & 1\end{vmatrix}=\sqrt{\frac{u}{k}}.$$

Thus,

$$|J|=\sqrt{\frac{u}{k}}$$

and so the joint probability density function of T and U is

$$g(t,u)=\frac{\sqrt{u}}{\sqrt{2\pi k}\,2^{k/2}\Gamma\left(\frac{k}{2}\right)}u^{(k/2)-1}e^{-\left[(u/k)t^2+u\right]/2}. \tag{9-18}$$

Now, since $v>0$ we must require that $u>0$, and since $-\infty<z<\infty$, then $-\infty<t<\infty$. On rearranging equation 9-18 we have

$$g(t,u)=\frac{1}{\sqrt{2\pi\, k}\,2^{k/2}\Gamma\left(\frac{k}{2}\right)}u^{(k-1)/2}e^{-(u/2)\left[\left(t^2/k\right)+1\right]},\qquad 0<u<\infty, -\infty<t<\infty,$$

and since $f(t)=\int_0^\infty g(t,u)\,du$, we obtain

$$f(t) = \frac{1}{\sqrt{2\pi k}\,2^{k/2}\,\Gamma\left(\frac{k}{2}\right)} \int_0^\infty u^{(k-1)/2} e^{-(u/2)\left[\left(t^2/k\right)+1\right]} du$$

$$= \frac{\Gamma[(k+1)/2]}{\sqrt{\pi k}\,\Gamma\left(\frac{k}{2}\right)} \cdot \frac{1}{\left[\left(t^2/k\right)+1\right]^{(k+1)/2}}, \qquad -\infty < t < \infty.$$

Primarily because of historical usage, many authors make no distinction between the random variable T and the symbol t. The mean and variance of the t distribution are $\mu = 0$ and $\sigma^2 = k/(k-2)$ for $k > 2$, respectively. Several t distributions are shown in Fig. 9-3. The general appearance of the t distribution is similar to the standard normal distribution, in that both distributions are symmetric and unimodal, and the maximum ordinate value is reached at the mean $\mu = 0$. However, the t distribution has heavier tails than the normal; that is, it has more probability further out. As the number of degrees of freedom $k \to \infty$, the limiting form of the t distribution is the standard normal distribution. In visualizing the t distribution it is sometimes useful to know that the ordinate of the density at the mean $\mu = 0$ is approximately four to five times larger than the ordinate at the 5th and 95th percentiles. For example, with 10 degrees of freedom for t this ratio is 4.8, with 20 degrees of freedom this factor is 4.3, and with 30 degrees of freedom this factor is 4.1. By comparison, for the normal distribution, this factor is 3.9.

The percentage points of the t distribution are given in Table IV of the Appendix. Let $t_{\alpha,k}$ be the percentage point or value of the t random variable with k degrees of freedom such that

$$P\{T \geq t_{\alpha,k}\} = \int_{t_{\alpha,k}}^{\infty} f(t)\,dt = \alpha.$$

This percentage point is illustrated in Fig. 9-4. Note that since the t distribution is symmetric about zero, we find $t_{1-\alpha,k} = -t_{\alpha,k}$. This relationship is useful, since Table IV gives only *upper-tail* percentage points, that is, values of $t_{\alpha,k}$ for $\alpha \leq 0.50$. To illustrate the use of the table, note that

$$P\{T \geq t_{0.05,10}\} = P\{T \geq 1.812\} = 0.05.$$

Thus, the upper 5% point of the t distribution with 10 degrees of freedom is $t_{0.05,10} = 1.812$. Similarly, the lower-tail point $t_{0.95,10} = -t_{0.05,10} = -1.812$.

Example 9-7

As an example of a random variable that follows the t distribution, suppose that $X_1, X_2, \ldots, X_n$ is a random sample from a normal distribution with mean μ and variance σ^2, and let $\overline{X}$ and S^2 denote the sample mean and variance. Consider the statistic

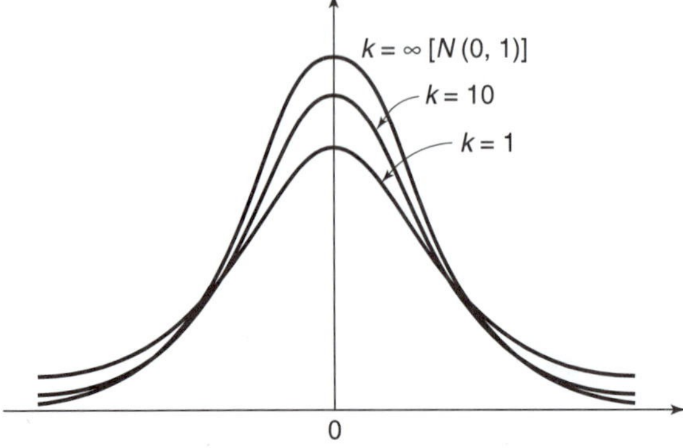

Figure 9-3 Several t distributions.

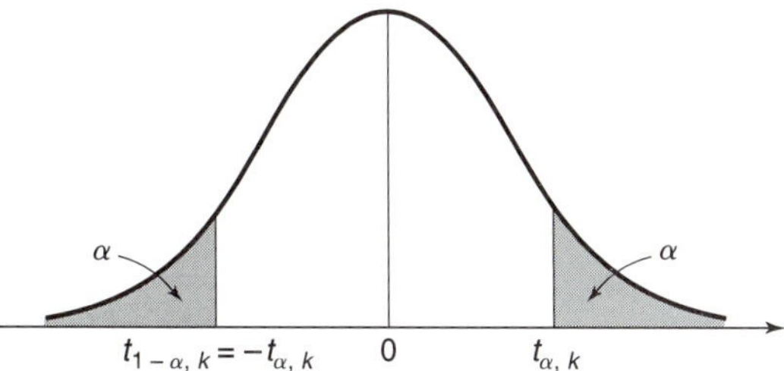

Figure 9-4 Percentage points of the *t* distribution.

$$\frac{\overline{X}-\mu}{S/\sqrt{n}}. \tag{9-19}$$

Dividing both the numerator and denominator of equation 9-19 by σ, we obtain

$$\frac{\dfrac{\overline{X}-\mu}{\sigma}}{S\ \sqrt{n}} = \frac{\dfrac{\overline{X}-\mu}{\sigma/\sqrt{n}}}{\sqrt{S^2/\sigma^2}}.$$

Since $(\overline{X}-\mu)/(\sigma/\sqrt{n}) \sim N(0,1)$ and $S^2/\sigma^2 \sim \chi^2_{n-1}/(n-1)$, and since $\overline{X}$ and S^2 are independent, we see from Theorem 9-3 that

$$T = \frac{\overline{X}-\mu}{S/\sqrt{n}} \tag{9-20}$$

follows a *t* distribution with $v = n - 1$ degrees of freedom. In Chapters 10 and 11 we will use the random variable in equation 9-20 to construct confidence intervals and test hypotheses about the mean of a normal distribution.

9-5 THE *F* DISTRIBUTION

A very useful sampling distribution is the *F* distribution.

Theorem 9-4

Let *W* and *Y* be independent chi-square random variables with *u* and *v* degrees of freedom, respectively. Then the ratio

$$F = \frac{W/u}{Y/v}$$

has the probability density function

$$h(f) = \frac{\Gamma\left(\dfrac{u+v}{2}\right)\left(\dfrac{u}{v}\right)^{u/2} f^{(u/2)-1}}{\Gamma\left(\dfrac{u}{2}\right)\Gamma\left(\dfrac{v}{2}\right)\left[\dfrac{u}{v}f+1\right]^{(u+v)/2}}, \qquad 0 < f < \infty, \tag{9-21}$$

and is said to follow the *F* distribution with *u* degrees of freedom in the numerator and *v* degrees of freedom in the denominator. It is usually abbreviated $F_{u,v}$.

Proof Since *W* and *Y* are independent, their joint probability density distribution is

$$f(w,y)=\frac{w^{(u/2)-1}y^{(v/2)-1}}{2^{u/2}\Gamma\left(\frac{u}{2}\right)2^{v/2}\Gamma\left(\frac{v}{2}\right)}e^{-(w+y)/2}, \qquad 0<w,y<\infty.$$

Proceeding as in Section 4-10, define the new random variable $M = Y$. The inverse solutions of $f = (w/u)/(y/u)$ and $m = y$ are

$$w=\frac{umf}{v}$$

and

$$y=m.$$

Therefore, the Jacobian

$$J=\begin{vmatrix}\frac{um}{v} & \frac{uf}{v}\\ 0 & 1\end{vmatrix}=\frac{u}{v}m.$$

Thus, the joint probability density function is given by

$$g(f,m)=\frac{\frac{u}{v}\left(\frac{u}{v}fm\right)^{(u/2)-1}m^{v/2}}{2^{u/2}\Gamma\left(\frac{u}{2}\right)2^{v/2}\Gamma\left(\frac{v}{2}\right)}e^{-(m/2)((u/v)f+1)}, \qquad 0<f,m<\infty,$$

and since $h(f) = \int_0^\infty g(f, m)\, dm$, we obtain equation 9-21, completing the proof.

The mean and variance of the F distribution are $\mu = v/(v-2)$ for $v > 2$, and

$$\sigma^2=\frac{2v^2(u+v-2)}{u(v-2)^2(v-4)}, \qquad v>4.$$

Several F distributions are shown in Fig. 9-5. The F random variable is nonnegative and the distribution is skewed to the right. The F distribution looks very similar to the chi-square distribution in Fig. 9-1; however, the parameters u and v provide extra flexibility regarding shape.

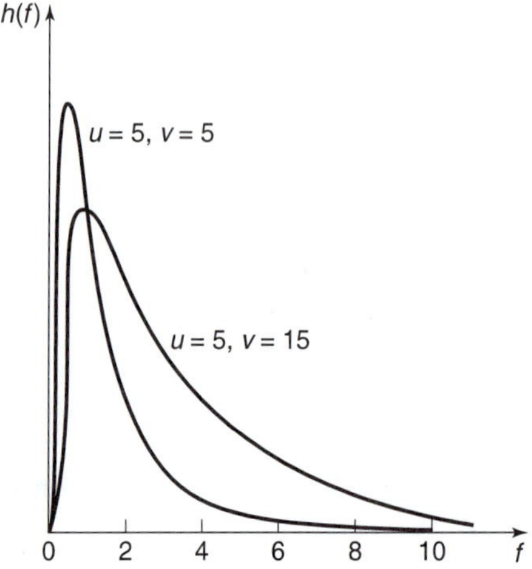

Figure 9-5 The F distribution.

The percentage points of the *F* distribution are given in Table V of the Appendix. Let $F_{\alpha,u,v}$ be the percentage point of the *F* distribution with *u* and *v* degrees of freedom, such that the probability that the random variable *F* exceeds this value is

$$P\{F \geq F_{\alpha,u,v}\} = \int_{F_{\alpha,u,v}}^{\infty} h(f)\,df = \alpha.$$

This is illustrated in Fig. 9-6. For example, if $u = 5$ and $v = 10$, we find from Table V of the Appendix that

$$P\{F \geq F_{0.05,5,10}\} = P\{F \geq 3.33\} = 0.05.$$

That is, the upper 5% point of $F_{5,10}$ is $F_{0.05,5,10} = 3.33$. Table V contains only upper-tail percentage points (values of $F_{\alpha,u,v}$ for $\alpha \leq 0.50$). The lower-tail percentage points $F_{1-\alpha,u,v}$ can be found as follows:

$$F_{1-\alpha,u,v} = \frac{1}{F_{\alpha,v,u}}. \tag{9-22}$$

For example, to find the lower-tail percentage point $F_{0.95,5,10}$, note that

$$F_{0.95,5,10} = \frac{1}{F_{0.05,10,5}} = \frac{1}{4.74} = 0.211.$$

Example 9-8

As an example of a statistic that follows the *F* distribution, suppose we have two normal populations with variances σ_1^2 and σ_2^2. Let independent random samples of sizes n_1 and n_2 be taken from popula tions 1 and 2, respectively, and let S_1^2 and S_2^2 be the sample variances. Then the ratio

$$F = \frac{S_1^2/\sigma_1^2}{S_2^2/\sigma_2^2} \tag{9-23}$$

has an *F* distribution with $n_1 - 1$ numerator degrees of freedom and $n_2 - 1$ denominator degrees of freedom. This follows directly from the facts that $(n_1 - 1)S_1^2/\sigma_1^2 \sim \chi^2_{n_1-1}$ and $(n_2 - 1)S_2^2/\sigma_2^2 \sim \chi^2_{n_2-1}$ and from Theorem 9-4. The random variable in equation 9-23 plays a key role in Chapters 10 and 11, where we address the problems of confidence interval estimation and hypothesis testing about the variances of two independent normal populations.

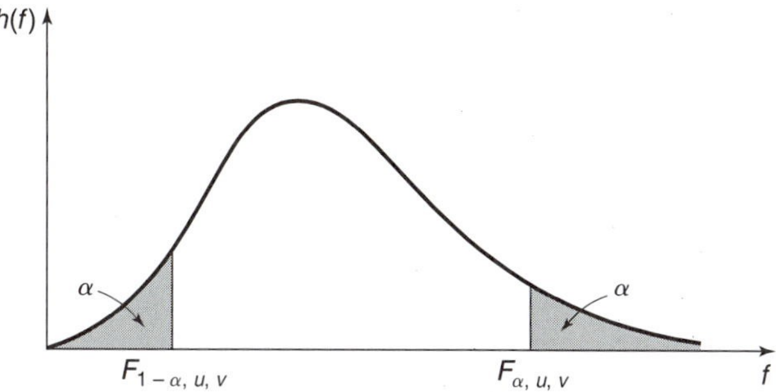

Figure 9-6 Upper and lower percentage points of the *F* distribution.

9-6 SUMMARY

This chapter has presented the concept of random sampling and introduced sampling distributions. In repeated sampling from a population, sample statistics of the sort discussed in Chapter 2 vary from sample to sample, and the probability distribution of such statistics (or functions of the statistics) is called the sampling distribution. The normal, chi-square, Student t, and F distributions have been presented in this chapter and will be employed extensively in later chapters to describe sampling variation.

9-7 EXERCISES

9-1. Suppose that a random variable is normally distributed with mean μ and variance σ^2. Draw a random sample of five observations. What is the joint density function of the sample?

9-2. Transistors have a life that is exponentially distributed with parameter λ. A random sample of n transistors is taken. What is the joint density function of the sample?

9-3. Suppose that X is uniformly distributed on the interval from 0 to 1. Consider a random sample of size 4 from X. What is the joint density function of the sample?

9-4. A lot consists of N transistors, and of these M ($M \le N$) are defective. We randomly select two transistors without replacement from this lot and determine whether they are defective or nondefective. The random variable

$$X_i = \begin{cases} 1 & \text{if the } i\text{th transistor is nondefective,} \\ 0 & \text{if the } i\text{th transistor is defective,} \end{cases} \qquad i = 1, 2.$$

Determine the joint probability function for X_1 and X_2. What are the marginal probability functions for X_1 and X_2? Are X_1 and X_2 independent random variables?

9-5. A population of power supplies for a personal computer has an output voltage that is normally distributed with a mean of 5.00 V and a standard deviation of 0.10 V. A random sample of eight power supplies is selected. Specify the sampling distribution of $\overline{X}$.

9-6. Consider the power supply problem described in Exercise 9-5. What is the standard error of $\overline{X}$?

9-7. Consider the power supply problem described in Exercise 9-5. Suppose that the population standard deviation is unknown. How would you obtain the estimated standard error?

9-8. A procurement specialist has purchased 25 resistors from vendor 1 and 30 resistors from vendor 2. Let $X_{11}, X_{12}, \ldots, X_{1,25}$ represent the vendor 1 observed resistances assumed normally and independently distributed with a mean of 100 Ω and a standard deviation of 1.5 Ω. Similarly, let $X_{21}, X_{22}, \ldots, X_{2,30}$ represent the vendor 2 observed resistances assumed normally and independently distributed with a mean of 105 Ω and a standard deviation of 2.0 Ω. What is the sampling distribution of $\overline{X}_1 - \overline{X}_2$?

9-9. Consider the resistor problem in Exercise 9-8. Find the standard error of $\overline{X}_1 - \overline{X}_2$.

9-10. Consider the resistor problem in Exercise 9-8. If we could not assume that resistance is normally distributed, what could be said about the sampling distribution of $\overline{X}_1 - \overline{X}_2$?

9-11. Suppose that independent random samples of sizes n_1 and n_2 are taken from two normal populations with means μ_1 and μ_2 and variances σ_1^2 and σ_2^2, respectively. If $\overline{X}_1$ and $\overline{X}_2$ are the sample means, find the sampling distribution of the statistic

$$\frac{\overline{X}_1 - \overline{X}_2 - (\mu_1 - \mu_2)}{\sqrt{(\sigma_1^2/n_1) + (\sigma_2^2/n_2)}}.$$

9-12. A manufacturer of semiconductor devices takes a random sample of 100 chips and tests them, classifying each chip as defective or nondefective. Let $X_i = 0(1)$ if the ith chip is nondefective (defective). The sample fraction defective is

$$\hat{p} = \frac{X_1 + X_2 + \cdots + X_{100}}{100}$$

What is the sampling distribution of $\hat{p}$?

9-13. For the semiconductor problem in Exercise 9-12, find the standard error of $\hat{p}$. Also find the estimated standard error of $\hat{p}$.

9-14. Develop the moment-generating function of the chi-square distribution.

9-15. Derive the mean and variance of the chi-square random variable with u degrees of freedom.

9-16. Derive the mean and variance of the t distribution.

9-17. Derive the mean and variance of the F distribution.

9-18. Order Statistics. Let $X_1, X_2, \ldots, X_n$ be a random sample of size n from X, a random variable having distribution function $F(x)$. Rank the elements in order of increasing numerical magnitude, resulting in $X_{(1)}$, $X_{(2)}, \ldots, X_{(n)}$, where $X_{(1)}$ is the smallest sample element ($X_{(1)} = \min\{X_1, X_2, \ldots, X_n\}$) and $X_{(n)}$ is the largest sample element ($X_{(n)} = \max\{X_1, X_2, \ldots, X_n\}$). $X_{(i)}$ is called the ith order statistic. Often, the distribution of some of the order statistics is of interest, particularly the minimum and maximum sample values, $X_{(1)}$ and $X_{(n)}$, respectively. Prove that the distribution functions of $X_{(1)}$ and $X_{(n)}$, denoted respectively by $F_{X_{(1)}}(t)$ and $F_{X_{(n)}}(t)$, are

$$F_{X_{(1)}}(t) = 1 - [1 - F(t)]^n,$$
$$F_{X_{(n)}}(t) = [F(t)]^n.$$

Prove that if X is continuous with probability distribution $f(x)$, then the probability distributions of $X_{(1)}$ and $X_{(n)}$ are

$$f_{X_{(1)}}(t) = n[1 - F(t)]^{n-1} f(t),$$
$$f_{X_{(n)}}(t) = n[F(t)]^{n-1} f(t).$$

9-19. Continuation of Exercise 9-18. Let $X_1, X_2, \ldots, X_n$ be a random sample of a Bernoulli random variable with parameter p. Show that

$$P(X_{(n)} = 1) = 1 - (1 - p)^n,$$
$$P(X_{(1)} = 0) = 1 - p^n.$$

Use the results of Exercise 9-18.

9-20. Continuation of Exercise 9-18. Let $X_1, X_2, \ldots, X_n$ be a random sample of a normal random variable with mean μ and variance σ^2. Using the results of Exercise 9-18, derive the density functions of $X_{(1)}$ and $X_{(n)}$.

9-21. Continuation of Exercise 9-18. Let $X_1, X_2, \ldots, X_{(n)}$ be a random sample of an exponential random variable with parameter λ. Derive the distribution functions and probability distributions for $X_{(1)}$ and $X_{(n)}$. Use the results of Exercise 9-18.

9-22. Let $X_1, X_2, \ldots, X_n$ be a random sample of a continuous random variable. Find

$$E[F(X_{(n)})]$$

and

$$E[F(X_{(1)})].$$

9-23. Using Table III of the Appendix, find the following values:

(a) $\chi^2_{0.95,8}$.
(b) $\chi^2_{0.50,12}$.
(c) $\chi^2_{0.025,20}$.
(d) $\chi^2_{\alpha,10}$ such that $P\{\chi^2_{10} \le \chi^2_{\alpha,10}\} = 0.975$.

9-24. Using Table IV of the Appendix, find the following values:

(a) $t_{0.25,10}$.
(b) $t_{0.25,20}$.
(c) $t_{\alpha,10}$ such that $P\{t_{10} \le t_{\alpha,10}\} = 0.95$.

9-25. Using Table V of the Appendix, find the following values:

(a) $F_{0.25,4,9}$.
(b) $F_{0.05,15,10}$.
(c) $F_{0.95,6,8}$.
(d) $F_{0.90,24,24}$.

9-26. Let $F_{1-\alpha,u,v}$ denote a lower-tail point ($\alpha \le 0.50$) of the $F_{u,v}$ distribution. Prove that $F_{1-\alpha,u,v} = 1/F_{\alpha,v,u}$.

Chapter 10

Parameter Estimation

Statistical inference is the process by which information from sample data is used to draw conclusions about the population from which the sample was selected. The techniques of statistical inference can be divided into two major areas: *parameter estimation* and *hypothesis testing*. This chapter treats parameter estimation, and hypothesis testing is presented in Chapter 11.

As an example of a parameter estimation problem, suppose that civil engineers are analyzing the compressive strength of concrete. There is a natural variability in the strength of each individual concrete specimen. Consequently, the engineers are interested in estimating the average strength for the population consisting of this type of concrete. They may also be interested in estimating the variability of compressive strength in this population. We present methods for obtaining point estimates of parameters such as the population mean and variance and we also discuss methods for obtaining certain kinds of interval estimates of parameters called confidence intervals.

10-1 POINT ESTIMATION

A point estimate of a population parameter is a single numerical value of a statistic that corresponds to that parameter. That is, the point estimate is a unique selection for the value of an unknown parameter. More precisely, if X is a random variable with probability distribution $f(x)$, characterized by the unknown parameter θ, and if $X_1, X_2, \ldots, X_n$ is a random sample of size n from X, then the statistic $\hat{\theta} = h(X_1, X_2, \ldots, X_n)$ corresponding to θ is called the *estimator* of θ. Note that the estimate $\hat{\theta}$ is a random variable, because it is a function of sample data. After the sample has been selected, $\hat{\theta}$ takes on a particular numerical value called the point estimate of θ.

As an example, suppose that the random variable X is normally distributed with unknown mean μ and known variance σ^2. The sample mean $\overline{X}$ is a point estimator of the unknown population mean μ. That is, $\hat{\mu} = \overline{X}$. After the sample has been selected, the numerical value $\bar{x}$ is the point estimate of μ. Thus, if $x_1 = 2.5$, $x_2 = 3.1$, $x_3 = 2.8$, and $x_4 = 3.0$, then the point estimate of μ is

$$\bar{x} = \frac{2.5 + 3.1 + 2.8 + 3.0}{4} = 2.85.$$

Similarly, if the population variance σ^2 is also unknown, a point estimator for σ^2 is the sample variance S^2 and the numerical value $s^2 = 0.07$ calculated from the sample data is the point estimate of σ^2.

Estimation problems occur frequently in engineering. We often need to estimate the following parameters:

- The mean μ of a single population

- The variance σ^2 (or standard deviation σ) of a single population
- The proportion p of items in a population that belong to a class of interest
- The difference between means of two populations, $\mu_1 - \mu_2$
- The difference between two population proportions, $p_1 - p_2$

Reasonable point estimates of these parameters are as follows:

- For μ, the estimate is $\hat{\mu} = \overline{X}$, the sample mean
- For σ^2, the estimate is $\hat{\sigma}^2 = S^2$, the sample variance
- For p, the estimate is $\hat{p} = X/n$, the sample proportion, where X is the number of items in a random sample of size n that belong to the class of interest
- For $\mu_1 - \mu_2$, the estimate is $\hat{\mu}_1 - \hat{\mu}_2 = \overline{X}_1 - \overline{X}_2$, the difference between the sample means of two independent random samples
- For $p_1 - p_2$, the estimate is $\hat{p}_1 - \hat{p}_2$, the difference between two sample proportions computed from two independent random samples

There may be several different potential point estimators for a parameter. For example, if we wish to estimate the mean of a random variable, we might consider the sample mean, the sample median, or perhaps the average of the smallest and largest observations in the sample as point estimators. In order to decide which point estimator of a particular parameter is the best one to use, we need to examine their statistical properties and develop some criteria for comparing estimators.

10-1.1 Properties of Estimators

A desirable property of an estimator is that it should be "close" in some sense to the true value of the unknown parameter. Formally, we say that $\hat{\Theta}$ is an *unbiased* estimator of the parameter θ if

$$E(\hat{\Theta}) = \theta. \tag{10-1}$$

That is, $\hat{\Theta}$ is an unbiased estimator of θ if "on the average" its values are equal to θ. Note that this is equivalent to requiring that the mean of the sampling distribution of $\hat{\Theta}$ be equal to θ.

Example 10-1

Suppose that X is a random variable with mean μ and variance σ^2. Let $X_1, X_2, \ldots, X_n$ be a random sample of size n from X. Show that the sample mean $\overline{X}$ and sample variance S^2 are unbiased estimators of μ and σ^2, respectively. Consider

$$\begin{aligned} E(\overline{X}) &= E\left(\frac{\sum_{i=1}^{n} X_i}{n}\right) \\ &= \frac{1}{n}\sum_{i=1}^{n} E(X_i), \end{aligned}$$

and since $E(X_i) = \mu$, for all $i = 1, 2, \ldots, n$,

$$E(\overline{X}) = \frac{1}{n}\sum_{i=1}^{n}\mu = \mu.$$

Therefore, the sample mean $\overline{X}$ is an unbiased estimator of the population mean μ. Now consider

$$\begin{aligned} E(S^2) &= E\left[\frac{\sum_{i=1}^{n}(X_i - \overline{X})^2}{n-1}\right] \\ &= \frac{1}{n-1}E\sum_{i=1}^{n}(X_i - \overline{X})^2 \\ &= \frac{1}{n-1}E\sum_{i=1}^{n}(X_i^2 + \overline{X}^2 - 2\overline{X}X_i) \\ &= \frac{1}{n-1}E\left(\sum_{i=1}^{n}X_i^2 - n\overline{X}^2\right) \\ &= \frac{1}{n-1}\left[\sum_{i=1}^{n}E(X_i^2) - nE(\overline{X}^2)\right]. \end{aligned}$$

However, since $E(X_i^2) = \mu^2 + \sigma^2$ and $E(\overline{X}^2) = \mu^2 + \sigma^2/n$, we have

$$\begin{aligned} E(S^2) &= \frac{1}{n-1}\left[\sum_{i=1}^{n}(\mu^2 + \sigma^2) - n(\mu^2 + \sigma^2/n)\right] \\ &= \frac{1}{n-1}(n\mu^2 + n\sigma^2 - n\mu^2 - \sigma^2) \\ &= \sigma^2. \end{aligned}$$

Therefore, the sample variance S^2 is an unbiased estimator of the population variance σ^2. However, the sample standard deviation S is a biased estimator of the population standard deviation σ. For large samples this bias is negligible.

The mean square error of an estimator $\hat{\theta}$ is defined as

$$MSE(\hat{\theta}) = E(\hat{\theta} - \theta)^2. \tag{10-2}$$

The mean square error can be rewritten as follows:

$$\begin{aligned} MSE(\hat{\theta}) &= E[\hat{\theta} - E(\hat{\theta})]^2 + [\theta - E(\hat{\theta})]^2 \\ &= V(\hat{\theta}) + (\text{bias})^2. \end{aligned} \tag{10-3}$$

That is, the mean square error of $\hat{\theta}$ is equal to the variance of the estimator plus the squared bias. If $\hat{\theta}$ is an unbiased estimator of θ, the mean square error of $\hat{\theta}$ is equal to the variance of $\hat{\theta}$.

The mean square error is an important criterion for comparing two estimators. Let $\hat{\theta}_1$ and $\hat{\theta}_2$ be two estimators of the parameter θ, and let $MSE(\hat{\theta}_1)$ and $MSE(\hat{\theta}_2)$ be the mean square errors of $\hat{\theta}_1$ and $\hat{\theta}_2$. Then the relative efficiency of $\hat{\theta}_2$ to $\hat{\theta}_1$ is defined as

$$\frac{MSE(\hat{\theta}_1)}{MSE(\hat{\theta}_2)}.$$

If this relative efficiency is less than one, we would conclude that $\hat{\theta}_1$ is a more efficient estimator of θ than is $\hat{\theta}_2$, in the sense that it has smaller mean square error. For example, suppose that we wish to estimate the mean μ of a population. We have a random sample of

n observations $X_1, X_2, \ldots, X_n$, and we wish to compare two possible estimators for μ: the sample mean $\overline{X}$ and a single observation from the sample, say X_i. Note that both $\overline{X}$ and X_i are unbiased estimators of μ; consequently, the mean square error of both estimators is simply the variance. For the sample mean, we have $MSE(\overline{X}) = V(\overline{X}) = \sigma^2/n$, where σ^2 is the population variance; for an individual observation, we have $MSE(X_i) = V(X_i) = \sigma^2$. Therefore, the relative efficiency of X_i to $\overline{X}$ is

$$\frac{MSE(\overline{X})}{MSE(X_i)} = \frac{\sigma^2/n}{\sigma^2} = \frac{1}{n}.$$

Since $(1/n) < 1$ for sample sizes $n \geq 2$, we would conclude that the sample mean is a better estimator of μ than a single observation X_i.

Within the class of unbiased estimators, we would like to find the estimator that has the smallest variance. Such an estimator is called a minimum variance unbiased estimator. Figure 10-1 shows the probability distribution of two unbiased estimators $\hat{\theta}_1$ and $\hat{\theta}_2$, with $\hat{\theta}_1$ having smaller variance than $\hat{\theta}_2$. The estimator $\hat{\theta}_1$ is more likely than $\hat{\theta}_2$ to produce an estimate that is close to the true value of the unknown parameter θ.

It is possible to obtain a lower bound on the variance of all unbiased estimators of θ. Let $\hat{\theta}$ be an unbiased estimator of the parameter θ, based on a random sample of n observations, and let $f(x, \theta)$ denote the probability distribution of the random variable X. Then a lower bound on the variance of $\hat{\theta}$ is[1]

$$V(\hat{\theta}) \geq \frac{1}{nE\left[\frac{d}{d\theta}\ln f(X,\theta)\right]^2}. \qquad (10\text{-}4)$$

This inequality is called the Cramér–Rao lower bound. If an unbiased estimator $\hat{\theta}$ satisfies equation 10-4 as an equality, it is the minimum variance unbiased estimator of θ.

Example 10-2

We will show that the sample mean $\overline{X}$ is the minimum variance unbiased estimator of the mean of a normal distribution with known variance.

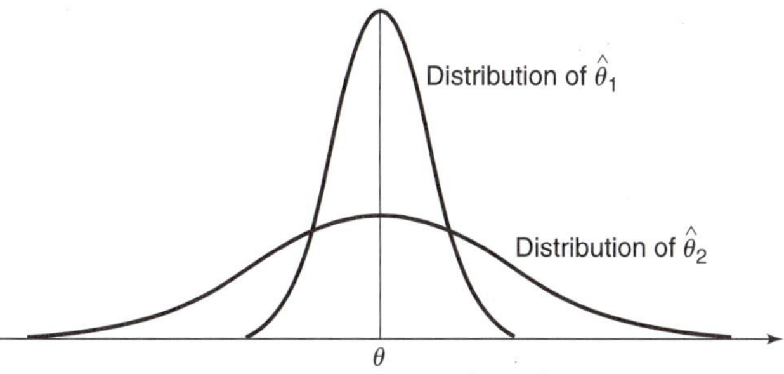

Figure 10-1 The probability distribution of two unbiased estimators, $\hat{\theta}_1$ and $\hat{\theta}_2$.

[1]Certain conditions on the function $f(X, \theta)$ are required for obtaining the Cramér–Rao inequality (for example, see Tucker 1962). These conditions are satisfied by most of the standard probability distributions.

From Example 10-1 we observe that $\overline{X}$ is an unbiased estimator of μ. Note that

$$\ln f(X,\mu) = \ln\left\{\left(\sigma\sqrt{2\pi}\right)^{-1} \exp\left[-\frac{1}{2}\left(\frac{X-\mu}{\sigma}\right)^2\right]\right\}$$

$$= -\ln\left(\sigma\sqrt{2\pi}\right) - \frac{1}{2}\left(\frac{X-\mu}{\sigma}\right)^2.$$

Substituting into equation 10-4 we obtain

$$V\left(\overline{X}\right) \geq \frac{1}{nE\left\{\frac{d}{d\mu}\left[-\ln\left(\sigma\sqrt{2\pi}\right) - \frac{1}{2}\left(\frac{X-\mu}{\sigma}\right)^2\right]\right\}^2}$$

$$= \frac{1}{nE\left[\frac{X-\mu}{\sigma^2}\right]^2}$$

$$= \frac{1}{\frac{nE(X-\mu)^2}{\sigma^4}}$$

$$= \frac{1}{\frac{n\sigma^2}{\sigma^4}}$$

$$= \frac{\sigma^2}{n}.$$

Since we know that, in general, the variance of the sample mean is $V(\overline{X}) = \sigma^2/n$, we see that $V(\overline{X})$ satisfies the Cramér–Rao lower bound as an equality. Therefore $\overline{X}$ is the minimum variance unbiased estimator of μ for the normal distribution where σ^2 is known.

Sometimes we find that biased estimators are preferable to unbiased estimators because they have smaller mean square error. That is, we can reduce the variance of the estimator considerably by introducing a relatively small amount of bias. So long as the reduction in variance is greater than the squared bias, an improved estimator in the mean square error sense will result. For example, Fig. 10-2 shows the probability distribution of a biased estimator $\hat{\theta}_1$ with smaller variance than the unbiased estimator $\hat{\theta}_2$. An estimate based on $\hat{\theta}_1$ would more likely be close to the true value of θ than would an estimate based on $\hat{\theta}_2$. We will see an application of biased estimation in Chapter 15.

An estimator $\hat{\theta}^*$ that has a mean square error that is less than or equal to the mean square error of any other estimator $\hat{\theta}$, for all values of the parameter θ, is called an *optimal* estimator of θ.

Another way to define the closeness of an estimator $\hat{\theta}$ to the parameter θ is in terms of *consistency.* If $\hat{\theta}_n$ is an estimator of θ based on a random sample of size n, we say that $\hat{\theta}_n$ is consistent for θ if, for $\varepsilon > 0$,

$$\lim_{n\to\infty} P\left(\left|\hat{\theta}_n - \theta\right| < \varepsilon\right) = 1. \tag{10-5}$$

Consistency is a large-sample property, since it describes the limiting behavior of the estimator $\hat{\theta}$ as the sample size tends to infinity. It is usually difficult to prove that an estimator

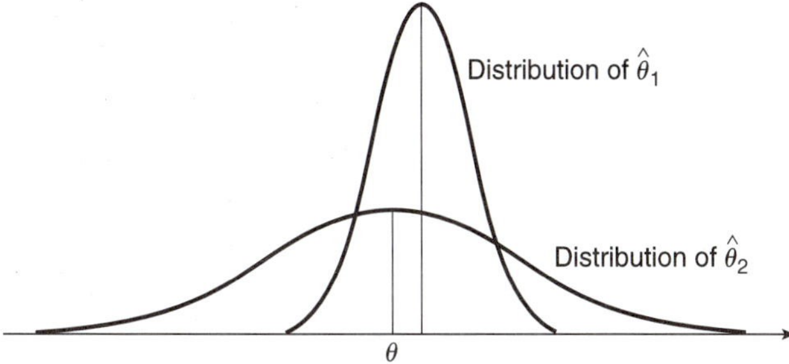

Figure 10-2 A biased estimator, $\hat{\theta}_1$, that has smaller variance than the unbiased estimator, $\hat{\theta}_2$.

is consistent using the definition of equation 10-5. However, estimators whose mean square error (or variance, if the estimator is unbiased) tends to zero as the sample size approaches infinity are consistent. For example, $\overline{X}$ is a consistent estimator of the mean of a normal distribution, since $\overline{X}$ is unbiased and $\lim_{n\to\infty} V(\overline{X}) = \lim_{n\to\infty} (\sigma^2/n) = 0$.

10-1.2 The Method of Maximum Likelihood

One of the best methods for obtaining a point estimator is the method of maximum likelihood. Suppose that X is a random variable with probability distribution $f(x, \theta)$, where θ is a single unknown parameter. Let $X_1, X_2, \ldots, X_n$ be the observed values in a random sample of size n. Then the *likelihood function* of the sample is

$$L(\theta) = f(x_1, \theta) \cdot f(x_2, \theta) \cdot \cdots \cdot f(x_n, \theta). \tag{10-6}$$

Note that the likelihood function is now a function of only the unknown parameter θ. The *maximum likelihood estimator* (MLE) of θ is the value of θ that maximizes the likelihood function $L(\theta)$. Essentially, the maximum likelihood estimator is the value of θ that maximizes the probability of occurrence of the sample results.

Example 10-3

Let X be a Bernoulli random variable. The probability mass function is

$$\begin{aligned} p(x) &= p^x(1-p)^{1-x}, & x &= 0, 1, \\ &= 0, & &\text{otherwise,} \end{aligned}$$

where p is the parameter to be estimated. The likelihood function of a sample of size n would be

$$L(p) = \prod_{i=1}^{n} p^{x_i}(1-p)^{1-x_i} = p^{\sum_{i=1}^{n} x_i}(1-p)^{n-\sum_{i=1}^{n} x_i}.$$

We observe that if $\hat{p}$ maximizes $L(p)$ then $\hat{p}$ also maximizes $\ln L(p)$ since the logarithm is a monotonically increasing function. Therefore,

$$\ln L(p) = \sum_{i=1}^{n} x_i \ln p + \left(n - \sum_{i=1}^{n} x_i\right) \ln(1-p).$$

Now

$$\frac{d \ln L(p)}{dp} = \frac{\sum_{i=1}^{n} x_i}{p} - \frac{\left(n - \sum_{i=1}^{n} x_i\right)}{1-p}.$$

Equating this zero and solving for p yields the MLE $\hat{p}$, as

$$\hat{p} = \frac{1}{n} \sum_{i=1}^{n} X_i = \overline{X},$$

an intuitively pleasing answer. Of course, one should also perform a second derivative test, but we have foregone that here.

Example 10-4

Let X be normally distributed with unknown mean μ and known variance σ^2. The likelihood function of a sample of size n is

$$\begin{aligned} L(\mu) &= \prod_{i=1}^{n} \frac{1}{\sigma\sqrt{2\pi}} e^{-(x_i-\mu)^2/2\sigma^2} \\ &= \frac{1}{\left(2\pi\sigma^2\right)^{n/2}} e^{-\left(1/2\sigma^2\right)\sum_{i=1}^{n}(x_i-\mu)^2}. \end{aligned}$$

Now

$$\ln L(\mu) = -(n/2)\ln\left(2\pi\sigma^2\right) - \left(2\sigma^2\right)^{-1} \sum_{i=1}^{n} (x_i - \mu)^2$$

and

$$\frac{d \ln L(\mu)}{d\mu} = \left(\sigma^2\right)^{-1} \sum_{i=1}^{n} (x_i - \mu).$$

Equating this last result to zero and solving for μ yields

$$\hat{\mu} = \frac{1}{n} \sum_{i=1}^{n} X_i = \overline{X}$$

as the MLE of μ.

It may not always be possible to use calculus methods to determine the maximum of $L(\theta)$. This is illustrated in the following example.

Example 10-5

Let X be uniformly distributed on the interval 0 to a. The likelihood function of a random sample X_1, X_2, …, X_n of size n is

$$L(a) = \prod_{i=1}^{n} \frac{1}{a} = \frac{1}{a^n}.$$

Note that the slope of this function is not zero anywhere, so we cannot use calculus methods to find the maximum likelihood estimator $\hat{a}$. However, notice that the likelihood function increases as a decreases. Therefore, we would maximize $L(a)$ by setting $\hat{a}$ to the smallest value that it could reasonably assume. Clearly, a can be no smaller than the largest sample value, so we would use the largest observation as $\hat{a}$. Thus, $\hat{a} = \max_i X_i$ is the MLE for a.

The method of maximum likelihood can be used in situations where there are several unknown parameters, say $\theta_1, \theta_2, \ldots, \theta_k$, to estimate. In such cases, the likelihood function is a function of the k unknown parameters $\theta_1, \theta_2, \ldots, \theta_k$ and the maximum likelihood estimators $\{\hat{\theta}_i\}$ would be found by equating the k first partial derivatives $\partial L(\theta_1, \theta_2, \ldots, \theta_k)/\partial\theta_i$, $i = 1, 2, \ldots, k$, to zero and solving the resulting system of equations.

Example 10-6

Let X be normally distributed with mean μ and variance σ^2, where both μ and σ^2 are unknown. Find the maximum likelihood estimators of μ and σ^2. The likelihood function for a random sample of size n is

$$L(\mu,\sigma^2) = \prod_{i=1}^{n} \frac{1}{\sigma\sqrt{2\pi}} e^{-(x_i-\mu)^2/2\sigma^2}$$
$$= \frac{1}{(2\pi\sigma^2)^{n/2}} e^{-(1/2\sigma^2)\sum_{i=1}^{n}(x_i-\mu)^2}$$

and

$$\ln L(\mu,\sigma^2) = -\frac{n}{2}\ln(2\pi\sigma^2) - \frac{1}{2\sigma^2}\sum_{i=1}^{n}(x_i-\mu)^2.$$

Now

$$\frac{\partial \ln L(\mu,\sigma^2)}{\partial\mu} = \frac{1}{\sigma^2}\sum_{i=1}^{n}(x_i-\mu) = 0,$$
$$\frac{\partial \ln L(\mu,\sigma^2)}{\partial(\sigma^2)} = \frac{-n}{2\sigma^2} + \frac{1}{2\sigma^4}\sum_{i=1}^{n}(x_i-\mu)^2 = 0.$$

The solutions to the above equations yield the maximum likelihood estimators

$$\hat{\mu} = \frac{1}{n}\sum_{i=1}^{n} X_i = \overline{X}$$

and

$$\hat{\sigma}^2 = \frac{1}{n}\sum_{i=1}^{n}(X_i - \overline{X})^2,$$

which is closely related to the unbiased sample variance S^2. Namely, $\hat{\sigma}^2 = ((n-1)/n)S^2$.

Maximum likelihood estimators are not necessarily unbiased (see the maximum likelihood estimator of σ^2 in Example 10-6), but they usually may be easily modified to make them unbiased. Further, the bias approaches zero for large samples. In general, maximum likelihood estimators have good large-sample or *asymptotic* properties. Specifically, they are asymptotically normally distributed, unbiased, and have a variance that approaches the Cramér–Rao lower bound for large n. More precisely, if $\hat{\theta}$ is the maximum likelihood estimator for θ, then $\sqrt{n}(\hat{\theta}-\theta)$ is normally distributed with mean zero and variance

$$V\left[\sqrt{n}(\hat{\theta}-\theta)\right] = V(\sqrt{n}\hat{\theta}) \simeq \frac{1}{E\left[\frac{d}{d\theta}\ln f(X,\theta)\right]^2}$$

for large n. Maximum likelihood estimators are also consistent. In addition, they possess the invariance property; that is, if $\hat{\theta}$ is the maximum likelihood estimator of θ and $u(\theta)$ is a function of θ that has a single-valued inverse, then the maximum likelihood estimator of $u(\theta)$ is $u(\hat{\theta})$.

It can be shown graphically that the maximum of the likelihood will occur at the value of the maximum likelihood estimator. Consider a sample of size $n = 10$ from a normal distribution:

14.15, 32.07, 32.30, 25.01, 21.86, 23.70, 25.92, 25.19, 22.59, 26.47.

Assume that the population variance is known to be 4. The MLE for the mean, μ, of a normal distribution has been shown to be $\overline{X}$. For this set of data, $\overline{x} = 25$. Figure 10-3 displays the log-likelihood for various values of the mean. Notice that the maximum value of the log-likelihood function occurs at approximately $\overline{x} = 25$. Sometimes, the likelihood function is relatively flat in the region around the maximum. This may be due to the size of the sample taken from the population. A small sample size can lead to a fairly flat log likelihood, implying less precision in the estimate of the parameter of interest.

10-1.3 The Method of Moments

Suppose that X is either a continuous random variable with probability density $f(x; \theta_1, \theta_2, \ldots, \theta_k)$ or a discrete random variable with distribution $p(x; \theta_1, \theta_2, \ldots, \theta_k)$ characterized by k unknown parameters. Let $X_1, X_2, \ldots, X_n$ be a random sample of size n from X, and define the first k sample moments about the origin as

$$m_t' = \frac{1}{n}\sum_{i=1}^{n} X_i^t, \qquad t = 1, 2, \ldots, k. \tag{10-7}$$

The first k population moments about the origin are

$$\begin{aligned} \mu_t' = E\left(X^t\right) &= \int_{-\infty}^{\infty} x^t f\left(x;\theta_1,\theta_2,\ldots,\theta_k\right)dx, \qquad t = 1,2,\ldots,k, \qquad X \text{ continuous,} \\ &= \sum_{x\in R_x} x^t p\left(x;\theta_1,\theta_2,\ldots,\theta_k\right), \qquad t = 1,2,\ldots,k, \qquad X \text{ discrete.} \end{aligned} \tag{10-8}$$

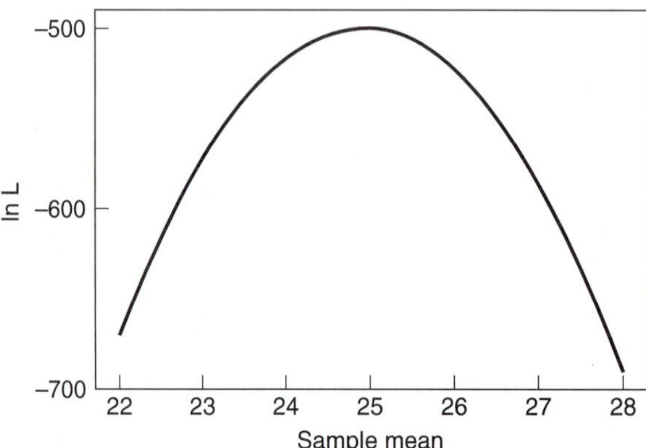

Figure 10-3 Log likelihood for various means.

The population moments $\{\mu_t'\}$ will, in general, be functions of the k unknown parameters $\{\theta_i\}$. Equating sample moments and population moments will yield k simultaneous equations in k unknowns (the θ_i); that is,

$$\mu_t' = m_t', \qquad t = 1, 2, \ldots, k. \tag{10-9}$$

The solution to equation 10-9, denoted $\hat{\theta}_1, \hat{\theta}_2, \ldots, \hat{\theta}_k$, yields the moment estimators of $\theta_1, \theta_2, \ldots, \theta_k$.

Example 10-7

Let $X \sim N(\mu, \sigma^2)$ where μ and σ^2 are unknown. To derive estimators for μ and σ^2 by the method of moments, recall that for the normal distribution

$$\mu_1' = \mu,$$
$$\mu_2' = \sigma^2 + \mu^2.$$

The sample moments are $m_1' = (1/n)\sum_{i=1}^{n} X_i$ and $m_2' = (1/n)\sum_{i=1}^{n} X_i^2$. From equation 10-9 we obtain

$$\mu = \frac{1}{n}\sum_{i=1}^{n} X_i,$$

$$\sigma^2 + \mu^2 = \frac{1}{n}\sum_{i=1}^{n} X_i^2,$$

which have the solution

$$\hat{\mu} = \frac{1}{n}\sum_{i=1}^{n} X_i = \overline{X},$$

$$\hat{\sigma}^2 = \frac{1}{n}\left(\sum_{i=1}^{n} X_i^2 - n\overline{X}^2\right) = \frac{1}{n}\sum_{i=1}^{n}\left(X_i - \overline{X}\right)^2.$$

Example 10-8

Let X be uniformly distributed on the interval $(0, a)$. To find an estimator of a by the method of moments, we note that the first population moment about zero is

$$\mu_1' = \int_0^a x\frac{1}{a}\,dx = \frac{a}{2}.$$

The first sample moment is just $\overline{X}$. Therefore,

$$\hat{a} = 2\overline{X},$$

or the moment estimator of a is just twice the sample mean.

The method of moments often yields estimators that are reasonably good. In Example 10-7, for instance, the moment estimators are identical to the maximum likelihood estimators. In general, moment estimators are asymptotically normally distributed (approximately) and consistent. However, their variance may be larger than the variance of estimators derived by other methods, such as the method of maximum likelihood. Occasionally, the method of moments yields estimators that are very poor, as in Example 10-8. The estimator in that example does not always generate an estimate that is compatible with our knowledge of the situation. For example, if our sample observations were $x_1 = 60$, $x_2 = 10$, and $x_3 = 5$, then $\hat{a} = 50$, which is unreasonable, since we know that $a \geq 60$.

10-1.4 Bayesian Inference

In the preceding chapters we made an extensive study of the use of probability. Until now, we have interpreted these probabilities in the frequency sense; that is, they refer to an experiment that can be repeated an indefinite number of times, and if the probability of occurrence of an event A is 0.6, then we would expect A to occur in about 60% of the experimental trials. This frequency interpretation of probability is often called the objectivist or classical viewpoint.

Bayesian inference requires a different interpretation of probability, called the subjective viewpoint. We often encounter subjective probabilistic statements, such as "There is a 30% chance of rain today." Subjective statements measure a person's "degree of belief" concerning some event, rather than a frequency interpretation. Bayesian inference requires us to make use of subjective probability to measure our degree of belief about a state of nature. That is, we must specify a probability distribution to describe our degree of belief about an unknown parameter. This procedure is totally unlike anything we have discussed previously. Until now, parameters have been treated as unknown constants. Bayesian inference requires us to think of parameters as *random variables.*

Suppose we let $f(\theta)$ be the probability distribution of the parameter or state of nature θ. The distribution $f(\theta)$ summarizes our objective information about θ prior to obtaining sample information. Obviously, if we are reasonably certain about the value of θ, we will choose $f(\theta)$ with a small variance, while if we are less certain about θ, $f(\theta)$ will be chosen with a larger variance. We call $f(\theta)$ the *prior distribution* of θ.

Now consider the distribution of the random variable X. The distribution of X we denote $f(x|\theta)$ to indicate that the distribution depends on the unknown parameter θ. Suppose we take a random sample from X, say $X_1, X_2, \ldots, X_n$. The joint density of *likelihood* of the sample is

$$f(x_1, x_2, \ldots, x_n|\theta) = f(x_1|\theta)f(x_2|\theta) \cdots f(x_n|\theta).$$

We define the *posterior distribution* of θ as the conditional distribution of θ, given the sample results. This is just

$$f(\theta|x_1, x_2, \ldots, x_n) = \frac{f(x_1, x_2, \ldots, x_n; \theta)}{f(x_1, x_2, \ldots, x_n)}. \tag{10-10}$$

The joint distribution of the sample and θ in the numerator of equation 10-10 is the product of the prior distribution of θ and the likelihood, or

$$f(x_1, x_2, \ldots, x_n; \theta) = f(\theta) \cdot f(x_1, x_2, \ldots, x_n|\theta).$$

The denominator of equation 10-10, which is the marginal distribution of the sample, is just a normalizing constant obtained by

$$f(x_1, x_2, \ldots, x_n) = \begin{cases} \int_{-\infty}^{\infty} f(\theta) f(x_1, x_2, \ldots, x_n|\theta)\,d\theta, & x \text{ continuous,} \\ \sum_{\theta} f(\theta) f(x_1, x_2, \ldots, x_n|\theta), & x \text{ discrete.} \end{cases} \tag{10-11}$$

Consequently, we may write the posterior distribution of θ as

$$f(\theta|x_1, x_2, \ldots, x_n) = \frac{f(\theta) f(x_1, x_2, \ldots, x_n|\theta)}{f(x_1, x_2, \ldots, x_n)}. \tag{10-12}$$

We note that Bayes' theorem has been used to transform or update the prior distribution to the posterior distribution. The posterior distribution reflects our degree of belief about θ given the sample information. Furthermore, the posterior distribution is proportional to the product of the prior distribution and the likelihood, the constant of proportionality being the normalizing constant $f(x_1, x_2, \ldots, x_n)$.

Thus, the posterior density for θ expresses our degree of belief about the value of θ given the result of the sample.

Example 10-9

The time to failure of a transistor is known to be exponentially distributed with parameter λ. For a random sample of n transistors, the joint density of the sample elements, given λ, is

$$f(x_1, x_2, \ldots, x_n | \lambda) = \lambda^n e^{-\lambda \sum_{i=1}^{n} x_i}.$$

Suppose we feel that the prior distribution for λ is also exponential,

$$\begin{aligned} f(\lambda) &= ke^{-k\lambda}, \qquad && \lambda > 0, \\ &= 0, && \text{otherwise,} \end{aligned}$$

where k would be chosen depending on the exact knowledge or degree of belief we have about the value of λ. The joint density of the sample and λ is

$$f(x_1, x_2, \ldots, x_n; \lambda) = k\lambda^n e^{-\lambda(\sum x_i + k)}.$$

and the marginal density of the sample is

$$\begin{aligned} f(x_1, x_2, \ldots, x_n) &= \int_0^\infty k\lambda^n e^{-\lambda(\sum x_i + k)} d\lambda \\ &= \frac{k\Gamma(n+1)}{(\sum x_i + k)^{n+1}}. \end{aligned}$$

Therefore, the posterior density for λ, by equation 10-12, is

$$f(\lambda | x_1, x_2, \ldots, x_n) = \frac{1}{\Gamma(n+1)} (\sum x_i + k)^{n+1} \lambda^n e^{-(\sum x_i + k)},$$

and we see that the posterior density for λ is a gamma distribution with parameters $n + 1$ and $\Sigma X_i + k$.

10-1.5 Applications to Estimation

In this section, we discuss the application of Bayesian inference to the problem of estimating an unknown parameter of a probability distribution. Let $X_1, X_2, \ldots, X_n$ be a random sample of the random variable X having density $f(x|\theta)$. We want to obtain a point estimate of θ. Let $f(\theta)$ be the prior distribution for θ and let $\ell(\hat{\theta}; \theta)$ be the *loss function*. The loss function is a penalty function reflecting the "payment" we must make for misidentifying θ by a realization of its point estimator $\hat{\theta}$. Common choices for $\ell(\hat{\theta}; \theta)$ are $(\hat{\theta} - \theta)^2$ and $|\hat{\theta} - \theta|^2$. Generally, the less accurate a realization of $\hat{\theta}$ is, the more we must pay. In conjunction with a particular loss function, the *risk* is defined as the expected value of the loss function with respect to the random variables $X_1, X_2, \ldots, X_n$ comprising $\hat{\theta}$. In other words, the risk is

$$\begin{aligned} R(d; \theta) &= E\left[\ell(\hat{\theta}; \theta)\right] \\ &= \int_{-\infty}^{\infty} \int_{-\infty}^{\infty} \cdots \int_{-\infty}^{\infty} \ell\{d(x_1, x_2, \ldots, x_n); \theta\} f(x_1, x_2, \ldots, x_n | \theta)\, dx_1 dx_2 \cdots dx_n, \end{aligned}$$

where the function $d(x_1, x_2, \ldots, x_n)$, an alternate notation for the estimator $\hat{\theta}$, is simply a function of the observations. Since θ is considered to be a random variable, the risk is itself a random variable. We would like to find the function d that minimizes the *expected* risk. We write the expected risk as

$$B(d) = E[R(d;\theta)] = \int_{-\infty}^{\infty} R(d;\theta) f(\theta)\, d\theta$$
$$= \int_{-\infty}^{\infty} \left\{ \int_{-\infty}^{\infty} \cdots \int_{-\infty}^{\infty} \ell\{d(x_1, x_2, \ldots, x_n); \theta\} f(x_1, x_2, \ldots, x_n | \theta)\, dx_1 dx_2 \cdots dx_n \right\} f(\theta)\, d\theta. \tag{10-13}$$

We define the *Bayes estimator* of the parameter θ to be the function d of the sample X_1, X_2, …, X_n that minimizes the expected risk. On interchanging the order of integration in equation 10-13 we obtain

$$B(d) = \int_{-\infty}^{\infty} \cdots \int_{-\infty}^{\infty} \left\{ \int_{-\infty}^{\infty} \ell\{d(x_1, x_2, \ldots, x_n); \theta\} f(x_1, x_2, \ldots, x_n | \theta) f(\theta)\, d\theta \right\} dx_1 dx_2 \cdots dx_n. \tag{10-14}$$

The function B will be minimized if we can find a function d that minimizes the quantity within the large braces in equation 10-14 for every set of the x values. That is, the Bayes estimator of θ is a function d of the x_i that minimizes

$$\int_{-\infty}^{\infty} \ell\{d(x_1, x_2, \ldots, x_n); \theta\} f(x_1, x_2, \ldots, x_n | \theta) f(\theta)\, d\theta$$
$$= \int_{-\infty}^{\infty} \ell(\hat{\theta}; \theta) f(x_1, x_2, \ldots, x_n; \theta)\, d\theta$$
$$= f(x_1, x_2, \ldots, x_n) \int_{-\infty}^{\infty} \ell(\hat{\theta}; \theta) f(\theta | x_1, x_2, \ldots, x_n)\, d\theta. \tag{10-15}$$

Thus, the Bayes estimator of θ is the value $\hat{\theta}$ that minimizes

$$\int_{-\infty}^{\infty} \ell(\hat{\theta}; \theta) f(\theta | x_1, x_2, \ldots, x_n)\, d\theta. \tag{10-16}$$

If the loss function $\ell(\hat{\theta}; \theta)$ is the squared-error loss $(\hat{\theta} - \theta)^2$, then we may show that the Bayes estimator of θ, say $\hat{\theta}$, is the mean of the posterior density for θ (refer to Exercise 10-78).

Example 10-10

Consider the situation in Example 10-9, where it was shown that if the random variable X is exponentially distributed with parameter λ, and if the prior distribution for λ is exponential with parameter k, then the posterior distribution for λ is a gamma distribution, with parameters $n + 1$ and $\sum_{i=1}^{n} X_i + k$. Therefore, if a squared-error loss function is assumed, the Bayes estimator for λ is the mean of this gamma distribution,

$$\hat{\lambda} = \frac{n+1}{\sum_{i=1}^{n} X_i + k}.$$

Suppose that in the time-to-failure problem in Example 10-9, a reasonable exponential prior distribution for λ has parameter $k = 140$. This is equivalent to saying that the prior estimate for λ is 0.07142. A random sample of size $n = 10$ yields $\sum_{i=1}^{10} x_i = 1500$. The Bayes estimate of λ is

$$\hat{\lambda} = \frac{n+1}{\sum_{i=1}^{10} x_i + k} = \frac{10+1}{1500+140} = 0.06707.$$

We may compare this with the results that would have been obtained by classical methods. The maximum likelihood estimator of the parameter λ in an exponential distribution is

$$\lambda^* = \frac{n}{\sum_{i=1}^{n} X_i}.$$

Consequently, the maximum likelihood estimate of λ, based on the foregoing sample data, is

$$\lambda^* = \frac{n}{\sum_{i=1}^{n} x_i} = \frac{10}{1500} = 0.06667.$$

Note that the results produced by the two methods differ somewhat. The Bayes estimate is slightly closer to the prior estimate than is the maximum likelihood estimate.

Example 10-11

Let $X_1, X_2, \ldots, X_n$ be a random sample from the normal density with mean μ and variance 1, where μ is unknown. Assume that the prior density for μ is normal with mean 0 and variance 1; that is,

$$f(\mu) = \frac{1}{\sqrt{2\pi}} e^{-(1/2)\mu^2} \qquad -\infty < \mu < \infty.$$

The joint conditional density of the sample given μ is

$$\begin{aligned} f(x_1, x_2 \ldots, x_n | \mu) &= \frac{1}{(2\pi)^{n/2}} e^{-(1/2)\Sigma(x_i - \mu)^2} \\ &= \frac{1}{(2\pi)^{n/2}} e^{-(1/2)\left(\Sigma x_i^2 - 2\mu \Sigma x_i + n\mu^2\right)}. \end{aligned}$$

Thus, the joint density of the sample and μ is

$$f(x_1, x_2 \ldots, x_n; \mu) = \frac{1}{(2\pi)^{(n+1)/2}} \exp\left\{-\frac{1}{2}\left[\Sigma x_i^2 + (n+1)\mu^2 - 2\mu n\bar{x}\right]\right\}.$$

The marginal density of the sample is

$$f(x_1, x_2 \ldots, x_n) = \frac{1}{(2\pi)^{(n+1)/2}} \exp\left\{-\frac{1}{2}\Sigma x_i^2\right\} \int_{-\infty}^{\infty} \exp\left\{-\frac{1}{2}\left[(n+1)\mu^2 - 2\mu n\bar{x}\right]\right\} d\mu.$$

By completing the square in the exponent under the integral, we obtain

$$\begin{aligned} f(x_1, x_2 \ldots, x_n) &= \frac{1}{(2\pi)^{n/2}} \exp\left[-\frac{1}{2}\left(\Sigma x_i^2 - \frac{n^2\bar{x}^2}{n+1}\right)\right] \times \left[\frac{1}{(2\pi)^{1/2}} \int_{-\infty}^{\infty} \exp\left[-\frac{1}{2}(n+1)\left(\mu - \frac{n\bar{x}}{n+1}\right)^2\right] d\mu\right] \\ &= \frac{1}{(n+1)^{1/2}(2\pi)^{n/2}} \exp\left[-\frac{1}{2}\left(\Sigma x_i^2 - \frac{n^2\bar{x}^2}{n+1}\right)\right] \end{aligned}$$

using the fact that the integral is $(2\pi)^{1/2}/(n+1)^{1/2}$ (since a normal density has to integrate to 1). Now the posterior density for μ is

$$\begin{aligned} f(\mu | x_1, x_2 \ldots, x_n) &= \frac{(2\pi)^{-(n+1)/2} \exp\left\{-\frac{1}{2}\left[\Sigma x_i^2 + (n+1)\mu^2 - 2n\bar{x}\mu\right]\right\}}{(2\pi)^{-n/2}(n+1)^{-1/2} \exp\left\{-\frac{1}{2}\left(\Sigma x_i^2 - \frac{n^2\bar{x}^2}{n+1}\right)\right\}} \\ &= \frac{(n+1)^{1/2}}{(2\pi)^{1/2}} \exp\left\{-\frac{1}{2}(n+1)\left[\mu^2 - \frac{2n\bar{x}\mu}{n+1} + \frac{n^2\bar{x}^2}{(n+1)^2}\right]\right\} \\ &= \frac{(n+1)^{1/2}}{(2\pi)^{1/2}} \exp\left\{-\frac{1}{2}(n+1)\left[\mu - \frac{n\bar{x}}{n+1}\right]^2\right\}. \end{aligned}$$

Therefore, the posterior density for μ is a normal density with mean $n\bar{X}/(n+1)$ and variance $(n+1)^{-1}$. If the loss function $\ell(\hat{\mu}; \mu)$ is squared error, the Bayes estimator of μ is

$$\hat{\mu} = \frac{n\overline{X}}{n+1} = \frac{\sum_{i=1}^{n} X_i}{n+1}.$$

There is a relationship between the Bayes estimator for a parameter and the maximum likelihood estimator of the same parameter. For large sample sizes the two are nearly equivalent. In general, the difference between the two estimators is small compared to $1/\sqrt{n}$. In practical problems, a moderate sample size will produce approximately the same estimate by either the Bayes or the maximum likelihood method, if the sample results are consistent with the assumed prior information. If the sample results are inconsistent with the prior assumptions, then the Bayes estimate may differ considerably from the maximum likelihood estimate. In these circumstances, if the sample results are accepted as being correct, the prior information must be incorrect. The maximum likelihood estimate would then be the better estimate to use.

If the sample results do not agree with the prior information, the Bayes estimator will tend to produce an estimate that is between the maximum likelihood estimate and the prior assumptions. If there is more inconsistency between the prior information and the sample, there will be a greater difference between the two estimates. For an illustration of this, refer to Example 10-10.

10-1.6 Precision of Estimation: The Standard Error

When we report the value of a point estimate, it is usually necessary to give some idea of its precision. The *standard error* is the usual measure of precision employed. If $\hat{\theta}$ is an estimator of θ, then the *standard error of* $\hat{\theta}$ is just the standard deviation of $\hat{\theta}$, or

$$\sigma_{\hat{\theta}} = \sqrt{V(\hat{\theta})}. \tag{10-17}$$

If $\sigma_{\hat{\theta}}$ involves any unknown parameters, then if we substitute estimates of these parameters into equation 10-17, we obtain the *estimated standard error of* $\hat{\theta}$, say $\hat{\sigma}_{\hat{\theta}}$. A small standard error implies that a relatively precise estimate has been reported.

Example 10-12

An article in the *Journal of Heat Transfer* (Trans. ASME, Ses. C, 96, 1974, p. 59) describes a method of measuring the thermal conductivity of Armco iron. Using a temperature of 100°F and a power input of 550 W, the following 10 measurements of thermal conductivity (in Btu/hr–ft–°F) were obtained:

$$41.60,\ 41.48,\ 42.34,\ 41.95,\ 41.86,$$
$$42.18,\ 41.72,\ 42.26,\ 41.81,\ 42.04.$$

A point estimate of mean thermal conductivity at 100°F and 550 W is the sample mean, or

$$\bar{x} = 41.924 \text{ Btu/hr–ft–°F}.$$

The standard error of the sample mean is $\sigma_{\bar{x}} = \sigma/\sqrt{n}$, and since σ is unknown, we may replace it with the sample standard deviation to obtain the estimated standard error of $\bar{x}$,

$$\hat{\sigma}_{\bar{x}} = \frac{s}{\sqrt{n}} = \frac{0.284}{\sqrt{10}} = 0.0898.$$

Notice that the standard error is about 0.2% of the sample mean, implying that we have obtained a relatively precise point estimate of thermal conductivity.

When the distribution of $\hat{\theta}$ is unknown or complicated, the standard error of $\hat{\theta}$ may be difficult to estimate using standard statistical theory. In this case, a computer-intensive technique called the *bootstrap* can be used. Efron and Tibshirani (1993) provide an excellent introduction to the bootstrap technique.

Suppose that the standard error of $\hat{\theta}$ is denoted $\sigma_{\hat{\theta}}$. Further, assume the population probability density function is given by $f(x;\theta)$. A bootstrap estimate of $\sigma_{\hat{\theta}}$ can be easily constructed.

1. Given a random sample from $f(x;\hat{\theta})$, $x_1, x_2, \ldots, x_n$, estimate θ, denoted by $\hat{\theta}$.
2. Using the estimate $\hat{\theta}$, generate a sample of size n from the distribution $f(x;\hat{\theta})$. This is the bootstrap sample.
3. Using the bootstrap sample, estimate θ. This estimate we denote $\hat{\theta}_i^*$.
4. Generate B bootstrap samples to obtain bootstrap estimates, $\hat{\theta}_i^*$, for $i = 1, 2, \ldots, B$ (B = 100 or 200 is often used).
5. Let $\overline{\theta}^* = \sum_{i=1}^{B} \hat{\theta}_i^* \Big/ B$ represent the sample mean of the bootstrap estimates.
6. The bootstrap standard error of $\overline{\theta}^*$ is found with the usual standard deviation formula:

$$S_{\hat{\theta}} = \sqrt{\frac{\sum_{i=1}^{B}\left(\hat{\theta}_i^* - \overline{\theta}^*\right)^2}{B-1}}.$$

In the literature, $B - 1$ is often replaced by B; for large values of B, however, there is little practical difference in the estimate obtained.

Example 10-13

The failure times X of an electronic component are known to follow an exponential distribution with unknown parameter λ. A random sample of ten components resulted in the following failure times (in hours):

195.2, 201.4, 183.0, 175.1, 205.1, 191.7, 188.6, 173.5, 200.8, 210.0.

The mean of the exponential distribution is given by $E(X) = 1/\lambda$. It is also known that $E(\overline{X}) = 1/\lambda$. A reasonable estimate for λ then is $\hat{\lambda} = 1/\overline{X}$. From the sample data, we find $\overline{X} = 192.44$, resulting in $\hat{\lambda} = 1/192.44 = 0.00520$. $B = 100$ bootstrap samples of size $n = 10$ were generated using Minitab® with $f(x; 0.00520) = 0.00520e^{-0.00520x}$. Some of the bootstrap estimates are shown in Table 10-1.

The average of the bootstrap estimates is found to be $\overline{\lambda}^* = \sum_{i=1}^{100} \hat{\lambda}_i^* \Big/ 100 = 0.00551$. The standard error of the estimate is

$$S_{\hat{\theta}} = \sqrt{\frac{\sum_{i=1}^{B}\left(\hat{\lambda}_i^* - \overline{\lambda}^*\right)^2}{B-1}} = \sqrt{\frac{\sum_{i=1}^{100}\left(\hat{\lambda}_i^* - 0.00551\right)^2}{100-1}} = 0.00169.$$

Table 10-1 Bootstrap Estimates for Example 10-13

Sample	Sample Mean, $\overline{x}_i^*$	$\hat{\lambda}_i^*$
1	243.407	0.00411
2	153.821	0.00650
3	126.554	0.00790
⋮		
100	204.390	0.00489

10-2 SINGLE-SAMPLE CONFIDENCE INTERVAL ESTIMATION

In many situations, a point estimate does not provide enough information about the parameter of interest. For example, if we are interested in estimating the mean compression strength of concrete, a single number may not be very meaningful. An interval estimate of the form $L \leq \mu \leq U$ might be more useful. The end points of this interval will be random variables, since they are functions of sample data.

In general, to construct an interval estimator of the unknown parameter θ, we must find two statistics, L and U, such that

$$P\{L \leq \theta \leq U\} = 1 - \alpha. \tag{10-18}$$

The resulting interval

$$L \leq \theta \leq U \tag{10-19}$$

is called a $100(1 - \alpha)\%$ *confidence interval* for the unknown parameter θ. L and U are called the lower- and upper-*confidence limits*, respectively, and $1 - \alpha$ is called the *confidence coefficient.* The interpretation of a confidence interval is that if many random samples are collected and a $100(1 - \alpha)\%$ confidence interval on θ is computed from each sample, then $100(1 - \alpha)\%$ of these intervals will contain the true value of θ. The situation is illustrated in Fig. 10-4 which shows several $100(1 - \alpha)\%$ confidence intervals for the mean μ of a distribution. The dots at the center of each interval indicate the point estimate of μ (in this case $\overline{X}$). Notice that one of the 15 intervals fails to contain the true value of μ. If this were a 95% confidence level, in the long run, only 5% of the intervals would fail to contain μ.

Now in practice, we obtain only one random sample and calculate one confidence interval. Since this interval either will or will not contain the true value of θ, it is not reasonable to attach a probability level to this specific event. The appropriate statement would be that θ lies in the observed interval $[L, U]$ with confidence $100(1 - \alpha)$. This statement has

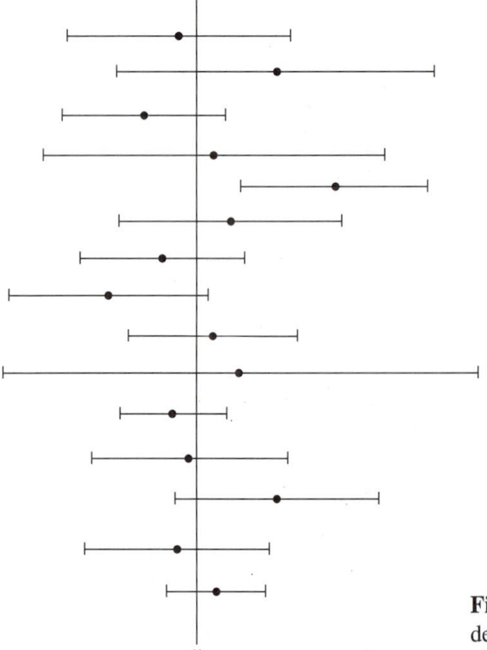

Figure 10-4 Repeated construction of a confidence interval for μ.

a frequency interpretation; that is, we do not know if the statement is true for this specific sample, but the *method* used to obtain the interval $[L, U]$ yields correct statements $100(1-\alpha)\%$ of the time.

The confidence interval in equation 10-19 might be more properly called a *two-sided confidence interval*, as it specifies both a lower and an upper limit on θ. Occasionally, a *one-sided* confidence interval might be more appropriate. A one-sided $100(1-\alpha)\%$ lower-confidence interval on θ is given by the interval

$$L \leq \theta, \tag{10-20}$$

where the lower-confidence limit L is chosen so that

$$P\{L \leq \theta\} = 1 - \alpha. \tag{10-21}$$

Similarly, a one-sided $100(1-\alpha)\%$ upper-confidence interval on θ is given by the interval

$$\theta \leq U, \tag{10-22}$$

where the upper-confidence limit U is chosen so that

$$P\{\theta \leq U\} = 1 - \alpha. \tag{10-23}$$

The length of the observed two-sided confidence interval is an important measure of the quality of the information obtained from the sample. The half-interval length $\theta - L$ or $U - \theta$ is called the *accuracy* of the estimator. The longer the confidence interval, the more confident we are that the interval actually contains the true value of θ. On the other hand, the longer the interval, the less information we have about the true value of θ. In an ideal situation, we obtain a relatively short interval with high confidence.

10-2.1 Confidence Interval on the Mean of a Normal Distribution, Variance Known

Let X be a normal random variable with unknown mean μ and known variance σ^2, and suppose that a random sample of size n, $X_1, X_2, \ldots, X_n$, is taken. A $100(1-\alpha)\%$ confidence interval on μ can be obtained by considering the sampling distribution of the sample mean $\overline{X}$. In Section 9-3 we noted that the sampling distribution of $\overline{X}$ is normal if X is normal and approximately normal if the conditions of the Central Limit Theorem are met. The mean of $\overline{X}$ is μ and the variance is σ^2/n. Therefore, the distribution of the statistic

$$Z = \frac{\overline{X} - \mu}{\sigma/\sqrt{n}}$$

is taken to be a standard normal distribution.

The distribution of $Z = (\overline{X} - \mu)/(\sigma/\sqrt{n})$ is shown in Fig. 10-5. From examination of this figure we see that

$$P\{-Z_{\alpha/2} \leq Z \leq Z_{\alpha/2}\} = 1 - \alpha$$

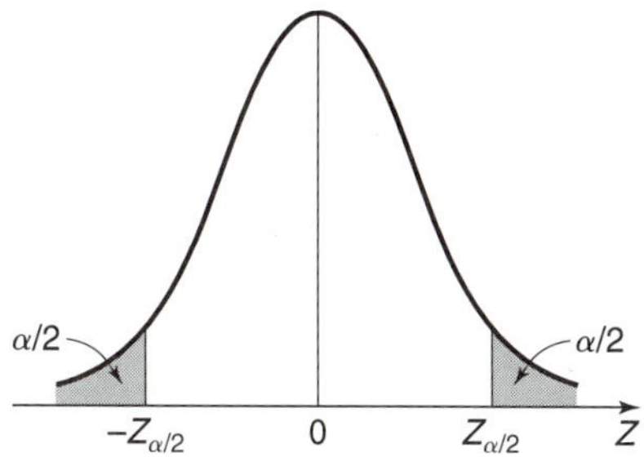

Figure 10-5 The distribution of Z.

or

$$P\left\{-Z_{\alpha/2} \le \frac{\overline{X}-\mu}{\sigma/\sqrt{n}} \le Z_{\alpha/2}\right\} = 1-\alpha.$$

This can be rearranged as

$$P\left\{\overline{X} - Z_{\alpha/2}\,\sigma/\sqrt{n} \le \mu \le \overline{X} + Z_{\alpha/2}\,\sigma/\sqrt{n}\right\} = 1-\alpha. \qquad (10\text{-}24)$$

Comparing equations 10-24 and 10-18, we see that the $100(1-\alpha)\%$ two-sided confidence interval on μ is

$$\overline{X} - Z_{\alpha/2}\,\sigma/\sqrt{n} \le \mu \le \overline{X} + Z_{\alpha/2}\,\sigma/\sqrt{n}. \qquad (10\text{-}25)$$

Example 10-14

Consider the thermal conductivity data in Example 10-12. Suppose that we want to find a 95% confidence interval on the mean thermal conductivity of Armco iron. Suppose we know that the standard deviation of thermal conductivity at 100°F and 550 W is $\sigma = 0.10$ Btu/hr–ft–°F. If we assume that thermal conductivity is normally distributed (or that the conditions of the Central Limit Theorem are met), then we can use equation 10-25 to construct the confidence interval. A 95% interval implies that $1-\alpha = 0.95$, so $\alpha = 0.05$, and from Table II in the Appendix $Z_{\alpha/2} = Z_{0.05/2} = Z_{0.025} = 1.96$. The lower confidence limit is

$$\begin{aligned} L &= \bar{x} - Z_{\alpha/2}\,\sigma/\sqrt{n} \\ &= 41.924 - 1.96(0.10)/\sqrt{10} \\ &= 41.924 - 0.062 \\ &= 41.862 \end{aligned}$$

and the upper confidence limit is

$$\begin{aligned} U &= \bar{x} + Z_{\alpha/2}\,\sigma/\sqrt{n} \\ &= 41.924 + 1.96(0.10)/\sqrt{10} \\ &= 41.924 + 0.062 \\ &= 41.986. \end{aligned}$$

Thus the 95% two-sided confidence interval is

$$41.862 \le \mu \le 41.986.$$

This is our interval of reasonable values for mean thermal conductivity at 95% confidence.

Confidence Level and Precision of Estimation

Notice that in the previous example our choice of the 95% level of confidence was essentially arbitrary. What would have happened if we had chosen a higher level of confidence, say 99%? In fact, doesn't it seem reasonable that we would want the higher level of confidence? At $\alpha = 0.01$, we find $Z_{\alpha/2} = Z_{0.01/2} = Z_{0.005} = 2.58$, while for $\alpha = 0.05$, $Z_{0.025} = 1.96$. Thus, the length of the 95% confidence interval is

$$2\left(1.96\sigma/\sqrt{n}\right) = 3.92\sigma/\sqrt{n},$$

whereas the length of the 99% confidence interval is

$$2\left(2.58\sigma/\sqrt{n}\right) = 5.15\sigma/\sqrt{n}.$$

The 99% confidence interval is longer than the 95% confidence interval. This is why we have a higher level of confidence in the 99% confidence interval. Generally, for a fixed sample size n and standard deviation σ, the higher the confidence level, the longer the resulting confidence interval.

Since the *length* of the confidence interval measures the *precision* of estimation, we see that precision is inversely related to the confidence level. As noted earlier, it is highly desirable to obtain a confidence interval that is short enough for decision-making purposes and that also has adequate confidence. One way to achieve this is by choosing the sample size n to be large enough to give a confidence interval of specified length with prescribed confidence.

Choice of Sample Size

The accuracy of the confidence interval in equation 10-25 is $Z_{\alpha/2}\sigma/\sqrt{n}$. This means that in using $\bar{x}$ to estimate μ, the error $E = |\bar{x} - \mu|$ is less than $Z_{\alpha/2}\sigma/\sqrt{n}$, with confidence $100(1 - \alpha)$. This is shown graphically in Fig. 10-6. In situations where the sample size can be controlled, we can choose n to be $100(1 - \alpha)\%$ confident that the error in estimating μ is less than a specified error E. The appropriate sample size is

$$n = \left(\frac{Z_{\alpha/2}\sigma}{E}\right)^2. \tag{10-26}$$

If the right-hand side of equation 10-26 is not an integer, it must be rounded up. Notice that $2\,E$ is the length of the resulting confidence interval.

To illustrate the use of this procedure, suppose that we wanted the error in estimating the mean thermal conductivity of Armco iron in Example 10-14 to be less than 0.05 Btu/hr –ft–°F, with 95% confidence. Since $\sigma = 0.10$ and $Z_{0.025} = 1.96$, we may find the required sample size from equation 10-26 to be

$$n = \left(\frac{Z_{\alpha/2}\sigma}{E}\right)^2 = \left[\frac{(1.96)0.10}{0.05}\right]^2 = 15.37 = 16.$$

Notice how, in general, the sample size behaves as a function of the length of the confidence interval $2\,E$, the confidence level $100(1 - \alpha)\%$, and the standard deviation σ as follows:

- As the desired length of the interval $2\,E$ decreases, the required sample size n increases for a fixed value of σ and specified confidence.
- As σ increases, the required sample size n increases for a fixed length $2\,E$ and specified confidence.
- As the level of confidence increases, the required sample size n increases for fixed length $2\,E$ and standard deviation σ.

E = error = $|\bar{x} - \mu|$

$L = \bar{x} - Z_{\alpha/2}\,\sigma/\sqrt{n}$ $\quad \bar{x} \quad \mu \quad$ $U = \bar{x} + Z_{\alpha/2}\,\sigma/\sqrt{n}$

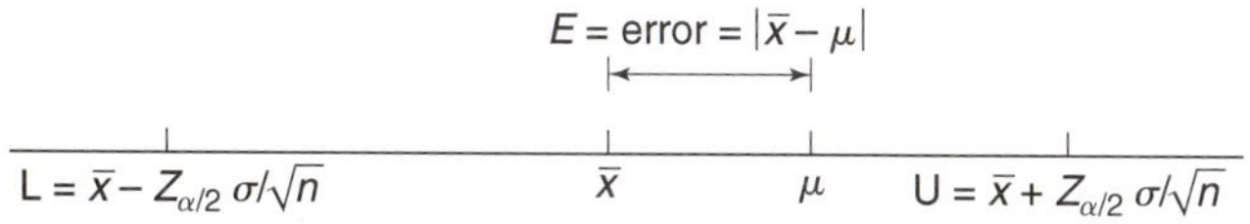

Figure 10-6 Error in estimating μ with $\bar{x}$.

One-Sided Confidence Intervals

It is also possible to obtain one-sided confidence intervals for μ by setting either $L = -\infty$ or $U = \infty$ and replacing $Z_{\alpha/2}$ by Z_α. The $100(1 - \alpha)\%$ upper-confidence interval for μ is

$$\mu \leq \overline{X} + Z_\alpha \sigma \sqrt{n}, \tag{10-27}$$

and the $100(1 - \alpha)\%$ lower-confidence interval for μ is

$$\overline{X} - Z_\alpha \sigma \sqrt{n} \leq \mu. \tag{10-28}$$

10-2.2 Confidence Interval on the Mean of a Normal Distribution, Variance Unknown

Suppose that we wish to find a confidence interval on the mean of a distribution but the variance is unknown. Specifically, a random sample of size n, $X_1, X_2, \ldots, X_n$ is available, and $\overline{X}$ and S^2 are the sample mean and sample variance, respectively. One possibility would be to replace σ in the confidence interval formulas for μ with known variance (equations 10-25, 10-27, and 10-28) with the sample standard deviation s. If the sample size, n, is relatively large, say $n > 30$, then this is an acceptable procedure. Consequently, we often call the confidence intervals in Sections 10-2.1 and 10-2.2 *large-sample confidence intervals*, because they are approximately valid even if the unknown population variances are replaced by the corresponding sample variances.

When sample sizes are small, this approach will not work, and we must use another procedure. To produce a valid confidence interval, we must make a stronger assumption about the underlying population. The usual assumption is that the underlying population is *normally* distributed. This leads to confidence intervals based on the t distribution. Specifically, let $X_1, X_2, \ldots, X_n$ be a random sample from a normal distribution with unknown mean μ and unknown variance σ^2. In Section 9-4 we noted that the sampling distribution of the statistic

$$t = \frac{\overline{X} - \mu}{S/\sqrt{n}}$$

is the t distribution with $n - 1$ degrees of freedom. We now show how the confidence interval on μ is obtained.

The distribution of $t = (\overline{X} - \mu)/(S/\sqrt{n})$ is shown in Fig. 10-7. Letting $t_{\alpha/2,n-1}$ be the upper $\alpha/2$ percentage point of the t distribution with $n - 1$ degrees of freedom, we observe from Fig. 10-7 that

$$P\{-t_{\alpha/2,n-1} \leq t \leq t_{\alpha/2,n-1}\} = 1 - \alpha$$

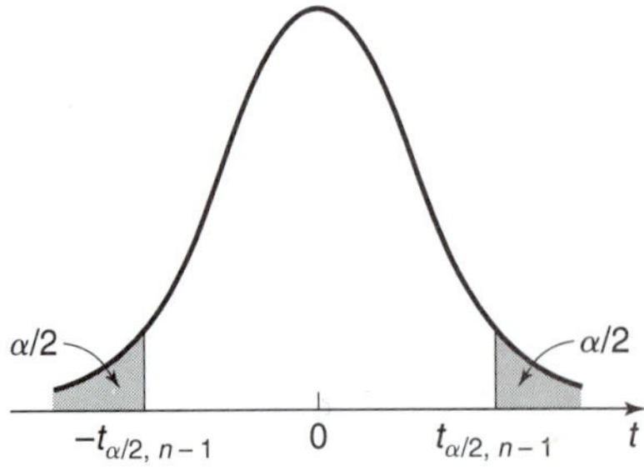

Figure 10-7 The t distribution.

or

$$P\left\{-t_{\alpha/2,n-1} \le \frac{\bar{X}-\mu}{S/\sqrt{n}} \le t_{\alpha/2,n-1}\right\} = 1-\alpha.$$

Rearranging this last equation yields

$$P\left\{\bar{X} - t_{\alpha/2,n-1}\, S/\sqrt{n} \le \mu \le \bar{X} + t_{\alpha/2,n-1}\, S/\sqrt{n}\right\} = 1-\alpha. \tag{10-29}$$

Comparing equations 10-29 and 10-18, we see that a $100(1-\alpha)\%$ two-sided confidence interval on μ is

$$\bar{X} - t_{\alpha/2,n-1}\, S/\sqrt{n} \le \mu \le \bar{X} + t_{\alpha/2,n-1}\, S/\sqrt{n}. \tag{10-30}$$

A $100(1-\alpha)\%$ lower-confidence interval on μ is given by

$$\bar{X} - t_{\alpha,n-1}\, S\ \sqrt{n} \le \mu, \tag{10-31}$$

and a $100(1-\alpha)\%$ upper-confidence interval on μ is

$$\mu \le \bar{X} + t_{\alpha,n-1}\, S\ \sqrt{n}. \tag{10-32}$$

Remember that these procedures assume that we are sampling from a normal population. This assumption is important for small samples. Fortunately, the normality assumption holds in many practical situations. When it does not, we must use *distribution-free* or *nonparametric* confidence intervals. Nonparametric methods are discussed in Chapter 16. However, when the population is normal, the t-distribution intervals are the shortest possible $100(1-\alpha)\%$ confidence intervals, and are therefore superior to the nonparametric methods.

Selecting the sample size n required to give a confidence interval of required length is not as easy as in the known σ case, because the length of the interval depends on the value of σ (unknown before the data is collected) and on n. Furthermore, n enters the confidence interval through both $1/\sqrt{n}$ and $t_{\alpha/2,n-1}$. Consequently, the required n must be determined through trial and error.

Example 10-15

An article in the *Journal of Testing and Evaluation* (Vol. 10, No. 4, 1982, p. 133) presents the following 20 measurements on residual flame time (in seconds) of treated specimens of children's nightwear:

9.85, 9.93, 9.75, 9.77, 9.67,
9.87, 9.67, 9.94, 9.85, 9.75,
9.83, 9.92, 9.74, 9.99, 9.88,
9.95, 9.95, 9.93, 9.92, 9.89.

We wish to find a 95% confidence interval on the mean residual flame time. The sample mean and standard deviation are

$$\bar{x} = 9.8475,$$
$$s = 0.0954.$$

From Table IV of the Appendix we find $t_{0.025,19} = 2.093$. The lower and upper 95% confidence limits are

$$\begin{aligned} L &= \bar{x} - t_{\alpha/2,n-1}\, s/\sqrt{n} \\ &= 9.8475 - 2.093(0.0954)/\sqrt{20} \\ &= 9.8029 \text{ seconds.} \end{aligned}$$

and

$$\begin{aligned} U &= \bar{x} + t_{\alpha/2,n-1}\, s/\sqrt{n} \\ &= 9.8475 + 2.093(0.0954)/\sqrt{20} \\ &= 9.8921 \text{ seconds.} \end{aligned}$$

Therefore the 95% confidence interval is

$$9.8029 \text{ sec} \leq \mu \leq 9.8921 \text{ sec}$$

We are 95% confident that the mean residual flame time is between 9.8025 and 9.8921 seconds.

10-2.3 Confidence Interval on the Variance of a Normal Distribution

Suppose that X is normally distributed with unknown mean μ and unknown variance σ^2. Let $X_1, X_2, \ldots, X_n$ be a random sample of size n, and let S^2 be the sample variance. It was shown in Section 9-3 that the sampling distribution of

$$\chi^2 = \frac{(n-1)S^2}{\sigma^2}$$

is chi-square with $n - 1$ degrees of freedom. This distribution is shown in Fig. 10-8.

To develop the confidence interval, we note from Fig. 10-8 that

$$P\{\chi^2_{1-\alpha/2,\, n-1} \leq \chi^2 \leq \chi^2_{\alpha/2,\, n-1}\} = 1 - \alpha$$

or

$$P\left\{\chi^2_{1-\alpha/2,n-1} \leq \frac{(n-1)S^2}{\sigma^2} \leq \chi^2_{\alpha/2,n-1}\right\} = 1 - \alpha.$$

This last equation can be rearranged to yield

$$P\left\{\frac{(n-1)S^2}{\chi^2_{\alpha/2,n-1}} \leq \sigma^2 \leq \frac{(n-1)S^2}{\chi^2_{1-\alpha/2,n-1}}\right\} = 1 - \alpha. \tag{10-33}$$

Comparing equations 10-33 and 10-18, we see that a $100(1 - \alpha)\%$ two-sided confidence interval for σ^2 is

$$\frac{(n-1)S^2}{\chi^2_{\alpha/2,n-1}} \leq \sigma^2 \leq \frac{(n-1)S^2}{\chi^2_{1-\alpha/2,n-1}}. \tag{10-34}$$

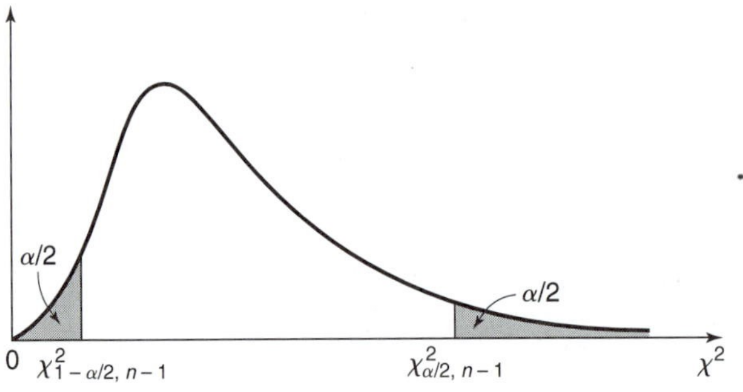

Figure 10-8 The χ^2 distribution.

To find a 100(1 − α)% lower-confidence interval on σ^2, set $U = \infty$ and replace $\chi^2_{\alpha/2,n-1}$ with $\chi^2_{\alpha,n-1}$, giving

$$\frac{(n-1)S^2}{\chi^2_{\alpha,n-1}} \leq \sigma^2. \tag{10-35}$$

The 100(1 − α)% upper-confidence interval is found by setting $L = 0$ and replacing $\chi^2_{1-\alpha/2,n-1}$ with $\chi^2_{1-\alpha,n-1}$, resulting in

$$\sigma^2 \leq \frac{(n-1)S^2}{\chi^2_{1-\alpha,n-1}}. \tag{10-36}$$

Example 10-16

A manufacturer of soft drink beverages is interested in the uniformity of the machine used to fill cans. Specifically, it is desirable that the standard deviation σ of the filling process be less than 0.2 fluid ounces; otherwise there will be a higher than allowable percentage of cans that are underfilled. We will assume that fill volume is approximately normally distributed. A random sample of 20 cans result in a sample variance of $S^2 = 0.0225$ (fluid ounces)^2. A 95% upper-confidence interval is found from equation 10-36 as follows:

$$\sigma^2 \leq \frac{(n-1)S^2}{\chi^2_{0.95,19}},$$

or

$$\sigma^2 \leq \frac{(19)0.0225}{10.117} = 0.0423 \text{ (fluid ounces)}^2.$$

This last statement may be converted into a confidence interval on the standard deviation σ by taking the square root of both sides, resulting in

$$\sigma \leq 0.21 \text{ fluid ounces.}$$

Therefore, at the 95% level of confidence, the data do not support the claim that the process standard deviation is less than 0.20 fluid ounces.

10-2.4 Confidence Interval on a Proportion

It is often necessary to construct a 100(1 − α)% confidence interval on a proportion. For example, suppose that a random sample of size n has been taken from a large (possibly infinite) population, and $X(\leq n)$ observations in this sample belong to a class of interest. Then $\hat{p} = X/n$ is the point estimator of the proportion of the population that belongs to this class. Note that n and p are the parameters of a binomial distribution. Furthermore, in Section 7-5 we saw that the sampling distribution of $\hat{p}$ is approximately normal with mean p and variance $p(1 - p)/n$, if p is not too close to either 0 or 1, and if n is relatively large. Thus, the distribution of

$$Z = \frac{\hat{p} - p}{\sqrt{\dfrac{p(1-p)}{n}}}$$

is approximately standard normal.

To construct the confidence interval on p, note that

$$P\{-Z_{\alpha/2} \leq Z \leq Z_{\alpha/2}\} \simeq 1 - \alpha$$

or

$$P\left\{-Z_{\alpha/2} \le \frac{\hat{p}-p}{\sqrt{\frac{p(1-p)}{n}}} \le Z_{\alpha/2}\right\} \simeq 1-\alpha.$$

This may be rearranged as

$$P\left\{\hat{p}-Z_{\alpha/2}\sqrt{\frac{p(1-p)}{n}} \le p \le \hat{p}+Z_{\alpha/2}\sqrt{\frac{p(1-p)}{n}}\right\} \simeq 1-\alpha. \tag{10-37}$$

We recognize the quantity $\sqrt{p(1-p)/n}$ as the standard error of the point estimator $\hat{p}$. Unfortunately, the upper and lower limits of the confidence interval obtained from equation 10-37 would contain the unknown parameter p. However, a satisfactory solution is to replace p by $\hat{p}$ in the standard error, giving an *estimated standard error.* Therefore,

$$P\left\{\hat{p}-Z_{\alpha/2}\sqrt{\frac{\hat{p}(1-\hat{p})}{n}} \le p \le \hat{p}+Z_{\alpha/2}\sqrt{\frac{\hat{p}(1-\hat{p})}{n}}\right\} \simeq 1-\alpha, \tag{10-38}$$

and the approximate $100(1-\alpha)\%$ two-sided confidence interval on p is

$$\hat{p}-Z_{\alpha/2}\sqrt{\frac{\hat{p}(1-\hat{p})}{n}} \le p \le \hat{p}+Z_{\alpha/2}\sqrt{\frac{\hat{p}(1-\hat{p})}{n}}. \tag{10-39}$$

An approximate $100(1-\alpha)\%$ lower-confidence interval is

$$\hat{p}-Z_{\alpha}\sqrt{\frac{\hat{p}(1-\hat{p})}{n}} \le p, \tag{10-40}$$

and an approximate $100(1-\alpha)\%$ upper-confidence interval is

$$p \le \hat{p}+Z_{\alpha}\sqrt{\frac{\hat{p}(1-\hat{p})}{n}}. \tag{10-41}$$

Example 10-17

In a random sample of 75 axle shafts, 12 have a surface finish that is rougher than the specifications will allow. Therefore, a point estimate of the proportion p of shafts in the population that exceed the roughness specifications is $\hat{p} = x/n = 12/75 = 0.16$. A 95% two-sided confidence interval for p is computed from equation 10-39 as

$$\hat{p}-Z_{0.025}\sqrt{\frac{\hat{p}(1-\hat{p})}{n}} \le p \le \hat{p}+Z_{0.025}\sqrt{\frac{\hat{p}(1-\hat{p})}{n}}$$

or

$$0.16-1.96\sqrt{\frac{0.16(0.84)}{75}} \le p \le 0.16+1.96\sqrt{\frac{0.16(0.84)}{75}},$$

which simplifies to

$$0.08 \le p \le 0.24.$$

Define the error in estimating p by $\hat{p}$ as $E = |p - \hat{p}|$. Note that we are approximately $100(1 - \alpha)\%$ confident that this error is less than $Z_{\alpha/2}\sqrt{p(1-p)/n}$. Therefore, in situations where the sample size can be selected, we may choose n to be $100(1 - \alpha)\%$ confident that the error is less than some specified value E. The appropriate sample size is

$$n = \left(\frac{Z_{\alpha/2}}{E}\right)^2 p(1-p). \tag{10-42}$$

This function is relatively flat from $p = 0.3$ to $p = 0.7$. An estimate of p is required to use equation 10-42. If an estimate $\hat{p}$ from a previous sample is available, it could be substituted for p in equation 10-42, or perhaps a subjective estimate could be made. If these alternatives are unsatisfactory, a preliminary sample could be taken, $\hat{p}$ computed, and then equation 10-42 used to determine how many additional observations are required to estimate p with the desired accuracy. The sample size from equation 10-42 will always be a maximum for $p = 0.5$ [that is, $p(1 - p) = 0.25$], and this can be used to obtain an upper bound on n. In other words, we are *at least* $100(1 - \alpha)\%$ confident that the error in estimating p using $\hat{p}$ is less than E if the sample size is

$$n = \left(\frac{Z_{\alpha/2}}{E}\right)^2 (0.25).$$

In order to maintain at least a $100(1 - \alpha)\%$ level of confidence the value for n is always rounded up to the next integer.

Example 10-18

Consider the data in Example 10-17. How large a sample is required if we want to be 95% confident that the error in using $\hat{p}$ to estimate p is less than 0.05? Using $\hat{p} = 0.16$ as an initial estimate of p, we find from equation 10-42 that the required sample size is

$$\left(\frac{Z_{0.025}}{E}\right)^2 \hat{p}(1-\hat{p}) = \left(\frac{1.96}{0.05}\right)^2 0.16(0.84) = 207.$$

We note that the procedures developed in this section depend on the normal approximation to the binomial. In situations where this approximation is inappropriate, particularly cases where n is small, other methods must be used. Tables of the binomial distribution could be used to obtain a confidence interval for p. If n is large but p is small, then the Poisson approximation to the binomial could be used to construct confidence intervals. These procedures are illustrated by Duncan (1986).

Agresti and Coull (1998) present an alternative form of a confidence interval on the population proportion, p, based on a large-sample hypothesis test on p (see Chapter 11 of this text). Agresti and Coull show that the upper and lower limits of an approximate $100(1 - \alpha)\%$ confidence interval on p are

$$\frac{\hat{p} + \dfrac{Z_{a/2}^2}{2n} \pm z_{a/2}\sqrt{\dfrac{\hat{p}(1-\hat{p})}{n} + \dfrac{Z_{a/2}^2}{4n^2}}}{1 + \dfrac{Z_{a\,2}^2}{n}}.$$

The authors refer to this as the *score* confidence interval. One-sided confidence intervals can be constructed simply by replacing $Z_{\alpha/2}$ with Z_α.

To illustrate this confidence interval, reconsider Example 10-17, which discusses the surface finish of a shaft with $n = 75$ and $\hat{p} = 0.16$. The lower and upper limits of a 95% confidence interval using the approach of Agresti and Coull are

$$\frac{\hat{p}+\frac{Z_{\alpha/2}^2}{2n}\pm Z_{\alpha/2}\sqrt{\frac{\hat{p}(1-\hat{p})}{n}+\frac{Z_{\alpha/2}^2}{4n^2}}}{1+\frac{Z_{\alpha 2}^2}{n}}=\frac{0.16+\frac{(1.96)^2}{2(75)}\pm 1.96\sqrt{\frac{0.16(0.84)}{75}+\frac{(1.96)^2}{4(75)^2}}}{1+\frac{(1.96)^2}{75}}$$

$$=\frac{0.186\pm 0.087}{1.051}$$

$$=0.177\pm 0.083.$$

The resulting lower- and upper-confidence limits are 0.094 and 0.260, respectively.

Agresti and Coull argue that the more complicated confidence interval has several advantages over the standard large-sample interval (given in equation 10-39). One advantage is that their confidence interval tends to maintain the stated level of confidence better than the standard large-sample interval. Another advantage is that the lower-confidence limit will always be non-negative. The large-sample confidence interval can result in negative lower–confidence limits, which the practitioner will generally then set to 0. A method which can report a negative lower limit on a parameter that is inherently non-negative (such as a proportion, p) is often considered an inferior method. Lastly, the requirements that p not be close to 0 or 1 and n be relatively large are not requirements for the approach suggested by Agresti and Coull. In other words, their approach results in an appropriate confidence interval for any combination of n and p.

10-3 TWO-SAMPLE CONFIDENCE INTERVAL ESTIMATION

10-3.1 Confidence Interval on the Difference between Means of Two Normal Distributions, Variances Known

Consider two independent random variables X_1 with unknown mean μ_1 and known variance σ_1^2 and X_2 with unknown mean μ_2 and known variance σ_2^2. We wish to find a $100(1 - \alpha)\%$ confidence interval on the difference in means $\mu_1 - \mu_2$. Let $X_{11}, X_{12}, \ldots, X_{1n_1}$ be a random sample of n_1 observations from X_1, and $X_{21}, X_{22}, \ldots, X_{2n_2}$ be a random sample of n_2 observations from X_2. If $\overline{X}_1$ and $\overline{X}_2$ are the sample means, the statistic

$$Z=\frac{\overline{X}_1-\overline{X}_2-(\mu_1-\mu_2)}{\sqrt{\frac{\sigma_1^2}{n_1}+\frac{\sigma_2^2}{n_2}}}$$

is standard normal if X_1 and X_2 are normal or approximately standard normal if the conditions of the Central Limit Theorem apply, respectively. From Fig. 10-5, this implies that

$$P\{-Z_{\alpha/2}\le Z\le Z_{\alpha/2}\}=1-\alpha$$

or

$$P\left\{-Z_{\alpha/2}\le\frac{\overline{X}_1-\overline{X}_2-(\mu_1-\mu_2)}{\sqrt{\frac{\sigma_1^2}{n_1}+\frac{\sigma_2^2}{n_2}}}\le Z_{\alpha/2}\right\}=1-\alpha.$$

This can be rearranged as

$$P\left\{\bar{X}_1 - \bar{X}_2 - Z_{\alpha/2}\sqrt{\frac{\sigma_1^2}{n_1}+\frac{\sigma_2^2}{n_2}} \le \mu_1 - \mu_2 \right.$$

$$\left. \le \bar{X}_1 - \bar{X}_2 + Z_{\alpha/2}\sqrt{\frac{\sigma_1^2}{n_1}+\frac{\sigma_2^2}{n_2}}\right\} = 1-\alpha. \tag{10-43}$$

Comparing equations 10-43 and 10-18, we note that the $100(1-\alpha)\%$ confidence interval for $\mu_1 - \mu_2$ is

$$\bar{X}_1 - \bar{X}_2 - Z_{\alpha/2}\sqrt{\frac{\sigma_1^2}{n_1}+\frac{\sigma_2^2}{n_2}} \le \mu_1 - \mu_2 \le \bar{X}_1 - \bar{X}_2 + Z_{\alpha/2}\sqrt{\frac{\sigma_1^2}{n_1}+\frac{\sigma_2^2}{n_2}}. \tag{10-44}$$

One-sided confidence intervals on $\mu_1 - \mu_2$ may also be obtained. A $100(1-\alpha)\%$ upper-confidence interval on $\mu_1 - \mu_2$ is

$$\mu_1 - \mu_2 \le \bar{X}_1 - \bar{X}_2 + Z_{\alpha}\sqrt{\frac{\sigma_1^2}{n_1}+\frac{\sigma_2^2}{n_2}}, \tag{10-45}$$

and a $100(1-\alpha)\%$ lower-confidence interval is

$$\bar{X}_1 - \bar{X}_2 - Z_{\alpha}\sqrt{\frac{\sigma_1^2}{n_1}+\frac{\sigma_2^2}{n_2}} \le \mu_1 - \mu_2. \tag{10-46}$$

Example 10-19

Tensile strength tests were performed on two different grades of aluminum spars used in manufacturing the wing of a commercial transport aircraft. From past experience with the spar manufacturing process and the testing procedure, the standard deviations of tensile strengths are assumed to be known. The data obtained are shown in Table 10-2.

If μ_1 and μ_2 denote the true mean tensile strengths for the two grades of spars, then we may find a 90% confidence interval on the difference in mean strength $\mu_1 - \mu_2$ as follows:

$$\begin{aligned} L &= \bar{x}_1 - \bar{x}_2 - Z_{\alpha/2}\sqrt{\frac{\sigma_1^2}{n_1}+\frac{\sigma_2^2}{n_2}} \\ &= 87.6 - 74.5 - 1.645\sqrt{\frac{(1.0)^2}{10}+\frac{(1.5)^2}{12}} \\ &= 13.1 - 0.88 \\ &= 12.22 \text{ kg/mm}^2, \end{aligned}$$

Table 10-2 Tensile Strength Test Result for Aluminum Spars

Spar Grade	Sample Size	Sample Mean Tensile Strength (kg / mm^2)	Standard Deviation (kg / mm^2)
1	$n_1 = 10$	$\bar{x}_1 = 87.6$	$\sigma_1 = 1.0$
2	$n_2 = 12$	$\bar{x}_2 = 74.5$	$\sigma_2 = 1.5$

$$
\begin{aligned}
U &= \bar{x}_1 - \bar{x}_2 + Z_{\alpha/2}\sqrt{\frac{\sigma_1^2}{n_1}+\frac{\sigma_2^2}{n_2}} \\
&= 87.6 - 74.5 + 1.645\sqrt{\frac{(1.0)^2}{10}+\frac{(1.5)^2}{12}} \\
&= 13.1 + 0.88 \\
&= 13.98 \text{ kg/mm}^2.
\end{aligned}
$$

Therefore the 90% confidence interval on the difference in mean tensile strength is

$$12.22 \text{ kg/mm}^2 \le \mu_1 - \mu_2 \le 13.98 \text{ kg/mm}^2.$$

We are 90% confident that the mean tensile strength of grade 1 aluminum exceeds that of grade 2 aluminum by between 12.22 and 13.98 kg/mm^2.

If the standard deviations σ_1 and σ_2 are known (at least approximately), and if the sample sizes n_1 and n_2 are equal ($n_1 = n_2 = n$, say), then we can determine the sample size required so that the error in estimating $\mu_1 - \mu_2$ using $\overline{X}_1 - \overline{X}_2$ will be less than E at $100(1 - \alpha)\%$ confidence. The required sample size from each population is

$$n = \left(\frac{Z_{\alpha/2}}{E}\right)^2 \left(\sigma_1^2 + \sigma_2^2\right). \qquad (10\text{-}47)$$

Remember to round up if n is not an integer.

10-3.2 Confidence Interval on the Difference between Means of Two Normal Distributions, Variances Unknown

We now extend the results of Section 10-2.2 to the case of two populations with unknown means and variances, and we wish to find confidence intervals on the difference in means $\mu_1 - \mu_2$. If the sample sizes n_1 and n_2 both exceed 30, then the normal known-variances distribution intervals in Section 10-3.1 can be used. However, when small samples are taken, we must assume that the underlying populations are normally distributed with unknown variances and base the confidence intervals on the t distribution.

Case I. $\sigma_1^2 = \sigma_2^2 = \sigma^2$

Consider two independent normal random variables, say X_1 with mean μ_1 and variance σ_1^2, and X_2 with mean μ_2 and variance σ_2^2. Both the means μ_1 and μ_2 and the variances σ_1^2 and σ_2^2 are unknown. However, suppose it is reasonable to assume that both variances are equal; that is, $\sigma_1^2 = \sigma_2^2 = \sigma^2$. We wish to find a $100(1 - \alpha)\%$ confidence interval on the difference in means $\mu_1 - \mu_2$.

Random samples of size n_1 and n_2 are taken on X_1 and X_2, respectively. Let the sample means be denoted $\overline{X}_1$ and $\overline{X}_2$ and the sample variances be denoted S_1^2 and S_2^2. Since both S_1^2 and S_2^2 are estimates of the common variance σ^2, we may obtain a combined (or "pooled") estimator of σ^2:

$$S_p^2 = \frac{(n_1 - 1)S_1^2 + (n_2 - 1)S_2^2}{n_1 + n_2 - 2}. \qquad (10\text{-}48)$$

To develop the confidence interval for $\mu_1 - \mu_2$, note that the distribution of the statistic

$$t = \frac{\overline{X}_1 - \overline{X}_2 - (\mu_1 - \mu_2)}{S_p\sqrt{\dfrac{1}{n_1} + \dfrac{1}{n_2}}}$$

is the t distribution with $n_1 + n_2 - 2$ degrees of freedom. Therefore,

$$P\{-t_{\alpha/2,n_1+n_2-2} \le t \le t_{\alpha/2,n_1+n_2-2}\} = 1 - \alpha$$

or

$$P\left\{-t_{\alpha/2,n_1+n_2-2} \le \frac{\overline{X}_1 - \overline{X}_2 - (\mu_1 - \mu_2)}{S_p\sqrt{\dfrac{1}{n_1} + \dfrac{1}{n_2}}} \le t_{\alpha/2,n_1+n_2-2}\right\} = 1 - \alpha.$$

This may be rearranged as

$$P\left\{\overline{X}_1 - \overline{X}_2 - t_{\alpha/2,n_1+n_2-2}S_p\sqrt{\frac{1}{n_1} + \frac{1}{n_2}}\right.$$

$$\left.\le \mu_1 - \mu_2 \le \overline{X}_1 - \overline{X}_2 + t_{\alpha/2,n_1+n_2-2}S_p\sqrt{\frac{1}{n_1} + \frac{1}{n_2}}\right\} = 1 - \alpha. \tag{10-49}$$

Therefore, a $100(1 - \alpha)\%$ two-sided confidence interval for the difference in means $\mu_1 - \mu_2$ is

$$\overline{X}_1 - \overline{X}_2 - t_{\alpha/2,n_1+n_2-2}S_p\sqrt{\frac{1}{n_1} + \frac{1}{n_2}}$$

$$\le \mu_1 - \mu_2 \le \overline{X}_1 - \overline{X}_2 + t_{\alpha/2,n_1+n_2-2}S_p\sqrt{\frac{1}{n_1} + \frac{1}{n_2}}. \tag{10-50}$$

A one-sided $100(1 - \alpha)\%$ lower-confidence interval on $\mu_1 - \mu_2$ is

$$\overline{X}_1 - \overline{X}_2 - t_{\alpha,n_1+n_2-2}S_p\sqrt{\frac{1}{n_1} + \frac{1}{n_2}} \le \mu_1 - \mu_2, \tag{10-51}$$

and a one-sided $100(1 - \alpha)\%$ upper-confidence interval on $\mu_1 - \mu_2$ is

$$\mu_1 - \mu_2 \le \overline{X}_1 - \overline{X}_2 + t_{\alpha,n_1+n_2-2}S_p\sqrt{\frac{1}{n_1} + \frac{1}{n_2}}. \tag{10-52}$$

Example 10-20

In a batch chemical process used for etching printed circuit boards, two different catalysts are being compared to determine whether they require different emersion times for removal of identical quantities of photoresist material. Twelve batches were run with catalyst 1, resulting in a sample mean emersion time of $\overline{x}_1 = 24.6$ minutes and a sample standard deviation of $s_1 = 0.85$ minutes. Fifteen batches were run with catalyst 2, resulting in a mean emersion time of $\overline{x}_2 = 22.1$ minutes and a standard deviation of $s_2 = 0.98$ minutes. We will find a 95% confidence interval on the difference in means

$\mu_1 - \mu_2$, assuming that the standard deviations (or variances) of the two populations are equal. The pooled estimate of the common variance is found using equation 10-48 as follows:

$$\begin{aligned} s_p^2 &= \frac{(n_1-1)s_1^2 + (n_2-1)s_2^2}{n_1+n_2-2} \\ &= \frac{11(0.85)^2 + 14(0.98)^2}{12+15-2} \\ &= 0.8557. \end{aligned}$$

The pooled standard deviation is $s_p = \sqrt{0.8557} = 0.925$. Since $t_{\alpha/2,n_1+n_2-2} = t_{0.025,25} = 2.060$, we may calculate the 95% lower- and upper-confidence limits as

$$\begin{aligned} L &= \bar{x}_1 - \bar{x}_2 - t_{\alpha/2,n_1+n_2-2}s_p\sqrt{\frac{1}{n_1}+\frac{1}{n_2}} \\ &= 24.6 - 22.1 - 2.060(0.925)\sqrt{\frac{1}{12}+\frac{1}{15}} \\ &= 1.76 \text{ minutes} \end{aligned}$$

and

$$\begin{aligned} U &= \bar{x}_1 - \bar{x}_2 + t_{\alpha/2,n_1+n_2-2}s_p\sqrt{\frac{1}{n_1}+\frac{1}{n_2}} \\ &= 24.6 - 22.1 + 2.060(0.925)\sqrt{\frac{1}{12}+\frac{1}{15}} \\ &= 3.24 \text{ minutes.} \end{aligned}$$

That is, the 95% confidence interval on the difference in mean emersion times is

$$1.76 \text{ minutes} \le \mu_1 - \mu_2 \le 3.24 \text{ minutes.}$$

We are 95% confident that catalyst 1 requires an emersion time that is between 1.76 minutes and 3.24 minutes longer than that required by catalyst 2.

Case II. $\sigma_1^2 \neq \sigma_2^2$

In many situations it is not reasonable to assume that $\sigma_1^2 = \sigma_2^2$. When this assumption is unwarranted, one may still find a $100(1-\alpha)\%$ confidence interval for $\mu_1 - \mu_2$ using the fact that the statistic

$$t^* = \frac{\bar{X}_1 - \bar{X}_2 - (\mu_1 - \mu_2)}{\sqrt{S_1^2/n_1 + S_2^2/n_2}}$$

is distributed approximately as t with degrees of freedom given by

$$v = \frac{\left(S_1^2/n_1 + S_2^2/n_2\right)^2}{\dfrac{\left(S_1^2/n_1\right)^2}{n_1+1} + \dfrac{\left(S_2^2/n_2\right)^2}{n_2+1}} - 2. \tag{10-53}$$

Consequently, an approximate $100(1-\alpha)\%$ two-sided confidence interval for $\mu_1 - \mu_2$, when $\sigma_1^2 \neq \sigma_2^2$, is

$$\bar{X}_1 - \bar{X}_2 - t_{\alpha/2,v}\sqrt{\frac{S_1^2}{n_1}+\frac{S_2^2}{n_2}} \le \mu_1 - \mu_2 \le \bar{X}_1 - \bar{X}_2 + t_{\alpha/2,v}\sqrt{\frac{S_1^2}{n_1}+\frac{S_2^2}{n_2}}. \tag{10-54}$$

Upper (lower) one-sided confidence limits may be found by replacing the lower (upper)-confidence limit with $-\infty(\infty)$ and changing $\alpha/2$ to α.

10-3.3 Confidence Interval on $\mu_1 - \mu_2$ for Paired Observations

In Sections 10-3.1 and 10-3.2 we developed confidence intervals for the difference in means where two independent random samples were selected from the two populations of interest. That is, n_1 observations were selected at random from the first population and a completely independent sample of n_2 observations was selected at random from the second population. There are also a number of experimental situations where there are only n different *experimental units* and the data are collected in *pairs*; that is, two observations are made on each unit.

For example, the journal *Human Factors* (1962, p. 375) reports a study in which 14 subjects were asked to park two cars having substantially different wheelbases and turning radii. The time in seconds was recorded for each car and subject, and the resulting data are shown in Table 10-3. Notice that each subject is the "experimental unit" referred to earlier. We wish to obtain a confidence interval on the difference in mean time to park the two cars, say $\mu_1 - \mu_2$.

In general, suppose that the data consist of n pairs $(X_{11}, X_{21}), (X_{12}, X_{22}), \ldots,(X_{1n}, X_{2n})$. Both X_1 and X_2 are assumed to be normally distributed with mean μ_1 and μ_2, respectively. The random variables within *different pairs* are *independent.* However, because there are two measurements on the same experimental unit, the two measurements *within the same pair* may not be independent. Consider the n differences $D_1 = X_{11} - X_{21}, D_2 = X_{12} - X_{22}, \ldots, D_n = X_{1n} - X_{2n}$. Now the mean of the differences D, say μ_D, is

$$\mu_D = E(D) = E(X_1 - X_2) = E(X_1) - E(X_2) = \mu_1 - \mu_2,$$

because the expected value of $X_1 - X_2$ is the difference in expected values regardless of whether X_1 and X_2 are independent. Consequently, we can construct a confidence interval for $\mu_1 - \mu_2$ just by finding a confidence interval on μ_D. Since the differences D_i are normally and independently distributed, we can use the t-distribution procedure described in Section

Table 10-3 Time in Seconds to Parallel Park Two Automobiles

	Automobile		
Subject	1	2	Difference
1	37.0	17.8	19.2
2	25.8	20.2	5.6
3	16.2	16.8	−0.6
4	24.2	41.4	−17.2
5	22.0	21.4	0.6
6	33.4	38.4	−5.0
7	23.8	16.8	7.0
8	58.2	32.2	26.0
9	33.6	27.8	5.8
10	24.4	23.2	1.2
11	23.4	29.6	−6.2
12	21.2	20.6	0.6
13	36.2	32.2	4.0
14	29.8	53.8	−24.0

10-2.2 to find the confidence interval on μ_D. By analogy with equation 10-30, the 100(1 − α)% confidence interval on $\mu_D = \mu_1 - \mu_2$ is

$$\bar{D} - t_{\alpha/2,n-1} S_D/\sqrt{n} \le \mu_D \le \bar{D} + t_{\alpha/2,n-1} S_D/\sqrt{n}, \tag{10-55}$$

where $\bar{D}$ and S_D are the sample mean and sample standard deviation of the differences D_i, respectively. This confidence interval is valid for the case where $\sigma_1^2 \ne \sigma_2^2$, because S_D^2 estimates $\sigma_D^2 = V(X_1 - X_2)$. Also, for large samples (say $n \ge 30$ pairs), the assumption of normality is unnecessary.

Example 10-21

We now return to the data in Table 10-3 concerning the time for $n = 14$ subjects to parallel park two cars. From the column of observed differences d_i we calculate $\bar{d} = 1.21$ and $s_d = 12.68$. The 90% confidence interval for $\mu_D = \mu_1 - \mu_2$ is found from equation 10-55 as follows:

$$\bar{d} - t_{0.05,13}\, s_d \;\; \sqrt{n} \le \mu_D \le \bar{d} + t_{0.05,13}\, s_d \;\; \sqrt{n},$$

$$1.21 - 1.771(12.68)/\sqrt{14} \le \mu_D \le 1.21 + 1.771(12.68)/\sqrt{14},$$

$$-4.79 \le \mu_D \le 7.21.$$

Notice that the confidence interval on μ_D includes zero. This implies that at the 90% level of confidence, the data do not support the claim that the two cars have different mean parking times μ_1 and μ_2. That is, the value $\mu_D = \mu_1 - \mu_2 = 0$ is not inconsistent with the observed data.

Note that when pairing data, degrees of freedom are lost in comparison to the two-sample confidence intervals, but typically a gain in precision of estimation is achieved because S_d is smaller than S_p.

10-3.4 Confidence Interval on the Ratio of Variances of Two Normal Distributions

Suppose that X_1 and X_2 are independent normal random variables with unknown means μ_1 and μ_2 and unknown variances σ_1^2 and σ_2^2, respectively. We wish to find a 100(1 − α)% con fidence interval on the ratio σ_1^2/σ_2^2. Let two random samples of sizes n_1 and n_2 be taken on X_1 and X_2, and let S_1^2 and S_2^2 denote the sample variances. To find the confidence interval, we note that the sampling distribution of

$$F = \frac{S_2^2/\sigma_2^2}{S_1^2/\sigma_1^2}$$

is F with $n_2 - 1$ and $n_1 - 1$ degrees of freedom. This distribution is shown in Fig. 10-9.

From Fig. 10-9, we see that

$$P\{F_{1-\alpha/2,n_2-1,n_1-1} \le F \le F_{\alpha/2,n_2-1,n_1-1}\} = 1 - \alpha$$

or

$$P\left\{F_{1-\alpha/2,n_2-1,n_1-1} \le \frac{S_2^2/\sigma_2^2}{S_1^2/\sigma_1^2} \le F_{\alpha/2,n_2-1,n_1-1}\right\} = 1 - \alpha.$$

Hence

$$P\left\{\frac{S_1^2}{S_2^2} F_{1-\alpha/2,n_2-1,n_1-1} \le \frac{\sigma_1^2}{\sigma_2^2} \le \frac{S_1^2}{S_2^2} F_{\alpha/2,n_2-1,n_1-1}\right\} = 1 - \alpha. \tag{10-56}$$

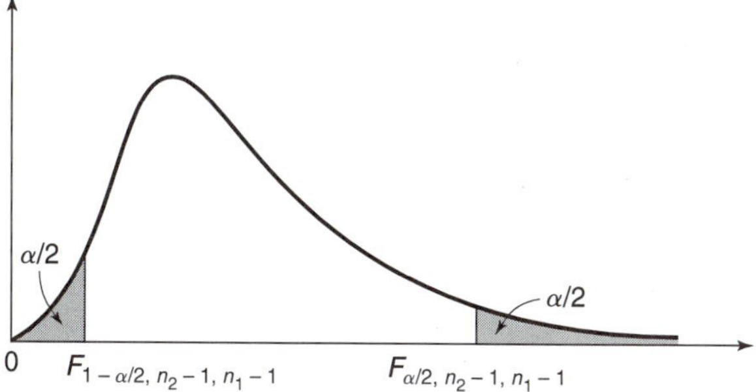

Figure 10-9 The distribution of F_{n_2-1,n_1-1}.

Comparing equations 10-56 and 10-18, we see that a 100(1 – α)% two-sided confidence interval for σ_1^2/σ_2^2 is

$$\frac{S_1^2}{S_2^2} F_{1-\alpha/2,n_2-1,n_1-1} \le \frac{\sigma_1^2}{\sigma_2^2} \le \frac{S_1^2}{S_2^2} F_{\alpha/2,n_2-1,n_1-1}, \tag{10-57}$$

where the lower 1 – $\alpha/2$ tail point of the F_{n_2-1,n_1-1} distribution is given by (see equation 9-22).

$$F_{1-\alpha/2,n_2-1,n_1-1} = \frac{1}{F_{\alpha/2,n_1-1,n_2-1}}. \tag{10-58}$$

We may also construct one-sided confidence intervals. A 100(1 – α)% lower-confidence limit on σ_1^2/σ_2^2 is

$$\frac{S_1^2}{S_2^2} F_{1-\alpha,n_2-1,n_1-1} \le \frac{\sigma_1^2}{\sigma_2^2}, \tag{10-59}$$

while a 100(1 – α)% upper-confidence interval on σ_1^2/σ_2^2 is

$$\frac{\sigma_1^2}{\sigma_2^2} \le \frac{S_1^2}{S_2^2} F_{\alpha,n_2-1,n_1-1}. \tag{10-60}$$

Example 10-22

Consider the batch chemical etching process described in Example 10-20. Recall that two catalysts are being compared to measure their effectiveness in reducing emersion times for printed circuit boards. $n_1 = 12$ batches were run with catalyst 1 and $n_2 = 15$ batches were run with catalyst 2, yielding $s_1 = 0.85$ minutes and $s_2 = 0.98$ minutes. We will find a 90% confidence interval on the ratio of variances σ_1^2/σ_2^2. From equation 10-57, we find that

$$\frac{s_1^2}{s_2^2} F_{0.95,14,11} \le \frac{\sigma_1^2}{\sigma_2^2} \le \frac{s_1^2}{s_2^2} F_{0.05,14,11},$$

$$\frac{(0.85)^2}{(0.98)^2} 0.39 \le \frac{\sigma_1^2}{\sigma_2^2} \le \frac{(0.85)^2}{(0.98)^2} 2.74,$$

or

$$0.29 \le \frac{\sigma_1^2}{\sigma_2^2} \le 2.06,$$

using the fact that $F_{0.95,14,11} = 1/F_{0.05,11,14} = 1/2.58 = 0.39$. Since this confidence interval includes unity, we could not claim that the standard deviations of the emersion times for the two catalysts are different at the 90% level of confidence.

10-3.5 Confidence Interval on the Difference between Two Proportions

If there are two proportions of interest, say p_1 and p_2, it is possible to obtain a $100(1-\alpha)\%$ confidence interval on their difference, $p_1 - p_2$. If two independent samples of size n_1 and n_2 are taken from infinite populations so that X_1 and X_2 are independent, binomial random variables with parameters (n_1, p_1) and (n_2, p_2), respectively, where X_1 represents the number of sample observations from the first population that belong to a class of interest and X_2 represents the number of sample observations from the second population that belong to the class of interest, then $\hat{p}_1 = X_1/n_1$ and $\hat{p}_2 = X_2/n_2$ are independent estimators of p_1 and p_2, respectively. Furthermore, under the assumption that the normal approximation to the binomial applies, the statistic

$$Z = \frac{\hat{p}_1 - \hat{p}_2 - (p_1 - p_2)}{\sqrt{\dfrac{p_1(1-p_1)}{n_1} + \dfrac{p_2(1-p_2)}{n_2}}}$$

is distributed approximately as standard normal. Using an approach analogous to that of the previous section, it follows that an approximate $100(1-\alpha)\%$ two-sided confidence interval for $p_1 - p_2$ is

$$\hat{p}_1 - \hat{p}_2 - Z_{\alpha/2}\sqrt{\frac{\hat{p}_1(1-\hat{p}_1)}{n_1} + \frac{\hat{p}_2(1-\hat{p}_2)}{n_2}}$$

$$\le p_1 - p_2 \le \hat{p}_1 - \hat{p}_2 + Z_{\alpha/2}\sqrt{\frac{\hat{p}_1(1-\hat{p}_1)}{n_1} + \frac{\hat{p}_2(1-\hat{p}_2)}{n_2}}. \qquad (10\text{-}61)$$

An approximate $100(1-\alpha)\%$ lower-confidence interval for $p_1 - p_2$ is

$$\hat{p}_1 - \hat{p}_2 - Z_{\alpha}\sqrt{\frac{\hat{p}_1(1-\hat{p}_1)}{n_1} + \frac{\hat{p}_2(1-\hat{p}_2)}{n_2}} \le p_1 - p_2, \qquad (10\text{-}62)$$

and an approximate $100(1-\alpha)\%$ upper-confidence interval for $p_1 - p_2$ is

$$p_1 - p_2 \le \hat{p}_1 - \hat{p}_2 + Z_{\alpha}\sqrt{\frac{\hat{p}_1(1-\hat{p}_1)}{n_1} + \frac{\hat{p}_2(1-\hat{p}_2)}{n_2}}. \qquad (10\text{-}63)$$

Example 10-23

Consider the data in Example 10-17. Suppose that a modification is made in the surface finishing process and subsequently a second random sample of 85 axle shafts is obtained. The number of defective shafts in this second sample is 10. Therefore, since $n_1 = 75$, $\hat{p}_1 = 0.16$, $n_2 = 85$, and $\hat{p}_2 = 10/85 = 0.12$, we can obtain an approximate 95% confidence interval on the difference in the proportions of defectives produced under the two processes from equation 10-61 as

$$\hat{p}_1 - \hat{p}_2 - Z_{0.025}\sqrt{\frac{\hat{p}_1(1-\hat{p}_1)}{n_1} + \frac{\hat{p}_2(1-\hat{p}_2)}{n_2}}$$

$$\le p_1 - p_2 \le \hat{p}_1 - \hat{p}_2 + Z_{0.025}\sqrt{\frac{\hat{p}_1(1-\hat{p}_1)}{n_1} + \frac{\hat{p}_2(1-\hat{p}_2)}{n_2}}$$

or

$$0.16-0.12-1.96\sqrt{\frac{0.16(0.84)}{75}+\frac{0.12(0.88)}{85}}$$
$$\le p_1-p_2\le 0.16-0.12+1.96\sqrt{\frac{0.16(0.84)}{75}+\frac{0.12(0.88)}{85}}.$$

This simplifies to

$$-0.07\le p_1-p_2\le 0.15.$$

This interval includes zero, so, based on the sample data, it seems unlikely that the changes made in the surface finish process have reduced the proportion of defective axle shafts being produced.

10-4 APPROXIMATE CONFIDENCE INTERVALS IN MAXIMUM LIKELIHOOD ESTIMATION

If the method of maximum likelihood is used for parameter estimation, the asymptotic properties of these estimators may be used to obtain approximate confidence intervals. Let $\hat{\theta}$ be the maximum likelihood estimator of θ. For large samples, $\hat{\theta}$ is approximately normally distributed with mean θ and variance $V(\hat{\theta})$ given by the Cramér–Rao lower bound (equation 10-4). Therefore, an approximate 100(1 – α)% confidence interval for θ is

$$\hat{\theta}-Z_{\alpha/2}[V(\hat{\theta})]^{1/2}\le\theta\le\hat{\theta}+Z_{\alpha/2}[V(\hat{\theta})]^{1/2}. \tag{10-64}$$

Usually, the $V(\hat{\theta})$ is a function of the unknown parameter θ. In these cases, replace θ with $\hat{\theta}$.

Example 10-24

Recall Example 10-3, where it was shown that the maximum likelihood estimator of the parameter p of a Bernoulli distribution is $\hat{p}=(1/n)\sum_{i=1}^{n}X_i=\bar{X}$. Using the Cramér–Rao lower bound, we may verify that the lower bound for the variance of $\hat{p}$ is

$$V(\hat{p})\ge\frac{1}{nE\left[\frac{d}{dp}\ln\left[p^X(1-p)^{1-X}\right]\right]^2}$$
$$=\frac{1}{nE\left[\frac{X}{p}-\frac{(1-X)}{(1-p)}\right]^2}$$
$$=\frac{1}{nE\left[\frac{X^2}{p^2}+\frac{(1-X)^2}{(1-p)^2}-2\frac{X(1-X)}{p(1-p)}\right]}.$$

For the Bernoulli distribution, we observe that $E(X)=p$ and $E(X^2)=p$. Therefore, this last expression simplifies to

$$V(\hat{p})\ge\frac{1}{n\left[\frac{1}{p}+\frac{1}{(1-p)}\right]}=\frac{p(1-p)}{n}.$$

This result should not be surprising, since we know directly that for the Bernoulli distribution, $V(\bar{X}) = V(X_i)/n = p(1-p)/n$. In any case, replacing p in $V(\hat{p})$ by $\hat{p}$, the approximate 100(1 – α)% confidence interval for p is found from equation 10–64 to be

$$\hat{p} - Z_{\alpha/2}\sqrt{\frac{\hat{p}(1-\hat{p})}{n}} \le p \le \hat{p} + Z_{\alpha/2}\sqrt{\frac{\hat{p}(1-\hat{p})}{n}}.$$

10-5 SIMULTANEOUS CONFIDENCE INTERVALS

Occasionally it is necessary to construct several confidence intervals on more than one parameter, and we wish the probability to be $(1 - \alpha)$ that *all* such confidence intervals simultaneously produce correct statements. For example, suppose that we are sampling from a normal population with unknown mean and variance, and we wish to construct confidence intervals for μ and σ^2 such that the probability is $(1 - \alpha)$ that both intervals simultaneously yield correct conclusions. Since $\overline{X}$ and S^2 are independent, we could ensure this result by constructing $100(1 - \alpha)^{1/2}\%$ confidence intervals for each parameter separately, and both intervals would simultaneously produce correct conclusions with probability $(1 - \alpha)^{1/2}(1 - \alpha)^{1/2} = (1 - \alpha)$.

If the sample statistics on which the confidence intervals are based are not independent random variables, then the confidence intervals are not independent, and other methods must be used. In general, suppose that m confidence intervals are required. The Bonferroni inequality states that

$$P\{\text{all } m \text{ statements are simultaneously correct}\} \equiv 1-\alpha \ge 1-\sum_{i=1}^{m}\alpha_i, \qquad (10\text{-}65)$$

where $1 - \alpha_i$ is the confidence level used in the ith confidence interval. In practice, we select a value for the simultaneous confidence level $1 - \alpha$, and then choose the individual α_i such that $\sum_{i=1}^{m}\alpha_i = \alpha$. Usually, we set $\alpha_i = \alpha/m$.

As an illustration, suppose we wished to construct two confidence intervals on the means of two normal distributions such that we are at least 90% confident that both statements are simultaneously correct. Therefore, since $1 - \alpha = 0.90$, we have $\alpha = 0.10$, and since two confidence intervals are required, each of these should be constructed with $\alpha_i = \alpha/2 = 0.10/2 = 0.05$, $i = 1, 2$. That is, two individual 95% confidence intervals on μ_1 and μ_2 will *simultaneously* lead to correct statements with probability at least 0.90.

10-6 BAYESIAN CONFIDENCE INTERVALS

Previously, we presented Bayesian techniques for point estimation. In this section, we will present the Bayesian approach to constructing confidence intervals.

We may use Bayesian methods to construct interval estimates of parameters that are similar to confidence intervals. If the posterior density for θ has been obtained, we can construct an interval, usually centered at the posterior mean, that contains $100(1 - \alpha)\%$ of the posterior probability. Such an interval is called the $100(1 - \alpha)\%$ Bayes interval for the unknown parameter θ.

While in many cases the Bayes interval estimate for θ will be quite similar to a classical confidence interval with the same confidence coefficient, the interpretation of the two is very different. A confidence interval is an interval that, before the sample is taken, will include the unknown θ with probability $1 - \alpha$. That is, the classical confidence interval relates to the relative frequency of an interval including θ. On the other hand, a Bayes interval is an interval that contains $100(1 - \alpha)\%$ of the posterior probability for θ. Since the posterior probability density measures a degree of belief about θ given the sample results, the Bayes interval provides a subjective degree of belief about θ rather than a frequency interpretation. The Bayes interval estimate of θ is affected by the sample results but is not completely determined by them.

Example 10-25

Suppose that the random variable X is normally distributed with mean μ and variance 4. The value of μ is unknown, but a reasonable prior density would be normal with mean 2 and variance 1. That is,

$$f(x_1, x_2, \ldots, x_n | \mu) = \frac{1}{(8\pi)^{n/2}} e^{-(1/8)\Sigma(x_i - \mu)^2}$$

and

$$f(\mu) = \frac{1}{\sqrt{2\pi}} e^{-(1/2)(\mu-2)^2}.$$

We can show that the posterior density for μ is

$$f(\mu | x_1, x_2, \ldots, x_n) = \frac{1}{\sqrt{2\pi}}\left(\frac{n}{4}+1\right)^{1/2} \exp\left\{-\frac{1}{2}\left(\frac{n}{4}+1\right)\left[\mu - \frac{\left(\frac{n\bar{x}}{4}+2\right)}{\left(\frac{n}{4}+1\right)}\right]^2\right\}$$

$$= \frac{1}{\sqrt{2\pi}}\left(\frac{n}{4}+1\right)^{1/2} \exp\left\{-\frac{1}{2}\left(\frac{n}{4}+1\right)\left(\mu - \frac{n\bar{x}+8}{n+4}\right)^2\right\}$$

using the methods of Section 10-1.4. Thus, the posterior distribution for μ is normal with mean $(n\bar{X} + 8)/(n + 4)$ and variance $4/(n + 4)$. A 95% Bayes interval for μ, which is symmetric about the posterior mean, would be

$$\frac{n\bar{X}+8}{n+4} - Z_{0.025}\frac{2}{\sqrt{n+4}} \le \mu \le \frac{n\bar{X}+8}{n+4} + Z_{0.025}\frac{2}{\sqrt{n+4}} \tag{10-66}$$

If a random sample of size 16 is taken and we find that $\bar{x} = 2.5$, equation 10-66 reduces to

$$1.52 \le \mu \le 3.28.$$

If we ignore the prior information, the classical confidence interval for μ is

$$1.52 \le \mu \le 3.48.$$

We see that the Bayes interval is slightly shorter than the classical confidence interval, because the prior information is equivalent to a slight increase in the sample size if no prior knowledge was assumed.

10-7 BOOTSTRAP CONFIDENCE INTERVALS

In Section 10-1.6 we introduced the bootstrap technique for estimating the standard error of a parameter, θ. The bootstrap technique can also be used to construct a confidence interval on θ.

For an arbitrary parameter θ, general $100(1 - \alpha)\%$ lower and upper limits are, respectively,

$$L = \hat{\theta} - 100(1 - \alpha/2) \text{ percentile of } (\hat{\theta} - \theta),$$
$$U = \hat{\theta} - 100(\alpha/2) \text{ percentile of } (\hat{\theta} - \theta).$$

Bootstrap samples can be generated to estimate the values of L and U.

Suppose B bootstrap samples are generated and $\hat{\theta}_1^*, \hat{\theta}_2^*, \ldots, \hat{\theta}_B^*$ and $\overline{\theta}^*$ are calculated. From these estimates, we then compute the differences $\hat{\theta}_1^* - \overline{\theta}^*, \hat{\theta}_2^* - \overline{\theta}^*, \ldots, \hat{\theta}_B^* - \overline{\theta}^*$, arrange the differences in increasing order, and find the necessary percentiles $100(1 - \alpha/2)$ and $100(\alpha/2)$ for L and U. For example, if $B = 200$ and a 90% confidence interval is desired, then the $100(1 - 0.10/2) = 95$th percentile and the $100(0.10/2) = 5$th percentile would be the 190th difference and the 10th difference, respectively.

Example 10-26

An electronic device consists of four components. The time to failure for each component follows an exponential distribution and the components are identical to and independent of one another. The electronic device will fail only after all four components have failed. The times to failure for the electronic components have been collected for 15 such devices. The total times to failure are

78.7778, 13.5260, 6.8291, 47.3746, 16.2033, 27.5387, 28.2515, 38.5826,
35.4363, 80.2757, 50.3861, 81.3155, 42.2532, 33.9970, 57.4312.

It is of interest to construct a 90% confidence interval on the exponential parameter λ. By definition, the sum of r independent and identically distributed exponential random variables follows a gamma distribution and is defined as gamma(r, λ). Therefore, $r = 4$, but λ needs to be estimated. A bootstrap estimate for λ can be found using the technique given in Section 10-1.6. Using the time-to-failure data above, we find the average time to failure to be $\bar{x} = 42.545$. The mean of a gamma distribution is $E(X) = r/\lambda$ and λ is calculated for each bootstrap sample. Running Minitab® for $B = 100$ bootstraps, we found that the bootstrap estimate $\overline{\lambda}^* = 0.0949$. Using the bootstrap estimates for each sample, the differences can be calculated and some of the calculations are shown in Table 10-4.

When the 100 differences are arranged in increasing order, the 5th percentile and the 95th percentiles turn out to be –0.0205 and 0.0232, respectively. Therefore, the resulting confidence limits are

$$L = 0.0949 - 0.0232 = 0.0717,$$
$$U = 0.0949 - (-0.0205) = 0.1154.$$

We are approximately 90% confident that the true value of λ lies between 0.0717 and 0.1154. Figure 10-10 displays the histogram of the bootstrap estimates $\hat{\lambda}_i^*$ while Fig. 10-11 depicts the differences $\hat{\lambda}_i^* - \overline{\lambda}^*$. The bootstrap estimates are reasonable when the estimator is unbiased and the standard error is approximately constant.

Table 10-4 Bootstrap Estimates for Example 10-26

Sample	$\hat{\lambda}_i^*$	$\hat{\lambda}_i^* - \overline{\lambda}^*$
1	0.087316	–0.0075392
2	0.090689	–0.0041660
3	0.096664	0.0018094
⋮		
100	0.090193	–0.0046623

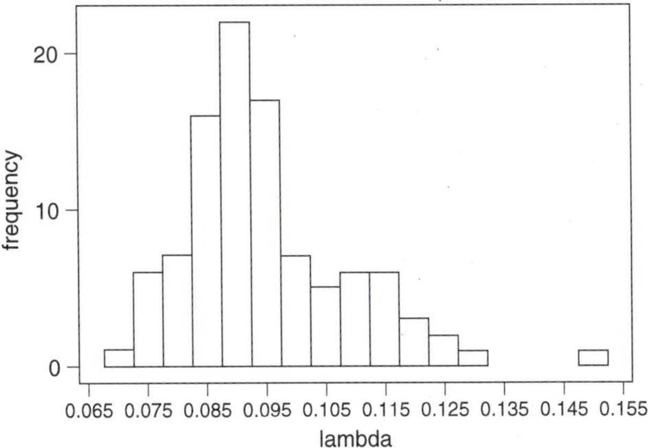

Figure 10-10 Histogram of the bootstrap estimates $\hat{\lambda}_i^*$.

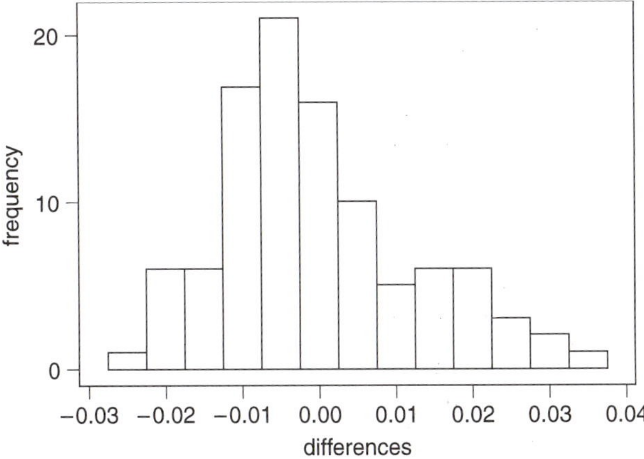

Figure 10-11 Histogram of the differences $\hat{\lambda}_i^* - \overline{\lambda}^*$

10-8 OTHER INTERVAL ESTIMATION PROBLEMS

10-8.1 Prediction Intervals

So far in this chapter we have presented interval estimators on population parameters, such as the mean, μ. There are many situations where the practitioner would like to predict a single future observation for the random variable of interest instead of predicting or estimating the average of this random variable. A *prediction interval* can be constructed for any single observation at some future time.

Consider a given random sample of size n, $X_1, X_2, \ldots, X_n$, from a normal population with mean μ and variance σ^2. Let the sample average be denoted by $\overline{X}$. Suppose we wish to predict the future observation X_{n+1}. Since $\overline{X}$ is the point predictor for this observation, the prediction error is given by $X_{n+1} - \overline{X}$. The expected value and variance of the prediction error are

$$E(X_{n+1} - \overline{X}) = E(X_{n+1}) - E(\overline{X}) = \mu - \mu = 0$$

and

$$\begin{aligned}\operatorname{Var}(X_{n+1}-\overline{X}) &= \operatorname{Var}(X_{n+1})+\operatorname{Var}(\overline{X})\\ &= \sigma^2+\frac{\sigma^2}{n}\\ &= \sigma^2\left(1+\frac{1}{n}\right).\end{aligned}$$

Since X_{n+1} and $\overline{X}$ are independent, normally distributed random variables, the prediction error is also normally distributed and

$$Z=\frac{(X_{n+1}-\overline{X})-0}{\sqrt{\sigma^2\left(1+\frac{1}{n}\right)}}=\frac{X_{n+1}-\overline{X}}{\sqrt{\sigma^2\left(1+\frac{1}{n}\right)}}$$

and is standard normal. If σ^2 is unknown, it can be estimated by the sample variance, S^2, and then

$$T=\frac{X_{n+1}-\overline{X}}{\sqrt{S^2\left(1+\frac{1}{n}\right)}}$$

follows the t distribution with $n-1$ degrees of freedom.

Following the usual procedure for constructing confidence intervals, the two-sided $100(1-\alpha)\%$ prediction interval is

$$-t_{\alpha/2,n-1}\leq\frac{X_{n+1}-\overline{X}}{\sqrt{S^2\left(1+\frac{1}{n}\right)}}\leq t_{\alpha/2,n-1}.$$

By rearranging the inequality we obtain the final form for the two-sided $100(1-\alpha)\%$ prediction interval:

$$\overline{X}-t_{\alpha/2,n-1}\sqrt{S^2\left(1+\frac{1}{n}\right)}\leq X_{n+1}\leq\overline{X}+t_{\alpha/2,n-1}\sqrt{S^2\left(1+\frac{1}{n}\right)}. \qquad (10\text{-}67)$$

The lower one-sided $100(1-\alpha)\%$ prediction interval on X_{n+1} is given by

$$\overline{X}-t_{\alpha,n-1}\sqrt{S^2\left(1+\frac{1}{n}\right)}\leq X_{n+1}. \qquad (10\text{-}68)$$

The upper one-sided $100(1-\alpha)\%$ prediction interval on X_{n+1} is given by

$$X_{n+1}\leq\overline{X}+t_{\alpha,n-1}\sqrt{S^2\left(1+\frac{1}{n}\right)}. \qquad (10\text{-}69)$$

Example 10-27

Maximum forces experienced by a transport aircraft for an airline on a particular route for 10 flights are (in units of gravity, g)

1.15, 1.23, 1.56, 1.69, 1.71, 1.83, 1.83, 1.85, 1.90, 1.91.

The sample average and sample standard deviation are calculated to be $\bar{x} = 1.666$ and $s = 0.273$, respectively. It may be of importance to predict the next maximum force experienced by the aircraft. Since $t_{0.025,9} = 2.262$, the 95% prediction interval on X_{11} is

$$1.666 - t_{\alpha/2,n-1}\sqrt{(0.273)^2\left(1+\frac{1}{10}\right)} \le X_{11} \le 1.666 + t_{\alpha/2,n-1}\sqrt{(0.273)^2\left(1+\frac{1}{10}\right)},$$

$$1.018 \le X_{11} \le 2.314.$$

10-8.2 Tolerance Intervals

As presented earlier in this chapter, confidence intervals are the intervals in which we expect the true population parameter, such as μ, to lie. In contrast, *tolerance intervals* are intervals in which we expect a *percentage* of the population values to lie.

Suppose that X is a normally distributed random variable with mean μ and variance σ^2. We would expect approximately 95% of all values of X to be contained within the interval $\mu \pm 1.645\sigma$. But what if μ and σ are unknown and must be estimated? Using the point estimates $\bar{x}$ and s for a sample of size n, we can construct the interval $\bar{x} \pm 1.645s$. Unfortunately, due to the variability in estimating μ and σ the resulting interval may contain less than 95% of the values. In this particular instance, a value larger than 1.645 will be needed to guarantee 95% coverage when using point estimates for the population parameters. We can construct an interval that will contain the stated percentage of population values and be relatively confident in the result. For example, we may want to be 90% confident that the resulting interval covers at least 95% of the population values. This type of interval is referred to as a *tolerance interval* and can be constructed easily for various confidence levels.

In general, for $0 < q < 100$, the two-sided tolerance interval for covering at least q% of the values from a normal population with $100(1 - \alpha)$% confidence is $\bar{x} \pm ks$. The value k is a constant tabulated for various combinations of q and $100(1 - \alpha)$. Values of k are given in Table XIV of the Appendix for $q = 90$, 95, and 99 and for $100(1 - \alpha) = 90$, 95, and 99.

The lower one-sided tolerance interval for covering at least q% of the values from a normal population with $100(1 - \alpha)$% confidence is $\bar{x} - ks$. The upper one-sided tolerance interval for covering at least q% of the values from a normal population with $100(1 - \alpha)$% confidence is $\bar{x} + ks$. Various values of k for one-sided tolerance intervals were calculated using the technique given in Odeh and Owens (1980) and are provided in Table XIV of the Appendix.

Example 10-28

Reconsider the maximum forces for the transport aircraft in Example 10-27. A two-sided tolerance interval is desired that would cover 99% of all maximum forces with 95% confidence. From Table XIV (Appendix), with $1 - \alpha = 0.95$, $q = 0.99$, and $n = 10$, we find that $k = 4.433$. The sample average and sample standard deviation were calculated as $\bar{x} = 1.666$ and $s = 0.273$, respectively. The resulting tolerance interval is then

$$1.666 \pm 4.433(0.273)$$

or

$$(0.456, 2.876).$$

Therefore, we conclude that we are 95% confident that at least 99% of all maximum forces would lie between 0.456 g and 2.876 g.

It is possible to construct *nonparametric* tolerance intervals that are based on the extreme values in a random sample of size n from any continuous population. If P is the minimum proportion of the population contained between the largest and smallest observation with confidence $1 - \alpha$, then it can be shown that

$$nP^{n-1} - (n-1)P^n = \alpha.$$

Further, the required n is approximately

$$n = \frac{1}{2} + \frac{1+P}{1-P} \cdot \frac{\chi^2_{\alpha,4}}{4}. \tag{10-70}$$

Thus, in order to be 95% certain that at least 90% of the population will be included between the extreme values of the sample, we require a sample of size

$$n = \frac{1}{2} + \frac{1.9}{0.1} \cdot \frac{9.488}{4} \simeq 46.$$

Note that there is a fundamental difference between confidence limits and tolerance limits. Confidence limits (and thus confidence intervals) are used to estimate a parameter of a population, while tolerance limits (and tolerance intervals) are used to indicate the limits between which we can expect to find a proportion of a population. As n approaches infinity, the length of a confidence interval approaches zero, while tolerance limits approach the corresponding quantiles for the population.

10-9 SUMMARY

This chapter has introduced the point and interval estimation of unknown parameters. A number of methods of obtaining point estimators were discussed, including the method of maximum likelihood and the method of moments. The method of maximum likelihood usually leads to estimators that have good statistical properties. Confidence intervals were derived for a variety of parameter estimation problems. These intervals have a frequency interpretation. The two-sided confidence intervals developed in Sections 10-2 and 10-3 are summarized in Table 10-5. In some instances, one-sided confidence intervals may be appropriate. These may be obtained by setting one confidence limit in the two-sided confidence interval equal to the lower (or upper) limit of a feasible region for the parameter, and using α instead of $\alpha/2$ as the probability level on the remaining upper (or lower) confidence limit. Confidence intervals using a bootstrapping technique were introduced. Tolerance intervals were also presented. Approximate confidence intervals in maximum likelihood estimation and simultaneous confidence intervals were also briefly introduced.

Table 10-5 Summary of Confidence Interval Procedures

Problem Type	Point Estimator	Two-Sided 100(1 – α)% Confidence Interval
Mean μ of a normal distribution, variance σ^2 known	$\overline{X}$	$\overline{X} - Z_{\alpha/2}\,\sigma/\sqrt{n} \le \mu \le \overline{X} + Z_{\alpha/2}\,\sigma/\sqrt{n}$
Difference in means of two normal distributions μ_1 and μ_2, variances σ_1^2 and σ_2^2 known	$\overline{X}_1 - \overline{X}_2$	$\overline{X}_1 - \overline{X}_2 - Z_{\alpha/2}\sqrt{\frac{\sigma_1^2}{n_1} + \frac{\sigma_2^2}{n_2}} \le \mu_1 - \mu_2 \le \overline{X}_1 - \overline{X}_2 + Z_{\alpha/2}\sqrt{\frac{\sigma_1^2}{n_1} + \frac{\sigma_2^2}{n_2}}$
Mean μ of a normal distribution, variance σ^2 unknown	$\overline{X}$	$\overline{X} - t_{\alpha/2,n-1}\,S/\sqrt{n} \le \mu \le \overline{X} + t_{\alpha/2,n-1}\,S/\sqrt{n}$
Difference in means of two normal distributions $\mu_1 - \mu_2$, variance $\sigma_1^2 = \sigma_2^2$ unknown	$\overline{X}_1 - \overline{X}_2$	$\overline{X}_1 - \overline{X}_2 - t_{\alpha/2,n_1+n_2-2}S_p\sqrt{\frac{1}{n_1} + \frac{1}{n_2}} \le \mu_1 - \mu_2 \le \overline{X}_1 - \overline{X}_2 + t_{\alpha/2,n_1+n_2-2}S_p\sqrt{\frac{1}{n_1} + \frac{1}{n_2}}$, where $S_p = \sqrt{\frac{(n_1-1)S_1^2 + (n_2-1)S_2^2}{n_1+n_2-2}}$
Difference in means of two normal distributions for paired samples $\mu_D = \mu_1 - \mu_2$	$\overline{D}$	$\overline{D} - t_{\alpha/2,n-1}\,S_D/\sqrt{n} \le \mu_D \le \overline{D} + t_{\alpha/2,n-1}\,S_D/\sqrt{n}$
Variance σ^2 of a normal distribution	S^2	$\frac{(n-1)S^2}{\chi^2_{\alpha/2,n-1}} \le \sigma^2 \le \frac{(n-1)S^2}{\chi^2_{1-\alpha/2,n-1}}$
Ratio of the variances σ_1^2/σ_2^2 of two normal distributions	$\frac{S_1^2}{S_2^2}$	$\frac{S_1^2}{S_2^2}F_{1-\alpha/2,n_2-1,n_1-1} \le \frac{\sigma_1^2}{\sigma_2^2} \le \frac{S_1^2}{S_2^2}F_{\alpha/2,n_2-1,n_1-1}$
Proportion or parameter of a binomial distribution p	$\hat{p}$	$\hat{p} - Z_{\alpha/2}\sqrt{\frac{\hat{p}(1-\hat{p})}{n}} \le p \le \hat{p} + Z_{\alpha/2}\sqrt{\frac{\hat{p}(1-\hat{p})}{n}}$
Difference in two proportions or two binomial parameters $p_1 - p_2$	$\hat{p}_1 - \hat{p}_2$	$\hat{p}_1 - \hat{p}_2 - Z_{\alpha/2}\sqrt{\frac{\hat{p}_1(1-\hat{p}_1)}{n_1} + \frac{\hat{p}_2(1-\hat{p}_2)}{n_2}} \le p_1 - p_2 \le \hat{p}_1 - \hat{p}_2 + Z_{\alpha/2}\sqrt{\frac{\hat{p}_1(1-\hat{p}_1)}{n_1} + \frac{\hat{p}_2(1-\hat{p}_2)}{n_2}}$

10-10 EXERCISES

10-1. Suppose we have a random sample of size $2n$ from a population denoted X, and $E(X) = \mu$ and $V(X) = \sigma^2$. Let

$$\bar{X}_1 = \frac{1}{2n}\sum_{i=1}^{2n} X_i \quad \text{and} \quad \bar{X}_2 = \frac{1}{n}\sum_{i=1}^{n} X_i$$

be two estimators of μ. Which is the better estimator of μ? Explain your choice.

10-2. Let $X_1, X_2, \ldots, X_7$ denote a random sample from a population having mean μ and variance σ^2. Consider the following estimators of μ:

$$\hat{\theta}_1 = \frac{X_1 + X_2 + \cdots + X_7}{7},$$

$$\hat{\theta}_2 = \frac{2X_1 - X_6 + X_4}{2}.$$

Is either estimator unbiased? Which estimator is "better"? In what sense is it better?

10-3. Suppose that $\hat{\theta}_1$ and $\hat{\theta}_2$ are estimators of the parameter θ. We know that $E(\hat{\theta}_1) = \theta$, $E(\hat{\theta}_2) = \theta/2$, $V(\hat{\theta}_1) = 10$, and $V(\hat{\theta}_2) = 4$. Which estimator is "better"? In what sense is it better?

10-4. Suppose that $\hat{\theta}_1$, $\hat{\theta}_2$, and $\hat{\theta}_3$ are estimators of θ. We know that $E(\hat{\theta}_1) = E(\hat{\theta}_2) = \theta$, $E(\hat{\theta}_3) \neq \theta$, $V(\hat{\theta}_1) = 12$, $V(\hat{\theta}_2) = 10$, and $E(\hat{\theta}_3 - \theta)^2 = 6$. Compare these three estimators. Which do you prefer? Why?

10-5. Let three random samples of sizes $n_1 = 10$, $n_2 = 8$, and $n_3 = 6$ be taken from a population with mean μ and variance σ^2. Let S_1^2, S_2^2, and S_3^2 be the sample variances. Show that

$$S^2 = \frac{10S_1^2 + 8S_2^2 + 6S_3^2}{24}$$

is an unbiased estimator of σ^2.

10-6. Best Linear Unbiased Estimators. An estimator $\hat{\theta}$ is called a linear estimator if it is a linear combination of the observations in the sample. $\hat{\theta}$ is called a best linear unbiased estimator if, of all linear functions of the observations, it both is unbiased and has minimum variance. Show that the sample mean $\bar{X}$ is the best linear unbiased estimator of the population mean μ.

10-7. Find the maximum likelihood estimator of the parameter c of the Poisson distribution, based on a random sample of size n.

10-8. Find the estimator of c in the Poisson distribution by the method of moments, based on a random sample of size n.

10-9. Find the maximum likelihood estimator of the parameter λ in the exponential distribution, based on a random sample of size n.

10-10. Find the estimator of λ in the exponential distribution by the method of moments, based on a random sample of size n.

10-11. Find moment estimators of the parameters r and λ of the gamma distribution, based on a random sample of size n.

10-12. Let X be a geometric random variable with parameter p. Find an estimator of p by the method of moments, based on a random sample of size n.

10-13. Let X be a geometric random variable with parameter p. Find the maximum likelihood estimator of p, based on a random sample of size n.

10-14. Let X be a Bernoulli random variable with parameter p. Find an estimator of p by the method of moments, based on a random sample of size n.

10-15. Let X be a binomial random variable with parameters n (known) and p. Find an estimator of p by the method of moments, based on a random sample of size N.

10-16. Let X be a binomial random variable with parameters n and p, both unknown. Find estimators of n and p by the method of moments, based on a random sample of size N.

10-17. Let X be a binomial random variable with parameters n (unknown) and p. Find the maximum likelihood estimator of p, based on a random sample of size N.

10-18. Set up the likelihood function for a random sample of size n from a Weibull distribution. What difficulties would be encountered in obtaining the maximum likelihood estimators of the three parameters of the Weibull distribution?

10-19. Prove that if $\hat{\theta}$ is an unbiased estimator of θ, and if $\lim_{n\to\infty} V(\hat{\theta}) = 0$, then $\hat{\theta}$ is a consistent estimator of θ.

10-20. Let X be a random variable with mean μ and variance σ^2. Given two random samples of sizes n_1 and n_2 with sample means $\bar{X}_1$ and $\bar{X}_2$, respectively, show that

$$\bar{X} = a\bar{X}_1 + (1 - a)\bar{X}_2, \qquad 0 < a < 1,$$

is an unbiased estimator of μ. Assuming $\bar{X}_1$ and $\bar{X}_2$ to be independent, find the value of a that minimizes the variance of $\bar{X}$.

10-21. Suppose that the random variable X has the probability distribution

$$\begin{aligned} f(x) &= (\gamma + 1)x^\gamma, && 0 < x < 1, \\ &= 0, && \text{otherwise.} \end{aligned}$$

Let $X_1, X_2, \ldots, X_n$ be a random sample of size n. Find the maximum likelihood estimator of γ.

10-22. Let X have the truncated (on the left at x) exponential distribution

$$f(x) = \lambda \exp[-\lambda(x - x_\ell)], \qquad x > x_\ell > 0,$$
$$= 0, \qquad \text{otherwise.}$$

Let $X_1, X_2, \ldots, X_n$ be a random sample of size n. Find the maximum likelihood estimator of λ.

10-23. Assume that λ in the previous exercise is known but x_ℓ is unknown. Obtain the maximum likelihood estimator of x_ℓ.

10-24. Let X be a random variable with mean μ and variance σ^2, and let $X_1, X_2, \ldots, X_n$ be a random sample of size n from X. Show that the estimator $G = K\sum_{i=1}^{n-1}\left(X_{i+1} - X_i\right)^2$ is unbiased for an appropriate choice for K. Find the appropriate value for K.

10-25. Let X be a normally distributed random variable with mean μ and variance σ^2. Assume that σ^2 is known and μ unknown. The prior density for μ is assumed to be normal with mean μ_0 and variance σ_0^2. Determine the posterior density for μ, given a random sample of size n from X.

10-26. Let X be normally distributed with known mean μ and unknown variance σ^2. Assume that the prior density for $1/\sigma^2$ is a gamma distribution with parameters $m + 1$ and $m\sigma_0^2$. Determine the posterior density for $1/\sigma^2$, given a random sample of size n from X.

10-27. Let X be a geometric random variable with parameter p. Suppose we assume a beta distribution with parameters a and b as the prior density for p. Determine the posterior density for p, given a random sample of size n from X.

10-28. Let X be a Bernoulli random variable with parameter p. If the prior density for p is a beta distribution with parameters a and b, determine the posterior density for p, given a random sample of size n from X.

10-29. Let X be a Poisson random variable with parameter λ. The prior density for λ is a gamma distribution with parameters $m + 1$ and $(m + 1)/\lambda_0$. Determine the posterior density for λ, given a random sample of size n from X.

10-30. Suppose that $X \sim N(\mu, 40)$, and let the prior density for μ be $N(4, 8)$. For a random sample of size 25, the value $\bar{x} = 4.85$ is obtained. What is the Bayes estimate of μ, assuming a squared-error loss?

10-31. A process manufactures printed circuit boards. A locating notch is drilled a distance X from a component hole on the board. The distance is a random variable $X \sim N(\mu, 0.01)$. The prior density for μ is uniform between 0.98 and 1.20 inches. A random sample of size 4 produces the value $\bar{x} = 1.05$. Assuming a squared-error loss, determine the Bayes estimate of μ.

10-32. The time between failures of a milling machine is exponentially distributed with parameter λ. Suppose we assume an exponential prior on λ with a mean of 3000 hours. Two machines are observed and the average time between failures is $\bar{x} = 3135$ hours. Assuming a squared-error loss, determine the Bayes estimate of λ.

10-33. The weight of boxes of candy is normally dis tributed with mean μ and variance $\frac{1}{10}$. It is reasonable to assume a prior density for μ that is normal with a mean of 10 pounds and a variance of $\frac{1}{25}$. Determine the Bayes estimate of μ given that a sample of size 25 produces $\bar{x} = 10.05$ pounds. If boxes that weigh less than 9.95 pounds are defective, what is the probability that defective boxes will be produced?

10-34. The number of defects that occur on a silicon wafer used in integrated circuit manufacturing is known to be a Poisson random variable with parameter λ. Assume that the prior density for λ is exponential with a parameter of 0.25. A total of 45 defects were observed on 10 wafers. Set up an integral that defines a 95% Bayes interval for λ. What difficulties would you encounter in evaluating this integral?

10-35. The random variable X has a density function

$$f(x|\theta) = \frac{2x}{\theta^2}, \qquad 0 < x < \theta,$$

and the prior density for θ is

$$f(\theta) = 1, \qquad 0 < \theta < 1.$$

(a) Find the posterior density for θ assuming $n = 1$.

(b) Find the Bayes estimator for θ assuming the loss function $\ell(\hat{\theta}; \theta) = \theta^2(\theta - \theta)^2$ and $n = 1$.

10-36. Let X follow the Bernoulli distribution with parameter p. Assume a reasonable prior density for p to be

$$f(p) = 6p(1 - p), \qquad 0 \le p \le 1,$$
$$= 0, \qquad \text{otherwise.}$$

If the loss function is squared error, find the Bayes estimator of p if one observation is available. If the loss function is

$$\ell(\hat{p}; p) = 2(\hat{p} - p)^2,$$

find the Bayes estimator of p for $n = 1$.

10-37. Consider the confidence interval for μ with known standard deviation σ:

$$\bar{X} - Z_{\alpha_2}\sigma\sqrt{n} \le \mu \le \bar{X} + Z_{\alpha_1}\sigma\sqrt{n},$$

where $\alpha_1 + \alpha_2 = \alpha$. Let $\alpha = 0.05$ and find the interval for $\alpha_1 = \alpha_2 = \alpha/2 = 0.025$. Now find the interval for the case $\alpha_1 = 0.01$ and $\alpha_2 = 0.04$. Which interval is shorter? Is there any advantage to a "symmetric" confidence interval?

10-38. When $X_1, X_2, \ldots, X_n$ are independent Poisson random variables, each with parameter λ, and when n is relatively large, the sample mean $\overline{X}$ is approximately normal with mean λ and variance λ/n.

(a) What is the distribution of the statistic

$$\frac{\overline{X}-\lambda}{\sqrt{\lambda/n}}?$$

(b) Use the results of (a) to find a $100(1-\alpha)\%$ confidence interval for λ.

10-39. A manufacturer produces piston rings for an automobile engine. It is known that ring diameter is approximately normally distributed and has standard deviation $\sigma = 0.001$ mm. A random sample of 15 rings has a mean diameter of $\bar{x} = 74.036$ mm.

(a) Construct a 99% two-sided confidence interval on the mean piston ring diameter.

(b) Construct a 95% lower-confidence limit on the mean piston ring diameter.

10-40. The life in hours of a 75-W light bulb is known to be approximately normally distributed, with a standard deviation $\sigma = 25$ hours. A random sample of 20 bulbs has a mean life of $\bar{x} = 1014$ hours.

(a) Construct a 95% two-sided confidence interval on the mean life.

(b) Construct a 95% lower-confidence interval on the mean life.

10-41. A civil engineer is analyzing the compressive strength of concrete. Compressive strength is approximately normally distributed with a variance $\sigma^2 = 1000\ (\text{psi})^2$. A random sample of 12 specimens has a mean compressive strength of $\bar{x} = 3250$ psi.

(a) Construct a 95% two-sided confidence interval on mean compressive strength.

(b) Construct a 99% two-sided confidence interval on mean compressive strength. Compare the width of this confidence interval with the width of the one found in part (a).

10-42. Suppose that in Exercise 10-40 we wanted to be 95% confident that the error in estimating the mean life is less than 5 hours. What sample size should be used?

10-43. Suppose that in Exercise 10-40 we wanted the total width of the confidence interval on mean life to be 8 hours. What sample size should be used?

10-44. Suppose that in Exercise 10-41 it is desired to estimate the compressive strength with an error that is less than 15 psi. What sample size is required?

10-45. Two machines are used to fill plastic bottles with dishwashing detergent. The standard deviations of fill volume are known to be $\sigma_1 = 0.15$ fluid ounces and $\sigma_2 = 0.18$ fluid ounces for the two machines, respectively. Two random samples of $n_1 = 12$ bottles from machine 1 and $n_2 = 10$ bottles from machine 2 are selected, and the sample mean fill volumes are $\bar{x}_1 = 30.87$ fluid ounces and $\bar{x}_2 = 30.68$ fluid ounces.

(a) Construct a 90% two-sided confidence interval on the mean difference in fill volume.

(b) Construct a 95% two-sided confidence interval on the mean difference in fill volume. Compare the width of this interval to the width of the interval in part (a).

(c) Construct a 95% upper-confidence interval on the mean difference in fill volume.

10-46. The burning rates of two different solid-fuel rocket propellants are being studied. It is known that both propellants have approximately the same standard deviation of burning rate; that is, $\sigma_1 = \sigma_2 = 3$ cm/s. Two random samples of $n_1 = 20$ and $n_2 = 20$ specimens are tested, and the sample mean burning rates are $\bar{x}_1 = 18$ cm/s and $\bar{x}_2 = 24$ cm/s. Construct a 99% confidence interval on the mean difference in burning rate.

10-47. Two different formulations of a lead-free gasoline are being tested to study their road octane numbers. The variance of road octane number for formulation 1 is $\sigma_1^2 = 1.5$ and for formulation 2 it is $\sigma_2^2 = 1.2$. Two random samples of size $n_1 = 15$ and $n_2 = 20$ are tested, and the mean road octane numbers observed are $\bar{x}_1 = 89.6$ and $\bar{x}_2 = 92.5$. Construct a 95% two-sided confidence interval on the difference in mean road octane number.

10-48. The compressive strength of concrete is being tested by a civil engineer. He tests 16 specimens and obtains the following data:

2216	2237	2249	2204
2225	2301	2281	2263
2318	2255	2275	2295
2250	2238	2300	2217

(a) Construct a 95% two-sided confidence interval on the mean strength.

(b) Construct a 95% lower-confidence interval on the mean strength.

(c) Construct a 95% two-sided confidence interval on the mean strength assuming that $\sigma = 36$. Compare this interval with the one from part (a).

(d) Construct a 95% two-sided prediction interval for a single compressive strength.

(e) Construct a two-sided tolerance interval that would cover 99% of all compressive strengths with 95% confidence.

10-49. An article in *Annual Reviews Material Research* (2001, p. 291) presents bond strengths for various energetic materials (explosives, propellants, and pyrotechnics). Bond strengths for 15 such materials are shown below. Construct a two-sided 95% confidence interval on the mean bond strength.

323, 312, 300, 284, 283, 261, 207, 183,
180, 179, 174, 167, 167, 157, 120.

10-50. The wall thickness of 25 glass 2-liter bottles was measured by a quality-control engineer. The sample mean was $\bar{x} = 4.05$ mm and the sample standard deviation was $s = 0.08$ mm. Find a 90% lower-confidence interval on the mean wall thickness.

10-51. An industrial engineer is interested in estimating the mean time required to assemble a printed circuit board. How large a sample is required if the engineer wishes to be 95% confident that the error in estimating the mean is less than 0.25 minutes? The standard deviation of assembly time is 0.45 minutes.

10-52. A random sample of size 15 from a normal population has mean $\bar{x} = 550$ and variance $s^2 = 49$. Find the following:

(a) A 95% two-sided confidence interval on μ.

(b) A 95% lower-confidence interval on μ.

(c) A 95% upper-confidence interval on μ.

(d) A 95% two-sided prediction interval for a single observation.

(e) A two-sided tolerance interval that would cover 90% of all observations with 99% confidence.

10-53. An article in *Computers in Cardiology* (1993, p. 317) presents the results of a heart stress test in which the stress is induced by a particular drug. The heart rates (in beats per minute) of nine male patients after the drug is administered are recorded. The average heart rate was found to be $\bar{x} = 102.9$ (bpm) with a sample standard deviation of $s = 13.9$ (bpm). Find a 90% confidence interval on the mean heart rate after the drug is administered.

10-54. Two independent random samples of sizes $n_1 = 18$ and $n_2 = 20$ are taken from two normal populations. The sample means are $\bar{x}_1 = 200$ and $\bar{x}_2 = 190$. We know that the variances are $\sigma_1^2 = 15$ and $\sigma_2^2 = 12$. Find the following:

(a) A 95% two-sided confidence interval on $\mu_1 - \mu_2$.

(b) A 95% lower-confidence interval on $\mu_1 - \mu_2$.

(c) A 95% upper-confidence interval on $\mu_1 - \mu_2$.

10-55. The output voltage from two different types of transformers is being investigated. Ten transformers of each type are selected at random and the voltage measured. The sample means are $\bar{x}_1 = 12.13$ volts and $\bar{x}_2 = 12.05$ volts. We know that the variances of output voltage for the two types of transformers are $\sigma_1^2 = 0.7$ and $\sigma_2^2 = 0.8$, respectively. Construct a 95% two-sided confidence interval on the difference in mean voltage.

10-56. Random samples of size 20 were drawn from two independent normal populations. The sample means and standard deviations were $\bar{x}_1 = 22.0$, $s_1 = 1.8$, $\bar{x}_2 = 21.5$, and $s_2 = 1.5$. Assuming that $\sigma_1^2 = \sigma_2^2$, find the following:

(a) A 95% two-sided confidence interval on $\mu_1 - \mu_2$.

(b) A 95% upper-confidence interval on $\mu_1 - \mu_2$.

(c) A 95% lower-confidence interval on $\mu_1 - \mu_2$.

10-57. The diameter of steel rods manufactured on two different extrusion machines is being investigated. Two random samples of sizes $n_1 = 15$ and $n_2 = 18$ are selected, and the sample means and sample variances are $\bar{x}_1 = 8.73$, $s_1^2 = 0.30$, $x_2 = 8.68$, and $s_2^2 = 0.34$, respectively. Assuming that $\sigma_1^2 = \sigma_2^2$, construct a 95% two-sided confidence interval on the difference in mean rod diameter.

10-58. Random samples of sizes $n_1 = 15$ and $n_2 = 10$ are drawn from two independent normal populations. The sample means and variances are $\bar{x}_1 = 300$, $s_1^2 = 16$, $\bar{x}_2 = 325$, $s_2^2 = 49$. Assuming that $\sigma_1^2 \neq \sigma_2^2$, construct a 95% two-sided confidence interval on $\mu_1 - \mu_2$.

10-59. Consider the data in Exercise 10-48. Construct the following:

(a) A 95% two-sided confidence interval on σ^2.

(b) A 95% lower-confidence interval on σ^2.

(c) A 95% upper-confidence interval on σ^2.

10-60. Consider the data in Exercise 10-49. Construct the following:

(a) A 99% two-sided confidence interval on σ^2.

(b) A 99% lower-confidence interval on σ^2.

(c) A 99% upper-confidence interval on σ^2.

10-61. Construct a 95% two-sided confidence interval on the variance of the wall thickness data in Exercise 10-50.

10-62. In a random sample of 100 light bulbs, the sample standard deviation of bulb life was found to be 12.6 hours. Compute a 90% upper-confidence interval on the variance of bulb life.

10-63. Consider the data in Exercise 10-56. Construct a 95% two-sided confidence interval on the ratio of the population variances σ_1^2/σ_2^2.

10-64. Consider the data in Exercise 10-57. Construct the following:

(a) A 90% two-sided confidence interval on σ_1^2/σ_2^2.

(b) A 95% two-sided confidence interval on σ_1^2/σ_2^2. Compare the width of this interval with the width of the interval in part (a).

(c) A 90% lower-confidence interval on σ_1^2/σ_2^2.
(d) A 90% upper-confidence interval on σ_1^2/σ_2^2.

10-65. Construct a 95% two-sided confidence interval on the ratio of the variances σ_1^2/σ_2^2 using the data in Exercise 10-58.

10-66. Of 400 randomly selected motorists, 48 were found to be uninsured. Construct a 95% two-sided confidence interval on the uninsured rate for motorists.

10-67. How large a sample would be required in Exercise 10-66 to be 95% confident that the error in estimating the uninsured rate for motorists is less than 0.03?

10-68. A manufacturer of electronic calculators is interested in estimating the fraction of defective units produced. A random sample of 8000 calculators contains 18 defectives. Compute a 99% upper-confidence interval on the fraction defective.

10-69. A study is to be conducted of the percentage of homeowners who own at least two television sets. How large a sample is required if we wish to be 99% confident that the error in estimating this quantity is less than 0.01?

10-70. A study is conducted to determine whether there is a significant difference in union membership based on sex of the person. A random sample of 5000 factory-employed men were polled, and of this group, 785 were members of a union. A random sample of 3000 factory-employed women were also polled, and of this group, 327 were members of a union. Construct a 99% confidence interval on the difference in proportions $p_1 - p_2$.

10-71. The fraction of defective product produced by two production lines is being analyzed. A random sample of 1000 units from line 1 has 10 defectives, while a random sample of 1200 units from line 2 has 25 defectives. Find a 99% confidence interval on the difference in fraction defective produced by the two lines.

10-72. The results of a study on powered wheelchair driving performance were presented in the *Proceedings of the IEEE 24th Annual Northeast Bioengineering Conference* (1998, p. 130). In this study, the effects of two types of joysticks, force sensing (FSJ) and position sensing (PSJ), on power wheelchair control were investigated. Each of 10 subjects was asked to test both joysticks. One response of interest is the time (in seconds) to complete a predetermined course. Data typical of this type of experiment are as follows:

Subject	PSJ	FSJ
1	25.9	33.4
2	30.2	37.4
3	33.7	48.0
4	27.6	30.5
5	33.3	27.8
6	34.6	27.5
7	33.1	36.9
8	30.6	31.1
9	30.5	27.1
10	25.4	38.0

Find a 95% confidence interval on the difference in mean completion times. Is there any indication that one joystick is preferable?

10-73. The manager of a fleet of automobiles is testing two brands of radial tires. He assigns one tire of each brand at random to the two rear wheels of eight cars and runs the cars until the tires wear out. The data (in kilometers) are shown below:

Car	Brand 1	Brand 2
1	36,925	34,318
2	45,300	42,280
3	36,240	35,500
4	32,100	31,950
5	37,210	38,015
6	48,360	47,800
7	38,200	37,810
8	33,500	33,215

Find a 95% confidence interval on the difference in mean mileage. Which brand do you prefer?

10-74. Consider the data in Exercise 10-50. Find confidence intervals on μ and σ^2 such that we are at least 90% confident that both intervals simultaneously lead to correct conclusions.

10-75. Consider the data in Exercise 10-56. Suppose that a random sample of size $n_3 = 15$ is obtained from a third normal population, with $\bar{x}_3 = 20.5$ and $s_3 = 1.2$. Find two-sided confidence intervals on $\mu_1 - \mu_2$, $\mu_1 - \mu_3$, and $\mu_2 - \mu_3$ such that the probability is at least 0.95 that all three intervals simultaneously lead to correct conclusions.

10-76. A random variable X is normally distributed with mean μ and variance $\sigma^2 = 10$. The prior density for μ is uniform between 6 and 12. A random sample of size 16 yields $\bar{x} = 8$. Construct a 90% Bayes interval for μ. Could you reasonably accept the hypothesis that $\mu = 9$?

10-77. Let X be a normally distributed random variable with mean $\mu = 5$ and unknown variance σ^2. The prior density for $1/\sigma^2$ is a gamma distribution with parameters $r = 3$ and $\lambda = 1.0$. Determine the posterior density for $1/\sigma^2$. If a random sample of size 10 yields $\Sigma(x_i - 4)^2 = 4.92$, determine the Bayes estimate of $1/\sigma^2$ assuming a squared-error loss. Set up an integral that defines a 90% Bayes interval for $1/\sigma^2$.

10-78. Prove that if a squared-error loss function is used, the Bayes estimator of θ is the mean of the posterior distribution for θ.

Chapter 11

Tests of Hypotheses

Many problems require that we decide whether to accept or reject a statement about some parameter. The statement is usually called a hypothesis, and the decision-making procedure about the hypothesis is called hypothesis testing. This is one of the most useful aspects of statistical inference, since many types of decision problems can be formulated as hypothesis-testing problems. This chapter will develop hypothesis-testing procedures for several important situations.

11-1 INTRODUCTION

11-1.1 Statistical Hypotheses

A statistical hypothesis is a statement about the probability distribution of a random variable. Statistical hypotheses often involve one or more parameters of this distribution. For example, suppose that we are interested in the mean compressive strength of a particular type of concrete. Specifically, we are interested in deciding whether or not the mean compressive strength (say μ) is 2500 psi. We may express this formally as

$$\begin{aligned} H_0&: \mu = 2500 \text{ psi}, \\ H_1&: \mu \neq 2500 \text{ psi}. \end{aligned} \tag{11-1}$$

The statement H_0: $\mu = 2500$ psi in equation 11-1 is called the *null hypothesis*, and the statement H_1: $\mu \neq 2500$ psi is called the *alternative hypothesis.* Since the alternative hypothesis specifies values of μ that could be either greater than 2500 psi or less than 2500 psi, it is called a *two-sided alternative hypothesis.* In some situations, we may wish to formulate a *one-sided alternative hypothesis*, as in

$$\begin{aligned} H_0&: \mu = 2500 \text{ psi}, \\ H_1&: \mu > 2500 \text{ psi}. \end{aligned} \tag{11-2}$$

It is important to remember that hypotheses are always statements about the population or distribution under study, not statements about the sample. The value of the population parameter specified in the null hypothesis (2500 psi in the above example) is usually determined in one of three ways. First, it may result from past experience or knowledge of the process, or even from prior experimentation. The objective of hypothesis testing then is usually to determine whether the experimental situation has changed. Second, this value may be determined from some theory or model regarding the process under study. Here the objective of hypothesis testing is to verify the theory or model. A third situation arises when the value of the population parameter results from external considerations, such as design or engineering specifications, or from contractual obligations. In this situation, the usual objective of hypothesis testing is conformance testing.

We are interested in making a decision about the truth or falsity of a hypothesis. A procedure leading to such a decision is called a *test of a hypothesis.* Hypothesis-testing procedures rely on using the information in a random sample from the population of interest. If this information is consistent with the hypothesis, then we would conclude that the hypothesis is true; however, if this information is inconsistent with the hypothesis, we would conclude that the hypothesis is false.

To test a hypothesis, we must take a random sample, compute an appropriate test statistic from the sample data, and then use the information contained in this test statistic to make a decision. For example, in testing the null hypothesis concerning the mean compressive strength of concrete in equation 11-1, suppose that a random sample of 10 concrete specimens is tested and the sample mean $\bar{x}$ is used as a test statistic. If $\bar{x} > 2550$ psi or if $\bar{x} < 2450$ psi, we will consider the mean compressive strength of this particular type of concrete to be different from 2500 psi. That is, we would *reject* the null hypothesis H_0: $\mu = 2500$. Rejecting H_0 implies that the alternative hypothesis, H_1, is true. The set of all possible values of $\bar{x}$ that are either greater than 2550 psi or less than 2450 psi is called the *critical region* or *rejection region* for the test. Alternatively, if 2450 psi $\le \bar{x} \le$ 2550 psi, then we would *accept* the null hypothesis H_0: $\mu = 2500$. Thus, the interval [2450 psi, 2550 psi] is called the *acceptance region* for the test. Note that the boundaries of the critical region, 2450 psi and 2550 psi (often called the *critical values* of the test statistic), have been determined somewhat arbitrarily. In subsequent sections we will show how to construct an appropriate test statistic to determine the critical region for several hypothesis-testing situations.

11-1.2 Type I and Type II Errors

The decision to accept or reject the null hypothesis is based on a test statistic computed from the data in a random sample. When a decision is made using the information in a random sample, this decision is subject to error. Two kinds of errors may be made when testing hypotheses. If the null hypothesis is rejected when it is true, then a type I error has been made. If the null hypothesis is accepted when it is false, then a type II error has been made. The situation is described in Table 11-1.

The probabilities of occurrence of type I and type II errors are given special symbols:

$$\alpha = P\{\text{type I error}\} = P\{\text{reject } H_0 | H_0 \text{ is true}\}, \tag{11-3}$$

$$\beta = P\{\text{type II error}\} = P\{\text{accept } H_0 | H_0 \text{ is false}\}. \tag{11-4}$$

Sometimes it is more convenient to work with the *power* of the test, where

$$\text{Power} = 1 - \beta = P\{\text{reject } H_0 | H_0 \text{ is false}\}. \tag{11-5}$$

Note that the power of the test is the probability that a false null hypothesis is correctly rejected. Because the results of a test of a hypothesis are subject to error, we cannot "prove" or "disprove" a statistical hypothesis. However, it is possible to design test procedures that control the error probabilities α and β to suitably small values.

Table 11-1 Decisions in Hypothesis Testing

	H_0 is True	H_0 is False
Accept H_0	No error	Type II error
Reject H_0	Type I error	No error

The probability of type I error α is often called the *significance level* or *size* of the test. In the concrete-testing example, a type I error would occur if the sample mean $\bar{x} > 2550$ psi or if $\bar{x} < 2450$ psi when in fact the true mean compressive strength $\mu = 2500$ psi. Generally, the type I error probability is controlled by the location of the critical region. Thus, it is usually easy in practice for the analyst to set the type I error probability at (or near) any desired value. Since the probability of wrongly rejecting H_0 is directly controlled by the decision maker, rejection of H_0 is always a *strong conclusion.* Now suppose that the null hypothesis H_0: $\mu = 2500$ psi is false. That is, the true mean compressive strength μ is some value other than 2500 psi. The probability of type II error is not a constant but depends on the true mean compressive strength of the concrete. If μ denotes the true mean compressive strength, then $\beta(\mu)$ denotes the type II error probability corresponding to μ. The function $\beta(\mu)$ is evaluated by finding the probability that the test statistic (in this case $\bar{x}$) falls in the acceptance region given a particular value of μ. We define the *operating characteristic curve* (or OC curve) of a test as the plot of $\beta(\mu)$ against μ. An example of an operating characteristic curve for the concrete-testing problem is shown in Fig. 11-1. From this curve, we see that the type II error probability depends on the extent to which H_0: $\mu = 2500$ psi is false. For example, note that $\beta(2700) < \beta(2600)$. Thus we can think of the type II error probability as a measure of the ability of the test procedure to detect a particular deviation from the null hypothesis H_0. Small deviations are harder to detect than large ones. We also observe that since this is a two-sided alternative hypothesis, the operating characteristic curve is symmetric; that is, $\beta(2400) = \beta(2600)$. Furthermore, when $\mu = 2500$ the probability of type II error $\beta = 1 - \alpha$.

The probability of type II error is also a function of sample size, as illustrated in Fig. 11-2. From this figure, we see that for a given value of the type I error probability α and a given value of mean compressive strength the type II error probability decreases as the sample size n increases. That is, a specified deviation of the true mean from the value specified in the null hypothesis is easier to detect for larger sample sizes than for smaller ones. The effect of the type I error probability α on the type II error probability β for a given sample size n is illustrated in Fig. 11-3. Decreasing α causes β to increase, and increasing α causes β to decrease.

Because the type II error probability β is a function of both the sample size and the extent to which the null hypothesis H_0 is false, it is customary to think of the decision to accept H_0 as a *weak conclusion*, unless we know that β is acceptably small. Therefore, rather than saying we "accept H_0," we prefer the terminology "*fail to reject* H_0." Failing to reject H_0 implies that we have not found sufficient evidence to reject H_0, that is, to make a strong statement. Thus failing to reject H_0 does not necessarily mean that there is a high

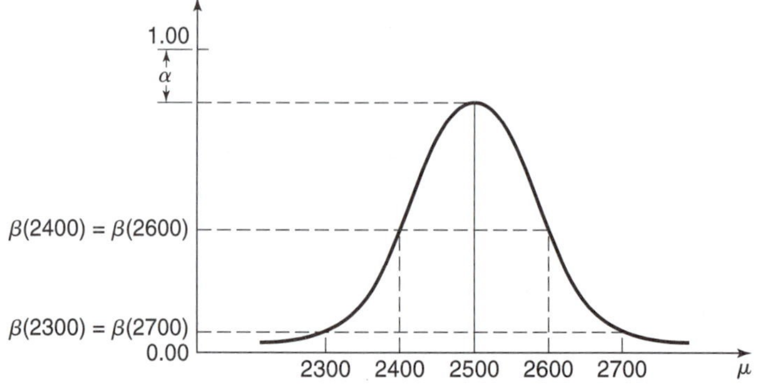

Figure 11-1 Operating characteristic curve for the concrete testing example.

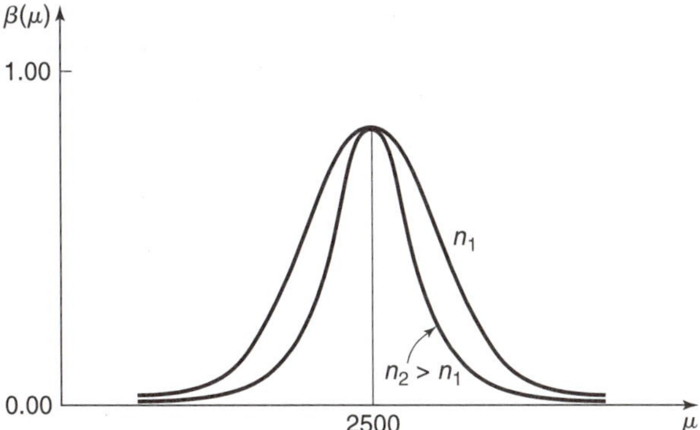

Figure 11-2 The effect of sample size on the operating characteristic curve.

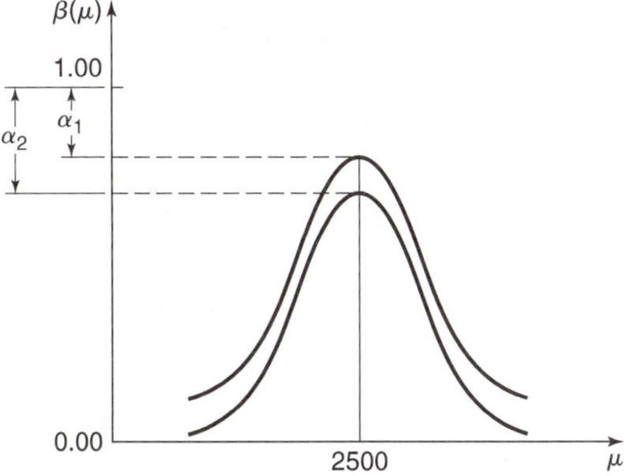

Figure 11-3 The effect of type I error on the operating characteristic curve.

probability that H_0 is true. It may imply that more data are required to reach a strong conclusion. This can have important implications for the formulation of hypotheses.

11-1.3 One-Sided and Two-Sided Hypotheses

Because rejecting H_0 is always a strong conclusion while failing to reject H_0 can be a weak conclusion unless β is known to be small, we usually prefer to construct hypotheses such that the statement about which a strong conclusion is desired is in the alternative hypothesis, H_1. Problems for which a two-sided alternative hypothesis is appropriate do not really present the analyst with a choice of formulation. That is, if we wish to test the hypothesis that the mean of a distribution μ equals some arbitrary value, say μ_0, and if it is important to detect values of the true mean μ that could be either greater than μ_0 or less than μ_0, then one must use the two-sided alternative in

$$H_0\colon \mu = \mu_0,$$
$$H_1\colon \mu \neq \mu_0.$$

Many hypothesis-testing problems naturally involve a one-sided alternative hypothesis. For example, suppose that we want to reject H_0 only when the true value of the mean exceeds μ_0. The hypotheses would be

$$H_0\colon \mu = \mu_0, \qquad\qquad (11\text{-}6)$$
$$H_1\colon \mu > \mu_0.$$

This would imply that the critical region is located in the upper tail of the distribution of the test statistic. That is, if the decision is to be based on the value of the sample mean $\bar{x}$, then we would reject H_0 in equation 11-6 if $\bar{x}$ is too large. The operating characteristic curve for the test for this hypothesis is shown in Fig. 11-4, along with the operating characteristic curve for a two-sided test. We observe that when the true mean μ exceeds μ_0 (i.e., when the alternative hypothesis, H_1: $\mu > \mu_0$ is true), the one-sided test is superior to the two-sided test in the sense that it has a steeper operating characteristic curve. When the true mean $\mu = \mu_0$, both the one-sided and two-sided tests are equivalent. However, when the true mean μ is less than μ_0, the two operating characteristic curves differ. If $\mu < \mu_0$, the two-sided test has a higher probability of detecting this departure from μ_0 than the one-sided test. This is intuitively appealing, as the one-sided test is designed assuming either that μ cannot be less than μ_0 or, if μ is less than μ_0, that it is desirable to accept the null hypothesis.

In effect there are two different models that can be used for the one-sided alternative hypothesis. For the case where the alternative hypothesis is H_1: $\mu > \mu_0$, these two models are

$$\begin{aligned} H_0&: \mu = \mu_0, \\ H_1&: \mu > \mu_0 \end{aligned} \tag{11-7}$$

and

$$\begin{aligned} H_0&: \mu \le \mu_0, \\ H_1&: \mu > \mu_0. \end{aligned} \tag{11-8}$$

In equation 11-7, we are assuming that μ cannot be less than μ_0, and the operating characteristic curve is undefined for values of $\mu < \mu_0$. In equation 11-8, we are assuming that μ can be less than μ_0 and that in such a situation it would be desirable to accept H_0. Thus for equation 11-8 the operating characteristic curve is defined for all values of $\mu \le \mu_0$. Specifically, if $\mu \le \mu_0$, we have $\beta(\mu) = 1 - \alpha(\mu)$, where $\alpha(\mu)$ is the significance level as a function of μ. For situations in which the model of equation 11-8 is appropriate, we define the significance level of the test as the maximum value of the type I error probability α; that is, the value of α at $\mu = \mu_0$. In situations where one-sided alternative hypotheses are appropriate, we will usually write the null hypothesis with an equality; for example, H_0: $\mu = \mu_0$. This will be interpreted as including the cases H_0: $\mu \le \mu_0$ or H_0: $\mu \ge \mu_0$, as appropriate.

In problems where one-sided test procedures are indicated, analysts occasionally experience difficulty in choosing an appropriate formulation of the alternative hypothesis. For example, suppose that a soft drink beverage bottler purchases 10-ounce nonreturnable bottles from a glass company. The bottler wants to be sure that the bottles exceed the specification on mean internal pressure or bursting strength, which for 10-ounce bottles is 200 psi.

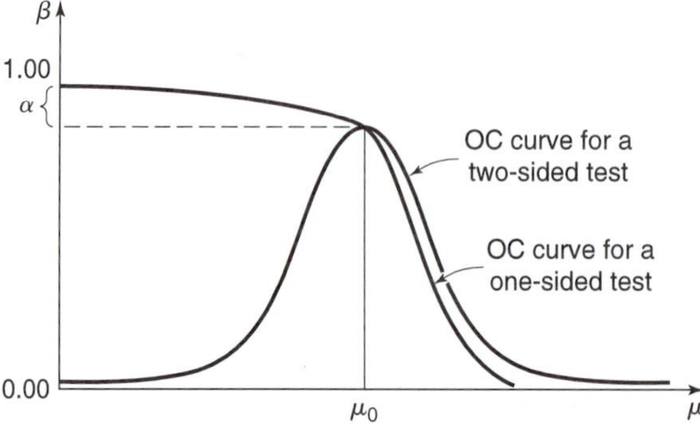

Figure 11-4 Operating characteristic curves for two-sided and one-sided tests.

The bottler has decided to formulate the decision procedure for a specific lot of bottles as a hypothesis problem. There are two possible formulations for this problem, either

$$H_0\colon \mu \leq 200 \text{ psi},$$
$$H_1\colon \mu > 200 \text{ psi} \tag{11-9}$$

or

$$H_0\colon \mu \geq 200 \text{ psi},$$
$$H_1\colon \mu < 200 \text{ psi}. \tag{11-10}$$

Consider the formulation in equation 11-9. If the null hypothesis is rejected, the bottles will be judged satisfactory; while if H_0 is not rejected, the implication is that the bottles do not conform to specifications and should not be used. Because rejecting H_0 is a strong conclusion, this formulation forces the bottle manufacturer to "demonstrate" that the mean bursting strength of the bottles exceeds the specification. Now consider the formulation in equation 11-10. In this situation, the bottles will be judged satisfactory *unless* H_0 is rejected. That is, we would conclude that the bottles are satisfactory unless there is strong evidence to the contrary.

Which formulation is correct, equation 11-9 or equation 11-10? The answer is "it depends." For equation 11-9, there is some probability that H_0 will be accepted (i.e., we would decide that the bottles are not satisfactory) even though the true mean is slightly greater than 200 psi. This formulation implies that we want the bottle manufacturer to *demonstrate* that the product meets or exceeds our specifications. Such a formulation could be appropriate if the manufacturer has experienced difficulty in meeting specifications in the past, or if product safety considerations force us to hold tightly to the 200 psi specification. On the other hand, for the formulation of equation 11-10 there is some probability that H_0 will be accepted and the bottles judged satisfactory even though the true mean is slightly less than 200 psi. We would conclude that the bottles are unsatisfactory only when there is strong evidence that the mean does not exceed 200 psi; that is, when $H_0\colon \mu \geq 200$ psi is rejected. This formulation assumes that we are relatively happy with the bottle manufacturer's past performance and that small deviations from the specification of $\mu \geq 200$ psi are not harmful.

In formulating one-sided alternative hypotheses, we should remember that rejecting H_0 is always a strong conclusion, and consequently, we should put the statement about which it is important to make a strong conclusion in the alternative hypothesis. Often this will depend on our point of view and experience with the situation.

11-2 TESTS OF HYPOTHESES ON A SINGLE SAMPLE

11-2.1 Tests of Hypotheses on the Mean of a Normal Distribution, Variance Known

Statistical Analysis

Suppose that the random variable X represents some process or population of interest. We assume that the distribution of X is either normal or that, if it is nonnormal, the conditions of the Central Limit Theorem hold. In addition, we assume that the mean μ of X is unknown but that the variance σ^2 is known. We are interested in testing the hypothesis

$$H_0\colon \mu = \mu_0,$$
$$H_1\colon \mu \neq \mu_0, \tag{11-11}$$

where μ_0 is a specified constant.

A random sample of size n, $X_1, X_2, \ldots, X_n$, is available. Each observation in this sample has unknown mean μ and known variance σ^2. The test procedure for H_0: $\mu = \mu_0$ uses the test statistic

$$Z_0 = \frac{\overline{X} - \mu_0}{\sigma/\sqrt{n}}. \tag{11-12}$$

If the null hypothesis H_0: $\mu = \mu_0$ is true, then $E(\overline{X}) = \mu_0$, and it follows that the distribution of Z_0 is $N(0, 1)$. Consequently, if H_0: $\mu = \mu_0$ is true, the probability is $1 - \alpha$ that a value of the test statistic Z_0 falls between $-Z_{\alpha/2}$ and $Z_{\alpha/2}$, where $Z_{\alpha/2}$ is the percentage point of the standard normal distribution such that $P\{Z \geq Z_{\alpha/2}\} = \alpha/2$ (i.e., $Z_{\alpha/2}$ is the $100(1-\alpha/2)$ percentage point of the standard normal distribution). The situation is illustrated in Fig. 11-5. Note that the probability is α that a value of the test statistic Z_0 would fall in the region $Z_0 > Z_{\alpha/2}$ or $Z_0 < -Z_{\alpha/2}$ when H_0: $\mu = \mu_0$ is true. Clearly, a sample producing a value of the test statistic that falls in the tails of the distribution of Z_0 would be unusual if H_0: $\mu = \mu_0$ is true; it is also an indication that H_0 is false. Thus, we should reject H_0 if either

$$Z_0 > Z_{\alpha/2} \tag{11-13a}$$

or

$$Z_0 < -Z_{\alpha/2} \tag{11-13b}$$

and fail to reject H_0 if

$$-Z_{\alpha/2} \leq Z_0 \leq Z_{\alpha/2}. \tag{11-14}$$

Equation 11-14 defines the *acceptance region* for H_0 and equation 11-13 defines the *critical region* or *rejection region.* The type I error probability for this test procedure is α.

Example 11-1

The burning rate of a rocket propellant is being studied. Specifications require that the mean burning rate must be 40 cm/s. Furthermore, suppose that we know that the standard deviation of the burning rate is approximately 2 cm/s. The experimenter decides to specify a type I error probability $\alpha = 0.05$, and he will base the test on a random sample of size $n = 25$. The hypotheses we wish to test are

$$H_0\text{: } \mu = 40 \text{ cm/s},$$
$$H_1\text{: } \mu \neq 40 \text{ cm/s}.$$

Twenty-five specimens are tested, and the sample mean burning rate obtained is $\bar{x} = 41.25$ cm/s. The value of the test statistic in equation 11-12 is

$$\begin{aligned} Z_0 &= \frac{\bar{x} - \mu_0}{\sigma/\sqrt{n}} \\ &= \frac{41.25 - 40}{2/\sqrt{25}} = 3.125. \end{aligned}$$

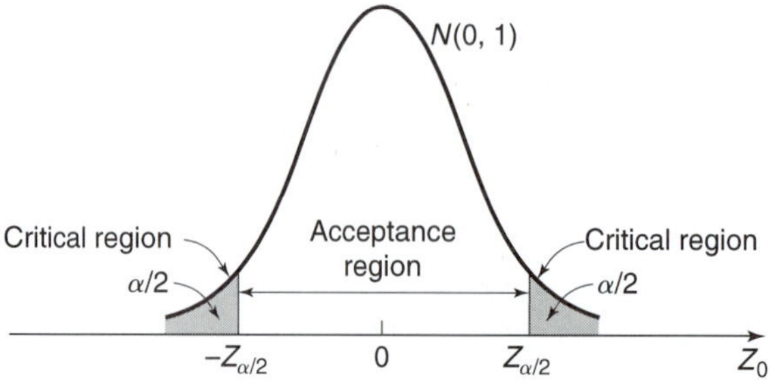

Figure 11-5 The distribution of Z_0 when H_0: $\mu = \mu_0$ is true.

Since $\alpha = 0.05$, the boundaries of the critical region are $Z_{0.025} = 1.96$ and $-Z_{0.025} = -1.96$, and we note that Z_0 falls in the critical region. Therefore, H_0 is rejected, and we conclude that the mean burning rate is not equal to 40 cm/s.

Now suppose that we wish to test the one-sided alternative, say

$$H_0: \mu = \mu_0,$$
$$H_1: \mu > \mu_0. \tag{11-15}$$

(Note that we could also write H_0: $\mu \le \mu_0$.) In defining the critical region for this test, we observe that a negative value of the test statistic Z_0 would never lead us to conclude that H_0: $\mu = \mu_0$ is false. Therefore, we would place the critical region in the upper tail of the $N(0, 1)$ distribution and reject H_0 on values of Z_0 that are too large. That is, we would reject H_0 if

$$Z_0 > Z_\alpha. \tag{11-16}$$

Similarly, to test

$$H_0: \mu = \mu_0,$$
$$H_1: \mu < \mu_0, \tag{11-17}$$

we would calculate the test statistic Z_0 and reject H_0 on values of Z_0 that are too small. That is, the critical region is in the lower tail of the $N(0, 1)$ distribution, and we reject H_0 if

$$Z_0 < -Z_\alpha. \tag{11-18}$$

Choice of Sample Size

In testing the hypotheses of equations 11-11, 11-15, and 11-17, the type I error probability α is directly selected by the analyst. However, the probability of type II error β depends on the choice of sample size. In this section, we will show how to select the sample size in order to arrive at a specified value of β.

Consider the two-sided hypothesis

$$H_0: \mu = \mu_0,$$
$$H_1: \mu \neq \mu_0.$$

Suppose that the null hypothesis is false and that the true value of the mean is $\mu = \mu_0 + \delta$, say, where $\delta > 0$. Now since H_1 is true, the distribution of the test statistic Z_0 is

$$Z_0 \sim N\left(\frac{\delta\sqrt{n}}{\sigma}, 1\right). \tag{11-19}$$

The distribution of the test statistic Z_0 under both the null hypothesis H_0 and the alternative hypothesis H_1 is shown in Fig. 11-6. From examining this figure, we note that if H_1 is true, a type II error will be made only if $-Z_{\alpha/2} \le Z_0 \le Z_{\alpha/2}$ where $Z_0 \sim N\left(\delta\sqrt{n}/\sigma, 1\right)$. That is, the probability of the type II error β is the probability that Z_0 falls between $-Z_{\alpha/2}$ and $Z_{\alpha/2}$ *given that H_1 is true.* This probability is shown as the shaded portion of Fig. 11-6. Expressed mathematically, this probability is

$$\beta = \Phi\left(Z_{\alpha/2} - \frac{\delta\sqrt{n}}{\sigma}\right) - \Phi\left(-Z_{\alpha/2} - \frac{\delta\sqrt{n}}{\sigma}\right), \tag{11-20}$$

where $\Phi(z)$ denotes the probability to the left of z on the standard normal distribution. Note that equation 11-20 was obtained by evaluating the probability that Z_0 falls in the interval

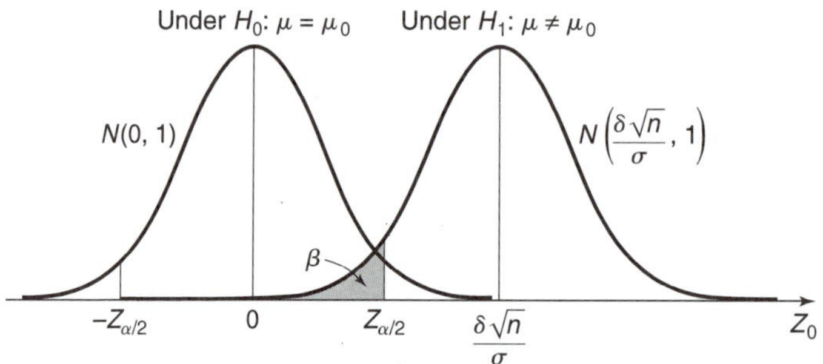

Figure 11-6 The distribution of Z_0 under H_0 and H_1.

$[-Z_{\alpha/2}, Z_{\alpha/2}]$ on the distribution of Z_0 when H_1 is true. These two points were standardized to produce equation 11-20. Furthermore, note that equation 11-20 also holds if $\delta < 0$, due to the symmetry of the normal distribution.

While equation 11-20 could be used to evaluate the type II error, it is more convenient to use the operating characteristic curves in Charts VI*a* and VI*b* of the Appendix. These curves plot β as calculated from equation 11-20 against a parameter d for various example sizes n. Curves are provided for both $\alpha = 0.05$ and $\alpha = 0.01$. The parameter d is defined as

$$d = \frac{|\mu - \mu_0|}{\sigma} = \frac{|\delta|}{\sigma}. \tag{11-21}$$

We have chosen d so that one set of operating characteristic curves can be used for all problems regardless of the values of μ_0 and σ. From examining the operating characteristic curves or equation 11-20 and Fig. 11-6 we note the following:

1. The further the true value of the mean μ from μ_0, the smaller the probability of type II error β for a given n and α. That is, we see that for a specified sample size and α, large differences in the mean are easier to detect than small ones.
2. For a given δ and α, the probability of type II error β decreases as n increases. That is, to detect a specified difference in the mean δ, we may make the test more powerful by increasing the sample size.

Example 11-2

Consider the rocket propellant problem in Example 11-1. Suppose that the analyst is concerned about the probability of a type II error if the true mean burning rate is $\mu = 41$ cm/s. We may use the operating characteristic curves to find β. Note that $\delta = 41 - 40 = 1$, $n = 25$, $\sigma = 2$, and $\alpha = 0.05$. Then

$$d = \frac{|\mu - \mu_0|}{\sigma} = \frac{|\delta|}{\sigma} = \frac{1}{2}$$

and from Chart VI*a* (Appendix), with $n = 25$, we find that $\beta = 0.30$. That is, if the true mean burning rate is $\mu = 41$ cm/s, then there is approximately a 30% chance that this will not be detected by the test with $n = 25$.

Example 11-3

Once again, consider the rocket propellant problem in Example 11-1. Suppose that the analyst would like to design the test so that if the true mean burning rate differs from 40 cm/s by as much as 1 cm/s, the test will detect this (i.e., reject H_0: $\mu = 40$) with a high probability, say 0.90. The operating characteristic curves can be used to find the sample size that will give such a test. Since $d = |\mu - \mu_0|/\sigma = 1/2$, $\alpha = 0.05$, and $\beta = 0.10$, we find from Chart VI*a* (Appendix) that the required sample size is $n = 40$, approximately.

In general, the operating characteristic curves involve three parameters: β, δ, and n. Given any two of these parameters, the value of the third can be determined. There are two typical applications of these curves.

1. For a given n and δ, find β. This was illustrated in Example 11-2. This kind of problem is often encountered when the analyst is concerned about the sensitivity of an experiment already performed, or when sample size is restricted by economic or other factors.
2. For a given β and δ, find n. This was illustrated in Example 11-3. This kind of problem is usually encountered when the analyst has the opportunity to select the sample size at the outset of the experiment.

Operating characteristic curves are given in Charts VI*c* and VI*d* (Appendix) for the one-sided alternatives. If the alternative hypothesis is H_1: $\mu > \mu_0$, then the abscissa scale on these charts is

$$d = \frac{\mu - \mu_0}{\sigma}. \tag{11-22}$$

When the alternative hypothesis is H_1: $\mu < \mu_0$, the corresponding abscissa scale is

$$d = \frac{\mu_0 - \mu}{\sigma}. \tag{11-23}$$

It is also possible to derive formulas to determine the appropriate sample size to use to obtain a particular value of β for a given δ and α. These formulas are alternatives to using the operating characteristic curves. For the two-sided alternative hypothesis, we know from equation 11-20 that

$$\beta = \Phi\left(Z_{\alpha/2} - \frac{\delta\sqrt{n}}{\sigma}\right) - \Phi\left(-Z_{\alpha/2} - \frac{\delta\sqrt{n}}{\sigma}\right),$$

or if $\delta > 0$,

$$\beta \simeq \Phi\left(Z_{\alpha/2} - \frac{\delta\sqrt{n}}{\sigma}\right), \tag{11-24}$$

since $\Phi\left(-Z_{\alpha/2} - \delta\sqrt{n}/\sigma\right) \simeq 0$ when δ is positive. From equation 11-24, we take normal inverses to obtain

$$-Z_\beta \simeq Z_{\alpha/2} - \frac{\delta\sqrt{n}}{\sigma}$$

or

$$n \simeq \frac{\left(Z_{\alpha/2} + Z_\beta\right)^2 \sigma^2}{\delta^2}. \tag{11-25}$$

This approximation is good when $\Phi\left(-Z_{\alpha/2} - \delta\sqrt{n}/\sigma\right)$ is small compared to β. For either of the one-sided alternative hypotheses in equation 11-15 or equation 11-17, the sample size required to produce a specified type II error with probability β given δ and α is

$$n = \frac{\left(Z_\alpha + Z_\beta\right)^2 \sigma^2}{\delta^2}. \tag{11-26}$$

Example 11-4

Returning to the rocket propellant problem of Example 11-3, we note that $\sigma = 2$, $\delta = 41 - 40 = 1$, $\alpha = 0.05$, and $\beta = 0.10$. Since $Z_{\alpha/2} = Z_{0.025} = 1.96$ and $Z_\beta = Z_{0.10} = 1.28$, the sample size required to detect this departure from H_0: $\mu = 40$ is found in equation 11-25 to be

$$n \simeq \frac{\left(Z_{\alpha/2} + Z_\beta\right)^2 \sigma^2}{\delta^2} = \frac{(1.96 + 1.28)^2 2^2}{1^2} = 42,$$

which is in close agreement with the value determined from the operating characteristic curve. Note that the approximation is good, since $\Phi\left(-Z_{\alpha/2} - \delta\sqrt{n}/\sigma\right) = \Phi\left(-1.96 - (1)\sqrt{42}/2\right) = -5.20 \simeq 0$ which is small relative to β.

The Relationship Between Tests of Hypotheses and Confidence Intervals

There is a close relationship between the test of a hypothesis about a parameter θ and the confidence interval for θ. If $[L, U]$ is a $100(1 - \alpha)\%$ confidence interval for the parameter θ, then the test of size α of the hypothesis

$$H_0\colon \theta = \theta_0,$$
$$H_1\colon \theta \neq \theta_0$$

will lead to rejection of H_0 if and only if θ_0 is not in the interval $[L, U]$. As an illustration, consider the rocket propellant problem in Example 11-1. The null hypothesis H_0: $\mu = 40$ was rejected using $\alpha = 0.05$. The 95% two-sided confidence interval on μ for these data may be computed from equation 10-25 as $40.47 \leq \mu \leq 42.03$. That is, the interval $[L, U]$ is [40.47, 42.03], and since $\mu_0 = 40$ is not included in this interval, the null hypothesis H_0: $\mu = 40$ is rejected.

Large Sample Test with Unknown Variance

Although we have developed the test procedure for the null hypothesis H_0: $\mu = \mu_0$ assuming that σ^2 is known, in many practical situations σ^2 will be unknown. In general, if $n \geq 30$, then the sample variance S^2 can be substituted for σ^2 in the test procedures with little harmful effect. Thus, while we have given a test for known σ^2, it can be converted easily in to a *large-sample* test procedure for unknown σ^2. Exact treatment of the case where σ^2 is unknown and n is small involves the use of the t distribution and is deferred until Section 11-2.2.

P-Values

Computer software packages are frequently used for statistical hypothesis testing. Most of these programs calculate and report the probability that the test statistic will take on a value at least as extreme as the observed value of the statistic when H_0 is true. This probability is usually called a *P-value.* It represents the smallest level of significance that would lead to rejection of H_0. Thus, if $P = 0.04$ is reported in the computer output, the null hypothesis H_0 would be rejected at the level $\alpha = 0.05$ but not at the level $\alpha = 0.01$. Generally, if P is less than or equal to α, we would reject H_0, whereas if P exceeds α we would fail to reject H_0.

It is customary to call the test statistic (and the data) *significant* when the null hypothesis H_0 is rejected, so we may think of the P-value as the smallest level α at which the data are significant. Once the P-value is known, the decision maker can determine for himself or herself how significant the data are without the data analyst formally imposing a preselected level of significance.

It is not always easy to compute the exact P-value of a test. However, for the foregoing normal distribution tests it is relatively easy. If Z_0 is the computed value of the test statistic, then the P-value is

$$P = \begin{cases} 2\left[1 - \Phi\left(\left|Z_0\right|\right)\right] & \text{for a two-tailed test,} \\ 1 - \Phi\left(Z_0\right) & \text{for an upper-tail test,} \\ \Phi\left(Z_0\right) & \text{for a lower-tail test.} \end{cases}$$

To illustrate, consider the rocket propellant problem in Example 11-1. The computed value of the test statistic is $Z_0 = 3.125$ and since the alternative hypothesis is two-tailed, the P-value is

$$P = 2[1 - \Phi(3.125)] = 0.0018.$$

Thus, H_0: $\mu = 40$ would be rejected at any level of significance where $\alpha \geq P = 0.0018$. For example, H_0 would be rejected if $\alpha = 0.01$, but it would not be rejected if $\alpha = 0.001$.

Practical Versus Statistical Significance

In Chapters 10 and 11, we present confidence intervals and tests of hypotheses for both single-sample and two-sample problems. In hypothesis testing, we have discussed the statistical significance when the null hypothesis is rejected. What has not been discussed is the *practical* significance of rejecting the null hypothesis. In hypothesis testing, the goal is to make a decision about a claim or belief. The decision as to whether or not the null hypothesis is rejected in favor of the alternative is based on a sample taken from the population of interest. If the null hypothesis is rejected, we say there is statistically significant evidence against the null hypothesis in favor of the alternative. Results that are statistically significant (by rejection of the null hypothesis) do not necessarily imply practically significant results.

To illustrate, suppose that the average temperature on a single day throughout a particular state is hypothesized to be $\mu = 63$ degrees. Suppose that $n = 50$ locations within the state had an average temperature of $\bar{x} = 62$ degrees and standard deviation of 0.5 degrees. If we were to test the hypothesis H_0: $\mu = 63$ against H_1: $\mu \neq 63$, we would get a resulting P-value of approximately 0 and we would reject the null hypothesis. Our conclusion would be that the true average temperature is not 63 degrees. In other words, we have illustrated a statistically significant difference between the hypothesized value and the sample average obtained from the data. But is this a practical difference? That is, is 63 degrees *different* from 62 degrees? Very few investigators would actually conclude that this difference is practical. In other words, statistical significance does not imply practical significance.

The size of the sample under investigation has a direct influence on the power of the test and the practical significance. As the sample size increases, even the smallest differences between the hypothesized value and the sample value may be detected by the hypothesis test. Therefore, care must be taken when interpreting the results of a hypothesis test when the sample sizes are large.

11-2.2 Tests of Hypotheses on the Mean of a Normal Distribution, Variance Unknown

When testing hypotheses about the mean μ of a population when σ^2 is unknown, we can use the test procedures discussed in Section 11-2.1 provided that the sample size is large ($n \geq 30$, say). These procedures are approximately valid regardless of whether or not the underlying population is normal. However, when the sample size is small and σ^2 is unknown, we must make an assumption about the form of the underlying distribution in order to obtain a test procedure. A reasonable assumption in many cases is that the underlying distribution is normal.

Many populations encountered in practice are quite well approximated by the normal distribution, so this assumption will lead to a test procedure of wide applicability. In fact, moderate departure from normality will have little effect on the test validity. When the assumption is unreasonable, we can either specify another distribution (exponential, Weibull, etc.) and use some general method of test construction to obtain a valid procedure, or we could use one of the nonparametric tests that are valid for any underlying distribution (see Chapter 16).

Statistical Analysis

Suppose that X is a normally distributed random variable with unknown mean μ and variance σ^2. We wish to test the hypothesis that μ equals a constant μ_0. Note that this situation is similar to that treated in Section 11-2.1, except that now *both* μ *and* σ^2 are unknown. Assume that a random sample of size n, say $X_1, X_2,\ldots, X_n$, is available, and let $\overline{X}$ and S^2 be the sample mean and variance, respectively.

Suppose that we wish to test the two-sided alternative

$$H_0: \mu = \mu_0,$$
$$H_1: \mu \neq \mu_0. \tag{11-27}$$

The test procedure is based on the statistic

$$t_0 = \frac{\overline{X} - \mu_0}{S/\sqrt{n}}, \tag{11-28}$$

which follows the t distribution with $n - 1$ degrees of freedom if the null hypothesis H_0: $\mu = \mu_0$ is true. To test H_0: $\mu = \mu_0$ in equation 11-27, the test statistic t_0 in equation 11-28 is calculated, and H_0 is rejected if either

$$t_0 > t_{\alpha/2,\, n-1} \tag{11-29a}$$

or

$$t_0 < -t_{\alpha/2,\, n-1}, \tag{11-29b}$$

where $t_{\alpha/2,\, n-1}$ and $-t_{\alpha/2,\, n-1}$ are the upper and lower $\alpha/2$ percentage points of the t distribution with $n - 1$ degrees of freedom.

For the one-sided alternative hypothesis

$$H_0: \mu = \mu_0,$$
$$H_1: \mu > \mu_0, \tag{11-30}$$

we calculate the test statistic t_0 from equation 11-28 and reject H_0 if

$$t_0 > t_{\alpha,\, n-1}. \tag{11-31}$$

For the other one-sided alternative,

$$\begin{aligned} H_0&: \mu = \mu_0, \\ H_1&: \mu < \mu_0, \end{aligned} \tag{11-32}$$

we would reject H_0 if

$$t_0 < -t_{\alpha,\, n-1}. \tag{11-33}$$

Example 11-5

The breaking strength of a textile fiber is a normally distributed random variable. Specifications require that the mean breaking strength should equal 150 psi. The manufacturer would like to detect any significant departure from this value. Thus, he wishes to test

$$\begin{aligned} H_0&: \mu = 150 \text{ psi}, \\ H_1&: \mu \neq 150 \text{ psi}. \end{aligned}$$

A random sample of 15 fiber specimens is selected and their breaking strengths determined. The sample mean and variance are computed from the sample data as $\bar{x} = 152.18$ and $s^2 = 16.63$. Therefore, the test statistic is

$$t_0 = \frac{\bar{x} - \mu_0}{s/\sqrt{n}} = \frac{152.18 - 150}{\sqrt{16.63/15}} = 2.07.$$

The type I error is specified as $\alpha = 0.05$. Therefore $t_{0.025,14} = 2.145$ and $-t_{0.025,14} = -2.145$, and we would conclude that there is not sufficient evidence to reject the hypothesis that $\mu = 150$ psi.

Choice of Sample Size

The type II error probability for tests on the mean of a normal distribution with unknown variance depends on the distribution of the test statistic in equation 11-28 when the null hypothesis H_0: $\mu = \mu_0$ is false. When the true value of the mean is $\mu = \mu_0 + \delta$, note that the test statistic can be written as

$$\begin{aligned} t_0 &= \frac{\bar{X} - \mu_0}{S/\sqrt{n}} \\ &= \frac{\dfrac{\left[\bar{X} - (\mu_0 + \delta)\right]\sqrt{n}}{\sigma} + \dfrac{\delta\sqrt{n}}{\sigma}}{S\,\sigma} \\ &= \frac{Z + \dfrac{\delta\sqrt{n}}{\sigma}}{W}. \end{aligned} \tag{11-34}$$

The distributions of Z and W in equation 11-34 are $N(0, 1)$ and $\sqrt{\chi^2_{n-1}/(n-1)}$, respectively, and Z and W are independent random variables. However, $\delta\sqrt{n}/\sigma$ is a nonzero constant, so that the numerator of equation 11-34 is a $N\left(\delta\sqrt{n}/\sigma,\ 1\right)$ random variable. The resulting distribution is called the *noncentral t* distribution with $n - 1$ degrees of freedom and noncentrality parameter $\delta\sqrt{n}/\sigma$. Note that if $\delta = 0$, then the noncentral t distribution reduces to the usual or central t distribution. In any case, the type II error of the two-sided alternative (for example) would be

$$\beta = P\{-t_{\alpha/2,\,n-1} \le t_0 \le t_{\alpha/2,\,n-1} | \delta \ne 0\}$$

$$= P\{-t_{\alpha/2,\,n-1} \le t_0' \le t_{\alpha/2,\,n-1}\},$$

where t_0' denotes the noncentral t random variable. Finding the type II error for the t-test involves finding the probability contained between two points on the noncentral t distribution.

The operating characteristic curves in Charts VIe, VIf, VIg, and VIh (Appendix) plot β against a parameter d for various sample sizes n. Curves are provided for both the two-sided and one-sided alternatives and for $\alpha = 0.05$ or $\alpha = 0.01$. For the two-sided alternative in equation 11-27, the abscissa scale factor d on Charts VIe and VIf is defined as

$$d = \frac{|\mu - \mu_0|}{\sigma} = \frac{|\delta|}{\sigma} \tag{11-35}$$

For the one-sided alternatives, if rejection is desired, i.e., $\mu > \mu_0$, as in equation 11-30, we use Charts VIg and VIh with

$$d = \frac{\mu - \mu_0}{\sigma} = \frac{\delta}{\sigma}, \tag{11-36}$$

while if rejection is desired, i.e., $\mu < \mu_0$, as in equation 11-32,

$$d = \frac{\mu_0 - \mu}{\sigma} = \frac{\delta}{\sigma}. \tag{11-37}$$

We note that d depends on the unknown parameter σ^2. There are several ways to avoid this difficulty. In some cases, we may use the results of a previous experiment or prior information to make a rough initial estimate of σ^2. If we are interested in examining the operating characteristic after the data has been collected, we could use the sample variance s^2 to estimate σ^2. If analysts do not have any previous experience on which to draw in estimating σ^2, they can define the difference in the mean δ that they wish to detect relative to σ. For example, if one wishes to detect a small difference in the mean, one might use a value of $d = |\delta|/\sigma \le 1$ (say) whereas if one is interested in detecting only moderately large differences in the mean, one might select $d = |\delta|/\sigma = 2$ (say). That is, it is the value of the ratio $|\delta|/\sigma$ that is important in determining sample size, and if it is possible to specify the relative size of the difference in means that we are interested in detecting, then a proper value of d can usually be selected.

Example 11-6

Consider the fiber-testing problem in Example 11-5. If the breaking strength of this fiber differs from 150 psi by as much as 2.5 psi, the analyst would like to reject the null hypothesis H_0: $\mu = 150$ psi with a probability of at least 0.90. Is the sample size $n = 15$ adequate to ensure that the test is this sensitive? If we use the sample standard deviation $s = \sqrt{16.63} = 4.08$ to estimate σ, then $d = |\delta|/\sigma = 2.5/4.08 = 0.61$. By referring to the operating characteristic curves in Chart VIe, with $d = 0.61$ and $n = 15$, we find $\beta \simeq 0.45$. Thus, the probability of rejecting H_0: $\mu = 150$ psi if the true mean differs from this value by ± 2.5 psi is $1 - \beta = 1 - 0.45 = 0.55$, approximately, and we would conclude that a sample size of

$n = 15$ is not adequate. To find the sample size required to give the desired degree of protection, enter the operating characteristic curves in Chart VI*e* with $d = 0.61$ and $\beta = 0.10$, and read the corresponding sample size as $n = 35$, approximately.

11-2.3 Tests of Hypotheses on the Variance of a Normal Distribution

There are occasions when tests assessing the variance or standard deviation of a population are needed. In this section we present two procedures, one based on the assumption of normality and the other one a large-sample test.

Test Procedures for a Normal Population

Suppose that we wish to test the hypothesis that the variance σ^2 of a normal distribution equals a specified value, say σ_0^2. Let $X \sim N(\mu, \sigma^2)$, where μ and σ^2 are unknown, and let X_1, X_2, …, X_n be a random sample of n observations from this population. To test

$$\begin{aligned} H_0&: \sigma^2 = \sigma_0^2, \\ H_1&: \sigma^2 \neq \sigma_0^2, \end{aligned} \tag{11-38}$$

we use the test statistic

$$\chi_0^2 = \frac{(n-1)S^2}{\sigma_0^2}, \tag{10-39}$$

where S^2 is the sample variance. Now if H_0: $\sigma^2 = \sigma_0^2$ is true, then the test statistic χ_0^2 follows the chi-square distribution with $n - 1$ degrees of freedom. Therefore, H_0: $\sigma^2 = \sigma_0^2$ would be rejected if

$$\chi_0^2 > \chi_{\alpha/2,\, n-1}^2 \tag{11-40a}$$

or if

$$\chi_0^2 < \chi_{1-\alpha/2,\, n-1}^2, \tag{11-40b}$$

where $\chi_{\alpha/2,\, n-1}^2$ and $\chi_{1-\alpha/2,\, n-1}^2$ are the upper and lower $\alpha/2$ percentage points of the chi-square distribution with $n - 1$ degrees of freedom.

The same test statistic is used for the one-sided alternatives. For the one-sided hypothesis

$$\begin{aligned} H_0&: \sigma^2 = \sigma_0^2, \\ H_1&: \sigma^2 > \sigma_0^2, \end{aligned} \tag{11-41}$$

we would reject H_0 if

$$\chi_0^2 > \chi_{\alpha,\, n-1}^2. \tag{11-42}$$

For the other one-sided hypothesis,

$$\begin{aligned} H_0&: \sigma^2 = \sigma_0^2, \\ H_1&: \sigma^2 < \sigma_0^2, \end{aligned} \tag{11-43}$$

we would reject H_0 if

$$\chi_0^2 < \chi_{1-\alpha,\, n-1}^2. \tag{11-44}$$

Example 11-7

Consider the machine described in Example 10-16, which is used to fill cans with a soft drink beverage. If the variance of the fill volume exceeds 0.02 (fluid ounces)^2, then an unacceptably large percentage of the cans will be underfilled. The bottler is interested in testing the hypothesis

$$H_0\colon \sigma^2 = 0.02,$$
$$H_1\colon \sigma^2 > 0.02.$$

A random sample of $n = 20$ cans yields a sample variance of $s^2 = 0.0225$. Thus, the test statistic is

$$\chi_0^2 = \frac{(n-1)s^2}{\sigma_0^2} = \frac{(19)0.0225}{0.02} = 21.38$$

If we choose $\alpha = 0.05$, we find that $\chi^2_{0.05,19} = 30.14$, and we would conclude that there is no strong evidence that the variance of fill volume exceeds 0.02 (fluid ounces)^2.

Choice of Sample Size

Operating characteristic curves for the χ^2 tests are provided in Charts VI*i* through VI*n* (Appendix) for $\alpha = 0.05$ and $\alpha = 0.01$. For the two-sided alternative hypothesis of equation 11-38, Charts VI*i* and VI*j* plot β against an abscissa parameter,

$$\lambda - \frac{\sigma}{\sigma_0}, \tag{11-45}$$

for various sample sizes n, where σ denotes the true value of the standard deviation. Charts VI*k* and VI*l* are for the one-sided alternative $H_1\colon \sigma^2 > \sigma_0^2$, while Charts VI*m* and VI*n* are for the other one-sided alternative $H_1\colon \sigma^2 < \sigma_0^2$. In using these charts, we think of σ as the value of the standard deviation that we want to detect.

Example 11-8

In Example 11-7, find the probability of rejecting $H_0\colon \sigma^2 = 0.02$ if the true variance is as large as $\sigma^2 = 0.03$. Since $\sigma = \sqrt{0.03} = 0.1732$ and $\sigma_0 = \sqrt{0.02} = 0.1414$ the abscissa parameter is

$$\lambda = \frac{\sigma}{\sigma_0} = \frac{0.1732}{0.1414} = 1.23.$$

From Chart VI*k*, with $\lambda = 1.23$ and $n = 20$, we find that $\beta \simeq 0.60$. That is, there is only about a 40% chance that $H_0\colon \sigma^2 = 0.02$ will be rejected if the variance is really as large as $\sigma^2 = 0.03$. To reduce β, a larger sample size must be used. From the operating characteristic curve, we note that to reduce β to 0.20 a sample size of 75 is necessary.

A Large-Sample Test Procedure

The chi-square test procedure prescribed above is rather sensitive to the normality assumption. Consequently, it would be desirable to develop a procedure that does not require this assumption. When the underlying population is not necessarily normal but n is large (say $n \geq 35$ or 40), then we can use the following result: if $X_1, X_2, \ldots, X_n$ is a random sample from a population with variance σ^2, the sample standard deviation S is approximately normal with mean $E(S) \simeq \sigma$ and variance $V(S) \simeq \sigma^2/2n$, if n is large.

Then the distribution of

$$Z_0 = \frac{S - \sigma}{\sigma/\sqrt{2n}} \tag{11-46}$$

is approximately standard normal.

To test

$$\begin{aligned} H_0&: \sigma^2 = \sigma_0^2, \\ H_1&: \sigma^2 \neq \sigma_0^2, \end{aligned} \tag{11-47}$$

substitute σ_0 for σ in equation 11-46. Thus, the test statistic is

$$Z_0 = \frac{S - \sigma_0}{\sigma_0/\sqrt{2n}}, \tag{11-48}$$

and we would reject H_0 if $Z_0 > Z_{\alpha/2}$ or if $Z_0 < -Z_{\alpha/2}$. The same test statistic would be used for the one-sided alternatives. If we are testing

$$\begin{aligned} H_0&: \sigma^2 = \sigma_0^2, \\ H_1&: \sigma^2 > \sigma_0^2, \end{aligned} \tag{11-49}$$

we would reject H_0 if $Z_0 > Z_\alpha$, while if we are testing

$$\begin{aligned} H_0&: \sigma^2 = \sigma_0^2, \\ H_1&: \sigma^2 < \sigma_0^2, \end{aligned} \tag{11-50}$$

we would reject H_0 if $Z_0 < -Z_\alpha$.

Example 11-9

An injection-molded plastic part is used in a graphics printer. Before agreeing to a long-term contract, the printer manufacturer wants to be sure using $\alpha = 0.01$ that the supplier can produce parts with a standard deviation of length of at most 0.025 mm. The hypotheses to be tested are

$$\begin{aligned} H_0&: \sigma^2 = 6.25 \times 10^{-4}, \\ H_1&: \sigma^2 < 6.25 \times 10^{-4}, \end{aligned}$$

since $(0.025)^2 = 0.000625$. A random sample of $n = 50$ parts is obtained, and the sample standard deviation is $s = 0.021$ mm. The test statistic is

$$Z_0 = \frac{s - \sigma_0}{\sigma_0/\sqrt{2n}} = \frac{0.021 - 0.025}{0.025/\sqrt{100}} = -1.60.$$

Since $-Z_{0.01} = -2.33$ and the observed value of Z_0 is not smaller than this critical value, H_0 is not rejected. That is, the evidence from the supplier's process is not strong enough to justify a long-term contract.

11-2.4 Tests of Hypotheses on a Proportion

Statistical Analysis

In many engineering and management problems, we are concerned with a random variable that follows the binomial distribution. For example, consider a production process that manufactures items that are classified as either acceptable or defective. It is usually reasonable to model the occurrence of defectives with the binomial distribution, where the binomial parameter p represents the proportion of defective items produced.

We will consider testing

$$H_0: p = p_0,$$
$$H_1: p \neq p_0. \tag{11-51}$$

An approximate test based on the normal approximation to the binomial will be given. This approximate procedure will be valid as long as p is not extremely close to zero or 1, and if the sample size is relatively large. Let X be the number of observations in a random sample of size n that belongs to the class associated with p. Then, if the null hypothesis H_0: $p = p_0$ is true, we have $X \sim N(np_0, np_0(1-p_0))$, approximately. To test H_0: $p = p_0$ calculate the test statistic

$$Z_0 = \frac{X - np_0}{\sqrt{np_0(1-p_0)}} \tag{11-52}$$

and reject H_0: $p = p_0$ if

$$Z_0 > Z_{\alpha/2} \quad \text{or} \quad Z_0 < -Z_{\alpha/2}. \tag{11-53}$$

Critical regions for the one-sided alternative hypotheses would be located in the usual manner.

Example 11-10

A semiconductor firm produces logic devices. The contract with their customer calls for a fraction defective of no more than 0.05. They wish to test

$$H_0: p = 0.05,$$
$$H_1: p > 0.05.$$

A random sample of 200 devices yields six defectives. The test statistic is

$$Z_0 = \frac{x - np_0}{\sqrt{np_0(1-p_0)}} = \frac{6 - 200(0.05)}{\sqrt{200(0.05)(0.95)}} = -1.30$$

Using $\alpha = 0.05$, we find that $Z_{0.05} = 1.645$, and so we cannot reject the null hypothesis that $p = 0.05$.

Choice of Sample Size

It is possible to obtain closed-form equations for the β error for the tests in this section. The β error for the two-sided alternative H_1: $p \neq p_0$ is approximately

$$\beta \simeq \Phi\left(\frac{p_0 - p + Z_{\alpha/2}\sqrt{p_0(1-p_0)/n}}{\sqrt{p(1-p)/n}}\right) - \Phi\left(\frac{p_0 - p - Z_{\alpha/2}\sqrt{p_0(1-p_0)/n}}{\sqrt{p(1-p)/n}}\right). \tag{11-54}$$

If the alternative is H_1: $p < p_0$, then

$$\beta \simeq 1-\Phi\left(\frac{p_0 - p - Z_\alpha\sqrt{p_0(1-p_0)/n}}{\sqrt{p(1-p)/n}}\right), \tag{11-55}$$

whereas if the alternative is H_1: $p > p_0$, then

$$\beta \simeq \Phi\left(\frac{p_0 - p + Z_\alpha\sqrt{p_0(1-p_0)/n}}{\sqrt{p(1-p)/n}}\right). \tag{11-56}$$

These equations can be solved to find the sample size n that gives a test of level α that has a specified β risk. The sample size equations are

$$n = \left(\frac{Z_{\alpha/2}\sqrt{p_0(1-p_0)} + Z_\beta\sqrt{p(1-p)}}{p - p_0}\right)^2 \tag{11-57}$$

for the two-sided alternative and

$$n = \left(\frac{Z_\alpha\sqrt{p_0(1-p_0)} + Z_\beta\sqrt{p(1-p)}}{p - p_0}\right)^2 \tag{11-58}$$

for the one-sided alternatives.

Example 11-11

For the situation described in Example 11-10, suppose that we wish to find the β error of the test if $p = 0.07$. Using equation 11-56, the β error is

$$\begin{aligned}\beta &\simeq \Phi\left(\frac{0.05 - 0.07 + 1.645\sqrt{(0.05)(0.95)/200}}{\sqrt{(0.07)(0.93)/200}}\right)\\ &= \Phi(0.30)\\ &= 0.6179.\end{aligned}$$

This type II error probability is not as small as one might like, but $n = 200$ is not particularly large and 0.07 is not very far from the null value $p_0 = 0.05$. Suppose that we want the β error to be no larger than 0.10 if the true value of the fraction defective is as large as $p = 0.07$. The required sample size would be found from equation 11-58 as

$$\begin{aligned}n &= \left(\frac{1.645\sqrt{(0.05)(0.95)} + 1.28\sqrt{(0.07)(0.93)}}{0.07 - 0.05}\right)^2\\ &= 1174,\end{aligned}$$

which is a very large sample size. However, notice that we are trying to detect a very small deviation from the null value $p_0 = 0.05$.

11-3 TESTS OF HYPOTHESES ON TWO SAMPLES

11-3.1 Tests of Hypotheses on the Means of Two Normal Distributions, Variances Known

Statistical Analysis

Suppose that there are two populations of interest, say X_1 and X_2. We assume that X_1 has unknown mean μ_1 and known variance σ_1^2, and that X_2 has unknown mean μ_2 and known variance σ_2^2. We will be concerned with testing the hypothesis that the means μ_1 and μ_2 are equal. It is assumed either that the random variables X_1 and X_2 are normally distributed or, if they are nonnormal, that the conditions of the Central Limit Theorem apply.

Consider first the two-sided alternative hypothesis

$$\begin{aligned} H_0&: \mu_1 = \mu_2, \\ H_1&: \mu_1 \neq \mu_2. \end{aligned} \tag{11-59}$$

Suppose that a random sample of size n_1 is drawn from X_1, say $X_{11}, X_{12},\ldots, X_{1n_1}$, and that a second random sample of size n_2 is drawn from X_2, say $X_{21}, X_{22},\ldots, X_{2n_2}$. It is assumed that the $\{X_{1j}\}$ are independently distributed with mean μ_1 and variance σ_1^2, that the $\{X_{2j}\}$ are independently distributed with mean μ_2 and variance σ_2^2, and that the two samples $\{X_{1j}\}$ and $\{X_{2j}\}$ are independent. The test procedure is based on the distribution of the difference in sample means, say $\overline{X}_1 - \overline{X}_2$. In general, we know that

$$\overline{X}_1 - \overline{X}_2 \sim N\left(\mu_1 - \mu_2, \frac{\sigma_1^2}{n_1} + \frac{\sigma_2^2}{n_2}\right).$$

Thus, if the null hypothesis H_0: $\mu_1 = \mu_2$ is true, the test statistic

$$Z_0 = \frac{\overline{X}_1 - \overline{X}_2}{\sqrt{\dfrac{\sigma_1^2}{n_1} + \dfrac{\sigma_2^2}{n_2}}} \tag{11-60}$$

follows the $N(0, 1)$ distribution. Therefore, the procedure for testing H_0: $\mu_1 = \mu_2$ is to calculate the test statistic Z_0 in equation 11-60 and reject the null hypothesis if

$$Z_0 > Z_{\alpha/2} \tag{11-61a}$$

or

$$Z_0 < -Z_{\alpha/2}. \tag{11-61b}$$

The one-sided alternative hypotheses are analyzed similarly. To test

$$\begin{aligned} H_0&: \mu_1 = \mu_2, \\ H_1&: \mu_1 > \mu_2, \end{aligned} \tag{11-62}$$

the test statistic Z_0 in equation 11-60 is calculated, and H_0: $\mu_1 = \mu_2$ is rejected if

$$Z_0 > Z_{\alpha}. \tag{11-63}$$

To test the other one-sided alternative hypothesis,

$$\begin{aligned} H_0&: \mu_1 = \mu_2, \\ H_1&: \mu_1 < \mu_2, \end{aligned} \tag{11-64}$$

use the test statistic Z_0 in equation 11-60 and reject H_0: $\mu_1 = \mu_2$ if

$$Z_0 < -Z_{\alpha}. \tag{11-65}$$

Example 11-12

The plant manager of an orange juice canning facility is interested in comparing the performance of two different production lines in her plant. As line number 1 is relatively new, she suspects that its output in number of cases per day is greater than the number of cases produced by the older line 2. Ten days of data are selected at random for each line, for which it is found that $\bar{x}_1 = 824.9$ cases per day and $\bar{x}_2 = 818.6$ cases per day. From experience with operating this type of equipment it is known that $\sigma_1^2 = 40$ and $\sigma_2^2 = 50$. We wish to test

$$H_0: \mu_1 = \mu_2,$$
$$H_1: \mu_1 > \mu_2.$$

The value of the test statistic is

$$Z_0 = \frac{\bar{x}_1 - \bar{x}_2}{\sqrt{\dfrac{\sigma_1^2}{n_1} + \dfrac{\sigma_2^2}{n_2}}} = \frac{824.9 - 818.6}{\sqrt{\dfrac{40}{10} + \dfrac{50}{10}}} = 2.10.$$

Using $\alpha = 0.05$ we find that $Z_{0.05} = 1.645$, and since $Z_0 > Z_{0.05}$, we would reject H_0 and conclude that the mean number of cases per day produced by the new production line is greater than the mean number of cases per day produced by the old line.

Choice of Sample Size

The operating characteristic curves in Charts VI*a*, VI*b*, VI*c*, and VI*d* (Appendix) may be used to evaluate the type II error probability for the hypotheses in equations 11-59, 11-62, and 11-64. These curves are also useful in sample size determination. Curves are provided for $\alpha = 0.05$ and $\alpha = 0.01$. For the two-sided alternative hypothesis in equation 11-59, the abscissa scale of the operating characteristic curves in Charts VI*a* and VI*b* is d, where

$$d = \frac{|\mu_1 - \mu_2|}{\sqrt{\sigma_1^2 + \sigma_2^2}} = \frac{|\delta|}{\sqrt{\sigma_1^2 + \sigma_2^2}}, \tag{11-66}$$

and one must choose equal sample sizes, say $n = n_1 = n_2$. The one-sided alternative hypotheses require the use of Charts VI*c* and VI*d*. For the one-sided alternative $H_1: \mu_1 > \mu_2$ in equation 11-62, the abscissa scale is

$$d = \frac{\mu_1 - \mu_2}{\sqrt{\sigma_1^2 + \sigma_2^2}} = \frac{\delta}{\sqrt{\sigma_1^2 + \sigma_2^2}}, \tag{11-67}$$

where $n = n_1 = n_2$. The other one-sided alternative hypothesis, $H_1: \mu_1 < \mu_2$, requires that d be defined as

$$d = \frac{\mu_2 - \mu_1}{\sqrt{\sigma_1^2 + \sigma_2^2}} = \frac{\delta}{\sqrt{\sigma_1^2 + \sigma_2^2}} \tag{11-68}$$

and $n = n_1 = n_2$.

It is not unusual to encounter problems where the costs of collecting data differ substantially between the two populations, or where one population variance is much greater than the other. In those cases, one often uses unequal sample sizes. If $n_1 \neq n_2$, the operating characteristic curves may be entered with an *equivalent* value of n computed from

$$n = \frac{\sigma_1^2 + \sigma_2^2}{\sigma_1^2/n_1 + \sigma_2^2/n_2}. \tag{11-69}$$

If $n_1 \neq n_2$, and their values are fixed in advance, then equation 11-69 is used directly to calculate n, and the operating characteristic curves are entered with a specified d to obtain β. If we are given d and it is necessary to determine n_1 and n_2 to obtain a specified β, say β^*, then one guesses at trial values of n_1 and n_2, calculates n in equation 11-69, enters the curves with the specified value of d, and finds β. If $\beta = \beta^*$, then the trial values of n_1 and n_2 are satisfactory. If $\beta \neq \beta^*$, then adjustments to n_1 and n_2 are made and the process is repeated.

Example 11-13

Consider the orange juice production line problem in Example 11-12. If the true difference in mean production rates were 10 cases per day, find the sample sizes required to detect this difference with a probability of 0.90. The appropriate value of the abscissa parameter is

$$d = \frac{\mu_1 - \mu_2}{\sqrt{\sigma_1^2 + \sigma_2^2}} = \frac{10}{\sqrt{40+50}} = 1.05,$$

and since $\alpha = 0.05$, we find from Chart VI*c* that $n = n_1 = n_2 = 8$.

It is also possible to derive formulas for the sample size required to obtain a specified β for a given δ and α. These formulas occasionally are useful supplements to the operating characteristic curves. For the two-sided alternative hypothesis, the sample size $n_1 = n_2 = n$ is

$$n \simeq \frac{\left(Z_{\alpha/2} + Z_\beta\right)^2 \left(\sigma_1^2 + \sigma_2^2\right)}{\delta^2}. \tag{11-70}$$

This approximation is valid when $\Phi\left(-Z_{\alpha/2} - \delta\sqrt{n} \big/ \sqrt{\sigma_1^2 + \sigma_2^2}\right)$ is small compared to β. For a one-sided alternative, we have $n_1 = n_2 = n$, where

$$n = \frac{\left(Z_\alpha + Z_\beta\right)^2 \left(\sigma_1^2 + \sigma_2^2\right)}{\delta^2}. \tag{11-71}$$

The derivations of equations 11-70 and 11-71 closely follow the single-sample case in Section 11-2. To illustrate the use of these equations, consider the situation in Example 11-13. We have a one-sided alternative with $\alpha = 0.05$, $\delta = 10$, $\sigma_1^2 = 40$, $\sigma_2^2 = 50$, and $\beta = 0.10$. Thus $Z_\alpha = Z_{0.05} = 1.645$, $Z_\beta = Z_{0.10} = 1.28$, and the required sample size is found from equation 11-71 to be

$$n = \frac{\left(Z_\alpha + Z_\beta\right)^2 \left(\sigma_1^2 + \sigma_2^2\right)}{\delta^2} = \frac{(1.645 + 1.28)^2 (40 + 50)}{10^2} = 8,$$

which agrees with the results obtained in Example 11-13.

11-3.2 Tests of Hypotheses on the Means of Two Normal Distributions, Variances Unknown

We now consider tests of hypotheses on the equality of the means μ_1 and μ_2 of two normal distributions where the variances σ_1^2 and σ_2^2 are unknown. A t statistic will be used to test these hypotheses. As noted in Section 11-2.2, the normality assumption is required to develop the test procedure, but moderate departures from normality do not adversely affect the procedure. There are two different situations that must be treated. In the first case, we assume that the variances of the two normal distributions are unknown but equal; that is, $\sigma_1^2 = \sigma_2^2 = \sigma^2$. In the second, we assume that σ_1^2 and σ_2^2 are unknown and not necessarily equal.

Case 1: $\sigma_1^2 = \sigma_2^2 = \sigma^2$ Let X_1 and X_2 be two independent normal populations with unknown means μ_1 and μ_2, and unknown but equal variances $\sigma_1^2 = \sigma_2^2 = \sigma^2$. We wish to test

$$\begin{aligned} H_0&: \mu_1 = \mu_2, \\ H_1&: \mu_1 \neq \mu_2. \end{aligned} \tag{11-72}$$

Suppose that $X_{11}, X_{12}, \ldots, X_{1n_1}$ is a random sample of n_1 observations from X_1, and $X_{21}, X_{22}, \ldots, X_{2n_2}$ is a random sample of n_2 observations from X_2. Let $\overline{X}_1$, $\overline{X}_2$, S_1^2, and S_2^2 be the sample means and sample variances, respectively. Since both S_1^2 and S_2^2 estimate the common variance σ^2, we may combine them to yield a single estimate, say

$$S_p^2 = \frac{(n_1 - 1)S_1^2 + (n_2 - 1)S_2^2}{n_1 + n_2 - 2}. \tag{11-73}$$

This combined or "pooled" estimator was introduced in Section 10-3.2. To test H_0: $\mu_1 = \mu_2$ in equation 11-72, compute the test statistic

$$t_0 = \frac{\overline{X}_1 - \overline{X}_2}{S_p \sqrt{\dfrac{1}{n_1} + \dfrac{1}{n_2}}}. \tag{11-74}$$

If H_0: $\mu_1 = \mu_2$ is true, t_0 is distributed as $t_{n_1 + n_2 - 2}$. Therefore, if

$$t_0 > t_{\alpha/2,\, n_1 + n_2 - 2} \tag{11-75a}$$

or if

$$t_0 < -t_{\alpha/2,\, n_1 + n_2 - 2} \tag{11-75b}$$

we reject H_0: $\mu_1 = \mu_2$.

The one-sided alternatives are treated similarly. To test

$$\begin{aligned} H_0&: \mu_1 = \mu_2, \\ H_1&: \mu_1 > \mu_2, \end{aligned} \tag{11-76}$$

compute the test statistic t_0 in equation 11-74 and reject H_0: $\mu_1 = \mu_2$ if

$$t_0 > t_{\alpha,\, n_1 + n_2 - 2}. \tag{11-77}$$

For the other one-sided alternative

$$\begin{aligned} H_0&: \mu_1 = \mu_2, \\ H_1&: \mu_1 < \mu_2, \end{aligned} \tag{11-78}$$

calculate the test statistic t_0 and reject H_0: $\mu_1 = \mu_2$ if

$$t_0 < -t_{\alpha,\, n_1 + n_2 - 2}. \tag{11-79}$$

The two-sample t-test given in this section is often called the *pooled* t-test, because the sample variances are combined or pooled to estimate the common variance. It is also known as the *independent* t-test, because the two normal populations are assumed to be independent.

Example 11-14

Two catalysts are being analyzed to determine how they affect the mean yield of a chemical process. Specifically, catalyst 1 is currently in use, but catalyst 2 is acceptable. Since catalyst 2 is cheaper, if it does not change the process yield, it should be adopted. Suppose we wish to test the hypotheses

$$H_0\colon \mu_1 = \mu_2,$$
$$H_1\colon \mu_1 \neq \mu_2.$$

Pilot plant data yields $n_1 = 8$, $\bar{x}_1 = 91.73$, $s_1^2 = 3.89$, $n_2 = 8$, $\bar{x}_2 = 93.75$, and $s_2^2 = 4.02$. From equation 11-73, we find

$$s_p^2 = \frac{(n_1-1)s_1^2 + (n_2-1)s_2^2}{n_1+n_2-2} = \frac{(7)3.89 + 7(4.02)}{8+8-2} = 3.96.$$

The test statistic is

$$t_0 = \frac{\bar{x}_1 - \bar{x}_2}{s_p\sqrt{\frac{1}{n_1} + \frac{1}{n_2}}} = \frac{91.73 - 93.75}{1.99\sqrt{\frac{1}{8} + \frac{1}{8}}} = -2.03.$$

Using $\alpha = 0.05$ we find that $t_{0.025,\,14} = 2.145$ and $-t_{0.025,\,14} = -2.145$, and, consequently, $H_0\colon \mu_1 = \mu_2$ cannot be rejected. That is, we do not have strong evidence to conclude that catalyst 2 results in a mean yield that differs from the mean yield when catalyst 1 is used.

Case 2: $\sigma_1^2 \neq \sigma_2^2$ In some situations, we cannot reasonably assume that the unknown variances σ_1^2 and σ_2^2 are equal. There is not an exact t statistic available for testing $H_0\colon \mu_1 = \mu_2$ in this case. However, the statistic

$$t_0^* = \frac{\bar{X}_1 - \bar{X}_2}{\sqrt{\frac{S_1^2}{n_1} + \frac{S_2^2}{n_2}}} \tag{11-80}$$

is distributed approximately as t with degrees of freedom given by

$$v = \frac{\left(\frac{S_1^2}{n_1} + \frac{S_2^2}{n_2}\right)^2}{\frac{\left(S_1^2/n_1\right)^2}{n_1+1} + \frac{\left(S_2^2/n_2\right)^2}{n_2+1}} - 2 \tag{11-81}$$

if the null hypothesis $H_0\colon \mu_1 = \mu_2$ is true. Therefore, if $\sigma_1^2 \neq \sigma_2^2$, the hypotheses of equations 11-72, 11-76, and 11-78 are tested as before, except that t_0^* is used as the test statistic and $n_1 + n_2 - 2$ is replaced by v in determining the degrees of freedom for the test. This general problem is often called the Behrens–Fisher problem.

Example 11-15

A manufacturer of video display units is testing two microcircuit designs to determine whether they produce equivalent current flow. Development engineering has obtained the following data:

Design 1	$n_1 = 15$	$\bar{x}_1 = 24.2$	$s_1^2 = 10$
Design 2	$n_2 = 10$	$\bar{x}_2 = 23.9$	$s_2^2 = 20$

We wish to test

$$H_0: \mu_1 = \mu_2,$$
$$H_1: \mu_1 \neq \mu_2,$$

where both populations are assumed to be normal, but we are unwilling to assume that the unknown variances σ_1^2 and σ_2^2 are equal. The test statistic is

$$t_0^* = \frac{\bar{x}_1 - \bar{x}_2}{\sqrt{\frac{s_1^2}{n_1} + \frac{s_2^2}{n_2}}} = \frac{24.2 - 23.9}{\sqrt{\frac{10}{15} + \frac{20}{10}}} = 0.184.$$

The degrees of freedom on t_0^* are found from equation 11-81 to be

$$v = \frac{\left(\frac{s_1^2}{n_1} + \frac{s_2^2}{n_2}\right)^2}{\frac{\left(s_1^2/n_1\right)^2}{n_1+1} + \frac{\left(s_2^2/n_2\right)^2}{n_2+1}} - 2 = \frac{\left(\frac{10}{15} + \frac{20}{10}\right)^2}{\frac{(10/15)^2}{16} + \frac{(20/10)^2}{11}} - 2 = 16.$$

Using $\alpha = 0.10$, we find that $t_{\alpha/2,v} = t_{0.05,16} = 1.746$. Since $|t_0^*| < t_{0.05,16}$, we cannot reject $H_0: \mu_1 = \mu_2$.

Choice of Sample Size

The operating characteristic curves in Charts VI*e*, VI*f*, VI*g*, and VI*h* (Appendix) are used to evaluate the type II error for the case where $\sigma_1^2 = \sigma_2^2 = \sigma^2$. Unfortunately, when $\sigma_1^2 \neq \sigma_2^2$, the distribution of t_0^* is unknown if the null hypothesis is false, and no operating characteristic curves are available for this case.

For the two-sided alternative in equation 11-72, when $\sigma_1^2 = \sigma_2^2 = \sigma^2$ and $n_1 = n_2 = n$, Charts VI*e* and VI*f* are used with

$$d = \frac{|\mu_1 - \mu_2|}{2\sigma} = \frac{|\delta|}{2\sigma}. \tag{11-82}$$

To use these curves, they must be entered with the sample size $n^* = 2n - 1$. For the one-sided alternative hypothesis of equation 11-76, we use Charts VI*g* and VI*h* and define

$$d = \frac{\mu_1 - \mu_2}{2\sigma} = \frac{\delta}{2\sigma}, \tag{11-83}$$

whereas for the other one-sided alternative hypothesis of equation 11-78, we use

$$d = \frac{\mu_2 - \mu_1}{2\sigma} = \frac{\delta}{2\sigma}. \tag{11-84}$$

It is noted that the parameter d is a function of σ, which is unknown. As in the single-sample t-test (Section 11-2.2), we may have to rely on a prior estimate of σ, or use a subjective estimate. Alternatively, we could define the differences in the mean that we wish to detect relative to σ.

Example 11-16

Consider the catalyst experiment in Example 11-14. Suppose that if catalyst 2 produces a yield that differs from the yield of catalyst 1 by 3.0% we would like to reject the null hypothesis with a probability of at least 0.85. What sample size is required? Using $s_p = 1.99$ as a rough estimate of the common standard deviation σ, we have $d = |\delta|/2\sigma = |3.00|/(2)(1.99) = 0.75$. From Chart VI*e* (Appendix) with $d = 0.75$ and $\beta = 0.15$, we find $n^* = 20$, approximately. Therefore, since $n^* = 2n - 1$,

$$n = \frac{n^*+1}{2} = \frac{20+1}{2} = 10.5 \simeq 11 \text{ (say)},$$

and we would use sample sizes of $n_1 = n_2 = n = 11$.

11-3.3 The Paired *t*-Test

A special case of the two-sample *t*-tests occurs when the observations on the two populations of interest are collected in pairs. Each pair of observations, say (X_{1j}, X_{2j}), is taken under homogeneous conditions, but these conditions may change from one pair to another. For example, suppose that we are interested in comparing two different types of tips for a hardness-testing machine. This machine presses the tip into a metal specimen with a known force. By measuring the depth of the depression caused by the tip, the hardness of the specimen can be determined. If several specimens were selected at random, half tested with tip 1, half tested with tip 2, and the pooled or independent *t*-test in Section 11-3.2 applied, the results of the test could be invalid. That is, the metal specimens could have been cut from bar stock that was produced in different heats, or they may not be homogeneous, which is another way hardness might be affected; then the observed differences between mean hardness readings for the two tip types also include hardness differences between specimens.

The correct experimental procedure is to collect the data in *pairs*; that is, to take two hardness readings of each specimen, one with each tip. The test procedure would then consist of analyzing the *differences* between hardness readings of each specimen. If there is no difference between tips, then the mean of the differences should be zero. This test procedure is called the *paired t-test.*

Let $(X_{11}, X_{21}), (X_{12}, X_{22}), \ldots, (X_{1n}, X_{2n})$ be a set of n *paired* observations, where we assume that $X_1 \sim N(\mu_1, \sigma_1^2)$ and $X_2 \sim N(\mu_2, \sigma_2^2)$. Define the differences between each pair of observations as $D_j = X_{1j} - X_{2j}$, $j = 1, 2, \ldots, n$.

The D_j are normally distributed with mean

$$\mu_D = E(X_1 - X_2) = E(X_1) - E(X_2) = \mu_1 - \mu_2,$$

so testing hypotheses about the equality of μ_1 and μ_2 can be accomplished by performing a one-sample *t*-test on μ_D. Specifically, testing H_0: $\mu_1 = \mu_2$ against H_1: $\mu_1 \neq \mu_2$ is equivalent to testing

$$\begin{aligned} H_0&: \mu_D = 0, \\ H_1&: \mu_D \neq 0. \end{aligned} \tag{11-85}$$

The appropriate test statistic for equation 11-85 is

$$t_0 = \frac{\overline{D}}{S_D/\sqrt{n}}, \tag{11-86}$$

where

$$\overline{D} = \frac{1}{n}\sum_{j=1}^{n} D_j \tag{11-87}$$

and

$$S_D^2 = \frac{\sum_{j=1}^{n} D_j^2 - \frac{1}{n}\left(\sum_{j=1}^{n} D_j\right)^2}{n-1} \tag{11-88}$$

are the sample mean and variance of the differences. We would reject H_0: $\mu_D = 0$ (implying that $\mu_1 \neq \mu_2$) if $t_0 > t_{\alpha/2,\, n-1}$ or if $t_0 < -t_{\alpha/2,\, n-1}$. One-sided alternatives would be treated similarly.

Example 11-17

An article in the *Journal of Strain Analysis* (Vol. 18, No. 2, 1983) compares several methods for predicting the shear strength for steel plate girders. Data for two of these methods, the Karlsruhe and Lehigh procedures, when applied to nine specific girders, are shown in Table 11-2. We wish to determine if there is any difference (on the average) between the two methods.

The sample average and standard deviation of the differences d_j are $\bar{d} = 0.2739$ and $s_d = 0.1351$, so the test statistic is

$$t_0 = \frac{\bar{d}}{s_d/\sqrt{n}} = \frac{0.2739}{0.1351/\sqrt{9}} = 6.08.$$

For the two-sided alternative H_1: $\mu_0 \neq 0$ and $\alpha = 0.1$, we would fail to reject only if $|t_0| < t_{0.05,\, 8} = 1.86$. Since $t_0 > t_{0.05,\, 8}$, we conclude that the two strength prediction methods yield different results. Specifically, the Karlsruhe method produces, on average, higher strength predictions than does the Lehigh method.

Paired Versus Unpaired Comparisons Sometimes in performing a comparative experiment, the investigator can choose between the paired analysis and the two-sample (or unpaired) *t*-test. If *n* measurements are to be made on each population, the two-sample *t* statistic is

$$t_0 = \frac{\bar{X}_1 - \bar{X}_2}{S_p\sqrt{\frac{1}{n} + \frac{1}{n}}},$$

Table 11-2 Strength Predictions for Nine Steel Plate Girders (Predicted Load/Observed Load)

Girder	Karlsruhe Method	Lehigh Method	Difference d_j
S1 / 1	1.186	1.061	0.125
S2 / 1	1.151	0.992	0.159
S3 / 1	1.322	1.063	0.259
S4 / 1	1.339	1.062	0.277
S5 / 1	1.200	1.065	0.135
S2 / 1	1.402	1.178	0.224
S2 / 2	1.365	1.037	0.328
S2 / 3	1.537	1.086	0.451
S2 / 4	1.559	1.052	0.507

which is compared to $t_{\alpha/2,\,2n-2}$, and of course, the paired t statistic is

$$t_0 = \frac{\overline{D}}{S_D/\sqrt{n}},$$

which is compared to $t_{\alpha/2,\,n-1}$. Notice that since

$$\begin{aligned}\overline{D} &= \frac{1}{n}\sum_{j=1}^{n} D_j = \frac{1}{n}\sum_{j=1}^{n}\left(X_{1j} - X_{2j}\right) = \frac{1}{n}\sum_{j=1}^{n} X_{1j} - \frac{1}{n}\sum_{j=1}^{n} X_{2j} \\ &= \overline{X}_1 - \overline{X}_2,\end{aligned}$$

the numerators of both statistics are identical. However, the denominator of the two-sample t-test is based on the assumption that X_1 and X_2 are *independent.* In many paired experiments, there is a strong positive correlation between X_1 and X_2. That is,

$$\begin{aligned}V\left(\overline{D}\right) &= V\left(\overline{X}_1 - \overline{X}_2\right) \\ &= V\left(\overline{X}_1\right) + V\left(\overline{X}_2\right) - 2\text{Cov}\left(\overline{X}_1, \overline{X}_2\right) \\ &= \frac{2\sigma^2(1-\rho)}{n},\end{aligned}$$

assuming that both populations X_1 and X_2 have identical variances. Furthermore, S_D^2/n estimates the variance of $\overline{D}$. Now, whenever there is positive correlation within the pairs, the denominator for the paired t-test will be smaller than the denominator of the two-sample t-test. This can cause the two-sample t-test to considerably understate the significance of the data if it is incorrectly applied to paired samples.

Although pairing will often lead to a smaller value of the variance of $\overline{X}_1 - \overline{X}_2$, it does have a disadvantage. Namely, the paired t-test leads to a loss of $n - 1$ degrees of freedom in comparison to the two-sample t-test. Generally, we know that increasing the degrees of freedom of a test increases the power against any fixed alternative values of the parameter.

So how do we decide to conduct the experiment—should we pair the observations or not? Although there is no general answer to this question, we can give some guidelines based on the above discussion. They are as follows:

1. If the experimental units are relatively homogenous (small σ) and the correlation between pairs is small, the gain in precision due to pairing will be offset by the loss of degrees of freedom, so an independent-samples experiment should be used.
2. If the experimental units are relatively heterogeneous (large σ) and there is large positive correlation between pairs, the paired experiment should be used.

The rules still require judgment in their implementation, because σ and ρ are usually not known precisely. Furthermore, if the number of degrees of freedom is large (say 40 or 50), then the loss of $n - 1$ of them for pairing may not be serious. However, if the number of degrees of freedom is small (say 10 or 20), then losing half of them is potentially serious if not compensated for by an increased precision from pairing.

11-3.4 Tests for the Equality of Two Variances

We now present tests for comparing two variances. Following the approach in Section 11-2.3, we present tests for normal populations and large-sample tests that may be applied to nonnormal populations.

Test Procedure for Normal Populations

Suppose that two independent populations are of interest, say $X_1 \sim N(\mu_1, \sigma_1^2)$ and $X_2 \sim N(\mu_2, \sigma_2^2)$, where μ_1, σ_1^2, μ_2, and σ_2^2 are unknown. We wish to test hypotheses about the equality of the two variances, say H_0: $\sigma_1^2 = \sigma_2^2$. Assume that two random samples of size n_1 from population 1 and of size n_2 from population 2 are available, and let S_1^2 and S_2^2 be the sample variances. To test the two-sided alternative

$$H_0: \sigma_1^2 = \sigma_2^2, \qquad H_1: \sigma_1^2 \neq \sigma_2^2, \tag{11-89}$$

we use the fact that the statistic

$$F_0 = \frac{S_1^2}{S_2^2} \tag{11-90}$$

is distributed as F, with $n_1 - 1$ and $n_2 - 1$ degrees of freedom, if the null hypothesis H_0: $\sigma_1^2 = \sigma_2^2$ is true. Therefore, we would reject H_0 if

$$F_0 > F_{\alpha/2,\, n_1-1,\, n_2-1} \tag{11-91a}$$

or if

$$F_0 < F_{1-\alpha/2,\, n_1-1,\, n_2-1}, \tag{11-91b}$$

where $F_{\alpha/2,\, n_1-1,\, n_2-1}$ and $F_{1-\alpha/2,\, n_1-1,\, n_2-1}$ are the upper and lower $\alpha/2$ percentage points of the F distribution with $n_1 - 1$ and $n_2 - 1$ degrees of freedom. Table V (Appendix) gives only the upper tail points of F, so to find $F_{1-\alpha/2,\, n_1-1,\, n_2-1}$ we must use

$$F_{1-\alpha/2, n_1-1, n_2-1} = \frac{1}{F_{\alpha/2, n_2-1, n_1-1}}. \tag{11-92}$$

The same test statistic can be used to test one-sided alternative hypotheses. Since the notation X_1 and X_2 is arbitrary, let X_1 denote the population that may have the largest variance. Therefore, the one-sided alternative hypothesis is

$$H_0: \sigma_1^2 = \sigma_2^2, \qquad H_1: \sigma_1^2 > \sigma_2^2. \tag{11-93}$$

If

$$F_0 > F_{\alpha,\, n_1-1,\, n_2-1}, \tag{11-94}$$

we would reject H_0: $\sigma_1^2 = \sigma_2^2$.

Example 11-18

Chemical etching is used to remove copper from printed circuit boards. X_1 and X_2 represent process yields when two different concentrations are used. Suppose that we wish to test

$$H_0: \sigma_1^2 = \sigma_2^2, \qquad H_1: \sigma_1^2 \neq \sigma_2^2.$$

Two samples of sizes $n_1 = n_2 = 8$ yield $s_1^2 = 3.89$ and $s_2^2 = 4.02$, and

$$F_0 = \frac{s_1^2}{s_2^2} = \frac{3.89}{4.02} = 0.97.$$

If $\alpha = 0.05$, we find that $F_{0.025,7,7} = 4.99$ and $F_{0.975,7,7} = (F_{0.025,7,7})^{-1} = (4.99)^{-1} = 0.20$. Therefore, we cannot reject H_0: $\sigma_1^2 = \sigma_2^2$, and we can conclude that there is no strong evidence that the variance of the yield is affected by the concentration.

Choice of Sample Size

Charts VI*o*, VI*p*, VI*q*, and VI*r* (Appendix) provide operating characteristic curves for the F-test for $\alpha = 0.05$, and $\alpha = 0.01$, assuming that $n_1 = n_2 = n$. Charts VI*o* and VI*p* are used with the two-sided alternative of equation 11-89. They plot β against the abscissa parameter

$$\lambda = \frac{\sigma_1}{\sigma_2} \tag{11-95}$$

for various $n_1 = n_2 = n$. Charts VI*q* and VI*r* are used for the one-sided alternative of equation 11-93.

Example 11-19

For the chemical process yield analyses problem in Example 11-18, suppose that one of the concentrations affected the variance of the yield so that one of the variances was four times the other and we wished to detect this with probability at least 0.80. What sample size should be used? Note that if one variance is four times the other, then

$$\lambda = \frac{\sigma_1}{\sigma_2} = 2.$$

By referring to Chart VI*o*, with $\beta = 0.20$ and $\lambda = 2$, we find that a sample size of $n_1 = n_2 = 20$, approximately, is necessary.

A Large-Sample Test Procedure

When both sample sizes n_1 and n_2 are large, a test procedure that does not require the normality assumption can be developed. The test is based on the result that the sample standard deviations S_1 and S_2 have approximate normal distributions with means σ_1 and σ_2, respectively, and variances $\sigma_1^2/2n_1$ and $\sigma_2^2/2n_2$, respectively. To test

$$H_0: \sigma_1^2 = \sigma_2^2, \qquad H_1: \sigma_1^2 \neq \sigma_2^2, \tag{11-96}$$

we would use the test statistic

$$Z_0 = \frac{S_1 - S_2}{S_p\sqrt{\dfrac{1}{2n_1} + \dfrac{1}{2n_2}}}, \tag{11-97}$$

where S_p is the pooled estimator of the common standard deviation σ. This statistic has an approximate standard normal distribution when $\sigma_1^2 = \sigma_2^2$. We would reject H_0 if $Z_0 > Z_{\alpha/2}$ or if $Z_0 < -Z_{\alpha/2}$. Rejection regions for the one-sided alternatives have the same form as in other two-sample normal tests.

11-3.5 Tests of Hypotheses on Two Proportions

The tests of Section 11-2.4 can be extended to the case where there are two binomial parameters of interest, say p_1 and p_2, and we wish to test that they are equal. That is, we wish to test

$$H_0: p_1 = p_2, \qquad H_1: p_1 \neq p_2. \tag{11-98}$$

We will present a large-sample procedure based on the normal approximation to the binomial and then outline one possible approach for small sample sizes.

Large-Sample Test for $H_0: p_1 = p_2$

Suppose that the two random samples of sizes n_1 and n_2 are taken from two populations, and let X_1 and X_2 represent the number of observations that belong to the class of interest in samples 1 and 2, respectively. Furthermore, suppose that the normal approximation to the binomial applies to each population, so that the estimators of the population proportions $\hat{p}_1 = X_1/n_1$ and $\hat{p}_2 = X_2/n_2$ have approximate normal distributions. Now, if the null hypothesis H_0: $p_1 = p_2$ is true, then using the fact that $p_1 = p_2 = p$, the random variable

$$Z = \frac{\hat{p}_1 - \hat{p}_2}{\sqrt{p(1-p)\left[\frac{1}{n_1} + \frac{1}{n_2}\right]}}$$

is distributed approximately $N(0, 1)$. An estimate of the common parameter p is

$$\hat{p} = \frac{X_1 + X_2}{n_1 + n_2}.$$

The test statistic for H_0: $p_1 = p_2$ is then

$$Z_0 = \frac{\hat{p}_1 - \hat{p}_2}{\sqrt{\hat{p}(1-\hat{p})\left[\frac{1}{n_1} + \frac{1}{n_2}\right]}}. \tag{11-99}$$

If

$$Z_0 > Z_{\alpha/2} \quad \text{or} \quad Z_0 < -Z_{\alpha/2}, \tag{11-100}$$

the null hypothesis is rejected.

Example 11-20

Two different types of fire control computers are being considered for use by the U.S. Army in six-gun 105-mm batteries. The two computer systems are subjected to an operational test in which the total number of hits of the target are counted. Computer system 1 gave 250 hits out of 300 rounds, while computer system 2 gave 178 hits out of 260 rounds. Is there reason to believe that the two computer systems differ? To answer this question, we test

$$H_0: p_1 = p_2,$$
$$H_1: p_1 \neq p_2.$$

Note that $\hat{p}_1 = 250/300 = 0.8333$, $\hat{p}_2 = 178/260 = 0.6846$, and

$$\hat{p} = \frac{x_1 + x_2}{n_1 + n_2} = \frac{250 + 178}{300 + 260} = 0.7643.$$

The value of the test statistic is

$$Z_0 = \frac{\hat{p}_1 - \hat{p}_2}{\sqrt{\hat{p}(1-\hat{p})\left[\frac{1}{n_1} + \frac{1}{n_2}\right]}} = \frac{0.8333 - 0.6846}{\sqrt{0.7643(0.2357)\left[\frac{1}{300} + \frac{1}{260}\right]}} = 4.13.$$

If we use $\alpha = 0.05$, then $Z_{0.025} = 1.96$ and $-Z_{0.025} = -1.96$, and we would reject H_0, concluding that there is a significant difference in the two computer systems.

Choice of Sample Size

The computation of the β error for the foregoing test is somewhat more involved than in the single-sample case. The problem is that the denominator of Z_0 is an estimate of the standard deviation of $\hat{p}_1 - \hat{p}_2$ under the assumption that $p_1 = p_2 = p$. When H_0: $p_1 = p_2$ is false, the standard deviation of $\hat{p}_1 - \hat{p}_2$ is

$$\sigma_{\hat{p}_1 - \hat{p}_2} = \sqrt{\frac{p_1(1-p_1)}{n_1} + \frac{p_2(1-p_2)}{n_2}}. \tag{11-101}$$

If the alternative hypothesis is two-sided, the β risk turns out to be approximately

$$\beta \simeq \Phi\left(\frac{Z_{\alpha/2}\sqrt{\bar{p}\bar{q}(1/n_1 + 1/n_2)} - (p_1 - p_2)}{\sigma_{\hat{p}_1 - \hat{p}_2}}\right) - \Phi\left(\frac{-Z_{\alpha/2}\sqrt{\bar{p}\bar{q}(1/n_1 + 1/n_2)} - (p_1 - p_2)}{\sigma_{\hat{p}_1 - \hat{p}_2}}\right), \tag{11-102}$$

where

$$\bar{p} = \frac{n_1 p_1 + n_2 p_2}{n_1 + n_2},$$

$$\bar{q} = \frac{n_1(1-p_1) + n_2(1-p_2)}{n_1 + n_2},$$

and $\sigma_{\hat{p}_1 - \hat{p}_2}$ is given by equation 11-101. If the alternative hypothesis is H_1: $p_1 > p_2$, then

$$\beta \simeq \Phi\left(\frac{Z_{\alpha}\sqrt{\bar{p}\bar{q}(1/n_1 + 1/n_2)} - (p_1 - p_2)}{\sigma_{\hat{p}_1 - \hat{p}_2}}\right), \tag{11-103}$$

and if the alternative hypothesis is H_1: $p_1 < p_2$, then

$$\beta \simeq 1 - \Phi\left(\frac{-Z_{\alpha}\sqrt{\bar{p}\bar{q}(1/n_1 + 1/n_2)} - (p_1 - p_2)}{\sigma_{\hat{p}_1 - \hat{p}_2}}\right). \tag{11-104}$$

For a specified pair of values p_1 and p_2 we can find the sample sizes $n_1 = n_2 = n$ required to give the test of size α that has specified type II error β. For the two-sided alternative the common sample size is approximately

$$n \simeq \frac{\left(Z_{\alpha/2}\sqrt{(p_1 + p_2)(q_1 + q_2)/2} + Z_{\beta}\sqrt{p_1 q_1 + p_2 q_2}\right)^2}{(p_1 - p_2)^2}, \tag{11-105}$$

where $q_1 = 1 - p_1$ and $q_2 = 1 - p_2$. For the one-sided alternatives, replace $Z_{\alpha/2}$ in equation 11-105 with Z_{α}.

Small-Sample Test for $H_0: p_1 = p_2$

Most problems involving the comparison of proportions p_1 and p_2 have relatively large sample sizes, so the procedure based on the normal approximation to the binomial is widely used in practice. However, occasionally, a small-sample-size problem is encountered. In such cases, the Z-tests are inappropriate and an alternative procedure is required. In this section we describe a procedure based on the hypergeometric distribution.

Suppose that X_1 and X_2 are the number of successes in two random samples of sizes n_1 and n_2, respectively. The test procedure requires that we view the total number of successes as fixed at the value $X_1 + X_2 = Y$. Now consider the hypotheses

$$H_0: p_1 = p_2,$$
$$H_1: p_1 > p_2.$$

Given that $X_1 + X_2 = Y$, large values of X_1 support H_1, whereas small or moderate values of X_1 support H_0. Therefore, we will reject H_0 whenever X_1 is sufficiently large.

Since the combined sample of $n_1 + n_2$ observations contains $X_1 + X_2 = Y$ total successes, if $H_0: p_1 = p_2$, the successes are no more likely to be concentrated in the first sample than in the second. That is, all the ways in which the $n_1 + n_2$ responses can be divided into one sample of n_1 responses and a second sample of n_2 responses are equally likely. The number of ways of selecting X_1 successes for the first sample leaving $Y - X_1$ successes for the second is

$$\binom{Y}{X_1}\binom{n_1 + n_2 - Y}{n_1 - X_1}.$$

Because outcomes are equally likely, the probability of there being exactly X_1 successes in sample 1 is determined by the ratio of the number of sample 1 outcomes having X_1 successes to the total number of outcomes, or

$$P\left(X_1 = x_1 \middle| Y \text{ success in } n_1 + n_2 \text{ responses}\right) = \frac{\binom{Y}{x_1}\binom{n_1 + n_2 - Y}{n_1 - x_1}}{\binom{n_1 + n_2}{n_1}}, \tag{11-106}$$

given that H_0: $p_1 = p_2$ is true. We recognize equation 11-106 as a hypergeometric distribution.

To use equation 11-106 for hypothesis testing, we would compute the probability of finding a value of X_1 at least as extreme as the observed value of X_1. Note that this probability is a P-value. If this P-value is sufficiently small, then the null hypothesis is rejected. This approach could also be applied to lower-tailed and two-tailed alternatives.

Example 11-21

Insulating cloth used in printed circuit boards is manufactured in large rolls. The manufacturer is trying to improve the process *yield*, that is, the number of defect-free rolls produced. A sample of 10 rolls contains exactly four defect-free rolls. From analysis of the defect types, manufacturing engineering suggests several changes in the process. Following implementation of these changes, another sample of 10 rolls yields 8 defect-free rolls. Do the data support the claim that the new process is better than the old one, using $\alpha = 0.10$?

To answer this question, we compute the P-value. In our example, $n_1 = n_2 = 10$, $y = 8 + 4 = 12$, and the observed value of $x_1 = 8$. The values of x_1 that are more extreme than 8 are 9 and 10. Therefore

$$P(X_1 = 8 | 12 \text{ successes}) = \frac{\binom{12}{8}\binom{8}{2}}{\binom{20}{10}} = 0.0750,$$

$$P(X_1 = 9 | 12 \text{ successes}) = \frac{\binom{12}{9}\binom{8}{1}}{\binom{20}{10}} = 0.0095,$$

$$P(X_1 = 10 | 12 \text{ successes}) = \frac{\binom{12}{10}\binom{8}{0}}{\binom{20}{10}} = 0.0003.$$

The P-value is $P = 0.0750 + 0.0095 + 0.0003 = 0.0848$. Thus, at the level $\alpha = 0.10$, the null hypothesis is rejected and we conclude that the engineering changes have improved the process yield.

This test procedure is sometimes called the Fisher–Irwin test. Because the test depends on the assumption that $X_1 + X_2$ is fixed at some value, some statisticians have argued against use of the test when $X_1 + X_2$ is not actually fixed. Clearly $X_1 + X_2$ is not fixed by the sampling procedure in our example. However, because there are no other better competing procedures, the Fisher–Irwin test is often used whether or not $X_1 + X_2$ is actually fixed in advance.

11-4 TESTING FOR GOODNESS OF FIT

The hypothesis-testing procedures that we have discussed in previous sections are for problems in which the form of the density function of the random variable is known, and the hypotheses involve the parameters of the distribution. Another kind of hypothesis is often encountered: we do not know the probability distribution of the random variable under study, say X, and we wish to test the hypothesis that X follows a particular probability distribution. For example, we might wish to test the hypothesis that X follows the normal distribution.

In this section, we describe a formal goodness-of-fit test procedure based on the chi-square distribution. We also describe a very useful graphical technique called probability plotting. Finally, we give some guidelines useful in selecting the form of the population distribution.

The Chi-Square Goodness-of-Fit Test

The test procedure requires a random sample of size n of the random variable X, whose probability density function is unknown. These n observations are arrayed in a frequency histogram having k class intervals. Let O_i be the observed frequency in the ith class interval. From the hypothesized probability distribution we compute the expected frequency in the ith class interval, denoted E_i. The test statistic is

$$\chi_0^2 = \sum_{i=1}^{k} \frac{(O_i - E_i)^2}{E_i}. \tag{11-107}$$

It can be shown that χ_0^2 approximately follows the chi-square distribution with $k - p - 1$ degrees of freedom, where p represents the number of parameters of the hypothesized distribution estimated by sample statistics. This approximation improves as n increases. We would reject the hypothesis that X conforms to the hypothesized distribution if $\chi_0^2 > \chi^2_{\alpha, k-p-1}$.

One point to be noted in the application of this test procedure concerns the magnitude of the expected frequencies. If these expected frequencies are too small, then χ_0^2 will not reflect the departure of observed from expected, but only the smallest of the expected frequencies. There is no general agreement regarding the minimum value of expected frequencies, but values of 3, 4, and 5 are widely used as minimal. Should an expected frequency be too small, it can be combined with the expected frequency in an adjacent class interval. The corresponding observed frequencies would then be combined also, and k would be reduced by 1. Class intervals are not required to be of equal width.

We now give three examples of the test procedure.

Example 11-22

A Completely Specified Distribution A computer scientist has developed an algorithm for generating pseudorandom integers over the interval 0–9. He codes the algorithm and generates 1000 pseudorandom digits. The data are shown in Table 11-3. Is there evidence that the random number generator is working correctly?

If the random number generator is working correctly, then the values 0–9 should follow the *discrete uniform* distribution, which implies that each of the integers should occur about 100 times. That is, the expected frequencies $E_i = 100$, for $i = 0, 1, \ldots, 9$. Since these expected frequencies can be determined without estimating any parameters from the sample data, the resulting chi-square goodness-of-fit test will have $k - p - 1 = 10 - 0 - 1 = 9$ degrees of freedom.

The observed value of the test statistic is

$$\begin{aligned}\chi_0^2 &= \sum_{i=1}^{k} \frac{(O_i - E_i)^2}{E_i} \\ &= \frac{(94-100)^2}{100} + \frac{(93-100)^2}{100} + \cdots + \frac{(94-100)^2}{100} \\ &= 3.72.\end{aligned}$$

Since $\chi^2_{0.05,9} = 16.92$ we are unable to reject the hypothesis that the data come from a discrete uniform distribution. Therefore, the random number generator seems to be working satisfactorily.

Example 11-23

A Discrete Distribution The number of defects in printed circuit boards is hypothesized to follow a Poisson distribution. A random sample of $n = 60$ printed boards have been collected, and the number of defects observed. The following data result:

Table 11-3 Data for Example 11-22

	0	1	2	3	4	5	6	7	8	9	Total n
Observed frequencies, O_i	94	93	112	101	104	95	100	99	108	94	1000
Expected frequencies, E_i	100	100	100	100	100	100	100	100	100	100	1000

Number of Defects	Observed Frequency
0	32
1	15
2	9
3	4

The mean of the assumed Poisson distribution in this example is unknown and must be estimated from the sample data. The estimate of the mean number of defects per board is the sample average; that is $(32 \cdot 0 + 15 \cdot 1 + 9 \cdot 2 + 4 \cdot 3)/60 = 0.75$. From the cumulative Poisson distribution with parameter 0.75 we may compute the expected frequencies as $E_i = np_i$, where p_i is the theoretical, hypothesized probability associated with the ith class interval and n is the total number of observations. The appropriate hypotheses are

$$H_0: p(x) = \frac{e^{-0.75}(0.75)^x}{x!}, \qquad x = 0, 1, 2, \ldots,$$

$$H_1: p(x) \text{ is not Poisson with } \lambda = 0.75.$$

We may compute the expected frequencies as follows:

Number of Failures	Probability	Expected Frequency
0	0.472	28.32
1	0.354	21.24
2	0.133	7.98
≥ 3	0.041	2.46

The expected frequencies are obtained by multiplying the sample size times the respective probabilities. Since the expected frequency in the last cell is less than 3, we combine the last two cells:

Number of Failures	Observed Frequency	Expected Frequency
0	32	28.32
1	15	21.24
≥ 2	13	10.44

The test statistic (which will have $k - p - 1 = 3 - 1 - 1 = 1$ degree of freedom) becomes

$$\chi_0^2 = \frac{(32 - 28.32)^2}{28.32} + \frac{(15 - 21.24)^2}{21.24} + \frac{(13 - 10.44)^2}{10.44} = 2.94,$$

and since $\chi^2_{0.05,1} = 3.84$, we cannot reject the hypothesis that the occurrence of defects follows a Poisson distribution with mean 0.75 defects per board.

Example 11-24

A Continuous Distribution A manufacturing engineer is testing a power supply used in a word processing work station. He wishes to determine whether output voltage is adequately described by a normal distribution. From a random sample of $n = 100$ units he obtains sample estimates of the mean and standard deviation $\bar{x} = 12.04$ V and $s = 0.08$ V.

A common practice in constructing the class intervals for the frequency distribution used in the chi-square goodness-of-fit test is to choose the cell boundaries so that the expected frequencies $E_i = np_i$ are equal for all cells. To use this method, we want to choose the cell boundaries $a_0, a_1, \ldots, a_k$ for the k cells so that all the probabilities

$$p_i = P(a_{i-1} \le X \le a_i) = \int_{a_{i-1}}^{a_i} f(x)\,dx$$

are equal. Suppose we decide to use $k = 8$ cells. For the standard normal distribution the intervals that divide the scale into eight equally likely segments are [0, 0.32), [0.32, 0.675), [0.675, 1.15), [1.15, ∞), and their four "mirror image" intervals on the other side of zero. Denoting these standard normal endpoints by $a_0, a_1, \ldots, a_8$, it is a simple matter to calculate the endpoints that are necessary for the general normal problem at hand; namely, we define the new class interval endpoints by the transformation $a_i' = \bar{x} + sa_i$, $i = 0, 1, \ldots, 8$. For example, the sixth interval's right endpoint is

$$a_6' = \bar{x} + sa_6 = 12.04 + (0.08)(0.675) = 12.094.$$

For each interval, $p_i = \frac{1}{8} = 0.125$, so the expected cell frequencies are $E_i = np_i = 100\ (0.125) = 12.5$. The complete table of observed and expected frequencies is given in Table 11-4.

The computed value of the chi-square statistic is

$$\begin{aligned}\chi_0^2 &= \sum_{i=1}^{8} \frac{(O_i - E_i)^2}{E_i} \\ &= \frac{(10-12.5)^2}{12.5} + \frac{(14-12.5)^2}{12.5} + \cdots + \frac{(14-12.5)^2}{12.5} \\ &= 1.12.\end{aligned}$$

Since two parameters in the normal distribution have been estimated, we would compare $\chi_0^2 = 1.12$ to a chi-square distribution with $k - p - 1 = 8 - 2 - 1 = 5$ degrees of freedom. Using $\alpha = 0.10$, we see that $\chi_{0.1,5}^2 = 13.36$, and so we conclude that there is no reason to believe that output voltage is not normally distributed.

Probability Plotting

Graphical methods are also useful when selecting a probability distribution to describe data. Probability plotting is a graphical method for determining whether the data conform to a hypothesized distribution based on a subjective visual examination of the data. The general procedure is very simple and can be performed quickly. Probability plotting requires special graph paper, known as *probability* paper, that has been designed for the hypothesized distribution. Probability paper is widely available for the normal, lognormal, Weibull, and various chi-square and gamma distributions. To construct a probability plot, the observations in the sample are first ranked from smallest to largest. That is, the sample

Table 11-4 Observed and Expected Frequencies

Class Interval	Observed Frequency, O_i	Expected Frequency, E_i
$x < 11.948$	10	12.5
$11.948 \le x < 11.986$	14	12.5
$11.986 \le x < 12.014$	12	12.5
$12.014 \le x < 12.040$	13	12.5
$12.040 \le x < 12.066$	11	12.5
$12.066 \le x < 12.094$	12	12.5
$12.094 \le x < 12.132$	14	12.5
$12.132 \le x$	14	12.5
	100	100

$X_1, X_2, \ldots, X_n$ is arranged as $X_{(1)}, X_{(2)}, \ldots, X_{(n)}$, where $X_{(j)} \le X_{(j+1)}$. The ordered observations $X_{(j)}$ are then plotted against their observed cumulative frequency $(j - 0.5)/n$ on the appropriate probability paper. If the hypothesized distribution adequately describes the data, the plotted points will fall approximately along a straight line; if the plotted points deviate significantly from a straight line, then the hypothesized model is not appropriate. Usually, the determination of whether or not the data plot as a straight line is subjective.

Example 11-25

To illustrate probability plotting, consider the following data:

–0.314, 1.080, 0.863, –0.179, –1.390, –0.563, 1.436, 1.153, 0.504, –0.801.

We hypothesize that these data are adequately modeled by a normal distribution. The observations are arranged in ascending order and their cumulative frequencies $(j - 0.5)/n$ calculated as follows:

j	$X_{(j)}$	$(j - 0.5)/n$
1	–1.390	0.05
2	–0.801	0.15
3	–0.563	0.25
4	–0.314	0.35
5	–0.179	0.45
6	0.504	0.55
7	0.863	0.65
8	1.080	0.75
9	1.153	0.85
10	1.436	0.95

The pairs of values $X_{(j)}$ and $(j - 0.5)/n$ are now plotted on normal probability paper. This plot is shown in Fig. 11-7. Most normal probability paper plots $100(j - 0.5)/n$ on the right vertical scale and $100[1 - (j - 0.5)/n]$ on the left vertical scale, with the variable value plotted on the horizontal scale. We have chosen to plot $X_{(j)}$ versus $100(j - 0.5)/n$ on the right vertical in Fig. 11-7. A straight line,

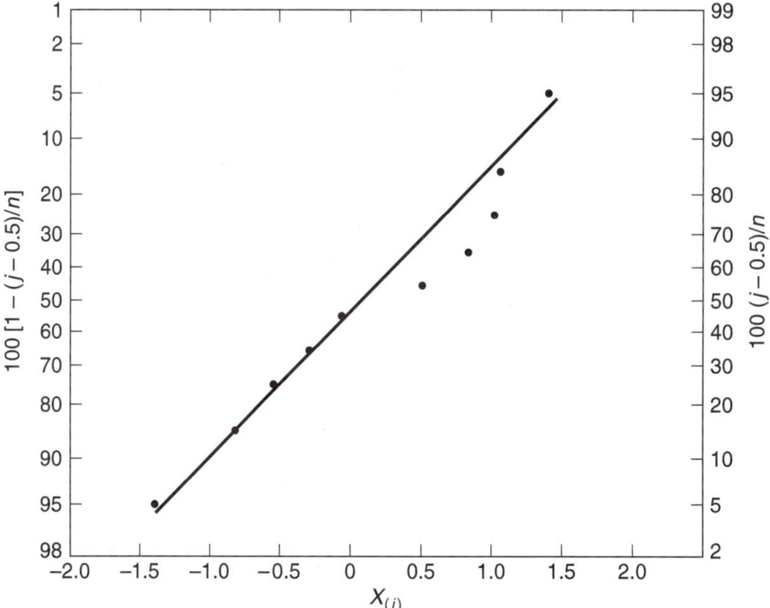

Figure 11-7 Normal probability plot.

chosen subjectively, has been drawn through the plotted points. In drawing the straight line, one should be influenced more by the points near the middle than the extreme points. Since the points fall generally near the line, we conclude that a normal distribution describes the data.

We can obtain an estimate of the mean and standard deviation directly from the normal probability plot. We see from the straight line in Fig. 11-7 that the mean is estimated as the 50th percentile of the sample, or $\hat{\mu} = 0.10$, approximately, and the standard deviation is estimated as the difference between the 84th and 50th percentiles, or $\hat{\sigma} = 0.95 - 0.10 = 0.85$, approximately.

A normal probability plot can also be constructed on ordinary graph paper by plotting the standardized normal scores Z_j against $X_{(j)}$, where the standardized normal scores satisfy

$$\frac{j-0.5}{n} = P(Z \le Z_j) = \Phi(Z_j).$$

For example, if $(j-0.5)/n = 0.05$, then $\Phi(Z_j) = 0.05$ implies that $Z_j = 1.64$. To illustrate, consider the data from Example 11-25. In the table below we have shown the standardized normal scores in the last column:

j	$X_{(j)}$	$(j-0.5)/n$	Z_j
1	−1.390	0.05	−1.64
2	−0.801	0.15	−1.04
3	−0.563	0.25	−0.67
4	−0.314	0.35	−0.39
5	−0.179	0.45	−0.13
6	0.504	0.55	0.13
7	0.863	0.65	0.39
8	1.080	0.75	0.67
9	1.153	0.85	1.04
10	1.436	0.95	1.64

Figure 11-8 presents the plot of Z_j versus $X_{(j)}$. This normal probability plot is equivalent to the one in Fig. 11-7. Many software packages will construct probability plots for various distributions. For a Minitab® example, see Section 11-6.

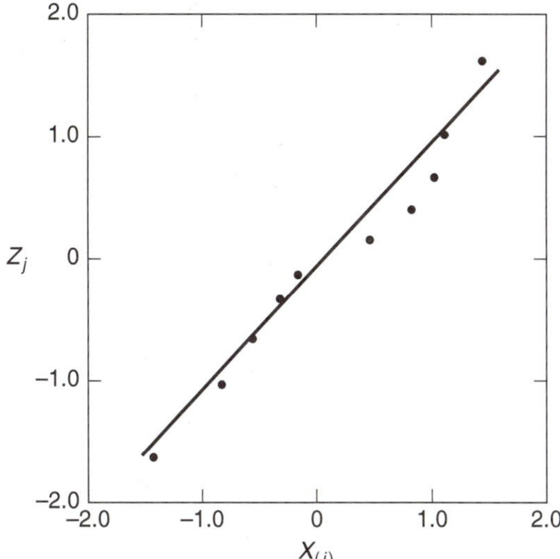

Figure 11-8 Normal probability plot.

Selecting the Form of a Distribution

The choice of the distribution hypothesized to fit the data is important. Sometimes analysts can use their knowledge of the physical phenomena to choose a distribution to model the data. For example, in studying the circuit board defect data in Example 11-23, a Poisson distribution was hypothesized to describe the data, because failures are an "event per unit" phenomena, and such phenomena are often well modeled by a Poisson distribution. Sometimes previous experience can suggest the choice of distribution.

In situations where there is no previous experience or theory to suggest a distribution that describes the data, analysts must rely on other methods. Inspection of a frequency histogram can often suggest an appropriate distribution. One may also use the display in Fig. 11-9 to assist in selecting a distribution that describes the data. When using Fig. 11-9, note that the β_2 axis increases downward. This figure shows the regions in the β_1, β_2 plane for several standard probability distributions, where

$$\sqrt{\beta_1} = \frac{E(X-\mu)^3}{\left(\sigma^2\right)^{3/2}}$$

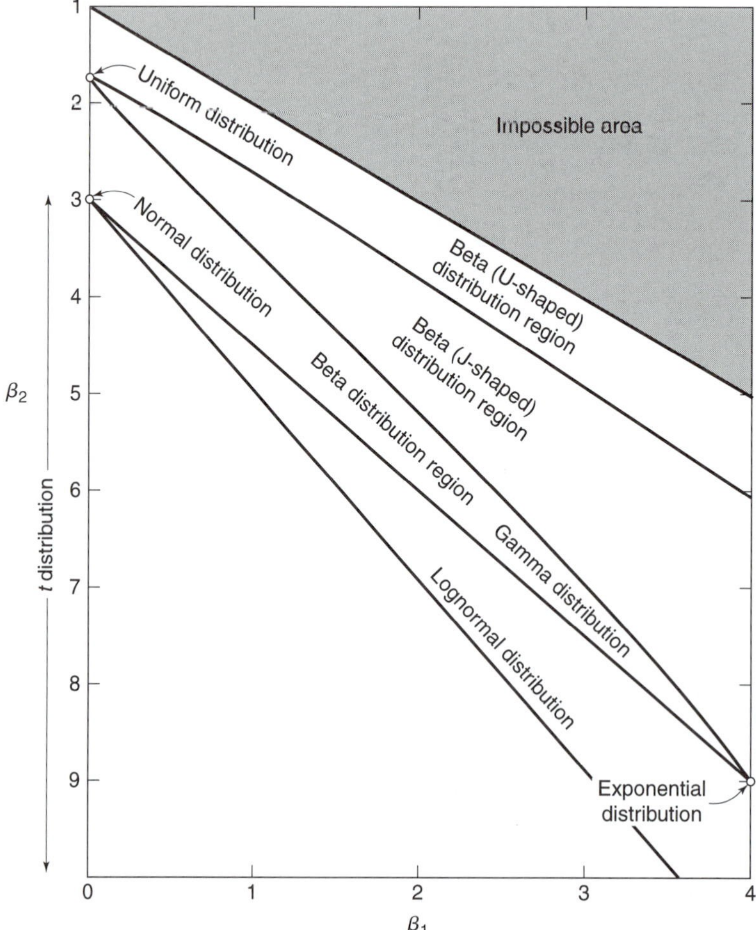

Figure 11-9 Regions in the β_1, β_2 plane for various standard distributions. (Adapted from G. J. Hahn and S. S. Shapiro, *Statistical Models in Engineering*, John Wiley & Sons, New York, 1967; used with permission of the publisher and Professor E. S. Pearson, University of London.)

is a standardized measure of skewness and

$$\beta_2 = \frac{E(X-\mu)^4}{\sigma^4}$$

is a standardized measure of kurtosis (or peakedness). To use Fig. 11-9, calculate the sample estimates of β_1 and β_2, say

$$\sqrt{\hat{\beta}_1} = \frac{M_3}{(M_2)^{3/2}}$$

and

$$\hat{\beta}_2 = \frac{M_4}{M_2^2},$$

where

$$M_j = \frac{1}{n}\sum_{i=1}^{n}(X_i - \overline{X})^j \qquad j = 1, 2, 3, 4,$$

and plot the point $\hat{\beta}_1, \hat{\beta}_2$. If this plotted point falls reasonably close to a point, line, or area that corresponds to one of the distributions given in the figure, then this distribution is a logical candidate to model the data.

From inspecting Fig. 11-9 we note that all normal distributions are represented by the point $\beta_1 = 0$ and $\beta_2 = 3$. This is reasonable, since all normal distributions have the same shape. Similarly, the exponential and uniform distributions are represented by a single point in the β_1, β_2 plane. The gamma and lognormal distributions are represented by lines, because their shapes depend on their parameter values. Note that these lines are close together, which may explain why some data sets are modeled equally well by either distribution. We also observe that there are regions of the β_1, β_2 plane for which none of the distributions in Fig. 11-9 is appropriate. Other, more general distributions, such as the Johnson or Pearson families of distributions, may be required in these cases. Procedures for fitting these families of distributions and figures similar to Fig. 11-9 are given in Hahn and Shapiro (1967).

11-5 CONTINGENCY TABLE TESTS

Many times, the n elements of a sample from a population may be classified according to two different criteria. It is then of interest to know whether the two methods of classification are statistically independent; for example, we may consider the population of graduating engineers and we may wish to determine whether starting salary is independent of academic disciplines. Assume that the first method of classification has r levels and that the second method of classification has c levels. We will let O_{ij} be the observed frequency for level i of the first classification method and for level j of the second classification method. The data would, in general, appear as in Table 11-5. Such a table is commonly called an *$r \times c$ contingency table.*

We are interested in testing the hypothesis that the row and column methods of classification are independent. If we reject this hypothesis, we conclude there is some *interaction* between the two criteria of classification. The exact test procedures are difficult to obtain, but an approximate test statistic is valid for large n. Assume the O_{ij} to be multinomial random variables and p_{ij} to be the probability that a randomly selected element falls in the ijth

Table 11-5 An $r \times c$ Contingency Table

Row	Column 1	2	...	c
1	O_{11}	O_{12}	...	O_{1c}
2	O_{21}	O_{22}	...	O_{2c}
⋮	⋮	⋮		⋮
r	O_{r1}	O_{r2}	...	O_{rc}

cell, given that the two classifications are independent. Then $p_{ij} = u_i v_j$, where u_i is the probability that a randomly selected element falls in row class i and v_j is the probability that a randomly selected element falls in column class j. Now, assuming independence, the maximum likelihood estimators of u_i and v_j are

$$\hat{u}_i = \frac{1}{n}\sum_{j=1}^{c} O_{ij},$$

$$\hat{v}_j = \frac{1}{n}\sum_{i=1}^{r} O_{ij}. \tag{11-108}$$

Therefore, assuming independence, the expected number of each cell is

$$E_{ij} = n\hat{u}_i\hat{v}_j = \frac{1}{n}\sum_{m=1}^{c} O_{im}\sum_{k=1}^{r} O_{kj}. \tag{11-109}$$

Then, for large n, the statistic

$$\chi_0^2 = \sum_{i=1}^{r}\sum_{j=1}^{c}\frac{\left(O_{ij} - E_{ij}\right)^2}{E_{ij}} \sim \chi^2_{(r-1)(c-1)}, \tag{11-110}$$

approximately, and we would reject the hypothesis of independence if $\chi_0^2 > \chi^2_{\alpha,(r-1)(c-1)}$.

Example 11-26

A company has to choose among three pension plans. Management wishes to know whether the preference for plan is independent of job classification. The opinions of a random sample of 500 employees are shown in Table 11-6. We may compute $\hat{u}_1 = (340/500) = 0.68$, $\hat{u}_2 = (160/500) = 0.32$, $\hat{v}_1 = (200/500) = 0.40$, $\hat{v}_2 = (200/500) = 0.40$, and $\hat{v}_3 = (100/500) = 0.20$. The expected frequencies may be computed from equation 11-109. For example, the expected number of salaried workers favoring pension plan 1 is

$$E_{11} = n\hat{u}_1\hat{v}_1 = 500(0.68)(0.40) = 136.$$

Table 11-6 Observed Data for Example 11-26

	Pension Plan 1	2	3	Total
Salaried workers	160	140	40	340
Hourly workers	40	60	60	160
Totals	200	200	100	500

Table 11-7 Expected Frequencies for Example 11-26

	Pension Plan			
	1	2	3	Total
Salaried workers	136	136	68	340
Hourly workers	64	64	32	160
Totals	200	200	100	500

The expected frequencies are shown in Table 11-7. The test statistic is computed from equation 11-110 as follows:

$$\chi_0^2 = \sum_{i=1}^{2}\sum_{j=1}^{3}\frac{\left(O_{ij} - E_{ij}\right)^2}{E_{ij}}$$

$$= \frac{(160-136)^2}{136} + \frac{(140-136)^2}{136} + \frac{(40-68)^2}{68} + \frac{(40-64)^2}{64} + \frac{(60-64)^2}{64} + \frac{(60-32)^2}{32} = 49.63.$$

Since $\chi^2_{0.05,2} = 5.99$, we reject the hypothesis of independence and conclude that the preference for pension plans is not independent of job classification.

Using the two-way contingency table to test independence between two variables of classification in a sample from a single population of interest is only one application of contingency table methods. Another common situation occurs when there are r populations of interest and each population is divided into the same c categories. A sample is then taken from the ith population and the counts entered in the appropriate columns of the ith row. In this situation we want to investigate whether or not the proportions in the c categories are the same for all populations. The null hypothesis in this problem states that the populations are *homogeneous* with respect to the categories. For example, when there are only two categories, such as success and failure, defective and nondefective, and so on, then the test for homogeneity is really a test of the equality of r binomial parameters. Calculation of expected frequencies, determination of degrees of freedom, and computation of the chi-square statistic for the test for homogeneity are identical to the test for independence.

11-6 SAMPLE COMPUTER OUTPUT

There are many statistical packages available that can be used to construct confidence intervals, carry out tests of hypotheses, and determine sample size. In this section we present results for several problems using Minitab®.

Example 11-27

A study was conducted on the tensile strength of a particular fiber under various temperatures. The results of the study (given in MPa) are

226, 237, 272, 245, 428, 298, 345, 201, 327, 301, 317, 395, 332, 238, 367.

Suppose it is of interest to determine if the mean tensile strength is greater than 250 MPa. That is, test

$$H_0: \mu = 250,$$
$$H_1: \mu > 250.$$

A normal probability plot was constructed for the tensile strength and is given in Fig. 11-10. The normality assumption appears to be satisfied. The population variance for tensile strength is assumed to be unknown, and as a result, a single-sample t-test will be used for this problem.

The results from Minitab® for hypothesis testing and confidence interval on the mean are

```
Test of mu = 250 vs mu > 250

Variable         N           Mean     StDev    SE Mean
TS              15          301.9      65.9       17.0

Variable     95.0%    Lower Bound         T          P
TS                          272.0      3.05      0.004
```

The P-value is reported as 0.004, leading us to reject the null hypothesis and conclude that the mean tensile strength is greater than 250 MPa. The lower one-sided 95% confidence interval is given as $272 < \mu$.

Example 11-28

Reconsider Example 11-17, comparing two methods for predicting the shear strength for steel plate girders. The Minitab® output for the paired t-test using $\alpha = 0.10$ is

```
Paired T for Karlsruhe - Lehigh

               N       Mean      StDev    SE Mean
Karlsruhe      9     1.3401     0.1460     0.0487
Lehigh         9     1.0662     0.0494     0.0165
Difference     9     0.2739     0.1351     0.0450

90% CI for mean difference:  (0.1901, 0.3576)
T-Test of mean difference = 0 (vs not = 0): T-Value = 6.08 P-Value =
0.000
```

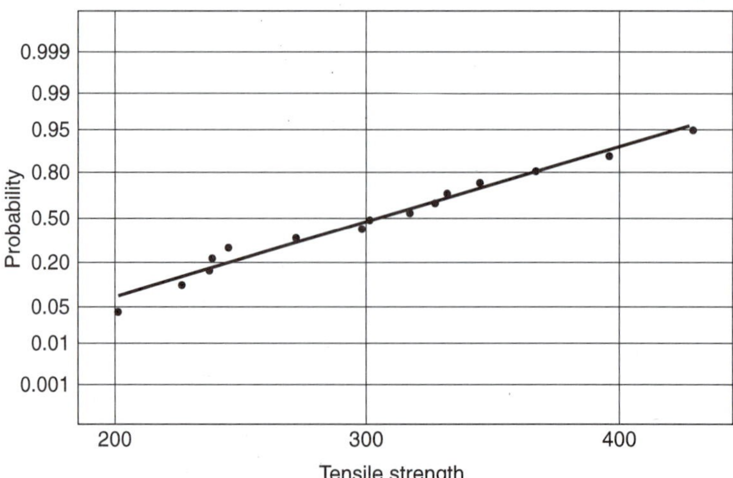

Figure 11-10 Normal probability plot for Example 11-27.

The results of the Minitab® output are in agreement with the results found in Example 11-17. Minitab® also provides the appropriate confidence interval for the problem. Using $\alpha = 0.10$, the level of confidence is 0.90; the 90% confidence interval on the difference between the two methods is (0.1901, 0.3576). Since the confidence interval does not contain zero, we also conclude that there is a significant difference between the two methods.

Example 11-29

The number of airline flights canceled is recorded for all airlines for each day of service. The number of flights recorded and the number of these flights that were canceled on a single day in March 2001 are provided below for two major airlines.

Airline	# of Flights	# of Canceled Flights	Proportion
American Airlines	2128	115	$\hat{p}_1 = 0.054$
America West Airlines	635	49	$\hat{p}_2 = 0.077$

Is there a significant difference in the proportion of canceled flights for the two airlines? The hypotheses of interest are H_0: $p_1 = p_2$ versus H_1: $p_1 \neq p_2$. A two-sample test on proportions and a two-sided confidence interval on proportions are

```
Sample        X        N     Sample p
1           115     2128     0.054041
2            49      635     0.077165

Estimate for p(1) - p(2):  -0.0231240
95% CI for p(1) - p(2):  (-0.0459949, -0.000253139)
Test for p(1) - p(2) = 0 (vs not = 0):  Z = -1.98 P-Value = 0.048
```

The *P*-value is given as 0.048, indicating that there is a significant difference between the proportions of flights canceled for American and America West airlines at a 5% level of significance. The 95% confidence interval of (–0.0460, –0.0003) indicates that America West airlines had a statistically significant higher proportion of canceled flights than American Airlines for the single day.

Example 11-30

The mean compressive strength for a particular high-strength concrete is hypothesized to be $\mu = 20$ (MPa). It is known that the standard deviation of compressive strength is $\sigma = 1.3$ MPa. A group of engineers wants to determine the number of concrete specimens that will be needed in the study to detect a decrease in the mean compressive strength of two standard deviations. If the average compressive strength is actually less than $\mu - 2\sigma$, they want to be confident of correctly detecting this significant difference. In other words, the test of interest would be H_0: $\mu = 20$ versus H_1: $\mu < 20$. For this study, the significance level is set at $\alpha = 0.05$ and the power of the test is $1 - \beta = 0.99$. What is the minimum number of concrete specimens that should be used in this study? For a difference of 2σ or 2.6 MPa, $\alpha = 0.05$, and $1 - \beta = 0.99$, the minimum sample size can be found using Minitab®. The resulting output is

```
Testing mean = null (versus < null)
Calculating power for mean = null + difference
Alpha = 0.05 Sigma = 1.3

                Sample   Target    Actual
Difference        Size    Power     Power
      -2.6           6   0.9900    0.9936
```

The minimum number of specimens to be used in the study should be $n = 6$ in order to attain the desired power and level of significance.

Example 11-31

A manufacturer of rubber belts wishes to inspect and control the number of nonconforming belts produced on line. The proportion of nonconforming belts that is acceptable is $p = 0.01$. For practical purposes, if the proportion increases to $p = 0.035$ or greater, the manufacturer wants to detect this change. That is, the test of interest would be H_0: $p = 0.01$ versus H_1: $p > 0.01$. If the acceptable level of significance is $\alpha = 0.05$ and the power is $1 - \beta = 0.95$, how many rubber belts should be selected for inspection? For $\alpha = 0.05$ and $1 - \beta = 0.95$, the appropriate sample size can be determined using Minitab®. The output is

```
Testing proportion = 0.01 (versus > 0.01)
Alpha = 0.05

 Alternative   Sample    Target    Actual
  Proportion     Size     Power     Power
    3.50E-02      348    0.9500    0.9502
```

Therefore, to adequately detect a significant change in the proportion of nonconforming rubber belts, random samples of at least $n = 348$ would be needed.

11-7 SUMMARY

This chapter has introduced hypothesis testing. Procedures for testing hypotheses on means and variances are summarized in Table 11-8. The chi-square goodness of fit test was introduced to test the hypothesis that an empirical distribution follows a particular probability law. Graphical methods are also useful in goodness-of-fit testing, particularly when sample sizes are small. Two-way contingency tables for testing the hypothesis that two methods of classification of a sample are independent were also introduced. Several computer examples were also presented.

11-8 EXERCISES

11-1. The breaking strength of a fiber used in manufacturing cloth is required to be at least 160 psi. Past experience has indicated that the standard deviation of breaking strength is 3 psi. A random sample of four specimens is tested and the average breaking strength is found to be 158 psi.

(a) Should the fiber be judged acceptable with $\alpha = 0.05$?

(b) What is the probability of accepting H_0: $\mu \leq 160$ if the fiber has a true breaking strength of 165 psi?

11-2. The yield of a chemical process is being studied. The variance of yield is known from previous experience with this process to be 5 (units of σ^2 = percentage^2). The past five days of plant operation have resulted in the following yields (in percentages): 91.6, 88.75, 90.8, 89.95, 91.3.

(a) Is there reason to believe the yield is less than 90%?

(b) What sample size would be required to detect a true mean yield of 85% with probability 0.95?

11-3. The diameters of bolts are known to have a standard deviation of 0.0001 inch. A random sample of 10 bolts yields an average diameter of 0.2546 inch.

(a) Test the hypothesis that the true mean diameter of bolts equals 0.255 inch, using $\alpha = 0.05$.

(b) What size sample would be necessary to detect a true mean bolt diameter of 0.2552 inch with a probability of at least 0.90?

11-4. Consider the data in Exercise 10-39.

(a) Test the hypothesis that the mean piston ring diameter is 74.035 mm. Use $\alpha = 0.01$.

(b) What sample size is required to detect a true mean diameter of 74.030 with a probability of at least 0.95?

Table 11-8 Summary of Hypothesis Testing Procedures on Means and Variances

Null Hypothesis	Test Statistic	Alternative Hypothesis	Criteria for Rejection	OC Curve Parameter
H_0: $\mu = \mu_0$, σ^2 known	$Z_0 = \dfrac{\overline{X} - \mu_0}{\sigma/\sqrt{n}}$	H_1: $\mu \neq \mu_0$	$\lvert Z_0 \rvert > Z_{\alpha/2}$	$d = \lvert \mu - \mu_0 \rvert / \sigma$
		H_1: $\mu > \mu_0$	$Z_0 > Z_\alpha$	$d = (\mu - \mu_0)/\sigma$
		H_1: $\mu < \mu_0$	$Z_0 < -Z_\alpha$	$d = (\mu_0 - \mu)/\sigma$
H_0: $\mu = \mu_0$, σ^2 unknown	$t_0 = \dfrac{\overline{X} - \mu_0}{S/\sqrt{n}}$	H_1: $\mu \neq \mu_0$	$\lvert t_0 \rvert > t_{\alpha/2,\, n-1}$	$d = \lvert \mu - \mu_0 \rvert / \sigma$
		H_1: $\mu > \mu_0$	$t_0 > t_{\alpha,\, n-1}$	$d = (\mu - \mu_0)/\sigma$
		H_1: $\mu < \mu_0$	$t_0 < -t_{\alpha,\, n-1}$	$d = (\mu_0 - \mu)/\sigma$
H_0: $\mu_1 = \mu_2$, σ_1^2 and σ_2^2 known	$Z_0 = \dfrac{\overline{X}_1 - \overline{X}_2}{\sqrt{\dfrac{\sigma_1^2}{n_1} + \dfrac{\sigma_2^2}{n_2}}}$	H_1: $\mu_1 \neq \mu_2$	$\lvert Z_0 \rvert > Z_{\alpha/2}$	$d = \lvert \mu_1 - \mu_2 \rvert \sqrt{\sigma_1^2 + \sigma_2^2}$
		H_1: $\mu_1 > \mu_2$	$Z_0 > Z_\alpha$	$d = (\mu_1 - \mu_2)/\sqrt{\sigma_1^2 + \sigma_2^2}$
		H_1: $\mu_1 < \mu_2$	$Z_0 < -Z_\alpha$	$d = (\mu_2 - \mu_1)/\sqrt{\sigma_1^2 + \sigma_2^2}$
H_0: $\mu_1 = \mu_2$, $\sigma_1^2 = \sigma_2^2 = \sigma^2$ unknown	$t_0 = \dfrac{\overline{X}_1 - \overline{X}_2}{S_p \sqrt{\dfrac{1}{n_1} + \dfrac{1}{n_2}}}$	H_1: $\mu_1 \neq \mu_2$	$\lvert t_0 \rvert > t_{\alpha/2,\, n_1 + n_2 - 2}$	$d = \lvert \mu_1 - \mu_2 \rvert / 2\sigma$
		H_1: $\mu_1 > \mu_2$	$t_0 > t_{\alpha,\, n_1 + n_2 - 2}$	$d = (\mu_1 - \mu_2)/2\sigma$
		H_1: $\mu_1 < \mu_2$	$t_0 < -t_{\alpha,\, n_1 + n_2 - 2}$	$d = (\mu_2 - \mu_1)/2\sigma$
H_0: $\mu_1 = \mu_2$, $\sigma_1^2 \neq \sigma_2^2$ unknown	$t_0 = \dfrac{\overline{X}_1 - \overline{X}_2}{\sqrt{\dfrac{S_1^2}{n_1} + \dfrac{S_2^2}{n_2}}}$	H_1: $\mu_1 \neq \mu_2$	$\lvert t_0 \rvert > t_{\alpha/2,\, v}$	—
		H_1: $\mu_1 > \mu_2$	$t_0 > t_{\alpha,\, v}$	—
		H_1: $\mu_1 < \mu_2$	$t_0 < -t_{\alpha,\, v}$	—
	$v = \dfrac{\left(\dfrac{S_1^2}{n_1} + \dfrac{S_2^2}{n_2}\right)^2}{\dfrac{\left(S_1^2/n_1\right)^2}{n_1 + 1} + \dfrac{\left(S_2^2/n_2\right)^2}{n_2 + 1}} - 2$			
H_0: $\sigma^2 = \sigma_0^2$,	$\chi_0^2 = \dfrac{(n-1)S^2}{\sigma_0^2}$	H_1: $\sigma^2 \neq \sigma_0^2$	$\chi_0^2 > \chi_{\alpha/2,\, n-1}^2$ or $\chi_0^2 < \chi_{1-\alpha/2,\, n-1}^2$	$\lambda = \sigma/\sigma_0$
		H_1: $\sigma^2 > \sigma_0^2$	$\chi_0^2 > \chi_{\alpha, n-1}^2$	$\lambda = \sigma/\sigma_0$
		H_1: $\sigma^2 < \sigma_0^2$	$\chi_0^2 < \chi_{1-\alpha,\, n-1}^2$	$\lambda = \sigma/\sigma_0$
H_0: $\sigma_1^2 = \sigma_2^2$	$F_0 = S_1^2/S_2^2$	H_1:$\sigma_1^2 \neq \sigma_2^2$	$F_0 > F_{\alpha/2,\, n_1 - 1,\, n_2 - 1}$ or $F_0 < F_{1-\alpha/2,\, n_1 - 1,\, n_2 - 1}$	$\lambda = \sigma_1/\sigma_2$
		H_1:$\sigma_1^2 > \sigma_2^2$	$F_0 > F_{\alpha,\, n_1 - 1,\, n_2 - 1}$	$\lambda = \sigma_1/\sigma_2$

11-5. Consider the data in Exercise 10-40. Test the hypothesis that the mean life of the light bulbs is 1000 hours. Use $\alpha = 0.05$.

11-6. Consider the data in Exercise 10-41. Test the hypothesis that mean compressive strength equals 3500 psi. Use $\alpha = 0.01$.

11-7. Two machines are used for filling plastic bottles with a net volume of 16.0 ounces. The filling processes can be assumed normal, with standard deviations $\sigma_1 = 0.015$ and $\sigma_2 = 0.018$. Quality engineering suspects that both machines fill to the same net volume, whether or not this volume is 16.0 ounces. A random sample is taken from the output of each machine.

Machine 1		Machine 2	
16.03	16.01	16.02	16.03
16.04	15.96	15.97	16.04
16.05	15.98	15.96	16.02
16.05	16.02	16.01	16.01
16.02	15.99	15.99	16.00

(a) Do you think that quality engineering is correct? Use $\alpha = 0.05$.

(b) Assuming equal sample sizes, what sample size should be used to assure that $\beta = 0.05$ if the true difference in means is 0.075? Assume that $\alpha = 0.05$.

(c) What is the power of the test in (a) for a true difference in means of 0.075?

11-8. The film development department of a local store is considering the replacement of its current film-processing machine. The time in which it takes the machine to completely process a roll of film is important. A random sample of 12 rolls of 24-exposure color film is selected for processing by the current machine. The average processing time is 8.1 minutes, with a sample standard deviation of 1.4 minutes. A random sample of 10 rolls of the same type of film is selected for testing in the new machine. The average processing time is 7.3 minutes, with a sample standard deviation of 0.9 minutes. The local store will not purchase the new machine unless the processing time is more than 2 minutes shorter than the current machine. Based on this information, should they purchase the new machine?

11-9. Consider the data in Exercise 10-45. Test the hypothesis that both machines fill to the same volume. Use $\alpha = 0.10$.

11-10. Consider the data in Exercise 10-46. Test H_0: $\mu_1 = \mu_2$ against H_1:$\mu_1 > \mu_2$, using $\alpha = 0.05$.

11-11. Consider the gasoline road octane number data in Exercise 10-47. If formulation 2 produces a higher road octane number than formulation 1, the manufacturer would like to detect this. Formulate and test an appropriate hypothesis, using $\alpha = 0.05$.

11-12. The lateral deviation in yards of a certain type of mortar shell is being investigated by the propellant manufacturer. The following data have been observed.

Round	Deviation	Round	Deviation
1	11.28	6	–9.48
2	–10.42	7	6.25
3	–8.51	8	10.11
4	1.95	9	–8.65
5	6.47	10	–0.68

Test the hypothesis that the mean lateral deviation of these mortar shells is zero. Assume that lateral deviation is normally distributed.

11-13. The shelf life of a photographic film is of interest to the manufacturer. The manufacturer observes the following shelf life for eight units chosen at random from the current production. Assume that shelf life is normally distributed.

108 days	128 days
134	163
124	159
116	134

(a) Is there any evidence that the mean shelf life is greater than or equal to 125 days?

(b) If it is important to detect a ratio of δ/σ of 1.0 with a probability 0.90, is the sample size sufficient?

11-14. The titanium content of an alloy is being studied in the hope of ultimately increasing the tensile strength. An analysis of six recent heats chosen at random produces the following titanium contents.

8.0%	7.7%
9.9	11.6
9.9	14.6

Is there any evidence that the mean titanium content is greater than 9.5%?

11-15. An article in the *Journal of Construction Engineering and Management* (1999, p. 39) presents some data on the number of work hours lost per day on a construction project due to weather-related incidents. Over 11 workdays, the following lost work hours were recorded.

8.8	8.8
12.5	12.2
5.4	13.3
12.8	6.9
9.1	2.2
14.7	

Assuming work hours are normally distributed, is there any evidence to conclude that the mean number of work hours lost per day is greater than 8 hours?

11-16. The percentage of scrap produced in a metal finishing operation is hypothesized to be less than 7.5%. Several days were chosen at random and the percentages of scrap were calculated.

5.51%	7.32%
6.49	8.81
6.46	8.56
5.37	7.46

(a) In your opinion, is the true scrap rate less than 7.5%?

(b) If it is important to detect a ratio of $\delta/\sigma = 1.5$ with a probability of at least 0.90, what is the minimum sample size that can be used?

(c) For $\delta/\sigma = 2.0$, what is the power of the above test?

11-17. Suppose that we must test the hypotheses

$$H_0: \mu \geq 15,$$
$$H_1: \mu < 15,$$

where it is known that $\sigma^2 = 2.5$. If $\alpha = 0.05$ and the true mean is 12, what sample size is necessary to assure a type II error of 5%?

11-18. An engineer desires to test the hypothesis that the melting point of an alloy is 1000°C. If the true melting point differs from this by more than 20°C he must change the alloy's composition. If we assume that the melting point is a normally distributed random variable, $\alpha = 0.05$, $\beta = 0.10$, and $\sigma = 10$°C, how many observations should be taken?

11-19. Two methods for producing gasoline from crude oil are being investigated. The yields of both processes are assumed to be normally distributed. The following yield data have been obtained from the pilot plant.

Process	Yields (%)					
1	24.2	26.6	25.7	24.8	25.9	26.5
2	21.0	22.1	21.8	20.9	22.4	22.0

(a) Is there reason to believe that process 1 has a greater mean yield? Use $\alpha = 0.01$. Assume that both variances are equal.

(b) Assuming that in order to adopt process 1 it must produce a mean yield that is at least 5% greater than that of process 2, what are your recommendations?

(c) Find the power of the test in part (a) if the mean yield of process 1 is 5% greater than that of process 2.

(d) What sample size is required for the test in part (a) to ensure that the null hypothesis will be rejected with a probability of 0.90 if the mean yield of process 1 exceeds the mean yield of process 2 by 5%?

11-20. An article that appeared in the *Proceedings of the 1998 Winter Simulation Conference* (1998, p. 1079) discusses the concept of validation for traffic simulation models. The stated purpose of the study is to design and modify the facilities (roadways and control devices) to optimize efficiency and safety of traffic flow. Part of the study compares speed observed at various intersections and speed simulated by a model being tested. The goal is to determine whether the simulation model is representative of the actual observed speed. Field data is collected at a particular location and then the simulation model is implemented. Fourteen speeds (ft/sec) are measured at a particular location. Fourteen observations are simulated using the proposed model. The data are:

Field		Model	
53.33	57.14	47.40	58.20
53.33	57.14	49.80	59.00
53.33	61.54	51.90	60.10
55.17	61.54	52.20	63.40
55.17	61.54	54.50	65.80
55.17	69.57	55.70	71.30
57.14	69.57	56.70	75.40

Assuming the variances are equal, conduct a test of hypothesis test to determine whether there is a significant difference between the field data and the model simulated data. Use $\alpha = 0.05$.

11-21. The following are the burning times (in minutes) of flares of two different types.

Type 1		Type 2	
63	82	64	56
81	68	72	63
57	59	83	74
66	75	59	82
82	73	65	82

(a) Test the hypothesis that the two variances are equal. Use $\alpha = 0.05$.

(b) Using the results of (a), test the hypothesis that the mean burning times are equal.

11-22. A new filtering device is installed in a chemical unit. Before its installation, a random sample yielded the following information about the percentage of impurity: $\bar{x}_1 = 12.5$, $s_1^2 = 101.17$, and $n_1 = 8$. After installation, a random sample yielded $\bar{x}_2 = 10.2$, $s_2^2 = 94.73$, $n_2 = 9$.

(a) Can you conclude that the two variances are equal?

(b) Has the filtering device reduced the percentage of impurity significantly?

11-23. Suppose that two random samples were drawn from normal populations with equal variances. The sample data yields $\bar{x}_1 = 20.0$, $n_1 = 10$, $\Sigma(x_{1i} - \bar{x}_1)^2 = 1480$, $\bar{x}_2 = 15.8$, $n_2 = 10$, and $\Sigma(x_{2i} - \bar{x}_2)^2 = 1425$.

(a) Test the hypothesis that the two means are equal. Use $\alpha = 0.01$.

(b) Find the probability that the null hypothesis in (a) will be rejected if the true difference in means is 10.

(c) What sample size is required to detect a true difference in means of 5 with probability at least 0.80 if it is known at the start of the experiment that a rough estimate of the common variance is 150?

11-24. Consider the data in Exercise 10-56.

(a) Test the hypothesis that the means of the two normal distributions are equal. Use $\alpha = 0.05$ and assume that $\sigma_1^2 = \sigma_2^2$.

(b) What sample size is required to detect a difference in means of 2.0 with a probability of at least 0.85?

(c) Test the hypothesis that the variances of the two distributions are equal. Use $\alpha = 0.05$.

(d) Find the power of the test in (c) if the variance of a population is four times the other.

11-25. Consider the data in Exercise 10-57. Assuming that $\sigma_1^2 = \sigma_2^2$, test the hypothesis that the mean rod diameters do not differ. Use $\alpha = 0.05$.

11-26. A chemical company produces a certain drug whose weight has a standard deviation of 4 mg. A new method of producing this drug has been proposed, although some additional cost is involved. Management will authorize a change in production technique only if the standard deviation of the weight in the new process is less than 4 mg. If the standard deviation of weight in the new process is as small as 3 mg, the company would like to switch production methods with a probability of at least 0.90. Assuming weight to be normally distributed and $\alpha = 0.05$, how many observations should be taken? Suppose the researchers choose $n = 10$ and obtain the data below. Is this a good choice for n? What should be their decision?

16.628 grams	16.630 grams
16.622	16.631
16.627	16.624
16.623	16.622
16.618	16.626

11-27. A manufacturer of precision measuring instruments claims that the standard deviation in the use of the instrument is 0.00002 inch. An analyst, who is unaware of the claim, uses the instrument eight times and obtains a sample standard deviation of 0.00005 inch.

(a) Using $\alpha = 0.01$, is the claim justified?

(b) Compute a 99% confidence interval for the true variance.

(c) What is the power of the test if the true standard deviation equals 0.00004?

(d) What is the smallest sample size that can be used to detect a true standard deviation of 0.00004 with a probability at least of 0.95? Use $\alpha = 0.01$.

11-28. The standard deviation of measurements made by a special thermocouple is supposed to be 0.005 degree. If the standard deviation is as great as 0.010, we wish to detect it with a probability of at least 0.90. Use $\alpha = 0.01$. What sample size should be used? If this sample size is used and the sample standard deviation $s = 0.007$, what is your conclusion, using $\alpha = 0.01$? Construct a 95% upper-confidence interval for the true variance.

11-29. The manufacturer of a power supply is interested in the variability of output voltage. He has tested 12 units, chosen at random, with the following results:

5.34	5.65	4.76
5.00	5.55	5.54
5.07	5.35	5.44
5.25	5.35	4.61

(a) Test the hypothesis that $\sigma^2 = 0.5$. Use $\alpha = 0.05$.

(b) If the true value of $\sigma^2 = 1.0$, what is the probability that the hypothesis in (a) will be rejected?

11-30. For the data in Exercise 11-7, test the hypothesis that the two variances are equal, using $\alpha = 0.01$. Does the result of this test influence the manner in which a test on means would be conducted? What sample size is necessary to detect $\sigma_1^2/\sigma_2^2 = 2.5$, with a probability of at least 0.90?

11-31. Consider the following two samples, drawn from two normal populations.

Sample 1	Sample 2
4.34	1.87
5.00	2.00
4.97	2.00
4.25	1.85
5.55	2.11
6.55	2.31
6.37	2.28
5.55	2.07
3.76	1.76
—	1.91
—	2.00

Is there evidence to conclude that the variance of population 1 is greater than the variance of population 2? Use $\alpha = 0.01$. Find the probability of detecting $\sigma_1^2/\sigma_2^2 = 4.0$.

11-32. Two machines produce metal parts. The variance of the weight of these parts is of interest. The following data have been collected.

Machine 1	Machine 2
$n_1 = 25$	$n_2 = 30$
$\bar{x}_1 = 0.984$	$\bar{x}_2 = 0.907$
$s_1^2 = 13.46$	$s_2^2 = 9.65$

(a) Test the hypothesis that the variances of the two machines are equal. Use $\alpha = 0.05$.

(b) Test the hypothesis that the two machines produce parts having the same mean weight. Use $\alpha = 0.05$.

11-33. In a hardness test, a steel ball is pressed into the material being tested at a standard load. The diameter of the indentation is measured, which is related to the hardness. Two types of steel balls are available, and their performance is compared on 10 specimens. Each specimen is tested twice, once with each ball. The results are given below:

Ball x	75	46	57	43	58	32	61	56	34	65
Ball y	52	41	43	47	32	49	52	44	57	60

Test the hypothesis that the two steel balls give the same expected hardness measurement. Use $\alpha = 0.05$.

11-34. Two types of exercise equipment, A and B, for handicapped individuals are often used to determine the effect of the particular exercise on heart rate (in beats per minute). Seven subjects participated in a study to determine whether the two types of equipment have the same effect on heart rate. The results are given in the table below.

Subject	A	B
1	162	161
2	163	187
3	140	199
4	191	206
5	160	161
6	158	160
7	155	162

Conduct an appropriate test of hypothesis to determine whether there is a significant difference in heart rate due to the type of equipment used.

11-35. An aircraft designer has theoretical evidence that painting the airplane reduces its speed at a specified power and flap setting. He tests six consecutive airplanes from the assembly line before and after painting. The results are shown below.

	Top Speed (mph)	
Airplane	Painted	Not Painted
1	286	289
2	285	286
3	279	283
4	283	288
5	281	283
6	286	289

Do the data support the designer's theory? Use $\alpha = 0.05$.

11-36. An article in the *International Journal of Fatigue* (1998, p. 537) discusses the bending fatigue resistance of gear teeth when using a particular prestressing or *presetting* process. Presetting of a gear tooth is obtained by applying and then removing a single overload to the machine element. To determine significant differences in fatigue resistance due to presetting, fatigue data were paired. A "preset" tooth and a "nonpreset" tooth were paired if they were present on the same gear. Eleven pairs were formed and the fatigue life measured for each. (The final response of interest is $\ln[(\text{fatigue life}) \times 10^{-3}]$.)

Pair	Preset Tooth	Nonpreset Tooth
1	3.813	2.706
2	4.025	2.364
3	3.042	2.773
4	3.831	2.558
5	3.320	2.430
6	3.080	2.616
7	2.498	2.765
8	2.417	2.486
9	2.462	2.688
10	2.236	2.700
11	3.932	2.810

Conduct a test of hypothesis to determine whether presetting significantly increases the fatigue life of gear teeth. Use $\alpha = 0.10$.

11-37. Consider the data in Exercise 10-66. Test the hypothesis that the uninsured rate is 10%. Use $\alpha = 0.05$.

11-38. Consider the data in Exercise 10-68. Test the hypothesis that the fraction of defective calculators produced is 2.5%.

11-39. Suppose that we wish to test the hypothesis H_0: $\mu_1 = \mu_2$ against the alternative H_1: $\mu_1 \neq \mu_2$, where both variances σ_1^2 and σ_2^2 are known. A total of $n_1 + n_2 = N$ observations can be taken. How should these observations be allocated to the two populations to maximize the probability that H_0 will be rejected if H_1 is true, and $\mu_1 - \mu_2 = \delta \neq 0$?

11-40. Consider the union membership study described in Exercise 10-70. Test the hypothesis that the proportion of men who belong to a union does not differ from the proportion of women who belong to a union. Use $\alpha = 0.05$.

11-41. Using the data in Exercise 10-71, determine whether it is reasonable to conclude that production line 2 produced a higher fraction of defective product than line 1. Use $\alpha = 0.01$.

11-42. Two different types of injection-molding machines are used to form plastic parts. A part is considered defective if it has excessive shrinkage or is discolored. Two random samples, each of size 500, are selected, and 32 defective parts are found in the sample from machine 1, while 21 defective parts are found in the sample from machine 2. Is it reasonable to conclude that both machines produce the same fraction of defective parts?

11-43. Suppose that we wish to test H_0: $\mu_1 = \mu_2$ against H_1: $\mu_1 \neq \mu_2$, where σ_1^2 and σ_2^2 are known. The total sample size N is fixed, but the allocation of observations to the two populations such that $n_1 + n_2 = N$ is to be made on the basis of cost. If the costs of sampling for populations 1 and 2 are C_1 and C_2, respectively, find the minimum cost sample sizes that provide a specified variance for the difference in sample means.

11-44. A manufacturer of a new pain relief tablet would like to demonstrate that her product works twice as fast as her competitor's product. Specifically, she would like to test

$$H_0\colon \mu_1 = 2\mu_2,$$
$$H_1\colon \mu_1 > 2\mu_2,$$

where μ_1 is the mean absorption time of the competitive product and μ_2 is the mean absorption time of the new product. Assuming that the variances σ_1^2 and σ_2^2 are known, suggest a procedure for testing this hypothesis.

11-45. Derive an expression similar to equation 11-20 for the β error for the test on the variance of a normal distribution. Assume that the two-sided alternative is specified.

11-46. Derive an expression similar to equation 11-20 for the β error for the test of the equality of the variances of two normal distributions. Assume that the two-sided alternative is specified.

11-47. The number of defective units found each day by an in-circuit functional tester in a printed circuit board assembly process is shown below.

Number of Defectives per Day	Times Observed
0–10	6
11–15	11
16–20	16
21–25	28
26–30	22
31–35	19
36–40	11
41–45	4

(a) It is reasonable to conclude that these data come from a normal distribution? Use a chi-square goodness-of-fit test.

(b) Plot the data on normal probability paper. Does an assumption of normality seem justified?

11-48. Defects on wafer surfaces in integrated circuit fabrication are unavoidable. In a particular process the following data were collected.

Number of Defects i	Number of Wafers with i Defects
0	4
1	13
2	34
3	56
4	70
5	70
6	58
7	42
8	25
9	15
10	9
11	3
12	1

Does the assumption of a Poisson distribution seem appropriate as a probability model for this process?

11-49. A pseudorandom number generator is designed so that integers 0 through 9 have an equal probability of occurrence. The first 10,000 numbers are as follows:

0	1	2	3	4	5	6	7	8	9
967	1008	975	1022	1003	989	1001	981	1043	1011

Does this generator seem to be working properly?

11-50. The cycle time of an automatic machine has been observed and recorded.

Sec	2.1	2.11	2.12	2.13	2.14	2.15	2.16	2.17	2.18	2.19	2.2
Freq	16	28	41	74	149	256	137	82	40	19	11

(a) Does the normal distribution seem to be a reasonable probability model for the cycle time? Use the chi-square goodness-of-fit test.

(b) Plot the data on normal probability paper. Does the assumption of normality seem reasonable?

11-51. A soft drink bottler is studying the internal pressure strength of 1-liter glass nonreturnable bottles. A random sample of 16 bottles is tested and the pressure strengths obtained. The data are shown below. Plot these data on normal probability paper.

Does it seem reasonable to conclude that pressure strength is normally distributed?

226.16 psi	211.14 psi
202.20	203.62
219.54	188.12
193.73	224.39
208.15	221.31
195.45	204.55
193.71	202.21
200.81	201.63

11-52. A company operates four machines for three shifts each day. From production records, the following data on the number of breakdowns are collected.

	Machines			
Shift	*A*	*B*	*C*	*D*
1	41	20	12	16
2	31	11	9	14
3	15	17	16	10

Test the hypothesis that breakdowns are independent of the shift.

11-53. Patients in a hospital are classified as surgical or medical. A record is kept of the number of times patients require nursing service during the night and whether these patients are on Medicare or not. The data are as follows:

	Patient Category	
Medicare	Surgical	Medical
Yes	46	52
No	36	43

Test the hypothesis that calls by surgical–medical patients are independent of whether the patients are receiving Medicare.

11-54. Grades in a statistics course and an operations research course taken simultaneously were as follows for a group of students.

Statistics	Operations Research Grade			
Grade	*A*	*B*	*C*	Other
A	25	6	17	13
B	17	16	15	6
C	18	4	18	10
Other	10	8	11	20

Are the grades in statistics and operations research related?

11-55. An experiment with artillery shells yields the following data on the characteristics of lateral deflections and ranges. Would you conclude that deflection and range are independent?

	Lateral Deflection		
Range (yards)	Left	Normal	Right
0 – 1,999	6	14	8
2,000 – 5,999	9	11	4
6,000 – 11,999	8	17	6

11-56. A study is being made of the failures of an electronic component. There are four types of failures possible and two mounting positions for the device. The following data have been taken.

	Failure Type			
Mounting Position	*A*	*B*	*C*	*D*
1	22	46	18	9
2	4	17	6	12

Would you conclude that the type of failure is independent of the mounting position?

11-57. An article in *Research in Nursing and Health* (1999, p. 263) summarizes data collected from a previous study (*Research in Nursing and Health*, 1998, p. 285) on the relationship between physical activity and socio-economic status of 1507 Caucasian women. The data are given in the table below.

	Physical Activity	
Socio-economic Status	Inactive	Active
Low	216	245
Medium	226	409
High	114	297

Test the hypothesis that physical activity is independent of socio-economic status.

11-58. Fabric is graded into three classifications: *A*, *B*, and *C*. The results below were obtained from five looms. Is fabric classification independent of the loom?

	Number of Pieces of Fabric in Fabric Classification		
Loom	*A*	*B*	*C*
1	185	16	12
2	190	24	21
3	170	35	16
4	158	22	7
5	185	22	15

11-59. An article in the *Journal of Marketing Research* (1970, p. 36) reports a study of the relationship between facility conditions at gasoline stations and the aggressiveness of their gasoline marketing policy. A sample of 441 gasoline stations was investigated with the results shown below obtained. Is there evidence that gasoline pricing strategy and facility conditions are independent?

	Condition		
Policy	Substandard	Standard	Modern
Aggressive	24	52	58
Neutral	15	73	86
Nonaggressive	17	80	36

11-60. Consider the injection molding process described in Exercise 11-42.

(a) Set up this problem as a 2×2 contingency table and perform the indicated statistical analysis.

(b) State clearly the hypothesis being tested. Are you testing homogeneity or independence?

(c) Is this procedure equivalent to the test procedure used in Exercise 11-42?

Chapter 12

Design and Analysis of Single-Factor Experiments: The Analysis of Variance

Experiments are a natural part of the engineering and management decision-making process. For example, suppose that a civil engineer is investigating the effect of curing methods on the mean compressive strength of concrete. The experiment would consist of making up several test specimens of concrete using each of the proposed curing methods and then testing the compressive strength of each specimen. The data from this experiment could be used to determine which curing method should be used to provide maximum compressive strength.

If there are only two curing methods of interest, the experiment could be *designed* and *analyzed* using the methods discussed in Chapter 11. That is, the experimenter has a *single factor* of interest—curing methods—and there are only two *levels* of the factor. If the experimenter is interested in determining which curing method produces the maximum compressive strength, then the number of specimens to test can be determined using the operating characteristic curves in Chart VI (Appendix), and the t-test can be used to determine whether the two means differ.

Many single-factor experiments require more than two levels of the factor to be considered. For example, the civil engineer may have five different curing methods to investigate. In this chapter we introduce the analysis of variance for dealing with more than two levels of a single factor. In Chapter 13, we show how to design and analyze experiments with several factors.

12-1 THE COMPLETELY RANDOMIZED SINGLE-FACTOR EXPERIMENT

12-1.1 An Example

A manufacturer of paper used for making grocery bags is interested in improving the tensile strength of the product. Product engineering thinks that tensile strength is a function of the hardwood concentration in the pulp, and that the range of hardwood concentrations of practical interest is between 5% and 20%. One of the engineers responsible for the study decides to investigate four levels of hardwood concentration: 5%, 10%, 15%, and 20%. She also decides to make up six test specimens at each concentration level, using a pilot plant. All 24 specimens are tested on a laboratory tensile tester in random order. The data from this experiment are shown in Table 12-1.

This is an example of a completely randomized single-factor experiment with four levels of the factor. The levels of the factor are sometimes called *treatments*. Each treatment

Table 12-1 Tensile Strength of Paper (psi)

Hardwood Concentration (%)	Observations 1	2	3	4	5	6	Totals	Averages
5	7	8	15	11	9	10	60	10.00
10	12	17	13	18	19	15	94	15.67
15	14	18	19	17	16	18	102	17.00
20	19	25	22	23	18	20	127	21.17
							383	15.96

has six observations, or *replicates*. The role of *randomization* in this experiment is extremely important. By randomizing the order of the 24 runs, the effect of any nuisance variable that may affect the observed tensile strength is approximately balanced out. For example, suppose that there is a warm-up effect on the tensile tester; that is, the longer the machine is on, the greater the observed tensile strength. If the 24 runs are made in order of increasing hardwood concentration (i.e., all six 5% concentration specimens are tested first, followed by all six 10% concentration specimens, etc.), then any observed differences due to hardwood concentration could also be due to the warm-up effect.

It is important to graphically analyze the data from a designed experiment. Figure 12-1 presents box plots of tensile strength at the four hardwood concentration levels. This plot indicates that changing the hardwood concentration has an effect on tensile strength; specifically, higher hardwood concentrations produce higher observed tensile strength. Furthermore, the distribution of tensile strength at a particular hardwood level is reasonably symmetric, and the variability in tensile strength does not change dramatically as the hardwood concentration changes.

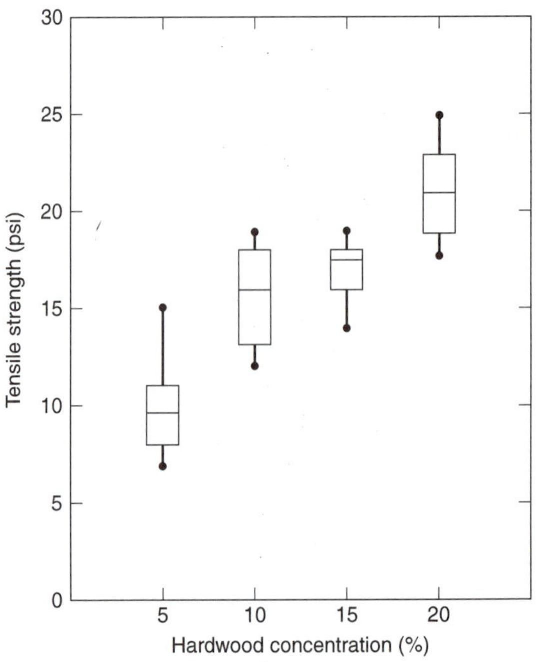

Figure 12-1 Box plots of hardwood concentration data.

Graphical interpretation of the data is always a good idea. Box plots show the variability of the observations *within* a treatment (factor level) and the variability *between* treatments. We now show how the data from a single-factor randomized experiment can be analyzed statistically.

12-1.2 The Analysis of Variance

Suppose we have a different levels of a single factor (treatments) that we wish to compare. The observed response for each of the a treatments is a random variable. The data would appear as in Table 12-2. An entry in Table 12-2, say y_{ij}, represents the jth observation taken under treatment i. We initially consider the case where there is an equal number of observations n on each treatment.

We may describe the observations in Table 12-2 by the linear statistical model,

$$y_{ij} = \mu + \tau_i + \epsilon_{ij} \begin{cases} i = 1, 2, \ldots, a, \\ j = 1, 2, \ldots, n, \end{cases} \tag{12-1}$$

where y_{ij} is the (ij)th observation, μ is a parameter common to all treatments, called the *overall mean*, τ_i is a parameter associated with the ith treatment, called the ith *treatment effect*, and ϵ_{ij} is a random error component. Note that y_{ij} represents both the random variable and its realization. We would like to test certain hypotheses about the treatment effects and to estimate them. For hypothesis testing, the model errors are assumed to be normally and independently distributed random variables with mean zero and variance σ^2 [abbreviated NID(0, σ^2)]. The variance σ^2 is assumed constant for all levels of the factor.

The model of equation 12-1 is called the *one-way-classification* analysis of variance, because only one factor is investigated. Furthermore, we will require that the observations be taken in random order so that the environment in which the treatments are used (often called the experimental units) is as uniform as possible. This is called a completely randomized experimental design. There are two different ways that the a factor levels in the experiment could have been chosen. First, the a treatments could have been specifically chosen by the experimenter. In this situation we wish to test hypotheses about the τ_i, and conclusions will apply only to the factor levels considered in the analysis. The conclusions cannot be extended to similar treatments that were not considered. Also, we may wish to estimate the τ_i. This is called the *fixed effects* model. Alternatively, the a treatments could be a random sample from a larger population of treatments. In this situation we would like to be able to extend the conclusions (which are based on the sample of treatments) to all treatments in the population, whether they were explicitly considered in the analysis or not. Here the τ_i are random variables, and knowledge about the particular ones investigated is relatively useless. Instead, we test hypotheses about the variability of the τ_i and try to estimate this variability. This is called the *random effects*, or *components of variance*, model.

Table 12-2 Typical Data for One-Way-Classification Analysis of Variance

Treatment	Observation				Totals	Averages
1	y_{11}	y_{12}	$\cdots$	y_{1n}	$y_{1\cdot}$	$\bar{y}_{1\cdot}$
2	y_{21}	y_{22}	$\cdots$	y_{2n}	$y_{2\cdot}$	$\bar{y}_{2\cdot}$
$\vdots$	$\vdots$	$\vdots$		$\vdots$	$\vdots$	$\vdots$
a	y_{a1}	y_{a2}	$\cdots$	y_{an}	$y_{a\cdot}$	$\bar{y}_{a\cdot}$

In this section we will develop the analysis of variance for the fixed-effects model, one-way classification. In the fixed-effects model, the treatment effects τ_i are usually defined as deviations from the overall mean, so that

$$\sum_{i=1}^{a} \tau_i = 0. \tag{12-2}$$

Let $y_{i\cdot}$ represent the total of the observations under the ith treatment and $\bar{y}_{i\cdot}$ represent the average of the observations under the ith treatment. Similarly, let $y_{\cdot\cdot}$ represent the grand total of all observations and $\bar{y}_{\cdot\cdot}$ represent the grand mean of all observations. Expressed mathematically,

$$\begin{aligned} y_{i\cdot} &= \sum_{j=1}^{n} y_{ij}, \qquad \bar{y}_{i\cdot} = y_{i\cdot}/n, \qquad i = 1, 2, \ldots, a, \\ y_{\cdot\cdot} &= \sum_{i=1}^{a}\sum_{j=1}^{n} y_{ij}, \qquad \bar{y}_{\cdot\cdot} = y_{\cdot\cdot}/N, \end{aligned} \tag{12-3}$$

where $N = an$ is the total number of observations. Thus the "dot" subscript notation implies summation over the subscript that it replaces.

We are interested in testing the equality of the a treatment effects. Using equation 12-2, the appropriate hypotheses are

$$\begin{aligned} &H_0\colon \tau_1 = \tau_2 = \cdots = \tau_a = 0, \\ &H_1\colon \tau_i \neq 0 \text{ for at least one } i. \end{aligned} \tag{12-4}$$

That is, if the null hypothesis is true, then each observation is made up of the overall mean μ plus a realization of the random error ϵ_{ij}.

The test procedure for the hypotheses in equation 12-4 is called the analysis of variance. The name "analysis of variance" results from partitioning total variability in the data into its component parts. The total corrected sum of squares, which is a measure of total variability in the data, may be written as

$$\sum_{i=1}^{a}\sum_{j=1}^{n}\left(y_{ij} - \bar{y}_{\cdot\cdot}\right)^2 = \sum_{i=1}^{a}\sum_{j=1}^{n}\left[\left(\bar{y}_{i\cdot} - \bar{y}_{\cdot\cdot}\right) + \left(y_{ij} - \bar{y}_{i\cdot}\right)\right]^2 \tag{12-5}$$

or

$$\begin{aligned} \sum_{i=1}^{a}\sum_{j=1}^{n}\left(y_{ij} - \bar{y}_{\cdot\cdot}\right)^2 &= n\sum_{i=1}^{a}\left(\bar{y}_{i\cdot} - \bar{y}_{\cdot\cdot}\right)^2 + \sum_{i=1}^{a}\sum_{j=1}^{n}\left(y_{ij} - \bar{y}_{i\cdot}\right)^2 \\ &\quad + 2\sum_{i=1}^{a}\sum_{j=1}^{n}\left(\bar{y}_{i\cdot} - \bar{y}_{\cdot\cdot}\right)\left(y_{ij} - \bar{y}_{i\cdot}\right). \end{aligned} \tag{12-6}$$

Note that the cross-product term in equation 12-6 is zero, since

$$\sum_{j=1}^{n}\left(y_{ij} - \bar{y}_{i\cdot}\right) = y_{i\cdot} - n\bar{y}_{i\cdot} = y_{i\cdot} - n\left(y_{i\cdot}/n\right) = 0.$$

Therefore, we have

$$\sum_{i=1}^{a}\sum_{j=1}^{n}\left(y_{ij} - \bar{y}_{\cdot\cdot}\right)^2 = n\sum_{i=1}^{a}\left(\bar{y}_{i\cdot} - \bar{y}_{\cdot\cdot}\right)^2 + \sum_{i=1}^{a}\sum_{j=1}^{n}\left(y_{ij} - \bar{y}_{i\cdot}\right)^2. \tag{12-7}$$

Equation 12-7 shows that the total variability in the data, measured by the total corrected sum of squares, can be partitioned into a sum of squares of differences between treatment means and the grand mean and a sum of squares of differences of observations within treatments and the treatment mean. Differences between observed treatment means and the grand mean measure the differences between treatments, while differences of observations within a treatment from the treatment mean can be due only to random error. Therefore, we write equation 12-7 symbolically as

$$SS_T = SS_{\text{treatments}} + SS_E,$$

where SS_T is the total sum of squares, $SS_{\text{treatments}}$ is the sum of squares due to treatments (i.e., *between* treatments), and SS_E is the sum of squares due to error (i.e., *within* treatments). There are $an = N$ total observations; thus SS_T has $N - 1$ degrees of freedom. There are a levels of the factor, so $SS_{\text{treatments}}$ has $a - 1$ degrees of freedom. Finally, within any treatment there are n replicates providing $n - 1$ degrees of freedom with which to estimate the experimental error. Since there are a treatments, we have $a(n - 1) = an - a = N - a$ degrees of freedom for error.

Now consider the distributional properties of these sums of squares. Since we have assumed that the errors ϵ_{ij} are NID$(0, \sigma^2)$, the observations y_{ij} are NID$(\mu + \tau_i, \sigma^2)$. Thus SS_T/σ^2 is distributed as chi-square with $N - 1$ degrees of freedom, since SS_T is a sum of squares in normal random variables. We may also show that $SS_{\text{treatments}}/\sigma^2$ is chi-square with $a - 1$ degrees of freedom, if H_0 is true, and SS_E/σ^2 is chi-square with $N - a$ degrees of freedom. However, all three sums of squares are not independent, since $SS_{\text{treatments}}$ and SS_E add up to SS_T. The following theorem, which is a special form of one due to Cochran, is useful in developing the test procedure.

Theorem 12-1 (Cochran)

Let Z_i be NID(0, 1) for $i = 1, 2, \ldots, v$ and let

$$\sum_{i=1}^{v} Z_i^2 = Q_1 + Q_2 + \cdots + Q_s,$$

where $s < v$ and Q_i is chi-square with v_i degrees of freedom ($i = 1, 2, \ldots, s$). Then $Q_1, Q_2, \ldots, Q_s$ are independent chi-square random variables with $v_1, v_2, \ldots, v_s$ degrees of freedom, respectively, if and only if

$$v = v_1 + v_2 + \cdots + v_s.$$

Using this theorem, we note that the degrees of freedom for $SS_{\text{treatments}}$ and SS_E add up to $N - 1$, so that $SS_{\text{treatments}}/\sigma^2$ and SS_E/σ^2 are independently distributed chi-square random variables. Therefore, under the null hypothesis, the statistic

$$F_0 = \frac{SS_{\text{treatments}}/(a-1)}{SS_E/(N-a)} = \frac{MS_{\text{treatments}}}{MS_E} \tag{12-8}$$

follows the $F_{a-1, N-a}$ distribution. The quantities $MS_{\text{treatments}}$ and MS_E are *mean squares.*

The expected values of the mean squares are used to show that F_0 in equation 12-8 is an appropriate test statistic for H_0: $\tau_i = 0$ and to determine the criterion for rejecting this null hypothesis. Consider

$$E(MS_E) = E\left(\frac{SS_E}{N-a}\right) = \frac{1}{N-a} E\left[\sum_{i=1}^{a}\sum_{j=1}^{n}\left(y_{ij} - \bar{y}_{i\cdot}\right)^2\right]$$

$$= \frac{1}{N-a} E\left[\sum_{i=1}^{a}\sum_{j=1}^{n}\left(y_{ij}^2 - 2y_{ij}\bar{y}_{i\cdot} + \bar{y}_{i\cdot}^2\right)\right]$$

$$= \frac{1}{N-a} E\left[\sum_{i=1}^{a}\sum_{j=1}^{n} y_{ij}^2 - 2n\sum_{i=1}^{a}\bar{y}_{i\cdot}^2 + n\sum_{i=1}^{a}\bar{y}_{i\cdot}^2\right]$$

$$= \frac{1}{N-a} E\left[\sum_{i=1}^{a}\sum_{j=1}^{n} y_{ij}^2 - \frac{1}{n}\sum_{i=1}^{a} y_{i\cdot}^2\right].$$

Substituting the model, equation 12-1, into this equation we obtain

$$E(MS_E) = \frac{1}{N-a} E\left[\sum_{i=1}^{a}\sum_{j=1}^{n}\left(\mu + \tau_i + \epsilon_{ij}\right)^2 - \frac{1}{n}\sum_{i=1}^{a}\left(\sum_{j=1}^{n}\left(\mu + \tau_i + \epsilon_{ij}\right)\right)^2\right].$$

Now on squaring and taking the expectation of the quantities within brackets, we see that terms involving ϵ_{ij}^2 and $\sum_{j=1}^{n}\epsilon_{ij}^2$ are replaced by σ^2 and $n\sigma^2$, respectively, because $E(\epsilon_{ij}) = 0$. Furthermore, all cross products involving ϵ_{ij} have zero expectation. Therefore, after squaring, taking expectation, and noting that $\sum_{i=1}^{a}\tau_i = 0$, we have

$$E(MS_E) = \frac{1}{N-a} E\left[N\mu^2 + n\sum_{i=1}^{a}\tau_i^2 + N\sigma^2 - N\mu^2 - n\sum_{i=1}^{a}\tau_i^2 - a\sigma^2\right]$$

or

$$E(MS_E) = \sigma^2.$$

Using a similar approach, we may show that

$$E\left(MS_{\text{treatments}}\right) = \sigma^2 + \frac{n\sum_{i=1}^{a}\tau_i^2}{a-1}.$$

From the expected mean squares we see that MS_E is an unbiased estimator of σ^2. Also, under the null hypothesis, $MS_{\text{treatments}}$ is an unbiased estimator of σ^2. However, if the null hypothesis if false, then the expected value of $MS_{\text{treatments}}$ is greater than σ^2. Therefore, under the alternative hypothesis the expected value of the numerator of the test statistic (equation 12-8) is greater than the expected value of the denominator. Consequently, we should reject H_0 if the test statistic is large. This implies an upper-tail, one-tail critical region. Thus, we would reject H_0 if

$$F_0 > F_{\alpha,\, a-1,\, N-a}$$

where F_0 is computed from equation 12-8.

Efficient computational formulas for the sums of squares may be obtained by expanding and simplifying the definitions of $SS_{\text{treatments}}$ and SS_T in equation 12-7. This yields

$$SS_T = \sum_{i=1}^{a}\sum_{j=1}^{n} y_{ij}^2 - \frac{y_{\cdot\cdot}^2}{N} \tag{12-9}$$

and

$$SS_{\text{treatments}} = \sum_{i=1}^{a}\frac{y_{i\cdot}^2}{n} - \frac{y_{\cdot\cdot}^2}{N}. \tag{12-10}$$

The error sum of squares is obtained by subtraction:

$$SS_E = SS_T - SS_{\text{treatments}}. \tag{12-11}$$

The test procedure is summarized in Table 12-3. This is called an analysis-of-variance table.

Example 12-1

Consider the hardwood concentration experiment described in Section 12-1.1 We can use the analysis of variance to test the hypothesis that different hardwood concentrations do not affect the mean tensile strength of the paper. The sums of squares for analysis of variance are computed from equations 12-9, 12-10, and 12-11 as follows:

$$\begin{aligned}
SS_T &= \sum_{i=1}^{4}\sum_{j=1}^{6} y_{ij}^2 - \frac{y_{..}^2}{N} \\
&= (7)^2 + (8)^2 + \cdots + (20)^2 - \frac{(383)^2}{24} = 512.96, \\
SS_{\text{treatments}} &= \sum_{i=1}^{4} \frac{y_{i.}^2}{n} - \frac{y_{..}^2}{N} \\
&= \frac{(60)^2 + (94)^2 + (102)^2 + (127)^2}{6} - \frac{(383)^2}{24} = 382.79, \\
SS_E &= SS_T - SS_{\text{treatments}} \\
&= 512.96 - 382.79 = 130.17.
\end{aligned}$$

The analysis of variance is summarized in Table 12-4. Since $F_{0.01,3,20} = 4.94$, we reject H_0 and conclude that hardwood concentration in the pulp significantly affects the strength of the paper.

12-1.3 Estimation of the Model Parameters

It is possible to derive estimators for the parameters in the one-way analysis-of-variance model

$$y_{ij} = \mu + \tau_i + \epsilon_{ij}.$$

Table 12-3 Analysis of Variance for the One-Way-Classification Fixed-Effects Model

Source of Variation	Sum of Squares	Degrees of Freedom	Mean Square	F_0
Between treatments	$SS_{\text{treatments}}$	$a-1$	$MS_{\text{treatments}}$	$\frac{MS_{\text{treatments}}}{MS_E}$
Error (within treatments)	SS_E	$N-a$	MS_E	
Total	SS_T	$N-1$		

Table 12-4 Analysis of Variance for the Tensile Strength Data

Source of Variation	Sum of Squares	Degrees of Freedom	Mean Square	F_0
Hardwood concentration	382.79	3	127.60	19.61
Error	130.17	20	6.51	
Total	512.96	23		

An appropriate estimation criterion is to estimate μ and τ_i such that the sum of the squares of the errors or deviations ϵ_{ij} is a minimum. This method of parameter estimation is called the method of *least squares.* In estimating μ and τ_i by least squares, the normality assumption on the errors ϵ_{ij} is not needed. To find the least-squares estimators of μ and τ_i, we form the sum of squares of the errors

$$L=\sum_{i=1}^{a}\sum_{j=1}^{n}\epsilon_{ij}^2=\sum_{i=1}^{a}\sum_{j=1}^{n}\left(y_{ij}-\mu-\tau_i\right)^2 \tag{12-12}$$

and find values of μ and τ_i, say $\hat{\mu}$ and $\hat{\tau}_i$, that minimize L. The values $\hat{\mu}$ and $\hat{\tau}_i$ are the solutions to the $a + 1$ simultaneous equations

$$\left.\frac{\partial L}{\partial \mu}\right|_{\hat{\mu},\hat{\tau}_i}=0,$$

$$\left.\frac{\partial L}{\partial \tau_i}\right|_{\hat{\mu},\hat{\tau}_i}=0, \qquad i=1,2,\ldots,a.$$

Differentiating equation 12-12 with respect to μ and τ_i and equating to zero, we obtain

$$-2\sum_{i=1}^{a}\sum_{j=1}^{n}\left(y_{ij}-\hat{\mu}-\hat{\tau}_i\right)=0$$

and

$$-2\sum_{j=1}^{n}\left(y_{ij}-\hat{\mu}-\hat{\tau}_i\right)=0, \qquad i=1,2,\ldots,a.$$

After simplification these equations become

$$\begin{aligned}
N\hat{\mu}+n\hat{\tau}_1+n\hat{\tau}_2+\cdots+n\hat{\tau}_a&=y_{..}\,,\\
n\hat{\mu}+n\hat{\tau}_1 &= y_{1\cdot}\,,\\
n\hat{\mu}\qquad+n\hat{\tau}_2 &= y_{2\cdot}\,,\\
\vdots\qquad\qquad\vdots\qquad &\quad\vdots\\
n\hat{\mu}\qquad\qquad+n\hat{\tau}_a&=y_{a\cdot}\,.
\end{aligned} \tag{12-13}$$

Equations 12-13 are called the *least-squares normal equations.* Notice that if we add the last a normal equations we obtain the first normal equation. Therefore, the normal equations are not linearly independent, and there are no unique estimates for $\mu, \tau_1, \tau_2, \ldots, \tau_a$. One way to overcome this difficulty is to impose a constraint on the solution to the normal equations. There are many ways to choose this constraint. Since we have defined the treatment effects as deviations from the overall mean, it seems reasonable to apply the constraint

$$\sum_{i=1}^{a}\hat{\tau}_i=0. \tag{12-14}$$

Using this constraint, we obtain as the solution to the normal equations

$$\begin{aligned}
\hat{\mu}&=\bar{y}_{..},\\
\hat{\tau}_i&=\bar{y}_{i\cdot}-\bar{y}_{..}, \qquad i=1,2,\ldots,a.
\end{aligned} \tag{12-15}$$

This solution has considerable intuitive appeal, since the overall mean is estimated by the grand average of the observations and the estimate of any treatment effect is just the difference between the treatment average and the grand average.

This solution is obviously not unique because it depends on the constraint (equation 12-14) that we have chosen. At first this may seem unfortunate, because two different experimenters could analyze the same data and obtain different results if they apply different constraints. However, certain *functions* of the model parameter are estimated uniquely, regardless of the constraint. Some examples are $\tau_i - \tau_j$, which would be estimated by $\hat{\tau}_i - \hat{\tau}_j = \bar{y}_{i.} - \bar{y}_{j.}$, and $\mu + \tau_i$, which would be estimated by $\hat{\mu} + \hat{\tau}_i = \bar{y}_{i.}$. Since we are usually interested in differences in the treatment effects rather than their actual values, it causes no concern that the τ_i cannot be estimated uniquely. In general, any function of the model parameters that is a linear combination of the left-hand side of the normal equations can be estimated uniquely. Functions that are uniquely estimated, regardless of which constraint is used, are called *estimable* functions.

Frequently, we would like to construct a confidence interval for the ith treatment mean. The mean of the ith treatment is

$$\mu_i = \mu + \tau_i, \qquad i = 1, 2, \ldots, a.$$

A point estimator of μ_i would be $\hat{\mu}_i = \hat{\mu} + \hat{\tau}_i = \bar{y}_{i.}$ Now, if we assume that the errors are normally distributed, each $\bar{y}_i.$ is NID(μ_i, σ^2/n). Thus, if σ^2 were known, we could use the normal distribution to construct a confidence interval for μ_i. Using MS_E as an estimator of σ^2, we can base the confidence interval on the t distribution. Therefore, a 100(1 – α)% confidence interval on the ith treatment mean μ_i is

$$\left[\bar{y}_i.\pm t_{\alpha/2,N-a}\sqrt{MS_E/n}\right]. \tag{12-16}$$

A 100(1 – α)% confidence interval on the difference between any two treatment means, say $\mu_i - \mu_j$, is

$$\left[\bar{y}_i.-\bar{y}_j.\pm t_{\alpha/2,N-a}\sqrt{2\,MS_E/n}\right]. \tag{12-17}$$

Example 12-2

We can use the results given previously to estimate the mean tensile strengths at different levels of hardwood concentration for the experiment in Section 12-1.1. The mean tensile strength estimates are

$$\bar{y}_1. = \hat{\mu}_{5\%} = 10.00 \text{ psi},$$

$$\bar{y}_2. = \hat{\mu}_{10\%} = 15.67 \text{ psi},$$

$$\bar{y}_3. = \hat{\mu}_{15\%} = 17.00 \text{ psi},$$

$$\bar{y}_4. = \hat{\mu}_{20\%} = 21.17 \text{ psi}.$$

A 95% confidence interval on the mean tensile strength at 20% hardwood is found from equation 12-16 as follows:

$$\left[\bar{y}_i.\pm t_{\alpha/2,N-a}\sqrt{MS_E/n}\right],$$

$$\left[21.17\pm(2.086)\sqrt{6.51/6}\right],$$

$$\left[21.17\pm2.17\right].$$

The desired confidence interval is

$$19.00 \text{ psi} \leq \mu_{20\%} \leq 23.34 \text{ psi}.$$

Visual examination of the data suggests that mean tensile strength at 10% and 15% hardwood is similar. A confidence interval on the difference in means $\mu_{15\%} - \mu_{10\%}$ is

$$\left[\bar{y}_i. - \bar{y}_j \pm t_{\alpha/2,N-a}\sqrt{2MS_E/n}\right],$$

$$\left[17.00 - 15.67 \pm (2.086)\sqrt{2(6.51)/6}\right],$$

$$[1.33 \pm 3.07].$$

Thus, the confidence interval on $\mu_{15\%} - \mu_{10\%}$ is

$$-1.74 \leq \mu_{15\%} - \mu_{10\%} \leq 4.40.$$

Since the confidence interval includes zero, we would conclude that there is no difference in mean tensile strength at these two particular hardwood levels.

12-1.4 Residual Analysis and Model Checking

The one-way model analysis of variance assumes that the observations are normally and independently distributed, with the same variance in each treatment or factor level. These assumptions should be checked by examining the *residuals*. We define a residual as $e_{ij} = y_{ij} - \bar{y}_{i.}$, that is, the difference between an observation and the corresponding treatment mean. The residuals for the hardwood percentage experiment are shown in Table 12-5.

The normality assumption can be checked by plotting the residuals on normal probability paper. To check the assumption of equal variances at each factor level, plot the residuals against the factor levels and compare the spread in the residuals. It is also useful to plot the residuals against $\bar{y}_{i.}$ (sometimes called the *fitted value*); the variability in the residuals should not depend in any way on the value of $\bar{y}_{i.}$. When a pattern appears in these plots, it usually suggests the need for *transformation*, that is, analyzing the data in a different metric. For example, if the variability in the residuals increases with $\bar{y}_i.$, then a transformation such as $\log y$ or $\sqrt{y}$ should be considered. In some problems the dependency of residual scatter in $\bar{y}_i.$ is very important information. It may be desirable to select the factor level that results in maximum y; however, this level may also cause more variation in y from run to run.

The independence assumption can be checked by plotting the residuals against the time or run order in which the experiment was performed. A pattern in this plot, such as sequences of positive and negative residuals, may indicate that the observations are not independent. This suggests that time or run order is important, or that variables that change over time are important and have not been included in the experimental design.

A normal probability plot of the residuals from the hardwood concentration experiment is shown in Fig. 12-2. Figures 12-3 and 12-4 present the residuals plotted against the treat-

Table 12-5 Residuals for the Tensile Strength Experiment

Hardwood Concentration	Residuals					
5%	−3.00	−2.00	5.00	1.00	−1.00	0.00
10%	−3.67	1.33	−2.67	2.33	3.33	−0.67
15%	−3.00	1.00	2.00	0.00	−1.00	1.00
20%	−2.17	3.83	0.83	1.83	−3.17	−1.17

ment number and the fitted value $\bar{y}_{i\cdot}$. These plots do not reveal any model inadequacy or unusual problem with the assumptions.

12-1.5 An Unbalanced Design

In some single-factor experiments the number of observations taken under each treatment may be different. We then say that the design is *unbalanced.* The analysis of variance described earlier is still valid, but slight modifications must be made in the sums of squares formulas. Let n_i observations be taken under treatment $i\,(i = 1, 2,\ldots, a)$; and let the total

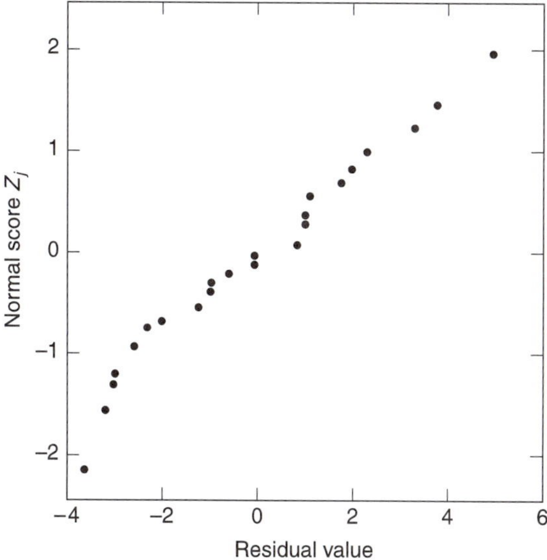

Figure 12-2 Normal probability plot of residuals from the hardwood concentration experiment.

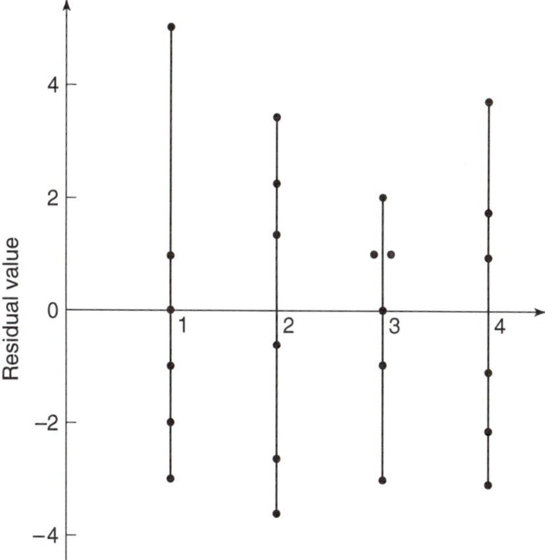

Figure 12-3 Plot of residuals vs. treatment.

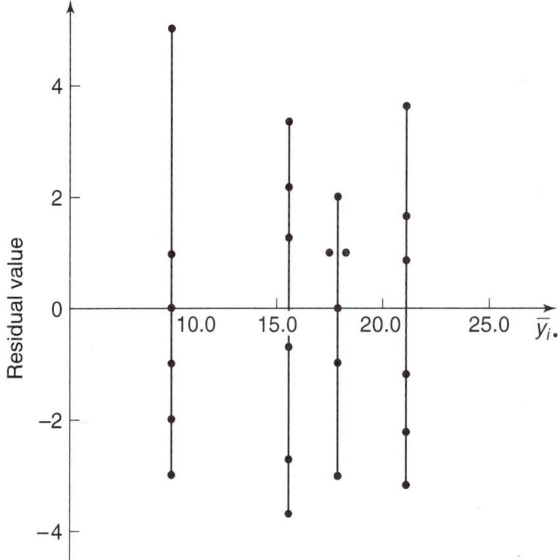

Figure 12-4 Plot of residuals vs. $\bar{y}_{i.}$.

number of observations $N = \sum_{i=1}^{a} n_i$. The computational formulas for SS_T and $SS_{\text{treatments}}$ become

$$SS_T = \sum_{i=1}^{a}\sum_{j=1}^{n_i} y_{ij}^2 - \frac{y_{..}^2}{N}$$

and

$$SS_{\text{treatments}} = \sum_{i=1}^{a} \frac{y_{i.}^2}{n_i} - \frac{y_{..}^2}{N}.$$

In solving the normal equations, the constraint $\sum_{i=1}^{a} n_i \hat{\tau}_i = 0$ is used. No other changes are required in the analysis of variance.

There are two important advantages in choosing a balanced design. First, the test statistic is relatively insensitive to small departures from the assumption of equality of variances if the sample sizes are equal. This is not the case for unequal sample sizes. Second, the power of the test is maximized if the samples are of equal size.

12-2 TESTS ON INDIVIDUAL TREATMENT MEANS

12-2.1 Orthogonal Contrasts

Rejecting the null hypothesis in the fixed-effects-model analysis of variance implies that there are differences between the a treatment means, but the exact nature of the differences is not specified. In this situation, further comparisons between groups of treatment means may be useful. The ith treatment mean is defined as $\mu_i = \mu + \tau_i$, and μ_i is estimated by $\bar{y}_{i}.$. Comparisons between treatment means are usually made in terms of the treatment totals $\{y_{i\cdot}\}$.

Consider the hardwood concentration experiment presented in Section 12-1.1. Since the hypothesis H_0: $\tau_i = 0$ was rejected, we know that some hardwood concentrations produce tensile strengths different from others, but which ones actually cause this difference?

We might suspect at the outset of the experiment that hardwood concentrations 3 and 4 produce the same tensile strength, implying that we would like to test the hypothesis

$$H_0: \mu_3 = \mu_4,$$
$$H_1: \mu_3 \neq \mu_4.$$

This hypothesis could be tested by using a linear combination of treatment totals, say

$$y_{3\cdot} - y_{4\cdot} = 0.$$

If we had suspected that the *average* of hardwood concentrations 1 and 3 did not differ from the *average* of hardwood concentrations 2 and 4, then the hypothesis would have been

$$H_0: \mu_1 + \mu_3 = \mu_2 + \mu_4,$$
$$H_1: \mu_1 + \mu_3 \neq \mu_2 + \mu_4,$$

which implies that the linear combination of treatment totals

$$y_{1\cdot} + y_{3\cdot} - y_{2\cdot} - y_{4\cdot} = 0.$$

In general, the comparison of treatment means of interest will imply a linear combination of treatment totals such as

$$C = \sum_{i=1}^{a} c_i y_{i\,.},$$

with the restriction that $\sum_{i=1}^{a} c_i = 0$. These linear combinations are called *contrasts*. The sum of squares for any contrast is

$$SS_C = \frac{\left(\sum_{i=1}^{a} c_i y_{i\,.}\right)^2}{n\sum_{i=1}^{a} c_i^2} \tag{12-18}$$

and has a single degree of freedom. If the design is unbalanced, then the comparison of treatment means requires that $\sum_{i=1}^{a} n_i c_i = 0$, and equation 12-18 becomes

$$SS_C = \frac{\left(\sum_{i=1}^{a} c_i y_{i\,.}\right)^2}{\sum_{i=1}^{a} n_i c_i^2}. \tag{12-19}$$

A contrast is tested by comparing its sum of squares to the mean square error. The resulting statistic would be distributed as F, with 1 and $N - a$ degrees of freedom.

A very important special case of the above procedure is that of *orthogonal contrasts.* Two contrasts with coefficients $\{c_i\}$ and $\{d_i\}$ are orthogonal if

$$\sum_{i=1}^{a} c_i d_i = 0$$

or, for an unbalanced design, if

$$\sum_{i=1}^{a} n_i c_i d_i = 0.$$

For a treatments a set of $a - 1$ orthogonal contrasts will partition the sum of squares due to treatments into $a - 1$ independent single-degree-of-freedom components. Thus, tests performed on orthogonal contrasts are independent.

There are many ways to choose the orthogonal contrast coefficients for a set of treatments. Usually, something in the nature of the experiment should suggest which comparisons will be of interest. For example, if there are $a = 3$ treatments, with treatment 1 a "control" and treatments 2 and 3 actual levels of the factor of interest to the experimenter, then appropriate orthogonal contrasts might be as follows:

Treatment	Orthogonal Contrasts	
1 (control)	–2	0
2 (level 1)	1	–1
3 (level 2)	1	1

Note that contrast 1 with $c_i = -2, 1, 1$ compares the average effect of the factor with the control while contrast 2 with $d_i = 0, -1, 1$ compares the two levels of the factor of interest.

Contrast coefficients must be chosen prior to running the experiment, for if these comparisons are selected after examining the data, most experimenters would construct tests that compare large observed differences in means. These large differences could be due to the presence of real effects or they could be due to random error. If experimenters always pick the largest differences to compare, they will inflate the type I error of the test, since it is likely that in an unusually high percentage of the comparisons selected the observed differences will be due to error.

Example 12-3

Consider the hardwood concentration experiment. There are four levels of hardwood concentration, and the possible sets of comparisons between these means and the associated orthogonal comparisons are

$$H_0: \mu_1 + \mu_4 = \mu_2 + \mu_3, \qquad C_1 = y_{1\cdot} - y_{2\cdot} - y_{3\cdot} + y_{4\cdot},$$

$$H_0: 3\mu_1 + \mu_2 = \mu_3 + 3\mu_4, \qquad C_2 = -3y_{1\cdot} - y_{2\cdot} + y_{3\cdot} + 3y_{4\cdot},$$

$$H_0: \mu_1 + 3\mu_3 = 3\mu_2 + \mu_4, \qquad C_3 = -y_{1\cdot} + 3y_{2\cdot} - 3y_{3\cdot} + y_{4\cdot}.$$

Notice that the contrast constants are orthogonal. Using the data from Table 12-1, we find the numerical values of the contrasts and the sums of squares as follows:

$$C_1 = 60 - 94 - 102 + 127 = -9, \qquad SS_{C_1} = \frac{(-9)^2}{6(4)} = 3.38,$$

$$C_2 = -3(60) - 94 + 102 + 3(127) = 209, \qquad SS_{C_2} = \frac{(209)^2}{6(20)} = 364.00,$$

$$C_3 = -60 + 3(94) - 3(102) + 127 = 43, \qquad SS_{C_3} = \frac{(43)^2}{6(20)} = 15.41.$$

These contrast sums of squares completely partition the treatment sum of squares; that is, $SS_{\text{treatments}} = SS_{C_1} + SS_{C_2} + SS_{C_3} = 382.79$. These tests on the contrasts are usually incorporated into the analysis of variance, such as shown in Table 12-6. From this analysis, we conclude that there are significant differences between hardwood concentrations 1, 2 vs. 3, 4, but that the average of 1 and 4 does not differ from the average of 2 and 3, nor does the average of 1 and 3 differ from the average of 2 and 4.

Table 12-6 Analysis of Variance for the Tensile Strength Data

Source of Variation	Sum of Squares	Degrees of Freedom	Mean Square	F_0
Hardwood concentration	382.79	3	127.60	19.61
C_1 (1, 4 vs. 2, 3)	(3.38)	(1)	3.38	0.52
C_2 (1, 2 vs. 3, 4)	(364.00)	(1)	364.00	55.91
C_3 (1, 3 vs. 2, 4)	(15.41)	(1)	15.41	2.37
Error	130.17	20	6.51	
Total	512.96	23		

12-2.2 Tukey's Test

Frequently, analysts do not know in advance how to construct appropriate orthogonal contrasts, or they may wish to test more than $a - 1$ comparisons using the same data. For example, analysts may want to test all possible pairs of means. The null hypotheses would then be H_0: $\mu_i = \mu_j$ for all $i \neq j$. If we test all possible pairs of means using t-tests, the probability of committing a type I error for the entire set of comparisons can be greatly increased. There are several procedures available that avoid this problem. Among the more popular of these procedures are the Newman–Keuls test [Newman (1939); Keuls (1952)], Duncan's multiple range test [Duncan (1955)], and Tukey's test [Tukey (1953)]. Here we describe Tukey's test.

Tukey's procedure makes use of another distribution, called the *Studentized range distribution.* The Studentized range statistic is

$$q = \frac{\bar{y}_{\max} - \bar{y}_{\min}}{\sqrt{MS_E/n}},$$

where $\bar{y}_{\max}$ is the largest sample mean and $\bar{y}_{\min}$ is the smallest sample mean out of p sample means. Let $q_\alpha(a, f)$ represent the upper α percentage point of q, where a is the number of treatments and f is the number of degrees of freedom for error. Two means, $\bar{y}_{i.}$ and $\bar{y}_{j.}$ $(i \neq j)$, are considered significantly different if

$$|\bar{y}_{i.} - \bar{y}_{j.}| > T_\alpha$$

where

$$T_\alpha = q_\alpha(a, f)\sqrt{\frac{MS_E}{n}}. \tag{12-20}$$

Table XII (Appendix) contains values of $q_\alpha(a, f)$ for $\alpha = 0.05$ and 0.01 and a selection of values for a and f. Tukey's procedure has the property that the overall significance level is exactly α for equal sample sizes and at most α for unequal sample sizes.

Example 12-4

We will apply Tukey's test to the hardwood concentration experiment. Recall that there are $a = 4$ means, $n = 6$, and $MS_E = 6.51$. The treatment means are

$$\bar{y}_{1\cdot} = 10.00 \text{ psi}, \quad \bar{y}_{2\cdot} = 15.67 \text{ psi}, \quad \bar{y}_{3\cdot} = 17.00 \text{ psi}, \quad \bar{y}_{4\cdot} = 21.17 \text{ psi}.$$

From Table XII (Appendix), with $\alpha = 0.05$, $a = 4$, and $f = 20$, we find $q_{0.05}(4, 20) = 3.96$.

Using Equation 12-20,

$$T_\alpha = q_{0.05}(4,20)\sqrt{\frac{MS_E}{n}} = 3.96\sqrt{\frac{6.51}{6}} = 4.12.$$

Therefore, we would conclude that two means are significantly different if

$$|\bar{y}_{i\cdot} - \bar{y}_{j\cdot}| > 4.12.$$

The differences in treatment averages are

$$|\bar{y}_{1\cdot} - \bar{y}_{2\cdot}| = |10.00 - 15.67| = 5.67,$$

$$|\bar{y}_{1\cdot} - \bar{y}_{3\cdot}| = |10.00 - 17.00| = 7.00,$$

$$|\bar{y}_{1\cdot} - \bar{y}_{4\cdot}| = |10.00 - 21.17| = 11.17,$$

$$|\bar{y}_{2\cdot} - \bar{y}_{3\cdot}| = |15.67 - 17.00| = 1.33,$$

$$|\bar{y}_{2\cdot} - \bar{y}_{4\cdot}| = |15.67 - 21.17| = 5.50,$$

$$|\bar{y}_{3\cdot} - \bar{y}_{4\cdot}| = |17.00 - 21.17| = 4.17.$$

From this analysis, we see significant differences between all pairs of means except 2 and 3. It may be of use to draw a graph of the treatment means, such as Fig. 12-5, with the means that are *not* different underlined.

Simultaneous confidence intervals can also be constructed on the differences in pairs of means using the Tukey approach. It can be shown that

$$P\left[\left(\bar{y}_{i.} - \bar{y}_{j.}\right) - q_\alpha(a,f)\sqrt{\frac{MS_E}{n}} \le \mu_i - \mu_j \le \left(\bar{y}_{i.} - \bar{y}_{j.}\right) + q_\alpha(a,f)\sqrt{\frac{MS_E}{n}}\right] = 1 - \alpha$$

when sample sizes are equal. This expression represents a 100(1 – α)% *simultaneous* confidence interval on all pairs of means $\mu_i - \mu_j$.

If the sample sizes are unequal, the 100(1 – α)% simultaneous confidence interval on all pairs of means $\mu_i - \mu_j$ is given by

$$P\left[\left(\bar{y}_{i\cdot} - \bar{y}_{j\cdot}\right) - \frac{q_\alpha(a,f)}{\sqrt{2}}\sqrt{MS_E\left(\frac{1}{n_i} + \frac{1}{n_j}\right)} \le \mu_i - \mu_j \le \left(\bar{y}_{i\cdot} - \bar{y}_{j\cdot}\right) + \frac{q_\alpha(a,f)}{\sqrt{2}}\sqrt{MS_E\left(\frac{1}{n_i} + \frac{1}{n_j}\right)}\right] = 1 - \alpha.$$

Interpretation of the confidence intervals is straightforward. If zero is contained in the interval, then there is no significant difference between the two means at the α significance level.

It should be noted that the significance level, α, in Tukey's multiple comparison procedure represents an *experimental error rate.* With respect to confidence intervals, α represents the probability that one or more of the confidence intervals on the pairwise differences will *not* contain the true difference for equal sample sizes (when sample sizes are unequal, this probability becomes at most α).

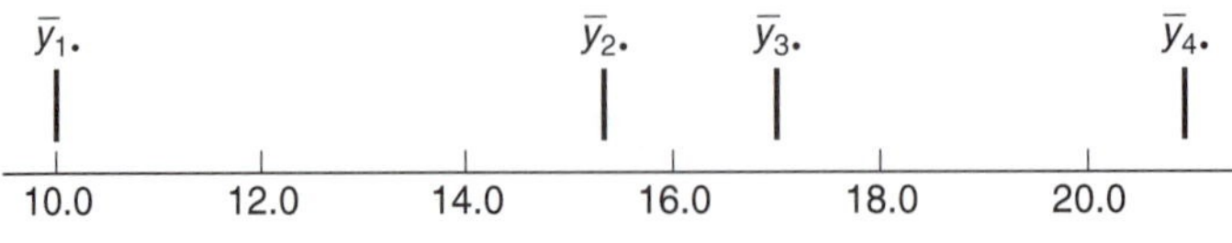

Figure 12-5 Results of Tukey's test.

12-3 THE RANDOM-EFFECTS MODEL

In many situations, the factor of interest has a large number of possible levels. The analyst is interested in drawing conclusions about the entire *population* of factor levels. If the experimenter randomly selects a of these levels from the population of factor levels, then we say that the factor is a *random* factor. Because the levels of the factor actually used in the experiment were chosen randomly, the conclusions reached will be valid about the entire population of factor levels. We will assume that the population of factor levels is either of infinite size or is large enough to be considered infinite.

The linear statistical model is

$$y_{ij} = \mu + \tau_i + \epsilon_{ij} \begin{cases} i = 1, 2, \ldots, a, \\ j = 1, 2, \ldots, n, \end{cases} \tag{12-21}$$

where τ_i and ϵ_{ij} are independent random variables. Note that the model is identical in structure to the fixed-effects case, but the parameters have a different interpretation. If the variance of τ_i is σ_τ^2, then the variance of any observation is

$$V(y_{ij}) = \sigma_\tau^2 + \sigma^2.$$

The variances σ_τ^2 and σ^2 are called *variance components*, and the model, equation 12-21, is called the *components-of-variance* or the *random-effects* model. To test hypotheses using this model, we require that the $\{\epsilon_{ij}\}$ are NID(0, σ^2), that $\{\tau_i\}$ are NID(0, σ_τ^2), and that τ_i and ϵ_{ij} are independent.The assumption that the $\{\tau_i\}$ are independent random variables implies that the usual assumption of $\sum_{i=1}^{a} \tau_i = 0$ from the fixed-effects model does not apply to the random-effects model.

The sum of squares identity

$$SS_T = SS_{\text{treatments}} + SS_E \tag{12-22}$$

still holds. That is, we partition the total variability in the observations into a component that measures variation between treatments ($SS_{\text{treatments}}$) and a component that measures variation within treatments (SS_E). However, instead of testing hypotheses about individual treatment effects, we test the hypotheses

$$H_0\colon \sigma_\tau^2 = 0,$$

$$H_1\colon \sigma_\tau^2 > 0.$$

If $\sigma_\tau^2 = 0$, all treatments are identical, but if $\sigma_\tau^2 > 0$, then there is variability between treatments. The quantity SS_E/σ^2 is distributed as chi-square with $N - a$ degrees of freedom, and under the null hypothesis, $SS_{\text{treatments}}/\sigma^2$ is distributed as chi-square with $a - 1$ degrees of freedom. Further, the random variables are independent of each other. Thus, under the null hypothesis, the ratio

$$F_0 = \frac{SS_{\text{treatments}}/(a-1)}{SS_E/(N-a)} = \frac{MS_{\text{treatments}}}{MS_E} \tag{12-23}$$

is distributed as F with $a - 1$ and $N - a$ degrees of freedom. By examining the expected mean squares we can determine the critical region for this statistic.

Consider

$$E(MS_{\text{treatments}}) = \frac{1}{a-1}E(SS_{\text{treatments}}) = \frac{1}{a-1}E\left[\sum_{i=1}^{a}\frac{y_{i.}^2}{n} - \frac{y_{..}^2}{N}\right]$$

$$= \frac{1}{a-1}E\left[\frac{1}{n}\sum_{i=1}^{a}\left(\sum_{j=1}^{n}(\mu+\tau_i+\epsilon_{ij})\right)^2 - \frac{1}{N}\left(\sum_{i=1}^{a}\sum_{j=1}^{n}(\mu+\tau_i+\epsilon_{ij})\right)^2\right].$$

If we square and take the expectation of the quantities in brackets, we see that terms involving τ_i^2 are replaced by σ_τ^2, as $E(\tau_i) = 0$. Also, terms involving $\sum_{j=1}^{n}\epsilon_{ij}^2$, $\sum_{i=1}^{a}\sum_{j=1}^{n}\epsilon_{ij}^2$, and $\sum_{i=1}^{a}\sum_{j=1}^{n}\tau_i^2$ are replaced by $n\sigma^2$, $an\sigma^2$, and $an\ \sigma_\tau^2$, respectively. Finally, all cross-product terms involving τ_i and ϵ_{ij} have zero expectation. This leads to

$$E(MS_{\text{treatments}}) = \frac{1}{a-1}\left[N\mu^2 + N\sigma_\tau^2 + a\sigma^2 - N\mu^2 - n\sigma_\tau^2 - \sigma^2\right]$$

or

$$E(MS_{\text{treatments}}) = \sigma^2 + n\sigma_\tau^2. \tag{12-24}$$

A similar approach will show that

$$E(MS_E) = \sigma^2. \tag{12-25}$$

From the expected mean squares, we see that if H_0 is true, both the numerator and the denominator of the test statistic, equation 12-23, are unbiased estimators of σ^2; whereas if H_1 is true, the expected value of the numerator is greater than the expected value of the denominator. Therefore, we should reject H_0 for values of F_0 that are too large. This implies an upper-tail, one-tail critical region, so we reject H_0 if $F_0 > F_{\alpha, a-1, N-a}$.

The computational procedure and analysis-of-variance table for the random-effects model are identical to the fixed-effects case. The conclusions, however, are quite different because they apply to the entire population of treatments.

We usually need to estimate the variance components (σ^2 and σ_τ^2) in the model. The procedure used to estimate σ^2 and σ_τ^2 is called the "analysis-of-variance method," because it uses the lines in the analysis-of-variance table. It does not require the normality assumption on the observations. The procedure consists of equating the expected mean squares to their observed values in the analysis-of-variance table and solving for the variance components. When equating observed and expected mean squares in the one-way-classification random-effects model, we obtain

$$MS_{\text{treatments}} = \sigma^2 + n\sigma_\tau^2$$

and

$$MS_E = \sigma^2.$$

Therefore, the estimators of the variance components are

$$\hat{\sigma}^2 = MS_E \tag{12-26}$$

and

$$\hat{\sigma}_\tau^2 = \frac{MS_{\text{treatments}} - MS_E}{n}. \tag{12-27}$$

For unequal sample sizes, replace n in equation 12-27 with

$$n_0 = \frac{1}{a-1}\left[\sum_{i=1}^{a} n_i - \frac{\sum_{i=1}^{a} n_i^2}{\sum_{i=1}^{a} n_i}\right].$$

Sometimes the analysis-of-variance method produces a negative estimate of a variance component. Since variance components are by definition nonnegative, a negative estimate of a variance component is unsettling. One course of action is to accept the estimate and use it as evidence that the true value of the variance component is zero, assuming that sampling variation led to the negative estimate. While this has intuitive appeal, it will disturb the statistical properties of other estimates. Another alternative is to reestimate the negative variance component with a method that always yields nonnegative estimates. Still another possibility is to consider the negative estimate as evidence that the assumed linear model is incorrect, requiring that a study of the model and its assumptions be made to find a more appropriate model.

Example 12-5

In his book *Design and Analysis of Experiments* (2001), D. C. Montgomery describes a single-factor experiment involving the random-effects model. A textile manufacturing company weaves a fabric on a large number of looms. The company is interested in loom-to-loom variability in tensile strength. To investigate this, a manufacturing engineer selects four looms at random and makes four strength determinations on fabric samples chosen at random for each loom. The data are shown in Table 12-7, and the analysis of variance is summarized in Table 12-8.

From the analysis of variance, we conclude that the looms in the plant differ significantly in their ability to produce fabric of uniform strength. The variance components are estimated by $\hat{\sigma}^2 = 1.90$ and

$$\hat{\sigma}_\tau^2 = \frac{29.73 - 1.90}{4} = 6.96.$$

Therefore, the variance of strength in the *manufacturing process* is estimated by

$$\begin{aligned}\widehat{V(y_{ij})} &= \hat{\sigma}_\tau^2 + \hat{\sigma}^2 \\ &= 6.96 + 1.90 \\ &= 8.86.\end{aligned}$$

Most of this variability is attributable to differences *between* looms.

Table 12-7 Strength Data for Example 12-5

Loom	Observations				Totals	Averages
	1	2	3	4		
1	98	97	99	96	390	97.5
2	91	90	93	92	366	91.5
3	96	95	97	95	383	95.8
4	95	96	99	98	388	97.0
					1527	95.4

Table 12-8 Analysis of Variance for the Strength Data

Source of Variation	Sum of Squares	Degrees of Freedom	Mean Square	F_0
Looms	89.19	3	29.73	15.68
Error	22.75	12	1.90	
Total	111.94	15		

This example illustrates an important application of analysis of variance—the isolation of different sources of variability in a manufacturing process. Problems of excessive variability in critical functional parameters or properties frequently arise in quality-improvement programs. For example, in the previous fabric-strength example, the process mean is estimated by $\bar{y}_{..} = 95.45$ psi and the process standard deviation is estimated by $\hat{\sigma}_y = \sqrt{\widehat{V(y_{ij})}} = \sqrt{8.86} = 2.98$ psi. If strength is approximately normally distributed, this would imply a distribution of strength in the outgoing product that looks like the normal distribution shown in Fig. 12-6*a*. If the lower specification limit (LSL) on strength is at 90 psi, then a substantial proportion of the process defective is *fallout*; that is, scrap or defective material that must be sold as second quality, and so on. This fallout is directly related to the excess variability resulting from *differences between looms.* Variability in loom performance could be caused by faulty setup, poor maintenance, inadequate supervision, poorly trained operators, and so forth. The engineer or manager responsible for quality improvement must identify and remove these sources of variability from the process. If he can do this, then strength variability will be greatly reduced, perhaps as low as $\hat{\sigma}_y = \hat{\sigma} = \sqrt{1.90} = 1.38$ psi, as shown in Fig. 12-6*b*. In this improved process, reducing the variability in strength has greatly reduced the fallout. This will result in lower cost, higher quality, a more satisfied customer, and enhanced competitive position for the company.

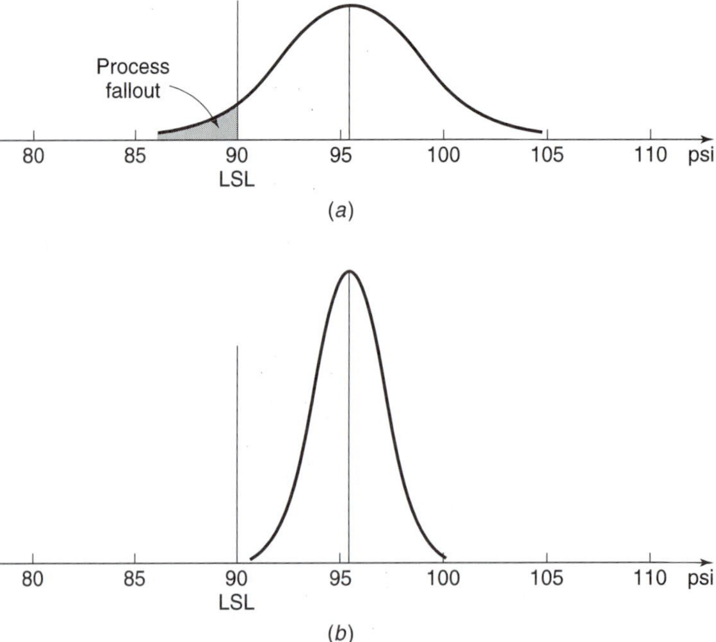

Figure 12-6 The distribution of fabric strength. (*a*) Current process, (*b*) Improved process.

12-4 THE RANDOMIZED BLOCK DESIGN

12-4.1 Design and Statistical Analysis

In many experimental problems it is necessary to design the experiment so that variability arising from nuisance variables can be controlled. As an example, recall the situation in Example 11-17, where two different procedures were used to predict the shear strength of steel plate girders. Because each girder has potentially different strength, and because this variability in strength was not of direct interest, we designed the experiment using the two methods on each girder and compared the difference in average strength readings to zero using the paired t-test. The paired t-test is a procedure for comparing two means when all experimental runs cannot be made under homogeneous conditions. Thus, the paired t-test reduces the noise in the experiment by blocking out a nuisance variable effect. The randomized block design is an extension of the paired t-test that is used in situations where the factor of interest has more than two levels.

As an example, suppose that we wish to compare the effect of four different chemicals on the strength of a particular fabric. It is known that the effect of these chemicals varies considerably from one fabric specimen to another. In this example, we have only one factor: chemical type. Therefore, we could select several pieces of fabric and compare all four chemicals within the relatively homogeneous conditions provided by each piece of fabric. This would remove any variation due to the fabric.

The general procedure for a randomized complete block design consists of selecting b blocks and running a complete replicate of the experiment in each block. A randomized complete block design for investigating a single factor with a levels would appear as in Fig. 12-7. There will be a observations (one per factor level) in each block, and the order in which these observations are run is randomly assigned within the block.

We will now describe the statistical analysis for a randomized block design. Suppose that a single factor with a levels is of interest, and the experiment is run in b blocks, as shown in Fig. 12-7. The observations may be represented by the linear statistical model,

$$y_{ij} = \mu + \tau_i + \beta_j + \epsilon_{ij} \begin{cases} i = 1, 2, \ldots, a, \\ j = 1, 2, \ldots, b, \end{cases} \tag{12-28}$$

where μ is an overall mean, τ_i is the effect of the ith treatment, β_j is the effect of the jth block, and ϵ_{ij} is the usual NID(0, σ^2) random error term. Treatments and blocks will be considered initially as fixed factors. Furthermore, the treatment and block effects are defined as deviations from the overall mean, so that $\sum_{i=1}^{a} \tau_i = 0$ and $\sum_{j=1}^{b} \beta_j = 0$. We are interested in testing the equality of the treatment effects. That is,

$$H_0: \tau_1 = \tau_2 = \cdots = \tau_a = 0,$$

$$H_1: \tau_i \neq 0 \text{ for at least one } i.$$

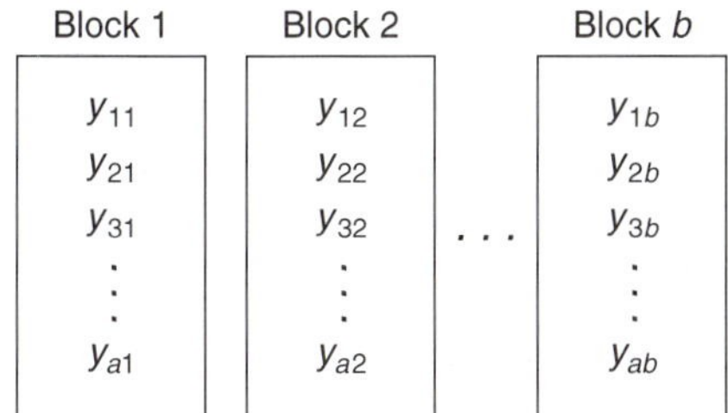

Figure 12-7 The randomized complete block design.

Let $y_{i.}$ be the total of all observations taken under treatment i, let $y._j$ be the total of all observations in block j, let $y..$ be the grand total of all observations, and let $N = ab$ be the total number of observations. Similarly, $\bar{y}_i.$ is the average of the observations taken under treatment i, $\bar{y}._j$ is the average of the observations in block j, and $\bar{y}..$ is the grand average of all observations. The total corrected sum of squares is

$$\sum_{i=1}^{a}\sum_{j=1}^{b}\left(y_{ij}-\bar{y}..\right)^2=\sum_{i=1}^{a}\sum_{j=1}^{b}\left[\left(\bar{y}_i.-\bar{y}..\right)+\left(\bar{y}._j-\bar{y}..\right)+\left(y_{ij}-\bar{y}_i.-\bar{y}._j+\bar{y}..\right)\right]^2. \quad (12\text{-}29)$$

Expanding the right-hand side of equation 12-29 and applying algebraic elbow grease yields

$$\begin{aligned}\sum_{i=1}^{a}\sum_{j=1}^{b}\left(y_{ij}-\bar{y}..\right)^2 &= b\sum_{i=1}^{a}\left(\bar{y}_i.-\bar{y}..\right)^2+a\sum_{j=1}^{b}\left(\bar{y}._j-\bar{y}..\right)^2 \\ &\quad +\sum_{i=1}^{a}\sum_{j=1}^{b}\left(y_{ij}-\bar{y}_i.-\bar{y}._j+\bar{y}..\right)^2\end{aligned} \quad (12\text{-}30)$$

or, symbolically,

$$SS_T = SS_{\text{treatments}} + SS_{\text{blocks}} + SS_E. \quad (12\text{-}31)$$

The degrees-of-freedom breakdown corresponding to equation 12-31 is

$$ab - 1 = (a - 1) + (b - 1) + (a - 1)(b - 1). \quad (12\text{-}32)$$

The null hypothesis of no treatment effects (H_0: $\tau_i = 0$) is tested by the F ratio, $MS_{\text{treatments}}/MS_E$. The analysis of variance is summarized in Table 12-9. Computing formulas for the sums of squares are also shown in this table. The same test procedure is used in cases where treatments and/or blocks are random.

Example 12-6

An experiment was performed to determine the effect of four different chemicals on the strength of a fabric. These chemicals are used as part of the permanent-press finishing process. Five fabric samples were selected, and a randomized block design was run by testing each chemical type once in random order on each fabric sample. The data are shown in Table 12-10.
The sums of squares for the analysis of variance are computed as follows:

Table 12-9 Analysis of Variance for Randomized Complete Block Design

Source of Variation	Sum of Squares	Degrees of Freedom	Mean Square	F_0
Treatments	$\sum_{i=1}^{a}\frac{y_{i.}^2}{b}-\frac{y_{..}^2}{ab}$	$a-1$	$\frac{SS_{\text{treatments}}}{a-1}$	$\frac{MS_{\text{treatments}}}{MS_E}$
Blocks	$\sum_{j=1}^{b}\frac{y_{.j}^2}{a}-\frac{y_{..}^2}{ab}$	$b-1$	$\frac{SS_{\text{blocks}}}{b-1}$	
Error	SS_E (by subtraction)	$(a-1)(b-1)$	$\frac{SS_E}{(a-1)(b-1)}$	
Total	$\sum_{i=1}^{a}\sum_{j=1}^{b}y_{ij}^2-\frac{y_{..}^2}{ab}$	$ab-1$		

Table 12-10 Fabric Strength Data — Randomized Block Design

Chemical Type	Fabric Sample 1	2	3	4	5	Row Totals, $y_{i\cdot}$	Row Averages, $\bar{y}_{i\cdot}$
1	1.3	1.6	0.5	1.2	1.1	5.7	1.14
2	2.2	2.4	0.4	2.0	1.8	8.8	1.76
3	1.8	1.7	0.6	1.5	1.3	6.9	1.38
4	3.9	4.4	2.0	4.1	3.4	17.8	3.56
Column totals, $y_{\cdot j}$	9.2	10.1	3.5	8.8	7.6	39.2	1.96
Column averages, $\bar{y}_{\cdot j}$	2.30	2.53	0.88	2.20	1.90	$(y_{\cdot\cdot})$	$(\bar{y}_{\cdot\cdot})$

$$
\begin{aligned}
SS_T &= \sum_{i=1}^{4}\sum_{j=1}^{5} y_{ij}^2 - \frac{y_{\cdot\cdot}^2}{ab} \\
&= (1.3)^2 + (1.6)^2 + \cdots + (3.4)^2 - \frac{(39.2)^2}{20} = 25.69, \\
SS_{\text{treatments}} &= \sum_{i=1}^{4} \frac{y_{i\cdot}^2}{b} - \frac{y_{\cdot\cdot}^2}{ab} \\
&= \frac{(5.7)^2 + (8.8)^2 + (6.9)^2 + (17.8)^2}{5} - \frac{(39.2)^2}{20} = 18.04, \\
SS_{\text{blocks}} &= \sum_{j=1}^{5} \frac{y_{\cdot j}^2}{a} - \frac{y_{\cdot\cdot}^2}{ab} \\
&= \frac{(9.2)^2 + (10.1)^2 + (3.5)^2 + (8.8)^2 + (7.6)^2}{4} - \frac{(39.2)^2}{20} = 6.69, \\
SS_E &= SS_T - SS_{\text{blocks}} - SS_{\text{treatments}} \\
&= 25.69 - 6.69 - 18.04 = 0.96.
\end{aligned}
$$

The analysis of variance is summarized in Table 12-11. We would conclude that there is a significant difference in the chemical types as far as their effect on fabric strength is concerned.

Table 12-11 Analysis of Variance for the Randomized Block Experiment

Source of Variation	Sum of Squares	Degrees of Freedom	Mean Square	F_0
Chemical type (treatments)	18.04	3	6.01	75.13
Fabric sample (blocks)	6.69	4	1.67	
Error	0.96	12	0.08	
Total	25.69	19		

Suppose an experiment is conducted as a randomized block design, and blocking was not really necessary. There are ab observations and $(a - 1)(b - 1)$ degrees of freedom for error. If the experiment had been run as a completely randomized single-factor design with b replicates, we would have had $a(b - 1)$ degrees of freedom for error. So, blocking has cost $a(b - 1) - (a - 1)(b - 1) = b - 1$ degrees of freedom for error. Thus, since the loss in error degrees of freedom is usually small, if there is a reasonable chance that block effects may be important, the experimenter should use the randomized block design.

For example, consider the experiment described in Example 12-6 as a one-way-classification analysis of variance. We would have 16 degrees of freedom for error. In the randomized block design there are 12 degrees of freedom for error. Therefore, blocking has cost only 4 degrees of freedom, a very small loss considering the possible gain in information that would be achieved if block effects are really important. As a general rule, when in doubt as to the importance of block effects, the experimenter should block and gamble that the block effect does exist. If the experimenter is wrong, the slight loss in the degrees of freedom for error will have a negligible effect, unless the number of degrees of freedom is very small. The reader should compare this discussion to the one at the end of Section 11-3.3.

12-4.2 Tests on Individual Treatment Means

When the analysis of variance indicates that a difference exists between treatment means, we usually need to perform some follow-up tests to isolate the specific differences. Any multiple comparison method, such as Tukey's test, could be used to do this.

Tukey's test presented in Section 12-2.2 can be used to determine differences between treatment means when blocking is involved simply by replacing n with the number of blocks b in equation 12-20. Keep in mind that the degrees of freedom for error have now changed. For the randomized block design, $f = (a - 1)(b - 1)$.

To illustrate this procedure, recall that the four chemical type means from Example 12-6 are

$$\bar{y}_{1.} = 1.14, \qquad \bar{y}_{2.} = 1.76, \qquad \bar{y}_{3.} = 1.38, \qquad \bar{y}_{4.} = 3.56,$$

$$T_\alpha = q_{0.05}(4,12)\sqrt{\frac{MS_E}{b}} = 4.20\sqrt{\frac{0.08}{5}} = 0.53.$$

Therefore, we would conclude that two means are significantly different if

$$|\bar{y}_{i\cdot} - \bar{y}_{j\cdot}| > 0.53.$$

The absolute values of the differences in treatment averages are

$$|\bar{y}_{1\cdot} - \bar{y}_{2\cdot}| = |1.14 - 1.76| = 0.62,$$

$$|\bar{y}_{1\cdot} - \bar{y}_{3\cdot}| = |1.14 - 1.38| = 0.24,$$

$$|\bar{y}_{1\cdot} - \bar{y}_{4\cdot}| = |1.14 - 3.56| = 2.42,$$

$$|\bar{y}_{2\cdot} - \bar{y}_{3\cdot}| = |1.76 - 1.38| = 0.38,$$

$$|\bar{y}_{2\cdot} - \bar{y}_{4\cdot}| = |1.76 - 3.56| = 1.80,$$

$$|\bar{y}_{3\cdot} - \bar{y}_{4\cdot}| = |1.38 - 3.56| = 2.18.$$

The results indicate chemical types 1 and 3 do not differ, and types 2 and 3 do not differ. Figure 12-8 represents the results graphically, where the underlined pairs do not differ.

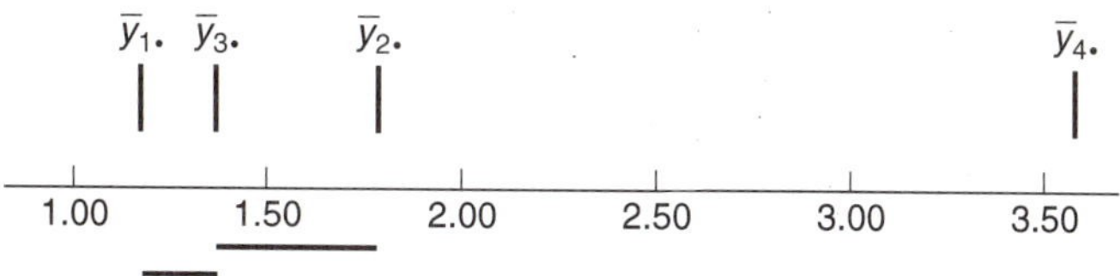

Figure 12-8 Results of Tukey's test.

12-4.3 Residual Analysis and Model Checking

In any designed experiment it is always important to examine the residuals and check for violations of basic assumptions that could invalidate the results. The residuals for the randomized block design are just the differences between the observed and fitted values

$$e_{ij} = y_{ij} - \hat{y}_{ij},$$

where the fitted values are

$$\hat{y}_{ij} = \bar{y}_{i\cdot} + \bar{y}_{\cdot j} - \bar{y}_{\cdot\cdot} \qquad (12\text{-}33)$$

The fitted value represents the estimate of the mean response when the ith treatment is run in the jth block. The residuals from the experiment from Example 12-6 are shown in Table 12-12.

Figures 12-9, 12-10, 12-11, and 12-12 present the important residual plots for the experiment. There is some indication that fabric sample (block) 3 has greater variability in strength when treated with the four chemicals than the other samples. Also, chemical type

Table 12-12 Residuals from the Randomized Block Design

Chemical Type	Fabric Sample				
	1	2	3	4	5
1	–0.18	–0.11	0.44	–0.18	0.02
2	0.10	0.07	–0.27	0.00	0.10
3	0.08	–0.24	0.30	–0.12	–0.02
4	0.00	0.27	–0.48	0.30	–0.10

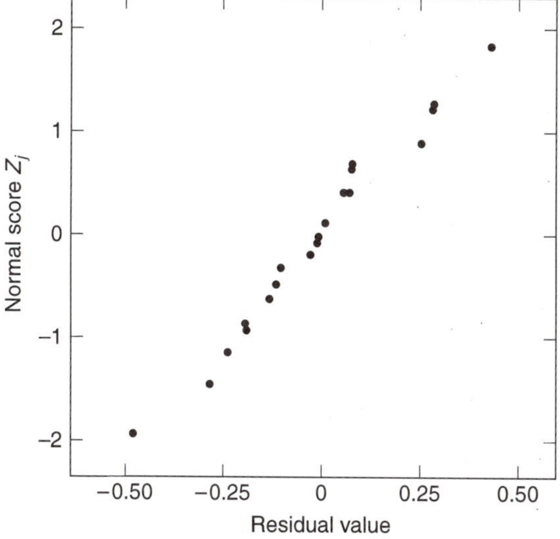

Figure 12-9 Normal probability plot of residuals from the randomized block design.

4, which provides the greatest strength, also has somewhat more variability in strength. Follow-up experiments may be necessary to confirm these findings if they are potentially important.

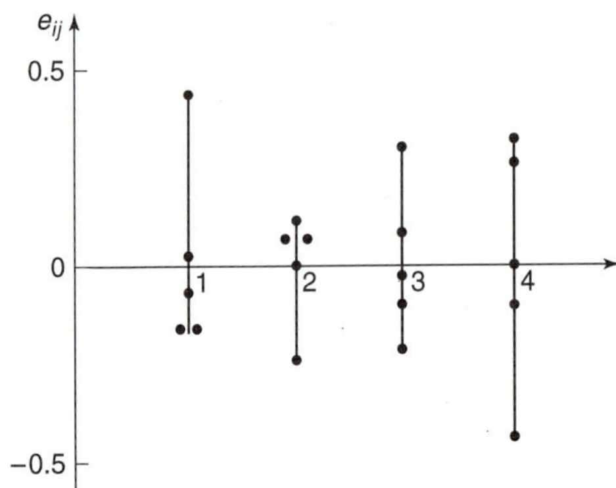

Figure 12-10 Residuals by treatment.

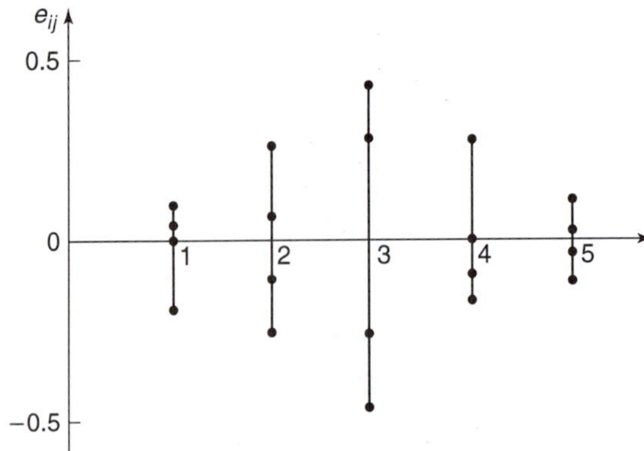

Figure 12-11 Residuals by block.

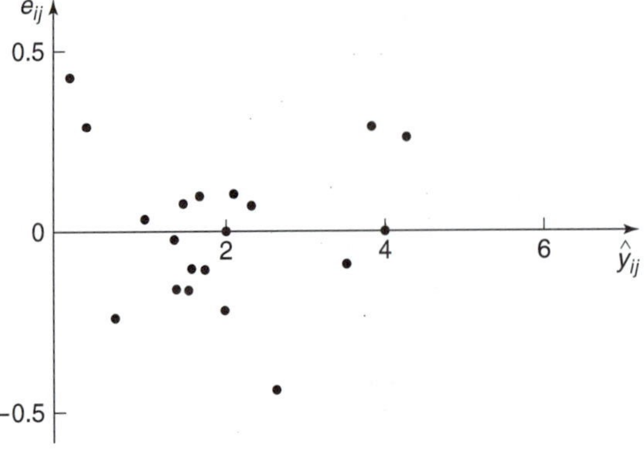

Figure 12-12 Residuals versus $\hat{y}_{ij}$.

12-5 DETERMINING SAMPLE SIZE IN SINGLE-FACTOR EXPERIMENTS

In any experimental design problem the choice of the sample size or number of replicates to use is important. Operating characteristic curves can be used to provide guidance in making this selection. Recall that the operating characteristic curve is a plot of the type II (β) error for various sample sizes against a measure of the difference in means that it is important to detect. Thus, if the experimenter knows how large a difference in means is of potential importance, the operating characteristic curves can be used to determine how many replicates are required to give adequate sensitivity.

We first consider sample size determination in a fixed-effects model for the case of equal sample size in each treatment. The power $(1 - \beta)$ of the test is

$$\begin{aligned} 1 - \beta &= P\{\text{Reject } H_0 | H_0 \text{ is false}\} \\ &= P\{F_0 > F_{\alpha, a-1, N-a} | H_0 \text{ is false}\}. \end{aligned} \tag{12-34}$$

To evaluate this probability statement, we need to know the distribution of the test statistic F_0 if the null hypothesis is false. It can be shown that if H_0 is false, the statistic $F_0 = MS_{\text{treatments}}/MS_E$ is distributed as a noncentral F random variable, with $a - 1$ and $N - a$ degrees of freedom and a noncentrality parameter δ. If $\delta = 0$, then the noncentral F distribution becomes the usual *central* F distribution.

The operating characteristic curves in Chart VII of the Appendix are used to calculate the power of the test for the fixed-effects model. These curves plot the probability of type II error (β) against Φ, where

$$\Phi^2 = \frac{n\sum_{i=1}^{a}\tau_i^2}{a\sigma^2}. \tag{12-35}$$

The parameter Φ^2 is related to the noncentrality parameter δ. Curves are available for $\alpha = 0.05$ and $\alpha = 0.01$ and for several values of degrees of freedom for numerator and denominator. In a completely randomized design, the symbol n in equation 12-35 is the number of replicates. In a randomized block design, replace n by the number of blocks.

In using the operating characteristic curves, we must define the difference in means that we wish to detect in terms of $\sum_{i=1}^{a}\tau_i^2$. Also, the error variance σ^2 is usually unknown. In such cases, we must choose ratios of $\sum_{i=1}^{a}\tau_i^2/\sigma^2$ that we wish to detect. Alternatively, if an estimate of σ^2 is available, one may replace σ^2 with this estimate. For example, if we were interested in the sensitivity of an experiment that has already been performed, we might use MS_E as the estimate of σ^2.

Example 12-7

Suppose that five means are being compared in a completely randomized experiment with $\alpha = 0.01$. The experimenter would like to know how many replicates to run if it is important to reject H_0 with a probability of at least 0.90 if $\sum_{i=1}^{5}\tau_{i=1}^2/\sigma^2 = 5.0$. The parameter Φ^2 is, in this case,

$$\Phi^2 = \frac{n\sum_{i=1}^{a}\tau_i^2}{a\sigma^2} = \frac{n}{5}(5) = n,$$

and the operating characteristic curve for $a - 1 = 5 - 1 = 4$ and $N - a = a(n - 1) = 5(n - 1)$ error degrees of freedom is shown in Chart VII (Appendix). As a first guess, try $n = 4$ replicates. This yields $\Phi^2 = 4$, $\Phi = 2$, and $5(3) = 15$ error degrees of freedom. Consequently, from Chart VII, we find that $\beta \simeq 0.38$.

Therefore, the power of the test is approximately $1 - \beta = 1 - 0.38 = 0.62$, which is less than the required 0.90, and so we conclude that $n = 4$ replicates are not sufficient. Proceeding in a similar manner, we can construct the following display.

n	Φ^2	Φ	$a(n-1)$	β	Power $(1-\beta)$
4	4	2.00	15	0.38	0.62
5	5	2.24	20	0.18	0.82
6	6	2.45	25	0.06	0.94

Thus, at least $n = 6$ replicates must be run in order to obtain a test with the required power.

The power of the test for the random-effects model is

$$\begin{aligned} 1 - \beta &= P\{\text{Reject } H_0 | H_0 \text{ is false}\} \\ &= P\{F_0 > F_{\alpha, a-1, N-a} | \sigma_\tau^2 > 0\}. \end{aligned} \tag{12-36}$$

Once again the distribution of the test statistic F_0 under the alternative hypothesis is needed. It can be shown that if H_1 is true ($\sigma_\tau^2 > 0$), the distribution of F_0 is central F, with $a - 1$ and $N - a$ degrees of freedom.

Since the power of the random-effects model is based on the central F distribution, we could use the tables of the F distribution in the Appendix to evaluate equation 12-36. However, it is much easier to evaluate the power of the test by using the operating characteristic curves in Chart VIII of the Appendix. These curves plot the probability of the type II error against λ, where

$$\lambda = \sqrt{1 + \frac{n\sigma_\tau^2}{\sigma^2}}. \tag{12-37}$$

In the randomized block design, replace n with b, the number of blocks. Since σ^2 is usually unknown, we may either use a prior estimate or define the value of σ_τ^2 that we are interested in detecting in terms of the ratio σ_τ^2/σ^2.

Example 12-8

Consider a completely randomized design with five treatments selected at random, with six observations per treatment and $\alpha = 0.05$. We wish to determine the power of the test if σ_τ^2 is equal to σ^2. Since $a = 5$, $n = 6$, and $\sigma_\tau^2 = \sigma^2$, we may compute

$$\lambda = \sqrt{1 + (6)1} = 2.646.$$

From the operating characteristic curve with $a - 1 = 4$, $N - a = 25$ degrees of freedom and $\alpha = 0.05$, we find that

$$\beta \simeq 0.20.$$

Therefore, the power is approximately 0.80.

12-6 SAMPLE COMPUTER OUTPUT

Many computer packages can be implemented to carry out the analysis of variance for the situations presented in this chapter. In this section, computer output from Minitab® is presented.

Computer Output for Hardwood Concentration Example

Reconsider Example 12-1, which investigates the effect of hardwood concentration on tensile strength. Using ANOVA in Minitab® provides the following output.

```
Analysis of Variance for TS
Source   DF        SS       MS        F        P
Concen    3    382.79   127.60    19.61    0.000
Error    20    130.17     6.51
Total    23    512.96

                                  Individual 95% CIs For Mean
                                  Based on Pooled StDev
Level     N      Mean    StDev    ---+------+------+------+-
5         6    10.000    2.828    (--*--)
10        6    15.667    2.805          (--*--)
15        6    17.000    1.789             (--*--)
20        6    21.167    2.639                 (--*--)
                                  ---+------+------+------+-
Pooled StDev =           2.551    10.0   15.0   20.0   25.0
```

The analysis of variance results are identical to those presented in Section 12-1.2. Minitab® also provides 95% confidence intervals for the means of each level of hardwood concentration using a pooled estimate of the standard deviation. Interpretation of the confidence intervals is straightforward. Factor levels with confidence intervals that do not overlap are said to be significantly different.

A better indicator of significant differences is provided by confidence intervals based on Tukey's test on pairwise differences, an option in Minitab®. The output provided is

```
Tukey's pairwise comparisons

Family error rate = 0.0500
Individual error rate = 0.0111

Critical value = 3.96

Intervals for (column level mean) - (row level mean)

               5          10          15

10        -9.791
          -1.542

15       -11.124      -5.458
          -2.876       2.791

20       -15.291      -9.624      -8.291
          -7.042      -1.376      -0.042
```

The (simultaneous) confidence intervals are easily interpreted. For example, the 95% confidence interval for the difference in mean tensile strength between 5% hardwood concentration and 10% hardwood concentration is (–9.791, –1.542). Since this confidence interval does not contain the value 0, we conclude there is a significant difference between 5% and 10% hardwood concentrations. The remaining confidence intervals are interpreted similarly. The results provided by Minitab® are identical to those found in Section 12-2.2.

12-7 SUMMARY

This chapter has introduced design and analysis methods for experiments with a single factor. The importance of randomization in single-factor experiments was emphasized. In a completely randomized experiment, all runs are made in random order to balance out the effects of unknown nuisance variables. If a known nuisance variable can be controlled, blocking can be used as a design alternative. The fixed-effects and random-effects models of analysis of variance were presented. The primary difference between the two models is the inference space. In the fixed-effects model inferences are valid only about the factor levels specifically considered in the analysis, while in the random-effects model the conclusions may be extended to the population of factor levels. Orthogonal contrasts and Tukey's test were suggested for making comparisons between factor level means in the fixed-effects experiment. A procedure was also given for estimating the variance components in a random-effects model. Residual analysis was introduced for checking the underlying assumptions of the analysis of variance.

12-8 EXERCISES

12-1. A study is conducted to determine the effect of cutting speed on the life (in hours) of a particular machine tool. Four levels of cutting speed are selected for the study with the following results:

Cutting Speed	Tool Life					
1	41	43	33	39	36	40
2	42	36	34	45	40	39
3	34	38	34	34	36	33
4	36	37	36	38	35	35

(a) Does cutting speed affect tool life? Draw comparative box plots and perform an analysis of variance.

(b) Plot average tool life against cutting speed and interpret the results.

(c) Use Tukey's test to investigate differences between the individual levels of cutting speed. Interpret the results.

(d) Find the residuals and examine them for model inadequacy.

12-2. In "Orthogonal Design for Process Optimization and Its Application to Plasma Etching" (*Solid State Technology*, May 1987), G. Z. Yin and D. W. Jillie describe an experiment to determine the effect of C_2F_6 flow rate on the uniformity of the etch on a silicon wafer used in integrated circuit manufacturing. Three flow rates are used in the experiment, and the resulting uniformity (in percent) for six replicates is as follows:

	Observations					
C_2F_6 Flow	1	2	3	4	5	6
125	2.7	4.6	2.6	3.0	3.2	3.8
160	4.9	4.6	5.0	4.2	3.6	4.2
200	4.6	3.4	2.9	3.5	4.1	5.1

(a) Does C_2F_6 flow rate affect etch uniformity? Construct box plots to compare the factor levels and perform the analysis of variance.

(b) Do the residuals indicate any problems with the underlying assumptions?

12-3. The compressive strength of concrete is being studied. Four different mixing techniques are being investigated. The following data have been collected:

Mixing Technique	Compressive Strength (psi)			
1	3129	3000	2865	2890
2	3200	3300	2975	3150
3	2800	2900	2985	3050
4	2600	2700	2600	2765

(a) Test the hypothesis that mixing techniques affect the strength of the concrete. Use $\alpha = 0.05$.

(b) Use Tukey's test to make comparisons between pairs of means. Estimate the treatment effects.

12-4. A textile mill has a large number of looms. Each loom is supposed to provide the same output of cloth per minute. To investigate this assumption, five looms are chosen at random and their output measured at different times. The following data are obtained:

Loom	Output (lb / min)				
1	4.0	4.1	4.2	4.0	4.1
2	3.9	3.8	3.9	4.0	4.0
3	4.1	4.2	4.1	4.0	3.9
4	3.6	3.8	4.0	3.9	3.7
5	3.8	3.6	3.9	3.8	4.0

(a) Is this a fixed- or random-effects experiment? Are the looms similar in output?

(b) Estimate the variability between looms.

(c) Estimate the experimental error variance.

(d) What is the probability of accepting H_0 if σ_τ^2 is four times the experimental error variance?

(e) Analyze the residuals from this experiment and check for model inadequacy.

12-5. An experiment was run to determine whether four specific firing temperatures affect the density of a certain type of brick. The experiment led to the following data:

Temperature (°F)	Density						
100	21.8	21.9	21.7	21.6	21.7	21.5	21.8
125	21.7	21.4	21.5	21.5	–	–	–
150	21.9	21.8	21.8	21.6	21.5	–	–
175	21.9	21.7	21.8	21.7	21.6	21.8	–

(a) Does the firing temperature affect the density of the bricks?

(b) Estimate the components in the model.

(c) Analyze the residuals from the experiment.

12-6. An electronics engineer is interested in the effect on tube conductivity of five different types of coating for cathode ray tubes used in a telecommunications system display device. The following conductivity data are obtained:

Coating Type	Conductivity			
1	143	141	150	146
2	152	149	137	143
3	134	133	132	127
4	129	127	132	129
5	147	148	144	142

(a) Is there any difference in conductivity due to coating type? Use $\alpha = 0.05$.

(b) Estimate the overall mean and the treatment effects.

(c) Compute a 95% interval estimate of the mean for coating type 1. Compute a 99% interval estimate of the mean difference between coating types 1 and 4.

(d) Test all pairs of means using Tukey's test, with $\alpha = 0.05$.

(e Assuming that coating type 4 is currently in use, what are your recommendations to the manufacturer? We wish to minimize conductivity.

12-7. The response time in milliseconds was determined for three different types of circuits used in an electronic calculator. The results are recorded here:

Circuit Type	Response Time				
1	19	22	20	18	25
2	20	21	33	27	40
3	16	15	18	26	17

(a) Test the hypothesis that the three circuit types have the same response time.

(b) Use Tukey's test to compare pairs of treatment means.

(c) Construct a set of orthogonal contrasts, assuming that at the outset of the experiment you suspected the response time of circuit type 2 to be different from the other two.

(d) What is the power of this test for detecting $\sum_{i=1}^{3} \tau_i^2 / \sigma^2 = 3.0$?

(e) Analyze the residuals from this experiment.

12-8. In "The Effect of Nozzle Design on the Stability and Performance of Turbulent Water Jets" (*Fire Safety Journal*, Vol. 4, August 1981), C. Theobald describes an experiment in which a shape factor was determined for several different nozzle designs at different levels of jet efflux velocity. Interest in this experiment focuses primarily on nozzle design, and velocity is a nuisance factor. The data are shown below:

Nozzle	Jet Efflux Velocity (m/s)					
Type	11.73	14.37	16.59	20.43	23.46	28.74
1	0.78	0.80	0.81	0.75	0.77	0.78
2	0.85	0.85	0.92	0.86	0.81	0.83
3	0.93	0.92	0.95	0.89	0.89	0.83
4	1.14	0.97	0.98	0.88	0.86	0.83
5	0.97	0.86	0.78	0.76	0.76	0.75

(a) Does nozzle type affect shape factor? Compare the nozzles using box plots and the analysis of variance.

(b) Use Tukey's test to determine specific differences between the nozzles. Does a graph of average (or standard deviation) of shape factor versus nozzle type assist with the conclusions?

(c) Analyze the residuals from this experiment.

12-9. In his book *Design and Analysis of Experiments* (2001), D. C. Montgomery describes an experiment to determine the effect of four chemical agents on the strength of a particular type of cloth. Due to possible variability from cloth to cloth, bolts of cloth are considered blocks. Five bolts are selected and all four chemicals in random order are applied to each bolt. The resulting tensile strengths are

	Bolt				
Chemical	1	2	3	4	5
1	73	68	74	71	67
2	73	67	75	72	70
3	75	68	78	73	68
4	73	71	75	75	69

(a) Is there any difference in tensile strength between the chemicals?

(b) Use Tukey's test to investigate specific differences between the chemicals.

(c) Analyze the residuals from this experiment.

12-10. Suppose that four normal populations have common variance $\sigma^2 = 25$ and means $\mu_1 = 50$, $\mu_2 = 60$, $\mu_3 = 50$, and $\mu_4 = 60$. How many observations should be taken on each population so that the probability of rejecting the hypothesis of equality of means is at least 0.90? Use $\alpha = 0.05$.

12-11. Suppose that five normal populations have common variance $\sigma^2 = 100$ and means $\mu_1 = 175$, $\mu_2 = 190$, $\mu_3 = 160$, $\mu_4 = 200$, and $\mu_5 = 215$. How many observations per population must be taken so that the probability of rejecting the hypothesis of equality of means is at least 0.95? Use $\alpha = 0.01$.

12-12. Consider testing the equality of the means of two normal populations where the variances are unknown but assumed equal. The appropriate test procedure is the two-sample t-test. Show that the two-sample t-test is equivalent to the one-way-classification analysis of variance.

12-13. Show that the variance of the linear combination $\sum_{i=1}^{a} c_i y_{i.}$ is $\sigma^2 \sum_{i=1}^{a} n_i c_i^2$.

12-14. In a fixed-effects model, suppose that there are n observations for each of four treatments. Let Q_1^2, Q_2^2, and Q_3^2 be single-degree-of-freedom components for the orthogonal contrasts. Prove that $SS_{\text{treatments}} = Q_1^2 + Q_2^2 + Q_3^2$.

12-15. Consider the data shown in Exercise 12-7.

(a) Write out the least squares normal equations for this problem, and solve them for $\hat{\mu}$ and $\hat{\tau}_i$, making the usual constraint ($\sum_{i=1}^{3} \hat{\tau}_i = 0$). Estimate $\tau_1 - \tau_2$.

(b Solve the equations in (a) using the constraint $\hat{\tau}_3 = 0$. Are the estimators $\hat{\tau}_i$ and $\hat{\mu}$ the same as you found in (a)? Why? Now estimate $\tau_1 - \tau_2$ and compare your answer with (a). What statement can you make about estimating contrasts in the τ_i?

(c) Estimate $\mu + \tau_1$, $2\tau_1 - \tau_2 - \tau_3$ and $\mu + \tau_1 + \tau_2$ using the two solutions to the normal equations. Compare the results obtained in each case.

Chapter 13

Design of Experiments with Several Factors

An experiment is just a test or a series of tests. Experiments are performed in all scientific and engineering disciplines and are a major part of the discovery and learning process. The conclusions that can be drawn from an experiment will depend, in part, on how the experiment was conducted and so the *design* of the experiment plays a major role in problem solution. This chapter introduces experimental design techniques useful when several factors are involved.

13-1 EXAMPLES OF EXPERIMENTAL DESIGN APPLICATIONS

Example 13-1

A Characterization Experiment A development engineer is working on a new process for soldering electronic components to printed circuit boards. Specifically, he is working with a new type of flow solder machine that he hopes will reduce the number of defective solder joints. (A flow solder machine preheats printed circuit boards and then moves them into contact with a wave of liquid solder. This machine makes all the electrical and most of the mechanical connections of the components to the printed circuit board. Solder defects require touchup or rework, which adds cost and often damages the boards.) The flow solder machine has several variables that the engineer can control. They are as follows:

1. Solder temperature
2. Preheat temperature
3. Conveyor speed
4. Flux type
5. Flux specific gravity
6. Solder wave depth
7. Conveyor angle

In addition to these controllable factors, there are several factors that cannot be easily controlled once the machine enters routine manufacturing, including the following:

1. Thickness of the printed circuit board
2. Types of components used on the board
3. Layout of the components of the board
4. Operator
5. Environmental factors
6. Production rate

Sometimes we call the uncontrollable factors *noise* factors. A schematic representation of the process is shown in Fig. 13-1.

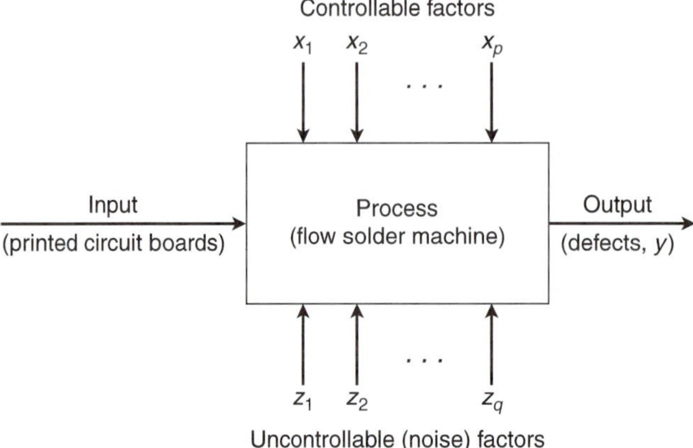

Figure 13-1 The flow solder experiment.

In this situation the engineer is interested in *characterizing* the flow solder machine; that is, he is interested in determining which factors (both controllable and uncontrollable) affect the occurrence of defects on the printed circuit boards. To accomplish this he can design an experiment that will enable him to estimate the magnitude and direction of the factor effects. Sometimes we call an experiment such as this a *screening* experiment. The information from this characterization study or screening experiment can be used to identify the critical factors, to determine the direction of adjustment for these factors to reduce the number of defects, and to assist in determining which factors should be carefully controlled during manufacturing to prevent high defect levels and erratic process performance.

Example 13-2

An Optimization Experiment In a characterization experiment, we are interested in determining *which* factors affect the response. A logical next step is to determine the region in the important factors that leads to an optimum response. For example, if the response is yield, we would look for a region of maximum yield, and if the response is cost, we would look for a region of minimum cost.

As an illustration, suppose that the yield of a chemical process is influenced by the operating temperature and the reaction time. We are currently operating the process at 155°F and 1.7 hours of reaction time and experiencing yields around 75%. Figure 13-2 shows a view of the time–temperature space from above. In this graph we have connected points of constant yield with lines. These lines are called *contours*, and we have shown the contours at 60%, 70%, 80%, 90%, and 95% yield. To locate the optimum, it is necessary to design an experiment that varies reaction time and temperature together. This design is illustrated in Fig. 13-2. The responses observed at the four points in the experiment (145°F, 1.2 hr), (145°F, 2.2 hr), (165°F, 1.2 hr), and (165°F, 2.2 hr) indicate that we should move in the general direction of increased temperature and lower reaction time to increase yield. A few additional runs could be performed in this direction to locate the region of maximum yield.

These examples illustrate only two potential applications of experimental design methods. In the engineering environment, experimental design applications are numerous. Some potential areas of use are as follows:

1. Process troubleshooting
2. Process development and optimization
3. Evaluation of material alternatives
4. Reliability and life testing
5. Performance testing

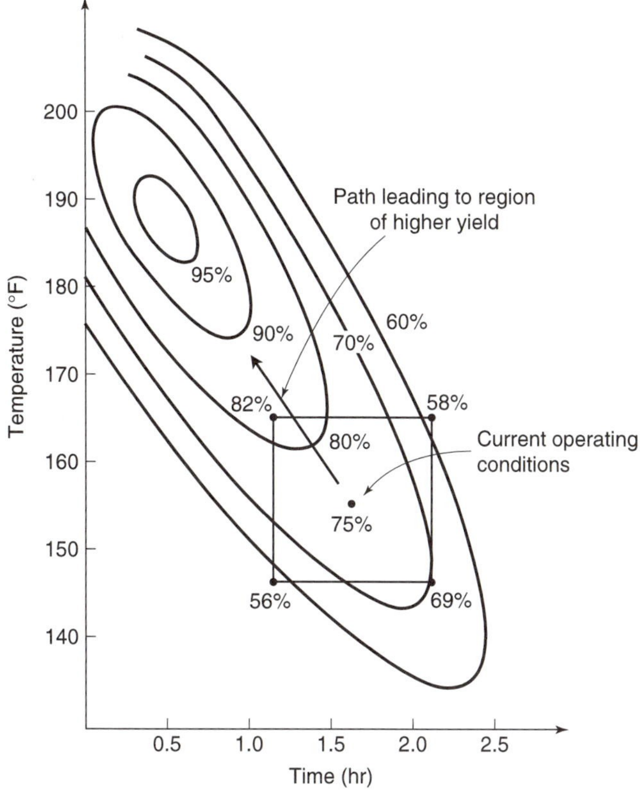

Figure 13-2 Contour plot of yield as a function of reaction time and reaction temperature, illustrating an optimization experiment.

6. Product design configuration
7. Component tolerance determination

Experimental design methods allow these problems to be solved efficiently during the early stages of the product cycle. This has the potential to dramatically lower overall product cost and reduce development lead time.

13-2 FACTORIAL EXPERIMENTS

When there are several factors of interest in an experiment, a *factorial design* should be used. These are designs in which factors are varied together. Specifically, by a factorial experiment we mean that in each complete trial or replicate of the experiment all possible combinations of the levels of the factors are investigated. Thus, if there are two factors, A and B, with a levels of factor A and b levels of factor B, then each replicate contains all ab treatment combinations.

The effect of a factor is defined as the change in response produced by a change in the level of the factor. This is called a *main effect* because it refers to the primary factors in the study. For example, consider the data in Table 13-1. The main effect of factor A is the difference between the average response at the first level of A and the average response at the second level of A, or

$$A = \frac{30+40}{2} - \frac{10+20}{2} = 20.$$

Table 13-1 A Factorial Experiment with Two Factors

	Factor B	
Factor A	B_1	B_2
A_1	10	20
A_2	30	40

That is, changing factor A from level 1 to level 2 causes an average response increase of 20 units. Similarly, the main effect of B is

$$B = \frac{20+40}{2} - \frac{10+30}{2} = 10.$$

In some experiments, the difference in response between the levels of one factor is not the same at all levels of the other factors. When this occurs, there is an *interaction* between the factors. For example, consider the data in Table 13-2. At the first level of factor B, the A effect is

$$A = 30 - 10 = 20,$$

and at the second level of factor B, the A effect is

$$A = 0 - 20 = -20.$$

Since the effect of A depends on the level chosen for factor B, there is interaction between A and B.

When an interaction is large, the corresponding main effects have little meaning. For example, by using the data in Table 13-2, we find the main effect of A to be

$$A = \frac{30+0}{2} - \frac{10+20}{2} = 0,$$

and we would be tempted to conclude that there is no A effect. However, when we examined the effects of A at *different levels of factor* B, we saw that this was not the case. The effect of factor A depends on the levels of factor B. Thus, knowledge of the AB interaction is more useful than knowledge of the main effects. A significant interaction can mask the significance of main effects.

The concept of interaction can be illustrated graphically. Figure 13-3 plots the data in Table 13-1 against the levels of A for both levels of B. Note that the B_1 and B_2 lines are roughly parallel, indicating that factors A and B do not interact significantly. Figure 13-4 plots the data in Table 13-2. In this graph, the B_1 and B_2 lines are not parallel, indicating the interaction between factors A and B. Such graphical displays are often useful in presenting the results of experiments.

An alternative to the factorial design that is (unfortunately) used in practice is to change the factors one at a time rather than to vary them simultaneously. To illustrate this one-factor-at-a-time procedure, consider the optimization experiment described in Example 13-2. The

Table 13-2 A Factorial Experiment with Interaction

	Factor B	
Factor A	B_1	B_2
A_1	10	20
A_2	30	0

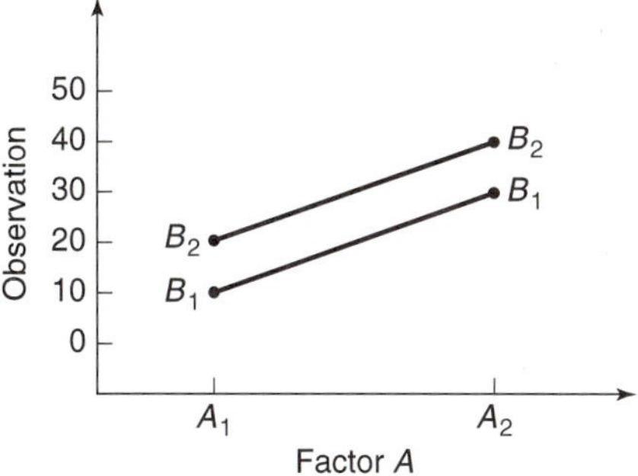

Figure 13-3 Factorial experiment, no interaction.

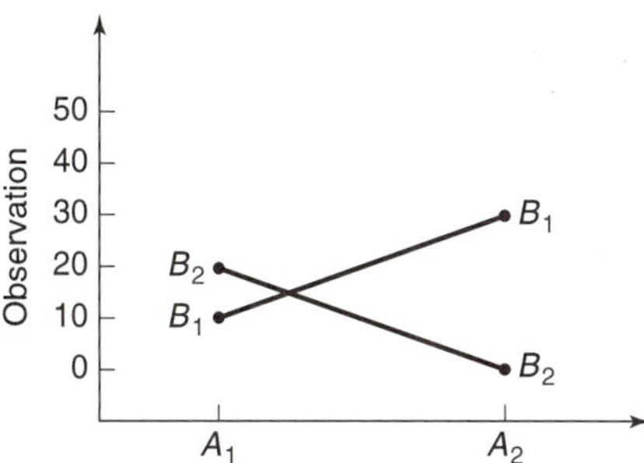

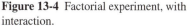

Figure 13-4 Factorial experiment, with interaction.

engineer is interested in finding the values of temperature and reaction time that maximize yield. Suppose that we fix temperature at 155°F (the current operating level) and perform five runs at different levels of time, say 0.5 hour, 1.0 hour, 1.5 hours, 2.0 hours, and 2.5 hours. The results of this series of runs are shown in Fig. 13-5. This figure indicates that maximum yield is achieved at about 1.7 hours of reaction time. To optimize temperature, the engineer fixes time at 1.7 hours (the apparent optimum) and performs five runs at different temperatures, say 140°F, 150°F, 160°F, 170°F, and 180°F. The results of this set of runs are plotted in Fig. 13-6. Maximum yield occurs at about 155°F. Therefore, we would conclude that running the process at 155°F and 1.7 hours is the best set of operating conditions, resulting in yields around 75%.

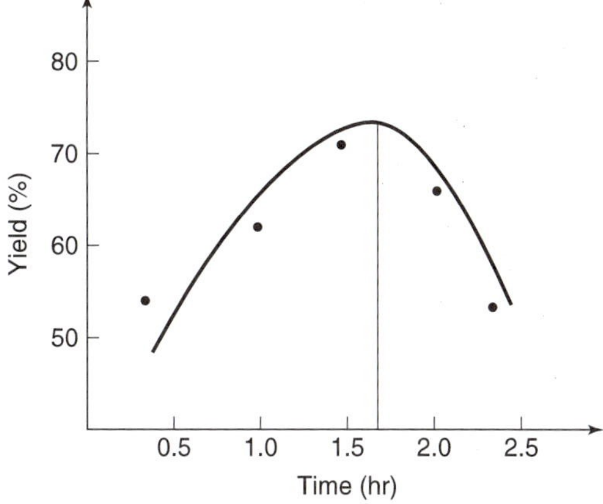

Figure 13-5 Yield versus reaction time with temperature constant at 155°F.

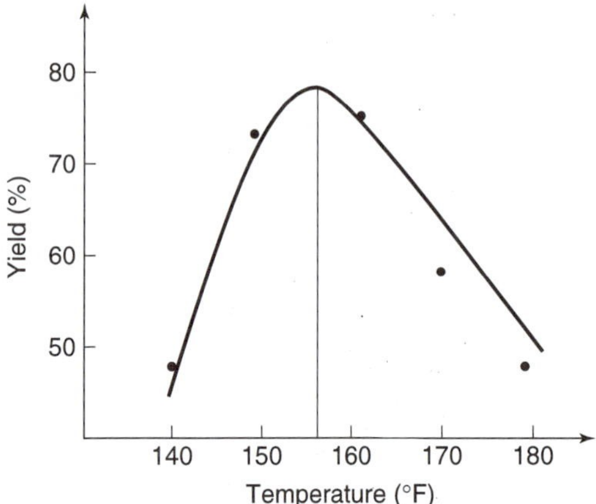

Figure 13-6 Yield versus temperature with reaction time constant at 1.7 hr.

Figure 13-7 displays the contour plot of yield as a function of temperature and time with the one-factor-at-a-time experiment shown on the contours. Clearly the one-factor-at-a-time design has failed dramatically here, as the true optimum is at least 20 yield points higher and occurs at much lower reaction times and higher temperatures. The failure to discover the shorter reaction times is particularly important as this could have significant impact on production volume or capacity, production planning, manufacturing cost, and total productivity.

The one-factor-at-a-time method has failed here because it fails to detect the interaction between temperature and time. Factorial experiments are the only way to detect inter-

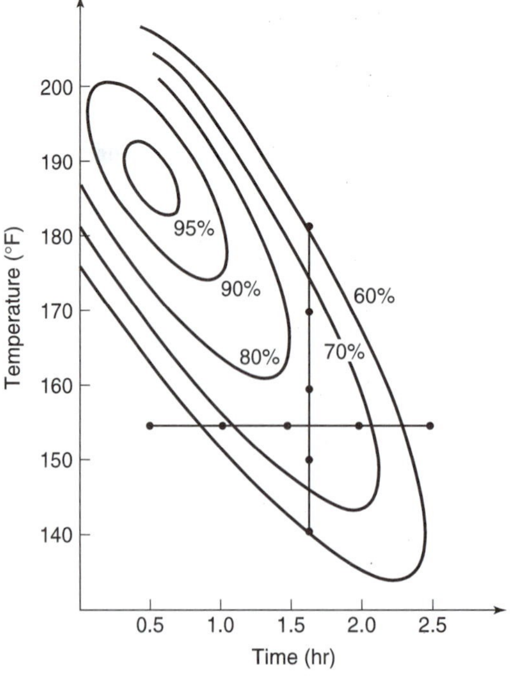

Figure 13-7 Optimization experiment using the one-factor-at-a-time method.

actions. Furthermore, the one-factor-at-a-time method is inefficient; it will require more experimentation than a factorial, and as we have just seen, there is no assurance that it will produce the correct results. The experiment shown in Fig. 13-2 that produced the information pointing to the region of the optimum is a simple example of a factorial experiment.

13-3 TWO-FACTOR FACTORIAL EXPERIMENTS

The simplest type of factorial experiment involves only two factors, say A and B. There are a levels of factor A and b levels of factor B. The two-factor factorial is shown in Table 13-3. Note that there are n *replicates* of the experiment and that each replicate contains all ab treatment combinations. The observation in the ijth cell in the kth replicate is denoted y_{ijk}. In collecting the data, the abn observations would be run in *random* order. Thus, like the single-factor experiment studied in Chapter 12, the two-factor factorial is a *completely randomized design.*

The observations may be described by the linear statistical model

$$y_{ijk} = \mu + \tau_i + \beta_j + (\tau\beta)_{ij} + \epsilon_{ijk} \begin{cases} i = 1, 2, \ldots, a, \\ j = 1, 2, \ldots, b, \\ k = 1, 2, \ldots, n, \end{cases} \tag{13-1}$$

where μ is the overall mean effect, τ_i is the effect of the ith level of factor A, β_j is the effect of the jth level of factor B, $(\tau\beta)_{ij}$ is the effect of the interaction between A and B, and ϵ_{ijk} is a NID(0, σ^2) (normal and independently distributed) random error component. We are interested in testing the hypotheses of no significant factor A effect, no significant factor B effect, and no significant AB interaction. As with the single-factor experiments of Chapter 12, the analysis of variance will be used to test these hypotheses. Since there are two factors under study, the procedure used is called the two-way analysis of variance.

13-3.1 Statistical Analysis of the Fixed-Effects Model

Suppose that factors A and B are fixed. That is, the a levels of factor A and the b levels of factor B are specifically chosen by the experimenter, and inferences are confined to these levels only. In this model, it is customary to define the effects τ_i, β_j, and $(\tau\beta)_{ij}$ as deviations from the mean, so that $\sum_{i=1}^{a} \tau_i = 0$, $\sum_{j=1}^{b} \beta_j = 0$, $\sum_{i=1}^{a} (\tau\beta)_{ij} = 0$, and $\sum_{j=1}^{b} (\tau\beta)_{ij} = 0$.

Let $y_{i..}$ denote the total of the observations under the ith level of factor A, let $y_{.j.}$ denote the total of the observations under the jth level of factor B, let $y_{ij.}$ denote the total of the observations in the ijth cell of Table 13-3, and let $y_{...}$ denote the grand total of all the

Table 13-3 Data Arrangement for a Two-Factor Factorial Design

	Factor B			
Factor A	1	2	...	b
1	$y_{111}, y_{112}, \ldots, y_{11n}$	$y_{121}, y_{122}, \ldots, y_{12n}$		$y_{1b1}, y_{1b2}, \ldots, y_{1bn}$
2	$y_{211}, y_{212}, \ldots, y_{21n}$	$y_{221}, y_{222}, \ldots, y_{22n}$		$y_{2b1}, y_{2b2}, \ldots, y_{2bn}$
⋮				
a	$y_{a11}, y_{a12}, \ldots, y_{a1n}$	$y_{a21}, y_{a22}, \ldots, y_{a2n}$		$y_{ab1}, y_{ab2}, \ldots, y_{abn}$

observations. Define $\bar{y}_{i\cdot\cdot}$, $\bar{y}_{\cdot j\cdot}$, $\bar{y}_{ij\cdot}$, and $\bar{y}_{\cdots}$ as the corresponding row, column, cell, and grand averages. That is,

$$
\begin{aligned}
y_{i..} &= \sum_{j=1}^{b}\sum_{k=1}^{n} y_{ijk}, & \bar{y}_{i..} &= \frac{y_{i..}}{bn}, & i &= 1,2,\ldots,a, \\
y_{.j.} &= \sum_{i=1}^{a}\sum_{k=1}^{n} y_{ijk}, & \bar{y}_{.j.} &= \frac{y_{.j.}}{an}, & j &= 1,2,\ldots,b, \\
y_{ij.} &= \sum_{k=1}^{n} y_{ijk}, & \bar{y}_{ij.} &= \frac{y_{ij.}}{n}, & i &= 1,2,\ldots,a, \quad j = 1,2,\ldots,b, \\
y_{...} &= \sum_{i=1}^{a}\sum_{j=1}^{b}\sum_{k=1}^{n} y_{ijk}, & \bar{y}_{...} &= \frac{y_{...}}{abn}.
\end{aligned}
\tag{13-2}
$$

The total corrected sum of squares may be written

$$
\begin{aligned}
\sum_{i=1}^{a}\sum_{j=1}^{b}\sum_{k=1}^{n}\left(y_{ijk}-\bar{y}_{...}\right)^2
&= \sum_{i=1}^{a}\sum_{j=1}^{b}\sum_{k=1}^{n}\Big[\left(\bar{y}_{i..}-\bar{y}_{...}\right)+\left(\bar{y}_{.j.}-\bar{y}_{...}\right) \\
&\qquad +\left(\bar{y}_{ij.}-\bar{y}_{i..}-\bar{y}_{.j.}+\bar{y}_{...}\right)+\left(y_{ijk}-\bar{y}_{ij.}\right)\Big]^2 \\
&= bn\sum_{i=1}^{a}\left(\bar{y}_{i..}-\bar{y}_{...}\right)^2 + an\sum_{j=1}^{b}\left(\bar{y}_{.j.}-\bar{y}_{...}\right)^2 \\
&\quad + n\sum_{i=1}^{a}\sum_{j=1}^{b}\left(\bar{y}_{ij.}-\bar{y}_{i..}-\bar{y}_{.j.}+\bar{y}_{...}\right)^2 + \sum_{i=1}^{a}\sum_{j=1}^{b}\sum_{k=1}^{n}\left(y_{ijk}-\bar{y}_{ij.}\right)^2.
\end{aligned}
\tag{13-3}
$$

Thus, the total sum of squares is partitioned into a sum of squares due to "rows," or factor A (SS_A), a sum of squares due to "columns," or factor B (SS_B), a sum of squares due to the interaction between A and B (SS_{AB}), and a sum of squares due to error (SS_E). Notice that there must be at least two replicates to obtain a nonzero error sum of squares.

The sum of squares identity in equation 13-3 may be written symbolically as

$$SS_T = SS_A + SS_B + SS_{AB} + SS_E. \tag{13-4}$$

There are $abn - 1$ total degrees of freedom. The main effects A and B have $a - 1$ and $b - 1$ degrees of freedom, while the interaction effect AB has $(a - 1)(b - 1)$ degrees of freedom. Within each of the ab cells in Table 13-3, there are $n - 1$ degrees of freedom between the n replicates, and observations in the same cell can differ only due to random error. Therefore, there are $ab(n - 1)$ degrees of freedom for error. The ratio of each sum of squares on the right-hand side of equation 13-4 to its degrees of freedom is a *mean square.*

Assuming that factors A and B are fixed, the expected values of the mean squares are

$$E(MS_A) = E\left(\frac{SS_A}{a-1}\right) = \sigma^2 + \frac{bn\sum_{i=1}^{a}\tau_i^2}{a-1},$$

$$E(MS_B) = E\left(\frac{SS_B}{b-1}\right) = \sigma^2 + \frac{an\sum_{j=1}^{b}\beta_i^2}{b-1},$$

$$E(MS_{AB}) = E\left(\frac{SS_{AB}}{(a-1)(b-1)}\right) = \sigma^2 + \frac{n\sum_{i=1}^{a}\sum_{j=1}^{b}(\tau\beta)_{ij}^2}{(a-1)(b-1)},$$

and

$$E(MS_E) = E\left(\frac{SS_E}{ab(n-1)}\right) = \sigma^2.$$

Therefore, to test H_0: $\tau_i = 0$ (no row factor effects), H_0: $\beta_j = 0$ (no column factor effects), and H_0: $(\tau\beta)_{ij} = 0$ (no interaction effects), we would divide the corresponding mean square by the mean square error. Each of these ratios will follow an F distribution with numerator degrees of freedom equal to the number of degrees of freedom for the numerator mean square and $ab(n-1)$ denominator degrees of freedom, and the critical region will be located in the upper tail. The test procedure is arranged in an analysis-of-variance table, such as is shown in Table 13-4.

Computational formulas for the sums of squares in equation 13-4 are obtained easily. The total sum of squares is computed from

$$SS_T = \sum_{i=1}^{a}\sum_{j=1}^{b}\sum_{k=1}^{n} y_{ijk}^2 - \frac{y_{\dots}^2}{abn}. \tag{13-5}$$

The sums of squares for main effects are

$$SS_A = \sum_{i=1}^{a}\frac{y_{i..}^2}{bn} - \frac{y_{\dots}^2}{abn} \tag{13-6}$$

and

$$SS_B = \sum_{j=1}^{b}\frac{y_{.j.}^2}{an} - \frac{y_{\dots}^2}{abn}. \tag{13-7}$$

We usually calculate the SS_{AB} in two steps. First, we compute the sum of squares between the ab cell totals, called the sum of squares due to "subtotals":

$$SS_{\text{subtotals}} = \sum_{i=1}^{a}\sum_{j=1}^{b}\frac{y_{ij.}^2}{n} - \frac{y_{\dots}^2}{abn}.$$

Table 13-4 The Analysis-of-Variance Table for the Two-Way-Classification Fixed-Effects Model

Source of Variation	Sum of Squares	Degrees of Freedom	Mean Square	F_0
A treatments	SS_A	$a-1$	$MS_A = \frac{SS_A}{a-1}$	$\frac{MS_A}{MS_E}$
B treatments	SS_B	$b-1$	$MS_B = \frac{SS_B}{b-1}$	$\frac{MS_B}{MS_E}$
Interaction	SS_{AB}	$(a-1)(b-1)$	$MS_{AB} = \frac{SS_{AB}}{(a-1)(b-1)}$	$\frac{MS_{AB}}{MS_E}$
Error	SS_E	$ab(n-1)$	$MS_E = \frac{SS_E}{ab(n-1)}$	
Total	SS_T	$abn-1$		

This sum of squares also contains SS_A and SS_B. Therefore, the second step is to compute SS_{AB} as

$$SS_{AB} = SS_{\text{subtotals}} - SS_A - SS_B. \tag{13-8}$$

The error sum of squares is found by subtraction as either

$$SS_E = SS_T - SS_{AB} - SS_A - SS_B \tag{13-9a}$$

or

$$SS_E = SS_T - SS_{\text{subtotals}}. \tag{13-9b}$$

Example 13-3

Aircraft primer paints are applied to aluminum surfaces by two methods: dipping and spraying. The purpose of the primer is to improve paint adhesion. Some parts can be primed using either application method and engineering is interested in learning whether three different primers differ in their adhesion properties. A factorial experiment is performed to investigate the effect of paint primer type and application method on paint adhesion. Three specimens are painted with each primer using each application method, a finish paint applied, and the adhesion force measured. The data from the experiment are shown in Table 13-5. The circled numbers in the cells are the cell totals $y_{ij\cdot}$. The sums of squares required to perform the analysis of variance are computed as follows:

$$SS_T = \sum_{i=1}^{a}\sum_{j=1}^{b}\sum_{k=1}^{n} y_{ijk}^2 - \frac{y_{\dots}^2}{abn}$$

$$= (4.0)^2 + (4.5)^2 + \cdots + (5.0)^2 - \frac{(89.8)^2}{18} = 10.72,$$

$$SS_{\text{types}} = \sum_{i=1}^{a} \frac{y_{i..}^2}{bn} - \frac{y_{\dots}^2}{abn}$$

$$= \frac{(28.7)^2 + (34.1)^2 + (27.0)^2}{6} - \frac{(89.8)^2}{18} = 4.58,$$

$$SS_{\text{methods}} = \sum_{j=1}^{b} \frac{y_{.j.}^2}{an} - \frac{y_{\dots}^2}{abn}$$

$$= \frac{(40.2)^2 + (49.6)^2}{9} - \frac{(89.8)^2}{18} = 4.91,$$

$$SS_{\text{interaction}} = \sum_{i=1}^{a}\sum_{j=1}^{b} \frac{y_{ij.}^2}{n} - \frac{y_{\dots}^2}{abn} - SS_{\text{types}} - SS_{\text{methods}}$$

$$= \frac{(12.8)^2 + (15.9)^2 + \ldots + (15.5)^2}{3} - \frac{(89.8)^2}{18} - 4.58 - 4.91 = 0.24,$$

and

$$SS_E = SS_T - SS_{\text{types}} - SS_{\text{method}} - SS_{\text{interaction}}$$

$$= 10.72 - 4.58 - 4.91 - 0.24 = 0.99.$$

The analysis of variance is summarized in Table 13-6. Since $F_{0.05,2,12} = 3.89$ and $F_{0.05,1,12} = 4.75$, we conclude that the main effects of primer type and application method affect adhesion force. Furthermore, since $1.5 < F_{0.05,2,12}$, there is no indication of interaction between these factors.

Table 13-5 Adhesion Force Data for Example 13-3

Primer Type	Application Method: Dipping	Application Method: Spraying	$y_{i..}$
1	4.0, 4.5, 4.3 (12.8)	5.4, 4.9, 5.6 (15.9)	28.7
2	5.6, 4.9, 5.4 (15.9)	5.8, 6.1, 6.3 (18.2)	34.1
3	3.8, 3.7, 4.0 (11.5)	5.5, 5.0, 5.0 (15.5)	27.0
$y_{.j.}$	40.2	49.6	$89.8 = y_{...}$

Table 13-6 Analysis of Variance for Example 13-3

Source of Variation	Sum of Squares	Degrees of Freedom	Mean Square	F_0
Primer types	4.581	2	2.291	27.86
Application methods	4.909	1	4.909	59.70
Interaction	0.241	2	0.121	1.47
Error	0.987	12	0.082	
Total	10.718	17		

A graph of the cell adhesion force averages $\bar{y}_{ij.}$ versus the levels of primer type for each application method is shown in Fig. 13-8. The absence of interaction is evident by the parallelism of the two lines. Furthermore, since a large response indicates greater adhesion force, we conclude that spraying is a superior application method and that primer type 2 is most effective.

Tests on Individual Means When both factors are fixed, comparisons between the individual means of either factor may be made using Tukey's test. When there is no interaction, these comparisons may be made using either the row averages $\bar{y}_{i..}$ or the column averages $\bar{y}_{.j.}$. However, when interaction is significant, comparisons between the means of one factor (say A) may be obscured by the AB interaction. In this case, we may apply Tukey's test to the means of factor A, with factor B set at a particular level.

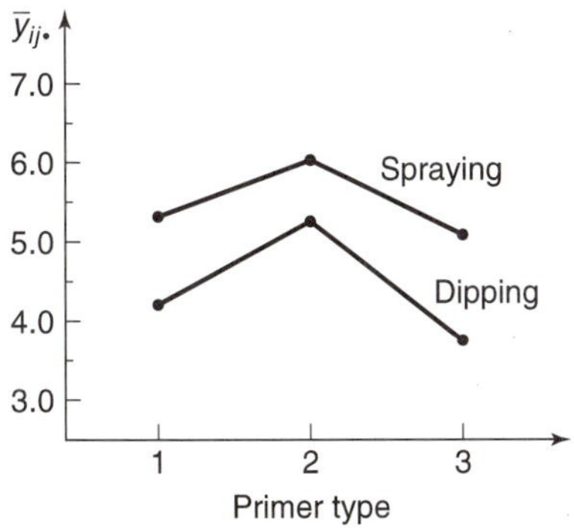

Figure 13-8 Graph of average adhesion force versus primer types for Example 13-3.

13-3.2 Model Adequacy Checking

Just as in the single-factor experiments discussed in Chapter 12, the residuals from a factorial experiment play an important role in assessing model adequacy. The residuals from a two-factor factorial experiment are

$$e_{ijk} = y_{ijk} - \bar{y}_{ij\cdot}\,.$$

That is, the residuals are just the difference between the observations and the corresponding cell averages.

Table 13-7 presents the residuals for the aircraft primer paint data in Example 13-3. The normal probability plot of these residuals is shown in Fig. 13-9. This plot has tails that do not fall exactly along a straight line passing through the center of the plot, indicating some potential problems with the normality assumption, but the deviation from normality does not appear severe. Figures 13-10 and 13-11 plot the residuals versus the levels of primer types and application methods, respectively. There is some indication that primer type 3 results in slightly lower variability in adhesion force than the other two primers. The graph of residuals versus fitted values $\hat{y}_{ijk} = \bar{y}_{ij\cdot}$ in Fig. 13-12 reveals no unusual or diagnostic pattern.

13-3.3 One Observation per Cell

In some cases involving a two-factor factorial experiment, we may have only one replicate, that is, only one observation per cell. In this situation there are exactly as many parameters

Table 13-7 Residuals for the Aircraft Primer Paint Experiment in Example 13-3

	Application Method	
Primer Type	Dipping	Spraying
1	−0.27, 0.23, 0.03	0.10, −0.40, 0.30
2	0.30, −0.40, 0.10	−0.27, 0.03, 0.23
3	−0.03, −0.13, 0.17	0.33, −0.17, −0.17

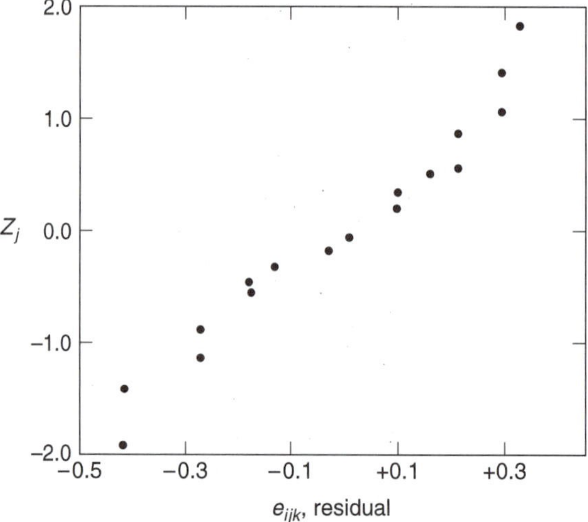

Figure 13-9 Normal probability plot of the residuals from Example 13-3.

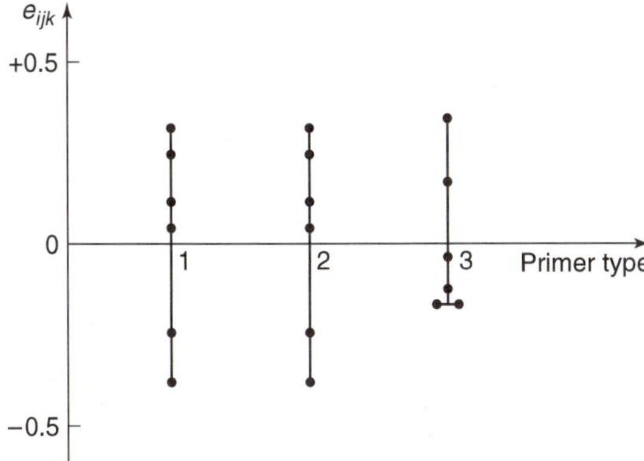

Figure 13-10 Plot of residuals versus primer type.

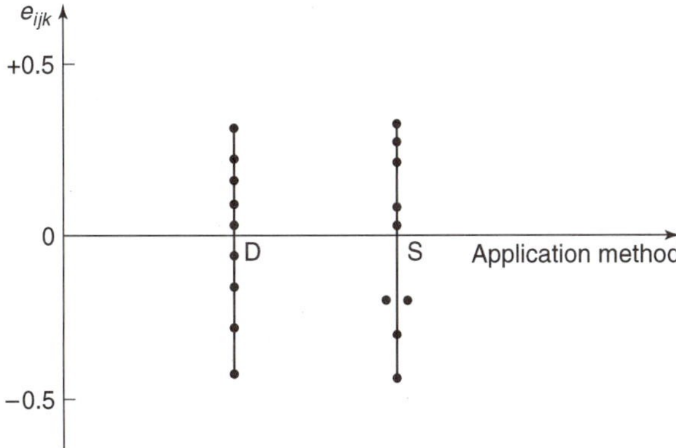

Figure 13-11 Plot of residuals versus application method.

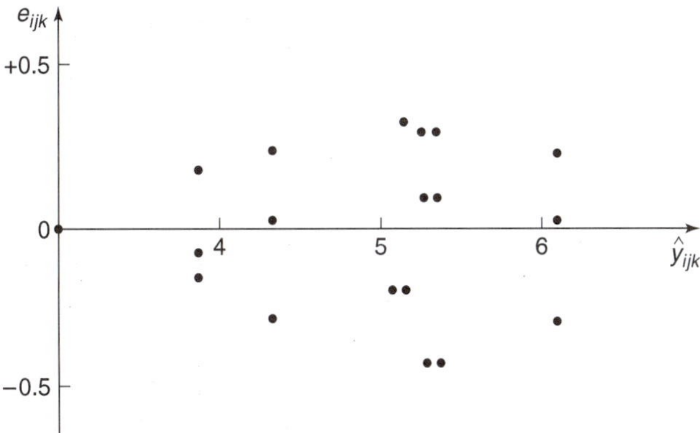

Figure 13-12 Plot of residuals versus predicted values $\hat{y}_{ijk\cdot} = \overline{y}_{ij\cdot}$.

in the analysis-of-variance model as there are observations, and the error degrees of freedom is zero. Thus, it is not possible to test a hypothesis about the main effects and interactions unless some additional assumptions are made. The usual assumption is to ignore the interaction effect and use the interaction mean square as an error mean square. Thus the analysis is equivalent to the analysis used in the randomized block design. This no-interaction assumption can be dangerous, and the experimenter should carefully examine the data and the residuals for indications that there really is interaction present. For more details, see Montgomery (2001).

13-3.4 The Random-Effects Model

So far we have considered the case where A and B are fixed factors. We now consider the situation in which the levels of both factors are selected at random from larger populations of factor levels, and we wish to extend our conclusions to the sampled population of factor levels. The observations are represented by the model

$$y_{ijk} = \mu + \tau_i + \beta_j + (\tau\beta)_{ij} + \epsilon_{ijk} \begin{cases} i = 1, 2, \ldots, a, \\ j = 1, 2, \ldots, b, \\ k = 1, 2, \ldots, n, \end{cases} \tag{13-10}$$

where the parameters τ_i, β_j, $(\tau\beta)_{ij}$, and ϵ_{ijk} are random variables. Specifically, we assume that τ_i is NID(0, σ_τ^2), β_j is NID(0, σ_β^2), $(\tau\beta)_{ij}$ is NID(0, $\sigma_{\tau\beta}^2$), and ϵ_{ijk} is NID(0, σ^2). The variance of any observation is

$$V(y_{ijk}) = \sigma_\tau^2 + \sigma_\beta^2 + \sigma_{\tau\beta}^2 + \sigma^2,$$

and σ_τ^2, σ_β^2, $\sigma_{\tau\beta}^2$, and σ^2 are called *variance components.* The hypotheses that we are interested in testing are H_0: $\sigma_\tau^2 = 0$, H_0: $\sigma_\beta^2 = 0$, and H_0: $\sigma_{\tau\beta}^2 = 0$. Notice the similarity to the one-way classification random-effects model.

The basic analysis of variance remains unchanged; that is, SS_A, SS_B, SS_{AB}, SS_T, and SS_E are all calculated as in the fixed-effects case. To construct the test statistics, we must examine the expected mean squares. They are

$$\begin{aligned} E(MS_A) &= \sigma^2 + n\sigma_{\tau\beta}^2 + bn\sigma_\tau^2, \\ E(MS_B) &= \sigma^2 + n\sigma_{\tau\beta}^2 + an\sigma_\beta^2, \\ E(MS_{AB}) &= \sigma^2 + n\sigma_{\tau\beta}^2, \end{aligned} \tag{13-11}$$

and

$$E(MS_E) = \sigma^2.$$

Note from the expected mean squares that the appropriate statistic for testing H_0: $\sigma_{\tau\beta}^2 = 0$ is

$$F_0 = \frac{MS_{AB}}{MS_E}, \tag{13-12}$$

since under H_0 both the numerator and denominator of F_0 have expectation σ^2, and only if H_0 is false is $E(MS_{AB})$ greater than $E(MS_E)$. The ratio F_0 is distributed as $F_{(a-1)(b-1),\, ab(n-1)}$. Similarly, for testing H_0: $\sigma_\tau^2 = 0$, we would use

$$F_0 = \frac{MS_A}{MS_{AB}}, \tag{13-13}$$

which is distributed as $F_{a-1,(a-1)(b-1)}$, and for testing H_0: $\sigma_\beta^2 = 0$, the statistic is

$$F_0 = \frac{MS_B}{MS_{AB}}, \tag{13-14}$$

which is distributed as $F_{b-1,(a-1)(b-1)}$. These are all upper-tail, one-tail tests. Notice that these test statistics are not the same as those used if both factors A and B are fixed. The expected mean squares are always used as a guide to test statistic construction.

The variance components may be estimated by equating the observed mean squares to their expected values and solving for the variance components. This yields

$$\begin{aligned}\hat{\sigma}^2 &= MS_E,\\ \hat{\sigma}^2_{\tau\beta} &= \frac{MS_{AB} - MS_E}{n},\\ \hat{\sigma}^2_{\beta} &= \frac{MS_B - MS_{AB}}{an},\\ \hat{\sigma}^2_{\tau} &= \frac{MS_A - MS_{AB}}{bn}.\end{aligned} \tag{13-15}$$

Example 13-4

Suppose that in Example 13-3, a large number of primers and several application methods could be used. Three primers, say 1, 2, and 3, were selected at random, as were the two application methods. The analysis of variance assuming the random effects model is shown in Table 13-8.

Notice that the first four columns in the analysis of variance table are exactly as in Example 13-3. Now, however, the F ratios are computed according to equations 13-12 through 13-14. Since $F_{0.05,2,12} = 3.89$, we conclude that interaction is not significant. Also, since $F_{0.05,2,2} = 19.0$ and $F_{0.05,1,2} = 18.5$, we conclude that both types and application methods significantly affect adhesion force, although primer type is just barely significant at $\alpha = 0.05$. The variance components may be estimated using equation 13-15 as follows:

$$\begin{aligned}\hat{\sigma}^2 &= 0.08,\\ \hat{\sigma}^2_{\tau\beta} &= \frac{0.12 - 0.08}{3} = 0.0133,\\ \hat{\sigma}^2_{\tau} &= \frac{2.29 - 0.12}{6} = 0.36,\\ \hat{\sigma}^2_{\beta} &= \frac{4.91 - 0.12}{9} = 0.53.\end{aligned}$$

Clearly, the two largest variance components are for primer types ($\hat{\sigma}^2_{\tau} = 0.36$) and application methods ($\hat{\sigma}^2_{\beta} = 0.53$).

13-3.5 The Mixed Model

Now suppose that one of the factors, A, is fixed and the other, B, is random. This is called the *mixed model* analysis of variance. The linear model is

$$y_{ijk} = \mu + \tau_i + \beta_j + (\tau\beta)_{ij} + \epsilon_{ijk} \begin{cases} i = 1, 2, \ldots, a, \\ j = 1, 2, \ldots, b, \\ k = 1, 2, \ldots, n. \end{cases} \tag{13-16}$$

Table 13-8 Analysis of Variance for Example 13-4

Source of Variation	Sum of Squares	Degrees of Freedom	Mean Square	F_0
Primer types	4.58	2	2.29	19.08
Application methods	4.91	1	4.91	40.92
Interaction	0.24	2	0.12	1.5
Error	0.99	12	0.08	
Total	10.72	17		

In this model, τ_i is a fixed effect defined such that $\sum_{i=1}^{a} \tau_i = 0$, β_j is a random effect, the interaction term $(\tau\beta)_{ij}$ is a random effect, and ϵ_{ijk} is a NID(0, σ^2) random error. It is also customary to assume that β_j is NID(0, σ_β^2) and that the interaction elements $(\tau\beta)_{ij}$ are normal random variables with mean zero and variance $[(a-1)/a]\sigma_{\tau\beta}^2$. The interaction elements are not all independent.

The expected mean squares in this case are

$$
\begin{aligned}
E(MS_A) &= \sigma^2 + n\sigma_{\tau\beta}^2 + \frac{bn\sum_{i=1}^{a}\tau_i^2}{a-1}, \\
E(MS_B) &= \sigma^2 + an\sigma_\beta^2, \\
E(MS_{AB}) &= \sigma^2 + n\sigma_{\tau\beta}^2,
\end{aligned}
\tag{13-17}
$$

and

$$E(MS_E) = \sigma^2.$$

Therefore, the appropriate test statistic for testing H_0: $\tau_i = 0$ is

$$F_0 = \frac{MS_A}{MS_{AB}}, \tag{13-18}$$

which is distributed as $F_{a-1,(a-1)(b-1)}$. For testing H_0: $\sigma_\beta^2 = 0$, the test statistic is

$$F_0 = \frac{MS_B}{MS_E}, \tag{13-19}$$

which is distributed as $F_{b-1, ab(n-1)}$. Finally, for testing H_0:$\sigma_{\tau\beta}^2 = 0$, we would use

$$F_0 = \frac{MS_{AB}}{MS_E}, \tag{13-20}$$

which is distributed as $F_{(a-1)(b-1), ab(n-1)}$.

The variance components σ_β^2, $\sigma_{\tau\beta}^2$, and σ^2 may be estimated by eliminating the first equation from equation 13-17, leaving three equations in three unknowns, the solutions of which are

$$
\begin{aligned}
\hat{\sigma}_\beta^2 &= \frac{MS_B - MS_E}{an}, \\
\hat{\sigma}_{\tau\beta}^2 &= \frac{MS_{AB} - MS_E}{n},
\end{aligned}
$$

and

$$\hat{\sigma}^2 = MS_E. \tag{13-21}$$

This general approach can be used to estimate the variance components in *any* mixed model. After eliminating the mean squares containing fixed factors, there will always be a set of equations remaining that can be solved for the variance components, Table 13-9 summarizes the analysis of variance for the two-factor mixed model.

Table 13-9 Analysis of Variance for the Two-Factor Mixed Model

Source of Variation	Sum of Squares	Degrees of Freedom	Mean Square	Expected Mean Square	F_0
Rows (A)	SS_A	$a-1$	MS_A	$\sigma^2 + n\sigma_{\tau\beta}^2 + bn\Sigma\tau_i^2 / (a-1)$	$\frac{MS_A}{MS_{AB}}$
Columns (B)	SS_B	$b-1$	MS_B	$\sigma^2 + an\sigma_\beta^2$	$\frac{MS_B}{MS_E}$
Interaction	SS_{AB}	$(a-1)(b-1)$	MS_{AB}	$\sigma^2 + n\sigma_{\tau\beta}^2$	$\frac{MS_{AB}}{MS_E}$
Error	SS_E	$ab(n-1)$	MS_E	σ^2	
Total	SS_T	$abn-1$			

13-4 GENERAL FACTORIAL EXPERIMENTS

Many experiments involve more than two factors. In this section we introduce the case where there are a levels of factor A, b levels of factor B, c levels of factor C, and so on, arranged in a factorial experiment. In general, there will be $abc \cdots n$ total observations, if there are n replicates of the complete experiment.

For example, consider the three-factor experiment with underlying model

$$y_{ijkl} = \mu + \tau_i + \beta_j + \gamma_k + (\tau\beta)_{ij} + (\tau\gamma)_{ik} + (\beta\gamma)_{jk} + (\tau\beta\gamma)_{ijk} + \epsilon_{ijkl} \begin{cases} i = 1,2,\ldots,a, \\ j = 1,2,\ldots,b, \\ k = 1,2,\ldots,c, \\ l = 1,2,\ldots,n. \end{cases} \tag{13-22}$$

Assuming that A, B, and C are fixed, the analysis of variance is shown in Table 13-10. Note that there must be at least two replicates ($n \geq 2$) to compute an error sum of squares. The F-tests on main effects and interactions follow directly from the expected mean squares.

Computing formulas for the sums of squares in Table 13-10 are easily obtained. The total sum of squares is, using the obvious "dot" notation,

$$SS_T = \sum_{i=1}^{a}\sum_{j=1}^{b}\sum_{k=1}^{c}\sum_{l=1}^{n} y_{ijkl}^2 - \frac{y_{....}^2}{abcn}. \tag{13-23}$$

The sum of squares for the main effects are computed from the totals for factors $A\,(y_{i...})$, $B(y_{.j..})$, and $C(y_{..k.})$ as follows:

$$SS_A = \sum_{i=1}^{a} \frac{y_{i...}^2}{bcn} - \frac{y_{....}^2}{abcn}, \tag{13-24}$$

$$SS_B = \sum_{j=1}^{b} \frac{y_{.j..}^2}{acn} - \frac{y_{....}^2}{abcn}, \tag{13-25}$$

$$SS_C = \sum_{k=1}^{c} \frac{y_{..k.}^2}{abn} - \frac{y_{....}^2}{abcn}. \tag{13-26}$$

To compute the two-factor interaction sums of squares, the totals for the $A \times B$, $A \times C$, and $B \times C$ cells are needed. It may be helpful to collapse the original data table into three two-way tables in order to compute these totals. The sums of squares are

$$\begin{aligned} SS_{AB} &= \sum_{i=1}^{a}\sum_{j=1}^{b} \frac{y_{ij..}^2}{cn} - \frac{y_{....}^2}{abcn} - SS_A - SS_B \\ &= SS_{\text{subtotals}(AB)} - SS_A - SS_B, \end{aligned} \tag{13-27}$$

$$\begin{aligned} SS_{AC} &= \sum_{i=1}^{a}\sum_{k=1}^{c} \frac{y_{i.k.}^2}{bn} - \frac{y_{....}^2}{abcn} - SS_A - SS_C \\ &= SS_{\text{subtotals}(AC)} - SS_A - SS_C, \end{aligned} \tag{13-28}$$

and

$$\begin{aligned} SS_{BC} &= \sum_{j=1}^{b}\sum_{k=1}^{c} \frac{y_{.jk.}^2}{an} - \frac{y_{....}^2}{abcn} - SS_B - SS_C \\ &= SS_{\text{subtotals}(BC)} - SS_B - SS_C. \end{aligned} \tag{13-29}$$

The three-factor interaction sum of squares is computed from the three-way cell totals $y_{ijk.}$ as

$$SS_{ABC} = \sum_{i=1}^{a}\sum_{j=1}^{b}\sum_{k=1}^{c} \frac{y_{ijk.}^2}{n} - \frac{y_{....}^2}{abcn} - SS_A - SS_B - S_C - SS_{AB} - SS_{AC} - SS_{BC} \tag{13-30a}$$

Table 13-10 The Analysis-of-Variance Table for the Three-Factor Fixed-Effects Model

Source of Variation	Sum of Squares	Degrees of Freedom	Mean Square	Expected Mean Squares	F_0
A	SS_A	$a-1$	MS_A	$\sigma^2 + \frac{bcn\Sigma\tau_i^2}{a-1}$	$\frac{MS_A}{MS_E}$
B	SS_B	$b-1$	MS_B	$\sigma^2 + \frac{acn\Sigma\beta_j^2}{b-1}$	$\frac{MS_B}{MS_E}$
C	SS_C	$c-1$	MS_C	$\sigma^2 + \frac{abn\Sigma\gamma_k^2}{c-1}$	$\frac{MS_C}{MS_E}$
AB	SS_{AB}	$(a-1)(b-1)$	MS_{AB}	$\sigma^2 + \frac{cn\Sigma\Sigma(\tau\beta)_{ij}^2}{(a-1)(b-1)}$	$\frac{MS_{AB}}{MS_E}$
AC	SS_{AC}	$(a-1)(c-1)$	MS_{AC}	$\sigma^2 + \frac{bn\Sigma\Sigma(\tau\gamma)_{ik}^2}{(a-1)(c-1)}$	$\frac{MS_{AC}}{MS_E}$
BC	SS_{BC}	$(b-1)(c-1)$	MS_{BC}	$\sigma^2 + \frac{an\Sigma\Sigma(\beta\gamma)_{jk}^2}{(b-1)(c-1)}$	$\frac{MS_{BC}}{MS_E}$
ABC	SS_{ABC}	$(a-1)(b-1)(c-1)$	MS_{ABC}	$\sigma^2 + \frac{n\Sigma\Sigma\Sigma(\tau\beta\gamma)_{ijk}^2}{(a-1)(b-1)(c-1)}$	$\frac{MS_{ABC}}{MS_E}$
Error	SS_E	$abc(n-1)$	MS_E	σ^2	
Total	SS_T	$abcn-1$			

$$= SS_{\text{subtotals}(ABC)} - SS_A - SS_B - SS_C - SS_{AB} - SS_{AC} - SS_{BC}. \tag{13-30b}$$

The error sum of squares may be found by subtracting the sum of squares for each main effect and interaction from the total sum of squares, or by

$$SS_E = SS_T - SS_{\text{subtotals}(ABC)}. \tag{13-31}$$

Example 13-5

A mechanical engineer is studying the surface roughness of a part produced in a metal-cutting operation. Three factors, feed rate (A), depth of cut (B), and tool angle (C), are of interest. All three factors have been assigned two levels, and two replicates of a factorial design are run. The coded data are shown in Table 13-11. The three-way cell totals $y_{ijk\cdot}$ are circled in this table.

The sums of squares are calculated as follows, using equations 13-23 to 13-31:

$$SS_T = \sum_{i=1}^{a}\sum_{j=1}^{b}\sum_{k=1}^{c}\sum_{l=1}^{n} y_{ijkl}^2 - \frac{y_{....}^2}{abcn} = 2051 - \frac{(177)^2}{16} = 92.9375,$$

$$SS_A = \sum_{i=1}^{a} \frac{y_{i...}^2}{bcn} - \frac{y_{....}^2}{abcn}$$

$$= \frac{(75)^2 + (102)^2}{8} - \frac{(177)^2}{16} = 45.5625,$$

$$SS_B = \sum_{j=1}^{b} \frac{y_{.j..}^2}{acn} - \frac{y_{....}^2}{abcn}$$

$$= \frac{(82)^2 + (95)^2}{8} - \frac{(177)^2}{16} = 10.5625,$$

$$SS_C = \sum_{k=1}^{c} \frac{y_{..k.}^2}{abn} - \frac{y_{....}^2}{abcn}$$

$$= \frac{(85)^2 + (92)^2}{8} - \frac{(177)^2}{16} = 3.0625,$$

$$SS_{AB} = \sum_{i=1}^{a}\sum_{j=1}^{b} \frac{y_{ij..}^2}{cn} - \frac{y_{....}^2}{abcn} - SS_A - SS_B$$

$$= \frac{(37)^2 + (38)^2 + (45)^2 + (57)^2}{4} - \frac{(177)^2}{16} - 45.5625 - 10.5625$$

$$= 7.5625,$$

$$SS_{AC} = \sum_{i=1}^{a}\sum_{k=1}^{c} \frac{y_{i.k.}^2}{bn} - \frac{y_{....}^2}{abcn} - SS_A - SS_C$$

$$= \frac{(36)^2 + (39)^2 + (49)^2 + (53)^2}{4} - \frac{(177)^2}{16} - 45.5625 - 3.0625$$

$$= 0.0625,$$

$$SS_{BC} = \sum_{j=1}^{b}\sum_{k=1}^{c} \frac{y_{.jk.}^2}{an} - \frac{y_{....}^2}{abcn} - SS_B - SS_C$$

$$= \frac{(38)^2 + (44)^2 + (47)^2 + (48)^2}{4} - \frac{(177)^2}{16} - 10.5625 - 3.0625$$

$$= 1.5625,$$

$$SS_{ABC} = \sum_{i=1}^{a}\sum_{j=1}^{b}\sum_{k=1}^{c}\frac{y_{ijk.}^2}{n} - \frac{y_{....}^2}{abcn} - SS_A - SS_B - SS_C - SS_{AB} - SS_{AC} - SS_{BC}$$

$$= \frac{(16)^2+(21)^2+\cdots+(30)^2}{2} - \frac{(177)^2}{16} - 45.5625 - 10.5625 - 3.0625 - 7.5625 - 0.0625 - 1.5625$$

$$= 5.0625,$$

$$SS_E = SS_T - SS_{\text{subtotals}(ABC)}$$

$$= 92.9375 - 73.4375 = 19.5000.$$

The analysis of variance is summarized in Table 13-12. Feed rate has a significant effect on surface finish ($\alpha < 0.01$), as does the depth of cut ($0.05 < \alpha < 0.10$). There is some indication of a mild interaction between these factors, as the F-test for the AB interaction is just less than the 10% critical value.

Obviously factorial experiments with three or more factors are complicated and require many runs, particularly if some of the factors have several (more than two) levels. This leads us to consider a class of factorial designs with all factors at two levels. These designs are extremely easy to set up and analyze, and as we will see, it is possible to greatly reduce the number of experimental runs through the technique of fractional replication.

Table 13-11 Coded Surface Roughness Data for Example 13-5

	Depth of Cut (B)				
	0.025 inch		0.040 inch		
Feed Rate (A)	Tool Angle (C)		Tool Angle (C)		
	15°	25°	15°	25°	$y_{i\cdots}$
20 in./min	9 7 (16)	11 10 (21)	9 11 (20)	10 8 (18)	75
30 in./min	10 12 (22)	10 13 (23)	12 15 (27)	16 14 (30)	102
$B \times C$ totals $y_{\cdot jk\cdot}$	38	44	47	48	$177 = y_{....}$

$A \times B$ Totals $y_{ij\cdot\cdot}$

A/B	0.025	0.040
20	37	38
30	45	57
$y_{.j..}$	82	95

$A \times C$ Totals $y_{i\cdot k\cdot}$

A/C	15	25
20	36	39
30	49	53
$y_{..k.}$	85	92

Table 13-12 Analysis of Variance for Example 13-5

Source of Variation	Sum of Squares	Degrees of Freedom	Mean Square	F_0
Feed rate (A)	45.5625	1	45.5625	18.69[a]
Depth of cut (B)	10.5625	1	10.5625	4.33[b]
Tool angle (C)	3.0625	1	3.0625	1.26
AB	7.5625	1	7.5625	3.10
AC	0.0625	1	0.0625	0.03
BC	1.5625	1	1.5625	0.64
ABC	5.0625	1	5.0625	2.08
Error	19.5000	8	2.4375	
Total	92.9375	15		

[a]Significant at 1%.

[b]Significant at 10%.

13-5 THE 2^k FACTORIAL DESIGN

There are certain special types of factorial designs that are very useful. One of these is a factorial design with k factors, each at two levels. Because each complete replicate of the design has 2^k runs or treatment combinations, the arrangement is called a 2^k factorial design. These designs have a greatly simplified statistical analysis, and they also form the basis of many other useful designs.

13-5.1 The 2^2 Design

The simplest type of 2^k design is the 2^2, that is, two factors, A and B, each at two levels. We usually think of these levels as the "low" and "high" levels of the factor. The 2^2 design is shown in Fig. 13-13. Note that the design can be represented geometrically as a square, with the $2^2 = 4$ runs forming the corners of the square. A special notation is used to represent the treatment combinations. In general a treatment combination is represented by a series of lowercase letters. If a letter is present, then the corresponding factor is run at the high level

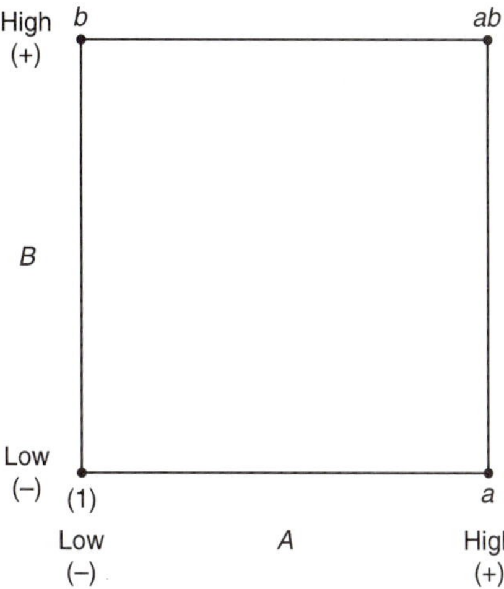

Figure 13-13 The 2^2 factorial design.

in that treatment combination; if it is absent, the factor is run at its low level. For example, treatment combination a indicates that factor A is at the high level and factor B is at the low level. The treatment combination with both factors at the low level is denoted (1). This notation is used throughout the 2^k design series. For example, the treatment combination in a 2^4 design with A and C at the high level and B and D at the low level is denoted ac.

The effects of interest in the 2^2 design are the main effects A and B and the two-factor interaction AB. Let (1), a, b, and ab also represent the totals of all n observations taken at these design points. It is easy to estimate the effects of these factors. To estimate the main effect of A we would average the observations on the right side of the square, where A is at the high level, and subtract from this the average of the observations on the left side of the square, where A is at the low level, or

$$\begin{aligned} A &= \frac{a+ab}{2n} - \frac{b+(1)}{2n} \\ &= \frac{1}{2n}\left[a+ab-b-(1)\right]. \end{aligned} \tag{13-32}$$

Similarly, the main effect of B is found by averaging the observations on the top of the square, where B is at the high level, and subtracting the average of the observations on the bottom of the square, where B is at the low level:

$$\begin{aligned} B &= \frac{b+ab}{2n} - \frac{a+(1)}{2n} \\ &= \frac{1}{2n}\left[b+ab-a-(1)\right]. \end{aligned} \tag{13-33}$$

Finally, the AB interaction is estimated by taking the difference in the diagonal averages in Fig. 13-13, or

$$\begin{aligned} AB &= \frac{ab+(1)}{2n} - \frac{a+b}{2n} \\ &= \frac{1}{2n}\left[ab+(1)-a-b\right]. \end{aligned} \tag{13-34}$$

The quantities in brackets in equations 13-32, 13-33, and 13-34 are called *contrasts*. For example, the A contrast is

$$\text{Contrast}_A = a + ab - b - (1).$$

In these equations, the contrast coefficients are always either +1 or −1. A table of plus and minus signs, such as Table 13-13, can be used to determine the sign of each treatment combination for a particular contrast. The column headings for Table 13-13 are the main effects A and B, AB interaction, and I, which represents the total. The row headings are the treatment combinations. Note that the signs in the AB column are the products of signs from

Table 13-13 Signs for Effects in the 2^2 Design

Treatment	Factorial Effect			
Combinations	I	A	B	AB
(1)	+	−	−	+
a	+	+	−	−
b	+	−	+	−
ab	+	+	+	+

columns A and B. To generate a contrast from this table, multiply the signs in the appropriate column of Table 13-13 by the treatment combinations listed in the rows and add.

To obtain the sums of squares for A, B, and AB, we can use equation 12-18, which expresses the relationship between a single-degree-of-freedom contrast and its sum of squares:

$$SS = \frac{(\text{Contrast})^2}{n\sum(\text{contrast coefficients})^2}. \tag{13-35}$$

Therefore, the sums of squares for A, B, and AB are

$$SS_A = \frac{[a + ab - b - (1)]^2}{4n},$$

$$SS_B = \frac{[b + ab - a - (1)]^2}{4n},$$

$$SS_{AB} = \frac{[ab + (1) - a - b]^2}{4n}.$$

The analysis of variance is completed by computing the total sum of squares SS_T (with $4n - 1$ degrees of freedom) as usual, and obtaining the error sum of squares SS_E [with $4(n - 1)$ degrees of freedom] by subtraction.

Example 13-6

An article in the *AT&T Technical Journal* (March/April, 1986, Vol. 65, p. 39) describes the application of two-level experimental designs to integrated circuit manufacturing. A basic processing step in this industry is to grow an epitaxial layer on polished silicon wafers. The wafers are mounted on a susceptor and positioned inside a bell jar. Chemical vapors are introduced through nozzles near the top of the jar. The susceptor is rotated and heat is applied. These conditions are maintained until the epitaxial layer is thick enough.

Table 13-14 presents the results of a 2^2 factorial design with $n = 4$ replicates using the factors A = deposition time and B = arsenic flow rate. The two levels of deposition time are $-$ = short and $+$ = long, and the two levels of arsenic flow rate are $-$ = 55% and $+$ = 59%. The response variable is epitaxial layer thickness (μm). We may find the estimates of the effects using equations 13-32, 13-33, and 13-34 as follows:

$$\begin{aligned} A &= \frac{1}{2n}[a + ab - b - (1)] \\ &= \frac{1}{2(4)}[59.299 + 59.156 - 55.686 - 56.081] = 0.836, \end{aligned}$$

Table 13-14 The 2^2 Design for the Epitaxial Process Experiment

Treatment	Design Factors				Thickness (μm)	
Combinations	A	B	AB	Thickness (μm)	Total	Average
(1)	$-$	$-$	$+$	14.037, 14.165, 13.972, 13.907	56.081	14.021
a	$+$	$-$	$-$	14.821, 14.757, 14.843, 14.878	59.299	14.825
b	$-$	$+$	$-$	13.880, 13.860, 14.032, 13.914	55.686	13.922
ab	$+$	$+$	$+$	14.888, 14.921, 14.415, 14.932	59.156	14.789

$$B = \frac{1}{2n}\left[b + ab - a - (1)\right]$$
$$= \frac{1}{2(4)}[55.686 + 59.156 - 59.299 - 56.081] = -0.067,$$
$$AB = \frac{1}{2n}\left[ab + (1) - a - b\right]$$
$$= \frac{1}{2(4)}[59.156 + 56.081 - 59.299 - 55.686] = 0.032.$$

The numerical estimates of the effects indicate that the effect of deposition time is large and has a positive direction (increasing deposition time increases thickness), since changing deposition time from low to high changes the mean epitaxial layer thickness by 0.836 μm. The effects of arsenic flow rate (B) and the AB interaction appear small.

The magnitude of these effects may be confirmed with the analysis of variance. The sums of squares for A, B, and AB are computed using equation 13-35:

$$SS = \frac{(\text{Contrast})^2}{n \cdot 4},$$
$$SS_A = \frac{\left[a + ab - b - (1)\right]^2}{16}$$
$$= \frac{[6.688]^2}{16}$$
$$= 2.7956,$$
$$SS_B = \frac{\left[b + ab - a - (1)\right]^2}{16}$$
$$= \frac{[-0.538]^2}{16}$$
$$= 0.0181,$$
$$SS_{AB} = \frac{\left[ab + (1) - a - b\right]^2}{16}$$
$$= \frac{[0.256]^2}{16}$$
$$= 0.0040.$$

The analysis of variance is summarized in Table 13-15. This confirms our conclusions obtained by examining the magnitude and direction of the effects; deposition time affects epitaxial layer thickness, and from the direction of the effect estimates we know that longer deposition times lead to thicker epitaxial layers.

Table 13-15 Analysis of Variance for the Epitaxial Process Experiment

Source of Variation	Sum of Squares	Degrees of Freedom	Mean Square	F_0
A (deposition time)	2.7956	1	2.7956	134.50
B (arsenic flow)	0.0181	1	0.0181	0.87
AB	0.0040	1	0.0040	0.19
Error	0.2495	12	0.0208	
Total	3.0672	15		

Residual Analysis It is easy to obtain the residuals from a 2^k design by fitting a regression model to the data. For the epitaxial process experiment, the regression model is

$$y = \beta_0 + \beta_1 x_1 + \epsilon,$$

since the only active variable is deposition time, which is represented by x_1. The low and high levels of deposition time are assigned the values $x_1 = -1$ and $x_1 = +1$, respectively. The fitted model is

$$\hat{y} = 14.389 + \left(\frac{0.836}{2}\right)x_1,$$

where the intercept $\hat{\beta}_0$ is the grand average of all 16 observations $(\bar{y})$ and the slope $\hat{\beta}_1$ is one-half the effect estimate for deposition time. The reason the regression coefficient is one-half the effect estimate is because regression coefficients measure the effect of a unit change in x_1 on the mean of y, and the effect estimate is based on a two-unit change (from –1 to +1).

This model can be used to obtain the predicted values at the four points in the design. For example, consider the point with low deposition time ($x_1 = -1$) and low arsenic flow rate. The predicted value is

$$\hat{y} = 14.389 + \left(\frac{0.836}{2}\right)(-1) = 13.971\ \mu\text{m},$$

and the residuals would be

$$e_1 = 14.037 - 13.971 = 0.066,$$

$$e_2 = 14.165 - 13.971 = 0.194,$$

$$e_3 = 13.972 - 13.971 = 0.001,$$

$$e_4 = 13.907 - 13.971 = -0.064.$$

It is easy to verify that for low deposition time ($x_1 = -1$) and high arsenic flow rate, $\hat{y} = 14.389 + (0.836/2)(-1) = 13.971\ \mu\text{m}$, the remaining predicted values and residuals are

$$e_5 = 13.880 - 13.971 = -0.091,$$

$$e_6 = 13.860 - 13.971 = -0.111,$$

$$e_7 = 14.032 - 13.971 = 0.061,$$

$$e_8 = 13.914 - 13.971 = -0.057,$$

that for high deposition time ($x_1 = +1$) and low arsenic flow rate, $\hat{y} = 14.389 + 0.836/2)(+1) = 14.807\ \mu\text{m}$, they are

$$e_9 = 14.821 - 14.807 = 0.014,$$

$$e_{10} = 14.757 - 14.807 = -0.050,$$

$$e_{11} = 14.843 - 14.807 = 0.036,$$

$$e_{12} = 14.878 - 14.807 = 0.071,$$

and that for high deposition time ($x_1 = +1$) and high arsenic flow rate, $\hat{y} = 14.389 + (0.836/2)(+1) = 14.807\ \mu\text{m}$, they are

$$e_{13} = 14.888 - 14.807 = 0.081,$$

$$e_{14} = 14.921 - 14.807 = 0.114,$$

$$e_{15} = 14.415 - 14.807 = -0.392,$$

$$e_{16} = 14.932 - 14.807 = 0.125.$$

A normal probability plot of these residuals is shown in Fig. 13-14. This plot indicates that one residual, $e_{15} = -0.392$, is an outlier. Examining the four runs with high deposition time and high arsenic flow rate reveals that observation $y_{15} = 14.415$ is considerably smaller than the other three observations at that treatment combination. This adds some additional evidence to the tentative conclusion that observation 15 is an outlier. Another possibility is that there are some process variables that affect the *variability* in epitaxial layer thickness, and if we could discover which variables produce this effect, then it might be possible to adjust these variables to levels that would minimize the variability in epitaxial layer thickness. This would have important implications in subsequent manufacturing stages. Figures 13-15 and 13-16 are plots of residuals versus deposition time and arsenic flow rate, respectively. Apart from the unusually large residual associated with y_{15}, there is no strong evidence that either deposition time or arsenic flow rate influences the variability in epitaxial layer thickness.

Figure 13-17 shows the estimated standard deviation of epitaxial layer thickness at all four runs in the 2^2 design. These standard deviations were calculated using the data in Table 13-14. Notice that the standard deviation of the four observations with A and B at the high level is considerably larger than the standard deviations at any of the other three design

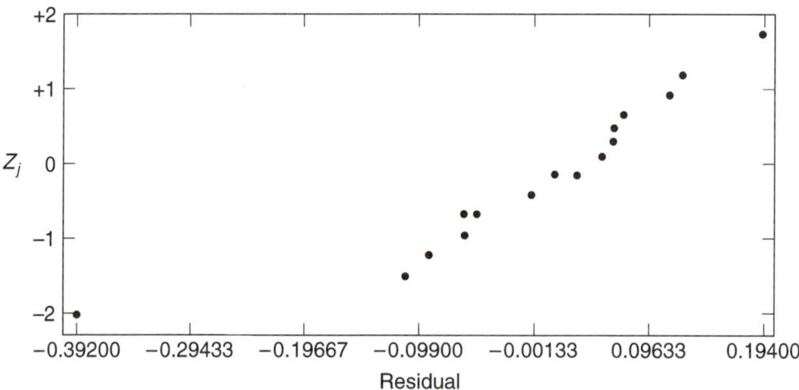

Figure 13-14 Normal probability plot of residuals for the epitaxial process experiment.

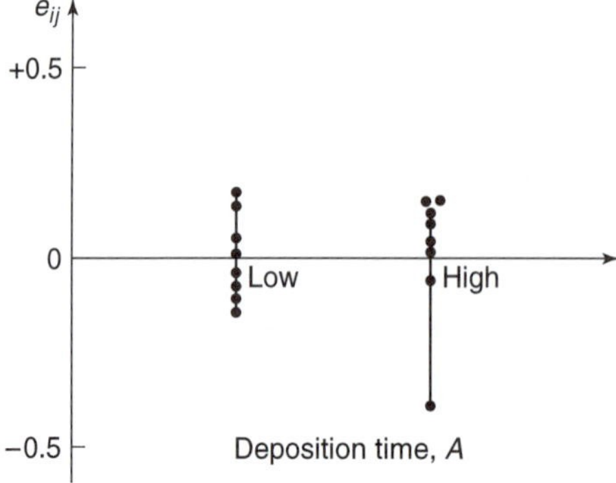

Figure 13-15 Plot of residuals versus deposition time.

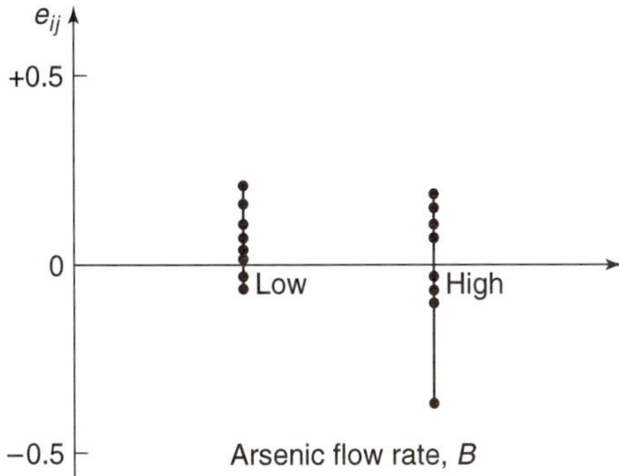

Figure 13-16 Plot of residuals versus arsenic flow rate.

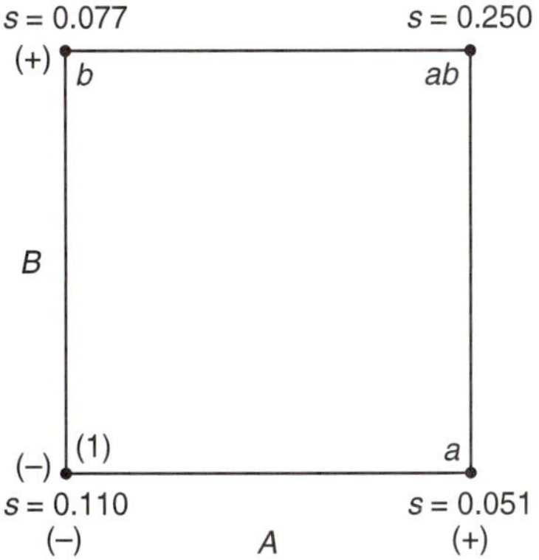

Figure 13-17 The estimated standard deviations of epitaxial layer thickness at the four runs in the 2^2 design.

points. Most of this difference is attributable to the unusually low thickness measurement associated with y_{15}. The standard deviation of the four observations with A and B at the low levels is also somewhat larger than the standard deviations at the remaining two runs. This could be an indication that there are other process variables not included in this experiment that affect the variability in epitaxial layer thickness. Another experiment to study this possibility, involving other process variables, could be designed and conducted (indeed, the original paper shows that there are two additional factors, unconsidered in this example, that affect process variability).

13-5.2 The 2^k Design for $k \geq 3$ Factors

The methods presented in the previous section for factorial designs with $k = 2$ factors each at two levels can be easily extended to more than two factors. For example, consider $k = 3$ factors, each at two levels. This design is a 2^3 factorial design, and it has eight treatment combinations. Geometrically, the design is a cube as shown in Fig. 13-18, with the eight runs forming the corners of the cube. This design allows three main effects to be estimated (A, B, and C) along with three two-factor interactions (AB, AC, and BC) and a three-factor interaction (ABC).

The main effects can be estimated easily. Remember that (1), a, b, ab, c, ac, bc, and abc represent the total of all n replicates at each of the eight treatment combinations in the design. Referring to the cube in Fig. 13-18, we would estimate the main effect of A by averaging the four treatment combinations on the right side of the cube, where A is at the high

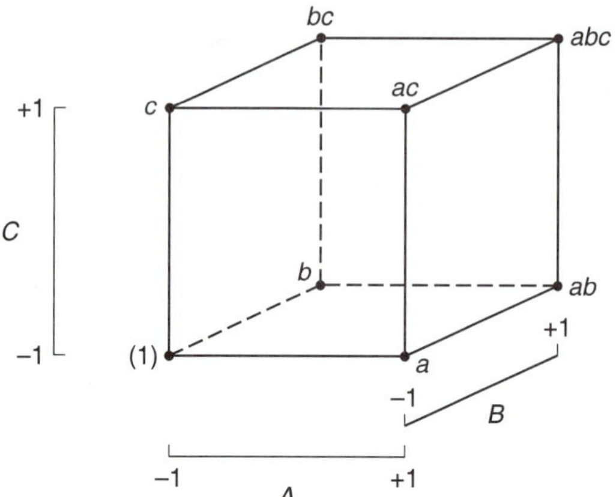

Figure 13-18 The 2^3 design.

level, and subtracting from that quantity the average of the four treatment combinations on the left side of the cube, where A is at the low level. This gives

$$A = \frac{1}{4n}[a + ab + ac + abc - b - c - bc - (1)]. \tag{13-36}$$

In a similar manner the effect of B is the average difference of the four treatment combinations in the back face of the cube and the four in the front, or

$$B = \frac{1}{4n}[b + ab + bc + abc - a - c - ac - (1)], \tag{13-37}$$

and the effect of C is the average difference between the four treatment combinations in the top face of the cube and the four in the bottom, or

$$C = \frac{1}{4n}[c + ac + bc + abc - a - b - ab - (1)]. \tag{13-38}$$

Now consider the two-factor interaction AB. When C is at the low level, AB is just the average difference in the A effect at the two levels of B, or

$$AB(C\text{ low}) = \frac{1}{2n}[ab - b] - \frac{1}{2n}[a - (1)].$$

Similarly, when C is at the high level, the AB interaction is

$$AB(C\text{ high}) = \frac{1}{2n}[abc - bc] - \frac{1}{2n}[ac - c].$$

The AB interaction is just the average of these two components, or

$$AB = \frac{1}{4n}[ab + (1) + abc + c - b - a - bc - ac]. \tag{13-39}$$

Using a similar approach, we can show that the AC and BC interaction effect estimates are as follows:

$$AC = \frac{1}{4n}[ac + (1) + abc + b - a - c - ab - bc], \tag{13-40}$$

$$BC = \frac{1}{4n}[bc + (1) + abc + a - b - c - ab - ac]. \tag{13-41}$$

The *ABC* interaction effect is the average difference between the *AB* interaction at the two levels of *C*. Thus

$$\begin{aligned} ABC &= \frac{1}{4n}\{[abc - bc] - [ac - c] - [ab - b] + [a - (1)]\} \\ &= \frac{1}{4n}[abc - bc - ac + c - ab + b + a - (1)]. \end{aligned} \tag{13-42}$$

The quantities in brackets in equations 13-36 through 13-42 are contrasts in the eight treatment combinations. These contrasts can be obtained from a table of plus and minus signs for the 2^3 design, shown in Table 13-16. Signs for the main effects (columns *A*, *B*, and *C*) are obtained by associating a plus with the high level of the factor and a minus with the low level. Once the signs for the main effects have been established, the signs for the remaining columns are found by multiplying the appropriate preceding columns row by row. For example, the signs in column *AB* are the product of the signs in columns *A* and *B*.

Table 13-16 has several interesting properties.

1. Except for the identity column *I*, each column has an equal number of plus and minus signs.
2. The sum of products of signs in any two columns is zero; that is, the columns in the table are *orthogonal.*
3. Multiplicating any column by column *I* leaves the column unchanged; that is, *I* is an *identity element.*
4. The product of any two columns yields a column in the table; for example, $A \times B = AB$ and $AB \times ABC = A^2B^2C = C$, since any column multiplied by itself is the identity column.

The estimate of any main effect or interaction is determined by multiplying the treatment combinations in the first column of the table by the signs in the corresponding main effect or interaction column, adding the result to produce a contrast, and then dividing the contrast by one-half the total number of runs in the experiment. Expressed mathematically,

$$\text{Effect} = \frac{\text{Contrast}}{n2^{k-1}}. \tag{13-43}$$

The sum of squares for any effect is

$$SS = \frac{(\text{Contrast})^2}{n2^k}. \tag{13-44}$$

Table 13-16 Signs for Effects in the 2^3 Design

Treatment	Factorial Effect							
Combinations	*I*	*A*	*B*	*AB*	*C*	*AC*	*BC*	*ABC*
(1)	+	−	−	+	−	+	+	−
a	+	+	−	−	−	−	+	+
b	+	−	+	−	−	+	−	+
ab	+	+	+	+	−	−	−	−
c	+	−	−	+	+	−	−	+
ac	+	+	−	−	+	+	−	−
bc	+	−	+	−	+	−	+	−
abc	+	+	+	+	+	+	+	+

Example 13-7

Consider the surface-roughness experiment described originally in Example 13-5. This is a 2^3 factorial design in the factors feed rate (A), depth of cut (B), and tool angle (C), with $n = 2$ replicates. Table 13-17 presents the observed surface-roughness data.

The main effects may be estimated using equations 13-36 through 13-42. The effect of A is, for example,

$$A = \frac{1}{4n}\left[a + ab + ac + abc - b - c - bc - (1)\right]$$
$$= \frac{1}{4(2)}[22 + 27 + 23 + 30 - 20 - 21 - 18 - 16]$$
$$= \frac{1}{8}[27] = 3.375,$$

and the sum of squares for A is found using equation 13-44:

$$SS_A = \frac{(\text{Contrast}_A)^2}{n2^k}$$
$$= \frac{(27)^2}{2(8)} = 45.5625.$$

It is easy to verify that the other effects are

$$B = 1.625,$$
$$C = 0.875,$$
$$AB = 1.375,$$
$$AC = 0.125,$$
$$BC = -0.625,$$
$$ABC = 1.125.$$

From examining the magnitude of the effects, clearly feed rate (factor A) is dominant, followed by depth of cut (B) and the AB interaction, although the interaction effect is relatively small. The analysis of variance is summarized in Table 13-18, and it confirms our interpretation of the effect estimates.

Table 13-17 Surface Roughness Data for Example 13-7

Treatment Combinations	Design Factors A	B	C	Roughness	Surface Totals
(1)	−1	−1	−1	9, 7	16
a	1	−1	−1	10, 12	22
b	−1	1	−1	9, 11	20
ab	1	1	−1	12, 5	27
c	−1	−1	1	11, 10	21
ac	1	−1	1	10, 13	23
bc	−1	1	1	10, 8	18
abc	1	1	1	16, 14	30

Table 13-18 Analysis of Variance for the Surface-Finish Experiment

Source of Variation	Sum of Squares	Degrees of Freedom	Mean Square	F_0
A	45.5625	1	45.5625	18.69
B	10.5625	1	10.5625	4.33
C	3.0625	1	3.0625	1.26
AB	7.5625	1	7.5625	3.10
AC	0.0625	1	0.0625	0.03
BC	1.5625	1	1.5625	0.64
ABC	5.0625	1	5.0625	2.08
Error	19.5000	8	2.4375	
Total	92.9375	15		

Other Methods for Judging Significance of Effects The analysis of variance is a formal way to determine which effects are nonzero. There are two other methods that are useful. In the first method, we can calculate the standard errors of the effects and compare the magnitudes of the effects to their standard errors. The second method uses normal probability plots to assess the importance of the effects.

The standard error of an effect is easy to find. If we assume that there are n replicates at each of the 2^k runs in the design, and if $y_{i1}, y_{i2}, \ldots, y_{in}$ are the observations at the ith run (design point), then

$$S_i^2 = \frac{1}{n-1}\sum_{j=1}^{n}\left(y_{ij} - \bar{y}_i\right)^2, \qquad i = 1, 2, \ldots, 2^k,$$

is an estimate of the variance at the ith run, where $\bar{y}_i = \sum_{j=1}^{n} y_{ij}/n$ is the sample mean of the n observations. The 2^k variance estimates can be pooled to give an overall variance estimate

$$S^2 = \frac{1}{2^k(n-1)}\sum_{i=1}^{2^k}\sum_{j=1}^{n}\left(y_{ij} - \bar{y}_i\right)^2, \tag{13-45}$$

where we have obviously assumed equal variances for each design point. This is also the variance estimate given by the mean square error from the analysis of variance procedure. Each effect estimate has variance given by

$$\begin{aligned} V(\text{Effect}) &= V\left[\frac{\text{Contrast}}{n2^{k-1}}\right] \\ &= \frac{1}{\left(n2^{k-1}\right)^2} V(\text{Contrast}). \end{aligned}$$

Each contrast is a linear combination of 2^k treatment totals, and each total consists of n observations. Therefore,

$$V(\text{Contrast}) = n2^k\sigma^2$$

and the variance of an effect is

$$\begin{aligned} V(\text{Effect}) &= \frac{1}{\left(n2^{k-1}\right)^2} n2^k\sigma^2 \\ &= \frac{1}{n2^{k-2}}\sigma^2. \end{aligned} \tag{13-46}$$

The estimated standard error of an effect would be found by replacing σ^2 with its estimate S^2 and taking the square root of equation 13-46.

To illustrate for the surface-roughness experiment, we find that $S^2 = 2.4375$ and the standard error of each estimated effect is

$$\begin{aligned} s.e.(\text{Effect}) &= \sqrt{\frac{1}{n2^{k-2}} S^2} \\ &= \sqrt{\frac{1}{2 \cdot 2^{3-2}}(2.4375)} \\ &= 0.78. \end{aligned}$$

Therefore two standard deviation limits on the effect estimates are

$$\begin{aligned} A&: \; 3.375 \pm 1.56, \\ B&: \; 1.625 \pm 1.56, \\ C&: \; 0.875 \pm 1.56, \\ AB&: \; 1.375 \pm 1.56, \\ AC&: \; 0.125 \pm 1.56, \\ BC&: \; -0.625 \pm 1.56, \\ ABC&: \; 1.125 \pm 1.56. \end{aligned}$$

These intervals are approximate 95% confidence intervals. They indicate that the two main effects, A and B, are important, but that the other effects are not, since the intervals for all effects except A and B include zero.

Normal probability plots can also be used to judge the significance of effects. We will illustrate that method in the next section.

Projection of 2^k Designs Any 2^k design will collapse or project into another 2^k design in fewer variables if one or more of the original factors are dropped. Sometimes this can provide additional insight into the remaining factors. For example, consider the surface-roughness experiment. Since factor C and all its interactions are negligible, we could eliminate factor C from the design. The result is to collapse the cube in Fig. 13-18 into a square in the $A-B$ plane—however, each of the four runs in the new design has four replicates. In general, if we delete h factors so that $r = k - h$ factors remain, the original 2^k design with n replicates will project into a 2^r design with $n2^h$ replicates.

Residual Analysis We may obtain the residuals from a 2^k design by using the method demonstrated earlier for the 2^2 design. As an example, consider the surface-roughness experiment. The three largest effects are A, B, and the AB interaction. The regression model used to obtain the predicted values is

$$\hat{y} = \hat{\beta}_0 + \hat{\beta}_1 x_1 + \hat{\beta}_2 x_2 + \hat{\beta}_{12} x_1 x_2,$$

where x_1 represents factor A, x_2 represents factor B, and x_1x_2 represents the AB interaction. The regression coefficients $\hat{\beta}_1, \hat{\beta}_2$, and $\hat{\beta}_{12}$ are estimated by one-half the corresponding effect estimates and $\hat{\beta}_0$ is the grand average. Thus

$$\hat{y} = 11.0625 + \left(\frac{3.375}{2}\right)x_1 + \left(\frac{1.625}{2}\right)x_2 + \left(\frac{1.375}{2}\right)x_1 x_2,$$

and the predicted values would be obtained by substituting the low and high levels of A and B into this equation. To illustrate, at the treatment combination where A, B, and C are all at the low level, the predicted value is

$$\begin{aligned}\hat{y} &= 11.0625 + \left(\frac{3.375}{2}\right)(-1) + \left(\frac{1.625}{2}\right)(-1) + \left(\frac{1.375}{2}\right)(-1)(-1) \\ &= 9.25.\end{aligned}$$

The observed values at this run are 9 and 7, so the residuals are $9 - 9.25 = -0.25$ and $7 - 9.25 = -2.25$. Residuals for the other seven runs are obtained similarly.

A normal probability plot of the residuals is shown in Fig. 13-19. Since the residuals lie approximately along a straight line, we do not suspect any severe nonnormality in the data. There are no indications of severe outliers. It would also be helpful to plot the residuals versus the predicted values and against each of the factors A, B, and C.

Yates' Algorithm for the 2^k Instead of using the table of plus and minus signs to obtain the contrasts for the effect estimates and the sums of squares, a simple tabular algorithm devised by Yates can be employed. To use Yates' algorithm, construct a table with the treatment combinations and the corresponding treatment totals recorded in *standard order.* By standard order, we mean that each factor is introduced one at a time by combining it with all factor levels above it. Thus for a 2^2, the standard order is (1), *a*, *b*, *ab*, while for a 2^3 it is (1), *a*, *b*, *ab*, *c*, *ac*, *bc*, *abc*, and for a 2^4 it is (1), *a*, *b*, *ab*, *c*, *ac*, *bc*, *abc*, *d*, *ad*, *bd*, *abd*, *cd*, *acd*, *bcd*, *abcd*. Then follow this four-step procedure:

1. Label the adjacent column [1]. Compute the entries in the top half of this column by adding the observations in adjacent pairs. Compute the entries in the bottom half of this column by changing the sign of the first entry in each pair of the original observations and adding the adjacent pairs.
2. Label the adjacent column [2]. Construct column [2] using the entries in column [1]. Follow the same procedure employed to generate column [1]. Continue this process until k columns have been constructed. Column [k] contains the contrasts designated in the rows.
3. Calculate the sums of squares for the effects by squaring the entries in column [k] and dividing by $n2^k$.
4. Calculate the effect estimates by dividing the entries in column [k] by $n2^{k-1}$.

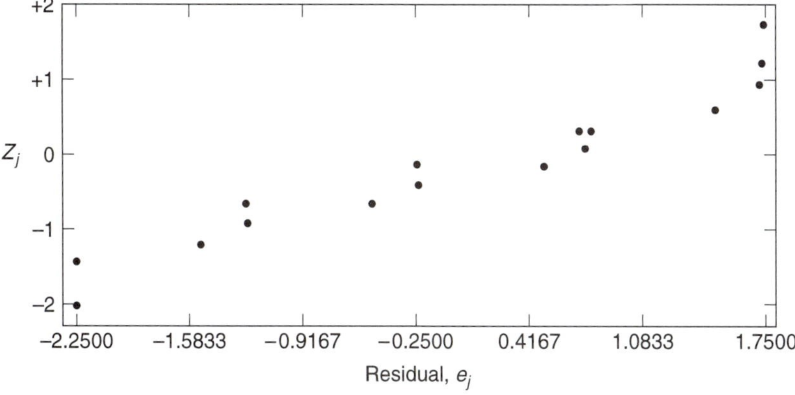

Figure 13-19 Normal probability plot of residuals from the surface-roughness experiment.

Table 13-19 Yates' Algorithm for the Surface-Roughness Experiment

Treatment Combinations	Response	[1]	[2]	[3]	Effect	Sum of Squares $[3]^2/n2^3$	Effect Estimates $[3]/n2^2$
(1)	16	38	85	177	Total	—	—
a	22	47	92	27	A	45.5625	3.375
b	20	44	13	13	B	10.5625	1.625
ab	27	48	14	11	AB	7.5625	1.375
c	21	6	9	7	C	3.0625	0.875
ac	23	7	4	1	AC	0.0625	0.125
bc	18	2	1	−5	BC	1.5625	−0.625
abc	30	12	10	9	ABC	5.0625	1.125

Example 13-8

Consider the surface-roughness experiment in Example 13-7. This is a 2^3 design with $n = 2$ replicates. The analysis of this data using Yates' algorithm is illustrated in Table 13-19. Note that the sums of squares computed from Yates' algorithm agree with the results obtained in Example 13-7.

13-5.3 A Single Replicate of the 2^k Design

As the number of factors in a factorial experiment grows, the number of effects that can be estimated grows also. For example, a 2^4 experiment has 4 main effects, 6 two factor interactions, 4 three-factor interactions, and 1 four-factor interaction, while a 2^6 experiment has six main effects, 15 two-factor interactions, 20 three-factor interactions, 15 four-factor interactions, 6 five-factor interactions, and 1 six-factor interaction. In most situations the *sparsity of effects principle* applies; that is, the system is usually dominated by the main effects and low-order interactions. Three-factor and higher interactions are usually negligible. Therefore, when the number of factors is moderately large, say $k \geq 4$ or 5, a common practice is to run only a single replicate of the 2^k design and then pool or combine the higher-order interactions as an estimate of error.

Example 13-9

An article in *Solid State Technology* ("Orthogonal Design for Process Optimization and its Application in Plasma Etching," May 1987, p. 127) describes the application of factorial designs in developing a nitride etch process on a single-wafer plasma etcher. The process uses C_2F_6 as the reactant gas. It is possible to vary the gas flow, the power applied to the cathode, the pressure in the reactor chamber, and the spacing between the anode and the cathode (gap). Several response variables would usually be of interest in this process, but in this example we will concentrate on etch rate for silicon nitride.

We will use a single replicate of a 2^4 design to investigate this process. Since it is unlikely that the three-factor and four-factor interactions are significant, we will tentatively plan to combine them as an estimate of error. The factor levels used in the design are shown here:

Design Factor

	A Gap	B Pressure	C C_2F_6 Flow	D Power
Level	(cm)	(mTorr)	(SCCM)	(w)
Low (−)	0.80	450	125	275
High (+)	1.20	550	200	325

Table 13-20 presents the data from the 16 runs of the 2^4 design. Table 13-21 is the table of plus and minus signs for the 2^4 design. The signs in the columns of this table can be used to estimate the factor effects. To illustrate, the estimate of factor A is

$$A = \frac{1}{8}\left[a + ab + ac + abc + ad + abd + acd + abcd - (1) - b - c - d - bc - bd - cd - bcd\right]$$

$$= \frac{1}{8}[669 + 650 + 642 + 635 + 749 + 868 + 860 + 729 - 550 - 604 - 633$$

$$- 601 - 1037 - 1052 - 1075 - 1063]$$

$$= -101.625.$$

Thus the effect of increasing the gap between the anode and the cathode from 0.80 cm to 1.20 cm is to decrease the etch rate by 101.625 Å/min.

It is easy to verify that the complete set of effect estimates is

$$\begin{aligned}
A &= -101.625, & D &= 306.125,\\
B &= -1.625, & AD &= -153.625,\\
AB &= -7.875, & BD &= -0.625,\\
C &= 7.375, & ABD &= 4.125,\\
AC &= -24.875, & CD &= -2.125,\\
BC &= -43.875, & ACD &= 5.625,\\
ABC &= -15.625, & BCD &= -25.375,\\
& & ABCD &= -40.125.
\end{aligned}$$

A very helpful method in judging the significance of factors in a 2^k experiment is to construct a normal probability plot of the effect estimates. If none of the effects is

Table 13-20 The 2^4 Design for the Plasma Etch Experiment

A	B	C	D	Etch Rate
(Gap)	(Pressure)	(C_2F_6 Flow)	(Power)	(Å/min)
−1	−1	−1	−1	550
1	−1	−1	−1	669
−1	1	−1	−1	604
1	1	−1	−1	650
−1	−1	1	−1	633
1	−1	1	−1	642
−1	1	1	−1	601
1	1	1	−1	635
−1	−1	−1	1	1037
1	−1	−1	1	749
−1	1	−1	1	1052
1	1	−1	1	868
−1	−1	1	1	1075
1	−1	1	1	860
−1	1	1	1	1063
1	1	1	1	729

Table 13-21 Contrast Constants for the 2^4 Design

	A	B	AB	C	AC	BC	ABC	D	AD	BD	ABD	CD	ACD	BCD	$ABCD$
(1)	–	–	+	–	+	+	–	–	+	+	–	+	–	–	+
a	+	–	–	–	–	+	+	–	–	+	+	+	+	–	–
b	–	+	–	–	+	–	+	–	+	–	+	+	–	+	–
ab	+	+	+	–	–	–	–	–	–	–	–	+	+	+	+
c	–	–	+	+	–	–	+	–	+	+	–	–	+	+	–
ac	+	–	–	+	+	–	–	–	–	+	+	–	–	+	+
bc	–	+	–	+	–	+	–	–	+	–	+	–	+	–	+
abc	+	+	+	+	+	+	+	–	–	–	–	–	–	–	–
d	–	–	+	–	+	+	–	+	–	–	+	–	+	+	–
ad	+	–	–	–	–	+	+	+	+	–	–	–	–	–	+
bd	–	+	–	–	+	–	+	+	–	+	–	–	+	–	+
abd	+	+	+	–	–	–	–	+	+	+	+	–	–	–	–
cd	–	–	+	+	–	–	+	+	–	–	+	+	–	–	+
acd	+	–	–	+	+	–	–	+	+	–	–	+	+	–	–
bcd	–	+	–	+	–	+	–	+	–	+	–	+	–	+	–
abcd	+	+	+	+	+	+	+	+	+	+	+	+	+	+	+

significant, then the estimates will behave like a random sample drawn from a normal distribution with zero mean, and the plotted effects will lie approximately along a straight line. Those effects that do not plot on the line are significant factors.

The normal probability plot of effect estimates from the plasma etch experiment is shown in Fig. 13-20. Clearly the main effects of A and D and the AD interaction are significant, as they fall far from the line passing through the other points. The analysis of variance summarized in Table 13-22 confirms these findings. Notice that in the analysis of variance we have pooled the three- and four-factor interactions to form the error mean square. If the normal probability plot had indicated that any of these interactions were important, they then should not be included in the error term.

Since $A = -101.625$, the effect of increasing the gap between the cathode and anode is to decrease the etch rate. However, $D = 306.125$, so applying higher power levels will increase the etch rate. Figure 13-21 is a plot of the AD interaction. This plot indicates that the effect of changing the gap width at low power settings is small, but that increasing the

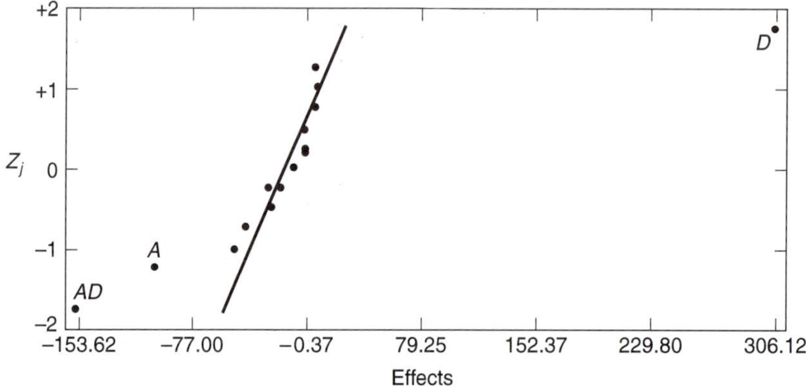

Figure 13-20 Normal probability plot of effects from the plasma etch experiment.

Table 13-22 Analysis of Variance for the Plasma Etch Experiment

Source of Variation	Sum of Squares	Degrees of Freedom	Mean Square	F_0
A	41,310.563	1	41,310.563	20.28
B	10.563	1	10.563	< 1
C	217.563	1	217.563	< 1
D	374,850.063	1	374,850.063	183.99
AB	248.063	1	248.063	< 1
AC	2,475.063	1	2,475.063	1.21
AD	94,402.563	1	99,402.563	48.79
BC	7,700.063	1	7,700.063	3.78
BD	1.563	1	1.563	< 1
CD	18.063	1	18.063	< 1
Error	10,186.815	5	2,037.363	
Total	531,420.938	15		

gap at high power settings dramatically reduces the etch rate. High etch rates are obtained at high power settings and narrow gap widths.

The residuals from the experiment can be obtained from the regression model

$$\hat{y} = 776.0625 - \left(\frac{101.625}{2}\right)x_1 + \left(\frac{306.125}{2}\right)x_4 - \left(\frac{153.625}{2}\right)x_1x_4.$$

For example, when A and D are both at the low level the predicted value is

$$\begin{aligned}\hat{y} &= 776.0625 - \left(\frac{101.625}{2}\right)(-1) + \left(\frac{306.125}{2}\right)(-1) - \left(\frac{153.625}{2}\right)(-1)(-1)\\ &= 597,\end{aligned}$$

and the four residuals at this treatment combination are

$$e_1 = 550 - 597 = -47,$$

$$e_2 = 604 - 597 = 7,$$

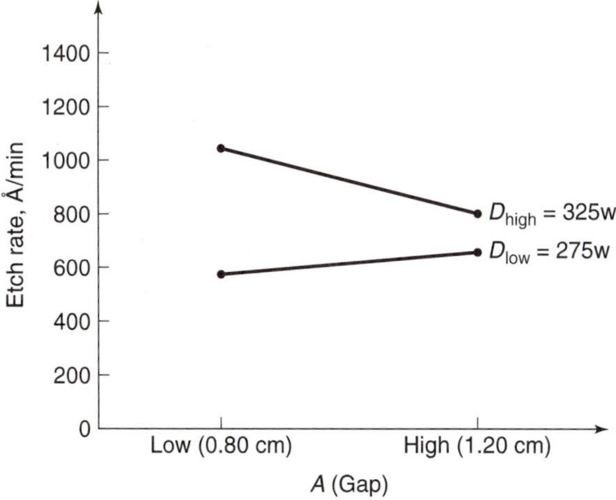

Figure 13-21 AD interaction from the plasma etch experiment.

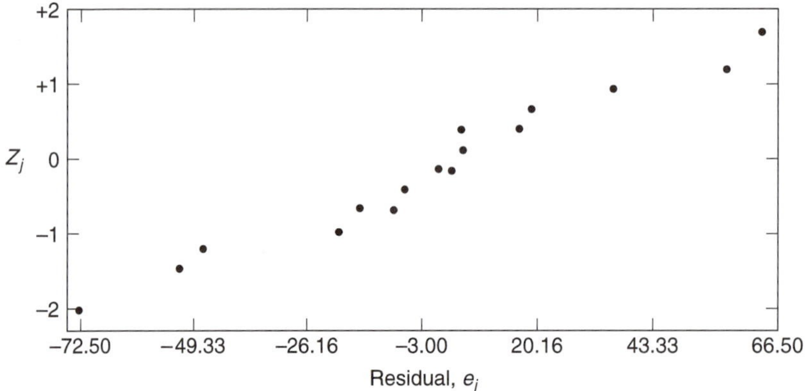

Figure 13-22 Normal probability plot of residuals from the plasma etch experiment.

$$e_3 = 638 - 597 = 41,$$

$$e_4 = 601 - 597 = 4.$$

The residuals at the other three treatment combinations (A high, D low), (A low, D high), and (A high, D high) are obtained similarly. A normal probability plot of the residuals is shown in Fig. 13-22. The plot is satisfactory.

13-6 CONFOUNDING IN THE 2^k DESIGN

It is often impossible to run a complete replicate of a factorial design under homogeneous experimental conditions. *Confounding* is a design technique for running a factorial experiment in blocks, where the block size is smaller than the number of treatment combinations in one complete replicate. The technique causes certain interaction effects to be indistinguishable from, or *confounded* with, blocks. We will illustrate confounding in the 2^k factorial design in 2^p blocks, where $p < k$.

Consider a 2^2 design. Suppose that each of the $2^2 = 4$ treatment combinations requires four hours of laboratory analysis. Thus, two days are required to perform the experiment. If days are considered as blocks, then we must assign two of the four treatment combinations to each day.

Consider the design shown in Fig. 13-23. Notice that block 1 contains the treatment combinations (1) and ab, and that block 2 contains a and b. The contrasts for estimating the main effects A and B are

$$\text{Contrast}_A = ab + a - b - (1),$$

$$\text{Contrast}_B = ab + b - a - (1).$$

Note that these contrasts are unaffected by blocking since in each contrast there is one plus and one minus treatment combination from each block. That is, any difference between block 1 and block 2 will cancel out. The contrast for the AB interaction is

$$\text{Contrast}_{AB} = ab + (1) - a - b.$$

Since the two treatment combinations with the plus sign, ab and (1), are in block 1 and the two with the minus sign, a and b, are in block 2, the block effect and the AB interaction are identical. That is, AB is confounded with blocks.

Block 1	Block 2
(1)	*a*
ab	*b*

Figure 13-23 The 2^2 design in two blocks.

Block 1	Block 2
(1)	*a*
ab	*b*
ac	*c*
bc	*abc*

Figure 13-24 The 2^3 design in two blocks, *ABC* confounded.

The reason for this is apparent from the table of plus and minus signs for the 2^2 design (Table 13-13). From this table, we see that all treatment combinations that have a plus on *AB* are assigned to block 1, while all treatment combinations that have a minus sign on *AB* are assigned to block 2.

This scheme can be used to confound any 2^k design in two blocks. As a second example, consider a 2^3 design, run in two blocks. Suppose we wish to confound the three-factor interaction *ABC* with blocks. From the table of plus and minus signs for the 2^3 design (Table 13-16), we assign the treatment combinations that are minus on *ABC* to block 1 and those that are plus on *ABC* to block 2. The resulting design is shown in Fig. 13-24.

There is a more general method of constructing the blocks. The method employs a *defining contrast*, say

$$L = \alpha_1 x_1 + \alpha_2 x_2 + \cdots + \alpha_k x_k, \tag{13-47}$$

where x_i is the level of the *i*th factor appearing in a treatment combination and α_i is the exponent appearing on the *i*th factor in the effect to be confounded. For the 2^k system, we have either $\alpha_i = 0$ or 1, and either $x_i = 0$ (low level) or $x_i = 1$ (high level). Treatment combinations that produce the same value of L (modulus 2) will be placed in the same block. Since the only possible values of L (mod 2) are 0 and 1, this will assign the 2^k treatment combinations to exactly two blocks.

As an example consider a 2^3 design with *ABC* confounded with blocks. Here x_1 corresponds to A, x_2 to B, x_3 to C, and $\alpha_1 = \alpha_2 = \alpha_3 = 1$. Thus, the defining contrast for *ABC* is

$$L = x_1 + x_2 + x_3.$$

To assign the treatment combinations to the two blocks, we substitute the treatment combinations into the defining contrast as follows:

$$\begin{aligned}
(1)&: L = 1(0) + 1(0) + 1(0) = 0 = 0 \text{ (mod 2)},\\
a&: L = 1(1) + 1(0) + 1(0) = 1 = 1 \text{ (mod 2)},\\
b&: L = 1(0) + 1(1) + 1(0) = 1 = 1 \text{ (mod 2)},\\
ab&: L = 1(1) + 1(1) + 1(0) = 2 = 0 \text{ (mod 2)},\\
c&: L = 1(0) + 1(0) + 1(1) = 1 = 1 \text{ (mod 2)},\\
ac&: L = 1(1) + 1(0) + 1(1) = 2 = 0 \text{ (mod 2)},\\
bc&: L = 1(0) + 1(1) + 1(1) = 2 = 0 \text{ (mod 2)},\\
abc&: L = 1(1) + 1(1) + 1(1) = 3 = 1 \text{ (mod 2)}.
\end{aligned}$$

Therefore, (1), *ab*, *ac*, and *bc* are run in block 1, and *a*, *b*, *c*, and *abc* are run in block 2. This is the same design shown in Fig. 13-24.

A shortcut method is useful in constructing these designs. The block containing the treatment combination (1) is called the *principal block.* Any element [except (1)] in the principal block may be generated by multiplying two other elements in the principal block modulus 2. For example, consider the principal block of the 2^3 design with ABC confounded, shown in Fig. 13-24. Note that

$$ab \cdot ac = a^2bc = bc,$$

$$ab \cdot bc = ab^2c = ac,$$

$$ac \cdot bc = abc^2 = ab.$$

Treatment combinations in the other block (or blocks) may be generated by multiplying one element in the new block by each element in the principal block modulus 2. For the 2^3 with ABC confounded, since the principal block is (1), ab, ac, and bc, we know that b is in the other block. Thus, the elements of this second block are

$$b \cdot (1) = b,$$

$$b \cdot ab = ab^2 = a,$$

$$b \cdot ac = abc,$$

$$b \cdot bc = b^2c = c.$$

Example 13-10

An experiment is performed to investigate the effects of four factors on the terminal miss distance of a shoulder-fired ground-to-air missile.

The four factors are target type (A), seeker type (B), target altitude (C), and target range (D). Each factor may be conveniently run at two levels, and the optical tracking system will allow terminal miss distance to be measured to the nearest foot. Two different gunners are used in the flight test, and since there may be differences between individuals, it was decided to conduct the 2^4 design in two blocks with $ABCD$ confounded. Thus, the defining contrast is

$$L = x_1 + x_2 + x_3 + x_4.$$

The experimental design and the resulting data are

Block 1	Block 2
(1) = 3	a = 7
ab = 7	b = 5
ac = 6	c = 6
bc = 8	d = 4
ad = 10	abc = 6
bd = 4	bcd = 7
cd = 8	acd = 9
abcd = 9	abd = 12

The analysis of the design by Yates' algorithm is shown in Table 13-23. A normal probability plot of the effects would reveal A (target type), D (target range), and AD to have large effects. A confirming analysis of variance, using three-factor interactions as error, is shown in Table 13-24.

It is possible to confound the 2^k design in four blocks of 2^{k-2} observations each. To construct the design, two effects are chosen to confound with blocks and their defining contrasts obtained. A third effect, the generalized interaction of the two initially chosen, is also

Table 13-23 Yates' Algorithm for the 2^4 Design in Example 13-10

Treatment Combinations	Response	[1]	[2]	[3]	[4]	Effect	Sum of Squares	Effect Estimate
(1)	3	10	22	48	111	Total	—	—
a	7	12	26	63	21	*A*	27.5625	2.625
b	5	12	30	4	5	*B*	1.5625	0.625
ab	7	14	33	17	−1	*AB*	0.0625	−0.125
c	6	14	6	4	7	*C*	3.0625	0.875
ac	6	16	−2	1	−19	*AC*	22.5625	−2.375
bc	8	17	14	−4	−3	*BC*	0.5625	−0.375
abc	6	16	3	3	−1	*ABC*	0.0625	−0.125
d	4	4	2	4	15	*D*	14.0625	1.375
ad	10	2	2	3	13	*AD*	10.5625	1.625
bd	4	0	2	−8	−3	*BD*	0.5625	−0.375
abd	12	−2	−1	−11	7	*ABD*	3.0625	0.875
cd	8	6	−2	0	−1	*CD*	0.0625	−0.125
acd	9	8	−2	−3	−3	*ACD*	0.5625	−0.375
bcd	7	1	2	0	−3	*BCD*	0.5625	−0.375
abcd	9	2	1	−1	−1	*ABCD*	0.0625	−0.125

Table 13-24 Analysis of Variance for Example 13-10

Source of Variation	Sum of Squares	Degrees of Freedom	Mean Square	F_0
Blocks (*ABCD*)	0.0625	1	0.0625	0.06
A	27.5625	1	27.5625	25.94
B	1.5625	1	1.5625	1.47
C	3.0625	1	3.0625	2.88
D	14.0625	1	14.0625	13.24
AB	0.0625	1	0.0625	0.06
AC	22.5625	1	22.5625	21.24
AD	10.5625	1	10.5625	9.94
BC	0.5625	1	0.5625	0.53
BD	0.5625	1	0.5625	0.53
CD	0.0625	1	0.0625	0.06
Error ($ABC + ABD + ACD + BCD$)	4.2500	4	1.0625	
Total	84.9375	15		

confounded with blocks. The generalized interaction of two effects is found by multiplying their respective columns.

For example, consider the 2^4 design in four blocks. If *AC* and *BD* are confounded with blocks, their generalized interaction is $(AC)(BD) = ABCD$. The design is constructed by using the defining contrasts for *AC* and *BD*:

$$L_1 = x_1 + x_3,$$

$$L_2 = x_3 + x_4.$$

It is easy to verify that the four blocks are

Block 1 $L_1 = 0, L_2 = 0$	Block 2 $L_1 = 1, L_2 = 0$	Block 3 $L_1 = 0, L_2 = 1$	Block 4 $L_1 = 1, L_2 = 1$
(1)	a	b	ab
ac	c	abc	bc
bd	abd	d	ad
$abcd$	bcd	acd	cd

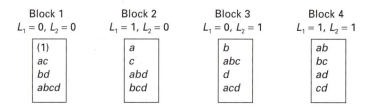

This general procedure can be extended to confounding the 2^k design in 2^p blocks, where $p < k$. Select p effects to be confounded, such that no effect chosen is a generalized interaction of the others. The blocks can be constructed from the p defining contrasts $L_1, L_2, \ldots, L_p$ associated with these effects. In addition, exactly $2^p - p - 1$ other effects are confounded with blocks, these being the generalized interaction of the original p effects chosen. Care should be taken so as not to confound effects of potential interest.

For more information on confounding refer to Montgomery (2001, Chapter 7). That book contains guidelines for selecting factors to confound with blocks so that main effects and low-order interactions are not confounded. In particular, the book contains a table of suggested confounding schemes for designs with up to seven factors and a range of block sizes, some as small as two runs.

13-7 FRACTIONAL REPLICATION OF THE 2^k DESIGN

As the number of factors in a 2^k increases, the number of runs required increases rapidly. For example, a 2^5 requires 32 runs. In this design, only 5 degrees of freedom correspond to main effects and 10 degrees of freedom correspond to two-factor interactions. If we can assume that certain high-order interactions are negligible, then a fractional factorial design involving fewer than the complete set of 2^k runs can be used to obtain information on the main effects and low-order interactions. In this section, we will introduce fractional replication of the 2^k design. For a more complete treatment, see Montgomery (2001, Chapter 8).

13-7.1 The One-Half Fraction of the 2^k Design

A one-half fraction of the 2^k design contains 2^{k-1} runs and is often called a 2^{k-1} fractional factorial design. As an example, consider the 2^{3-1} design, that is, a one-half fraction of the 2^3. The table of plus and minus signs for the 2^3 design is shown in Table 13-25. Suppose we select the four treatment combinations a, b, c, and abc as our one-half fraction. These

Table 13-25 Plus and Minus Signs for the 2^3 Factorial Design

Treatment	Factorial Effect							
Combinations	I	A	B	C	AB	AC	BC	ABC
a	+	+	−	−	−	−	+	+
b	+	−	+	−	−	+	−	+
c	+	−	−	+	+	−	−	+
abc	+	+	+	+	+	+	+	+
ab	+	+	+	−	+	−	−	−
ac	+	+	−	+	−	+	−	−
bc	+	−	+	+	−	−	+	−
(1)	+	−	−	−	+	+	+	−

treatment combinations are shown in the top half of Table 13-25. We will use both the conventional notation (a, b, c, ...) and the plus and minus notation for the treatment combinations. The equivalence between the two notations is as follows:

Notation 1	Notation 2
a	+ − −
b	− + −
c	− − +
abc	+ + +

Notice that the 2^{3-1} design is formed by selecting only those treatment combinations that yield a plus on the ABC effect. Thus ABC is called the *generator* of this particular fraction. Furthermore, the identity element I is also plus for the four runs, so we call

$$I = ABC$$

the defining relation for the design.

The treatment combinations in the 2^{3-1} designs yield three degrees of freedom associated with the main effects. From Table 13-25, we obtain the estimates of the main effects as

$$A = \frac{1}{2}[a - b - c + abc],$$

$$B = \frac{1}{2}[-a + b - c + abc],$$

$$C = \frac{1}{2}[-a - b + c + abc].$$

It is also easy to verify that the estimates of the two-factor interactions are

$$BC = \frac{1}{2}[a - b - c + abc],$$

$$AC = \frac{1}{2}[-a + b - c + abc],$$

$$AB = \frac{1}{2}[-a - b + c + abc].$$

Thus, the linear combination of observations in column A, say ℓ_A, estimates $A + BC$. Similarly, ℓ_B estimates $B + AC$, and ℓ_C estimates $C + AB$. Two or more effects that have this property are called *aliases*. In our 2^{3-1} design, A and BC are aliases, B and AC are aliases, and C and AB are aliases. Aliasing is the direct result of fractional replication. In many practical situations, it will be possible to select the fraction so that the main effects and low-order interactions of interest will be aliased with high-order interactions (which are probably negligible).

The alias structure for this design is found by using the defining relation $I = ABC$. Multiplying any effect by the defining relation yields the aliases for that effect. In our example, the alias of A

$$A = A \cdot ABC = A^2BC = BC,$$

since $A \cdot I = A$ and $A^2 = I$. The aliases of B and C are

$$B = B \cdot ABC = AB^2C = AC$$

and

$$C = C \cdot ABC = ABC^2 = AB.$$

Now suppose that we had chosen the other one-half fraction, that is, the treatment combinations in Table 13-25 associated with minus on ABC. The defining relation for this design is $I = -ABC$. The aliases are $A = -BC$, $B = -AC$, and $C = -AB$. Thus the estimates of A, B, and C with this fraction really estimate $A - BC$, $B - AC$, and $C - AB$. In practice, it usually does not matter which one-half fraction we select. The fraction with the plus sign in the defining relation is usually called the *principal fraction*, and the other fraction is usually called the *alternate fraction*.

Sometimes we use *sequences* of fractional factorial designs to estimate effects. For example, suppose we had run the principal fraction of the 2^{3-1} design. From this design we have the following effect estimates:

$$\ell_A = A + BC,$$

$$\ell_B = B + AC,$$

$$\ell_C = C + AB.$$

Suppose that we are willing to assume at this point that the two-factor interactions are negligible. If they are, then the 2^{3-1} design has produced estimates of the three main effects, A, B, and C. However, if after running the principal fraction we are uncertain about the interactions, it is possible to estimate them by running the *alternate* fraction. The alternate fraction produces the following effect estimates:

$$\ell_A' = A - BC,$$

$$\ell_B' = B - AC,$$

$$\ell_C' = C - AC.$$

If we combine the estimates from the two fractions, we obtain the following:

Effect i	From $\frac{1}{2}(\ell_i + \ell_i')$	From $\frac{1}{2}(\ell_i - \ell_i')$
$i = A$	$\frac{1}{2}(A + BC + A - BC) = A$	$\frac{1}{2}[A + BC - (A - BC)] = BC$
$i = B$	$\frac{1}{2}(B + AC + B - AC) = B$	$\frac{1}{2}[B + AC - (B - AC)] = AC$
$i = C$	$\frac{1}{2}(C + AB + C - AB) = C$	$\frac{1}{2}[C + AB - (C - AB)] = AB$

Thus by combining a sequence of two fractional factorial designs we can isolate both the main effects and the two-factor interactions. This property makes the fractional factorial design highly useful in experimental problems, as we can run sequences of small, efficient experiments, combine information across *several* experiments, and take advantage of learning about the process we are experimenting with as we go along.

A 2^{k-1} design may be constructed by writing down the treatment combinations for a full factorial with $k - 1$ factors and then adding the kth factor by identifying its plus and minus levels with the plus and minus signs of the highest-order interaction $\pm ABC \cdots (K - 1)$. Therefore, a 2^{3-1} fractional factorial is obtained by writing down the full 2^2 factorial and then equating factor C to the $\pm AB$ interaction. Thus, to obtain the principal fraction, we would use $C = +AB$ as follows:

Full 2^2		2^{3-1}, $I = +ABC$		
A	B	A	B	$C = AB$
−	−	−	−	+
+	−	+	−	−
−	+	−	+	−
+	+	+	+	+

To obtain the alternate fraction we would equate the last column to $C = -AB$.

Example 13-11

To illustrate the use of a one-half fraction, consider the plasma etch experiment described in Example 13-9. Suppose that we decide to use a 2^{4-1} design with $I = ABCD$ to investigate the four factors, gap (A), pressure (B), C_2F_6 flow rate (C), and power setting (D). This design would be constructed by writing down a 2^3 in the factors A, B, and C and then setting $D = ABC$. The design and the resulting etch rates are shown in Table 13-26.

In this design, the main effects are aliased with the three-factor interactions; note that the alias of A is

$$
\begin{aligned}
A \cdot I &= A \cdot ABCD, \\
&= A^2BCD, \\
&= BCD,
\end{aligned}
$$

and similarly

$$
\begin{aligned}
B &= ACD, \\
C &= ABD, \\
D &= ABC.
\end{aligned}
$$

The two-factor interactions are aliased with each other. For example, the alias of AB is CD:

$$
\begin{aligned}
AB \cdot I &= AB \cdot ABCD, \\
&= A^2B^2CD, \\
&= CD.
\end{aligned}
$$

The other aliases are

Table 13-26 The 2^{4-1} Design with Defining Relation $I = ABCD$

A	B	C	$D = ABC$	Treatment Combinations	Etch Rate
−	−	−	−	(1)	550
+	−	−	+	ad	749
−	+	−	+	bd	1052
+	+	−	−	ab	650
−	−	+	+	cd	1075
+	−	+	−	ac	642
−	+	+	−	bc	601
+	+	+	+	$abcd$	729

$$AC = BD,$$

$$AD = BC.$$

The estimates of the main effects and their aliases are found using the four columns of signs in Table 13-26. For example, from column A we obtain

$$\ell_A = A + BCD = \tfrac{1}{4}(-550 + 749 - 1052 + 650 - 1075 + 642 - 601 + 729)$$
$$= -127.00.$$

The other columns produce

$$\ell_B = B + ACD = 4.00,$$

$$\ell_C = C + ABD = 11.50,$$

and

$$\ell_D = D + ABC = 290.50.$$

Clearly ℓ_A and ℓ_D are large, and if we believe that the three-factor interactions are negligible, then the main effects A (gap) and D (power setting) significantly affect etch rate.

The interactions are estimated by forming the AB, AC, and AD columns and adding them to the table. The signs in the AB column are +, −, −, +, +, −, −, +, and this column produces the estimate

$$\ell_{AB} = AB + CD = \tfrac{1}{4}(550 - 749 - 1052 + 650 + 1075 - 642 - 601 + 729)$$
$$= -10.00.$$

From the AC and AD columns we find

$$\ell_{AC} = AC + BD = -25.50,$$

$$\ell_{AD} = AD + BC = -197.50.$$

The ℓ_{AD} estimate is large; the most straightforward interpretation of the results is that this is the AD interaction. Thus, the results obtained from the 2^{4-1} design agree with the full factorial results in Example 13-9.

Normality Probability Plots and Residuals The normal probability plot is very useful in assessing the significance of effects from a fractional factorial. This is particularly true when there are many effects to be estimated. Residuals can be obtained from a fractional factorial by the regression model method shown previously. These residuals should be plotted against the predicted values, against the levels of the factors, and on normal probability paper, as we have discussed before, both to assess the validity of the underlying model assumptions and to gain additional insight into the experimental situation.

Projection of the 2^{k-1} Design If one or more factors from a one-half fraction of a 2^k can be dropped, the design will project into a full factorial design. For example, Fig. 13-25 presents a 2^{3-1} design. Notice that this design will project into a full factorial in any two of the three original factors. Thus, if we think that at most two of the three factors are important, the 2^{3-1} design is an excellent design for identifying the significant factors. Sometimes we call experiments to identify a relatively few significant factors from a larger number of factors *screening experiments.* This projection property is highly useful in factor screening, as it allows negligible factors to be eliminated, resulting in a stronger experiment in the active factors that remain.

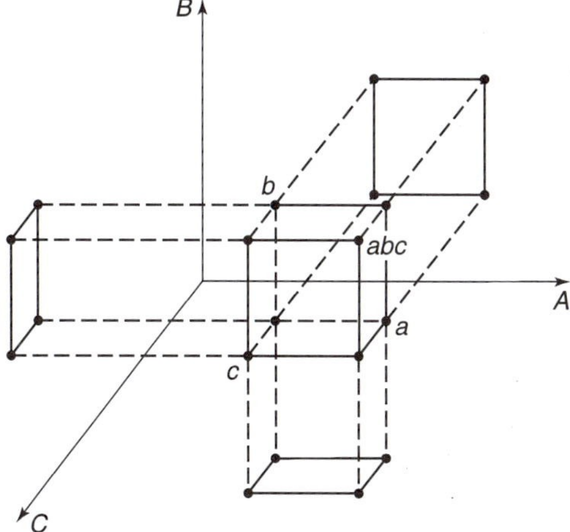

Figure 13-25 Projection of a 2^{3-1} design into three 2^2 designs.

In the 2^{4-1} design used in the plasma etch experiment in Example 13-11, we found that two of the four factors (B and C) could be dropped. If we eliminate these two factors, the remaining columns in Table 13-26 form a 2^2 design in the factors A and D, with two replicates. This design is shown in Fig. 13-26.

Design Resolution The concept of design resolution is a useful way to catalog fractional factorial designs according to the alias patterns they produce. Designs of resolution III, IV, and V are particularly important. The definitions of these terms and an example of each follow:

1. *Resolution III Designs.* These are designs in which no main effects are aliased with any other main effect, but main effects are aliased with two-factor interactions, and two-factor interactions may be aliased with each other. The 2^{3-1} design with $I = ABC$ is of resolution III. We usually employ a subscript Roman numeral to indicate design resolution; thus this one-half fraction is a 2^{3-1}_{III} design.
2. *Resolution IV Designs.* These are designs in which no main effect is aliased with any other main effect or two-factor interaction, but two-factor interactions are

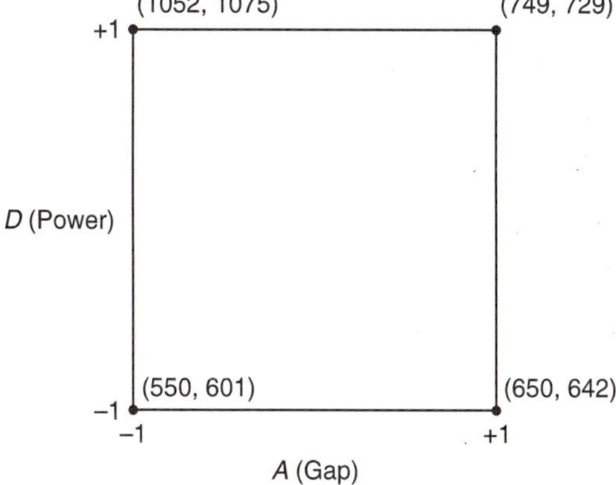

Figure 13-26 The 2^2 design obtained by dropping factors B and C from the plasma etch experiment.

aliased with each other. The 2^{4-1} design with $I = ABCD$ used in Example 13-11 is of resolution IV (2^{4-1}_{IV}).

3. *Resolution V Designs.* These are designs in which no main effect or two-factor interaction is aliased with any other main effect or two-factor interaction, but two-factor interactions are aliased with three-factor interactions. A 2^{5-1} design with $I = ABCDE$ is of resolution V (2^{5-1}_{V}).

Resolution III and IV designs are particularly useful in factor screening experiments. A resolution IV design provides very good information about main effects and will provide some information about two-factor interactions.

13-7.2 Smaller Fractions: The 2^{k-p} Fractional Factorial

Although the 2^{k-1} design is valuable in reducing the number of runs required for an experiment, we frequently find that smaller fractions will provide almost as much useful information at even greater economy. In general, a 2^k design may be run in a $1/2^p$ fraction called a 2^{k-p} fractional factorial design. Thus, a 1/4 fraction is called a 2^{k-2} fractional factorial design, a 1/8 fraction is called a 2^{k-3} design, and so on.

To illustrate a 1/4 fraction, consider an experiment with six factors and suppose that the engineer is interested primarily in main effects but would also like to get some information about the two-factor interactions. A 2^{6-1} design would require 32 runs and would have 31 degrees of freedom for estimation of effects. Since there are only six main effects and 15 two-factor interactions, the one-half fraction is inefficient—it requires too many runs. Suppose we consider a 1/4 fraction, or a 2^{6-2} design. This design contains 16 runs and, with 15 degrees of freedom, will allow estimation of all six main effects with some capability for examination of the two-factor interactions. To generate this design we would write down a 2^4 design in the factors A, B, C, and D, and then add two columns for E and F. To find the new columns we would select the two *design generators* $I = ABCE$ and $I = ACDF$. Thus column E would be found from $E = ABC$ and column F would be $F = ACD$, and also columns $ABCE$ and $ACDF$ are equal to the identity column. However, we know that the product of any two columns in the table of plus and minus signs for a 2^k is just another column in the table; therefore, the product of $ABCE$ and $ACDF$ or $ABCE\,(ACDF) = A^2BC^2DEF = BDEF$ is also an identity column. Consequently, the *complete defining relation* for the 2^{6-2} design is

$$I = ABCE = ACDF = BDEF.$$

To find the alias of any effect, simply multiply the effect by each *word* in the foregoing defining relation. The complete alias structure is

$$A = BCE = CDF = ABDEF,$$

$$B = ACE = DEF = ABCDF,$$

$$C = ABE = ADF = BCDEF,$$

$$D = ACF = BEF = ABCDE,$$

$$E = ABC = BDF = ACDEF,$$

$$F = ACD = BDE = ABCEF,$$

$$AB = CE = BCDF = ADEF,$$

$$AC = BE = DF = ABCDEF,$$

$$AD = CF = BCDE = ABEF,$$

$$AE = BC = CDEF = ABDF,$$

$$AF = CD = BCEF = ABDE,$$

$$BD = EF = ACDE = ABCF,$$

$$BF = DE = ABCD = ACEF$$

$$ABF = CEF = BCD = ADE,$$

$$CDE = ABD = AEF = CBF.$$

Notice that this is a resolution IV design; main effects are aliased with three-factor and higher interactions, and two-factor interactions are aliased with each other. This design would provide very good information on the main effects and give some idea about the strength of the two-factor interactions. For example, if the *AD* interaction appears significant, either *AD* and/or *CF* are significant. If *A* and/or *D* are significant main effects, but *C* and *F* are not, the experimenter may reasonably and tentatively attribute the significance to the *AD* interaction. The construction of the design is shown in Table 13-27.

The same principles can be applied to obtain even smaller fractions. Suppose we wish to investigate seven factors in 16 runs. This is a 2^{7-3} design (a 1/8 fraction). This design is constructed by writing down a 2^4 design in the factors *A, B, C,* and *D* and then adding three new columns. Reasonable choices for the three generators required are $I = ABCE$, $I = BCDF$, and $I = ACDG$. Therefore, the new columns are formed by setting $E = ABC$, $F = BCD$, and $G = ACD$. The complete defining relation is found by multiplying the generators together two at a time and then three at a time, resulting in

$$I = ABCE = BCDF = ACDG = ADEF = BDEG = ABFG = CEFG.$$

Notice that every main effect in this design will be aliased with three-factor and higher interactions and that two-factor interactions will be aliased with each other. Thus this is a resolution IV design.

For seven factors, we can reduce the number of runs even further. The 2^{7-4} design is an eight-run experiment accommodating seven variables. This is a 1/16 fraction and is

Table 13-27 Construction of the 2^{6-2} Design with Generators $I = ABCE$ and $I = ACDF$

A	*B*	*C*	*D*	*E* = *ABC*	*F* = *ACD*	
−	−	−	−	−	−	(1)
+	−	−	−	+	+	*aef*
−	+	−	−	+	−	*be*
+	+	−	−	−	+	*abf*
−	−	+	−	+	+	*cef*
+	−	+	−	−	−	*ac*
−	+	+	−	−	+	*bcf*
+	+	+	−	+	−	*abce*
−	−	−	+	−	+	*df*
+	−	−	+	+	−	*ade*
−	+	−	+	+	+	*bdef*
+	+	−	+	−	−	*abd*
−	−	+	+	+	−	*cde*
+	−	+	+	−	+	*acdf*
−	+	+	+	−	−	*bcd*
+	+	+	+	+	+	*abcdef*

obtained by first writing down a 2^3 design in the factors A, B, and C, and then forming the four new columns, from $I = ABD$, $I = ACE$, $I = BCF$, and $I = ABCG$. The design is shown in Table 13-28.

The complete defining relation is found by multiplying the generators together two, three, and finally four at a time, producing

$$I = ABD = ACE = BCF = ABCG = BCDE = ACDF = CDG = ABEF$$
$$= BEG = AFG = DEF = ADEG = CEFG = BDFG = ABCDEFG.$$

The alias of any main effect is found by multiplying that effect through each term in the defining relation. For example, the alias of A is

$$A = BD = CE = ABCF = BCG = ABCDE = CDF = ACDG$$
$$= BEF = ABEG = FG = ADEF = DEG = ACEFG = ABDFG$$
$$= BCDEFG.$$

This design is of resolution III, since the main effect is aliased with two-factor interactions. If we assume that all three-factor and higher interactions are negligible, the aliases of the seven main effects are

$$\ell_A = A + BD + CE + FG,$$
$$\ell_B = B + AD + CF + FG,$$
$$\ell_C = C + AE + BF + DG,$$
$$\ell_D = D + AB + CG + EF,$$
$$\ell_E = E + AC + BG + DF$$
$$\ell_F = F + BC + AG + DE,$$
$$\ell_G = G + CD + BE + AF.$$

This 2^{7-4}_{III} design is called a *saturated* fractional factorial, because all of the available degrees of freedom are used to estimate main effects. It is possible to combine sequences of these resolution III fractional factorials to separate the main effects from the two-factor interactions. The procedure is illustrated in Montgomery (2001, Chapter 8).

In constructing a fractional factorial design it is important to select the best set of design generators. Montgomery (2001) presents a table of optimum design generators for 2^{k-p} designs with up to 10 factors. The generators in this table will produce designs of maximum resolution for any specified combination of k and p. For more than 10 factors, a

Table 13-28 A 2^{7-4}_{III} Fractional Factorial Design

A	B	C	$D(=AB)$	$E(=AC)$	$F(=BC)$	$G(=ABC)$
−	−	−	+	+	+	−
+	−	−	−	−	+	+
−	+	−	−	+	−	+
+	+	−	+	−	−	−
−	−	+	+	−	−	+
+	−	+	−	+	−	−
−	+	+	−	−	+	−
+	+	+	+	+	+	+

resolution III design is recommended. These designs may be constructed by using the same method illustrated earlier for the 2^{7-4}_{III} design. For example, to investigate up to 15 factors in 16 runs, write down a 2^4 design in the factors *A, B, C*, and *D*, and then generate 11 new columns by taking the products of the original four columns two at a time, three at a time, and four at a time. The resulting design is a 2^{15-11}_{III} fractional factorial. These designs, along with other useful fractional factorials, are discussed by Montgomery (2001, Chapter 8).

13-8 SAMPLE COMPUTER OUTPUT

We provide Minitab® output for some of the examples presented in this chapter.

Sample Computer Output for Example 13-3

Reconsider Example 13-3, dealing with aircraft primer paints. The Minitab® results of the 3 × 3 factorial design with three replicates are

```
Analysis of Variance for Force

Source          DF        SS        MS        F        P
Type             2    4.5811    2.2906    27.86    0.000
Applicat         1    4.9089    4.9089    59.70    0.000
Type*Applicat    2    0.2411    0.1206     1.47    0.269
Error           12    0.9867    0.0822
Total           17   10.7178
```

The Minitab® results are in agreement with the results given in Table 13-6.

Sample Output for Example 13-7

Reconsider Example 13-7, dealing with surface roughness. The Minitab® results for the 2^3 design with two replicates are

```
Term        Effect       Coef    SE Coef        T        P
Constant              11.0625     0.3903    28.34    0.000
A           3.3750     1.6875     0.3903     4.32    0.003
B           1.6250     0.8125     0.3903     2.08    0.071
C           0.8750     0.4375     0.3903     1.12    0.295
A*B         1.3750     0.6875     0.3903     1.76    0.116
A*C         0.1250     0.0625     0.3903     0.16    0.877
B*C        -0.6250    -0.3125     0.3903    -0.80    0.446
A*B*C       1.1250     0.5625     0.3903     1.44    0.188

Analysis of Variance

Source               DF    Seq SS    Adj SS    Adj MS       F        P
Main Effects          3    59.187    59.187    19.729    8.09    0.008
2-Way Interactions    3     9.187     9.187     3.062    1.26    0.352
3-Way Interactions    1     5.062     5.062     5.062    2.08    0.188
Residual Error        8    19.500    19.500     2.437
Pure Error            8    19.500    19.500     2.438
Total                15    92.937
```

The output from Minitab® is slightly different from the results given in Example 13-7. *t*-tests on the main effects and interactions are provided in addition to the analysis of variance on the significance of main effects, two-factor interactions, and three-factor interactions. The

ANOVA results indicate that at least one of the main effects is significant, whereas no two-factor or three-factor interaction is significant.

13-9 SUMMARY

This chapter has introduced the design and analysis of experiments with several factors, concentrating on factorial designs and fractional factorials. Fixed, random, and mixed models were considered. The F-tests for main effects and interactions in these designs depend on whether the factors are fixed or random.

The 2^k factorial designs were also introduced. These are very useful designs in which all k factors appear at two levels. They have a greatly simplified method of statistical analysis. In situations where the design cannot be run under homogeneous conditions, the 2^k design can be confounded easily in 2^p blocks. This requires that certain interactions be confounded with blocks. The 2^k design also lends itself to fractional replication, in which only a particular subset of the 2^k treatment combinations are run. In fractional replication, each effect is aliased with one or more other effects. The general idea is to alias main effects and low-order interactions with higher-order interactions. This chapter discussed methods for construction of the 2^{k-p} fractional factorial designs, that is, a $1/2^p$ fraction of the 2^k design. These designs are particularly useful in industrial experimentation.

13-10 EXERCISES

13-1. An article in the *Journal of Materials Processing Technology* (2000, p. 113) presents results from an experiment involving tool wear estimation in milling. The objective is to minimize tool wear. Two factors of interest in the study were cutting speed (m/min) and depth of cut (mm). One response of interest is tool flank wear (mm). Three levels of each factor were selected and a factorial experiment with three replicates is run. Analyze the data and draw conclusions.

	Depth of Cut		
Cutting Speed	1	2	3
	0.170	0.198	0.217
12	0.185	0.210	0.241
	0.110	0.232	0.223
	0.178	0.215	0.260
15	0.210	0.243	0.289
	0.250	0.292	0.320
	0.212	0.250	0.285
18.75	0.238	0.282	0.325
	0.267	0.321	0.354

13-2. An engineer suspects that the surface finish of a metal part is influenced by the type of paint used and the drying time. He selects three drying times—20, 25, and 30 minutes—and randomly chooses two types of paint from several that are available. He conducts an experiment and obtains the data shown here. Analyze the data and draw conclusions. Estimate the variance components.

	Drying Time (min)		
Paint	20	25	30
1	74	73	78
	64	61	85
	50	44	92
2	92	98	66
	86	73	45
	68	88	85

13-3. Suppose that in Exercise 13-2 paint types were fixed effects. Compute a 95% interval estimate of the mean difference between the responses for paint type 1 and paint type 2.

13-4. The factors that influence the breaking strength of cloth are being studied. Four machines and three operators are chosen at random and an experiment is run using cloth from the same one-yard segment. The results are as follows:

	Machine			
Operator	1	2	3	4
A	109	110	108	110
	110	115	109	116
B	111	110	111	114
	112	111	109	112
C	109	112	114	111
	111	115	109	112

Test for interaction and main effects at the 5% level. Estimate the components of variance.

13-5. Suppose that in Exercise 13-4 the operators were chosen at random, but only four machines were available for the test. Does this influence the analysis or your conclusions?

13-6. A company employs two time-study engineers. Their supervisor wishes to determine whether the standards set by them are influenced by any interaction between engineers and operators. She selects three operators at random and conducts an experiment in which the engineers set standard times for the same job. She obtains the data shown here. Analyze the data and draw conclusions.

	Operator		
Engineer	1	2	3
1	2.59	2.38	2.40
	2.78	2.49	2.72
2	2.15	2.85	2.66
	2.86	2.72	2.87

13-7. An article in *Industrial Quality Control* (1956, p. 5) describes an experiment to investigate the effect of two factors (glass type and phosphor type) on the brightness of a television tube. The response variable measured is the current necessary (in microamps) to obtain a specified brightness level. The data are shown here. Analyze the data and draw conclusions, assuming that both factors are fixed.

	Phosphor Type		
Glass Type	1	2	3
1	280	300	290
	290	310	285
	285	295	290
2	230	260	220
	235	240	225
	240	235	230

13-8. Consider the tool wear data in Exercise 13-1. Plot the residuals from this experiment against the levels of cutting speed and against the depth of cut. Comment on the graphs obtained. What are the possible consequences of the information conveyed by the residual plots?

13-9. The percentage of hardwood concentration in raw pulp and the freeness and cooking time of pulp are being investigated for their effects on the strength of paper. Analyze the data shown in the following table, assuming that all three factors are fixed.

Percentage of Hardwood	Cooking Time 1.5 hours			Cooking Time 2.0 hours		
	Freeness			Freeness		
Concentration	400	500	650	400	500	650
10	96.6	97.7	99.4	98.4	99.6	100.6
	96.0	96.0	99.8	98.6	100.4	100.9
15	98.5	96.0	98.4	97.5	98.7	99.6
	97.2	96.9	97.6	98.1	98.0	99.0
20	97.5	95.6	97.4	97.6	97.0	98.5
	96.6	96.2	98.1	98.4	97.8	99.8

13-10. An article in *Quality Engineering* (1999, p. 357) presents the results of an experiment conducted to determine the effects of three factors on warpage in an injection-molding process. Warpage is defined as the nonflatness property in the product manufactured. This particular company manufactures plastic molded components for use in television sets, washing machines, and automobiles. The three factors of interest (each at two levels) are A = melt temperature, B = injection speed, and C = injection process. A complete 2^3 factorial design was carried out with replication. Two replicates are provided in the table below. Analyze the data from this experiment.

A	B	C	I	II
−1	−1	−1	1.35	1.40
1	−1	−1	2.15	2.20
−1	1	−1	1.50	1.50
1	1	−1	1.10	1.20
−1	−1	1	0.70	0.70
1	−1	1	1.40	1.35
−1	1	1	1.20	1.35
1	1	1	1.10	1.00

13-11. For the Warpage experiment in Exercise 13-10, obtain the residuals and plot them on normal probability paper. Also plot the residuals versus the predicted values. Comment on these plots.

13-12. Four factors are thought to possibly influence the taste of a soft drink beverage: type of sweetener (A), ratio of syrup to water (B), carbonation level (C), and temperature (D). Each factor can be run at two levels, producing a 2^4 design. At each run in the design, samples of the beverage are given to a test panel consisting of 20 people. Each tester assigns a point score from 1 to 10 to the beverage. Total score is the response variable, and the objective is to find a formulation that maximizes total score. Two replicates of this design are run, and the results shown here. Analyze the data and draw conclusions.

Treatment Combinations	Replicate I	Replicate II	Treatment Combinations	Replicate I	Replicate II
(1)	190	193	d	198	195
a	174	178	ad	172	176
b	181	185	bd	187	183
ab	183	180	abd	185	186
c	177	178	cd	199	190
ac	181	180	acd	179	175
bc	188	182	bcd	187	184
abc	173	170	$abcd$	180	180

13-13. Consider the experiment in Exercise 13-12. Plot the residuals against the levels of factors A, B, C, and D. Also construct a normal probability plot of the residuals. Comment on these plots.

13-14. Find the standard error of the effects for the experiment in Exercise 13-12. Using the standard errors as a guide, what factors appear significant?

13-15. The data shown here represent a single replicate of a 2^5 design that is used in an experiment to study the compressive strength of concrete. The factors are mix (A), time (B), laboratory (C), temperature (D), and drying time (E). Analyze the data, assuming that three-factor and higher interactions are negligible. Use a normal probability plot to assess the effects.

(1) = 700	d = 1000	e = 800	de = 1900
a = 900	ad = 1100	ae = 1200	ade = 1500
b = 3400	bd = 3000	be = 3500	bde = 4000
ab = 5500	abd = 6100	abe = 6200	$abde$ = 6500
c = 600	cd = 800	ce = 600	cde = 1500
ac = 1000	acd = 1100	ace = 1200	$acde$ = 2000
bc = 3000	bcd = 3300	bce = 3006	$bcde$ = 3400
abc = 5300	$abcd$ = 6000	$abce$ = 5500	$abcde$ = 6300

13-16. An experiment described by M. G. Natrella in the National Bureau of Standards *Handbook of Experimental Statistics* (No. 91, 1963) involves flame-testing fabrics after applying fire-retardant treatments. There are four factors: type of fabric (A), type of fire-retardant treatment (B), laundering condition (C—the low level is no laundering, the high level is after one laundering), and the method of conducting the flame test (D). All factors are run at two levels, and the response variable is the inches of fabric burned on a standard size test sample. The data are

(1) = 42	d = 40
a = 31	ad = 30
b = 45	bd = 50
ab = 29	abd = 25
c = 39	cd = 40
ac = 28	acd = 25
bc = 46	bcd = 50
abc = 32	$abcd$ = 23

(a) Estimate the effects and prepare a normal probability plot of the effects.

(b) Construct a normal probability plot of the residuals and comment on the results.

(c) Construct an analysis of variance table assuming that three- and four-factor interactions are negligible.

13-17. Consider the data from the first replicate of Exercise 13-10. Suppose that these observations could not all be run under the same conditions. Set up a design to run these observations in two blocks of four observations, each with ABC confounded. Analyze the data.

13-18. Consider the data from the first replicate of Exercise 13-12. Construct a design with two blocks of eight observations each, with $ABCD$ confounded. Analyze the data.

13-19. Repeat Exercise 13-18 assuming that four blocks are required. Confound ABD and ABC (and consequently CD) with blocks.

13-20. Construct a 2^5 design in four blocks. Select the effects to be confounded so that we confound the highest possible interactions with blocks.

13-21. An article in *Industrial and Engineering Chemistry* ("Factorial Experiments in Pilot Plant Studies," 1951, p. 1300) reports on an experiment to investigate the effects of temperature (A), gas throughput (B), and concentration (C) on the strength of product solution in a recirculation unit. Two blocks were used with ABC confounded, and the experiment was replicated twice. The data are as follows:

Replicate 1		Replicate 2	
Block 1	Block 2	Block 1	Block 2
(1) = 99	a = 18	(1) = 46	a = 18
ab = 52	b = 51	ab = − 47	b = 62
ac = 42	c = 108	ac = 22	c = 104
bc = 95	abc = 35	bc = 67	abc = 36

(a) Analyze the data from this experiment.

(b) Plot the residuals on normal probability paper and against the predicted values. Comment on the plots obtained.

(c) Comment on the efficiency of this design. Note that we have replicated the experiment twice, yet we have no information on the ABC interaction.

(d) Suggest a better design; specifically, one that would provide some information on *all* interactions.

13-22. R. D. Snee ("Experimenting with a Large Number of Variables," in *Experiments in Industry: Design, Analysis and Interpretation of Results*, by R. D. Snee, L. B. Hare, and J. B. Trout, Editors, ASQC, 1985) describes an experiment in which a 2^{5-1} design with $I = ABCDE$ was used to investigate the effects of five factors on the color of a chemical product. The factors are A = solvent/reactant, B = catalyst/reactant, C = temperature, D = reactant purity, and E = reactant pH. The results obtained are as follows:

$e = -0.63$	$d = 6.79$
$a = 2.51$	$ade = 6.47$
$b = -2.68$	$bde = 3.45$
$abe = 1.66$	$abd = 5.68$
$c = 2.06$	$cde = 5.22$
$ace = 1.22$	$acd = 4.38$
$bce = -2.09$	$bcd = 4.30$
$abc = 1.93$	$abcde = 4.05$

(a) Prepare a normal probability plot of the effects. Which factors are active?

(b) Calculate the residuals. Construct a normal probability plot of the residuals and plot the residuals versus the fitted values. Comment on the plots.

(c) If any factors are negligible, collapse the 2^{5-1} design into a full factorial in the active factors. Comment on the resulting design, and interpret the results.

13-23. An article in the *Journal of Quality Technology*, (Vol. 17, 1985, p. 198) describes the use of a replicated fractional factorial to investigate the effects of five factors on the free height of leaf springs used in an automotive application. The factors are A = furnace temperature, B = heating time, C = transfer time, D = hold down time, and E = quench oil temperature. The data are shown below.

A	B	C	D	E			
−	−	−	−	−	7.78,	7.78,	7.81
+	−	−	+	−	8.15,	8.18,	7.88
−	+	−	+	−	7.50,	7.56,	7.50
+	+	−	−	−	7.59,	7.56,	7.75
−	−	+	+	−	7.54,	8.00,	7.88
+	−	+	−	−	7.69,	8.09,	8.06
−	+	+	−	−	7.56,	7.52,	7.44
+	+	+	+	−	7.56,	7.81,	7.69
−	−	−	−	+	7.50,	7.25,	7.12
+	−	−	+	+	7.88,	7.88,	7.44
−	+	−	+	+	7.50,	7.56,	7.50
+	+	−	−	+	7.63,	7.75,	7.56
−	−	+	+	+	7.32,	7.44,	7.44
+	−	+	−	+	7.56,	7.69,	7.62
−	+	+	−	+	7.18,	7.18,	7.25
+	+	+	+	+	7.81,	7.50,	7.59

(a) What is the generator for this fraction? Write out the alias structure.

(b Analyze the data. What factors influence mean free height?

(c) Calculate the range of free height for each run. Is there any indication that any of these factors affects variability in free height?

(d) Analyze the residuals from this experiment and comment on your findings.

13-24. An article in *Industrial and Engineering Chemistry* ("More on Planning Experiments to Increase Research Efficiency," 1970, p. 60) uses a 2^{5-2} design to investigate the effects of A = condensation temperature, B = amount of material 1, C = solvent volume, D = condensation time, and E = amount of material 2 on yield. The results obtained are as follows:

e = 23.2, ad = 16.9, cd = 23.8, bde = 16.8,
ab = 15.5, bc = 16.2, ace = 23.4, $abcde$ = 18.1.

(a) Verify that the design generators used were $I = ACE$ and $I = BDE$.

(b Write down the complete defining relation and the aliases from this design.

(c) Estimate the main effects.

(d) Prepare an analysis of variance table. Verify that the AB and AD interactions are available to use as error.

(e) Plot the residuals versus the fitted values. Also construct a normal probability plot of the residuals. Comment on the results.

13-25. An article in *Cement and Concrete Research* (2001, p. 1213) describes an experiment to investigate the effects of four metal oxides on several cement properties. The four factors are all run at two levels and one response of interest is the mean bulk density (g/cm^3). The four factors and their levels are

Factor	Low Level (−1)	High Level (+1)
A: %Fe_2O_3	0	30
B: % ZnO	0	15
C: %PbO	0	2.5
D: %Cr_2O_3	0	2.5

Typical results from this type of experiment are given in the following table:

Run	A	B	C	D	Density
1	−1	−1	−1	1	2.001
2	1	−1	−1	−1	2.062
3	−1	1	−1	−1	2.019
4	1	1	−1	1	2.059
5	−1	−1	1	−1	1.990
6	1	−1	1	1	2.076
7	−1	1	1	1	2.038
8	1	1	1	−1	2.118

(a) What is the generator for this fraction?

(b) Analyze the data. What factors influence mean bulk density?

(c) Analyze the residuals from this experiment and comment on your findings.

13-26. Consider the 2^{6-2} design in Table 13-27. Suppose that after analyzing the original data, we find that factors C and E can be dropped. What type of 2^k design is left in the remaining variables?

13-27. Consider the 2^{6-2} design in Table 13-27. Suppose that after the original data analysis, we find that factors D and F can be dropped. What type of 2^k design is left in the remaining variables? Compare the results with Exercise 13-26. Can you explain why the answers are different?

13-28. Suppose that in Exercise 13-12 it was possible to run only a one-half fraction of the 2^4 design. Construct the design and perform the statistical analysis, use the data from replicate I.

13-29. Suppose that in Exercise 13-15 only a one-half fraction of the 2^5 design could be run. Construct the design and perform the analysis.

13-30. Consider the data in Exercise 13-15. Suppose that only a one-quarter fraction of the 2^5 design could be run. Construct the design and analyze the data.

13-31. Construct a 2^{6-3}_{III} fractional factorial design. Write down the aliases, assuming that only main effects and two-factor interactions are of interest.

Chapter 14

Simple Linear Regression and Correlation

In many problems there are two or more variables that are inherently related, and it is necessary to explore the nature of this relationship. Regression analysis is a statistical technique for modeling and investigating the relationship between two or more variables. For example, in a chemical process, suppose that the yield of product is related to the process operating temperature. Regression analysis can be used to build a model that expresses yield as a function of temperature. This model can then be used to predict yield at a given temperature level. It could also be used for process optimization or process control purposes.

In general, suppose that there is a single dependent variable, or *response* y, that is related to k independent, or *regressor*, variables, say $x_1, x_2, \ldots, x_k$. The response variable y is a random variable, while the regressor variables $x_1, x_2, \ldots, x_k$ are measured with negligible error. The x_j are called *mathematical* variables and are frequently controlled by the experimenter. Regression analysis can also be used in situations where $y, x_1, x_2, \ldots, x_k$ are jointly distributed random variables, such as when the data are collected as different measurements on a common experimental unit. The relationship between these variables is characterized by a mathematical model called a *regression equation.* More precisely, we speak of the regression of y on $x_1, x_2, \ldots, x_k$. This regression model is fitted to a set of data. In some instances, the experimenter will know the exact form of the true functional relationship between y and $x_1, x_2, \ldots, x_k$, say $y = \phi(x_1, x_2, \ldots, x_k)$. However, in most cases, the true functional relationship is unknown, and the experimenter will choose an appropriate function to approximate ϕ. A polynomial model is usually employed as the approximating function.

In this chapter, we discuss the case where only a single regressor variable, x, is of interest. Chapter 15 will present the case involving more than one regressor variable.

14-1 SIMPLE LINEAR REGRESSION

We wish to determine the relationship between a single regressor variable x and a response variable y. The regressor variable x is assumed to be a continuous mathematical variable, controllable by the experimenter. Suppose that the true relationship between y and x is a straight line, and that the observation y at each level of x is a random variable. Now, the expected value of y for each value of x is

$$E(y|x) = \beta_0 + \beta_1 x, \tag{14-1}$$

where the intercept β_0 and the slope β_1 are unknown constants. We assume that each observation, y, can be described by the model

$$y = \beta_0 + \beta_1 x + \epsilon, \tag{14-2}$$

where ϵ is a random error with mean zero and variance σ^2. The $\{\epsilon\}$ are also assumed to be uncorrelated random variables. The regression model of equation 14-2 involving only a single regressor variable x is often called the simple linear regression model.

Suppose that we have n pairs of observations, say (y_1, x_1), (y_2, x_2), …, (y_n, x_n). These data may be used to estimate the unknown parameters β_0 and β_1 in equation 14-2. Our estimation procedure will be the method of least squares. That is, we will estimate β_0 and β_1 so that the sum of squares of the deviations between the observations and the regression line is a minimum. Now using equation 14-2, we may write

$$y_i = \beta_0 + \beta_1 x_i + \epsilon_i, \qquad i = 1, 2, \ldots, n, \tag{14-3}$$

and the sum of squares of the deviations of the observations from the true regression line is

$$L = \sum_{i=1}^{n} \epsilon_i^2 = \sum_{i=1}^{n} (y_i - \beta_0 - \beta_1 x_i)^2. \tag{14-4}$$

The least-squares estimators of β_0 and β_1, say $\hat{\beta}_0$ and $\hat{\beta}_1$, must satisfy

$$\begin{aligned} \left.\frac{\partial L}{\partial \beta_0}\right|_{\hat{\beta}_0, \hat{\beta}_1} &= -2\sum_{i=1}^{n}\left(y_i - \hat{\beta}_0 - \hat{\beta}_1 x_i\right) = 0, \\ \left.\frac{\partial L}{\partial \beta_1}\right|_{\hat{\beta}_0, \hat{\beta}_1} &= -2\sum_{i=1}^{n}\left(y_i \quad \hat{\beta}_0 - \hat{\beta}_1 x_i\right)x_i = 0. \end{aligned} \tag{14-5}$$

Simplifying these two equations yields

$$\begin{aligned} n\hat{\beta}_0 + \hat{\beta}_1 \sum_{i=1}^{n} x_i &= \sum_{i=1}^{n} y_i, \\ \hat{\beta}_0 \sum_{i=1}^{n} x_i + \hat{\beta}_1 \sum_{i=1}^{n} x_i^2 &= \sum_{i=1}^{n} y_i x_i. \end{aligned} \tag{14-6}$$

Equations 14-6 are called the least-squares *normal equations.* The solution to the normal equations is

$$\hat{\beta}_0 = \bar{y} - \hat{\beta}_1 \bar{x}, \tag{14-7}$$

$$\hat{\beta}_1 = \frac{\sum_{i=1}^{n} y_i x_i - \frac{1}{n}\left(\sum_{i=1}^{n} y_i\right)\left(\sum_{i=1}^{n} x_i\right)}{\sum_{i=1}^{n} x_i^2 - \frac{1}{n}\left(\sum_{i=1}^{n} x_i\right)^2} \tag{14-8}$$

where $\bar{y} = (1/n)\sum_{i=1}^{n} y_i$ and $\bar{x} = (1/n)\sum_{i=1}^{n} x_i$. Therefore, equations 14-7 and 14-8 are the least-squares estimators of the intercept and slope, respectively. The fitted simple linear regression model is

$$\hat{y} = \hat{\beta}_0 + \hat{\beta}_1 x. \tag{14-9}$$

Notationally, it is convenient to give special symbols to the numerator and denominator of equation 14-8. That is, let

$$S_{xx} = \sum_{i=1}^{n}(x_i - \bar{x})^2 = \sum_{i=1}^{n} x_i^2 - \frac{1}{n}\left(\sum_{i=1}^{n} x_i\right)^2 \tag{14-10}$$

and

$$S_{xy} = \sum_{i=1}^{n} y_i(x_i - \bar{x}) = \sum_{i=1}^{n} x_i y_i - \frac{1}{n}\left(\sum_{i=1}^{n} x_i\right)\left(\sum_{i=1}^{n} y_i\right). \tag{14-11}$$

We call S_{xx} the corrected sum of squares of x and S_{xy} the corrected sum of cross products of x and y. The extreme right-hand sides of equations 14-10 and 14-11 are the usual computational formulas. Using this new notation, the least-squares estimator of the slope is

$$\hat{\beta}_1 = \frac{S_{xy}}{S_{xx}}. \tag{14-12}$$

Example 14-1

A chemical engineer is investigating the effect of process operating temperature on product yield. The study results in the following data:

Temperature, °C (x)	100	110	120	130	140	150	160	170	180	190
Yield, % (y)	45	51	54	61	66	70	74	78	85	89

These pairs of points are plotted in Fig. 14-1. Such a display is called a *scatter diagram*. Examination of this scatter diagram indicates that there is a strong relationship between yield and temperature, and the tentative assumption of the straight-line model $y = \beta_0 + \beta_1 x + \epsilon$ appears to be reasonable. The following quantities may be computed:

$$n = 10, \quad \sum_{i=1}^{10} x_i = 1450, \quad \sum_{i=1}^{10} y_i = 673, \quad \bar{x} = 145, \quad \bar{y} = 67.3,$$

$$\sum_{i=1}^{10} x_i^2 = 218{,}500, \quad \sum_{i=1}^{10} y_i^2 = 47{,}225, \quad \sum_{i=1}^{10} x_i y_i = 101{,}570.$$

From equations 14-10 and 14-11, we find

$$S_{xx} = \sum_{i=1}^{10} x_i^2 - \frac{1}{10}\left(\sum_{i=1}^{10} x_i\right)^2 = 218{,}500 - \frac{(1450)^2}{10} = 8250$$

and

$$S_{xy} = \sum_{i=1}^{10} x_i y_i - \frac{1}{10}\left(\sum_{i=1}^{10} x_i\right)\left(\sum_{i=1}^{10} y_i\right) = 101{,}570 - \frac{(1450)(673)}{10} = 3985.$$

Therefore, the least-squares estimates of the slope and intercept are

$$\hat{\beta}_1 = \frac{S_{xy}}{S_{xx}} = \frac{3985}{8250} = 0.483$$

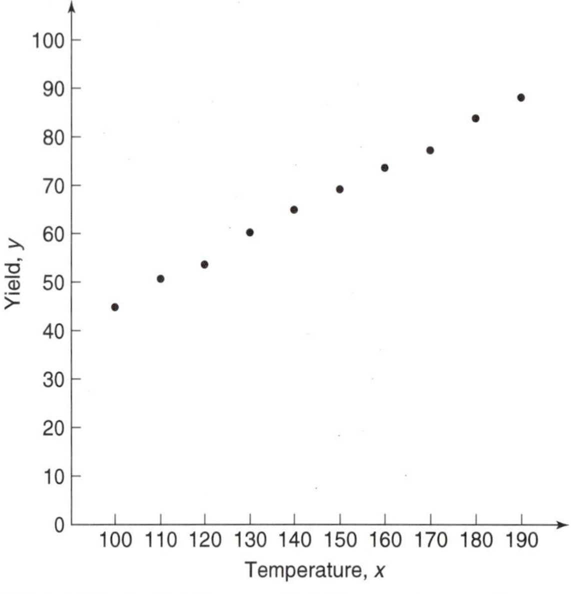

Figure 14-1 Scatter diagram of yield versus temperature.

and

$$\hat{\beta}_0 = \bar{y} - \hat{\beta}_1\bar{x} = 67.3 - (0.483)145 = -2.739.$$

The fitted simple linear regression model is

$$\hat{y} = -2.739 + 0.483x.$$

Since we have only tentatively assumed the straight-line model to be appropriate, we will want to investigate the adequacy of the model. The statistical properties of the least-squares estimators $\hat{\beta}_0$ and $\hat{\beta}_1$ are useful in assessing model adequacy. The estimators $\hat{\beta}_0$ and $\hat{\beta}_1$ are random variables, since they are just linear combinations of the y_i, and the y_i are random variables. We will investigate the bias and variance properties of these estimators. Consider first $\hat{\beta}_1$. The expected value of $\hat{\beta}_1$ is

$$\begin{aligned} E(\hat{\beta}_1) &= E\left(\frac{S_{xy}}{S_{xx}}\right) \\ &= \frac{1}{S_{xx}} E\left[\sum_{i=1}^{n} y_i(x_i - \bar{x})\right] \\ &= \frac{1}{S_{xx}} E\left[\sum_{i=1}^{n} (x_i - \bar{x})(\beta_0 + \beta_1 x_i + \epsilon_i)\right] \\ &= \frac{1}{S_{xx}} \left\{ E\left[\beta_0 \sum_{i=1}^{n} (x_i - \bar{x})\right] + E\left[\beta_1 \sum_{i=1}^{n} x_i(x_i - \bar{x})\right] \right. \\ &\qquad \left. + E\left[\sum_{i=1}^{n} \epsilon_i(x_i - \bar{x})\right] \right\} \end{aligned}$$

$$
\begin{aligned}
&= \frac{1}{S_{xx}} \beta_1 S_{xx} \\
&= \beta_1,
\end{aligned}
$$

since $\Sigma_{i=1}^{n}(x_i - \bar{x}) = 0$, $\Sigma_{i=1}^{n} x_i(x_i - \bar{x}) = S_{xx}$, and by assumption $E(\epsilon_i) = 0$. Thus, $\hat{\beta}_1$ is an *unbiased* estimator of the true slope β_1. Now consider the variance of $\hat{\beta}_1$. Since we have assumed that $V(\epsilon_i) = \sigma^2$, it follows that $V(y_i) = \sigma^2$, and

$$
\begin{aligned}
V(\hat{\beta}_1) &= V\left(\frac{S_{xy}}{S_{xx}}\right) \\
&= \frac{1}{S_{xx}^2} V\left[\sum_{i=1}^{n} y_i(x_i - \bar{x})\right].
\end{aligned}
\tag{14-13}
$$

The random variables $\{y_i\}$ are uncorrelated because the $\{\epsilon_i\}$ are uncorrelated. Therefore, the variance of the sum in equation 14-13 is just the sum of the variances, and the variance of each term in the sum, say $V[y_i(x_i - \bar{x})]$, is $\sigma^2(x_i - \bar{x})^2$. Thus,

$$
\begin{aligned}
V(\hat{\beta}_1) &= \frac{1}{S_{xx}^2} \sigma^2 \sum_{i=1}^{n} (x_i - \bar{x})^2 \\
&= \frac{\sigma^2}{S_{xx}}.
\end{aligned}
\tag{14-14}
$$

Using a similar approach, we can show that

$$
E(\hat{\beta}_0) = \beta_0 \quad \text{and} \quad V(\hat{\beta}_0) = \sigma^2\left[\frac{1}{n} + \frac{\bar{x}^2}{S_{xx}}\right].
\tag{14-15}
$$

Note that $\hat{\beta}_0$ is an unbiased estimator of β_0. The covariance of $\hat{\beta}_0$ and $\hat{\beta}_1$ is not zero; in fact, $\text{Cov}(\hat{\beta}_0, \hat{\beta}_1) = -\sigma^2\bar{x}/S_{xx}$.

It is usually necessary to obtain an estimate of σ^2. The difference between the observation y_i and the corresponding predicted value $\hat{y}_i$, say $e_i = y_i - \hat{y}_i$, is called a *residual.* The sum of the squares of the residuals, or the *error sum of squares*, would be

$$
\begin{aligned}
SS_E &= \sum_{i=1}^{n} e_i^2 \\
&= \sum_{i=1}^{n} (y_i - \hat{y}_i)^2.
\end{aligned}
\tag{14-16}
$$

A more convenient computing formula for SS_E may be found by substituting the fitted model $\hat{y}_i = \hat{\beta}_0 + \hat{\beta}_1 x_i$ into equation 14-16 and simplifying. The result is

$$
SS_E = \sum_{i=1}^{n} y_i^2 - n\bar{y}^2 - \hat{\beta}_1 S_{xy},
$$

and if we let $\Sigma_{i=1}^{n} y_i^2 - n\bar{y}^2 = \sum_{i=1}^{n}(y_i - \bar{y})^2 \equiv S_{yy}$, then we may write SS_E as

$$
SS_E = S_{yy} - \hat{\beta}_1 S_{xy}.
\tag{14-17}
$$

The expected value of the error sum of squares SS_E is $E(SS_E) = (n - 2)\sigma^2$. Therefore,

$$\hat{\sigma}^2 = \frac{SS_E}{n-2} \equiv MS_E \tag{14-18}$$

is an unbiased estimator of σ^2.

Regression analysis is widely used and frequently *misused.* There are several common abuses of regression that should be briefly mentioned. Care should be taken in selecting variables with which to construct regression models and in determining the form of the approximating function. It is quite possible to develop statistical relationships among variables that are completely unrelated in a practical sense. For example, one might attempt to relate the shear strength of spot welds with the number of boxes of computer paper used by the data processing department. A straight line may even appear to provide a good fit to the data, but the relationship is an unreasonable one on which to rely. A strong observed association between variables does not necessarily imply that a *causal* relationship exists between those variables. Designed experiments are the only way to determine causal relationships.

Regression relationships are valid only for values of the independent variable within the range of the original data. The linear relationship that we have tentatively assumed may be valid over the original range of x, but it may be unlikely to remain so as we encounter x values beyond that range. In other words, as we move beyond the range of values of x for which data were collected, we become less certain about the validity of the assumed model. Regression models are not necessarily valid for extrapolation purposes.

Finally, one occasionally feels that the model $y = \beta x + \epsilon$ is appropriate. The omission of the intercept from this model implies, of course, that $y = 0$ when $x = 0$. This is a very strong assumption that often is unjustified. Even when two variables, such as the height and weight of men, would seem to qualify for the use of this model, we would usually obtain a better fit by including the intercept, because of the limited range of data on the independent variable.

14-2 HYPOTHESIS TESTING IN SIMPLE LINEAR REGRESSION

An important part of assessing the adequacy of the simple linear regression model is testing statistical hypotheses about the model parameters and constructing certain confidence intervals. Hypothesis testing is discussed in this section, and Section 14-3 presents methods for constructing confidence intervals. To test hypotheses about the slope and intercept of the regression model, we must make the additional assumption that the error component ϵ_i is normally distributed. Thus, the complete assumptions are that the errors are NID(0, σ^2) (normal and independently distributed). Later we will discuss how these assumptions can be checked through *residual analysis.*

Suppose we wish to test the hypothesis that the slope equals a constant, say $\beta_{1,0}$. The appropriate hypotheses are

$$\begin{aligned} H_0&: \beta_1 = \beta_{1,0}, \\ H_1&: \beta_1 \neq \beta_{1,0}, \end{aligned} \tag{14-19}$$

where we have assumed a two-sided alternative. Now since the ϵ_i are NID(0, σ^2), it follows directly that the observations y_i are NID($\beta_0 + \beta_1 x_i$, σ^2). From equation 14-8 we observe that $\hat{\beta}_1$ is a linear combination of the observations y_i. Thus, $\hat{\beta}_1$ is a linear combination of independent normal random variables and, consequently, $\hat{\beta}_1$ is $N(\beta_1, \sigma^2/S_{xx})$, using the bias and variance properties of $\hat{\beta}_1$ from Section 14-1. Furthermore, $\hat{\beta}_1$ is independent of MS_E. Then, as a result of the normality assumption, the statistic

$$t_0 = \frac{\hat{\beta}_1 - \beta_{1,0}}{\sqrt{MS_E / S_{xx}}} \tag{14-20}$$

follows the t distribution with $n - 2$ degrees of freedom under H_0: $\beta_1 = \beta_{1,0}$. We would reject H_0: $\beta_1 = \beta_{1,0}$ if

$$|t_0| > t_{\alpha/2,\, n-2}, \tag{14-21}$$

where t_0 is computed from equation 14-20.

A similar procedure can be used to test hypotheses about the intercept. To test

$$H_0: \beta_0 = \beta_{0,0}, \\ H_1: \beta_0 \neq \beta_{0,0}, \tag{14-22}$$

we would use the statistic

$$t_0 = \frac{\hat{\beta}_0 - \beta_{0,0}}{\sqrt{MS_E\left[\dfrac{1}{n} + \dfrac{\bar{x}^2}{S_{xx}}\right]}} \tag{14-23}$$

and reject the null hypothesis if $|t_0| > t_{\alpha/2,\, n-2}$.

A very important special case of the hypothesis of equation 14-19 is

$$H_0: \beta_1 = 0, \\ H_1: \beta_1 \neq 0. \tag{14-24}$$

This hypothesis relates to the *significance of regression.* Failing to reject H_0: $\beta_1 = 0$ is equivalent to concluding that there is no linear relationship between x and y. This situation is illustrated in Fig. 14-2. Note that this may imply either that x is of little value in explaining the variation in y and that the best estimator of y for any x is $\hat{y} = \bar{y}$ (Fig. 14-2*a*) or that the true relationship between x and y is not linear (Fig. 14-2*b*). Alternatively, if H_0: $\beta_1 = 0$ is rejected, this implies that x is of value in explaining the variability in y. This is illustrated in Fig. 14-3. However, rejecting H_0:$\beta_1 = 0$ could mean either that the straight-line model is adequate (Fig. 14-3*a*) or that even though there is a linear effect of x, better results could be obtained with the addition of higher-order polynomial terms in x (Fig. 14-3*b*).

The test procedure for H_0:$\beta_1 = 0$ may be developed from two approaches. The first approach starts with the following partitioning of the total corrected sum of squares for y:

$$S_{yy} \equiv \sum_{i=1}^{n}(y_i - \bar{y})^2 = \sum_{i=1}^{n}(\hat{y}_i - \bar{y})^2 + \sum_{i=1}^{n}(y_i - \hat{y}_i)^2. \tag{14-25}$$

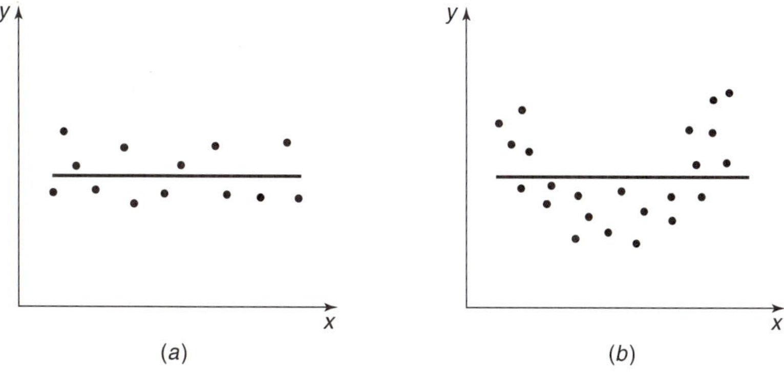

Figure 14-2 The hypothesis H_0: $\beta_1 = 0$ is not rejected.

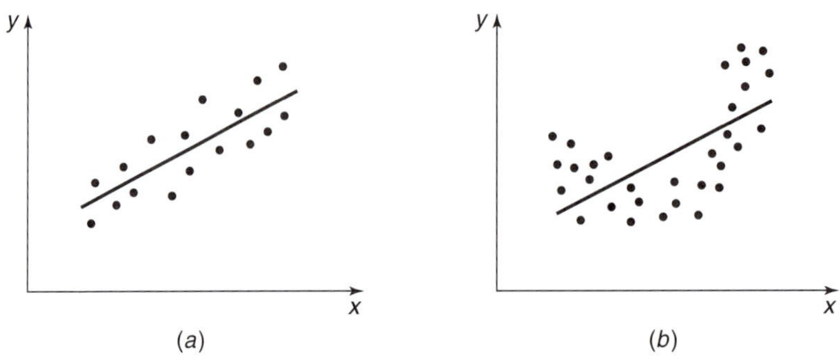

Figure 14-3 The hypothesis H_0: $\beta_1 = 0$ is rejected.

The two components of S_{yy} measure, respectively, the amount of variability in the y_i accounted for by the regression line, and the residual variation left unexplained by the regression line. We usually call $SS_E = \sum_{i=1}^{n}(y_i - \hat{y}_i)^2$ the *error* sum of squares and $SS_R = \sum_{i=1}^{n}(\hat{y}_i - \bar{y})^2$ the *regression sum* of squares. Thus, equation 14-25 may be written

$$S_{yy} = SS_R + SS_E. \tag{14-26}$$

Comparing equation 14-26 with equation 14-17, we note that the regression sum of squares SS_R is

$$SS_R = \hat{\beta}_1 S_{xy}. \tag{14-27}$$

S_{yy} has $n - 1$ degrees of freedom, and SS_R and SS_E have 1 and $n - 2$ degrees of freedom, respectively.

We may show that $E[SS_E/(n-2)] = \sigma^2$ and $E(SS_R) = \sigma^2 + \beta_1^2 S_{xx}$, and that SS_E and SS_R are independent. Thus, if H_0: $\beta_1 = 0$ is true, the statistic

$$F_0 = \frac{SS_R/1}{SS_E/(n-2)} = \frac{MS_R}{MS_E} \tag{14-28}$$

follows the $F_{1,n-2}$ distribution, and we would reject H_0 if $F_0 > F_{\alpha,1,n-2}$. The test procedure is usually arranged in an analysis of variance table (or ANOVA), such as Table 14-1.

The test for significance of regression may also be developed from equation 14-20 with $\beta_{1,0} = 0$, say

$$t_0 = \frac{\hat{\beta}_1}{\sqrt{MS_E/S_{xx}}}. \tag{14-29}$$

Squaring both sides of equation 14-29, we obtain

$$t_0^2 = \frac{\hat{\beta}_1^2 S_{xx}}{MS_E} = \frac{\hat{\beta}_1 S_{xy}}{MS_E} = \frac{MS_R}{MS_E}. \tag{14-30}$$

Table 14-1 Analysis of Variance for Testing Significance of Regression

Source of Variation	Sum of Squares	Degrees of Freedom	Mean Square	F_0
Regression	$SS_R = \hat{\beta}_1 S_{xy}$	1	MS_R	MS_R/MS_E
Error or residual	$SS_E = S_{yy} - \hat{\beta}_1 S_{xy}$	$n - 2$	MS_E	
Total	S_{yy}	$n - 1$		

Table 14-2 Testing for Significance of Regression, Example 14-2

Source of Variation	Sum of Squares	Degrees of Freedom	Mean Square	F_0
Regression	1924.87	1	1924.87	2138.74
Error	7.23	8	0.90	
Total	1932.10	9		

Note that t_0^2 in equation 14-30 is identical to F_0 in equation 14-28. It is true, in general, that the square of a t random variable with f degrees of freedom is an F random variable, with one and f degrees of freedom in the numerator and denominator, respectively. Thus, the test using t_0 is equivalent to the test based on F_0.

Example 14-2

We will test the model developed in Example 14-1 for significance of regression. The fitted model is $\hat{y} = -2.739 + 0.483x$, and S_{yy} is computed as

$$S_{yy} = \sum_{i=1}^{n} y_i^2 - \frac{1}{n}\left(\sum_{i=1}^{n} y_i\right)^2$$

$$= 47{,}225 - \frac{(673)^2}{10}$$

$$= 1932.10.$$

The regression sum of squares is

$$SS_R = \hat{\beta}_1 S_{xy} = (0.483)(3985) = 1924.87,$$

and the error sum of squares is

$$SS_E = S_{yy} - SS_R$$

$$= 1932.10 - 1924.87$$

$$= 7.23.$$

The analysis of variance for testing H_0: $\beta_1 = 0$ is summarized in Table 14-2. Noting that $F_0 = 2138.74 > F_{0.01,1,8} = 11.26$, we reject H_0 and conclude that $\beta_1 \neq 0$.

14-3 INTERVAL ESTIMATION IN SIMPLE LINEAR REGRESSION

In addition to point estimates of the slope and intercept, it is possible to obtain confidence interval estimates of these parameters. The width of these confidence intervals is a measure of the overall quality of the regression line. If the ϵ_i are normally and independently distributed, then

$$\left(\hat{\beta}_1 - \beta_1\right)\Big/\sqrt{MS_E/S_{xx}} \quad \text{and} \quad \left(\hat{\beta}_0 - \beta_0\right)\Big/\sqrt{MS_E\left[\frac{1}{n} + \frac{\bar{x}^2}{S_{xx}}\right]}$$

are both distributed as t with $n-2$ degrees of freedom. Therefore, a $100(1-\alpha)\%$ confidence interval on the slope β_1 is given by

$$\hat{\beta}_1 - t_{\alpha/2,n-2}\sqrt{\frac{MS_E}{S_{xx}}} \le \beta_1 \le \hat{\beta}_1 + t_{\alpha/2,n-2}\sqrt{\frac{MS_E}{S_{xx}}}. \tag{14-31}$$

Similarly, a $100(1-\alpha)\%$ confidence interval on the intercept β_0 is

$$\hat{\beta}_0 - t_{\alpha/2,n-2}\sqrt{MS_E\left[\frac{1}{n}+\frac{\bar{x}^2}{S_{xx}}\right]} \le \beta_0 \le \hat{\beta}_0 + t_{\alpha/2,n-2}\sqrt{MS_E\left[\frac{1}{n}+\frac{\bar{x}^2}{S_{xx}}\right]}. \tag{14-32}$$

Example 14-3

We will find a 95% confidence interval on the slope of the regression line using the data in Example 14-1. Recall that $\hat{\beta}_1 = 0.483$, $S_{xx} = 8250$, and $MS_E = 0.90$ (see Table 14-2). Then, from equation 14-31 we find

$$\hat{\beta}_1 - t_{0.025,8}\sqrt{\frac{MS_E}{S_{xx}}} \le \beta_1 \le \hat{\beta}_1 + t_{0.025,8}\sqrt{\frac{MS_E}{S_{xx}}}$$

or

$$0.483 - 2.306\sqrt{\frac{0.90}{8250}} \le \beta_1 \le 0.483 + 2.306\sqrt{\frac{0.90}{8250}}.$$

This simplifies to

$$0.459 \le \beta_1 \le 0.507.$$

A confidence interval may be constructed for the mean response at a specified x, say x_0. This is a confidence interval about $E(y|x_0)$ and is often called a confidence interval about the regression line. Since $E(y|x_0) = \beta_0 + \beta_1 x_0$, we may obtain a point estimate of $E(y|x_0)$ from the fitted model as

$$\widehat{E(y|x_0)} \equiv \hat{y}_0 = \hat{\beta}_0 + \hat{\beta}_1 x_0.$$

Now $\hat{y}_0$ is an unbiased point estimator of $E(y|x_0)$, since $\hat{\beta}_0$ and $\hat{\beta}_1$ are unbiased estimators of β_0 and β_1. The variance of $\hat{y}_0$ is

$$V(\hat{y}_0) = \sigma^2\left[\frac{1}{n} + \frac{(x_0-\bar{x})^2}{S_{xx}}\right],$$

and $\hat{y}_0$ is normally distributed, as $\hat{\beta}_0$ and $\hat{\beta}_1$ are normally distributed. Therefore, a $100(1-\alpha)\%$ confidence interval about the true regression line at $x = x_0$ may be computed from

$$\hat{y}_0 - t_{\alpha/2,n-2}\sqrt{MS_E\left(\frac{1}{n}+\frac{(x_0-\bar{x})^2}{S_{xx}}\right)} \le E(y|x_0) \le \hat{y}_0 + t_{\alpha/2,n-2}\sqrt{MS_E\left(\frac{1}{n}+\frac{(x_0-\bar{x})^2}{S_{xx}}\right)}. \tag{14-33}$$

The width of the confidence interval for $E(y|x_0)$ is a function of x_0. The interval width is a minimum for $x_0 = \bar{x}$ and widens as $|x_0 - \bar{x}|$ increases. This widening is one reason why using regression to extrapolate is ill-advised.

Example 14-4

We will construct a 95% confidence interval about the regression line for the data in Example 14-1. The fitted model is $\hat{y}_0 = -2.739 + 0.483x_0$, and the 95% confidence interval on $E(y|x_0)$ is found from equation 14-33 as

$$\left[\hat{y}_0 \pm 2.306\sqrt{0.90\left(\frac{1}{10}+\frac{(x_0-145)^2}{8250}\right)}\right].$$

The fitted values $\hat{y}_0$ and the corresponding 95% confidence limits for the points $x_0 = x_i$, $i = 1, 2, \ldots, 10$, are displayed in Table 14-3. To illustrate the use of this table, we may find the 95% confidence interval on the true mean process yield at $x_0 = 140$°C (say) as

$$64.88 - 0.71 \le E(y|x_0 = 140) \le 64.88 + 0.71$$

or

$$64.17 \le E(y|x_0 = 140) \le 65.49.$$

The fitted model and the 95% confidence interval about the regression line are shown in Fig. 14-4.

Table 14-3 Confidence Interval about the Regression Line, Example 14-4

x_0	100	110	120	130	140	150	160	170	180	190
$\hat{y}_0$	45.56	50.39	55.22	60.05	64.88	69.72	74.55	79.38	84.21	89.04
95% confidence limits	±1.30	±1.10	±0.93	±0.79	±0.71	±0.71	±0.79	±0.93	±1.10	±1.30

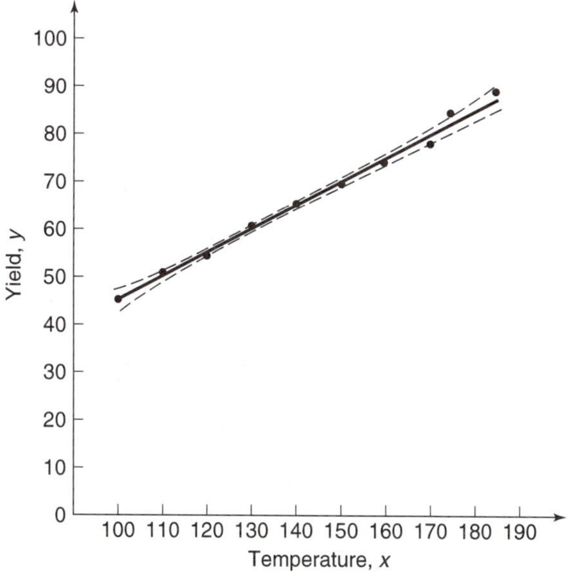

Figure 14-4 A 95% confidence interval about the regression line for Example 14-4.

14-4 PREDICTION OF NEW OBSERVATIONS

An important application of regression analysis is predicting new or future observations y corresponding to a specified level of the regressor variable x. If x_0 is the value of the regressor variable of interest, then

$$\hat{y}_0 = \hat{\beta}_0 + \hat{\beta}_1 x_0 \tag{14-34}$$

is the point estimate of the new or future value of the response y_0.

Now consider obtaining an interval estimate of this future observation y_0. This new observation is independent of the observations used to develop the regression model. Therefore, the confidence interval about the regression line, equation 14-33, is inappropriate, since it is based only on the data used to fit the regression model. The confidence interval about the regression line refers to the true mean response at $x = x_0$ (that is, a population parameter), not to future observations.

Let y_0 be the future observation at $x = x_0$, and let $\hat{y}_0$ given by equation 14-34 be the estimator of y_0. Note that the random variable

$$\psi = y_0 - \hat{y}_0$$

is normally distributed with mean zero and variance

$$\begin{aligned} V(\psi) &= V(y_0 - \hat{y}_0) \\ &= \sigma^2\left[1 + \frac{1}{n} + \frac{(x_0 - \bar{x})^2}{S_{xx}}\right], \end{aligned}$$

because y_0 is independent of $\hat{y}_0$. Thus, the $100(1 - \alpha)\%$ prediction interval on a future observations at x_0 is

$$\begin{aligned} &\hat{y}_0 - t_{\alpha/2,n-2}\sqrt{MS_E\left[1 + \frac{1}{n} + \frac{(x_0 - \bar{x})^2}{S_{xx}}\right]} \\ &\leq y_0 \leq \hat{y}_0 + t_{\alpha/2,n-2}\sqrt{MS_E\left[1 + \frac{1}{n} + \frac{(x_0 - \bar{x})^2}{S_{xx}}\right]}. \end{aligned} \tag{14-35}$$

Notice that the prediction interval is of minimum width at $x_0 = \bar{x}$ and widens as $|x_0 - \bar{x}|$ increases. By comparing equation 14-35 with equation 14-33, we observe that the prediction interval at x_0 is always wider than the confidence interval at x_0. This results because the prediction interval depends on both the error from the estimated model and the error associated with future observations (σ^2).

We may also find a $100(1 - \alpha)\%$ prediction interval on the *mean* of k future observations on the response at $x = x_0$. Let $\bar{y}_0$ be the mean of k future observations at $x = x_0$. The $100(1 - \alpha)\%$ prediction interval on $\bar{y}_0$ is

$$\begin{aligned} &\hat{y}_0 - t_{\alpha/2,n-2}\sqrt{MS_E\left[\frac{1}{k} + \frac{1}{n} + \frac{(x_0 - \bar{x})^2}{S_{xx}}\right]} \\ &\leq \bar{y}_0 \leq \hat{y}_0 + t_{\alpha/2,n-2}\sqrt{MS_E\left[\frac{1}{k} + \frac{1}{n} + \frac{(x_0 - \bar{x})^2}{S_{xx}}\right]}. \end{aligned} \tag{14-36}$$

To illustrate the construction of a prediction interval, suppose we use the data in Example 14-1 and find a 95% prediction interval on the next observation on the process yield at $x_0 = 160°\text{C}$. Using equation 14-35, we find that the prediction interval is

$$74.55 - 2.306\sqrt{0.90\left[1 + \frac{1}{10} + \frac{(160-145)^2}{8250}\right]}$$

$$\leq y_0 \leq 74.55 + 2.306\sqrt{0.90\left[1 + \frac{1}{10} + \frac{(160-145)^2}{8250}\right]},$$

which simplifies to

$$72.21 \leq y_0 \leq 76.89.$$

14-5 MEASURING THE ADEQUACY OF THE REGRESSION MODEL

Fitting a regression model requires several assumptions. Estimation of the model parameters requires the assumption that the errors are uncorrelated random variables with mean zero and constant variance. Tests of hypotheses and interval estimation require that the errors be normally distributed. In addition, we assume that the order of the model is correct; that is, if we fit a first-order polynomial, then we are assuming that the phenomenon actually behaves in a first-order manner.

The analyst should always consider the validity of these assumptions to be doubtful and conduct analyses to examine the adequacy of the model that has been tentatively entertained. In this section we discuss methods useful in this respect.

14-5.1 Residual Analysis

We define the residuals as $e_i = y_i - \hat{y}_i$, $i = 1, 2, \ldots, n$, where y_i is an observation and $\hat{y}_i$ is the corresponding estimated value from the regression model. Analysis of the residuals is frequently helpful in checking the assumption that the errors are NID$(0, \sigma^2)$ and in determining whether additional terms in the model would be useful.

As an approximate check of normality, the experimenter can construct a frequency histogram of the residuals or plot them on normal probability paper. It requires judgment to assess the nonnormality of such plots. One may also standardize the residuals by computing $d_i = e_i/\sqrt{MS_E}$, $i = 1, 2, \ldots, n$. If the errors are NID$(0, \sigma^2)$, then approximately 95% of the standardized residuals should fall in the interval $(-2, +2)$. Residuals far outside this interval may indicate the presence of an *outlier*, that is, an observation that is atypical of the rest of the data. Various rules have been proposed for discarding outliers. However, sometimes outliers provide important information about unusual circumstances of interest to the experimenter and should not be discarded. Therefore a detected outlier should be investigated first, then discarded if warranted. For further discussion of outliers, see Montgomery, Peck, and Vining (2001).

It is frequently helpful to plot the residuals (1) in time sequence (if known), (2) against the $\hat{y}_i$, and (3) against the independent variable x. These graphs will usually look like one of the four general patterns shown in Fig. 14-5. The pattern in Fig. 14-5*a* represents normality, while those in Figs. 14-5*b*, *c*, and d represent anomalies. If the residuals appear as in Fig. 14-5*b*, then the variance of the observations may be increasing with time or with the magnitude of the y_i or x_i. If a plot of the residuals against time has the appearance of Fig. 14-5*b*, then the variance of the observations is increasing with time. Plots against $\hat{y}_i$ and

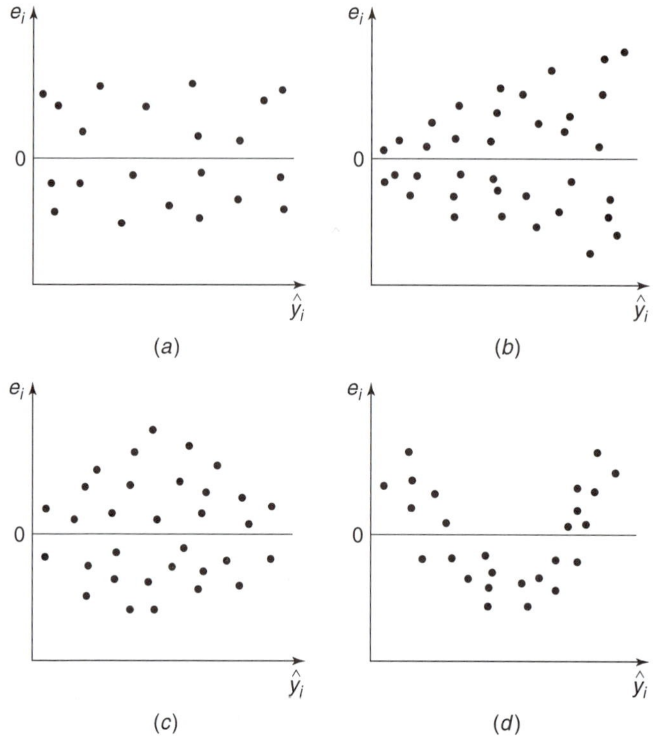

Figure 14-5 Patterns for residual plots. (*a*) Satisfactory, (*b*) funnel, (*c*) double bow, (*d*) nonlinear. [Adapted from Montgomery, Peck, and Vining (2001).]

x_i that look like Fig. 14-5*c* also indicate inequality of variance. Residual plots that look like Fig. 14.5*d* indicate model inadequacy; that is, higher-order terms should be added to the model.

Example 14-5

The residuals for the regression model in Example 14-1 are computed as follows:

$$e_1 = 45.00 - 45.56 = -0.56, \qquad e_6 = 70.00 - 69.72 = 0.28,$$
$$e_2 = 51.00 - 50.39 = 0.61, \qquad e_7 = 74.00 - 74.55 = -0.55,$$
$$e_3 = 54.00 - 55.22 = -1.22, \qquad e_8 = 78.00 - 79.38 = -1.38,$$
$$e_4 = 61.00 - 60.05 = 0.95, \qquad e_9 = 85.00 - 84.21 = 0.79,$$
$$e_5 = 66.00 - 64.88 = 1.12, \qquad e_{10} = 89.00 - 89.04 = -0.04.$$

These residuals are plotted on normal probability paper in Fig. 14-6. Since the residuals fall approximately along a straight line in Fig. 14-6, we conclude that there is no severe departure from normality. The residuals are also plotted against $\hat{y}_i$ in Fig. 14-7*a* and against x_i in Fig. 14-7*b*. These plots indicate no serious model inadequacies.

14-5.2 The Lack-of-Fit Test

Regression models are often fit to data when the true functional relationship is unknown. Naturally, we would like to know whether the order of the model tentatively assumed is correct. This section will describe a test for the validity of this assumption.

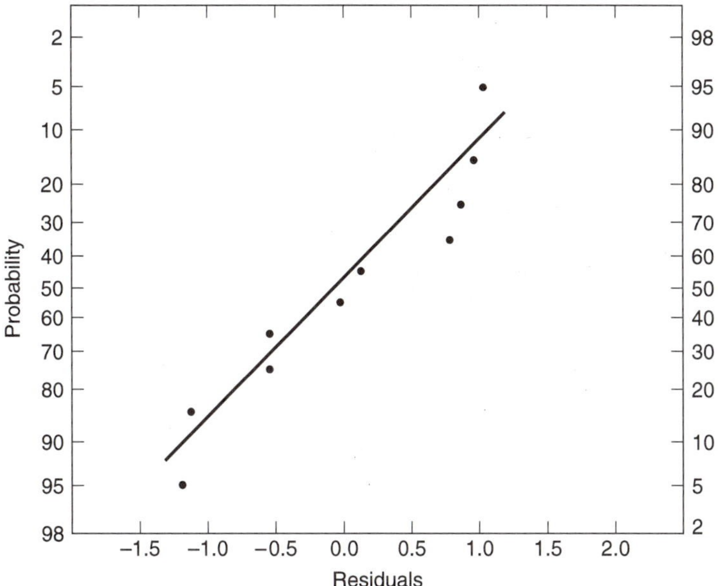

Figure 14-6 Normal probability plot of residuals.

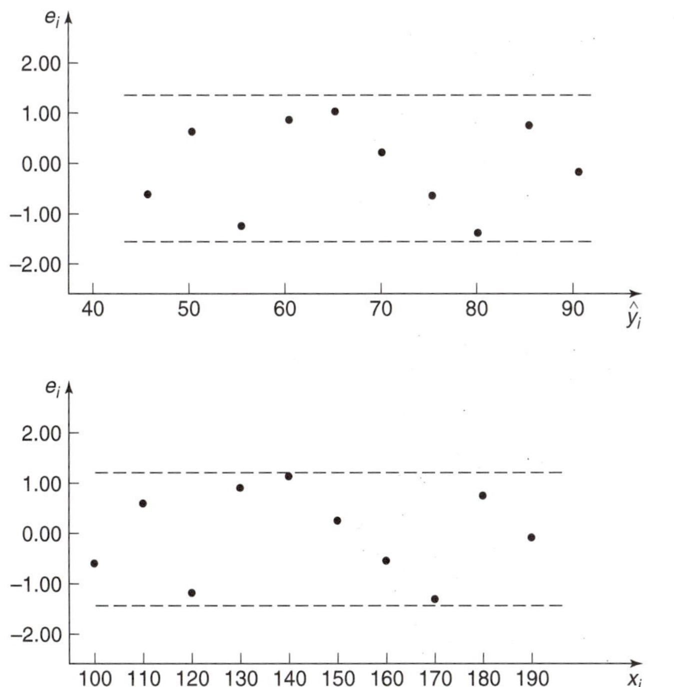

Figure 14-7 Residual plots for Example 14-5. (*a*) Plot against $\hat{y}_i$; (*b*) plot against x_i.

The danger of using a regression model that is a poor approximation of the true functional relationship is illustrated in Fig. 14-8. Obviously, a polynomial of degree two or greater should have been used in this situation.

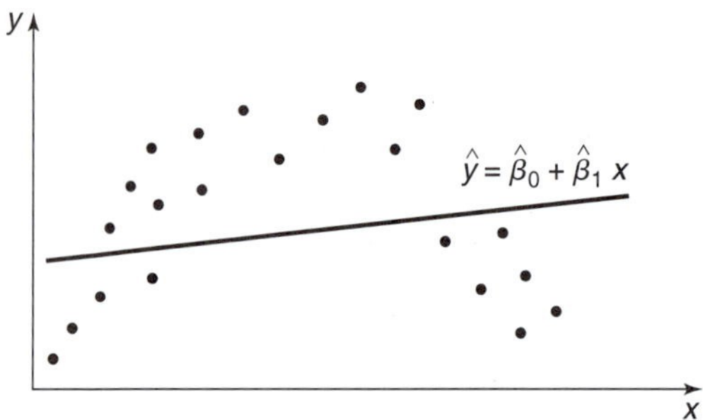

Figure 14-8 A regression model displaying lack of fit.

We present a test for the "goodness of fit" of the regression model. Specifically, the hypotheses we wish to test are

$$H_0\text{: The model adequately fits the data}$$

$$H_1\text{: The model does not fit the data}$$

The test involves partitioning the error or residual sum of squares into the two components

$$SS_E = SS_{PE} + SS_{LOF}$$

where SS_{PE} is the sum of squares attributable to "pure" error and SS_{LOF} is the sum of squares attributable to the lack of fit of the model. To compute SS_{PE} we must have repeated observations on y for at least one level of x. Suppose that we have n total observations such that

$$\begin{array}{ll} y_{11}, y_{12}, \ldots, y_{1n_1} & \text{repeated observations at } x_1, \\ y_{21}, y_{22}, \ldots, y_{2n_2} & \text{repeated observations at } x_2, \\ \vdots & \vdots \\ y_{m1}, y_{m2}, \ldots, y_{mn_m} & \text{repeated observations at } x_m. \end{array}$$

Note that there are m distinct levels of x. The contribution to the pure-error sum of squares at x_1 (say) would be

$$\sum_{u=1}^{n_1} \left(y_{1u} - \bar{y}_1\right)^2. \tag{14-37}$$

The total sum of squares for pure error would be obtained by summing equation 14-37 over all levels of x as

$$SS_{PE} = \sum_{i=1}^{m} \sum_{u=1}^{n_i} \left(y_{iu} - \bar{y}_i\right)^2. \tag{14-38}$$

There are $n_e = \sum_{i=1}^{m} 1(n_i - 1) = n - m$ degrees of freedom associated with the pure-error sum of squares. The sum of squares for lack of fit is simply

$$SS_{LOF} = SS_E - SS_{PE} \tag{14-39}$$

with $n - 2 - n_e = m - 2$ degrees of freedom. The test statistic for lack of fit would then be

$$F_0 = \frac{SS_{LOF}/(m-2)}{SS_{PE}/(n-m)} = \frac{MS_{LOF}}{MS_{PE}}, \tag{14-40}$$

and we would reject it if $F_0 > F_{\alpha, m-2, n-m}$.

This test procedure may be easily introduced into the analysis of variance conducted for the significance of regression. If the null hypothesis of model adequacy is rejected, then the model must be abandoned and attempts made to find a more appropriate model. If H_0 is not rejected, then there is no apparent reason to doubt the adequacy of the model, and MS_{PE} and MS_{LOF} are often combined to estimate σ^2.

Example 14-6

Suppose we have the following data:

x	1.0	1.0	2.0	3.3	3.3	4.0	4.0	4.0	4.7	5.0
y	2.3	1.8	2.8	1.8	3.7	2.6	2.6	2.2	3.2	2.0

x	5.6	5.6	5.6	6.0	6.0	6.5	6.9
y	3.5	2.8	2.1	3.4	3.2	3.4	5.0

We may compute $S_{yy} = 10.97$, $S_{xy} = 13.20$, $S_{xx} = 52.53$, $\bar{y} = 2.847$, and $\bar{x} = 4.382$. The regression model is $\hat{y} = 1.708 + 0.260x$, and the regression sum of squares is $SS_R = \hat{\beta}_1 S_{xy} = (0.260)(13.62) = 3.541$. The pure-error sum of squares is computed as follows:

Level of x	$\Sigma(y_i - \bar{y})^2$	Degrees of Freedom
1.0	0.1250	1
3.3	1.8050	1
4.0	0.1066	2
5.6	0.9800	2
6.0	0.0200	1
Total:	3.0366	7

The analysis of variance is summarized in Table 14-4. Since $F_{0.25,\,8,\,7} = 1.70$, we cannot reject the hypothesis that the tentative model adequately describes the data. We will pool lack-of-fit and pure-error mean squares to form the denominator mean square in the test for significance of regression. Also, since $F_{0.05,\,1,15} = 4.54$, we conclude that $\beta_1 \neq 0$.

In fitting a regression model to experimental data, a good practice is to use the lowest degree model that adequately describes the data. The lack-of-fit test may be useful in this

Table 14-4 Analysis of Variance for Example 14-6

Source of Variation	Sum of Squares	Degrees of Freedom	Mean Square	F_0
Regression	3.541	1	3.541	7.15
Residual	7.429	15	0.495	
(Lack of fit)	4.392	8	0.549	1.27
(Pure error)	3.037	7	0.434	
Total	10.970	16		

respect. However, it is always possible to fit a polynomial of degree $n - 1$ to n data points, and the experimenter should not consider using a model that is "saturated," that is, that has very nearly as many independent variables as observations on y.

14-5.3 The Coefficient of Determination

The quantity

$$R^2 = \frac{SS_R}{S_{yy}} = 1 - \frac{SS_E}{S_{yy}} \tag{14-41}$$

is called the coefficient of determination and is often used to judge the adequacy of a regression model. (We will see subsequently that in the case where x and y are jointly distributed random variables, R^2 is the square of the correlation coefficient between x and y.) Clearly $0 \leq R^2 \leq 1$. We often refer loosely to R^2 as the amount of variability in the data explained or accounted for by the regression model. For the data in Example 14-1, we have $R^2 = SS_R/S_{yy} = 1924.87/1932.10 = 0.9963$; that is, 99.63% of the variability in the data is accounted for by the model.

The statistic R^2 should be used with caution, since it is always possible to make R^2 unity simply by adding enough terms to the model. For example, we can obtain a "perfect" fit to n data points with a polynomial of degree $n - 1$. Also, R^2 will always increase if we add a variable to the model, but this does not necessarily mean the new model is superior to the old one. Unless the error sum of squares in the new model is reduced by an amount equal to the original error mean square, the new model will have a larger error mean square than the old one, because of the loss of one degree of freedom. Thus the new model will actually be worse than the old one.

There are several misconceptions about R^2. In general, R^2 does not measure the magnitude of the slope of the regression line. A large value of R^2 does not imply a steep slope. Furthermore, R^2 does not measure the appropriateness of the model, since it can be artificially inflated by adding higher-order polynomial terms. Even if y and x are related in a nonlinear fashion, R^2 will often be large. For example, R^2 for the regression equation in Fig. 14-3b will be relatively large, even though the linear approximation is poor. Finally, even though R^2 is large, this does not necessarily imply that the regression model will provide accurate predictions of future observations.

14-6 TRANSFORMATIONS TO A STRAIGHT LINE

We occasionally find that the straight-line regression model $y = \beta_0 + \beta_1 x + \epsilon$ is inappropriate because the true regression function is nonlinear. Sometimes this is visually determined from the scatter diagram, and sometimes we know in advance that the model is nonlinear because of prior experience or underlying theory. In some situations a nonlinear function can be expressed as a straight line by using a suitable transformation. Such nonlinear models are called *intrinsically linear.*

As an example of a nonlinear model that is intrinsically linear, consider the exponential function

$$y = \beta_0 e^{\beta_1 x} \epsilon.$$

This function is intrinsically linear, since it can be transformed to a straight line by a logarithmic transformation

$$\ln y = \ln \beta_0 + \beta_1 x + \ln \epsilon.$$

This transformation requires that the transformed error terms ln ϵ be normally and independently distributed with mean 0 and variance σ^2.

Another intrinsically linear function is

$$y = \beta_0 + \beta_1\left(\frac{1}{x}\right) + \epsilon.$$

By using the reciprocal transformation $z = 1/x$, the model is linearized to

$$y = \beta_0 + \beta_1 z + \epsilon.$$

Sometimes several transformations can be employed jointly to linearize a function. For example, consider the function

$$y = \frac{1}{\exp(\beta_0 + \beta_1 x + \epsilon)}.$$

Letting $y^* = 1/y$, we have the linearized form

$$\ln y^* = \beta_0 + \beta_1 x + \epsilon.$$

Several other examples of nonlinear models that are intrinsically linear are given by Daniel and Wood (1980).

14-7 CORRELATION

Our development of regression analysis thus far has assumed that x is a mathematical variable, measured with negligible error, and that y is a random variable. Many applications of regression analysis involve situations where *both* x and y are random variables. In these situations, it is usually assumed that the observations (y_i, x_i), $i = 1, 2, \ldots, n$, are jointly distributed random variables obtained from the distribution $f(y, x)$. For example, suppose we wish to develop a regression model relating the shear strength of spot welds to the weld diameter. In this example, weld diameter cannot be controlled. We would randomly select n spot welds and observe a diameter (x_i) and a shear strength (y_i) for each. Therefore, (y_i, x_i) are jointly distributed random variables.

We usually assume that the joint distribution of y_i and x_i is the bivariate normal distribution. That is,

$$f(y, x) = \frac{1}{2\pi\sigma_1\sigma_2\sqrt{1-\rho^2}} \exp\left\{-\frac{1}{2(1-\rho^2)}\left[\left(\frac{y-\mu_1}{\sigma_1}\right)^2 + \left(\frac{x-\mu_2}{\sigma_2}\right)^2 - 2\rho\left(\frac{y-\mu_1}{\sigma_1}\right)\left(\frac{x-\mu_2}{\sigma_2}\right)\right]\right\}. \tag{14-42}$$

where μ_1 and σ_1^2 are the mean and variance of y, μ_2 and σ_2^2 are the mean and variance of x, and ρ is the correlation coefficient between y and x. Recall from Chapter 4 that the correlation coefficient is defined as

$$\rho = \frac{\sigma_{12}}{\sigma_1\sigma_2},$$

where σ_{12} is the covariance between y and x.

The conditional distribution of y for a given value of x is (see Chapter 7)

$$f(y|x) = \frac{1}{\sqrt{2\pi}\sigma_{12}} \exp\left[-\frac{1}{2}\left(\frac{y - \beta_0 - \beta_1 x}{\sigma_{12}}\right)^2\right], \tag{14-43}$$

where

$$\beta_0 = \mu_1 - \mu_2 \rho \frac{\sigma_1}{\sigma_2}, \tag{14-44a}$$

$$\beta_1 = \frac{\sigma_1}{\sigma_2} \rho, \tag{14-44b}$$

and

$$\sigma_{12}^2 = \sigma_1^2(1 - \rho^2). \tag{14-44c}$$

That is, the conditional distribution of y given x is normal with mean

$$E(y|x) = \beta_0 + \beta_1 x \tag{14-45}$$

and variance σ_{12}^2. Note that the mean of the conditional distribution of y given x is a straight-line regression model. Furthermore, there is a relationship between the correlation coefficient ρ and the slope β_1. From equation 14-44b we see that if $\rho = 0$, then $\beta_1 = 0$, which implies that there is no regression of y on x. That is, knowledge of x does not assist us in predicting y.

The method of maximum likelihood may be used to estimate the parameters β_0 and β_1. It may be shown that the maximum likelihood estimators of these parameters are

$$\hat{\beta}_0 = \bar{y} - \hat{\beta}_1 \bar{x} \tag{14-46a}$$

and

$$\hat{\beta}_1 = \frac{\sum_{i=1}^{n} y_i (x_i - \bar{x})}{\sum_{i=1}^{n} (x_i - \bar{x})^2} = \frac{S_{xy}}{S_{xx}}. \tag{14-46b}$$

We note that the estimators of the intercept and slope in equation 14-46 are identical to those given by the method of least squares in the case where x was assumed to be a mathematical variable. That is, the regression model with y and x jointly normally distributed is equivalent to the model with x considered as a mathematical variable. This follows because the random variables y given x are independently and normally distributed with mean $\beta_0 + \beta_1 x$ and constant variance σ_{12}^2. These results will also hold for *any* joint distribution of y and x such that the conditional distribution of y given x is normal.

It is possible to draw inferences about the correlation coefficient ρ in this model. The estimator of ρ is the *sample correlation coefficient*

$$\begin{aligned} r &= \frac{\sum_{i=1}^{n} y_i (x_i - \bar{x})}{\left[\sum_{i=1}^{n} (x_i - \bar{x})^2 \sum_{i=1}^{n} (y_i - \bar{y})^2\right]^{1/2}} \\ &= \frac{S_{xy}}{\left[S_{xx} S_{yy}\right]^{1/2}}. \end{aligned} \tag{14-47}$$

Note that

$$\hat{\beta}_1 = \left(\frac{S_{yy}}{S_{xx}}\right)^{1/2} r, \tag{14-48}$$

so the slope $\hat{\beta}_1$ is just the sample correlation coefficient r multiplied by a scale factor that is the square root of the "spread" of the y values divided by the "spread" of the x values. Thus $\hat{\beta}_1$ and r are closely related, although they provide somewhat different information. The sample correlation coefficient r measures the linear association between y and x, while $\hat{\beta}_1$ measures the predicted change in the mean of y for a unit change in x. In the case of a mathematical variable x, r has no meaning because the magnitude of r depends on the choice of spacing for x. We may also write, from equation 14-48,

$$\begin{aligned} R^2 \equiv r^2 &= \hat{\beta}_1^2 \frac{S_{xx}}{S_{yy}} \\ &= \frac{\hat{\beta}_1 S_{xy}}{S_{yy}} \\ &= \frac{SS_R}{S_{yy}}, \end{aligned}$$

which we recognize from equation 14-41 as the coefficient of determination. That is, the coefficient of determination R^2 is just the square of the sample correlation coefficient between y and x.

It is often useful to test the hypothesis

$$\begin{aligned} &H_0: \rho = 0, \\ &H_1: \rho \neq 0. \end{aligned} \tag{14-49}$$

The appropriate test statistic for this hypothesis is

$$t_0 = \frac{r\sqrt{n-2}}{\sqrt{1-r^2}}, \tag{14-50}$$

which follows the t distribution with $n - 2$ degrees of freedom if H_0: $\rho = 0$ is true. Therefore, we would reject the null hypothesis if $|t_0| > t_{\alpha/2, n-2}$. This test is equivalent to the test of the hypothesis H_0: $\beta_1 = 0$ given in Section 14-2. This equivalence follows directly from equation 14-48.

The test procedure for the hypothesis

$$\begin{aligned} &H_0: \rho = \rho_0, \\ &H_1: \rho \neq \rho_0, \end{aligned} \tag{14-51}$$

where $\rho_0 \neq 0$, is somewhat more complicated. For moderately large samples (say $n \geq 25$) the statistic

$$Z = \text{arctanh } r = \frac{1}{2}\ln\frac{1+r}{1-r} \tag{14-52}$$

is approximately normally distributed with mean

$$\mu_Z = \text{arctanh } \rho = \frac{1}{2}\ln\frac{1+\rho}{1-\rho}$$

and variance

$$\sigma_Z^2 = (n-3)^{-1}.$$

Therefore, to test the hypothesis H_0: $\rho = \rho_0$, we may compute the statistic

$$Z_0 = (\text{arctanh } r - \text{arctanh } \rho_0)(n-3)^{1/2} \tag{14-53}$$

and reject H_0: $\rho = \rho_0$ if $|Z_0| > Z_{\alpha/2}$.

It is also possible to construct a $100(1-\alpha)\%$ confidence interval for ρ using the transformation in equation 14-52. The $100(1-\alpha)\%$ confidence interval is

$$\tanh\left(\text{arctanh } r - \frac{Z_{\alpha/2}}{\sqrt{n-3}}\right) \le \rho \le \tanh\left(\text{arctanh } r + \frac{Z_{\alpha/2}}{\sqrt{n-3}}\right), \tag{14-54}$$

where $\tanh u = (e^u - e^{-u})/(e^u + e^{-u})$.

Example 14-7

Montgomery, Peck, and Vining (2001) describe an application of regression analysis in which an engineer at a soft-drink bottler is investigating the product distribution and route service operations for vending machines. She suspects that the time required to load and service a machine is related to the number of cases of product delivered. A random sample of 25 retail outlets having vending machines is selected, and the in-outlet delivery time (in minutes) and volume of product delivered (in cases) is observed for each outlet. The data are shown in Table 14-5. We assume that delivery time and volume of product delivered are jointly normally distributed.

Using the data in Table 14-5, we may calculate

$$S_{yy} = 6105.9447, \qquad S_{xx} = 698.5600, \qquad \text{and} \qquad S_{xy} = 2027.7132.$$

The regression model is

$$\hat{y} = 5.1145 + 2.9027x.$$

The sample correlation coefficient between x and y is computed from equation 14-47 as

$$r = \frac{S_{xy}}{\left[S_{xx}S_{yy}\right]^{1/2}} = \frac{2027.7132}{\left[(698.5600)(6105.9447)\right]^{1/2}} = 0.9818.$$

Table 14-5 Data for Example 14-7

Observation	Delivery Time (y)	Number of Cases (x)	Observation	Delivery Time (y)	Number of Cases (x)
1	9.95	2	14	11.66	2
2	24.45	8	15	21.65	4
3	31.75	11	16	17.89	4
4	35.00	10	17	69.00	20
5	25.02	8	18	10.30	1
6	16.86	4	19	34.93	10
7	14.38	2	20	46.59	15
8	9.60	2	21	44.88	15
9	24.35	9	22	54.12	16
10	27.50	8	23	56.63	17
11	17.08	4	24	22.13	6
12	37.00	11	25	21.15	5
13	41.95	12			

Note that $R^2 = (0.9818)^2 = 0.9640$, or that approximately 96.40% of the variability in delivery time is explained by the linear relationship with delivery volume. To test the hypothesis

$$H_0\colon \rho = 0,$$

$$H_1\colon \rho \neq 0,$$

we can compute the test statistic of equation 14-50 as follows:

$$t_0 = \frac{r\sqrt{n-2}}{\sqrt{1-r^2}} = \frac{0.9818\sqrt{23}}{\sqrt{1-0.9640}} = 24.80.$$

Since $t_{0.025,23} = 2.069$, we reject H_0 and conclude that the correlation coefficient $\rho \neq 0$. Finally, we may construct an approximate 95% confidence interval on ρ from equation 14-54. Since arctanh r = arctanh 0.9818 = 2.3452, equation 14-54 becomes

$$\tanh\left(2.3452 - \frac{1.96}{\sqrt{22}}\right) \leq \rho \leq \tanh\left(2.3452 + \frac{1.96}{\sqrt{22}}\right),$$

which reduces to

$$0.9585 \leq \rho \leq 0.9921.$$

14-8 SAMPLE COMPUTER OUTPUT

Many of the procedures presented in this chapter can be implemented using statistical software. In this section, we present the Minitab® output for the data in Example 14-1.

Recall that Example 14-1 provides data on the effect of process operating temperature on product yield. The Minitab® output is

```
The regression equation is
Yield = - 2.74 + 0.483 Temp

Predictor       Coef      SE Coef        T         P
Constant      -2.739       1.546     -1.77     0.114
Temp         0.48303     0.01046     46.17     0.000

S = 0.9503   R-Sq = 99.6%   R-Sq(adj) = 99.6%

Analysis of Variance

Source            DF        SS        MS         F        P
Regression         1    1924.9    1924.9   2131.57    0.000
Residual Error     8       7.2       0.9
Total              9    1932.1
```

The regression equation is provided along with the results from the t-tests on the individual coefficients. The P-values indicate that the intercept does not appear to be significant (P-value = 0.114) while the regressor variable, temperature, is statistically significant (P-value $\simeq$ 0). The analysis of variance is also testing the hypothesis that $H_0\colon \beta_1 = 0$ and can be rejected (P-value $\simeq$ 0). Note also that T = 46.17 for temperature, and $t^2 = (46.17)^2 = 2131.67 \simeq F$. Aside from rounding, the computer results are in agreement with those found earlier in the chapter.

14-9 SUMMARY

This chapter has introduced the simple linear regression model and shown how least-squares estimates of the model parameters may be obtained. Hypothesis-testing procedures and confidence interval estimates of the model parameters have also been developed. Tests of hypotheses and confidence intervals require the assumption that the observations y are normally and independently distributed random variables. Procedures for testing model adequacy, including a lack-of-fit test and residual analysis, were presented. The correlation model was also introduced to deal with the case where x and y are jointly normally distributed. The equivalence of the regression model parameter estimation problem for the case where x and y are jointly normal to the case where x is a mathematical variable was also discussed. Procedures for obtaining point and interval estimates of the correlation coefficient and for testing hypotheses about the correlation coefficient were developed.

14-10 EXERCISES

14-1. Montgomery, Peck, and Vining (2001) present data concerning the performance of the 28 National Football League teams in 1976. It is suspected that the number of games won (y) is related to the number of yards gained rushing by an opponent (x). The data are shown below.

Teams	Games Won (y)	Yards Rushing by Opponent (x)
Washington	10	2205
Minnesota	11	2096
New England	11	1847
Oakland	13	1903
Pittsburgh	10	4757
Baltimore	11	1848
Los Angeles	10	1564
Dallas	11	1821
Atlanta	4	2577
Buffalo	2	2476
Chicago	7	1984
Cincinnati	10	1917
Cleveland	9	1761
Denver	9	1709
Detroit	6	1901
GreenBay	5	2288
Houston	5	2072
Kansas City	5	2861
Miami	6	2411
New Orleans	4	2289
New York Giants	3	2203
New York Jets	3	2592
Philadelphia	4	2053
St. Louis	10	1979
San Diego	6	2048
San Francisco	8	1786
Seattle	2	2876
Tampa Bay	0	2560

(a) Fit a linear regression model relating games won to yards gained by an opponent.
(b) Test for significance of regression.
(c) Find a 95% confidence interval for the slope.
(d) What percentage of total variability is explained by the model?
(e) Find the residuals and prepare appropriate residual plots.

14-2. Suppose we would like to use the model developed in Exercise 14-1 to predict the number of games a team will win if it can limit the opponents to 1800 yards rushing. Find a point estimate of the number of games won if the opponents gain only 1800 yards rushing. Find a 95% prediction interval on the number of games won.

14-3. *Motor Trend* magazine frequently presents performance data for automobiles. The table below presents data from the 1975 volume of *Motor Trend* concerning the gasoline milage performance and the engine displacement for 15 automobiles.

Automobile	Miles / Gallon (y)	Displacement (Cubic Inches) (x)
Apollo	18.90	350
Omega	17.00	350
Nova	20.00	250
Monarch	18.25	351
Duster	20.07	225
Jensen Conv.	11.20	440
Skyhawk	22.12	231
Monza	21.47	262
Corolla SR-5	30.40	96.9
Camaro	16.50	350
Eldorado	14.39	500
Trans Am	16.59	400
Charger SE	19.73	318
Cougar	13.90	351
Corvette	16.50	350

(a) Fit a regression model relating mileage performance to engine displacement.

(b) Test for significance of regression.

(c) What percentage of total variability in mileage is explained by the model?

(d) Find a 90% confidence interval on the mean mileage if the engine displacement is 275 cubic inches.

14-4. Suppose that we wish to predict the gasoline mileage from a car with a 275 cubic inch displacement engine. Find a point estimate, using the model developed in Exercise 14-3, and an appropriate 90% interval estimate. Compare this interval to the one obtained in Exercise 14-3d. Which one is wider, and why?

14-5. Find the residuals from the model in Exercise 14-3. Prepare appropriate residual plots and comment on model adequacy.

14-6. An article in *Technometrics* by S. C. Narula and J. F. Wellington ("Prediction, Linear Regression, and a Minimum Sum of Relative Errors," Vol. 19, 1977) presents data on the selling price and annual taxes for 27 houses. The data are shown below.

Sale Price / 1000	Taxes (Local, School, County) / 1000
25.9	4.9176
29.5	5.0208
27.9	4.5429
25.9	4.5573
29.9	5.0597
29.9	3.8910
30.9	5.8980
28.9	5.6039
35.9	5.8282
31.5	5.3003
31.0	6.2712
30.9	5.9592
30.0	5.0500
36.9	8.2464
41.9	6.6969
40.5	7.7841
43.9	9.0384
37.5	5.9894
37.9	7.5422
44.5	8.7951
37.9	6.0831
38.9	8.3607
36.9	8.1400
45.8	9.1416

(a) Fit a regression model relating sales price to taxes paid.

(b) Test for significance of regression.

(c) What percentage of the variability in selling price is explained by the taxes paid?

(d) Find the residuals for this model. Construct a normal probability plot for the residuals. Plot the residuals versus $\hat{y}$ and versus x. Does the model seem satisfactory?

14-7. The strength of paper used in the manufacture of cardboard boxes (y) is related to the percentage of hardwood concentration in the original pulp (x). Under controlled conditions, a pilot plant manufactures 16 samples, each from a different batch of pulp, and measures the tensile strength. The data are shown here.

y	101.4	117.4	117.1	106.2	131.9	146.9	146.8	133.9
x	1.0	1.5	1.5	1.5	2.0	2.0	2.2	2.4
y	111.3	123.0	125.1	145.2	134.3	144.5	143.7	146.9
x	2.5	2.5	2.8	2.8	3.0	3.0	3.2	3.3

(a) Fit a simple linear regression model to the data.

(b) Test for lack of fit and significance of regression.

(c) Construct a 90% confidence interval on the slope β_1.

(d) Construct a 90% confidence interval on the intercept β_0.

(e) Construct a 95% confidence interval on the true regression line at $x = 2.5$.

14-8. Compute the residuals for the regression model in Exercise 14-7. Prepare appropriate residual plots and comment on model adequacy.

14-9. The number of pounds of steam used per month by a chemical plant is thought to be related to the average ambient temperature for that month. The past year's usage and temperatures are shown in the following table.

Month	Temp.	Usage / 1000
Jan.	21	185.79
Feb.	24	214.47
Mar.	32	288.03
Apr.	47	424.84
May	50	454.58
June	59	539.03
July	68	621.55
Aug.	74	675.06
Sept.	62	562.03
Oct.	50	452.93
Nov.	41	369.95
Dec.	30	273.98

(a) Fit a simple linear regression model to the data.

(b) Test for significance of regression.

(c) Test the hypothesis that the slope $\beta_1 = 10$.

(d) Construct a 99% confidence interval about the true regression line at $x = 58$.

(e) Construct a 99% prediction interval on the steam usage in the next month having a mean ambient temperature of 58°.

14-10. Compute the residuals for the regression model in Exercise 14-9. Prepare appropriate residual plots and comment on model adequacy.

14-11. The percentage of impurity in oxygen gas produced by a distilling process is thought to be related to the percentage of hydrocarbon in the main condenser of the processor. One month's operating data are available, as shown in the table at the bottom of this page.

(a) Fit a simple linear regression model to the data.

(b) Test for lack of fit and significance of regression.

(c) Calculate R^2 for this model.

(d) Calculate a 95% confidence interval for the slope β_1.

14-12. Compute the residuals for the data in Exercise 14-11.

(a) Plot the residuals on normal probability paper and draw appropriate conclusions.

(b) Plot the residuals against $\hat{y}$ and x. Interpret these displays.

14-13. An article in *Transportation Research* (1999, p. 183) presents a study on world maritime employment. The purpose of the study was to determine a relationship between average manning level and the average size of the fleet. Manning level refers to the ratio of number of posts that must be manned by a seaman per ship (posts/ship). Data collected for ships of the United Kingdom over a 16-year period are

Average Size	Level
9154	20.27
9277	19.98
9221	20.28
9198	19.65
8705	18.81
8530	18.20
8544	18.05
7964	16.81
7440	15.56
6432	13.98
6032	14.51
5125	10.99
4418	12.83
4327	11.85
4133	11.33
3765	10.25

(a) Fit a linear regression model relating average manning level to average ship size.

(b) Test for significance of regression.

(c) Find a 95% confidence interval on the slope.

(d) What percentage of total variability is explained by the model?

(e) Find the residuals and construct appropriate residual plots.

14-14. The final averages for 20 randomly selected students taking a course in engineering statistics and a course in operations research at Georgia Tech are shown here. Assume that the final averages are jointly normally distributed.

Statistics	86	75	69	75	90	94	83	86	71	65
OR	80	81	75	81	92	95	80	81	76	72

Statistics	84	71	62	90	83	75	71	76	84	97
OR	85	72	65	93	81	70	73	72	80	98

(a) Find the regression line relating the statistics final average to the OR final average.

(b) Estimate the correlation coefficient.

(c) Test the hypothesis that $\rho = 0$.

(d) Test the hypothesis that $\rho = 0.5$.

(e) Construct a 95% confidence interval estimate of the correlation coefficient.

14-15. The weight and systolic blood pressure of 26 randomly selected males in the age group 25–30 are

Purity (%)	86.91	89.85	90.28	86.34	92.58	87.33	86.29	91.86	95.61	89.86
Hydrocarbon (%)	1.02	1.11	1.43	1.11	1.01	0.95	1.11	0.87	1.43	1.02

Purity (%)	96.73	99.42	98.66	96.07	93.65	87.31	95.00	96.85	85.20	90.56
Hydrocarbon (%)	1.46	1.55	1.55	1.55	1.40	1.15	1.01	0.99	0.95	0.98

shown in the following table. Assume that weight and blood pressure are jointly normally distributed.

Subject	Weight	Systolic BP
1	165	130
2	167	133
3	180	150
4	155	128
5	212	151
6	175	146
7	190	150
8	210	140
9	200	148
10	149	125
11	158	133
12	169	135
13	170	150
14	172	153
15	159	128
16	168	132
17	174	149
18	183	158
19	215	150
20	195	163
21	180	156
22	143	124
23	240	170
24	235	165
25	192	160
26	187	159

(a) Find a regression line relating systolic blood pressure to weight.
(b) Estimate the correlation coefficient.
(c) Test the hypothesis that $\rho = 0$.
(d) Test the hypothesis that $\rho = 0.6$.
(e) Construct a 95% confidence interval estimate of the correlation coefficient.

14-16. Consider the simple linear regression model $y = \beta_0 + \beta_1 x + \epsilon$. Show that $E(MS_R) = \sigma^2 + \beta_1^2 S_{xx}$.

14-17. Suppose that we have assumed the straight-line regression model

$$y = \beta_0 + \beta_1 x_1 + \epsilon$$

but that the response is affected by a second variable, x_2, such that the true regression function is

$$E(y) = \beta_0 + \beta_1 x_1 + \beta_2 x_2.$$

Is the estimator of the slope in the simple linear regression model unbiased?

14-18. Suppose that we are fitting a straight line and we wish to make the variance of the slope $\hat{\beta}_1$ as small as possible. Where should the observations x_i, $i = 1, 2, \ldots, n$, be taken so as to minimize $V(\hat{\beta}_1)$? Discuss the practical implications of this allocation of the x_i.

14-19. Weighted Least Squares. Suppose that we are fitting the straight line $y = \beta_0 + \beta_1 x + \epsilon$ but that the variance of the y values now depends on the level of x; that is,

$$V(y_i|x_i) = \sigma_i^2 = \frac{\sigma^2}{w_i}, \qquad i = 1, 2, \ldots, n,$$

where the w_i are unknown constants, often called *weights*. Show that the resulting least-squares normal equations are

$$\hat{\beta}_0 \sum_{i=1}^{n} w_i + \hat{\beta}_1 \sum_{i=1}^{n} w_i x_i = \sum_{i=1}^{n} w_i y_i,$$

$$\hat{\beta}_0 \sum_{i=1}^{n} w_i x_i + \hat{\beta}_1 \sum_{i=1}^{n} w_i x_i^2 = \sum_{i=1}^{n} w_i x_i y_i.$$

14-20. Consider the data shown below. Suppose that the relationship between y and x is hypothesized to be $y = (\beta_0 + \beta_1 x + \epsilon)^{-1}$. Fit an appropriate model to the data. Does the assumed model form seem appropriate?

x	10	15	18	12	9	8	11	6
y	0.17	0.13	0.09	0.15	0.20	0.21	0.18	0.24

14-21. Consider the weight and blood pressure data in Exercise 14-15. Fit a no-intercept model to the data, and compare it to the model obtained in Exercise 14-15. Which model is superior?

14-22. The following data, adapted from Montgomery, Peck, and Vining (2001), present the number of certified mental defectives per 10,000 of estimated population in the United Kingdom (y) and the number of radio receiver licenses issued (x) by the BBC (in millions) for the years 1924–1937. Fit a regression model relating y to x. Comment on the model. Specifically, does the existence of a strong correlation imply a cause-and-effect relationship?

Year	Number of Certified Mental Defectives per 10,000 of Estimated U.K. Population (y)	Number of Radio Receiver Licenses Issued (Millions) in the U.K. (x)
1924	8	1.350
1925	8	1.960
1926	9	2.270
1927	10	2.483
1928	11	2.730
1929	11	3.091
1930	12	3.674
1931	16	4.620
1932	18	5.497
1933	19	6.260
1934	20	7.012
1935	21	7.618
1936	22	8.131
1937	23	8.593

Chapter 15

Multiple Regression

Many regression problems involve more than one regressor variable. Such models are called *multiple regression models*. Multiple regression is one of the most widely used statistical techniques. This chapter presents the basic techniques of parameter estimation, confidence interval estimation, and model adequacy checking for multiple regression. We also introduce some of the special problems often encountered in the practical use of multiple regression, including model building and variable selection, autocorrelation in the errors, and multicollinearity or near-linear dependence among the regressors.

15-1 MULTIPLE REGRESSION MODELS

A regression model that involves more than one regressor variable is called a *multiple regression* model. As an example, suppose that the effective life of a cutting tool depends on the cutting speed and the tool angle. A multiple regression model that might describe this relationship is

$$y = \beta_0 + \beta_1 x_1 + \beta_2 x_2 + \varepsilon, \tag{15-1}$$

where y represents the tool life, x_1 represents the cutting speed, and x_2 represents the tool angle. This is a *multiple linear regression model* with two regressors. The term "linear" is used because equation 15-1 is a linear function of the unknown parameters β_0, β_1, and β_2. Note that the model describes a plane in the two-dimensional x_1, x_2 space. The parameter β_0 defines the intercept of the plane. We sometimes call β_1 and β_2 *partial* regression coefficients, because β_1 measures the expected change in y per unit change in x_1 when x_2 is held constant, and β_2 measures the expected change in y per unit change in x_2 when x_1 is held constant.

In general, the dependent variable or response y may be related to k independent variables. The model

$$y = \beta_0 + \beta_1 x_1 + \beta_2 x_2 + \cdots + \beta_k x_k + \varepsilon \tag{15-2}$$

is called a multiple linear regression model with k independent variables. The parameters β_j, $j = 0, 1, \ldots, k$, are called the regression coefficients. This model describes a hyperplane in the k-dimensional space of the regressor variables $\{x_j\}$. The parameter β_j represents the expected change in response y per unit change in x_j when all the remaining independent variables x_i $(i \neq j)$ are held constant. The parameters β_j, $j = 1, 2, \ldots, k$, are often called partial regression coefficients, because they describe the partial effect of one independent variable when the other independent variables in the model are held constant.

Multiple linear regression models are often used as approximating functions. That is, the true functional relationship between y and $x_1, x_2, \ldots, x_k$ is unknown, but over certain ranges of the independent variables the linear regression model is an adequate approximation.

Models that are more complex in appearance than equation 15-2 may often still be analyzed by multiple linear regression techniques. For example, consider the cubic polynomial model in one independent variable,

$$y = \beta_0 + \beta_1 x + \beta_2 x^2 + \beta_3 x^3 + \varepsilon. \tag{15-3}$$

If we let $x_1 = x$, $x_2 = x^2$, and $x_3 = x^3$, then equation 15-3 can be written

$$y = \beta_0 + \beta_1 x_1 + \beta_2 x_2 + \beta_3 x_3 + \varepsilon, \tag{15-4}$$

which is a multiple linear regression model with three regressor variables. Models that include interaction effects may also be analyzed by multiple linear regression methods. For example, suppose that the model is

$$y = \beta_0 + \beta_1 x_1 + \beta_2 x_2 + \beta_{12} x_1 x_2 + \varepsilon. \tag{15-5}$$

If we let $x_3 = x_1 x_2$ and $\beta_3 = \beta_{12}$, then equation 15-5 can be written

$$y = \beta_0 + \beta_1 x_1 + \beta_2 x_2 + \beta_3 x_3 + \varepsilon, \tag{15-6}$$

which is a linear regression model. In general, any regression model that is linear in the *parameters* (the β's) is a linear regression model, regardless of the shape of the surface that it generates.

15-2 ESTIMATION OF THE PARAMETERS

The method of least squares may be used to estimate the regression coefficients in equation 15-2. Suppose that $n > k$ observations are available, and let x_{ij} denote the ith observation or level of variable x_j. The data will appear as in Table 15-1. We assume that the error term ε in the model has $E(\varepsilon) = 0$, $V(\varepsilon) = \sigma^2$, and that the $\{\varepsilon_i\}$ are uncorrelated random variables.

We may write the model, equation 15-2, in terms of the observations,

$$\begin{aligned} y_i &= \beta_0 + \beta_1 x_{i1} + \beta_2 x_{i2} + \cdots + \beta_k x_{ik} + \varepsilon_i \\ &= \beta_0 + \sum_{j=1}^{k} \beta_j x_{ij} + \varepsilon_i, \qquad i = 1, 2, \ldots, n. \end{aligned} \tag{15-7}$$

The least-squares function is

$$\begin{aligned} L &= \sum_{i=1}^{n} \varepsilon_i^2 \\ &= \sum_{i=1}^{n} \left(y_i - \beta_0 - \sum_{j=1}^{k} \beta_j x_{ij} \right)^2 . \end{aligned} \tag{15-8}$$

Table 15-1 Data for Multiple Linear Regression

y	x_1	x_2	$\cdots$	x_k
y_1	x_{11}	x_{12}	$\cdots$	x_{1k}
y_2	x_{21}	x_{22}	$\cdots$	x_{2k}
.	.	.		.
.	.	.		.
.	.	.		.
y_n	x_{n1}	x_{n2}	$\cdots$	x_{nk}

The function L is to be minimized with respect to $\beta_0, \beta_1, \ldots, \beta_k$. The least-squares estimators of $\beta_0, \beta_1, \ldots, \beta_k$ must satisfy

$$\left.\frac{\partial L}{\partial \beta_0}\right|_{\hat{\beta}_0, \hat{\beta}_1, \ldots, \hat{\beta}_k} = -2\sum_{i=1}^{n}\left(y_i - \hat{\beta}_0 - \sum_{j=1}^{k}\hat{\beta}_j x_{ij}\right) = 0 \tag{15-9a}$$

and

$$\left.\frac{\partial L}{\partial \beta_j}\right|_{\hat{\beta}_0, \hat{\beta}_1, \ldots, \hat{\beta}_k} = -2\sum_{i=1}^{n}\left(y_i - \hat{\beta}_0 - \sum_{j=1}^{k}\hat{\beta}_j x_{ij}\right)x_{ij} = 0, \qquad j = 1, 2, \ldots, k. \tag{15-9b}$$

Simplifying equation 15-9, we obtain the least-squares normal equations

$$\begin{aligned}
n\hat{\beta}_0 + \hat{\beta}_1\sum_{i=1}^{n}x_{i1} + \hat{\beta}_2\sum_{i=1}^{n}x_{i2} + \cdots + \hat{\beta}_k\sum_{i=1}^{n}x_{ik} &= \sum_{i=1}^{n}y_i, \\
\hat{\beta}_0\sum_{i=1}^{n}x_{i1} + \hat{\beta}_1\sum_{i=1}^{n}x_{i1}^2 + \hat{\beta}_2\sum_{i=1}^{n}x_{i1}x_{i2} + \cdots + \hat{\beta}_k\sum_{i=1}^{n}x_{i1}x_{ik} &= \sum_{i=1}^{n}x_{i1}y_i, \\
\vdots \qquad\qquad \vdots \qquad\qquad \vdots \qquad\qquad \vdots \qquad \vdots \\
\hat{\beta}_0\sum_{i=1}^{n}x_{ik} + \hat{\beta}_1\sum_{i=1}^{n}x_{ik}x_{i1} + \hat{\beta}_2\sum_{i=1}^{n}x_{ik}x_{i2} + \cdots + \hat{\beta}_k\sum_{i=1}^{n}x_{ik}^2 &= \sum_{i=1}^{n}x_{ik}y_i.
\end{aligned} \tag{15-10}$$

Note that there are $p = k + 1$ normal equations, one for each of the unknown regression coefficients. The solution to the normal equations will be the least-squares estimators of the regression coefficients $\hat{\beta}_0, \hat{\beta}_1, \ldots, \hat{\beta}_k$.

It is simpler to solve the normal equations if they are expressed in matrix notation. We now give a matrix development of the normal equations that parallels the development of equation 15-10. The model in terms of the observations, equation 15-7, may be written in matrix notation,

$$\mathbf{y} = \mathbf{X}\boldsymbol{\beta} + \boldsymbol{\varepsilon},$$

where

$$\mathbf{y} = \begin{bmatrix} y_1 \\ y_2 \\ \vdots \\ y_n \end{bmatrix}, \qquad \mathbf{X} = \begin{bmatrix} 1 & x_{11} & x_{12} & \cdots & x_{1k} \\ 1 & x_{21} & x_{22} & \cdots & x_{2k} \\ \vdots & \vdots & \vdots & & \vdots \\ 1 & x_{n1} & x_{n2} & \cdots & x_{nk} \end{bmatrix},$$

$$\boldsymbol{\beta} = \begin{bmatrix} \beta_0 \\ \beta_1 \\ \vdots \\ \beta_k \end{bmatrix}, \qquad \text{and} \qquad \boldsymbol{\varepsilon} = \begin{bmatrix} \varepsilon_1 \\ \varepsilon_2 \\ \vdots \\ \varepsilon_n \end{bmatrix}.$$

In general, $\mathbf{y}$ is an $(n \times 1)$ vector of the observations, $\mathbf{X}$ is an $(n \times p)$ matrix of the levels of the independent variables, $\boldsymbol{\beta}$ is a $(p \times 1)$ vector of the regression coefficients, and $\boldsymbol{\varepsilon}$ is an $(n \times 1)$ vector of random errors.

We wish to find the vector of least-squares estimators, $\hat{\boldsymbol{\beta}}$, that minimizes

$$L = \sum_{i=1}^{n}\varepsilon_i^2 = \boldsymbol{\varepsilon}'\boldsymbol{\varepsilon} = (\mathbf{y} - \mathbf{X}\boldsymbol{\beta})'(\mathbf{y} - \mathbf{X}\boldsymbol{\beta}).$$

Note that L may be expressed as

$$\begin{aligned} L &= \mathbf{y}'\mathbf{y} - \boldsymbol{\beta}'\mathbf{X}'\mathbf{y} - \mathbf{y}'\mathbf{X}\boldsymbol{\beta} + \boldsymbol{\beta}'\mathbf{X}'\mathbf{X}\boldsymbol{\beta} \\ &= \mathbf{y}'\mathbf{y} - 2\boldsymbol{\beta}'\mathbf{X}'\mathbf{y} + \boldsymbol{\beta}'\mathbf{X}'\mathbf{X}\boldsymbol{\beta}, \end{aligned} \tag{15-11}$$

since $\boldsymbol{\beta}'\mathbf{X}'\mathbf{y}$ is a (1×1) matrix, hence a scalar, and its transpose $(\boldsymbol{\beta}'\mathbf{X}'\mathbf{y})' = \mathbf{y}'\mathbf{X}\boldsymbol{\beta}$ is the same scalar. The least-squares estimators must satisfy

$$\left.\frac{\partial L}{\partial \boldsymbol{\beta}}\right|_{\hat{\boldsymbol{\beta}}} = -2\mathbf{X}'\mathbf{y} + 2\mathbf{X}'\mathbf{X}\hat{\boldsymbol{\beta}} = \mathbf{0},$$

which simplifies to

$$\mathbf{X}'\mathbf{X}\hat{\boldsymbol{\beta}} = \mathbf{X}'\mathbf{y}. \tag{15-12}$$

Equations 15-12 are the least-squares normal equations. They are identical to equations 15-10. To solve the normal equations, multiply both sides of equation 15-12 by the inverse of $\mathbf{X}'\mathbf{X}$. Thus, the least-squares estimator of $\boldsymbol{\beta}$ is

$$\hat{\boldsymbol{\beta}} = (\mathbf{X}'\mathbf{X})^{-1}\mathbf{X}'\mathbf{y}. \tag{15-13}$$

It is easy to see that the matrix form of the normal equations is identical to the scalar form. Writing out equation 15-12 in detail we obtain

$$\begin{bmatrix} n & \sum_{i=1}^{n} x_{i1} & \sum_{i=1}^{n} x_{i2} & \cdots & \sum_{i=1}^{n} x_{ik} \\ \sum_{i=1}^{n} x_{i1} & \sum_{i=1}^{n} x_{i1}^2 & \sum_{i=1}^{n} x_{i1}x_{i2} & \cdots & \sum_{i=1}^{n} x_{i1}x_{ik} \\ \vdots & \vdots & \vdots & & \vdots \\ \sum_{i=1}^{n} x_{ik} & \sum_{i=1}^{n} x_{ik}x_{i1} & \sum_{i=1}^{n} x_{ik}x_{i2} & \cdots & \sum_{i=1}^{n} x_{ik}^2 \end{bmatrix} \begin{bmatrix} \hat{\beta}_0 \\ \hat{\beta}_1 \\ \vdots \\ \hat{\beta}_k \end{bmatrix} = \begin{bmatrix} \sum_{i=1}^{n} y_i \\ \sum_{i=1}^{n} x_{i1}y_i \\ \vdots \\ \sum_{i=1}^{n} x_{ik}y_i \end{bmatrix}.$$

If the indicated matrix multiplication is performed, the scalar form of the normal equations (that is, equation 15-10) will result. In this form it is easy to see that $\mathbf{X}'\mathbf{X}$ is a $(p \times p)$ symmetric matrix and $\mathbf{X}'\mathbf{y}$ is a $(p \times 1)$ column vector. Note the special structure of the $\mathbf{X}'\mathbf{X}$ matrix. The diagonal elements of $\mathbf{X}'\mathbf{X}$ are the sums of squares of the elements in the columns of $\mathbf{X}$, and the off-diagonal elements are the sums of cross products of the elements in the columns of $\mathbf{X}$. Furthermore, note that the elements of $\mathbf{X}'\mathbf{y}$ are the sums of cross products of the columns of $\mathbf{X}$ and the observations $\{y_i\}$.

The fitted regression model is

$$\hat{\mathbf{y}} = \mathbf{X}\hat{\boldsymbol{\beta}}. \tag{15-14}$$

In scalar notation, the fitted model is

$$\hat{y}_i = \hat{\beta}_0 + \sum_{j=1}^{k} \hat{\beta}_j x_{ij}, \qquad i = 1, 2, \ldots, n.$$

The difference between the observation y_i and the fitted value $\hat{y}_i$ is a residual, say $e_i = y_i - \hat{y}_i$. The $(n \times 1)$ vector of residuals is denoted

$$\mathbf{e} = \mathbf{y} - \hat{\mathbf{y}}. \tag{15-15}$$

Example 15-1

An article in the *Journal of Agricultural Engineering and Research* (2001, p. 275) describes the use of a regression model to relate the damage susceptibility of peaches to the height at which they are dropped (drop height, measured in mm) and the density of the peach (measured in g/cm^3). One goal of the analysis is to provide a predictive model for peach damage to serve as a guideline for harvesting and postharvesting operations. Data typical of this type of experiment is given in Table 15-2.

We will fit the multiple linear regression model

$$y = \beta_0 + \beta_1 x_1 + \beta_2 x_2 + \varepsilon$$

to these data. The **X** matrix and **y** vector for this model are

$$\mathbf{X} = \begin{bmatrix} 1 & 303.7 & 0.90 \\ 1 & 366.7 & 1.04 \\ 1 & 336.8 & 1.01 \\ 1 & 304.5 & 0.95 \\ 1 & 346.8 & 0.98 \\ 1 & 600.0 & 1.04 \\ 1 & 369.0 & 0.96 \\ 1 & 418.0 & 1.00 \\ 1 & 269.0 & 1.01 \\ 1 & 323.0 & 0.94 \\ 1 & 562.2 & 1.01 \\ 1 & 284.2 & 0.97 \\ 1 & 558.6 & 1.03 \\ 1 & 415.0 & 1.01 \\ 1 & 349.5 & 1.04 \\ 1 & 462.8 & 1.02 \\ 1 & 333.1 & 1.05 \\ 1 & 502.1 & 1.10 \\ 1 & 311.4 & 0.91 \\ 1 & 351.4 & 0.96 \end{bmatrix}, \quad \mathbf{y} = \begin{bmatrix} 3.62 \\ 7.27 \\ 2.66 \\ 1.53 \\ 4.91 \\ 10.36 \\ 5.26 \\ 6.09 \\ 6.57 \\ 4.24 \\ 8.04 \\ 3.46 \\ 8.50 \\ 9.34 \\ 5.55 \\ 8.11 \\ 7.32 \\ 12.58 \\ 0.15 \\ 5.23 \end{bmatrix}.$$

The $\mathbf{X}'\mathbf{X}$ matrix is

$$\mathbf{X}'\mathbf{X} = \begin{bmatrix} 1 & 1 & \cdots & 1 \\ 303.7 & 366.7 & \cdots & 351.4 \\ 0.90 & 1.04 & \cdots & 0.96 \end{bmatrix} \begin{bmatrix} 1 & 303.7 & 0.90 \\ 1 & 366.7 & 1.04 \\ \vdots & \vdots & \vdots \\ 1 & 351.4 & 0.96 \end{bmatrix}$$

$$= \begin{bmatrix} 20 & 7767.8 & 19.93 \\ 7767.8 & 3201646 & 7791.878 \\ 19.93 & 7791.878 & 19.9077 \end{bmatrix},$$

and the $\mathbf{X}'\mathbf{y}$ vector is

$$\mathbf{X}'\mathbf{y} = \begin{bmatrix} 1 & 1 & \cdots & 1 \\ 303.7 & 366.7 & \cdots & 351.4 \\ 0.90 & 1.04 & \cdots & 0.96 \end{bmatrix} \begin{bmatrix} 3.62 \\ 7.27 \\ \vdots \\ 5.23 \end{bmatrix} = \begin{bmatrix} 120.79 \\ 51129.17 \\ 122.70 \end{bmatrix}.$$

Table 15-2 Peach Damage Data for Example 15-1

Observation Number	Damage (mm), y	Drop Height (mm), x_1	Fruit Density (g/cm^3), x_2
1	3.62	303.7	0.90
2	7.27	366.7	1.04
3	2.66	336.8	1.01
4	1.53	304.5	0.95
5	4.91	346.8	0.98
6	10.36	600.0	1.04
7	5.26	369.0	0.96
8	6.09	418.0	1.00
9	6.57	269.0	1.01
10	4.24	323.0	0.94
11	8.04	562.2	1.01
12	3.46	284.2	0.97
13	8.50	558.6	1.03
14	9.34	415.0	1.01
15	5.55	349.5	1.04
16	8.11	462.8	1.02
17	7.32	333.1	1.05
18	12.58	502.1	1.10
19	0.15	311.4	0.91
20	5.23	351.4	0.96

The least-squares estimators are found from equation 15-13 to be

$$\hat{\boldsymbol{\beta}} = (\mathbf{X}'\mathbf{X})^{-1}\mathbf{X}'\mathbf{y},$$

or

$$\begin{bmatrix} \hat{\beta}_0 \\ \hat{\beta}_1 \\ \hat{\beta}_2 \end{bmatrix} = \begin{bmatrix} 20 & 7767.8 & 19.93 \\ 7767.8 & 3201646 & 7791.878 \\ 19.93 & 7791.878 & 19.9077 \end{bmatrix}^{-1} \begin{bmatrix} 120.79 \\ 51129.17 \\ 122.70 \end{bmatrix}$$

$$= \begin{bmatrix} 24.63666 & 0.005321 & -26.74679 \\ 0.005321 & 0.0000077 & -0.008353 \\ -26.74679 & -0.008353 & 30.096389 \end{bmatrix} \begin{bmatrix} 120.79 \\ 51129.17 \\ 122.70 \end{bmatrix}$$

$$= \begin{bmatrix} -33.831 \\ 0.01314 \\ 34.890 \end{bmatrix}.$$

Therefore, the fitted regression model is

$$\hat{y} = -33.831 + 0.01314x_1 + 34.890x_2.$$

Table 15-3 shows the fitted values of y and the residuals. The fitted values and residuals are calculated to the same accuracy as the original data.

Table 15-3 Observations, Fitted Values, and Residuals for Example 15-1

Observation Number	y_i	$\hat{y}_i$	$e_i = y_i - \hat{y}_i$
1	3.62	1.56	2.06
2	7.27	7.27	0.00
3	2.66	5.83	–3.17
4	1.53	3.31	–1.78
5	4.91	4.92	–0.01
6	10.36	10.34	0.02
7	5.26	4.51	0.75
8	6.09	6.55	–0.46
9	6.57	4.94	1.63
10	4.24	3.21	1.03
11	8.04	8.79	–0.75
12	3.46	3.75	–0.29
13	8.50	9.44	–0.94
14	9.34	6.86	2.48
15	5.55	7.05	–1.50
16	8.11	7.84	0.27
17	7.32	7.18	0.14
18	12.58	11.14	1.44
19	0.15	2.01	–1.86
20	5.23	4.28	0.95

The statistical properties of the least-squares estimator $\hat{\boldsymbol{\beta}}$ may be easily demonstrated. Consider first bias:

$$\begin{aligned} E(\hat{\boldsymbol{\beta}}) &= E\,[(\mathbf{X}'\mathbf{X})^{-1}\mathbf{X}'\mathbf{y}] \\ &= E\,[(\mathbf{X}'\mathbf{X})^{-1}\mathbf{X}'(\mathbf{X}\boldsymbol{\beta} + \boldsymbol{\varepsilon})] \\ &= E\,[(\mathbf{X}'\mathbf{X})^{-1}\mathbf{X}'\mathbf{X}\boldsymbol{\beta} + (\mathbf{X}'\mathbf{X})^{-1}\mathbf{X}'\boldsymbol{\varepsilon}] \\ &= \boldsymbol{\beta}, \end{aligned}$$

since $E(\boldsymbol{\varepsilon}) = \mathbf{0}$ and $(\mathbf{X}'\mathbf{X})^{-1}\mathbf{X}'\mathbf{X} = \mathbf{I}$. Thus $\hat{\boldsymbol{\beta}}$ is an unbiased estimator of β. The variance property of $\hat{\boldsymbol{\beta}}$ is expressed by the covariance matrix

$$\text{Cov}(\hat{\boldsymbol{\beta}}) = E\,\{[\hat{\boldsymbol{\beta}} - E(\hat{\boldsymbol{\beta}})][\hat{\boldsymbol{\beta}} - E(\hat{\boldsymbol{\beta}})]'\}.$$

The covariance matrix of $\hat{\boldsymbol{\beta}}$ is a $(p \times p)$ symmetric matrix whose jjth element is the variance of $\hat{\beta}_j$ and whose (i, j)th element is the covariance between $\hat{\beta}_i$ and $\hat{\beta}_j$. The covariance matrix of $\hat{\boldsymbol{\beta}}$ is

$$\text{Cov}(\hat{\boldsymbol{\beta}}) = \sigma^2(\mathbf{X}'\mathbf{X})^{-1}.$$

It is usually necessary to estimate σ^2. To develop this estimator, consider the sum of squares of the residuals, say

$$\begin{aligned} SS_E &= \sum_{i=1}^{n}(y_i - \hat{y}_i)^2 \\ &= \sum_{i=1}^{n} e_i^2 \\ &= \mathbf{e}'\mathbf{e}. \end{aligned}$$

Substituting $\mathbf{e} = \mathbf{y} - \hat{\mathbf{y}} = \mathbf{y} - \mathbf{X}\hat{\boldsymbol{\beta}}$, we have

$$\begin{aligned} SS_E &= (\mathbf{y} - \mathbf{X}\hat{\boldsymbol{\beta}})'(\mathbf{y} - \mathbf{X}\hat{\boldsymbol{\beta}}) \\ &= \mathbf{y}'\mathbf{y} - \hat{\boldsymbol{\beta}}'\mathbf{X}'\mathbf{y} - \mathbf{y}'\mathbf{X}\hat{\boldsymbol{\beta}} + \hat{\boldsymbol{\beta}}'\mathbf{X}'\mathbf{X}\hat{\boldsymbol{\beta}} \\ &= \mathbf{y}'\mathbf{y} - 2\hat{\boldsymbol{\beta}}'\mathbf{X}'\mathbf{y} + \hat{\boldsymbol{\beta}}'\mathbf{X}'\mathbf{X}\hat{\boldsymbol{\beta}}. \end{aligned}$$

Since $\mathbf{X}'\mathbf{X}\hat{\boldsymbol{\beta}} = \mathbf{X}'\mathbf{y}$, this last equation becomes

$$SS_E = \mathbf{y}'\mathbf{y} - \hat{\boldsymbol{\beta}}'\mathbf{X}'\mathbf{y}. \tag{15-16}$$

Equation 15-16 is called the *error* or *residual* sum of squares, and it has $n - p$ degrees of freedom associated with it. The mean square for error is

$$MS_E = \frac{SS_E}{n-p}. \tag{15-17}$$

It can be shown that the expected value of MS_E is σ^2; thus an unbiased estimator of σ^2 is given by

$$\hat{\sigma}^2 = MS_E. \tag{15-18}$$

Example 15-2

We will estimate the error variance σ^2 for the multiple regression problem in Example 15-1. Using the data in Table 15-2, we find

$$\mathbf{y}'\mathbf{y} = \sum_{i=1}^{20} y_i^2 = 904.60$$

and

$$\hat{\boldsymbol{\beta}}'\mathbf{X}'\mathbf{y} = \begin{bmatrix} -33.831 & 0.01314 & 34.890 \end{bmatrix} \begin{bmatrix} 120.79 \\ 51129.17 \\ 122.70 \end{bmatrix}$$

$$= 866.39.$$

Therefore, the error sum of squares is

$$\begin{aligned} SS_E &= \mathbf{y}'\mathbf{y} - \hat{\boldsymbol{\beta}}'\mathbf{X}'\mathbf{y} \\ &= 904.60 - 866.39 \\ &= 38.21. \end{aligned}$$

The estimate of σ^2 is

$$\hat{\sigma}^2 = \frac{SS_E}{n-p} = \frac{38.21}{20-3} = 2.247.$$

15-3 CONFIDENCE INTERVALS IN MULTIPLE LINEAR REGRESSION

It is often necessary to construct confidence interval estimates for the regression coefficients $\{\beta_j\}$. The development of a procedure for obtaining these confidence intervals requires that we assume the errors $\{\varepsilon_i\}$ to be normally and independently distributed with mean zero and variance σ^2. Therefore, the observations $\{y_i\}$ are normally and independently distributed with mean $\beta_0 + \sum_{j=1}^{k} \beta_j x_{ij}$ and variance σ^2. Since the least-squares estimator $\hat{\beta}$ is a linear combination of the observations, it follows that $\hat{\boldsymbol{\beta}}$ is normally distributed with mean vector $\boldsymbol{\beta}$ and covariance matrix $\sigma^2(\mathbf{X}'\mathbf{X})^{-1}$. Then each of the quantities

$$\frac{\hat{\beta}_j - \beta_j}{\sqrt{\hat{\sigma}^2 C_{jj}}}, \qquad j = 0, 1, \ldots, k, \tag{15-19}$$

is distributed as t with $n - p$ degrees of freedom, where C_{jj} is the jjth element of the $(\mathbf{X}'\mathbf{X})^{-1}$ matrix and $\hat{\sigma}^2$ is the estimate of the error variance, obtained from equation 15-18. Therefore, a $100(1 - \alpha)\%$ confidence interval for the regression coefficient β_j, $j = 0, 1, \ldots, k$, is

$$\hat{\beta}_j - t_{\alpha/2,n-p}\sqrt{\hat{\sigma}^2 C_{jj}} \le \beta_j \le \hat{\beta}_j + t_{\alpha/2,n-p}\sqrt{\hat{\sigma}^2 C_{jj}}. \tag{15-20}$$

Example 15-3

We will construct a 95% confidence interval on the parameter β_1 in Example 15-1. Note that the point estimate of β_1 is $\hat{\beta}_1 = 0.01314$, and the diagonal element of $(\mathbf{X}'\mathbf{X})^{-1}$ corresponding to β_1 is $C_{11} = 0.0000077$. The estimate of σ^2 was obtained in Example 15-2 as 2.247, and $t_{0.025,17} = 2.110$. Therefore, the 95% confidence interval on β_1 is computed from equation 15-20 as

$$0.01314 - (2.110)\sqrt{(2.247)(0.0000077)} \le \beta_1 \le 0.01314 + (2.110)\sqrt{(2.247)(0.0000077)},$$

which reduces to

$$0.00436 \le \beta_1 \le 0.0219.$$

We may also obtain a confidence interval on the mean response at a particular point, say $(x_{01}, x_{02}, \ldots, x_{0k})$. To estimate the mean response at this point define the vector

$$\mathbf{x}_0 = \begin{bmatrix} 1 \\ x_{01} \\ x_{02} \\ \vdots \\ x_{0k} \end{bmatrix}.$$

The estimated mean response at this point is

$$\hat{y}_0 = \mathbf{x}_0'\hat{\boldsymbol{\beta}}. \tag{15-21}$$

This estimator is unbiased, since $E(\hat{y}_0) = E(\mathbf{x}_0'\hat{\boldsymbol{\beta}}) = \mathbf{x}_0'\boldsymbol{\beta} = E(y_0)$, and the variance of $\hat{y}_0$ is

$$V(\hat{y}_0) = \sigma^2\mathbf{x}_0'(\mathbf{X}'\mathbf{X})^{-1}\mathbf{x}_0. \tag{15-22}$$

Therefore, a $100(1 - \alpha)\%$ confidence interval on the mean response at the point $(x_{01}, x_{02}, \ldots, x_{0k})$ is

$$\hat{y}_0 - t_{\alpha/2,n-p}\sqrt{\hat{\sigma}^2\mathbf{x}_0'(\mathbf{X}'\mathbf{X})^{-1}\mathbf{x}_0} \le E(y_0) \le \hat{y}_0 + t_{\alpha/2,n-p}\sqrt{\hat{\sigma}^2\mathbf{x}_0'(\mathbf{X}'\mathbf{X})^{-1}\mathbf{x}_0}. \tag{15-23}$$

Equation 15-23 is a confidence interval about the regression hyperplane. It is the multiple regression generalization of equation 14-33.

Example 15-4

The scientists conducting the experiment on damaged peaches in Example 15-1 would like to construct a 95% confidence interval on the mean damage for a peach dropped from a height of $x_1 = 325$ mm if its density is $x_2 = 0.98$ g/cm^3. Therefore,

$$\mathbf{x}_0 = \begin{bmatrix} 1 \\ 325 \\ 0.98 \end{bmatrix}.$$

The estimated mean response at this point is found from equation 15-21 to be

$$\hat{y}_0 = \mathbf{x}_0'\hat{\boldsymbol{\beta}} = [1 \quad 325 \quad 0.98]\begin{bmatrix} -33.831 \\ 0.01314 \\ 34.890 \end{bmatrix} = 4.63.$$

The variance of $\hat{y}_0$ is estimated by

$$\hat{\sigma}^2\mathbf{x}_0'(\mathbf{X}'\mathbf{X})^{-1}\mathbf{x}_0 = 2.247[1 \quad 325 \quad 0.98]\begin{bmatrix} 24.63666 & 0.005321 & -26.74679 \\ 0.005321 & 0.0000077 & -0.008353 \\ -26.74679 & -0.008353 & 30.096389 \end{bmatrix}\begin{bmatrix} 1 \\ 325 \\ 0.98 \end{bmatrix}$$

$$= 2.247(0.0718) = 0.1613.$$

Therefore, a 95% confidence interval on the mean damage at this point is found from equation 15-23 to be

$$4.63 - 2.110\sqrt{0.1613} \le E(y_0) \le 4.63 + 2.110\sqrt{0.1613},$$

which reduces to

$$3.78 \le E(y_0) \le 5.48.$$

15-4 PREDICTION OF NEW OBSERVATIONS

The regression model can be used to predict future observations on y corresponding to particular values of the independent variables, say $x_{01}, x_{02}, \ldots, x_{0k}$. If $\mathbf{x}_0' = [1, x_{01}, x_{02}, \ldots, x_{0k}]$, then a point estimate of the future observation y_0 at the point $(x_{01}, x_{02}, \ldots, x_{0k})$ is

$$\hat{y}_0 = \mathbf{x}_0'\hat{\boldsymbol{\beta}}. \qquad (15\text{-}24)$$

A $100(1 - \alpha)\%$ prediction interval for this future observation is

$$\hat{y}_0 - t_{\alpha/2,n-p}\sqrt{\hat{\sigma}^2\left(1 + \mathbf{x}_0'(\mathbf{X}'\mathbf{X})^{-1}\mathbf{x}_0\right)} \le y_0 \le \hat{y}_0 + t_{\alpha/2,n-p}\sqrt{\hat{\sigma}^2\left(1 + \mathbf{x}_0'(\mathbf{X}'\mathbf{X})^{-1}\mathbf{x}_0\right)}. \qquad (15\text{-}25)$$

This prediction interval is a generalization of the prediction interval for a future observation in simple linear regression, equation 14-35.

In predicting new observations and in estimating the mean response at a given point $(x_{01}, x_{02}, \ldots, x_{0k})$, one must be careful about extrapolating beyond the region containing the original observations. It is very possible that a model that fits well in the region of the original data will no longer fit well outside that region. In multiple regression it is often easy to inadvertently extrapolate, since the levels of the variables $(x_{i1}, x_{i2}, \ldots x_{ik})$, $i = 1, 2, \ldots, n$, jointly define the region containing the data. As an example, consider Fig. 15-1, which illus-

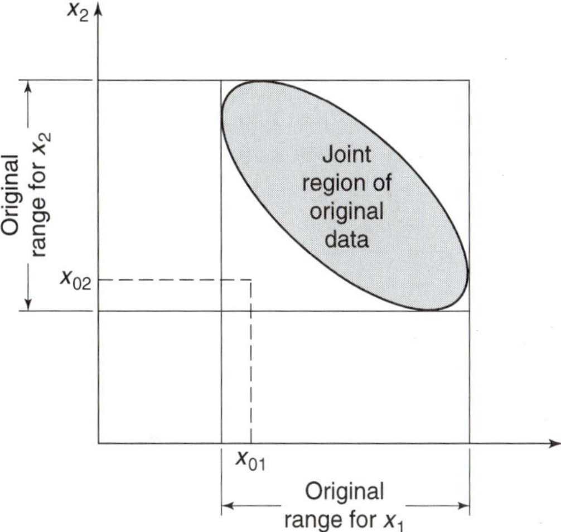

Figure 15-1 An example of extrapolation in multiple regression.

trates the region containing the observations for a two-variable regression model. Note that the point (x_{01}, x_{02}) lies within the ranges of both independent variables x_1 and x_2, but it is outside the region of the original observations. Thus, either predicting the value of a new observation or estimating the mean response at this point is an extrapolation of the original regression model.

Example 15-5

Suppose that the scientists in Example 15-1 wish to construct a 95% prediction interval on the damage on a peach that is dropped from a height of $x_1 = 325$ mm and has a density of $x_2 = 0.98$ g/cm^3. Note that $\mathbf{x}_0' = [1\ 325\ 0.98]$, and the point estimate of the damage is $\hat{y}_0 = \mathbf{x}_0'\boldsymbol{\beta} = 4.63$ mm. Also, in Example 15-4 we calculated $\mathbf{x}_0'(\mathbf{X}'\mathbf{X})^{-1}\mathbf{x}_0 = 0.0718$. Therefore, from equation 15-25 we have

$$4.63 - 2.110\sqrt{2.247(1 + 0.0718)} \le y_0 \le 4.63 + 2.110\sqrt{2.247(1 + 0.0718)},$$

and the 95% prediction interval is

$$1.36 \le y_0 \le 7.90.$$

15-5 HYPOTHESIS TESTING IN MULTIPLE LINEAR REGRESSION

In multiple linear regression problems, certain tests of hypotheses about the model parameters are useful in measuring model adequacy. In this section, we describe several important hypothesis-testing procedures. We continue to require the normality assumption on the errors, which was introduced in the previous section.

15-5.1 Test for Significance of Regression

The test for significance of regression is a test to determine whether there is a linear relationship between the dependent variable y and a subset of the independent variables x_1, x_2, …, x_k. The appropriate hypotheses are

$$H_0\text{: } \beta_1 = \beta_2 = \cdots = \beta_k = 0,$$
$$H_1\text{: } \beta_j \neq 0 \text{ for at least one } j. \tag{15-26}$$

Rejection of H_0: $\beta_j = 0$ implies that at least one of the independent variables $x_1, x_2, \ldots, x_k$ contributes significantly to the model. The test procedure is a generalization of the procedure used in simple linear regression. The total sum of squares S_{yy} is partitioned into a sum of squares due to regression and a sum of squares due to error, say

$$S_{yy} = SS_R + SS_{E,}$$

and if H_0: $\beta_j = 0$ is true, then $SS_R/\sigma^2 \sim \chi^2_k$, where the number of degrees of freedom for the χ^2 is equal to the number of regressor variables in the model. Also, we can show that $SS_E/\sigma^2 \sim \chi^2_{n-k-1}$, and SS_E and SS_R are independent. The test procedure for H_0: $\beta_j = 0$ is to compute

$$F_0 = \frac{SS_R/k}{SS_E/(n-k-1)} = \frac{MS_R}{MS_E} \tag{15-27}$$

and to reject H_0 if $F_0 > F_{\alpha,k,n-k-1}$. The procedure is usually summarized in an analysis of variance table such as Table 15-4.

A computational formula for SS_R may be found easily. We have derived a computational formula for SS_E in equation 15-16, that is,

$$SS_E = \mathbf{y}'\mathbf{y} - \hat{\boldsymbol{\beta}}'\mathbf{X}'\mathbf{y}.$$

Now since $S_{yy} = \sum_{i=1}^n y_i^2 - (\sum_{i=1}^n y_i)^2/n = \mathbf{y}'\mathbf{y} - (\sum_{i=1}^n y_i)^2/n$, we may rewrite the foregoing equation

$$SS_E = \mathbf{y}'\mathbf{y} - \frac{1}{n}\left(\sum_{i=1}^n y_i\right)^2 - \left[\hat{\boldsymbol{\beta}}'\mathbf{X}'\mathbf{y} - \frac{1}{n}\left(\sum_{i=1}^n y_i\right)^2\right]$$

or

$$SS_E = S_{yy} - SS_R.$$

Therefore, the regression sum of squares is

$$SS_R = \hat{\boldsymbol{\beta}}'\mathbf{X}'\mathbf{y} - \frac{1}{n}\left(\sum_{i=1}^n y_i\right)^2, \tag{15-28}$$

the error sum of squares is

$$SS_E = \mathbf{y}'\mathbf{y} - \hat{\boldsymbol{\beta}}'\mathbf{X}'\mathbf{y}, \tag{15-29}$$

and the total sum of squares is

$$S_{yy} = \mathbf{y}'\mathbf{y} - \frac{1}{n}\left(\sum_{i=1}^n y_i\right)^2. \tag{15-30}$$

Table 15-4 Analysis of Variance for Significance of Regression in Multiple Regression

Source of Variation	Sum of Squares	Degrees of Freedom	Mean Square	F_0
Regression	SS_R	k	MS_R	MS_R/MS_E
Error or residual	SS_E	$n-k-1$	MS_E	
Total	S_{yy}	$n-1$		

Example 15-6

We will test for significance of regression using the damaged peaches data from Example 15-1. Some of the numerical quantities required are calculated in Example 15-2. Note that

$$\begin{aligned} S_{yy} &= \mathbf{y}'\mathbf{y} - \frac{1}{n}\left(\sum_{i=1}^{n} y_i\right)^2 \\ &= 904.60 - \frac{(120.79)^2}{20} \\ &= 175.089, \\ SS_R &= \hat{\boldsymbol{\beta}}'\mathbf{X}'\mathbf{y} - \frac{1}{n}\left(\sum_{i=1}^{n} y_i\right)^2 \\ &= 866.39 - \frac{(120.79)^2}{20} \\ &= 136.88, \\ SS_E &= S_{yy} - SS_R \\ &= \mathbf{y}'\mathbf{y} - \hat{\boldsymbol{\beta}}'\mathbf{X}'\mathbf{y} \\ &= 38.21. \end{aligned}$$

The analysis of variance is shown in Table 15-5. To test H_0: $\beta_1 = \beta_2 = 0$, we calculate the statistic

$$F_0 = \frac{MS_R}{MS_E} = \frac{68.44}{2.247} = 30.46.$$

Since $F_0 > F_{0.05,2,17} = 3.59$, peach damage is related to drop height, fruit density, or both. However, we note that this does not necessarily imply that the relationship found is an appropriate one for predicting damage as a function of drop height or fruit density. Further tests of model adequacy are required.

Table 15-5 Test for Significance of Regression for Example 15-6

Source of Variation	Sum of Squares	Degrees of Freedom	Mean Square	F_0
Regression	136.88	2	68.44	30.46
Error	38.21	17	2.247	
Total	175.09	19		

15-5.2 Tests on Individual Regression Coefficients

We are frequently interested in testing hypotheses on the individual regression coefficients. Such tests would be useful in determining the value of each of the independent variables in the regression model. For example, the model might be more effective with the inclusion of additional variables, or perhaps with the deletion of one or more of the variables already in the model.

Adding a variable to a regression model always causes the sum of squares for regression to increase and the error sum of squares to decrease. We must decide whether the increase in the regression sum of squares is sufficient to warrant using the additional variable in the model. Furthermore, adding an unimportant variable to the model can actually increase the mean square error, thereby decreasing the usefulness of the model.

The hypotheses for testing the significance of any individual regression coefficient, say β_j, are

$$\begin{aligned} H_0&: \beta_j = 0, \\ H_1&: \beta_j \neq 0. \end{aligned} \tag{15-31}$$

If H_0: $\beta_j = 0$ is not rejected, then this indicates that x_j can possibly be deleted from the model. The test statistic for this hypothesis is

$$t_0 = \frac{\hat{\beta}_j}{\sqrt{\hat{\sigma}^2 C_{jj}}}, \tag{15-32}$$

where C_{jj} is the diagonal element of $(\mathbf{X}'\mathbf{X})^{-1}$ corresponding to $\hat{\beta}_j$. The null hypothesis H_0: $\beta_j = 0$ is rejected if $|t_0| > t_{\alpha/2,\, n-k-1}$. Note that this is really a partial or marginal test, because the regression coefficient $\hat{\beta}_j$ depends on all the other regressor variables $x_i (i \neq j)$ that are in the model. To illustrate the use of this test, consider the data in Example 15-1, and suppose that we want to test

$$\begin{aligned} H_0&: \beta_2 = 0, \\ H_1&: \beta_2 \neq 0. \end{aligned}$$

The main diagonal element of $(\mathbf{X}'\mathbf{X})^{-1}$ corresponding to $\hat{\beta}_2$ is $C_{22} = 30.096$, so the t statistic in equation 15-32 is

$$t_0 = \frac{\hat{\beta}_j}{\sqrt{\hat{\sigma}^2 C_{22}}} = \frac{34.89}{\sqrt{(2.247)(30.096)}} = 4.24.$$

Since $t_{0.025,22} = 2.110$, we reject H_0: $\beta_2 = 0$ and conclude that the variable x_2 (density) contributes significantly to the model. Note that this test measures the marginal or partial contribution of x_2 given that x_1 is in the model.

We may also examine the contribution to the regression sum of squares of a variable, say x_j, given that other variables x_i $(i \neq j)$ are included in the model. The procedure used to do this is called the general regression significance test, or the "extra sum of squares" method. This procedure can also be used to investigate the contribution of a *subset* of the regressor variables to the model. Consider the regression model with k regressor variables

$$\mathbf{y} = \mathbf{X}\boldsymbol{\beta} + \boldsymbol{\varepsilon},$$

where $\mathbf{y}$ is $(n \times 1)$, $\mathbf{X}$ is $(n \times p)$, $\boldsymbol{\beta}$ is $(p \times 1)$, $\boldsymbol{\varepsilon}$ is $(n \times 1)$, and $p = k + 1$. We would like to determine whether the subset of regressor variables $x_1, x_2, \ldots, x_r$ $(r < k)$ contributes significantly to the regression model. Let the vector of regression coefficients be partitioned as follows:

$$\boldsymbol{\beta} = \begin{bmatrix} \boldsymbol{\beta}_1 \\ \boldsymbol{\beta}_2 \end{bmatrix},$$

where $\boldsymbol{\beta}_1$ is $(r \times 1)$ and $\boldsymbol{\beta}_2$ is $[(p - r) \times 1]$. We wish to test the hypotheses

$$\begin{aligned} H_0&: \boldsymbol{\beta}_1 = \mathbf{0}, \\ H_1&: \boldsymbol{\beta}_1 \neq \mathbf{0}. \end{aligned} \tag{15-33}$$

The model may be written

$$\mathbf{y} = \mathbf{X}\boldsymbol{\beta} + \boldsymbol{\varepsilon} = \mathbf{X}_1\boldsymbol{\beta}_1 + \mathbf{X}_2\boldsymbol{\beta}_2 + \boldsymbol{\varepsilon}, \tag{15-34}$$

where $\mathbf{X}_1$ represents the columns of $\mathbf{X}$ associated with $\boldsymbol{\beta}_1$ and $\mathbf{X}_2$ represents the columns of $\mathbf{X}$ associated with $\boldsymbol{\beta}_2$.

For the *full* model (including both $\boldsymbol{\beta}_1$ and $\boldsymbol{\beta}_2$), we know that $\hat{\boldsymbol{\beta}} = (\mathbf{X}'\mathbf{X})^{-1}\mathbf{X}'\mathbf{y}$. Also, the regression sum of squares for all variables including the intercept is

$$SS_R(\boldsymbol{\beta}) = \hat{\boldsymbol{\beta}}'\mathbf{X}'\mathbf{y} \qquad (p \text{ degrees of freedom})$$

and

$$MS_E = \frac{\mathbf{y}'\mathbf{y} - \hat{\boldsymbol{\beta}}'\mathbf{X}'\mathbf{y}}{n - p}.$$

$SS_R(\boldsymbol{\beta})$ is called the regression sum of squares *due to* $\boldsymbol{\beta}$. To find the contribution of the terms in $\boldsymbol{\beta}_1$ to the regression, fit the model assuming the null hypothesis H_0: $\boldsymbol{\beta}_1 = \mathbf{0}$ to be true. The *reduced* model is found from equation 15-34 to be

$$\mathbf{y} = \mathbf{X}_2\boldsymbol{\beta}_2 + \boldsymbol{\varepsilon}. \tag{15-35}$$

The least-squares estimator of $\boldsymbol{\beta}_2$ is $\hat{\boldsymbol{\beta}}_2 = (\mathbf{X}_2'\mathbf{X}_2)^{-1}\mathbf{X}_2'\mathbf{y}$, and

$$SS_R(\boldsymbol{\beta}_2) = \hat{\boldsymbol{\beta}}_2'\mathbf{X}_2'\mathbf{y} \qquad (p - r \text{ degrees of freedom}). \tag{15-36}$$

The regression sum of squares due to $\boldsymbol{\beta}_1$ given that $\boldsymbol{\beta}_2$ is already in the model is

$$SS_R(\boldsymbol{\beta}_1|\boldsymbol{\beta}_2) = SS_R(\boldsymbol{\beta}) - SS_R(\boldsymbol{\beta}_2). \tag{15-37}$$

This sum of squares has r degrees of freedom. It is sometimes called the "extra sum of squares" due to $\boldsymbol{\beta}_1$. Note that $SS_R(\boldsymbol{\beta}_1|\boldsymbol{\beta}_2)$ is the increase in the regression sum of squares due to including the variables $x_1, x_2, \ldots, x_r$ in the model. Now $SS_R(\boldsymbol{\beta}_1|\boldsymbol{\beta}_2)$ is independent of MS_E, and the null hypothesis $\boldsymbol{\beta}_1 = \mathbf{0}$ may be tested by the statistic

$$F_0 = \frac{SS_R(\boldsymbol{\beta}_1|\boldsymbol{\beta}_2)/r}{MS_E}. \tag{15-38}$$

If $F_0 > F_{\alpha, r, n-p}$ we reject H_0, concluding that at least one of the parameters in $\boldsymbol{\beta}_1$ is not zero and, consequently, at least one of the variables $x_1, x_2, \ldots, x_r$ in $\mathbf{X}_1$ contributes significantly to the regression model. Some authors call the test in equation 15-38 a *partial* F-test.

The partial F-test is very useful. We can use it to measure the contribution of x_j as if it were the last variable added to the model by computing

$$SS_R(\beta_j|\beta_0, \beta_1, \ldots, \beta_{j-1}, \beta_{j+1}, \ldots, \beta_k).$$

This is the increase in the regression sum of squares due to adding x_j to a model that already includes $x_1, \ldots, x_{j-1}, x_{j+1}, \ldots, x_k$. Note that the partial F-test on a single variable x_j is equivalent to the t-test in equation 15-32. However, the partial F-test is a more general procedure in that we can measure the effect of sets of variables. In Section 15-11 we will show how the partial F-test plays a major role in *model building*, that is, in searching for the best set of regressor variables to use in the model.

Example 15-7

Consider the damaged peaches data in Example 15-1. We will investigate the contribution of the variable x_2 (density) to the model. That is, we wish to test

$$H_0: \beta_2 = 0,$$
$$H_1: \beta_2 \neq 0.$$

To test this hypothesis, we need the extra sum of squares due to β_2, or

$$\begin{aligned} SS_R(\beta_2|\beta_1, \beta_0) &= SS_R(\beta_1, \beta_2, \beta_0) - SS_R(\beta_1, \beta_0) \\ &= SS_R(\beta_1, \beta_2|\beta_0) - SS_R(\beta_1|\beta_0). \end{aligned}$$

In Example 15-6 we calculated

$$SS_R(\beta_1, \beta_2|\beta_0) = \hat{\boldsymbol{\beta}}'\mathbf{X}'\mathbf{y} - \frac{1}{n}\left(\sum_{i=1}^{n} y_i\right)^2 = 136.88 \qquad \text{(2 degrees of freedom)},$$

and if the model $y = \beta_0 + \beta_1 x_1 + \varepsilon$ is fit, we have

$$SS_R(\beta_1|\beta_0) = \hat{\beta}_1 S_{xy} = 96.21 \qquad \text{(1 degree of freedom)}.$$

Therefore, we have

$$\begin{aligned} SS_R(\beta_2|\beta_1, \beta_0) &= 136.88 - 96.21 \\ &= 40.67 \qquad \text{(1 degree of freedom)}. \end{aligned}$$

This is the increase in the regression sum of squares attributable to adding x_2 to a model already containing x_1. To test H_0: $\beta_2 = 0$, form the test statistic

$$F_0 = \frac{SS_R(\beta_2|\beta_1, \beta_0)/1}{MS_E} = \frac{40.67}{2.247} = 18.10.$$

Note that the MS_E from the *full* model, using both x_1 and x_2, is used in the denominator of the test statistic. Since $F_{0.05,1,17} = 4.45$, we reject H_0: $\beta_2 = 0$ and conclude that density (x_2) contributes significantly to the model.

Since this partial F-test involves a single variable, it is equivalent to the t-test. To see this, recall that the t-test on H_0: $\beta_2 = 0$ resulted in the test statistic $t_0 = 4.24$. Furthermore, recall that the square of a t random variable with v degrees of freedom is an F random variable with one and v degrees of freedom, and we note that $t_0^2 = (4.24)^2 = 17.98 \simeq F_0$.

15-6 MEASURES OF MODEL ADEQUACY

A number of techniques can be used to measure the adequacy of a multiple regression model. This section will present several of these techniques. Model validation is an important part of the multiple regression model building process. A good paper on this subject is Snee (1977) (see also Montgomery, Peck, and Vining, 2001).

15-6.1 The Coefficient of Multiple Determination

The coefficient of multiple determination R^2 is defined as

$$R^2 = \frac{SS_R}{S_{yy}} = 1 - \frac{SS_E}{S_{yy}}. \tag{15-39}$$

R^2 is a measure of the amount of reduction in the variability of y obtained by using the regressor variables $x_1, x_2, \ldots, x_k$. As in the simple linear regression case, we must have $0 \le R^2 \le 1$. However, as before, a large value of R^2 does not necessarily imply that the regression model is a good one. Adding a variable to the model will always increase R^2, regardless of whether the additional variable is statistically significant or not. Thus it is possible for models that have large values of R^2 to yield poor predictions of new observations or estimates of the mean response.

The positive square root of R^2 is the multiple correlation coefficient between y and the set of regressor variables $x_1, x_2, \ldots, x_k$. That is, R is a measure of the linear association between y and $x_1, x_2, \ldots, x_k$. When $k = 1$, this becomes the simple correlation between y and x.

Example 15-8

The coefficient of multiple determination for the regression model estimated in Example 15-1 is

$$R^2 = \frac{SS_R}{S_{yy}} = \frac{136.88}{175.09} = 0.782.$$

That is, about 78.2% of the variability in damage y is explained when the two regressor variables, drop height (x_1) and fruit density (x_2) are used. The model relating density to x_1 only was developed. The value of R^2 for this model turns out to be $R^2 = 0.549$. Therefore, adding the variable x_2 to the model has increased R^2 from 0.549 to 0.782.

Adjusted R^2

Some practitioners prefer to use the *adjusted coefficient of multiple determination*, adjusted R^2, defined as

$$R^2_{\text{adj}} = 1 - \frac{SS_E/(n-p)}{S_{yy}/(n-1)}. \tag{15-40}$$

The value $S_{yy}/(n-1)$ will be constant regardless of the number of variables in the model. $SS_E/(n-p)$ is the mean square for error, which will change with the addition or removal of terms (new regressor variables, interaction terms, higher-order terms) from the model. Therefore, R^2_{adj} will increase only if the addition of a new term significantly reduces the mean square for error. In other words, the R^2_{adj} will penalize adding terms to the model that are not significant in modeling the response. Interpretation of the adjusted coefficient of multiple determination is identical to that of R^2.

Example 15-9

We can calculate R^2_{adj} for the model fit in Example 15-1. From Example 15-6, we found that SS_E = 38.21 and S_{yy} = 175.09. The estimate R^2_{adj} is then

$$R^2_{\text{adj}} = 1 - \frac{38.21/(20-3)}{175.09/(20-1)} = 1 - \frac{2.247}{9.215}$$
$$= 0.756.$$

The adjusted R^2 will play a significant role in variable selection and model building later in this chapter.

15-6.2 Residual Analysis

The residuals from the estimated multiple regression model, defined by $e_i = y_i - \hat{y}_i$, play an important role in judging model adequacy, just as they do in simple linear regression. As noted in Section 14-5.1, there are several residual plots that are often useful. These are illustrated in Example 15-9. It is also helpful to plot the residuals against variables not presently in the model that are possible candidates for inclusion. Patterns in these plots, similar to those in Fig. 14-5, indicate that the model may be improved by adding the candidate variable.

Example 15-10

The residuals for the model estimated in Example 15-1 are shown in Table 15-3. These residuals are plotted on a normal probability plot in Fig. 15-2. No severe deviations from normality are obvious, although the smallest residual ($e_3 = -3.17$) does not fall near the remaining residuals. The standardized residual, $-3.17/\sqrt{2.247} = -2.11$, appears to be large and could indicate an unusual observation. The residuals are plotted against $\hat{y}$ in Fig. 15-3, and against x_1 and x_2 in Figs. 15-4 and 15-5, respectively. In Fig. 15-4, there is some indication that the assumption of constant variance may not be satisfied. Removal of the unusual observation may improve the model fit, but there is no indication of error in data collection. Therefore, the point will be retained. We will see subsequently (Example 15-16) that two other regressor variables are required to adequately model these data.

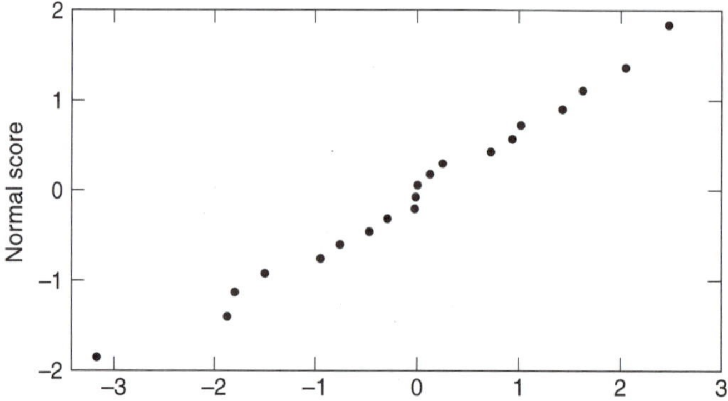

Figure 15-2 Normal probability plot of residuals for Example 15-10.

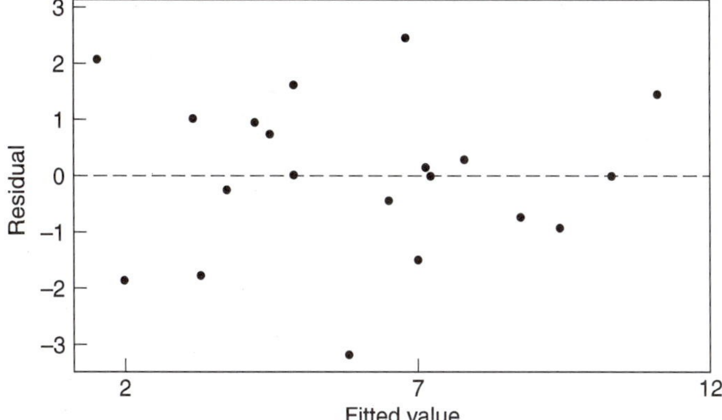

Figure 15-3 Plot of residuals against $\hat{y}$ for Example 15-10.

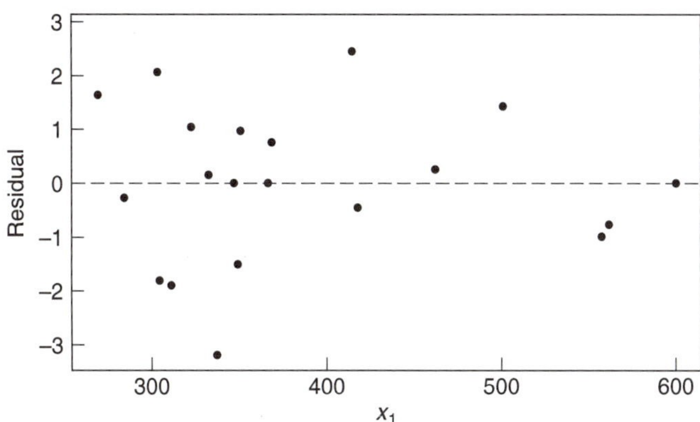

Figure 15-4 Plot of residuals against x_1 for Example 15-10.

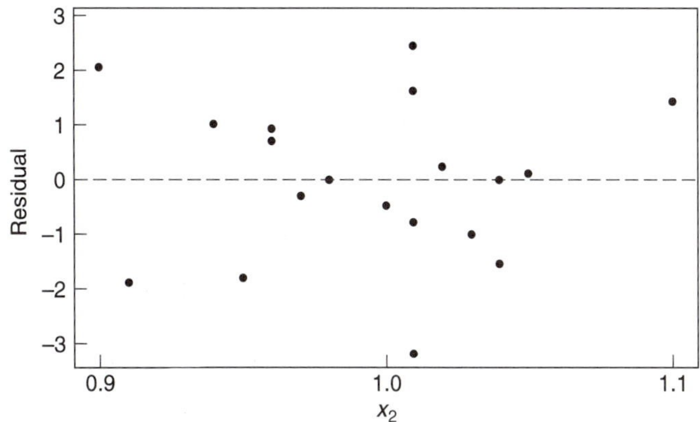

Figure 15-5 Plot of residuals against x_2 for Example 15-10.

15-7 POLYNOMIAL REGRESSION

The linear model $\mathbf{y} = \mathbf{X}\boldsymbol{\beta} + \boldsymbol{\varepsilon}$ is a general model that can be used to fit any relationship that is *linear* in the unknown parameters $\boldsymbol{\beta}$. This includes the important class of polynomial regression models. For example, the second-degree polynomial in one variable,

$$y = \beta_0 + \beta_1 x + \beta_{11} x^2 + \varepsilon, \tag{15-41}$$

and the second-degree polynomial in two variables,

$$y = \beta_0 + \beta_1 x_1 + \beta_2 x_2 + \beta_{11} x_1^2 + \beta_{22} x_2^2 + \beta_{12} x_1 x_2 + \varepsilon, \tag{15-42}$$

are linear regression models.

Polynomial regression models are widely used in cases where the response is curvilinear, because the general principles of multiple regression can be applied. The following example illustrates some of the types of analyses that can be performed.

Example 15-11

Sidewall panels for the interior of an airplane are formed in a 1500-ton press. The unit manufacturing cost varies with the production lot size. The data shown below give the average cost per unit (in hundreds of dollars) for this product (y) and the production lot size (x). The scatter diagram, shown in Fig. 15-6, indicates that a second-order polynomial may be appropriate.

y	1.81	1.70	1.65	1.55	1.48	1.40	1.30	1.26	1.24	1.21	1.20	1.18
x	20	25	30	35	40	50	60	65	70	75	80	90

We will fit the model

$$y = \beta_0 + \beta_1 x + \beta_{11} x^2 + \varepsilon.$$

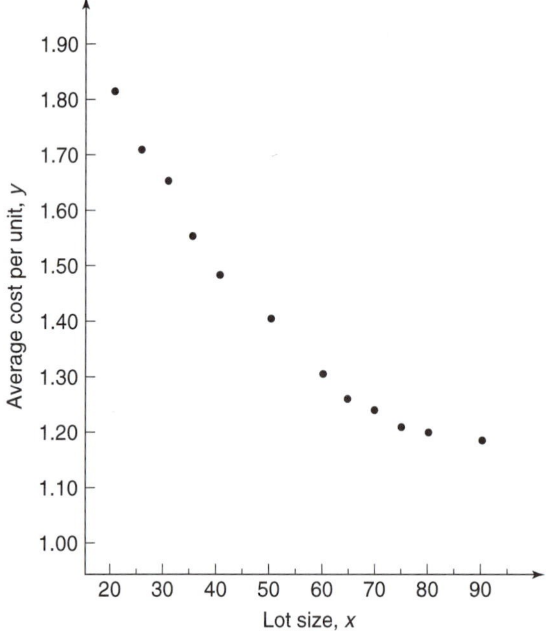

Figure 15-6 Data for Example 15-11.

The **y** vector, **X** matrix, and **β** vector are as follows:

$$\mathbf{y}=\begin{bmatrix}1.81\\1.70\\1.65\\1.55\\1.48\\1.40\\1.30\\1.26\\1.24\\1.21\\1.20\\1.18\end{bmatrix},\quad \mathbf{X}=\begin{bmatrix}1&20&400\\1&25&625\\1&30&900\\1&35&1225\\1&40&1600\\1&50&2500\\1&60&3600\\1&65&4225\\1&70&4900\\1&75&5625\\1&80&6400\\1&90&8100\end{bmatrix},\quad \boldsymbol{\beta}=\begin{bmatrix}\beta_0\\\beta_1\\\beta_{11}\end{bmatrix}.$$

Solving the normal equations $\mathbf{X}'\mathbf{X}\hat{\boldsymbol{\beta}}=\mathbf{X}'\mathbf{y}$ gives the fitted model

$$\hat{y}=2.1983-0.0225x+0.0001251x^2.$$

The test for significance of regression is shown in Table 15-6. Since $F_0=2171.07$ is significant at 1%, we conclude that at least one of the parameters β_1 and β_{11} is not zero. Furthermore, the standard tests for model adequacy reveal no unusual behavior.

In fitting polynomials, we generally like to use the lowest-degree model consistent with the data. In this example, it would seem logical to investigate dropping the quadratic term from the model. That is, we would like to test

$$H_0\colon \beta_{11}=0,$$
$$H_1\colon \beta_{11}\neq 0.$$

The general regression significance test can be used to test this hypothesis. We need to determine the "extra sum of squares" due to β_{11}, or

$$SS_R(\beta_{11}|\beta_1,\beta_0)=SS_R(\beta_1,\beta_{11}|\beta_0)-SS_R(\beta_1|\beta_0).$$

The sum of squares $SS_R(\beta_1,\beta_{11}|\beta_0)=0.5254$, from Table 15-6. To find $SS_R(\beta_1|\beta_0)$, we fit a simple linear regression model to the original data, yielding

$$\hat{y}=1.9004-0.0091x.$$

It can be easily verified that the regression sum of squares for this model is

$$SS_R(\beta_1|\beta_0)=0.4942.$$

Table 15-6 Test for Significance of Regression for the Second-Order Model in Example 15-11

Source of Variation	Sum of Squares	Degrees of Freedom	Mean Square	F_0
Regression	0.5254	2	0.2627	2171.07
Error	0.0011	9	0.000121	
Total	0.5265	11		

Table 15-7 Analysis of Variance of Example 15-11 Showing the Test for H_0: $\beta_{11} = 0$

Source of Variation	Sum of Squares	Degree of Freedom	Mean Square	F_0
Regression	$SS_R(\beta_1, \beta_{11}\|\beta_0) = 0.5254$	2	0.2627	2171.07
Linear	$SS_R(\beta_1\|\beta_0) = 0.4942$	1	0.4942	4084.30
Quadratic	$SS_R(\beta_{11}\|\beta_0, \beta_1) = 0.0312$	1	0.0312	257.85
Error	0.0011	9	0.000121	
Total	0.5265	11		

Therefore, the extra sum of squares due to β_{11}, given that β_1 and β_0 are in the model, is

$$\begin{aligned} SS_R(\beta_{11}|\beta_0, \beta_1) &= SS_R(\beta_1, \beta_{11}|\beta_0) - SS_R(\beta_1|\beta_0) \\ &= 0.5254 - 0.4942 \\ &= 0.0312. \end{aligned}$$

The analysis of variance, with the test of H_0: $\beta_{11} = 0$ incorporated into the procedure, is displayed in Table 15-7. Note that the quadratic term contributes significantly to the model.

15-8 INDICATOR VARIABLES

The regression models presented in previous sections have been based on *quantitative* variables, that is, variables that are measured on a numerical scale. For example, variables such as temperature, pressure, distance, and age are quantitative variables. Occasionally, we need to incorporate *qualitative* variables in a regression model. For example, suppose that one of the variables in a regression model is the operator who is associated with each observation y_i. Assume that only two operators are involved. We may wish to assign different levels to the two operators to account for the possibility that each operator may have a different effect on the response.

The usual method of accounting for the different levels of a qualitative variable is by using indicator variables. For instance, to introduce the effect of two different operators into a regression model, we could define an indicator variable as follows:

$$x = 0 \text{ if the observation is from operator 1,}$$

$$x = 1 \text{ if the observation is from operator 2.}$$

In general, a qualitative variable with t levels is represented by $t - 1$ indicator variables, which are assigned values of either 0 or 1. Thus, if there were *three* operators, the different levels would be accounted for by *two* indicator variables defined as follows:

x_1	x_2	
0	0	if the observation is from operator 1,
1	0	if the observation is from operator 2,
0	1	if the observation is from operator 3.

Indicator variables are also referred to as *dummy* variables. The following example illustrates some of the uses of indicator variables. For other applications, see Montgomery, Peck, and Vining (2001).

Example 15-12

(Adapted from Montgomery, Peck, and Vining, 2001). A mechanical engineer is investigating the surface finish of metal parts produced on a lathe and its relationship to the speed (in RPM) of the lathe. The data are shown in Table 15-8. Note that the data have been collected using two different types of cutting tools. Since it is likely that the type of cutting tool affects the surface finish, we will fit the model

$$y = \beta_0 + \beta_1 x_1 + \beta_2 x_2 + \varepsilon,$$

where y is the surface finish, x_1 is the lathe speed in RPM, and x_2 is an indicator variable denoting the type of cutting tool used; that is,

$$x_2 = \begin{cases} 0 \text{ for tool type 302,} \\ 1 \text{ for tool type 416.} \end{cases}$$

The parameters in this model may be easily interpreted. If $x_2 = 0$, then the model becomes

$$y = \beta_0 + \beta_1 x_1 + \varepsilon,$$

which is a straight-line model with slope β_1 and intercept β_0. However, if $x_2 = 1$, then the model becomes

$$y = \beta_0 + \beta_1 x_1 + \beta_2(1) + \varepsilon = \beta_0 + \beta_2 + \beta_1 x_1 + \varepsilon,$$

which is a straight-line model with slope β_1 and intercept $\beta_0 + \beta_2$. Thus, the model $y = \beta_0 + \beta_1 x + \beta_2 x_2 + \varepsilon$ implies that surface finish is linearly related to lathe speed and that the slope β_1 does not depend on the type of cutting tool used. However, the type of cutting tool does affect the intercept, and β_2 indicates the change in the intercept associated with a change in tool type from 302 to 416.

Table 15-8 Surface Finish Data for Example 15-12

Observation Number, i	Surface Finish, y_i	RPM	Type of Cutting Tool
1	45.44	225	302
2	42.03	200	302
3	50.10	250	302
4	48.75	245	302
5	47.92	235	302
6	47.79	237	302
7	52.26	265	302
8	50.52	259	302
9	45.58	221	302
10	44.78	218	302
11	33.50	224	416
12	31.23	212	416
13	37.52	248	416
14	37.13	260	416
15	34.70	243	416
16	33.92	238	416
17	32.13	224	416
18	35.47	251	416
19	33.49	232	416
20	32.29	216	416

The **X** matrix and **y** vector for this problem are as follows:

$$\mathbf{X}=\begin{bmatrix}1 & 225 & 0\\1 & 200 & 0\\1 & 250 & 0\\1 & 245 & 0\\1 & 235 & 0\\1 & 237 & 0\\1 & 265 & 0\\1 & 259 & 0\\1 & 221 & 0\\1 & 218 & 0\\1 & 224 & 1\\1 & 212 & 1\\1 & 248 & 1\\1 & 260 & 1\\1 & 243 & 1\\1 & 238 & 1\\1 & 224 & 1\\1 & 251 & 1\\1 & 232 & 1\\1 & 216 & 1\end{bmatrix},\qquad \mathbf{y}=\begin{bmatrix}45.44\\42.03\\50.10\\48.75\\47.92\\47.79\\52.26\\50.52\\45.58\\44.78\\33.50\\31.23\\37.52\\37.13\\34.70\\33.92\\32.13\\35.47\\33.49\\32.29\end{bmatrix}.$$

The fitted model is

$$\hat{y}=14.2762+0.1411x_1-13.2802x_2.$$

The analysis of variance for this model is shown in Table 15-9. Note that the hypothesis H_0: $\beta_1=\beta_2=0$ (significance of regression) is rejected. This table also contains the sum of squares

$$\begin{aligned}SS_R&=SS_R(\beta_1,\beta_2|\beta_0)\\&=SS_R(\beta_1|\beta_0)+SS_R(\beta_2|\beta_1,\beta_0),\end{aligned}$$

so a test of the hypothesis H_0: $\beta_2=0$ can be made. This hypothesis is also rejected, so we conclude that tool type has an effect on surface finish.

It is also possible to use indicator variables to investigate whether tool type affects *both* slope *and* intercept. Let the model be

$$y=\beta_0+\beta_1x_1+\beta_2x_2+\beta_3x_1x_2+\varepsilon,$$

Table 15-9 Analysis of Variance of Example 15-12

Source of Variation	Sum of Squares	Degrees of Freedom	Mean Square	F_0
Regression	1012.0595	2	506.0297	1103.69[a]
$SS_R(\beta_1\|\beta_0)$	(130.6091)	(1)	130.6091	284.87[a]
$SS_R(\beta_2\|\beta_1,\beta_0)$	(881.4504)	(1)	881.4504	1922.52[a]
Error	7.7943	17	0.4508	
Total	1019.8538	19		

[a]Significant at 1%.

where x_2 is the indicator variable. Now if tool type 302 is used, $x_2 = 0$, and the model is

$$y = \beta_0 + \beta_1 x_1 + \varepsilon.$$

If tool type 416 is used, $x_2 = 1$, and the model becomes

$$\begin{aligned} y &= \beta_0 + \beta_1 x_1 + \beta_2 + \beta_3 x_1 + \varepsilon \\ &= (\beta_0 + \beta_2) + (\beta_1 + \beta_3)x_1 + \varepsilon. \end{aligned}$$

Note that β_2 is the change in the intercept, and β_3 is the change in slope produced by a change in tool type.

Another method of analyzing these data set is to fit separate regression models to the data for each tool type. However, the indicator variable approach has several advantages. First, only one regression model must be estimated. Second, by pooling the data on both tool types, more degrees of freedom for error are obtained. Third, tests of both hypotheses on the parameters β_2 and β_3 are just special cases of the general regression significance test.

15-9 THE CORRELATION MATRIX

Suppose we wish to estimate the parameters in the model

$$y_i = \beta_0 + \beta_1 x_{i1} + \beta_2 x_{i2} + \varepsilon_i, \qquad i = 1, 2, \ldots, n. \tag{15-43}$$

We may rewrite this model with a transformed intercept β_0' as

$$y_i = \beta_0' + \beta_1(x_{i1} - \bar{x}_1) + \beta_2(x_{i2} - \bar{x}_2) + \varepsilon_i \tag{15-44}$$

or, since $\hat{\beta}_0' = \bar{y}$,

$$y_i - \bar{y} = \beta_1(x_{i1} - \bar{x}_1) + \beta_2(x_{i2} - \bar{x}_2) + \varepsilon_i. \tag{15-45}$$

The $\mathbf{X'X}$ matrix for this model is

$$\mathbf{X'X} = \begin{bmatrix} S_{11} & S_{12} \\ S_{12} & S_{22} \end{bmatrix}, \tag{15-46}$$

where

$$S_{kj} = \sum_{i=1}^{n} \left(x_{ik} - \bar{x}_k\right)\left(x_{ij} - \bar{x}_j\right), \qquad k, j = 1, 2. \tag{15-47}$$

It is possible to express this $\mathbf{X'X}$ matrix in correlation form. Let

$$r_{kj} = \frac{S_{kj}}{\left(S_{kk} S_{jj}\right)^{1/2}}, \qquad k, j = 1, 2, \tag{15-48}$$

and note that $r_{11} = r_{22} = 1$. Then the correlation form of the $\mathbf{X'X}$ matrix, equation 15-46, is

$$\mathbf{R} = \begin{bmatrix} 1 & r_{12} \\ r_{12} & 1 \end{bmatrix}. \tag{15-49}$$

The quantity r_{12} is the sample correlation between x_1 and x_2. We may also define the sample correlation between x_j and y as

$$r_{jy} = \frac{S_{jy}}{\left(S_{jj} S_{yy}\right)^{1/2}}, \qquad j = 1, 2, \tag{15-50}$$

where

$$S_{jy} = \sum_{u=1}^{n} \left(x_{uj} - \bar{x}_j\right)\left(y_u - \bar{y}\right), \qquad j = 1, 2, \tag{15-51}$$

is the corrected sum of cross products between x_j and y, and S_{yy} is the usual total corrected sum of squares of y.

These transformations result in a new regression model,

$$y_i^* = b_1 z_{i1} + b_2 z_{i2} + \varepsilon_i^*, \tag{15-52}$$

where

$$y_i^* = \frac{y_i - \bar{y}}{S_{yy}^{1/2}},$$

$$z_{ij} = \frac{x_{ij} - \bar{x}_j}{S_{jj}^{1/2}}, \qquad j = 1, 2.$$

The relationship between the parameters b_1 and b_2 in the new model, equation 15-52, and the parameters β_0, β_1, and β_2 in the original model, equation 15-43, is as follows:

$$\beta_1 = b_1\left(\frac{S_{yy}}{S_{11}}\right)^{1/2}, \tag{15-53}$$

$$\beta_2 = b_2\left(\frac{S_{yy}}{S_{22}}\right)^{1/2}, \tag{15-54}$$

$$\beta_0 = \bar{y} - \beta_1\bar{x}_1 - \beta_2\bar{x}_2. \tag{15-55}$$

The least-squares normal equations for the transformed model, equation 15-52, are

$$\begin{bmatrix} 1 & r_{12} \\ r_{12} & 1 \end{bmatrix}\begin{bmatrix} \hat{b}_1 \\ \hat{b}_2 \end{bmatrix} = \begin{bmatrix} r_{1y} \\ r_{2y} \end{bmatrix}. \tag{15-56}$$

The solution to equation 15-56 is

$$\begin{bmatrix} \hat{b}_1 \\ \hat{b}_2 \end{bmatrix} = \begin{bmatrix} 1 & r_{12} \\ r_{12} & 1 \end{bmatrix}^{-1}\begin{bmatrix} r_{1y} \\ r_{2y} \end{bmatrix}$$

$$= \frac{1}{1 - r_{12}^2}\begin{bmatrix} 1 & -r_{12} \\ -r_{12} & 1 \end{bmatrix}\begin{bmatrix} r_{1y} \\ r_{2y} \end{bmatrix}$$

or

$$\hat{b}_1 = \frac{r_{1y} - r_{12}r_{2y}}{1 - r_{12}^2}, \tag{15-57a}$$

$$\hat{b}_2 = \frac{r_{2y} - r_{12}r_{1y}}{1 - r_{12}^2}. \tag{15-57b}$$

The regression coefficients, equations 15-57, are usually called *standardized regression coefficients.* Many multiple regression computer programs use this transformation to reduce round-off errors in the $(\mathbf{X'X})^{-1}$ matrix. These round-off errors may be very serious if the original variables differ considerably in magnitude. Some of these computer programs also display both the original regression coefficients and the standardized coefficients. The standardized regression coefficients are dimensionless, and this may make it easier to compare regression coefficients in situations where the original variables x_j differ considerably in their units of measurement. In interpreting these standardized regression coefficients, however, we must remember that they are still partial regression coefficients (i.e., b_j shows the effect of z_j given that other z_i, $i \neq j$, are in the model). Furthermore, the $\hat{b}_j$ are affected by the spacing of the levels of the x_j. Consequently, we should not use the magnitude of the $\hat{b}_j$ as a measure of the importance of the regressor variables.

While we have explicitly treated only the case of two regressor variables, the results generalize. If there are k regressor variables $x_1, x_2, \ldots, x_k$, one may write the $\mathbf{X}'\mathbf{X}$ matrix in correlation form,

$$\mathbf{R} = \begin{bmatrix} 1 & r_{12} & r_{13} & \cdots & r_{1k} \\ r_{12} & 1 & r_{23} & \cdots & r_{2k} \\ r_{13} & r_{23} & 1 & \cdots & r_{3k} \\ & & \vdots & & \\ r_{1k} & r_{2k} & r_{3k} & \cdots & 1 \end{bmatrix}, \qquad (15\text{-}58)$$

where $r_{ij} = S_{ij}/(S_{ii}S_{jj})^{1/2}$ is the sample correlation between x_i and x_j and $S_{ij} = \sum_{u=1}^{n}(x_{ui} - \bar{x}_i)(x_{uj} - \bar{x}_j)$. The correlations between x_j and y are

$$\mathbf{g} = \begin{bmatrix} r_{1y} \\ r_{2y} \\ \vdots \\ r_{ky} \end{bmatrix}, \qquad (15\text{-}59)$$

where $r_{iy} = \sum_{u=1}^{n}(x_{ui} - \bar{x}_i)(y_u - \bar{y})$. The vector of standardized regression coefficients $\hat{\mathbf{b}}' = [\hat{b}_1, \hat{b}_2, \ldots, \hat{b}_k]$ is

$$\hat{\mathbf{b}} = \mathbf{R}^{-1}\mathbf{g}. \qquad (15\text{-}60)$$

The relationship between the standardized regression coefficients and the original regression coefficients is

$$\hat{\beta}_j = \hat{b}_j\left(\frac{S_{yy}}{S_{jj}}\right)^{1/2}, \qquad j = 1, 2, \ldots, k. \qquad (15\text{-}61)$$

Example 15-13

For the data in Example 15-1, we find

$$S_{yy} = 175.089, \qquad S_{11} = 184710.16,$$
$$S_{1y} = 4215.372, \qquad S_{22} = 0.047755,$$
$$S_{2y} = 2.33, \qquad S_{12} = 51.2873.$$

Therefore,

$$r_{12} = \frac{S_{12}}{(S_{11}S_{22})^{1/2}} = \frac{51.2873}{\sqrt{(184710.16)(0.047755)}} = 0.5460,$$

$$r_{1y} = \frac{S_{1y}}{\left(S_{11}S_{yy}\right)^{1/2}} = \frac{4215.372}{\sqrt{(184710.16)(175.089)}} = 0.7412,$$

$$r_{2y} = \frac{S_{2y}}{\left(S_{22}S_{yy}\right)^{1/2}} = \frac{2.33}{\sqrt{(0.047755)(175.089)}} = 0.8060,$$

and the correlation matrix for this problem is

$$\begin{bmatrix} 1 & 0.5460 \\ 0.5460 & 1 \end{bmatrix}.$$

From equation 15-56, the normal equations in terms of the standardized regression coefficients are

$$\begin{bmatrix} 1 & 0.5460 \\ 0.5460 & 1 \end{bmatrix} \begin{bmatrix} \hat{b}_1 \\ \hat{b}_2 \end{bmatrix} = \begin{bmatrix} 0.7412 \\ 0.8060 \end{bmatrix}.$$

Consequently, the standardized regression coefficients are

$$\begin{aligned} \begin{bmatrix} \hat{b}_1 \\ \hat{b}_2 \end{bmatrix} &= \begin{bmatrix} 1 & 0.5460 \\ 0.5460 & 1 \end{bmatrix}^{-1} \begin{bmatrix} 0.7412 \\ 0.8060 \end{bmatrix} \\ &= \begin{bmatrix} 1.424737 & -0.77791 \\ -0.77791 & 1.424737 \end{bmatrix} \begin{bmatrix} 0.7412 \\ 0.8060 \end{bmatrix} \\ &= \begin{bmatrix} 0.429022 \\ 0.571754 \end{bmatrix}. \end{aligned}$$

These standardized regression coefficients could also have been computed directly from either equation 15-57 or equation 15-61. Note that although $\hat{b}_2 > \hat{b}_1$, we should be cautious about concluding that the fruit density x_2 is more important than drop height (x_1), since $\hat{b}_1$ and $\hat{b}_2$ are still *partial* regression coefficients.

15-10 PROBLEMS IN MULTIPLE REGRESSION

There are a number of problems often encountered in the use of multiple regression. In this section, we briefly discuss three of these problem areas: the effect of multicollinearity on the regression model, the effect of outlying points in the x-space on the regression coefficients, and autocorrelation in the errors.

15-10.1 Multicollinearity

In most multiple regression problems, the independent or regressor variables x_j are intercorrelated. In situations which this intercorrelation is very large, we say that *multicollinearity* exists. Multicollinearity can have serious effects on the estimates of the regression coefficients and on the general applicability of the estimated model.

The effects of multicollinearity may be easily demonstrated. Consider a regression model with two regressor variables x_1 and x_2, and suppose that x_1 and x_2 have been "standardized," as in Section 15-9, so that the **X′X** matrix is in correlation form, as in equation 15-49. The model is

$$y_i = \beta_0 + \beta_1 x_{i1} + \beta_2 x_{i2} + \varepsilon_i, \quad i = 1, 2, \ldots, n.$$

The $(\mathbf{X}'\mathbf{X})^{-1}$ matrix for this model is

$$\mathbf{C} = (\mathbf{X}'\mathbf{X})^{-1} = \begin{bmatrix} 1/\left(1-r_{12}^2\right) & -r_{12}/\left(1-r_{12}^2\right) \\ -r_{12}/\left(1-r_{12}^2\right) & 1/\left(1-r_{12}^2\right) \end{bmatrix}$$

and the estimators of the parameters are

$$\hat{\beta}_1 = \frac{\mathbf{x}_1'\mathbf{y} - r_{12}\mathbf{x}_2'\mathbf{y}}{1-r_{12}^2},$$

$$\hat{\beta}_2 = \frac{\mathbf{x}_2'\mathbf{y} - r_{12}\mathbf{x}_1'\mathbf{y}}{1-r_{12}^2},$$

where r_{12} is the sample correlation between x_1 and x_2, and $\mathbf{x}_1'\mathbf{y}$ and $\mathbf{x}_2'\mathbf{y}$ are the elements of the $\mathbf{X}'\mathbf{y}$ vector.

Now, if multicollinearity is present, x_1 and x_2 are highly correlated, and $|r_{12}| \to 1$. In such a situation, the variances and covariances of the regression coefficients become very large, since $V(\hat{\beta}_j) = C_{jj}\sigma^2 \to \infty$ as $|r_{12}| \to 1$, and Cov $(\hat{\beta}_1, \hat{\beta}_2) = C_{12}\sigma^2 \to \pm\infty$ depending on whether $r_{12} \to \pm 1$. The large variances for $\hat{\beta}_j$ imply that the regression coefficients are very poorly estimated. Note that the effect of multicollinearity is to introduce a "near" linear dependency in the columns of the $\mathbf{X}$ matrix. As $r_{12} \to \pm 1$, this linear dependency becomes exact. Furthermore, if we assume that $\mathbf{x}_1'\mathbf{y} \to \mathbf{x}_2'\mathbf{y}$ as $|r_{12}| \to \pm 1$, then the estimates of the regression coefficients become equal in magnitude but opposite in sign; that is $\hat{\beta}_1 = -\hat{\beta}_2$, *regardless* of the true values of β_1 and β_2.

Similar problems occur when multicollinearity is present and there are more than two regressor variables. In general, the diagonal elements of the matrix $\mathbf{C} = (\mathbf{X}'\mathbf{X})^{-1}$ can be written

$$C_{jj} = \frac{1}{1-R_j^2}, \qquad j = 1, 2, \ldots, k, \tag{15-62}$$

where R_j^2 is the coefficient of multiple determination resulting from regressing x_j on the other $k-1$ regressor variables. Clearly, the stronger the linear dependency of x_j on the remaining regressor variables (and hence the stronger the multicollinearity), the larger the value of R_j^2 will be. We say that the variance of $\hat{\beta}_j$ is "inflated" by the quantity $(1 - R_j^2)^{-1}$. Consequently, we usually call

$$VIF\left(\hat{\beta}_j\right) = \frac{1}{1-R_j^2}, \qquad j = 1, 2, \ldots, k, \tag{15-63}$$

the *variance inflation factor* for $\hat{\beta}_j$. Note that these factors are the main diagonal elements of the inverse of the correlation matrix. They are an important measure of the extent to which multicollinearity is present.

Although the estimates of the regression coefficients are very imprecise when multicollinearity is present, the estimated equation may still be useful. For example, suppose we wish to predict new observations. If these predictions are required in the region of the x-space where the multicollinearity is in effect, then often satisfactory results will be obtained because while individual β_j may be poorly estimated, the function $\sum_{j=1}^{k}\beta_j x_{ij}$ may be estimated quite well. On the other hand, if the prediction of new observations requires extrapolation, then generally we would expect to obtain poor results. Extrapolation usually requires good estimates of the individual model parameters.

Multicollinearity arises for several reasons. It will occur when the analyst collects the data such that a constraint of the form $\sum_{j=1}^{k} a_j \mathbf{x}_j = \mathbf{0}$ holds among the columns of the $\mathbf{X}$ matrix (the a_j are constants, not all zero). For example, if four regressor variables are the components of a mixture, then such a constraint will always exist because the sum of the components is always constant. Usually, these constraints do not hold exactly, and the analyst does not know that they exist.

There are several ways to detect the presence of multicollinearity. Some of the more important of these are briefly discussed.

1. The variance inflation factors, defined in equation 15-63, are very useful measures of multicollinearity. The larger the variance inflation factor, the more severe the multicollinearity. Some authors have suggested that if any variance inflation factors exceed 10, then multicollinearity is a problem. Other authors consider this value too liberal and suggest that the variance inflation factors should not exceed 4 or 5.
2. The determinant of the correlation matrix may also be used as a measure of multicollinearity. The value of this determinant can range between 0 and 1. When the value of the determinant is 1, the columns of the $\mathbf{X}$ matrix are orthogonal (i.e., there is no intercorrelation between the regression variables), and when the value is 0, there is an exact linear dependency among the columns of $\mathbf{X}$. The smaller the value of the determinant, the greater the degree of multicollinearity.
3. The eigenvalues or characteristic roots of the correlation matrix provide a measure of multicollinearity. If $\mathbf{X}'\mathbf{X}$ is in correlation form, then the eigenvalues of $\mathbf{X}'\mathbf{X}$ are the roots of the equation
$$|\mathbf{X}'\mathbf{X} - \lambda\mathbf{I}| = 0.$$
One or more eigenvalues near zero implies that multicollinearity is present. If $\lambda_{\max}$ and $\lambda_{\min}$ denote the largest and smallest eigenvalues of $\mathbf{X}'\mathbf{X}$, then the ratio $\lambda_{\max}/\lambda_{\min}$ can also be used as a measure of multicollinearity. The larger the value of this ratio, the greater the degree of multicollinearity. Generally, if the ratio $\lambda_{\max}/\lambda_{\min}$ is less than 10, there is little problem with multicollinearity.
4. Sometimes inspection of the individual elements of the correlation matrix can be helpful in detecting multicollinearity. If an element $|r_{ij}|$ is close to 1, then x_i and x_j may be strongly multicollinear. However, when more than two regressor variables are involved in a multicollinear fashion, the individual r_{ij} are not necessarily large. Thus, this method will not always enable us to detect the presence of multicollinearity.
5. If the F-test for significance of regression is significant but tests on the individual regression coefficients are not significant, then multicollinearity may be present.

Several remedial measures have been proposed for resolving the problem of multicollinearity. Augmenting the data with new observations specifically designed to break up the approximate linear dependencies that currently exist is often suggested. However, sometimes this is impossible for economic reasons, or because of the physical constraints that relate the x_j. Another possibility is to delete certain variables from the model. This suffers from the disadvantage that one must discard the information contained in the deleted variables.

Since multicollinearity primarily affects the stability of the regression coefficients, it would seem that estimating these parameters by some method that is less sensitive to multicollinearity than ordinary least squares would be helpful. Several methods have been suggested for this. Hoerl and Kennard (1970a, b) have proposed ridge regression as an alternative to ordinary least squares. In ridge regression, the parameter estimates are obtained by solving

$$\boldsymbol{\beta}^*(l) = (\mathbf{X}'\mathbf{X} + l\mathbf{I})^{-1}\mathbf{X}'\mathbf{y}, \tag{15-64}$$

where $l > 0$ is a constant. Generally, values of l in the interval $0 \leq l \leq 1$ are appropriate. The ridge estimator $\boldsymbol{\beta}^*(l)$ is not an unbiased estimator of $\boldsymbol{\beta}$, as is the ordinary least-squares estimator $\hat{\boldsymbol{\beta}}$, but the mean square error of $\boldsymbol{\beta}^*(l)$ will be smaller than the mean square error of $\hat{\boldsymbol{\beta}}$. Thus ridge regression seeks to find a set of regression coefficients that is more "stable," in the sense of having a small mean square error. Since multicollinearity usually results in ordinary least-squares estimators that may have extremely large variances, ridge regression is suitable for situations where the multicollinearity problem exists.

To obtain the ridge regression estimator from equation 15-64, one must specify a value for the constant l. Of course, there is an "optimum" l for any problem, but the simplest approach is to solve equation 15-64 for several values of l in the interval $0 \leq l \leq 1$. Then a plot of the values of $\boldsymbol{\beta}^*(l)$ against l is constructed. This display is called the *ridge trace*. The appropriate value of l is chosen subjectively by inspection of the ridge trace. Typically, a value for l is chosen such that relatively stable parameter estimates are obtained. In general, the variance of $\boldsymbol{\beta}^*(l)$ is a decreasing function of l, while the squared bias $[\boldsymbol{\beta} - \boldsymbol{\beta}^*(l)]^2$ is an increasing function of l. Choosing the value of l involves trading off these two properties of $\boldsymbol{\beta}^*(l)$.

A good discussion of the practical use of ridge regression is in Marquardt and Snee (1975). Also, there are several other biased estimation techniques that have been proposed for dealing with multicollinearity. Several of these are discussed in Montgomery, Peck, and Vining (2001).

Example 15-14

(Based on an example in Hald, 1952.) The heat generated in calories per gram for a particular type of cement as a function of the quantities of four additives (z_1, z_2, z_3, and z_4) is shown in Table 15-10. We wish to fit a multiple linear regression model to these data.

Table 15-10 Data for Example 15-14

Observation Number	y	z_1	z_2	z_3	z_4
1	28.25	10	31	5	45
2	24.80	12	35	5	52
3	11.86	5	15	3	24
4	36.60	17	42	9	65
5	15.80	8	6	5	19
6	16.23	6	17	3	25
7	29.50	12	36	6	55
8	28.75	10	34	5	50
9	43.20	18	40	10	70
10	38.47	23	50	10	80
11	10.14	16	37	5	61
12	38.92	20	40	11	70
13	36.70	15	45	8	68
14	15.31	7	22	2	30
15	8.40	9	12	3	24

The data will be coded by defining a new set of regressor variables as

$$x_{ij} = \frac{z_{ij} - \bar{z}_j}{\sqrt{S_{jj}}}, \qquad i = 1, 2, \ldots, 15, \qquad j = 1, 2, 3, 4,$$

where $S_{jj} = \sum_{i=1}^{n} (z_{ij} - \bar{z}_j)^2$ is the corrected sum of squares of the levels of z_j. The coded data are shown in Table 15-11. This transformation makes the intercept orthogonal to the other regression coefficients, since the first column of the **X** matrix consists of ones. Therefore, the intercept in this model will always be estimated by $\bar{y}$. The (4×4) $\mathbf{X'X}$ matrix for the four coded variables is the correlation matrix

$$\mathbf{X'X} = \begin{bmatrix} 1.00000 & 0.84894 & 0.91412 & 0.93367 \\ 0.84894 & 1.00000 & 0.76899 & 0.97567 \\ 0.91412 & 0.76899 & 1.00000 & 0.86784 \\ 0.93367 & 0.97567 & 0.86784 & 1.00000 \end{bmatrix}.$$

This matrix contains several large correlation coefficients, and this may indicate significant multicollinearity. The inverse of $\mathbf{X'X}$ is

$$(\mathbf{X'X})^{-1} = \begin{bmatrix} 20.769 & 25.813 & -0.608 & -44.042 \\ 25.813 & 74.486 & 12.597 & -107.710 \\ -0.608 & 12.597 & 8.274 & -18.903 \\ -44.042 & -107.710 & -18.903 & 163.620 \end{bmatrix}.$$

The variance inflation factors are the main diagonal elements of this matrix. Note that three of the variance inflation factors exceed 10, a good indication that multicollinearity is present. The eigenvalues of $\mathbf{X'X}$ are $\lambda_1 = 3.657$, $\lambda_2 = 0.2679$, $\lambda_3 = 0.07127$, and $\lambda_4 = 0.004014$. Two of the eigenvalues, λ_3 and λ_4, are relatively close to zero. Also, the ratio of the largest to the smallest eigenvalue is

$$\frac{\lambda_{\max}}{\lambda_{\min}} = \frac{3.657}{0.004014} = 911.06,$$

which is considerably larger than 10. Therefore, since examination of the variance inflation factors and the eigenvalues indicates potential problems with multicollinearity, we will use ridge regression to estimate the model parameters.

Table 15-11 Coded Data for Example 15-14

Observation Number	y	x_1	x_2	x_3	x_4
1	28.25	−0.12515	0.00405	−0.09206	−0.05538
2	24.80	−0.02635	0.08495	−0.09206	0.03692
3	11.86	−0.37217	−0.31957	−0.27617	−0.33226
4	36.60	0.22066	0.22653	0.27617	0.20832
5	15.80	−0.22396	−0.50161	−0.09206	−0.39819
6	16.23	−0.32276	−0.27912	−0.27617	−0.31907
7	29.50	−0.02635	0.10518	0.00000	0.07647
8	28.75	−0.12515	0.06472	−0.09206	0.01055
9	43.20	0.27007	0.18608	0.36823	0.27425
10	38.47	0.51709	0.38834	0.36823	0.40609
11	10.14	0.17126	0.12540	−0.09206	0.15558
12	38.92	0.36887	0.18608	0.46029	0.27425
13	36.70	0.12186	0.28721	0.18411	0.24788
14	15.31	−0.27336	−0.17799	−0.36823	−0.25315
15	8.40	−0.17456	−0.38025	−0.27617	−0.33226

We solved Equation 15-64 for various values of l, and the results are summarized in Table 15-12. The ridge trace is shown in Fig. 15-7. The instability of the least-squares estimates $\beta_j^*(l = 0)$ is evident from inspection of the ridge trace. It is often difficult to choose a value of l from the ridge trace that simultaneously stabilizes the estimates of all regression coefficients. We will choose $l = 0.064$, which implies that the regression model is

$$\hat{y} = 25.53 - 18.0566x_1 + 17.2202x_2 + 36.0743x_3 + 4.7242x_4,$$

using $\hat{\beta}_0 = \bar{y} = 25.53$. Converting the model to the original variables z_j, we have

$$\hat{y} = 2.9913 - 0.8920z_1 + 0.3483z_2 + 3.3209z_3 - 0.0623z_4.$$

Table 15-12 Ridge Regression Estimates for Example 15-14

l	$\beta_1^*(l)$	$\beta_2^*(l)$	$\beta_3^*(l)$	$\beta_4^*(l)$
0.000	−28.3318	65.9996	64.0479	−57.2491
0.001	−31.0360	57.0244	61.9645	−44.0901
0.002	−32.6441	50.9649	60.3899	−35.3088
0.004	−34.1071	43.2358	58.0266	−24.3241
0.008	−34.3195	35.1426	54.7018	−13.3348
0.016	−31.9710	27.9534	50.0949	−4.5489
0.032	−26.3451	22.0347	43.8309	1.2950
0.064	−18.0566	17.2202	36.0743	4.7242
0.128	−9.1786	13.4944	27.9363	6.5914
0.256	−1.9896	10.9160	20.8028	7.5076
0.512	2.4922	9.2014	15.3197	7.7224

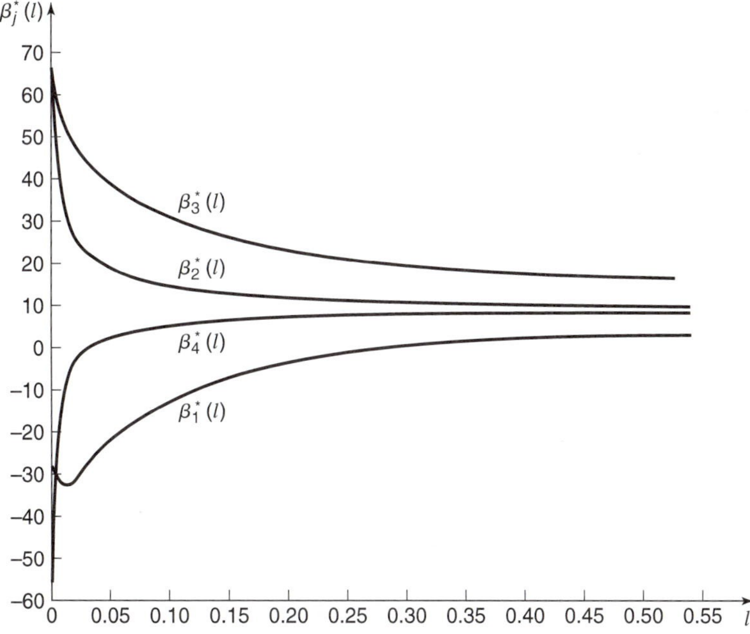

Figure 15-7 Ridge trace for Example 15-14.

15-10.2 Influential Observations in Regression

When using multiple regression we occasionally find that some small subset of the observations is unusually influential. Sometimes these influential observations are relatively far away from the vicinity where the rest of the data were collected. A hypothetical situation for two variables is depicted in Fig. 15-8, where one observation in x-space is remote from the rest of the data. The disposition of points in the x-space is important in determining the properties of the model. For example, the point (x_{i1}, x_{i2}) in Fig. 15-8 may be very influential in determining the estimates of the regression coefficients, the value of R^2, and the value of MS_E.

We would like to examine the data points used to build a regression model to determine if they control many model properties. If these influential points are "bad" points, or are erroneous in any way, then they should be eliminated. On the other hand, there may be nothing wrong with these points, but at least we would like to determine whether or not they produce results consistent with the rest of the data. In any event, even if an influential point is a valid one, if it controls important model properties, we would like to know this, since it could have an impact on the use of the model.

Montgomery, Peck, and Vining (2001) describe several methods for detecting influential observations. An excellent diagnostic is the Cook (1977, 1979) distance measure. This is a measure of the squared distance between the least squares estimate of $\boldsymbol{\beta}$ based on all n observations and the estimate $\hat{\boldsymbol{\beta}}_{(i)}$ based on removal of the ith point. The Cook distance measure is

$$D_i = \frac{\left(\hat{\boldsymbol{\beta}}_{(i)} - \hat{\boldsymbol{\beta}}\right)' \mathbf{X}'\mathbf{X}\left(\hat{\boldsymbol{\beta}}_{(i)} - \hat{\boldsymbol{\beta}}\right)}{pMS_E}, \qquad i = 1, 2, \ldots, n.$$

Clearly if the ith point is influential, its removal will result in $\hat{\boldsymbol{\beta}}_{(i)}$ changing considerably from the value $\hat{\boldsymbol{\beta}}$. Thus a large value of D_i implies that the ith point is influential. The statistic D_i is actually computed using

$$D_i = \frac{f_i^2}{p} \frac{h_{ii}}{(1 - h_{ii})}, \qquad i = 1, 2, \ldots, n, \tag{15-65}$$

where $f_i = e_i \Big/ \sqrt{MS_E(1 - h_{ii})}$ and h_{ii} is the ith diagonal element of the matrix

$$\mathbf{H} = \mathbf{X}(\mathbf{X}'\mathbf{X})^{-1}\mathbf{X}'.$$

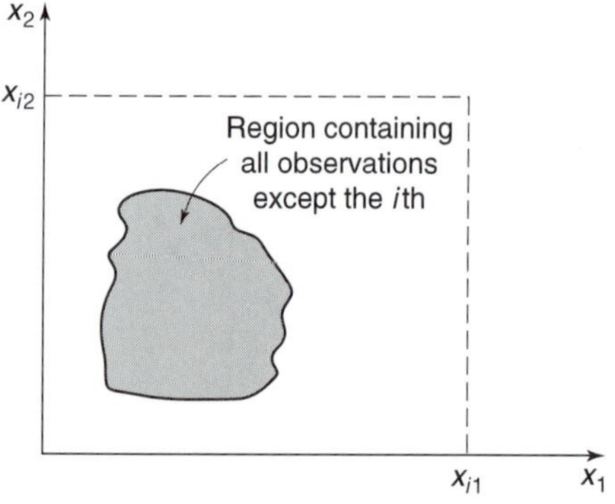

Figure 15-8 A point that is remote in x-space.

The **H** matrix is sometimes called the "hat" matrix, since

$$\begin{aligned}\hat{\mathbf{y}} &= \mathbf{X}\hat{\boldsymbol{\beta}} \\ &= \mathbf{X}(\mathbf{X}'\mathbf{X})^{-1}\mathbf{X}'\mathbf{y} \\ &= \mathbf{H}\mathbf{y}.\end{aligned}$$

Thus **H** is a projection matrix that transforms the observed values of **y** into a set of fitted values $\hat{\mathbf{y}}$.

From equation 15-65 we note that D_i is made up of a component that reflects how well the model fits the ith observation $y_i[e_i/\sqrt{MS_E(1-h_{ii})}]$ is called a *Studentized* residual, and it is a method of scaling residuals so that they have unit variance] and a component that measures how far that point is from the rest of the data [$h_{ii}/(1 - h_{ii})$ is the distance of the ith point from the centroid of the remaining $n - 1$ points]. A value of $D_i > 1$ would indicate that the point is influential. Either component of D_i (or both) may contribute to a large value.

Example 15-15

Table 15-13 lists the values of D_i for the damaged peaches data in Example 15-1. To illustrate the calculations, consider the first observation:

$$D_i = \frac{f_i^2}{p} \cdot \frac{h_{ii}}{(1-h_{ii})}$$

Table 15-13 Influence Diagnostics for the Damaged Peaches Data in Example 15-15

Observation (i)	h_{ii}	Cook's Distance Measure (D_i)
1	0.249	0.277
2	0.126	0.000
3	0.088	0.156
4	0.104	0.061
5	0.060	0.000
6	0.299	0.000
7	0.081	0.008
8	0.055	0.002
9	0.193	0.116
10	0.117	0.024
11	0.250	0.037
12	0.109	0.002
13	0.213	0.045
14	0.055	0.056
15	0.147	0.067
16	0.080	0.001
17	0.209	0.001
18	0.276	0.160
19	0.210	0.171
20	0.078	0.012

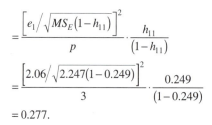

$$= \frac{\left[e_1 / \sqrt{MS_E(1-h_{11})}\right]^2}{p} \cdot \frac{h_{11}}{(1-h_{11})}$$

$$= \frac{\left[2.06 / \sqrt{2.247(1-0.249)}\right]^2}{3} \cdot \frac{0.249}{(1-0.249)}$$

$$= 0.277.$$

The values in Table 15-13 were calculated using Minitab®. The Cook distance measure D_i does not identify any potentially influential observations in the data, as no value of D_i exceeds unity.

15-10.3 Autocorrelation

The regression models developed thus far have assumed that the model error components ε_i are uncorrelated random variables. Many applications of regression analysis involve data for which this assumption may be inappropriate. In regression problems where the dependent and independent variables are time oriented or are time-series data, the assumption of uncorrelated errors is often untenable. For example, suppose we regressed the quarterly sales of a product against the quarterly point-of-sale advertising expenditures. Both variables are time series, and if they are positively correlated with other factors such as disposable income and population size, which are not included in the model, then it is likely that the error terms in the regression model are positively correlated over time. Variables that exhibit correlation over time are referred to as *autocorrelated* variables. Many regression problems in economics, business, and agriculture involve autocorrelated errors.

The occurrence of positively autocorrelated errors has several potentially serious consequences. The ordinary least-squares estimators of the parameters are affected in that they are no longer minimum variance estimators, although they are still unbiased. Furthermore, the mean square error MS_E may underestimate the error variance σ^2. Also, confidence intervals and tests of hypotheses, which are developed assuming uncorrelated errors, are not valid if autocorrelation is present.

There are several statistical procedures that can be used to determine whether the error terms in the model are uncorrelated. We will describe one of these, the Durbin-Watson test. This test assumes that the data are generated by the *first-order autoregressive model*

$$y_t = \beta_0 + \beta_1 x_t + \varepsilon_t, \qquad t = 1, 2, \ldots, n, \tag{15-66}$$

where t is the index of time and the error terms are generated according to the process

$$\varepsilon_t = \rho\varepsilon_{t-1} + a_t, \tag{15-67}$$

where $|\rho| < 1$ is an unknown parameter and a_t is a NID(0, σ^2) random variable. Equation 15-66 is a simple linear regression model, except for the errors, which are generated from equation 15-67. The parameter ρ in equation 15-67 is the autocorrelation coefficient. The Durbin–Watson test can be applied to the hypotheses

$$\begin{aligned} H_0&: \rho = 0, \\ H_1&: \rho > 0. \end{aligned} \tag{15-68}$$

Note that if H_0: $\rho = 0$ is not rejected, we are implying that there is no autocorrelation in the errors, and the ordinary linear regression model is appropriate.

To test H_0: $\rho = 0$, first fit the regression model by ordinary least squares. Then, calculate the Durbin–Watson test statistic

$$D = \frac{\sum_{t=2}^{n} (e_t - e_{t-1})^2}{\sum_{t=1}^{n} e_t^2}, \tag{15-69}$$

where e_t is the tth residual. For a suitable value of α, obtain the critical values $D_{\alpha, U}$ and $D_{\alpha, L}$ from Table 15-14. If $D > D_{\alpha, U}$, do not reject H_0: $\rho = 0$; but if $D < D_{\alpha, L}$, reject H_0: $\rho = 0$ and conclude that the errors are positively autocorrelated. If $D_{\alpha, L} \le D \le D_{\alpha, U}$, the test is

Table 15-14 Critical Values of the Durbin–Watson Statistic

Sample Size	Probability in Lower Tail (Significance Level = α)	k = Number of Regressors (Excluding the Intercept)									
		1		2		3		4		5	
		D_L	D_U	D_L	D_U	D_L	D_U	D_L	D_U	D_L	D_U
	0.01	0.81	1.07	0.70	1.25	0.59	1.46	0.49	1.70	0.39	1.96
15	0.025	0.95	1.23	0.83	1.40	0.71	1.61	0.59	1.84	0.48	2.09
	0.05	1.08	1.36	0.95	1.54	0.82	1.75	0.69	1.97	0.56	2.21
	0.01	0.95	1.15	0.86	1.27	0.77	1.41	0.63	1.57	0.60	1.74
20	0.025	1.08	1.28	0.99	1.41	0.89	1.55	0.79	1.70	0.70	1.87
	0.05	1.20	1.41	1.10	1.54	1.00	1.68	0.90	1.83	0.79	1.99
	0.01	1.05	1.21	0.98	1.30	0.90	1.41	0.83	1.52	0.75	1.65
25	0.025	1.13	1.34	1.10	1.43	1.02	1.54	0.94	1.65	0.86	1.77
	0.05	1.20	1.45	1.21	1.55	1.12	1.66	1.04	1.77	0.95	1.89
	0.01	1.13	1.26	1.07	1.34	1.01	1.42	0.94	1.51	0.88	1.61
30	0.025	1.25	1.38	1.18	1.46	1.12	1.54	1.05	1.63	0.98	1.73
	0.05	1.35	1.49	1.28	1.57	1.21	1.65	1.14	1.74	1.07	1.83
	0.01	1.25	1.34	1.20	1.40	1.15	1.46	1.10	1.52	1.05	1.58
40	0.025	1.35	1.45	1.30	1.51	1.25	1.57	1.20	1.63	1.15	1.69
	0.05	1.44	1.54	1.39	1.60	1.34	1.66	1.29	1.72	1.23	1.79
	0.01	1.32	1.40	1.28	1.45	1.24	1.49	1.20	1.54	1.16	1.59
50	0.025	1.42	1.50	1.38	1.54	1.34	1.59	1.30	1.64	1.26	1.69
	0.05	1.50	1.59	1.46	1.63	1.42	1.67	1.38	1.72	1.34	1.77
	0.01	1.38	1.45	1.35	1.48	1.32	1.52	1.28	1.56	1.25	1.60
60	0.025	1.47	1.54	1.44	1.57	1.40	1.61	1.37	1.65	1.33	1.69
	0.05	1.55	1.62	1.51	1.65	1.48	1.69	1.44	1.73	1.41	1.77
	0.01	1.47	1.52	1.44	1.54	1.42	1.57	1.39	1.60	1.36	1.62
80	0.025	1.54	1.59	1.52	1.62	1.49	1.65	1.47	1.67	1.44	1.70
	0.05	1.61	1.66	1.59	1.69	1.56	1.72	1.53	1.74	1.51	1.77
	0.01	1.52	1.56	1.50	1.58	1.48	1.60	1.45	1.63	1.44	1.65
100	0.025	1.59	1.63	1.57	1.65	1.55	1.67	1.53	1.70	1.51	1.72
	0.05	1.65	1.69	1.63	1.72	1.61	1.74	1.59	1.76	1.57	1.78

Source: Adapted from *Econometrics*, by R. J. Wonnacott and T. H. Wonnacott, John Wiley & Sons, New York, 1970, with permission of the publisher.

inconclusive. When the test is inconclusive, the implication is that more data must be collected. In many problems this is difficult to do.

To test for *negative* autocorrelation, that is, if the alternative hypothesis in equation 15-68 is H_1: $\rho < 0$, then use $D' = 4 - D$ as the test statistic, where D is defined in equation 15-69. If a two-sided alternative is specified, then use both of the one-sided procedures, noting that the type I error for the two-sided test is 2α, where α is the type I error for the one-sided tests.

The only effective remedial measure when autocorrelation is present is to build a model that accounts explicitly for the autocorrelative structure of the errors. For an introductory treatment of these methods, refer to Montgomery, Peck, and Vining (2001).

15-11 SELECTION OF VARIABLES IN MULTIPLE REGRESSION

15-11.1 The Model-Building Problem

An important problem in many applications of regression analysis is the selection of the set of independent or regressor variables to be used in the model. Sometimes previous experience or underlying theoretical considerations can help the analyst specify the set of independent variables. Usually, however, the problem consists of selecting an appropriate set of regressors from a set that quite likely includes all the important variables, but we are sure that not *all* these candidate variables are necessary to adequately model the response y.

In such a situation, we are interested in screening the candidate variables to obtain a regression model that contains the "best" subset of regressor variables. We would like the final model to contain enough regressor variables so that in the intended use of the model (prediction, for example) it will perform satisfactorily. On the other hand, to keep model maintenance costs to a minimum, we would like the model to use as few regressor variables as possible. The compromise between these conflicting objectives is often called finding the "best" regression equation. However, in most problems, there is no single regression model that is "best" in terms of the various evaluation criteria that have been proposed. A great deal of judgment and experience with the system being modeled is usually necessary to select an appropriate set of independent variables for a regression equation.

No algorithm will always produce a good solution to the variable selection problem. Most currently available procedures are search techniques. To perform satisfactorily, they require interaction with and judgment by the analyst. We now briefly discuss some of the more popular variable selection techniques.

15-11.2 Computational Procedures for Variable Selection

We assume that there are k candidate variables, $x_1, x_2, \ldots, x_k$, and a single dependent variable y. All models will include an intercept term β_0, so that the model with *all* variables included would have $k + 1$ terms. Furthermore, the functional form of each candidate variable (for example, $x_1 = 1/x_1$, $x_2 = \ln x_2$, etc.) is correct.

All Possible Regressions This approach requires that the analyst fit all the regression equations involving one candidate variable, all regression equations involving two candidate variables, and so on. Then these equations are evaluated according to some suitable criteria to select the "best" regression model. If there are k candidate variables, there are 2^k total equations to be examined. For example, if $k = 4$, there are $2^4 = 16$ possible regression equations, while if $k = 10$, there are $2^{10} = 1024$ possible regression equations. Hence, the number of equations to be examined increases rapidly as the number of candidate variables increases.

There are a number of criteria that may be used for evaluating and comparing the different regression models obtained. Perhaps the most commonly used criterion is based on the coefficient of multiple determination. Let R_p^2 denote the coefficient of determination for a regression model with p terms, that is, $p - 1$ candidate variables and an intercept term (note that $p \leq k + 1$). Computationally, we have

$$R_p^2 = \frac{SS_R(p)}{S_{yy}} = 1 - \frac{SS_E(p)}{S_{yy}}, \tag{15-70}$$

where $SS_R(p)$ and $SS_E(p)$ denote the regression sum of squares and the error sum of squares, respectively, for the p-variable equation. Now R_p^2 increases as p increases and is a maximum when $p = k + 1$. Therefore, the analyst uses this criterion by adding variables to the model up to the point where an additional variable is not useful in that it gives only a small increase in R_p^2. The general approach is illustrated in Fig. 15-9, which gives a hypothetical plot of R_p^2 against p. Typically, one examines a display such as this and chooses the number of variables in the model as the point at which the "knee" in the curve becomes apparent. Clearly, this requires judgment on the part of the analyst.

A second criterion is to consider the mean square error for the p-variable equation, say $MS_E(p) = SS_E(p)/(n - p)$. Generally, $MS_E(p)$ decreases as p increases, but this is not necessarily so. If the addition of a variable to the model with $p - 1$ terms does not reduce the error sum of squares in the new p term model by an amount equal to the error mean square in the old $p - 1$ term model, $MS_E(p)$ will *increase*, because of the loss of one degree of freedom for error. Therefore, a logical criterion is to select p as the value that minimizes $MS_E(p)$; or since $MS_E(p)$ is usually relatively flat in the vicinity of the minimum, we could choose p such that adding more variables to the model produces only very small reductions in $MS_E(p)$. The general procedure is illustrated in Fig. 15-10.

A third criterion is the C_p statistic, which is a measure of the total mean square error for the regression model. We define the total standardized mean square error as

$$\begin{aligned}\Gamma_p &= \frac{1}{\sigma^2}\sum_{i=1}^{n} E\left[\hat{y}_i - E(y_i)\right]^2 \\ &= \frac{1}{\sigma^2}\left[\sum_{i=1}^{n}\left\{E(y_i) - E(\hat{y}_i)\right\}^2 + \sum_{i=1}^{n} V(\hat{y}_i)\right] \\ &= \frac{1}{\sigma^2}\left[(\text{bias})^2 + \text{variance}\right].\end{aligned}$$

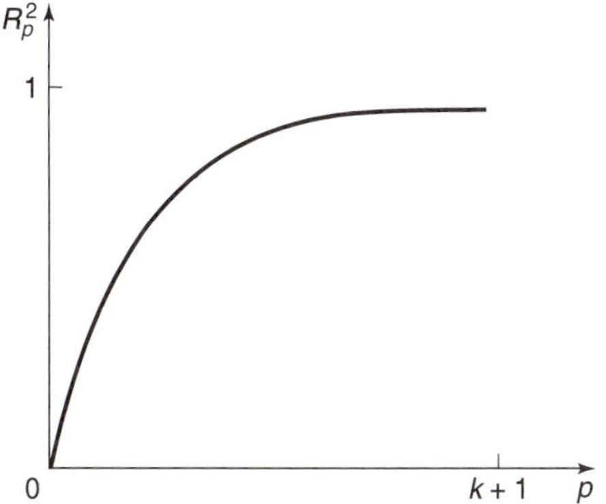

Figure 15-9 Plot of R_p^2 against p.

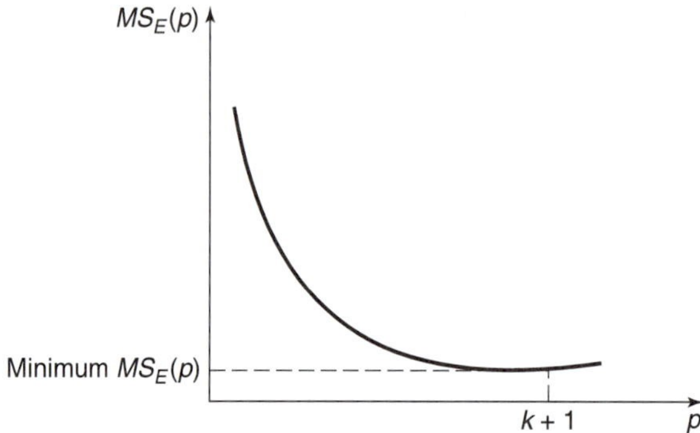

Figure 15-10 Plot of $MS_E(p)$ against p.

We use the mean square error from the *full* $k + 1$ term model as an estimate of σ^2; that is, $\hat{\sigma}^2 = MS_E(k + 1)$. An estimator of Γ_p is

$$C_p = \frac{SS_E(p)}{\hat{\sigma}^2} - n + 2p. \tag{15-71}$$

If the p-term model has negligible bias, then it can be shown that

$$E(C_p|\text{zero bias}) = p.$$

Therefore, the values of C_p for each regression model under consideration should be plotted against p. The regression equations that have negligible bias will have values of C_p that fall near the line $C_p = p$, while those with significant bias will have values of C_p that plot above this line. One then chooses as the "best" regression equation either a model with minimum C_p or a model with a slightly larger C_p that contains less bias than the minimum.

Another criterion is based on a modification of R_p^2 that accounts for the number of variables in the model. We presented this statistic in Section 15-6.1, the adjusted R^2 for the model fit in Example 15-1. This statistic is called the *adjusted* R_p^2 defined as

$$R_{adj}^2(p) = 1 - \frac{n-1}{n-p}\left(1 - R_p^2\right). \tag{15-72}$$

Note that $R_{adj}^2(p)$ may decrease as p increases if the decrease in $(n - 1)(1 - R_p^2)$ is not compensated for by the loss of one degree of freedom in $n - p$. The experimenter would usually select the regression model that has the maximum value of $R_{adj}^2(p)$. However, note that this is equivalent to the model that minimizes $MS_E(p)$, since

$$\begin{aligned}
R_{adj}^2(p) &= 1 - \left(\frac{n-1}{n-p}\right)\left(1 - R_p^2\right) \\
&= 1 - \left(\frac{n-1}{n-p}\right)\frac{SS_E(p)}{S_{yy}} \\
&= 1 - \left(\frac{n-1}{S_{yy}}\right)MS_E(p).
\end{aligned}$$

Example 15-16

The data in Table 15-15 are an expanded set of data for the damaged peach data in Example 15-1. There are now five candidate variables, drop height (x_1), fruit density (x_2), fruit height at impact point (x_3), fruit pulp thickness (x_4), and potential energy of the fruit before the impact (x_5).

Table 15-16 presents the results of running all possible regressions (except the trivial model with only an intercept) on these data. The values of R_p^2, $R_{adj}^2(p)$, $MS_E(p)$, and C_p are given in the table. A plot of the maximum R_p^2 for each subset of size p is shown in Fig. 15-11. Based on this plot there does not appear to be much gain in adding the fifth variable. The value of R_p^2 does not seem to increase significantly with the addition of x_5 over the four-variable model with the highest R_p^2 value. A plot of the minimum $MS_E(p)$ for each subset of size p is shown in Fig. 15-12. The best two-variable model is either (x_1, x_2) or (x_2, x_3); the best three-variable model is (x_1, x_2, x_3); the best four-variable model is either (x_1, x_2, x_3, x_4) or (x_1, x_2, x_3, x_5). There are several models with relatively small values of $MS_E(p)$, but either the three-variable model (x_1, x_2, x_3) or the four-variable model (x_1, x_2, x_3, x_4) would be superior to the other models based on the $MS_E(p)$ criterion. Further investigation will be necessary.

A C_p plot is shown in Fig. 15-13. Only the five-variable model has a $C_p \le p$ (specifically C_p = 6.0), but the C_p value for the four-variable model (x_1, x_2, x_3, x_4) is $C_p = 6.1732$. There appears to be insuffcient gain in the C_p value to justify including x_5. To illustrate the calculations, for this equation [for the model including (x_1, x_2, x_3, x_4)] we would find

$$\begin{aligned} C_p &= \frac{SS_E(p)}{\hat{\sigma}^2} - n + 2p \\ &= \frac{18.29715}{1.13132} - 20 + 2(5) = 6.1732, \end{aligned}$$

Table 15-15 Damaged Peach Data for Example 15-16

Observation	Delivery Time, y	Drop Height, x_1	Fruit Density, x_2	Fruit Height, x_3	Fruit Pulp Thickness, x_4	Potential Energy, x_5
1	3.62	303.7	0.90	26.1	22.3	184.5
2	7.27	366.7	1.04	18.0	21.5	185.2
3	2.66	336.8	1.01	39.0	22.9	128.4
4	1.53	304.5	0.95	48.5	20.4	173.0
5	4.91	346.8	0.98	43.1	18.7	139.6
6	10.36	600.0	1.04	21.0	17.0	146.5
7	5.26	369.0	0.96	12.7	20.4	155.5
8	6.09	418.0	1.00	46.0	18.1	129.2
9	6.57	269.0	1.01	2.6	21.5	154.6
10	4.24	323.0	0.94	6.9	24.4	152.8
11	8.04	562.2	1.01	27.3	19.5	199.6
12	3.46	284.2	0.97	30.6	20.2	177.5
13	8.50	558.6	1.03	37.7	22.6	210.0
14	9.34	415.0	1.01	26.1	17.1	165.1
15	5.55	349.5	1.04	48.0	21.5	195.3
16	8.11	462.8	1.02	32.8	23.7	171.0
17	7.32	333.1	1.05	29.2	21.9	163.9
18	12.58	502.1	1.10	4.0	16.9	140.8
19	0.15	311.4	0.91	39.2	26.0	154.1
20	5.23	351.4	0.96	36.3	23.5	194.6

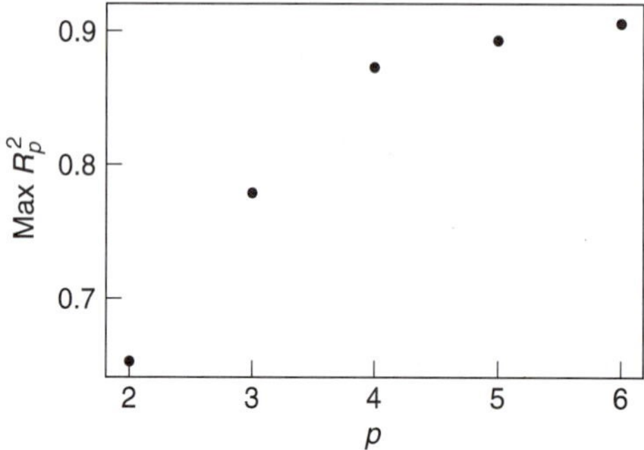

Figure 15-11 The maximum R_p^2 plot for Example 15-16.

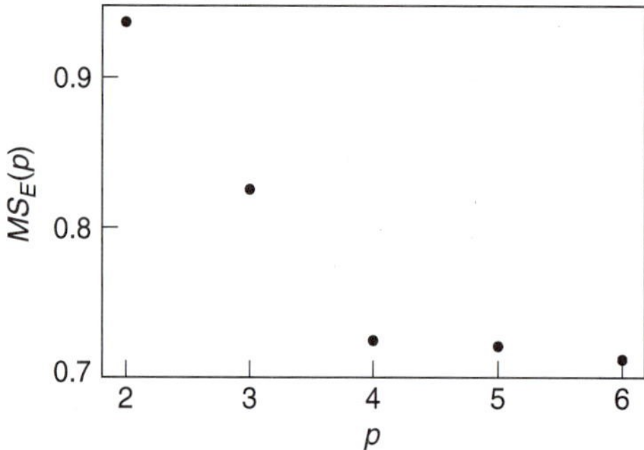

Figure 15-12 The $MS_E(p)$ plot for Example 15-16.

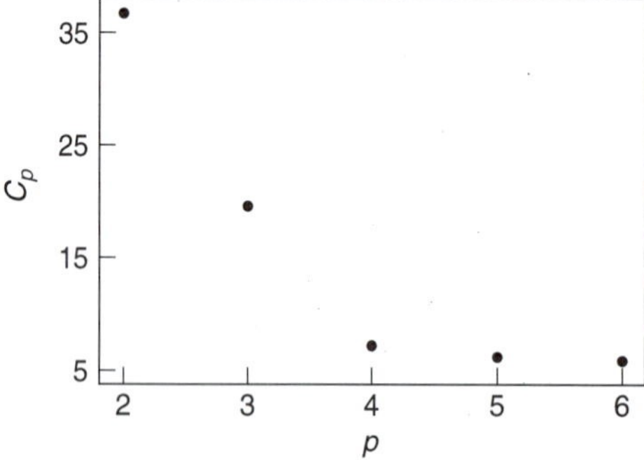

Figure 15-13 The C_p plot for Example 15-16.

noting that $\hat{\sigma}^2 = 1.13132$ is obtained from the full equation $(x_1, x_2, x_3, x_4, x_5)$. Since all other models (with the exclusion of the five-variable model) contain substantial bias, we would conclude on the basis of the C_p criterion that the best subset of the regressor variables is (x_1, x_2, x_3, x_4). Since this model also results in relatively small $MS_E(p)$ and a relatively high R_p^2, we would select it as the "best" regression equation. The final model is

$$\hat{y} = -19.9 + 0.0123x_1 + 27.3x_2 - 0.0655x_3 - 0.196x_4.$$

Keep in mind, though, that further analysis should be conducted on this model as well as other possible candidate models. With additional investigation, it is possible to discover an even better-fitting model. We will discuss this in more detail later in this chapter.

The all-possible-regressions approach requires considerable computational effort, even when k is moderately small. However, if the analyst is willing to look at something less than the estimated model and all its associated statistics, it possible to devise algorithms for all possible regressions that produce less information about each model but which are more efficient computationally. For example, suppose that we could efficiently calculate only the MS_E for each model. Since models with large MS_E are not likely to be selected as the best regression equations, we would then have only to examine in detail the models with small values of MS_E. There are several approaches to developing a computationally efficient algorithm for all possible regressions (for example, see Furnival and Wilson, 1974). Both Minitab® and SAS computer packages provide the Furnival and Wilson (1974) algorithm as an option. The SAS output is provided in Table 15-16.

Stepwise Regression This is probably the most widely used variable selection technique. The procedure iteratively constructs a sequence of regression models by adding or

Table 15-16 All Possible Regressions for the Data in Example 15-16

Number in Model p–1	R_p^2	$R_{adj}^2(p)$	C_p	$MS_E(p)$	Variables in Model
1	0.6530	0.6337	37.7047	3.37540	x_2
1	0.5495	0.5245	53.7211	4.38205	x_1
1	0.3553	0.3194	83.7824	6.27144	x_4
1	0.1980	0.1535	108.1144	7.80074	x_3
1	0.0021	–0.0534	138.4424	9.70689	x_5
2	0.7805	0.7547	19.9697	2.26063	x_1, x_2
2	0.7393	0.7086	26.3536	2.68546	x_2, x_3
2	0.7086	0.6744	31.0914	3.00076	x_2, x_4
2	0.7030	0.6681	31.9641	3.05883	x_1, x_3
2	0.6601	0.6201	38.6031	3.50064	x_2, x_5
2	0.6412	0.5990	41.5311	3.69550	x_1, x_4
2	0.5528	0.5002	55.2140	4.60607	x_1, x_5
2	0.4940	0.4345	64.3102	5.21141	x_3, x_4
2	0.4020	0.3316	78.5531	6.15925	x_4, x_5
2	0.2125	0.1199	107.8731	8.11045	x_3, x_5
3	0.8756	0.8523	7.2532	1.36135	x_1, x_2, x_3
3	0.8049	0.7683	18.1949	2.13501	x_1, x_2, x_4
3	0.7898	0.7503	20.5368	2.30060	x_2, x_3, x_4
3	0.7807	0.7396	21.9371	2.39961	x_1, x_2, x_5
3	0.7721	0.7294	23.2681	2.49372	x_1, x_3, x_4
3	0.7568	0.7112	25.6336	2.66098	x_2, x_3, x_5
3	0.7337	0.6837	29.2199	2.91456	$x_2, x_4\ x_5$
3	0.7032	0.6475	33.9410	3.24838	x_1, x_3, x_5
3	0.6448	0.5782	42.9705	3.88683	x_1, x_4, x_5
3	0.5666	0.4853	55.0797	4.74304	x_3, x_4, x_5
4	0.8955	0.8676	6.1732	1.21981	x_1, x_2, x_3, x_4
4	0.8795	0.8474	8.6459	1.40630	x_1, x_2, x_3, x_5
4	0.8316	0.7866	16.0687	1.96614	x_2, x_3, x_4, x_5
4	0.8103	0.7597	19.3611	2.21446	x_1, x_2, x_4, x_5
4	0.7854	0.7282	23.2090	2.50467	x_1, x_3, x_4, x_5
5	0.9095	0.8772	6.0000	1.13132	x_1, x_2, x_3, x_4, x_5

removing variables at each step. The criterion for adding or removing a variable at any step is usually expressed in terms of a partial F-test. Let F_{in} be the value of the F statistic for adding a variable to the model, and let F_{out} be the value of the F statistic for removing a variable from the model. We must have $F_{\text{in}} \geq F_{\text{out}}$, and usually $F_{\text{in}} = F_{\text{out}}$.

Stepwise regression begins by forming a one-variable model using the regressor variable that has the highest correlation with the response variable y. This will also be the variable producing the largest F statistic. If no F statistic exceeds F_{in}, the procedure terminates. For example, suppose that at this step x_1 is selected. At the second step the remaining $k - 1$ candidate variables are examined, and the variable for which the statistic

$$F_j = \frac{SS_R(\beta_j|\beta_1, \beta_0)}{MS_E(x_j, x_1)} \tag{15-73}$$

is a maximum is added to the equation, provided that $F_j > F_{\text{in}}$. In equation 15-73, $MS_E(x_j, x_1)$ denotes the mean square for error for the model containing both x_1 and x_j. Suppose that this procedure now indicates that x_2 should be added to the model. Now the stepwise regression algorithm determines whether the variable x_1 added at the first step should be removed. This is done by calculating the F statistic

$$F_1 = \frac{SS_R(\beta_1|\beta_2, \beta_0)}{MS_E(x_1, x_2)}. \tag{15-74}$$

If $F_1 < F_{\text{out}}$, the variable x_1 is removed.

In general, at each step the set of remaining candidate variables is examined and the variable with the largest partial F statistic is entered, provided that the observed value of F exceeds F_{in}. Then the partial F statistic for each variable in the model is calculated, and the variable with the smallest observed value of F is deleted if the observed $F < F_{\text{out}}$. The procedure continues until no other variables can be added to or removed from the model.

Stepwise regression is usually performed using a computer program. The analyst exercises control over the procedure by the choice of F_{in} and F_{out}. Some stepwise regression computer programs require that numerical values be specified for F_{in} and F_{out}. Since the number of degrees of freedom on MS_E depends on the number of variables in the model, which changes from step to step, a fixed value of F_{in} and F_{out} causes the type I and type II error rates to vary. Some computer programs allow the analyst to specify the type I error levels for F_{in} and F_{out}. However, the "advertised" significance level is not the true level, because the variable selected is the one that maximizes the partial F statistic at that stage. Sometimes it is useful to experiment with different values of F_{in} and F_{out} (or different advertised type I error rates) in several runs to see if this substantially affects the choice of the final model.

Example 15-17

We will apply stepwise regression to the damaged peaches data in Table 15-15. Minitab® output is provided in Fig. 15-14. From this figure, we see that variables x_1, x_2, and x_3 are significant; this is because the last column contains entries for only x_1, x_2, and x_3. Figure 15-15 provides the SAS computer output that will support the computations to be calculated next. Instead of specifying numerical values of F_{in} and F_{out}, we use an *advertised* type I error of $\alpha = 0.10$. The first step consists of building a simple linear regression model using the variable that gives the largest F statistic. This is x_2, and since

$$F_2 = \frac{SS_R(\beta_2|\beta_0)}{MS_E(x_2)} = \frac{114.32885}{3.37540} = 33.87 > F_{\text{in}} = F_{0.10,1,18} = 3.01,$$

```
Alpha-to-Enter:      0.1          Alpha-to-Remove        0.1

Response is y on 5 predictors, with N = 20

       Step            1             2             3
   Constant       -42.87        -33.83        -27.89

x2                  49.1          34.9          30.7
T-Value             5.82          4.23          4.71
P-Value            0.000         0.001         0.000

x1                              0.0131        0.0136
T-Value                           3.14          4.19
P-Value                          0.006         0.001

x3                                            -0.067
T-Value                                        -3.50
P-Value                                        0.003

S                   1.84          1.50          1.17
R-Sq               65.30         78.05         87.56
R-Sq (adj)         63.37         75.47         85.23
C-p                 37.7          20.0           7.3
```

Figure 15-14 Minitab® output for stepwise regression in Example 15-17.

x_2 is entered into the model.

The second step begins by finding the variable x_j that has the largest partial F statistic, given that x_2 is in the model. This is x_1, and since

$$F_1 = \frac{SS_R(\beta_1|\beta_2,\beta_0)}{MS_E(x_2,x_1)} = \frac{46.45691}{2.26063} = 20.55 > F_{\text{in}} = F_{0.10,1,17} = 3.03,$$

x_1 is added to the model. Now the procedure evaluates whether or not x_2 should be retained, given that x_1 is in the model. This involves calculating

$$F_2 = \frac{SS_R(\beta_2|\beta_1,\beta_0)}{MS_E(x_2,x_1)} = \frac{40.44627}{2.26063} = 17.89 > F_{\text{in}} = F_{0.10,1,17} = 3.03.$$

Therefore x_2 should be retained. Step 2 terminates with both x_1 and x_2 in the model.

The third step finds the next variable for entry as x_3. Since

$$F_3 = \frac{SS_R(\beta_3|\beta_2,\beta_1,\beta_0)}{MS_E(x_3,x_2,x_1)} = \frac{16.64910}{1.36135} = 12.23 > F_{\text{in}} = F_{0.10,1,16} = 3.05,$$

x_3 is added to the model. Partial F-tests on x_2 (given x_1 and x_3) and x_1 (given x_2 and x_3) indicate that these variables should be retained. Therefore, the third step concludes with the variables x_1, x_2, and x_3 in the model.

At the fourth step, neither of the remaining terms, x_4 or x_5, is significant enough to be included in the model. Therefore, the stepwise procedure terminates.

The stepwise regression procedure would conclude that the best model includes x_1, x_2, and x_3. The usual checks of model adequacy, such as residual analysis and C_p plots, should be applied to the equation. These results are similar to those found by all possible regressions, with the exception that x_4 was also considered a possible significant variable with all possible regressions.

Forward Selection This variable selection procedure is based on the principle that variables should be added to the model one at a time until no remaining candidate variables produce a significant increase in the regression sum of squares. That is, variables are added one at a time as long as $F > F_{in}$. Forward selection is a simplification of stepwise regression that omits the partial F-test for deleting variables from the model that have been added at previous steps. This is a potential weakness of forward selection; the procedure does not explore the effect that adding a variable at the current step has on variables added at earlier steps.

```
                           The REG Procedure
                         Dependent Variable: y
Forward Selection: Step 1 Variable x2 Entered: R-Square = 0.6530 and C(p) =
37.7047
Source              DF    Sum of Squares    Mean Square     F Value    Pr > F
Model                1         114.32885      114.32885       33.87    <.0001
Error               18          60.75725        3.37540
Corrected Total     19         175.08609
Variable   Parameter Estimate   Standard Error   Type II SS    F Value    Pr > F
Intercept        -42.87237             8.41429     87.62858      25.96    <.0001
x2                49.08366             8.43377    114.32885      33.87    <.0001
                      Bounds on condition number: 1, 1
-------------------------------------------------------------------------------
Forward Selection: Step 2 Variable x1 Entered: R-Square = 0.7805 and C(p) =
19.9697
Source              DF    Sum of Squares    Mean Square     F Value    Pr > F
Model                2         136.65541       68.32771       30.23    <.0001
Error               17          38.43068        2.26063
Corrected Total     19          175.0860
Variable   Parameter Estimate   Standard Error   Type II SS    F Value    Pr > F
Intercept        -33.83110             7.46286     46.45691      20.55    0.0003
x1                 0.01314             0.00418     22.32656       9.88    0.0059
x2                34.88963             8.24844     40.44627      17.89    0.0006
                 Bounds on condition number: 1.4282, 5.7129
-------------------------------------------------------------------------------
Forward Selection: Step 3 Variable x3 Entered: R-Square = 0.8756 and C(p) =
7.2532
Source              DF    Sum of Squares    Mean Square     F Value    Pr > F
Model                3         153.30451       51.10150       37.54    <.0001
Error               16          21.78159        1.36135
Corrected Total     19         175.08609
Variable   Parameter Estimate   Standard Error   Type II SS    F Value    Pr > F
Intercept        -27.89190             6.03518     29.07675      21.36    0.0003
x1                 0.01360             0.00325     23.87130      17.54    0.0007
x2                30.68486             6.51286     30.21859      22.40    0.0002
x3                -0.06701             0.01916     16.64910      12.23    0.0030
                 Bounds on condition number: 1.4786, 11.85
-------------------------------------------------------------------------------
No other variable met the 0.1000 significance level for entry into the model.
                       Summary of Forward Selection
       Variable    Number      Partial     Model
Step   Entered     Vars In     R-Square    R-Square    C(p)     F Value    Pr > F
  1      x2           1         0.6530      0.6530   37.7047     33.87     <.0001
  2      x1           2         0.1275      0.7805   19.9697      9.88     0.0059
  3      x3           3         0.0951      0.8756    7.2532     12.23     0.0030
```

Figure 15-15 SAS output for stepwise regression in Example 15-17.

Example 15-18

Application of the forward selection algorithm to the damaged peach data in Table 15-15 would begin by adding x_2 to the model. Then the variable that induces the largest partial F-test, given that x_2 is in the model, is added—this is variable x_1. The third step enters x_3, which produces the largest partial F statistic given that x_1 and x_2 are in the model. Since the partial F statistics for x_4 and x_5 are not significant, the procedure terminates. The SAS output for forward selection is given in Fig. 15-16. Note that forward selection leads to the same final model as stepwise regression. This is not always the case.

Backward Elimination This algorithm begins with all k candidate variables in the model. Then the variable with the smallest partial F statistic is deleted if this F statistic is insignificant, that is, if $F < F_{out}$. Next, the model with $k - 1$ variables is estimated, and the next variable for potential elimination is found. The algorithm terminates when no further variables can be deleted.

Example 15-19

To apply backward elimination to the data in Table 15-15, we begin by estimating the full model in all five variables. This model is $\hat{y} = -20.89732 + 0.01102x_1 + 27.37046x_2 - 0.06929x_3 - 0.25695x_4 + 0.01668x_5$.

The SAS computer output is given in Fig. 15-17. The partial F-tests for each variable are as follows:

$$F_1 = \frac{SS_R(\beta_1|\beta_2,\beta_3,\beta_4,\beta_5,\beta_0)}{MS_E} = \frac{13.65360}{1.13132} = 12.07,$$

$$F_2 = \frac{SS_R(\beta_2|\beta_1,\beta_3,\beta_4,\beta_5,\beta_0)}{MS_E} = \frac{21.73153}{1.13132} = 19.21,$$

$$F_3 = \frac{SS_R(\beta_3|\beta_1,\beta_2,\beta_4,\beta_5,\beta_0)}{MS_E} = \frac{17.37834}{1.13132} = 15.36,$$

$$F_4 = \frac{SS_R(\beta_4|\beta_1,\beta_2,\beta_3,\beta_5,\beta_0)}{MS_E} = \frac{5.25602}{1.13132} = 4.65,$$

$$F_5 = \frac{SS_R(\beta_5|\beta_1,\beta_2,\beta_3,\beta_4,\beta_0)}{MS_E} = \frac{2.45862}{1.13132} = 2.17.$$

The variable x_5 has the smallest F statistic, $F_5 = 2.17 < F_{out} = F_{0.10,1,14} = 3.10$; therefore, x_5 is removed from the model at step 1. The model is now fit with only the four remaining variables. In step 2, the F statistic for x_4 ($F_4 = 2.86$) is less than $F_{out} = F_{0.10,1,15} = 3.07$, therefore, x_4 is removed from the model. No remaining variables have F statistics less than the appropriate F_{out} values, and the procedure is terminated. The three-variable model (x_1, x_2, x_3) has all variables significant according to the partial F-test criterion. Note that backward elimination has resulted in the same model that was found by forward selection and stepwise regression. This may not always happen.

Some Comments on Final Model Selection We have illustrated several different approaches to the selection of variables in multiple linear regression. The final model obtained from any model-building procedure should be subjected to the usual adequacy checks, such as residual analysis and examination of the effects of outermost points. The analyst may also consider augmenting the original set of candidate variables with cross products, polynomial terms, or other transformations of the original variables that might improve the model.

```
                              The REG Procedure
                            Dependent Variable: y
Backward Elimination: Step 0 All Variables Entered: R-Square = 0.9095 and
C(p) = 6.0000
                              Sum of          Mean
Source              DF       Squares        Square     F Value       Pr > F
Model                5     159.24760      31.84952       28.15       <.0001
Error               14      15.83850       1.13132
Corrected Total     19     175.08609
                Parameter    Standard
Variable         Estimate       Error     Type II SS    F Value       Pr > F
Intercept       -20.89732     7.16035       9.63604        8.52       0.0112
x1                0.01102     0.00317      13.65360       12.07       0.0037
x2               27.37046     6.24496      21.73153       19.21       0.0006
x3               -0.06929     0.01768      17.37834       15.36       0.0015
x4               -0.25695     0.11921       5.25602        4.65       0.0490
x5                0.01668     0.01132       2.45862        2.17       0.1626
                 Bounds on condition number: 1.6438, 35.628
---------------------------------------------------------------------------

Backward Elimination: Step 1 Variable x5 Removed: R-Square = 0.8955 and
C(p) = 6.1732
                              Sum of          Mean
Source              DF       Squares        Square     F Value       Pr > F
Model                4     156.78898      39.19724       32.13       <.0001
Error               15      18.29712       1.21981
Corrected Total     19     175.08609
                Parameter    Standard
Variable         Estimate       Error     Type II SS    F Value       Pr > F
Intercept       -19.93175     7.40393       8.84010        7.25       0.0167
x1                0.01233     0.00316      18.51245       15.18       0.0014
x2               27.28797     6.48433      21.60246       17.71       0.0008
x3               -0.06549     0.01816      15.86299       13.00       0.0026
x4               -0.19641     0.11621       3.48447        2.86       0.1117
                  Bounds on condition number: 1.6358, 22.31
---------------------------------------------------------------------------

Backward Elimination: Step 2 Variable x4 Removed: R-Square = 0.8756 and
C(p) = 7.2532
                              Sum of          Mean
Source              DF       Squares        Square     F Value       Pr > F
Model                3     153.30451      51.10150       37.54       <.0001
Error               16      21.78159       1.36135
Corrected Total     19     175.08609
                Parameter    Standard
Variable         Estimate       Error     Type II SS    F Value       Pr > F
Intercept       -27.89190     6.03518      29.07675       21.36       0.0003
x1                0.01360     0.00325      23.87130       17.54       0.0007
x2               30.68486     6.51286      30.21859       22.20       0.0002
x3               -0.06701     0.01916      16.64910       12.23       0.0030
                  Bounds on condition number: 1.4786, 11.85
---------------------------------------------------------------------------

   All variables left in the model are significant at the 0.1000 level.
                      Summary of Backward Elimination
      Variable     Number      Partial      Model
Step   Entered     Vars In     R-Square     R-Square     C(p)     F Value    Pr > F
  1      x5           4         0.0140       0.8955     6.1732      2.17     0.1626
  2      x4           3         0.0199       0.8756     7.2532      2.86     0.1117
```

Figure 15-16 SAS output for forward selection in Example 15-18.

```
                                 The REG Procedure
                               Dependent Variable: y
Stepwise Selection: Step 1 Variable x2 Entered: R-Square = 0.6530 and C(p) =
37.7047
                               Sum of          Mean
Source                 DF     Squares         Square    F Value       Pr > F
Model                   1    114.32885     114.32885    33.87         <.0001
Error                  18     60.75725       3.37540
Corrected Total        19    175.08609
                Parameter    Standard
Variable         Estimate      Error        Type II SS    F Value      Pr > F
Intercept       -42.87237     8.41429       87.62858      25.96        <.0001
x2               49.08366     8.43377      114.32885      33.87        <.0001
                         Bounds on condition number: 1, 1
--------------------------------------------------------------------------------
Stepwise Selection: Step 2 Variable x1 Entered: R-Square = 0.7805 and C(p) =
19.9697
                               Sum of          Mean
Source                 DF     Squares         Square    F Value       Pr > F
Model                   2    136.65541      68.32771    30.23         <.0001
Error                  17     38.43068       2.26063
Corrected Total        19    175.08609
                Parameter    Standard
Variable         Estimate      Error        Type II SS    F Value      Pr > F
Intercept       -33.83110     7.46286       46.45691      20.55        0.0003
x1                0.01314     0.00418       22.32656       9.88        0.0059
x2               34.88963     8.24844       40.44627      17.89        0.0006
                    Bounds on condition number: 1.4282, 5.7129
--------------------------------------------------------------------------------
Stepwise Selection: Step 3 Variable x3 Entered: R-Square = 0.8756 and C(p) =
7.2532
                               Sum of          Mean
Source                 DF     Squares         Square    F Value       Pr > F
Model                   3    153.30451      51.10150    37.54         <.0001
Error                  16     21.78159       1.36135
Corrected Total        19    175.08609
                Parameter    Standard
Variable         Estimate      Error        Type II SS    F Value      Pr > F
Intercept       -27.89190     6.03518       29.07675      21.36        0.0003
x1                0.01360     0.00326       23.87130      17.54        0.0007
x2               30.68486     6.51286       30.21859      22.20        0.0002
x3               -0.06701     0.01916       16.64910      12.23        0.0030
                     Bounds on condition number: 1.4786, 11.85
--------------------------------------------------------------------------------
    All variables left in the model are significant at the 0.1000 level.
No other variable met the 0.1000 significance level for entry into the model.
                          Summary of Stepwise Selection
     Variable  Variable  Number   Partial    Model
Step Entered   Removed   Vars In  R-Square   R-Square    C(p)    F Value    Pr > F
  1     x2                  1      0.6530     0.6530   37.7047    33.87     <.0001
  2     x1                  2      0.1275     0.7805   19.9697     9.88     0.0059
  3     x3                  3      0.0951     0.8756    7.2532    12.23     0.0030
```

Figure 15-17 SAS output for backward elimination in Example 15-19.

A major criticism of variable selection methods, such as stepwise regression, is that the analyst may conclude that there is one "best" regression equation. This generally is not the case, because there are often several equally good regression models that can be used. One way to avoid this problem is to use several different model-building techniques and see if different models result. For example, we have found the same model for the damaged peach data by using stepwise regression, forward selection, and backward elimination. This is a good indication that the three-variable model is the best regression equation. Furthermore, there are variable selection techniques that are designed to find the best one-variable model, the best two-variable model, and so forth. For a discussion of these methods, and the variable selection problem in general, see Montgomery, Peck, and Vining (2001).

If the number of candidate regressors is not too large, the all-possible-regressions method is recommended. It is not distorted by multicollinearity among the regressors, as stepwise-type methods are.

15-12 SUMMARY

This chapter has introduced multiple linear regression, including least-squares estimation of the parameters, interval estimation, prediction of new observations, and methods for hypothesis testing. Various tests of model adequacy, including residual plots, have been discussed. It was shown that polynomial regression models can be handled by the usual multiple linear regression methods. Indicator variables were introduced for dealing with qualitative variables. It also was observed that the problem of multicollinearity, or intercorrelation between the regressor variables, can greatly complicate the regression problem and often leads to a regression model that may not predict new observations well. Several causes and remedial measures of this problem, including biased estimation techniques, were discussed. Finally, the variable selection problem in multiple regression was introduced. A number of model-building procedures, including all possible regressions, stepwise regression, forward selection, and backward elimination, were illustrated.

15-13 EXERCISES

15-1. Consider the damaged peach data in Table 15-15.

(a) Fit a regression model using x_1 (drop height) and x_4 (fruit pulp thickness) to these data.

(b) Test for significance of regression.

(c) Compute the residuals from this model. Analyze these residuals using the methods discussed in this chapter.

(d) How does this two-variable model compare with the two-variable model using x_1 and x_2 from Example 15-1?

15-2. Consider the damaged peach data in Table 15-15.

(a) Fit a regression model using x_1 (drop height), x_2 (fruit density), and x_3 (fruit height at impact point) to these data.

(b) Test for significance of regression.

(c) Compute the residuals from this model. Analyze these residuals using the methods discussed in this chapter.

15-3. Using the results of Exercise 15-1, find a 95% confidence interval on β_4.

15-4. Using the results of Exercise 15-2, find a 95% confidence interval on β_3.

15-5. The data in the table at the top of page 487 are the 1976 team performance statistics for the teams in the National Football League (*Source: The Sporting News*).

(a) Fit a multiple regression model relating the number of games won to the teams' passing yardage (x_2), the percentage of rushing plays (x_7), and the opponents' yards rushing (x_8).

(b) Construct the appropriate residual plots and comment on model adequacy.

(c) Test the significance of each variable to the model, using either the t-test or the partial F-test.

15-6. The table at the top of page 488 presents gasoline mileage performance for 25 automobiles (*Source: Motor Trend*, 1975).

(a) Fit a multiple regression model relating gasoline mileage to engine displacement (x_1) and number of carburetor barrels (x_6).

(b) Analyze the residuals and comment on model adequacy.

National Football League 1976 Team Performance

Team	y	x_1	x_2	x_3	x_4	x_5	x_6	x_7	x_8	x_9
Washington	10	2113	1985	38.9	64.7	+4	868	59.7	2205	1917
Minnesota	11	2003	2855	38.8	61.3	+3	615	55.0	2096	1575
New England	11	2957	1737	40.1	60.0	+14	914	65.6	1847	2175
Oakland	13	2285	2905	41.6	45.3	−4	957	61.4	1903	2476
Pittsburgh	10	2971	1666	39.2	53.8	+15	836	66.1	1457	1866
Baltimore	11	2309	2927	39.7	74.1	+8	786	61.0	1848	2339
Los Angeles	10	2528	2341	38.1	65.4	+12	754	66.1	1564	2092
Dallas	11	2147	2737	37.0	78.3	−1	761	58.0	1821	1909
Atlanta	4	1689	1414	42.1	47.6	−3	714	57.0	2577	2001
Buffalo	2	2566	1838	42.3	54.2	−1	797	58.9	2476	2254
Chicago	7	2363	1480	37.3	48.0	+19	984	67.5	1984	2217
Cincinnati	10	2109	2191	39.5	51.9	+6	700	57.2	1917	1758
Cleveland	9	2295	2229	37.4	53.6	−5	1037	58.8	1761	2032
Denver	9	1932	2204	35.1	71.4	+3	986	58.6	1709	2025
Detroit	6	2213	2140	38.8	58.3	+6	819	59.2	1901	1686
Green Bay	5	1722	1730	36.6	52.6	−19	791	54.4	2288	1835
Houston	5	1498	2072	35.3	59.3	−5	776	49.6	2072	1914
Kansas City	5	1873	2929	41.1	55.3	+10	789	54.3	2861	2496
Miami	6	2118	2268	38.2	69.6	+6	582	58.7	2411	2670
New Orleans	4	1775	1983	39.3	78.3	+7	901	51.7	2289	2202
New York Giants	3	1904	1792	39.7	38.1	−9	734	61.9	2203	1988
New York Jets	3	1929	1606	39.7	68.8	−21	627	52.7	2592	2324
Philadelphia	4	2080	1492	35.5	68.8	−8	722	57.8	2053	2550
St. Louis	10	2301	2835	35.3	74.1	+2	683	59.7	1979	2110
San Diego	6	2040	2416	38.7	50.0	0	576	54.9	2048	2628
San Francisco	8	2447	1638	39.9	57.1	−8	848	65.3	1786	1776
Seattle	2	1416	2649	37.4	56.3	−22	684	43.8	2876	2524
Tampa Bay	0	1503	1503	39.3	47.0	−9	875	53.5	2560	2241

y: Games won (per 14 game season).
x_1: Rushing yards (season).
x_2: Passing yards (season).
x_3: Punting average (yds / punt).
x_4: Field goal percentage (fgs made / fgs attempted).
x_5: Turnover differential (turnovers acquired / turnovers lost).
x_6: Penalty yards (season).
x_7: Percent rushing (rushing plays / total plays).
x_8: Opponents' rushing yards (season).
x_9: Opponents' passing yards (season).

(c) What is the value of adding x_6 to a model that already contains x_1?

15-7. The electric power consumed each month by a chemical plant is thought to be related to the average ambient temperature (x_1), the number of days in the month (x_2), the average product purity (x_3), and the tons of product produced (x_4). The past year's historical data is available and is presented in the table at the bottom of page 488.

Gasoline Mileage Performance for 25 Automobiles

Automobile	y	x_1	x_2	x_3	x_4	x_5	x_6	x_7	x_8	x_9	x_{10}	x_{11}
Apollo	18.90	350	165	260	8.0:1	2.56:1	4	3	200.3	69.9	3910	A
Nova	20.00	250	105	185	8.25:1	2.73:1	1	3	196.7	72.2	3510	A
Monarch	18.25	351	143	255	8.0:1	3.00:1	2	3	199.9	74.0	3890	A
Duster	20.07	225	95	170	8.4:1	2.76:1	1	3	194.1	71.8	3365	M
Jenson Conv.	11.20	440	215	330	8.2:1	2.88:1	4	3	184.5	69.0	4215	A
Skyhawk	22.12	231	110	175	8.0:1	2.56:1	2	3	179.3	65.4	3020	A
Scirocco	34.70	89.7	70	81	8.2:1	3.90:1	2	4	155.7	64.0	1905	M
Corolla SR-5	30.40	96.9	75	83	9.0:1	4.30:1	2	5	165.2	65.0	2320	M
Camaro	16.50	350	155	250	8.5:1	3.08:1	4	3	195.4	74.4	3885	A
Datsun B210	36.50	85.3	80	83	8.5:1	3.89:1	2	4	160.6	62.2	2009	M
Capri II	21.50	171	109	146	8.2:1	3.22:1	2	4	170.4	66.9	2655	M
Pacer	19.70	258	110	195	8.0:1	3.08:1	1	3	171.5	77.0	3375	A
Granada	17.80	302	129	220	8.0:1	3.00:1	2	3	199.9	74.0	3890	A
Eldorado	14.39	500	190	360	8.5:1	2.73:1	4	3	224.1	79.8	5290	A
Imperial	14.89	440	215	330	8.2:1	2.71:1	4	3	231.0	79.7	5185	A
Nova LN	17.80	350	155	250	8.5:1	3.08:1	4	3	196.7	72.2	3910	A
Starfire	23.54	231	110	175	8.0:1	2.56:1	2	3	179.3	65.4	3050	A
Cordoba	21.47	360	180	290	8.4:1	2.45:1	2	3	214.2	76.3	4250	A
Trans Am	16.59	400	185	NA	7.6:1	3.08:1	4	3	196	73.0	3850	A
Corolla E-5	31.90	96.9	75	83	9.0:1	4.30:1	2	5	165.2	61.8	2275	M
Mark IV	13.27	460	223	366	8.0:1	3.00:1	4	3	228	79.8	5430	A
Celica GT	23.90	133.6	96	120	8.4:1	3.91:1	2	5	171.5	63.4	2535	M
Charger SE	19.73	318	140	255	8.5:1	2.71:1	2	3	215.3	76.3	4370	A
Cougar	13.90	351	148	243	8.0:1	3.25:1	2	3	215.5	78.5	4540	A
Corvette	16.50	350	165	255	8.5:1	2.73:1	4	3	185.2	69.0	3660	A

y: Miles / gallon.
x_1: Displacement (cubic in.).
x_2: Horsepower (ft-lb).
x_3: Torque (ft-lb).
x_4: Compression ratio.
x_5: Rear axle ratio.
x_6: Carburetor (barrels).
x_7: No. of transmission speeds.
x_8: Overall length (in.).
x_9: Width (in.).
x_{10}: Weight (lbs).
x_{11}: Type of transmission (A—automatic, M—manual).

y	x_1	x_2	x_3	x_4	y	x_1	x_2	x_3	x_4
240	25	24	91	100	300	80	25	87	97
236	31	21	90	95	296	84	25	86	96
290	45	24	88	110	267	75	24	88	110
274	60	25	87	88	276	60	25	91	105
301	65	25	91	94	288	50	25	90	100
316	72	26	94	99	261	38	23	89	98

(a) Fit a multiple regression model to these data.

(b) Test for significance of regression.

(c) Use partial F statistics to test H_0: $\beta_3 = 0$ and H_0: $\beta_4 = 0$.

(d) Compute the residuals from this model. Analyze the residuals using the methods discussed in this chapter.

15-8. Hald (1952) reports data on the heat evolved in calories per gram of cement (y) for various amounts of four ingredients (x_1, x_2, x_3, x_4).

Observation Number	y	x_1	x_2	x_3	x_4
1	78.5	7	26	6	60
2	74.3	1	29	15	52
3	104.3	11	56	8	20
4	87.6	11	31	8	47
5	95.9	7	52	6	33
6	109.2	11	55	9	22
7	102.7	3	71	17	6
8	72.5	1	31	22	44
9	93.1	2	54	18	22
10	115.9	21	47	4	26
11	83.8	1	40	23	34
12	113.3	11	66	9	12
13	109.4	10	68	8	12

(a) Fit a multiple regression model to these data.

(b) Test for significance of regression.

(c) Test the hypothesis $\beta_4 = 0$ using the partial F-test.

(d) Compute the t statistics for each independent variable. What conclusions can you draw?

(e) Test the hypothesis $\beta_2 = \beta_3 = \beta_4 = 0$ using the partial F-test.

(f) Construct a 95% confidence interval estimate for β_2.

15-9. An article entitled "A Method for Improving the Accuracy of Polynomial Regression Analysis" in the *Journal of Quality Technology* (1971, p. 149) reported the following data on y = ultimate shear strength of a rubber compound (psi) and x = cure temperature (°F).

y	770	800	840	810	735	640	590	560
x	280	284	292	295	298	305	308	315

(a) Fit a second-order polynomial to this data.

(b) Test for significance of regression.

(c) Test the hypothesis that $\beta_{11} = 0$.

(d) Compute the residuals and test for model adequacy.

15-10. Consider the following data, which result from an experiment to determine the effect of x = test time in hours at a particular temperature on y = change in oil viscosity.

y	−4.42	−1.39	−1.55	−1.89	−2.43	−3.15	−4.05	−5.15	−6.43	−7.89
x	0.25	0.50	0.75	1.00	1.25	1.50	1.75	2.00	2.25	2.50

(a) Fit a second-order polynomial to the data.

(b) Test for significance of regression.

(c) Test the hypothesis that $\beta_{11} = 0$.

(d) Compute the residuals and check for model adequacy.

15-11. For many polynomial regression models we subtract $\bar{x}$ from each x value to produce a "centered" regressor $x' = x - \bar{x}$. Using the data from Exercise 15-9, fit the model $y = \beta_0^* + \beta_1^* x' + \beta_{11}^*(x')^2 + \varepsilon$. Use the results to estimate the coefficients in the uncentered model $y = \beta_0 + \beta_1 x + \beta_{11}x^2 + \varepsilon$.

15-12. Suppose that we use a standardized variable $x' = (x - \bar{x})/s_x$, where s_x is the standard deviation of x, in constructing a polynomial regression model. Using the data in Exercise 15-9 and the standardized variable approach, fit the model $y = \beta_0^* + \beta_1^* x' + \beta_{11}^*(x')^2 + \varepsilon$.

(a) What value of y do you predict when x = 285°F?

(b) Estimate the regression coefficients in the unstandardized model $y = \beta_0 + \beta_1 x + \beta_{11}x^2 + \varepsilon$.

(c) What can you say about the relationship between SS_E and R^2 for the standardized and unstandardized models?

(d) Suppose that $y' = (y - \bar{y})/s_y$ is used in the model along with x'. Fit the model and comment on the relationship between SS_E and R^2 in the standardized model and the unstandardized model.

15-13. The data shown at the bottom of this page were collected during an experiment to determine the change in thrust efficiency (%) (y) as the divergence angle of a rocket nozzle (x) changes.

(a) Fit a second-order model to the data.

(b) Test for significance of regression and lack of fit.

(c) Test the hypothesis that $\beta_{11} = 0$.

15-14. Discuss the hazards inherent in fitting polynomial models.

15-15. Consider the data in Example 15-12. Test the hypothesis that two different regression models (with

y	24.60	24.71	23.90	39.50	39.60	57.12	67.11	67.24	67.15	77.87	80.11	84.67
x	4.0	4.0	4.0	5.0	5.0	6.0	6.5	6.5	6.75	7.0	7.1	7.3

different slopes and intercepts) are required to adequately model the data.

15-16. Piecewise Linear Regression (I). Suppose that y is piecewise linearly related to x. That is, different linear relationships are appropriate over the intervals $-\infty < x \leq x^*$ and $x^* < x < \infty$. Show how indicator variables can be used to fit such a piecewise linear regression model, assuming that point x^* is known.

15-17. Piecewise Linear Regression (II). Consider the piecewise linear regression model described in Exercise 15-16. Suppose that at point x^* a discontinuity occurs in the regression function. Show how indicator variables can be used to incorporate the discontinuity into the model.

15-18. Piecewise Linear Regression (III). Consider the piecewise linear regression model described in Exercise 15-16. Suppose that point x^* is not known with certainty and must be estimated. Develop an approach that could be used to fit the piecewise linear regression model.

15-19. Calculate the standardized regression coefficients for the regression model developed in Exercise 15-1.

15-20. Calculate the standardized regression coefficients for the regression model developed in Exercise 15-2.

15-21. Find the variance inflation factors for the regression model developed in Example 15-1. Do they indicate that multicollinearity is a problem in this model?

15-22. Use the National Football League Team Performance data in Exercise 15-5 to build regression models using the following techniques:

(a) All possible regressions.

(b) Stepwise regression.

(c) Forward selection.

(d) Backward elimination.

(e) Comment on the various models obtained.

15-23. Use the gasoline mileage data in Exercise 15-6 to build regression models using the following techniques:

(a) All possible regressors.

(b) Stepwise regression.

(c) Forward selection.

(d) Backward elimination.

(e) Comment on the various models obtained.

15-24. Consider the Hald cement data in Exercise 15-8. Build regression models for the data using the following techniques:

(a) All possible regressions.

(b) Stepwise regression.

(c) Forward selection.

(d) Backward elimination.

15-25. Consider the Hald cement data in Exercise 15-8. Fit a regression model involving all four regressors and find the variance inflation factors. Is multicollinearity a problem in this model? Use ridge regression to estimate the coefficients in this model. Compare the ridge model to the models obtained in Exercise 15-25 using variable selection methods.

Chapter 16

Nonparametric Statistics

16-1 INTRODUCTION

Most of the hypothesis testing and confidence interval procedures in previous chapters are based on the assumption that we are working with random samples from normal populations. Fortunately, most of these procedures are relatively insensitive to slight departures from normality. In general, the t- and F-tests and t confidence intervals will have actual levels of significance or confidence levels that differ from the nominal or advertised levels chosen by the experimenter, although the difference between the actual and advertised levels is usually fairly small when the underlying population is not too different from the normal distribution. Traditionally, we have called these procedures *parametric* methods because they are based on a particular parametric family of distributions—in this case, the normal. Alternatively, sometimes we say that these procedures are not *distribution free* because they depend on the assumption of normality.

In this chapter we describe procedures called *nonparametric* or *distribution-free* methods and usually make no assumptions about the distribution of the underlying population, other than that it is continuous. These procedures have actual levels of significance α or confidence levels $100(1 - \alpha)\%$ for many different types of distributions. These procedures also have considerable appeal. One of their advantages is that the data need not be quantitative; it could be categorical (such as yes or no, defective or nondefective, etc.) or rank data. Another advantage is that nonparametric procedures are usually very quick and easy to perform.

The procedures described in this chapter are competitors of the parametric t- and F-procedures described earlier. Consequently, it is important to compare the performance of both parametric and nonparametric methods under the assumptions of both normal and nonnormal populations. In general, nonparametric procedures do not utilize all the information provided by the sample, and as a result a nonparametric procedure will be less efficient than the corresponding parametric procedure when the underlying population is normal. This loss of efficiency usually is reflected by a requirement for a larger sample size for the nonparametric procedure than would be required by the parametric procedure in order to achieve the same probability of type II error. On the other hand, this loss of efficiency is usually not large, and often the difference in sample size is very small. When the underlying distributions are not normal, then nonparametric methods have much to offer. They often provide considerable improvement over the normal-theory parametric methods.

16-2 THE SIGN TEST

16-2.1 A Description of the Sign Test

The sign test is used to test hypotheses about the median $\tilde{\mu}$ of a continuous distribution. Recall that the median of a distribution is a value of the random variable such that the probability is 0.5 that an observed value of X is less than or equal to the median, and the probability is 0.5 that an observed value of X is greater than or equal to the median. That is, $P(X \leq \tilde{\mu}) = P(X \geq \tilde{\mu}) = 0.5$.

Since the normal distribution is symmetric, the mean of a normal distribution equals the median. Therefore the sign test can be used to test hypotheses about the mean of a normal distribution. This is the same problem for which we used the t-test in Chapter 11. We will discuss the relative merits of the two procedures in Section 16-2.4. Note that while the t-test was designed for samples from a normal distribution, the sign test is appropriate for samples from any continuous distribution. Thus, the sign test is a nonparametric procedure.

Suppose that the hypotheses are

$$\begin{aligned} H_0&: \tilde{\mu} = \tilde{\mu}_0, \\ H_1&: \tilde{\mu} \neq \tilde{\mu}_0. \end{aligned} \tag{16-1}$$

The test procedure is as follows. Suppose that $X_1, X_2, \ldots, X_n$ is a random sample of n observations from the population of interest. Form the differences $(X_i - \tilde{\mu}_0)$, $i = 1, 2, \ldots, n$. Now if H_0: $\tilde{\mu} = \tilde{\mu}_0$ is true, any difference $X_i - \tilde{\mu}_0$ is equally likely to be positive or negative. Therefore let R^+ denote the number of these differences $(X_i - \tilde{\mu}_0)$ that are positive and let R^- denote the number of these differences that are negative, where $R = \min(R^+, R^-)$.

When the null hypothesis is true, R has a binomial distribution with parameters n and $p = 0.5$. Therefore, we would find a critical value, say R^*_α, from the binomial distribution that ensures that P(type I error) $= P$(reject H_0 when H_0 is true) $= \alpha$. A table of these critical values R^*_α is given in the Appendix, Table X. If the test statistic $R \leq R^*_\alpha$, then the null hypothesis H_0: $\tilde{\mu} - \tilde{\mu}_0$ should be rejected.

Example 16-1

Montgomery, Peck, and Vining (2001) report on a study in which a rocket motor is formed by binding an igniter propellant and a sustainer propellant together inside a metal housing. The shear strength of the bond between the two propellant types is an important characteristic. Results of testing 20 randomly selected motors are shown in Table 16-1. We would like to test the hypothesis that the median shear strength is 2000 psi.

The formal statement of the hypotheses of interest is

$$\begin{aligned} H_0&: \tilde{\mu} = 2000, \\ H_1&: \tilde{\mu} \neq 2000. \end{aligned}$$

The last two columns of Table 16-1 show the differences $(X_i - 2000)$ for $i = 1, 2, \ldots, 20$ and the corresponding signs. Note that $R^+ = 14$ and $R^- = 6$. Therefore $R = \min(R^+, R^-) = \min(14, 6) = 6$. From the Appendix, Table X, with $n = 20$, we find that the critical value for $\alpha = 0.05$ is $R^*_{0.05} = 5$. Therefore, since $R = 6$ is not less than or equal to the critical value $R^*_{0.05} = 5$, we cannot reject the null hypothesis that the median shear strength is 2000 psi.

We note that since R is a binomial random variable, we could test the hypothesis of interest by directly calculating a P-value from the binomial distribution. When H_0: $\tilde{\mu} = 2000$ is true, R has a binomial distribution with parameters $n = 20$ and $p = 0.5$. Thus the probability of observing six or fewer negative signs in a sample of 20 observations is

$$\begin{aligned} P(R \leq 6) &= \sum_{r=0}^{6} \binom{20}{r} (0.5)^r (0.5)^{20-r} \\ &= 0.058. \end{aligned}$$

Since the P-value is not less than the desired level of significance, we cannot reject the null hypothesis of $\tilde{\mu} = 2000$ psi.

Table 16-1 Propellant Shear Strength Data

Observation (i)	Shear Strength (X_i)	Differences ($X_i - 2000$)	Sign
1	2158.70	+158.70	+
2	1678.15	–321.85	–
3	2316.00	+316.00	+
4	2061.30	+61.30	+
5	2207.50	+207.50	+
6	1708.30	–291.70	–
7	1784.70	–215.30	–
8	2575.10	+575.00	+
9	2357.90	+357.90	+
10	2256.70	+256.70	+
11	2165.20	+165.20	+
12	2399.55	+399.55	+
13	1779.80	–220.20	–
14	2336.75	+336.75	+
15	1765.30	–234.70	–
16	2053.50	+53.50	+
17	2414.40	+414.40	+
18	2200.50	+200.50	+
19	2654.20	+654.20	+
20	1753.70	–246.30	–

Exact Significance Levels When a test statistic has a discrete distribution, such as R does in the sign test, it may be impossible to choose a critical value R_α^* that has a level of significance exactly equal to α. The usual approach is to choose R_α^* to yield an α as close to the advertised level α as possible.

Ties in the Sign Test Since the underlying population is assumed to be continuous, it is theoretically impossible to find a "tie," that is, a value of X_i exactly equal to $\tilde{\mu}_0$. However, this may sometimes happen in practice because of the way the data are collected. When ties occur, they should be set aside and the sign test applied to the remaining data.

One-Sided Alternative Hypotheses We can also use the sign test when a one-sided alternative hypothesis is appropriate. If the alternative is H_1: $\tilde{\mu} > \tilde{\mu}_0$, then reject H_0: $\tilde{\mu} = \tilde{\mu}_0$ if $R^- < R_\alpha^*$; if the alternative is H_1: $\tilde{\mu} < \tilde{\mu}_0$, then reject H_0: $\tilde{\mu} = \tilde{\mu}_0$ if $R^+ < R_\alpha^*$. The level of significance of a one-sided test is one-half the value shown in the Appendix, Table X. It is also possible to calculate a P-value from the binomial distribution for the one-sided case.

The Normal Approximation When $p = 0.5$, the binomial distribution is well approximated by a normal distribution when n is at least 10. Thus, since the mean of the binomial is np and the variance is $np(1 - p)$, the distribution of R is approximately normal, with mean $0.5n$ and variance $0.25n$ whenever n is moderately large. Therefore, in these cases the null hypothesis can be tested with the statistic

$$Z_0 = \frac{R - 0.5n}{0.5\sqrt{n}}. \tag{16-2}$$

The two-sided alternative would be rejected if $|Z_0| > Z_{\alpha/2}$, and the critical regions of the one-sided alternative would be chosen to reflect the sense of the alternative (if the alternative is $H_1: \tilde{\mu} > \tilde{\mu}_0$, reject H_0 if $Z_0 > Z_\alpha$, for example).

16-2.2 The Sign Test for Paired Samples

The sign test can also be applied to paired observations drawn from continuous populations. Let (X_{1j}, X_{2j}), $j = 1, 2, \ldots, n$, be a collection of paired observations from two continuous populations, and let

$$D_j = X_{1j} - X_{2j} \qquad j = 1, 2, \ldots, n,$$

be the paired differences. We wish to test the hypothesis that the two populations have a common median, that is, that $\tilde{\mu}_1 = \tilde{\mu}_2$. This is equivalent to testing that the median of the differences $\tilde{\mu}_d = 0$. This can be done by applying the sign test to the n differences D_j, as illustrated in the following example.

Example 16-2

An automotive engineer is investigating two different types of metering devices for an electronic fuel injection system to determine if they differ in their fuel mileage performance. The system is installed on 12 different cars, and a test is run with each metering system on each car. The observed fuel mileage performance data, corresponding differences, and their signs are shown in Table 16-2. Note that $R^+ = 8$ and $R^- = 4$. Therefore $R = \min(R^+, R^-) = \min(8, 4) = 4$. From the Appendix, Table X, with $n = 12$, we find the critical value for $\alpha = 0.05$ is $R^*_{0.05} = 2$. Since R is not less than the critical value $R^*_{0.05}$, we cannot reject the null hypothesis that the two metering devices produce the same fuel mileage performance.

16-2.3 Type II Error (β) for the Sign Test

The sign test will control the probability of type I error at an advertised level α for testing the null hypothesis H_0: $\tilde{\mu} = \tilde{\mu}_0$ for any continuous distribution. As with any hypothesis-testing procedure, it is important to investigate the type II error, β. The test should be able to effectively detect departures from the null hypothesis, and a good measure of this effec-

Table 16-2 Performance of Flow Metering Devices

	Metering Device			
Car	1	2	Difference, D_j	Sign
1	17.6	16.8	0.8	+
2	19.4	20.0	−0.6	−
3	19.5	18.2	1.3	+
4	17.1	16.4	0.7	+
5	15.3	16.0	−0.7	−
6	15.9	15.4	0.5	+
7	16.3	16.5	−0.2	−
8	18.4	18.0	0.4	+
9	17.3	16.4	0.9	+
10	19.1	20.1	−1.0	−
11	17.8	16.7	1.1	+
12	18.2	17.9	0.3	+

tiveness is the value of β for departures that are important. A small value of β implies an effective test procedure.

In determining β, it is important to realize that not only must a particular value of $\tilde{\mu}$, say $\tilde{\mu}_0 + \Delta$, be used, also the *form* of the underlying distribution will affect the calculations. To illustrate, suppose that the underlying distribution is normal with $\sigma = 1$ and we are testing the hypothesis that $\tilde{\mu} = 2$ (since $\tilde{\mu} = \mu$ in the normal distribution this is equivalent to testing that the mean equals 2). It is important to detect a departure from $\tilde{\mu} = 2$ to $\tilde{\mu} = 3$. The situation is illustrated graphically in Fig. 16-1*a*. When the alternative hypothesis is true (H_1: $\tilde{\mu} =$ 3), the probability that the random variable X exceeds the value 2 is

$$p = P(X > 2) = P(Z > -1) = 1 - \Phi(-1) = 0.8413.$$

Suppose we have taken samples of size 12. At the $\alpha = 0.05$ level, Appendix Table X indicates that we would reject H_0: $\tilde{\mu} = 2$ if $R \le R^*_{0.05} = 2$. Therefore, the β error is the probability that we do not reject H_0: $\tilde{\mu} = 2$ when in fact $\tilde{\mu} = 3$, or

$$\beta = 1 - \sum_{x=0}^{2} \binom{12}{x} (0.1587)^x (0.8413)^{12-x} = 0.2944.$$

If the distribution of X has been exponential rather than normal, then the situation would be as shown in Fig. 16-1*b*, and the probability that the random variable X exceeds the value $x = 2$ when $\tilde{\mu} = 3$ (note when the median of an exponential distribution is 3 the mean is 4.33) is

$$p = P(X > 2) = \int_2^{\infty} \frac{1}{4.33} e^{-\frac{1}{4.33}x} dx = e^{-\frac{2}{4.33}} = 0.6301.$$

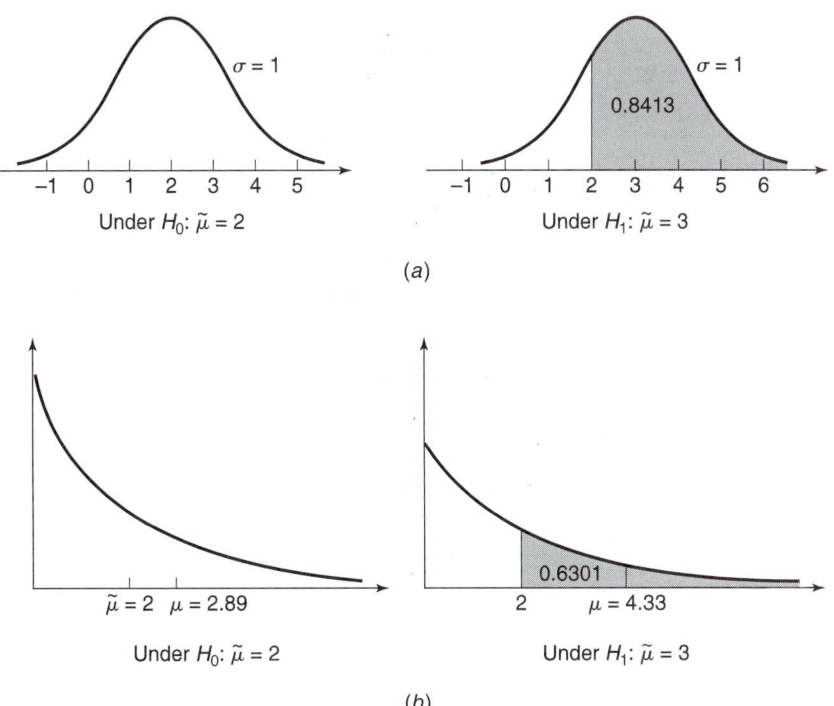

Figure 16-1 Calculation of β for the sign test. (*a*) Normal distributions, (*b*) exponential distributions.

The β error in this case is

$$\beta = 1 - \sum_{x=0}^{2} \binom{12}{x} (0.3699)^x (0.6301)^{12-x} = 0.8794.$$

Thus, the β error for the sign test depends not only on the alternative value of $\tilde{\mu}$ but on the area to the right of the value specified in the null hypothesis under the population probability distribution. This area is highly dependent on the shape of that particular probability distribution.

16-2.4 Comparison of the Sign Test and the *t*-Test

If the underlying population is normal, then either the sign test or the *t*-test could be used to test H_0: $\tilde{\mu} = \tilde{\mu}_0$. The *t*-test is known to have the smallest value of β possible among all tests that have significance level α, so it is superior to the sign test in the normal distribution case. When the population distribution is symmetric and nonnormal (but with finite mean $\mu = \tilde{\mu}$), then the *t*-test will have a β error that is smaller than β for the sign test, unless the distribution has very heavy tails compared with the normal. Thus, the sign test is usually considered a test procedure for the median rather than a serious competitor for the *t*-test. The Wilcoxon signed rank test in the next section is preferable to the sign test and compares well with the *t*-test for symmetric distributions.

16-3 THE WILCOXON SIGNED RANK TEST

Suppose that we are willing to assume that the population of interest is *continuous* and *symmetric.* As in the previous section, our interest focuses on the median $\tilde{\mu}$ (or equivalently, the mean μ, since $\tilde{\mu} = \mu$ for symmetric distributions). A disadvantage of the sign test in this situation is that it considers only the signs of the deviations $X_i - \tilde{\mu}_0$ and not their magnitudes. The Wilcoxon signed rank test is designed to overcome that disadvantage.

16-3.1 A Description of the Test

We are interested in testing H_0: $\mu = \mu_0$ against the usual alternatives. Assume that $X_1, X_2, \ldots, X_n$ is a random sample from a continuous and symmetric distribution with mean (and median) μ. Compute the differences $X_i - \mu_0$, $i = 1, 2, \ldots, n$. Rank the absolute differences $|X_i - \mu_0|$, $i = 1, 2, \ldots, n$, in ascending order, and then give the ranks the signs of their corresponding differences. Let R^+ be the sum of the positive ranks and R^- be the absolute value of the sum of the negative ranks, and let $R = \min(R^+, R^-)$. Appendix Table XI contains critical values of R, say R^*_α. If the alternative hypothesis is H_1: $\mu \neq \mu_0$, then if $R \leq R^*_\alpha$ the null hypothesis H_0: $\mu = \mu_0$ is rejected.

For one-sided tests, if the alternative is H_1: $\mu > \mu_0$, reject H_0: $\mu = \mu_0$ if $R^- < R^*_\alpha$; and if the alternative is H_1: $\mu < \mu_0$, reject H_0: $\mu = \mu_0$ if $R^+ < R^*_\alpha$. The significance level for one-sided tests is one-half the advertised level in Appendix Table XI.

Example 16-3

To illustrate the Wilcoxon signed rank test, consider the propellant shear strength data presented in Table 16-1. The signed ranks are

Observation	Difference, $X_i - 2000$	Signed Rank
16	+53.50	+1
4	+61.30	+2
1	+158.70	+3
11	+165.20	+4
18	+200.50	+5
5	+207.50	+6
7	−215.30	−7
13	−220.20	−8
15	−234.70	−9
20	−246.30	−10
10	+256.70	+11
6	−291.70	−12
3	+316.00	+13
2	−321.85	−14
14	+336.75	+15
9	+357.90	+16
12	+399.55	+17
17	+414.40	+18
8	+575.00	+19
19	+654.20	+20

The sum of the positive ranks is $R^+ = (+1 + 2 + 3 + 4 + 5 + 6 + 11 + 13 + 15 + 16 + 17 + 18 + 19 + 20) = 150$ and the sum of the negative ranks is $R^- = (7 + 8 + 9 + 10 + 12 + 14) = 60$. Therefore $R = \min(R^+, R^-) = \min(150, 60) = 60$. From Appendix Table XI, with $n = 20$ and $\alpha = 0.05$, we find the critical value $R^*_{0.05} = 52$. Since R exceeds R^*_α, we cannot reject the null hypothesis that the mean (or median, since the populations are assumed to be symmetric) shear strength is 2000 psi.

Ties in the Wilcoxon Signed Rank Test

Because the underlying population is continuous, ties are theoretically impossible, although they will sometimes occur in practice. If several observations have the same absolute magnitude, they are assigned the average of the ranks that they would receive if they differed slightly from one another.

16-3.2 A Large-Sample Approximation

If the sample size is moderately large, say $n > 20$, then it can be shown that R has approximately a normal distribution with mean

$$\mu_R = \frac{n(n+1)}{4}$$

and variance

$$\sigma_R^2 = \frac{n(n+1)(2n+1)}{24}.$$

Therefore, a test of H_0: $\mu = \mu_0$ can be based on the statistic

$$Z_0 = \frac{R - n(n+1)/4}{\sqrt{n(n+1)(2n+1)/24}}. \tag{16-3}$$

An appropriate critical region can be chosen from a table of the standard normal distribution.

16-3.3 Paired Observations

The Wilcoxon signed rank test can be applied to paired data. Let (X_{1j}, X_{2j}), $j = 1, 2, \ldots, n$, be a collection of paired observations from continuous distributions that differ only with respect to their means (it is *not* necessary that the distributions of X_1 and X_2 be symmetric). This assures that the distribution of the *differences* $D_j = X_{1j} - X_{2j}$ is *continuous* and *symmetric*.

To use the Wilcoxon signed rank test, the differences are first ranked in ascending order of their absolute values, and then the ranks are given the signs of the differences. Ties are assigned average ranks. Let R^+ be the sum of the positive ranks and R^- be the absolute value of the sum of the negative ranks, and let $R = \min(R^+, R^-)$. We reject the hypothesis of equality of means if $R \le R_\alpha^*$, where R_α^* is chosen from Appendix Table XI.

For one-sided tests, if the alternative is H_1: $\mu_1 > \mu_2$ (or H_1: $\mu_D > 0$), reject H_0 if $R^- < R_\alpha^*$; and if H_1: $\mu_1 < \mu_2$ (or H_1: $\mu_D < 0$), reject H_0 if $R^+ < R_\alpha^*$. Note that the significance level for one-sided tests is one-half the value given in Table XI.

Example 16-4

Consider the fuel metering device data examined in Example 16-2. The signed ranks are shown below.

Car	Difference	Signed Rank
7	−0.2	−1
12	0.3	2
8	0.4	3
6	0.5	4
2	−0.6	−5
4	0.7	6.5
5	−0.7	−6.5
1	0.8	8
9	0.9	9
10	−1.0	−10
11	1.1	11
3	1.3	12

Note that $R^+ = 55.5$ and $R^- = 22.5$; therefore, $R = \min(R^+, R^-) = \min(55.5, 22.5) = 22.5$. From Appendix Table XI, with $n = 12$ and $\alpha = 0.05$, we find the critical value $R_{0.05}^* = 13$. Since R exceeds $R_{0.05}^*$, we cannot reject the null hypothesis that the two metering devices produce the same mileage performance.

16-3.4 Comparison with the *t*-Test

When the underlying population is normal, either the *t*-test or the Wilcoxon signed rank test can be used to test hypotheses about μ. The *t*-test is the best test in such situations in the sense that it produces a minimum value of β for all tests with significance level α. However, since it is not always clear that the normal distribution is appropriate, and since there are many situations in which we know it to be inappropriate, it is of interest to compare the two procedures for both normal and nonnormal populations.

Unfortunately, such a comparison is not easy. The problem is that β for the Wilcoxon signed rank test is very difficult to obtain, and β for the *t*-test is difficult to obtain for non-normal distributions. Because type II error comparisons are difficult, other measures of comparison have been developed. One widely used measure is *asymptotic relative efficiency* (ARE). The ARE of one test relative to another is the limiting ratio of the sample sizes necessary to obtain identical error probabilities for the two procedures. For example, if the ARE of one test relative to a competitor is 0.5, then when sample sizes are large, the first test will require a sample twice as large as the second one to obtain similar error performance. While this does not tell us anything for small sample sizes, we can say the following:

1. For normal populations the ARE of the Wilcoxon signed rank test relative to the *t*-test is approximately 0.95.
2. For nonnormal populations, the ARE is at least 0.86, and in many cases it will exceed unity. When it exceeds unity, the Wilcoxon signed rank test requires a smaller sample size than does the *t*-test.

Although these are large-sample results, we generally conclude that the Wilcoxon signed rank test will never be much worse than the *t*-test, and in many cases where the population is nonnormal it may be superior. Thus the Wilcoxon signed rank test is a useful alternative to the *t*-test.

16-4 THE WILCOXON RANK-SUM TEST

Suppose that we have two independent continuous populations X_1 and X_2 with means μ_1 and μ_2. The distributions of X_1 and X_2 have the same shape and spread and differ only (possibly) in their means. The Wilcoxon rank-sum test can be used to test the hypothesis H_0: $\mu_1 = \mu_2$. Sometimes this procedure is called the Mann–Whitney test, although the Mann–Whitney test statistic is usually expressed in a different form.

16-4.1 A Description of the Test

Let $X_{11}, X_{12}, \ldots, X_{1n_1}$ and $X_{21}, X_{22}, \ldots, X_{2n_2}$ be two independent random samples from the continuous populations X_1 and X_2 described earlier. We assume that $n_1 \leq n_2$. Arrange all $n_1 + n_2$ observations in ascending order of magnitude and assign ranks to them. If two or more observations are tied (identical), then use the mean of the ranks that would have been assigned if the observations differed. Let R_1 be the sum of the ranks in the smaller X_1 sample, and define

$$R_2 = n_1(n_1 + n_2 + 1) - R_1. \tag{16-4}$$

Now if the two means do not differ, we would expect the sum of the ranks to be nearly equal for both samples. Consequently, if the sums of the ranks differ greatly, we would conclude that the means are not equal.

Appendix Table IX contains the critical values R^*_α of the rank sums for $\alpha = 0.05$ and $\alpha = 0.01$. Refer to Appendix Table IX, with the appropriate sample sizes n_1 and n_2. The null hypothesis H_0: $\mu_1 = \mu_2$ is rejected in favor of H_1: $\mu_1 \neq \mu_2$ if either R_1 or R_2 is less than or equal to the tabulated critical value R^*_α.

The procedure can also be used for one-sided alternatives. If the alternative is H_1: $\mu_1 < \mu_2$, then reject H_0 if $R_1 \leq R^*_\alpha$; while for H_0: $\mu_1 > \mu_2$, reject H_0 if $R_2 \leq R^*_\alpha$. For these one-sided tests the tabulated critical values R^*_α correspond to levels of significance of $\alpha = 0.025$ and $\alpha = 0.005$.

Example 16-5

The mean axial stress in tensile members used in an aircraft structure is being studied. Two alloys are being investigated. Alloy 1 is a traditional material and alloy 2 is a new aluminum–lithium alloy that is much lighter than the standard material. Ten specimens of each alloy type are tested, and the axial stress measured. The sample data are assembled in the following table:

Alloy 1		Alloy 2	
3238 psi	3254 psi	3261 psi	3248 psi
3195	3229	3187	3215
3246	3225	3209	3226
3190	3217	3212	3240
3204	3241	3258	3234

The data are arranged in ascending order and ranked as follows:

Alloy Number	Axial Stress	Rank
2	3187 psi	1
1	3190	2
1	3195	3
1	3204	4
2	3209	5
2	3212	6
2	3215	7
1	3217	8
1	3225	9
2	3226	10
1	3229	11
2	3234	12
1	3238	13
2	3240	14
1	3241	15
1	3246	16
2	3248	17
1	3254	18
2	3258	19
2	3261	20

The sum of the ranks for alloy 1 are

$$R_1 = 2 + 3 + 4 + 8 + 9 + 11 + 13 + 15 + 16 + 18 = 99,$$

and for alloy 2 they are

$$R_2 = n_1(n_1 + n_2 + 1) - R_1 = 10(10 + 10 + 1) - 99 = 111.$$

From Appendix Table IX, with $n_1 = n_2 = 10$ and $\alpha = 0.05$, we find that $R^*_{0.05} = 78$. Since neither R_1 nor R_2 is less than $R^*_{0.05}$, we cannot reject the hypothesis that both alloys exhibit the same mean axial stress.

16-4.2 A Large-Sample Approximation

When both n_1 and n_2 are moderately large, say greater than 8, the distribution of R_1 can be well approximated by the normal distribution with mean

$$\mu_{R_1} = \frac{n_1(n_1 + n_2 + 1)}{2}$$

and variance

$$\sigma^2_{R_1} = \frac{n_1 n_2(n_1 + n_2 + 1)}{12}.$$

Therefore, for n_1 and $n_2 > 8$ we could use

$$Z_0 = \frac{R_1 - \mu_{R_1}}{\sigma_{R_1}} \tag{16-5}$$

as a test statistic, and the appropriate critical region as $|Z_0| > Z_{\alpha/2}$, $Z_0 > Z_\alpha$, or $Z_0 < -Z_\alpha$, depending on whether the test is a two-tailed, upper-tail, or lower-tail test.

16-4.3 Comparison with the *t*-Test

In Section 16-3.4 we discussed the comparison of the *t*-test with the Wilcoxon signed rank test. The results for the two-sample problem are identical to the one-sample case; that is, when the normality assumption is correct the Wilcoxon rank-sum test is approximately 95% as efficient as the *t*-test in large samples. On the other hand, regardless of the form of the distributions, the Wilcoxon rank-sum test will always be at least 86% as efficient if the underlying distributions are very nonnormal. The efficiency of the Wilcoxon test relative to the *t*-test is usually high if the underlying distribution has heavier tails than the normal, because the behavior of the *t*-test is very dependent on the sample mean, which is quite unstable in heavy-tailed distributions.

16-5 NONPARAMETRIC METHODS IN THE ANALYSIS OF VARIANCE

16-5.1 The Kruskal–Wallis Test

The single-factor analysis of variance model developed in Chapter 12 for comparing *a* population means is

$$y_{ij} = \mu + \tau_i + \varepsilon_{ij} \begin{cases} i = 1, 2, \ldots, a, \\ j = 1, 2, \ldots, n_i. \end{cases} \tag{16-6}$$

In this model the error terms ε_{ij} are assumed to be normally and independently distributed with mean zero and variance σ^2. The assumption of normality led directly to the F-test described in Chapter 12. The Kruskal–Wallis test is a nonparametric alternative to the F-test; it requires only that the ε_{ij} have the same continuous distribution for all treatments $i = 1, 2, \ldots, a$.

Suppose that $N = \sum_{i=1}^{a} n_i$ is the total number of observations. Rank all N observations from smallest to largest and assign the smallest observation rank 1, the next smallest rank 2, …, and the largest observation rank N. If the null hypothesis

$$H_0\colon \mu_1 = \mu_2 = \cdots = \mu_a$$

is true, the N observations come from the same distribution, and all possible assignments of the N ranks to the a samples are equally likely, then we would expect the ranks 1, 2, …, N to be mixed throughout the a samples. If, however, the null hypothesis H_0 is false, then some samples will consist of observations having predominantly small ranks while other samples will consist of observations having predominantly large ranks. Let R_{ij} be the rank of observation y_{ij}, and let $R_{i\cdot}$ and $\bar{R}_{i\cdot}$ denote the total and average of the n_i ranks in the ith treatment. When the null hypothesis is true, then

$$E(R_{ij}) = \frac{N+1}{2}$$

and

$$E(\bar{R}_{i\cdot}) = \frac{1}{n_i}\sum_{j=1}^{n_i} E(R_{ij}) = \frac{N+1}{2}.$$

The Kruskal–Wallis test statistic measures the degree to which the actual observed average ranks $\bar{R}_{i\cdot}$ differ from their expected value $(N+1)/2$. If this difference is large, then the null hypothesis H_0 is rejected. The test statistic is

$$K = \frac{12}{N(N+1)}\sum_{i=1}^{a} n_i\left(\bar{R}_{i\cdot} - \frac{N+1}{2}\right)^2. \tag{16-7}$$

An alternate computing formula is

$$K = \frac{12}{N(N+1)}\sum_{i=1}^{a}\frac{R_{i\cdot}^2}{n_i} - 3(N+1). \tag{16-8}$$

We would usually prefer equation 16-8 to equation 16-7, as it involves the rank totals rather than the averages.

The null hypothesis H_0 should be rejected if the sample data generate a large value for K. The null distribution for K has been obtained by using the fact that under H_0 each possible assignment of ranks to the a treatments is equally likely. Thus we could enumerate all possible assignments and count the number of times each value of K occurs. This has led to tables of the critical values of K, although most tables are restricted to small sample sizes n_i. In practice, we usually employ the following large-sample approximation: Whenever H_0 is true and either

$$a = 3 \text{ and } n_i \geq 6 \qquad \text{for } i = 1, 2, 3$$

or

$$a > 3 \text{ and } n_i \geq 5 \qquad \text{for } i = 1, 2, \ldots, a,$$

then K has approximately a chi-square distribution with $a - 1$ degrees of freedom. Since large values of K imply that H_0 is false, we would reject H_0 if

$$K \geq \chi^2_{\alpha, a-1}.$$

The test has approximate significance level α.

Ties in the Kruskal–Wallis Test When observations are tied, assign an average rank to each of the tied observations. When there are ties, we should replace the test statistic in equation 16-8 with

$$K = \frac{1}{S^2}\left[\sum_{i=1}^{a}\frac{R_{i\cdot}^2}{n_i} - \frac{N(N+1)^2}{4}\right], \tag{16-9}$$

where n_i is the number of observations in the ith treatment, N is the total number of observations, and

$$S^2 = \frac{1}{N-1}\left[\sum_{i=1}^{a}\sum_{j=1}^{n_i} R_{ij}^2 - \frac{N(N+1)^2}{4}\right]. \tag{16-10}$$

Note that S^2 is just the variance of the ranks. When the number of ties is moderate, there will be little difference between equations 16-8 and 16-9, and the simpler form (equation 16-8) may be used.

Example 16-6

In *Design and Analysis of Experiments*, 5th Edition (John Wiley & Sons, 2001), D. C. Montgomery presents data from an experiment in which five different levels of cotton content in a synthetic fiber were tested to determine if cotton content has any effect on fiber tensile strength. The sample data and ranks from this experiment are shown in Table 16-3. Since there is a fairly large number of ties, we use equation 16-9 as the test statistic. From equation 16-10 we find

$$\begin{aligned} S^2 &= \frac{1}{N-1}\left[\sum_{i=1}^{a}\sum_{j=1}^{n_i} R_{ij}^2 - \frac{N(N+1)^2}{4}\right] \\ &= \frac{1}{24}\left[5497.79 - \frac{25(26)^2}{4}\right] \\ &= 53.03, \end{aligned}$$

Table 16-3 Data and Ranks for the Tensile Testing Experiment

Percentage of Cotton									
15		20		25		30		35	
y_{1j}	R_{1j}	y_{2j}	R_{2j}	y_{3j}	R_{3j}	y_{4j}	R_{4j}	y_{5j}	R_{5j}
7	2.0	12	9.5	14	11.0	19	20.5	7	2.0
7	2.0	17	14.0	18	16.5	25	25.0	10	5.0
15	12.5	12	9.5	18	16.5	22	23.0	11	7.0
11	7.0	18	16.5	19	20.5	19	20.5	15	12.5
9	4.0	18	16.5	19	20.5	23	24.0	11	7.0
$R_{i\cdot}$	27.5		66.0		85.0		113.0		33.5

and the test statistic is

$$K = \frac{1}{S^2}\left[\sum_{i=1}^{a}\frac{R_{i\cdot}^2}{n_i} - \frac{N(N+1)^2}{4}\right]$$
$$= \frac{1}{53.03}\left[5245.0 - \frac{25(26)^2}{4}\right]$$
$$= 19.25.$$

Since $K > \chi^2_{0.01,4} = 13.28$, we would reject the null hypothesis and conclude that treatments differ. This is the same conclusion given by the usual analysis of variance F-test.

16-5.2 The Rank Transformation

The procedure used in the previous section of replacing the observations by their ranks is called the *rank transformation.* It is a very powerful and widely useful technique. If we were to apply the ordinary F-test to the ranks rather than to the original data, we would obtain

$$F_0 = \frac{K/(a-1)}{(N-1-K)/(N-a)}$$

as the test statistic. Note that as the Kruskal–Wallis statistic K increases or decreases, F_0 also increases or decreases, so the Kruskal–Wallis test is nearly equivalent to applying the usual analysis of variance to the ranks.

The rank transformation has wide applicability in experimental design problems for which no nonparametric alternative to the analysis of variance exists. If the data are ranked and the ordinary F-test applied, an approximate procedure results, but one that has good statistical properties. When we are concerned about the normality assumption or the effect of outliers or "wild" values, we recommend that the usual analysis of variance be performed on both the original data and the ranks. When both procedures give similar results, the analysis of variance assumptions are probably satisfied reasonably well, and the standard analysis is satisfactory. When the two procedures differ, the rank transformation should be preferred since it is less likely to be distorted by nonnormality and unusual observations. In such cases, the experimenter may want to investigate the use of transformations for nonnormality and examine the data and the experimental procedure to determine whether outliers are present and if so, why they have occurred.

16-6 SUMMARY

This chapter has introduced nonparametric or distribution-free statistical methods. These procedures are alternatives to the usual parametric t- and F-tests when the normality assumption for the underlying population is not satisfied. The sign test can be used to test hypotheses about the median of a continuous distribution. It can also be applied to paired observations. The Wilcoxon signed rank test can be used to test hypotheses about the mean of a symmetric continuous distribution. It can also be applied to paired observations. The Wilcoxon signed rank test is a good alternative to the t-test. The two-sample hypothesis-testing problem on means of continuous symmetric distributions is approached using the Wilcoxon rank-sum test. This procedure compares very favorably with the two-sample t-test. The Kruskal–Wallis test is a useful alternative to the F-test in the analysis of variance.

16-7 EXERCISES

16-1. Ten samples were taken from a plating bath used in an electronics manufacturing process and the bath pH determined. The sample pH values are given below:

7.91, 7.85, 6.82, 8.01, 7.46, 6.95, 7.05, 7.35, 7.25, 7.42.

Manufacturing engineering believes that pH has a median value of 7.0. Do the sample data indicate that this statement is correct? Use the sign test to investigate this hypothesis.

16-2. The titanium content in an aircraft-grade alloy is an important determinant of strength. A sample of 20 test coupons reveals the following titanium contents (in percent):

8.32, 8.05, 8.93, 8.65, 8.25, 8.46, 8.52, 8.35, 8.36, 8.41, 8.42, 8.30, 8.71, 8.75, 8.60, 8.83, 8.50, 8.38, 8.29, 8.46.

The median titanium content should be 8.5%. Use the sign test to investigate this hypothesis.

16-3. The distribution time between arrivals in a telecommunication system is exponential, and the system manager wishes to test the hypothesis that H_0: $\tilde{\mu} = 3.5$ min versus H_1: $\tilde{\mu} > 3.5$ min.

(a) What is the value of the mean of the exponential distribution under H_0: $\tilde{\mu} = 3.5$?

(b) Suppose that we have taken a sample of $n = 10$ observations and we observe $R^- = 3$. Would the sign test reject H_0 at $\alpha = 0.05$?

(c) What is the type II error of this test if $\tilde{\mu} = 4.5$?

16-4. Suppose that we take a sample of $n = 10$ measurements from a normal distribution with $\sigma = 1$. We wish to test H_0: $\mu = 0$ against H_1: $\mu > 0$. The normal test statistic is $Z_0 = (\bar{X} - \mu_0)/\sigma/\sqrt{n}$, and we decide to use a critical region of 1.96 (that is, reject H_0 if $Z_0 \geq 1.96$).

(a) What is α for this test?

(b) What is β for this test if $\mu = 1$?

(c) If a sign test is used, specify the critical region that gives an α value consistent with α for the normal test.

(d) What is the β value for the sign test if $\mu = 1$? Compare this with the result obtained in part (b).

16-5. Two different types of tips can be used in a Rockwell hardness tester. Eight coupons from test ingots of a nickel-based alloy are selected, and each coupon is tested twice, once with each tip. The Rockwell C-scale hardness readings are shown next. Use the sign test to determine whether or not the two tips produce equivalent hardness readings.

Coupon	Tip 1	Tip 2
1	63	60
2	52	51
3	58	56
4	60	59
5	55	58
6	57	54
7	53	52
8	59	61

16-6. Testing for Trends. A turbocharger wheel is manufactured using an investment casting process. The shaft fits into the wheel opening, and this wheel opening is a critical dimension. As wheel wax patterns are formed, the hard tool producing the wax patterns wears. This may cause growth in the wheel-opening dimension. Ten wheel-opening measurements, in time order of production, are shown below:

4.00 (mm), 4.02, 4.03, 4.01, 4.00, 4.03, 4.04, 4.02, 4.03, 4.03.

(a) Suppose that p is the probability that observation X_{i+5} exceeds observation X_i. If there is no upward or downward trend, then X_{i+5} is no more or less likely to exceed X_i or lie below X_i. What is the value of p?

(b) Let V be the number of values of i for which $X_{i+5} > X_i$. If there is no upward or downward trend in the measurements, what is the probability distribution of V?

(c) Use the data above and the results of parts (a) and (b) to test H_0: there is no trend versus H_1: there is upward trend. Use $\alpha = 0.05$.

Note that this test is a modification of the sign test. It was developed by Cox and Stuart.

16-7. Consider the Wilcoxon signed rank test, and suppose that $n = 5$. Assume that H_0: $\mu = \mu_0$ is true.

(a) How many different sequences of signed ranks are possible? Enumerate these sequences.

(b) How many different values of R^+ are there? Find the probability associated with each value of R^+.

(c) Suppose that we define the critical region of the test to be R^*_α such that we would reject if $R^+ > R^*_\alpha$, and $R^*_\alpha = 13$. What is the approximate α level of this test?

(d) Can you see from this exercise how the critical values for the Wilcoxon signed rank test were developed? Explain.

16-8. Consider the data in Exercise 16-1, and assume that the distribution of pH is symmetric and continuous. Use the Wilcoxon signed rank test to test the hypothesis H_0: $\mu = 7$ against H_1: $\mu \neq 7$.

16-9. Consider the data in Exercise 16-2. Suppose that the distribution of titanium content is symmetric and continuous. Use the Wilcoxon signed rank test to test the hypotheses H_0: $\mu = 8.5$ versus H_1: $\mu \neq 8.5$.

16-10. Consider the data in Exercise 16-2. Use the large-sample approximation for the Wilcoxon signed rank test to test the hypotheses H_0: $\mu = 8.5$ versus H_1: $\mu \neq 8.5$. Assume that the distribution of titanium content is continuous and symmetric.

16-11. For the large-sample approximation to the Wilcoxon signed rank test, derive the mean and standard deviation of the test statistic used in the procedure.

16-12. Consider the Rockwell hardness test data in Exercise 16-5. Assume that both distributions are continuous and use the Wilcoxon signed rank test to test that the mean difference in hardness readings between the two tips is zero.

16-13. An electrical engineer must design a circuit to deliver the maximum amount of current to a display tube to achieve sufficient image brightness. Within his allowable design constraints, he has developed two candidate circuits and tests prototypes of each. The resulting data (in microamperes) is shown below:

Circuit 1: 251, 255, 258, 257, 250, 251, 254, 250, 248

Circuit 2: 250, 253, 249, 256, 259, 252, 260, 251

Use the Wilcoxon rank-sum test to test H_0: $\mu_1 = \mu_2$ against the alternative H_1: $\mu_1 > \mu_2$.

16-14. A consultant frequently travels from Phoenix, Arizona, to Los Angeles, California. He will use one of two airlines, United or Southwest. The number of minutes that his flight arrived late for the last six trips on each airline is shown below. Is there evidence that either airline has superior on-time arrival performance?

United	2	19	1	−2	8	0	(minutes late)
Southwest	20	4	8	8	−3	5	(minutes late)

16-15. The manufacturer of a hot tub is interested in testing two different heating elements for his product. The element that produces the maximum heat gain after 15 minutes would be preferable. He obtains 10 samples of each heating unit and tests each one. The heat gain after 15 minutes (in °F) is shown below. Is there any reason to suspect that one unit is superior to the other?

Unit 1 25, 27, 29, 31, 30, 26, 24, 32, 33, 38.

Unit 2 31, 33, 32, 35, 34, 29, 38, 35, 37, 30.

16-16. In *Design and Analysis of Experiments*, 5th Edition (John Wiley & Sons, 2001), D. C. Montgomery presents the results of an experiment to compare four different mixing techniques on the tensile strength of portland cement. The results are shown below. Is there any indication that mixing technique affects the strength?

Mixing Technique	Tensile Strength (lb / in.^{2})			
1	3129	3000	2865	2890
2	3200	3000	2975	3150
3	2800	2900	2985	3050
4	2600	2700	2600	2765

16-17. An article in the *Quality Control Handbook*, 3rd Edition (McGraw-Hill, 1962) presents the results of an experiment performed to investigate the effect of three different conditioning methods on the breaking strength of cement briquettes. The data are shown below. Is there any indication that conditioning method affects breaking strength?

Conditioning Method	Breaking Strength (lb / in.^{2})				
1	553	550	568	541	537
2	553	599	579	545	540
3	492	530	528	510	571

16-18. In *Statistics for Research* (John Wiley & Sons, 1983), S. Dowdy and S. Wearden present the results of an experiment to measure stress resulting from operating hand-held chain saws. The experimenters measured the kickback angle through which the saw is deflected when it begins to cut a 3-inch stock synthetic board. Shown below are deflection angles for five saws chosen at random from each of four different manufacturers. Is there any evidence that the manufacturers' products differ with respect to kickback angle?

Manufacturer	Kickback Angle				
A	42	17	24	39	43
B	28	50	44	32	61
C	57	45	48	41	54
D	29	40	22	34	30

Chapter 17

Statistical Quality Control and Reliability Engineering

The quality of the products and services used by our society has become a major consumer decision factor in many, if not most, businesses today. Regardless of whether the consumer is an individual, a corporation, a military defense program, or a retail store, the consumer is likely to consider quality of equal importance to cost and schedule. Consequently, quality improvement has become a major concern of many U.S. corporations. This chapter is about statistical quality control and reliability engineering methods, two sets of tools that are essential in quality-improvement activities.

17-1 QUALITY IMPROVEMENT AND STATISTICS

Quality means *fitness for use*. For example, we may purchase automobiles that we expect to be free of manufacturing defects and that should provide reliable and economical transportation, a retailer buys finished goods with the expectation that they are properly packaged and arranged for easy storage and display, or a manufacturer buys raw material and expects to process it with minimal rework or scrap. In other words, all consumers expect that the products and services they buy will meet their requirements, and those requirements define fitness for use.

Quality, or fitness for use, is determined through the interaction of *quality of design* and *quality of conformance*. By quality of design we mean the different grades or levels of performance, reliability, serviceability, and function that are the result of deliberate engineering and management decisions. By quality of conformance, we mean the systematic *reduction of variability* and *elimination of defects* until every unit produced is identical and defect free.

There is some confusion in our society about *quality improvement*; some people still think that it means gold plating a product or spending more money to develop a product or process. This thinking is wrong. Quality improvement means the systematic *elimination of waste*. Examples of waste include scrap and rework in manufacturing, inspection and test, errors on documents (such as engineering drawings, checks, purchase orders, and plans), customer complaint hotlines, warranty costs, and the time required to do things over again that could have been done right the first time. A successful quality-improvement effort can eliminate much of this waste and lead to lower costs, higher productivity, increased customer satisfaction, increased business reputation, higher market share, and ultimately higher profits for the company.

Statistical methods play a vital role in quality improvement. Some applications include the following:

1. In product design and development, statistical methods, including designed experiments, can be used to compare different materials and different components or ingredients, and to help in both system and component tolerance determination. This can significantly lower development costs and reduce development time.
2. Statistical methods can be used to determine the capability of a manufacturing process. Statistical process control can be used to systematically improve a process by reduction of variability.
3. Experiment design methods can be used to investigate improvements in the process. These improvements can lead to higher yields and lower manufacturing costs.
4. Life testing provides reliability and other performance data about the product. This can lead to new and improved designs and products that have longer useful lives and lower operating and maintenance costs.

Some of these applications have been illustrated in earlier chapters of this book. It is essential that engineers and managers have an in-depth understanding of these statistical tools in any industry or business that wants to be the high-quality, low-cost producer. In this chapter we give an introduction to the basic methods of statistical quality control and reliability engineering that, along with experimental design, form the basis of a successful quality-improvement effort.

17-2 STATISTICAL QUALITY CONTROL

The field of statistical quality control can be broadly defined as consisting of those statistical and engineering methods useful in the measurement, monitoring, control, and improvement of quality. In this chapter, a somewhat more narrow definition is employed. We will define statistical quality control as the statistical and engineering methods for process control.

Statistical quality control is a relatively new field, dating back to the 1920s. Dr. Walter A. Shewhart of the Bell Telephone Laboratories was one of the early pioneers of the field. In 1924, he wrote a memorandum showing a modern control chart, one of the basic tools of statistical process control. Harold F. Dodge and Harry G. Romig, two other Bell System employees, provided much of the leadership in the development of statistically based sampling and inspection methods. The work of these three men forms the basis of the modern field of statistical quality control. World War II saw the widespread introduction of these methods to U.S. industry. Dr. W. Edwards Deming and Dr. Joseph M. Juran have been instrumental in spreading statistical quality-control methods since World War II.

The Japanese have been particularly successful in deploying statistical quality-control methods and have used statistical methods to gain significant advantage relative to their competitors. In the 1970s American industry suffered extensively from Japanese (and other foreign) competition, and that has led, in turn, to renewed interest in statistical quality-control methods in the United States. Much of this interest focuses on *statistical process control* and *experimental design*. Many U.S. companies have begun extensive programs to implement these methods into their manufacturing, engineering, and other business organizations.

17-3 STATISTICAL PROCESS CONTROL

It is impossible to inspect quality into a product; the product must be built right the first time. This implies that the manufacturing process must be stable or repeatable and capable of operating with little variability around the target or nominal dimension. Online statistical process controls are powerful tools useful in achieving process stability and improving capability through the reduction of variability.

It is customary to think of statistical process control (SPC) as a set of problem-solving tools that may be applied to any process. The major tools of SPC are the following:

1. Histogram
2. Pareto chart
3. Cause-and-effect diagram
4. Defect-concentration diagram
5. Control chart
6. Scatter diagram
7. Check sheet

While these tools are an important part of SPC, they really constitute only the technical aspect of the subject. SPC is an *attitude*—a desire of all individuals in the organization for continuous improvement in quality and productivity by the systematic reduction of variability. The control chart is the most powerful of the SPC tools. We now give an introduction to several basic types of control charts.

17-3.1 Introduction to Control Charts

The basic theory of the control chart was developed by Walter Shewhart in the 1920s. To understand how a control chart works, we must first understand Shewhart's theory of variation. Shewhart theorized that all processes, however good, are characterized by a certain amount of variation if we measure with an instrument of sufficient resolution. When this variability is confined to *random* or *chance variation* only, the process is said to be in a state of *statistical control*. However, another situation may exist in which the process variability is also affected by some *assignable cause*, such as a faulty machine setting, operator error, unsatisfactory raw material, worn machine components, and so on.[1] These assignable causes of variation usually have an adverse effect on product quality, so it is important to have some systematic technique for detecting serious departures from a state of statistical control as soon after they occur as possible. Control charts are principally used for this purpose.

The power of the control chart lies in its ability to distinguish assignable causes from random variation. It is the job of the individual using the control chart to identify the underlying root cause responsible for the out-of-control condition, develop and implement an appropriate corrective action, and then follow up to ensure that the assignable cause has been eliminated from the process. There are three points to remember.

1. A state of statistical control is not a natural state for most processes.
2. The attentive use of control charts will result in the elimination of assignable causes, yielding an in-control process and reduced process variability.
3. The control chart is ineffective without the system to develop and implement corrective actions that attack the root causes of problems. Management and engineering involvement is usually necessary to accomplish this.

---

[1]Sometimes *common* cause is used instead of "random" or "chance cause," and *special* cause is used instead of "assignable cause."

We distinguish between control charts for measurements and control charts for attributes, depending on whether the observations on the quality characteristic are measurements or enumeration data. For example, we may choose to measure the diameter of a shaft, say with a micrometer, and utilize these data in conjunction with a control chart for measurements. On the other hand, we may judge each unit of product as either defective or nondefective and use the fraction of defective units found or the total number of defects in conjunction with a control chart for attributes. Obviously, certain products and quality characteristics lend themselves to analysis by either method, and a clear-cut choice between the two methods may be difficult.

A control chart, whether for measurements or attributes, consists of a *centerline*, corresponding to the average quality at which the process should perform when statistical control is exhibited, and two *control limits*, called the upper and lower control limits (UCL and LCL). A typical control chart is shown in Fig. 17-1. The control limits are chosen so that values falling between them can be attributed to chance variation, while values falling beyond them can be taken to indicate a lack of statistical control. The general approach consists of periodically taking a random sample from the process, computing some appropriate quantity, and plotting that quantity on the control chart. When a sample value falls outside the control limits, we search for some assignable cause of variation. However, even if a sample value falls between the control limits, a trend or some other systematic pattern may indicate that some action is necessary, usually to avoid more serious trouble. The samples should be selected in such a way that each sample is as homogeneous as possible and at the same time maximizes the opportunity for variation due to an assignable cause to be present. This is usually called the *rational subgroup* concept. Order of production and source (if more than one source exists) are commonly used bases for obtaining rational subgroups.

The ability to interpret control charts accurately is usually acquired with experience. It is necessary that the user be thoroughly familiar with both the statistical foundation of control charts and the nature of the production process itself.

17-3.2 Control Charts for Measurements

When dealing with a quality characteristic that can be expressed as a measurement, it is customary to exercise control over both the average value of the quality characteristic and its variability. Control over the average quality is exercised by the control chart for means, usually called the $\overline{X}$ chart. Process variability can be controlled by either a range (R) chart or a

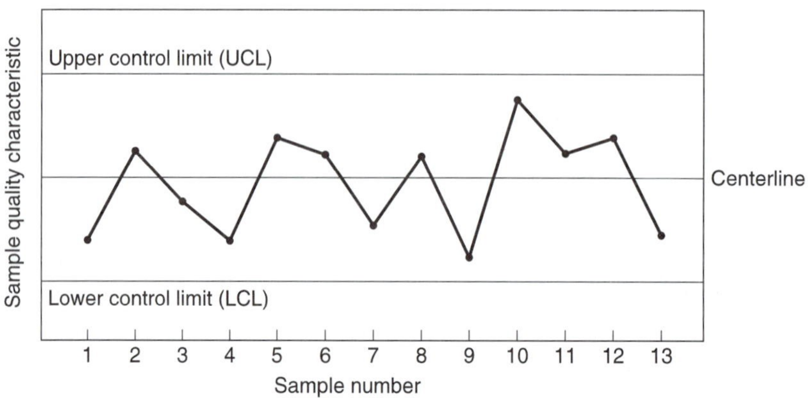

Figure 17-1 A typical control chart.

standard deviation chart, depending on how the population standard deviation is estimated. We will discuss only the R chart.

Suppose that the process mean and standard deviation, say μ and σ, are known, and, furthermore, that we can assume that the quality characteristic follows the normal distribution. Let $\overline{X}$ be the sample mean based on a random sample of size n from this process. Then the probability is $1 - \alpha$ that the mean of such random samples will fall between $\mu + Z_{a/2}\left(\sigma/\sqrt{n}\right)$ and $\mu - Z_{a/2}\left(\sigma/\sqrt{n}\right)$. Therefore, we could use these two values as the upper and lower control limits, respectively. However, we usually do not know μ and σ and they must be estimated. In addition, we may not be able to make the normality assumption. For these reasons, the probability limit $1 - \alpha$ is seldom used in practice. Usually $Z_{\alpha/2}$ is replaced by 3, and "3-sigma" control limits are used.

When μ and σ are unknown, we often estimate them on the basis of preliminary samples, taken when the process is thought to be in control. We recommend the use of at least 20–25 preliminary samples. Suppose k preliminary samples are available, each of size n. Typically, n will be 4, 5, or 6; these relatively small sample sizes are widely used and often arise from the construction of rational subgroups. Let the sample mean for the ith sample be $\overline{X}_i$. Then we estimate the mean of the population, μ, by the grand mean

$$\overline{\overline{X}} = \frac{1}{k}\sum_{i=1}^{k} \overline{X}_i. \tag{17-1}$$

Thus, we may take $\overline{\overline{X}}$ as the centerline on the $\overline{X}$ control chart.

We may estimate σ from either the standard deviations or the ranges of the k samples. Since it is more frequently used in practice, we confine our discussion to the range method. The sample size is relatively small, so there is little loss in efficiency in estimating σ from the sample ranges. The relationship between the range, R, of a sample from a normal population with known parameters and the standard deviation of that population is needed. Since R is a random variable, the quantity $W = R/\sigma$, called the relative range, is also a random variable. The parameters of the distribution of W have been determined for any sample size n. The mean of the distribution of W is called d_2, and a table of d_2 for various n is given in Table XIII of the Appendix. Let R_i be the range of the ith sample, and let

$$\overline{R} = \frac{1}{k}\sum_{i=1}^{k} R_i \tag{17-2}$$

be the average range. Then an estimate of σ would be

$$\hat{\sigma} = \frac{\overline{R}}{d_2}. \tag{17-3}$$

Therefore, we may use as our upper and lower control limits for the $\overline{X}$ chart

$$\begin{aligned} UCL &= \overline{\overline{X}} + \frac{3}{d_2\sqrt{n}}\overline{R}, \\ LCL &= \overline{\overline{X}} - \frac{3}{d_2\sqrt{n}}\overline{R}. \end{aligned} \tag{17-4}$$

We note that the quantity

$$A_2 = \frac{3}{d_2\sqrt{n}}$$

is a constant depending on the sample size, so it is possible to rewrite equations 17-4 as

$$UCL = \overline{\overline{X}} + A_2\overline{R},$$
$$LCL = \overline{\overline{X}} - A_2\overline{R}. \tag{17-5}$$

The constant A_2 is tabulated for various sample sizes in Table XIII of the Appendix.

The parameters of the R chart may also be easily determined. The centerline will obviously be $\overline{R}$. To determine the control limits, we need an estimate of σ_R, the standard deviation of R. Once again, assuming the process is in control, the distribution of the relative range, W, will be useful. The standard deviation of W, say σ_W, is a function of n, which has been determined. Thus, since

$$R = W\sigma,$$

we may obtain the standard deviation of R as

$$\sigma_R = \sigma_W\sigma.$$

As σ is unknown, we may estimate σ_R as

$$\hat{\sigma}_R = \sigma_W \frac{\overline{R}}{d_2},$$

and we would use as the upper and lower control limits on the R chart

$$UCL = \overline{R} + \frac{3\sigma_W}{d_2}\overline{R},$$
$$LCL = \overline{R} - \frac{3\sigma_W}{d_2}\overline{R}. \tag{17-6}$$

Setting $D_3 = 1 - 3\sigma_W/d_2$ and $D_4 = 1 + 3\sigma_W/d_2$, we may rewrite equation 17-6 as

$$UCL = D_4\overline{R}, \tag{17-7}$$
$$LCL = D_3\overline{R},$$

where D_3 and D_4 are tabulated in Table XIII of the Appendix.

When preliminary samples are used to construct limits for control charts, it is customary to treat these limits as trial values. Therefore, the k sample means and ranges should be plotted on the appropriate charts, and any points that exceed the control limits should be investigated. If assignable causes for these points are discovered, they should be eliminated and new limits for the control charts determined. In this way, the process may eventually be brought into statistical control and its inherent capabilities assessed. Other changes in process centering and dispersion may then be contemplated.

Example 17-1

A component part for a jet aircraft engine is manufactured by an investment casting process. The vane opening on this casting is an important functional parameter of the part. We will illustrate the use of $\overline{X}$ and R control charts to assess the statistical stability of this process. Table 17-1 presents 20 samples of five parts each. The values given in the table have been coded by using the last three digits of the dimension; that is, 31.6 should be 0.50316 inch.

The quantities $\overline{\overline{X}} = 33.33$ and $\overline{R} = 5.85$ are shown at the foot of Table 17-1. Notice that even though $\overline{X}$, $\overline{\overline{X}}$, R, and $\overline{R}$ are now realizations of random variables, we have still written them as

Table 17-1 Vane Opening Measurements

Sample Number	x_1	x_2	x_3	x_4	x_5	$\overline{X}$	R
1	33	29	31	32	33	31.6	4
2	35	33	31	37	31	33.2	6
3	35	37	33	34	36	35.0	4
4	30	31	33	34	33	32.2	4
5	33	34	35	33	34	33.8	2
6	38	37	39	40	38	38.4	3
7	30	31	32	34	31	31.6	4
8	29	39	38	39	39	36.8	10
9	28	34	35	36	43	35.2	15
10	39	33	32	34	32	34.0	7
11	28	30	28	32	31	29.8	4
12	31	35	35	35	34	34.0	4
13	27	32	34	35	37	33.0	10
14	33	33	35	37	36	34.8	4
15	35	37	32	35	39	35.6	7
16	33	33	27	31	30	30.8	6
17	35	34	34	30	32	33.0	5
18	32	33	30	30	33	31.6	3
19	25	27	34	27	28	28.2	9
20	35	35	36	33	30	33.8	6
						$\overline{\overline{X}} = 33.33$	$\overline{R} = 5.85$

uppercase letters. This is the usual convention in quality control, and it will always be clear from the context what the notation implies. The trial control limits are, for the $\overline{X}$ chart,

$$\overline{\overline{X}} \pm A_2\overline{R} = 33.3 \pm (0.577)(5.85) = 33.33 \pm 3.37,$$

or

$$UCL = 36.70,$$
$$LCL = 29.96.$$

For the R chart, the trial control limits are

$$UCL = D_4\overline{R} = (2.115)(5.85) = 12.37,$$
$$LCL = D_3\overline{R} = (0)(5.85) = 0.$$

The $\overline{X}$ and R control charts with these trial control limits are shown in Fig. 17-2. Notice that samples 6, 8, 11, and 19 are out of control on the $\overline{X}$ chart, and that sample 9 is out of control on the R chart. Suppose that all of these assignable causes can be traced to a defective tool in the wax-molding area. We should discard these five samples and recompute the limits for the $\overline{X}$ and R charts. These new revised limits are, for the $\overline{X}$ chart,

$$UCL = \overline{\overline{X}} + A_2\overline{R} = 32.90 + (0.577)(5.313) = 35.96,$$
$$LCL = \overline{\overline{X}} - A_2\overline{R} = 32.90 - (0.577)(5.313) = 29.84,$$

and, for the R chart, they are

$$UCL = D_4\overline{R} = (2.115)(5.067) = 10.71,$$
$$LCL = D_3\overline{R} = (0)(5.067) = 0.$$

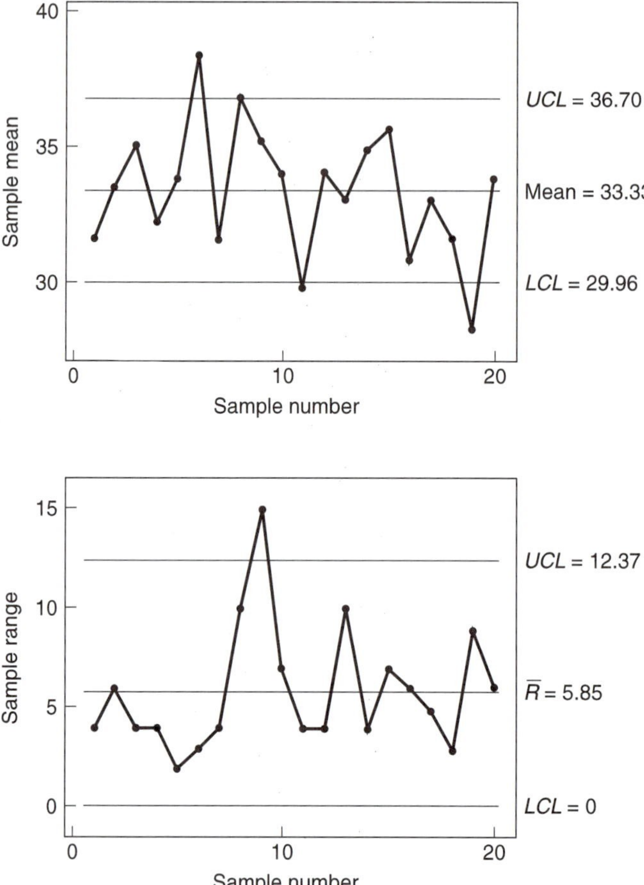

Figure 17-2 The $\overline{X}$ and R control charts for vane opening.

The revised control charts are shown in Fig. 17-3. Notice that we have treated the first 20 preliminary samples as *estimation data* with which to establish control limits. These limits can now be used to judge the statistical control of future production. As each new sample becomes available, the values of $\overline{X}$ and R should be computed and plotted on the control charts. It may be desirable to revise the limits periodically, even if the process remains stable. The limits should always be revised when process improvements are made.

Estimating Process Capability

It is usually necessary to obtain some information about the *capability* of the process, that is, about the performance of the process when it is operating in control. Two graphical tools, the *tolerance* chart (or tier chart) and the *histogram,* are helpful in assessing process capability. The tolerance chart for all 20 samples from the vane manufacturing process is shown in Fig. 17-4. The specifications on vane opening, 0.5030 ± 0.001 inch, are also shown on the chart. In terms of the coded data, the upper specification limit is $USL = 40$ and the lower specification limit is $LSL = 20$. The tolerance chart is useful in revealing patterns over time in the individual measurements, or it may show that a particular value of $\overline{X}$ or R was produced by one or two unusual observations in the sample. For example, note the two unusual observations in sample 9 and the single unusual observation in sample 8. Note also that it

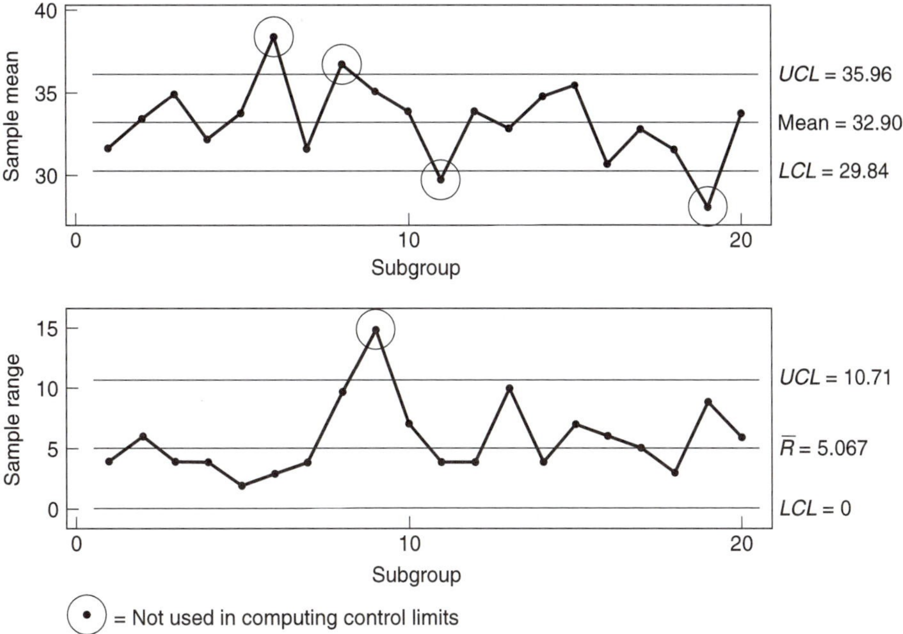

Figure 17-3 $\overline{X}$ and R control charts for vane opening, revised limits.

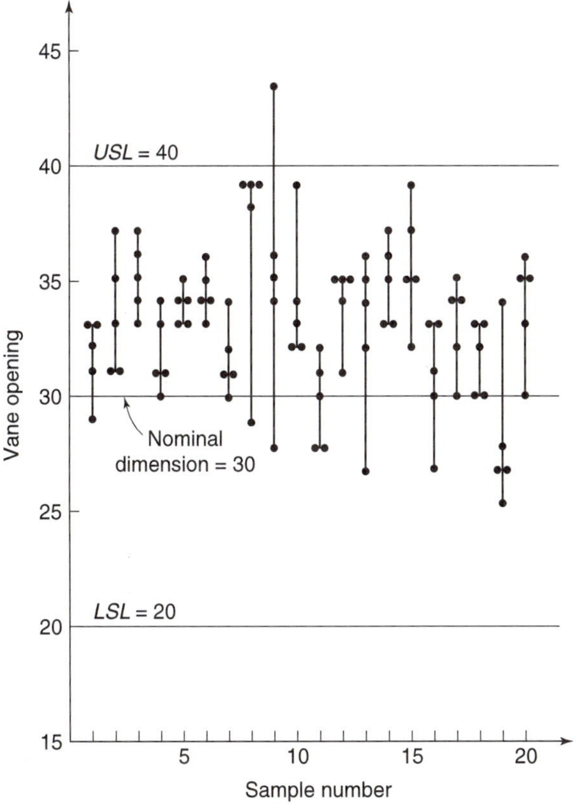

Figure 17-4 Tolerance diagram of vane openings.

is appropriate to plot the specification limits on the tolerance chart, since it is a chart of individual measurements. *It is never appropriate to plot specification limits on a control chart, or to use the specifications in determining the control limits.* Specification limits and control limits are unrelated. Finally, note from Fig. 17-4 that the process is running off center from the nominal dimension of 0.5030 inch.

The histogram for the vane opening measurements is shown in Fig. 17-5. The observations from samples 6, 8, 9, 11, and 19 have been deleted from this histogram. The general impression from examining this histogram is that the process is capable of meeting the specifications, but that it is running off center.

Another way to express process capability is in terms of the process capability ratio (*PCR*), defined as

$$PCR = \frac{USL - LSL}{6\sigma}. \tag{17-8}$$

Notice that the 6σ spread (3σ on either side of the mean) is sometimes called the *basic capability* of the process. The limits 3σ on either side of the process mean are sometimes called *natural* tolerance limits, as these represent limits that an in-control process should meet with most of the units produced. For the vane opening, we could estimate σ as

$$\hat{\sigma} = \frac{\overline{R}}{d_2} = \frac{4.8}{2.326} = 2.06.$$

Therefore, an estimator for the *PCR* is

$$\begin{aligned} PCR &= \frac{USL - LSL}{6\sigma} \\ &= \frac{40 - 20}{6(2.06)} \\ &= 1.62. \end{aligned}$$

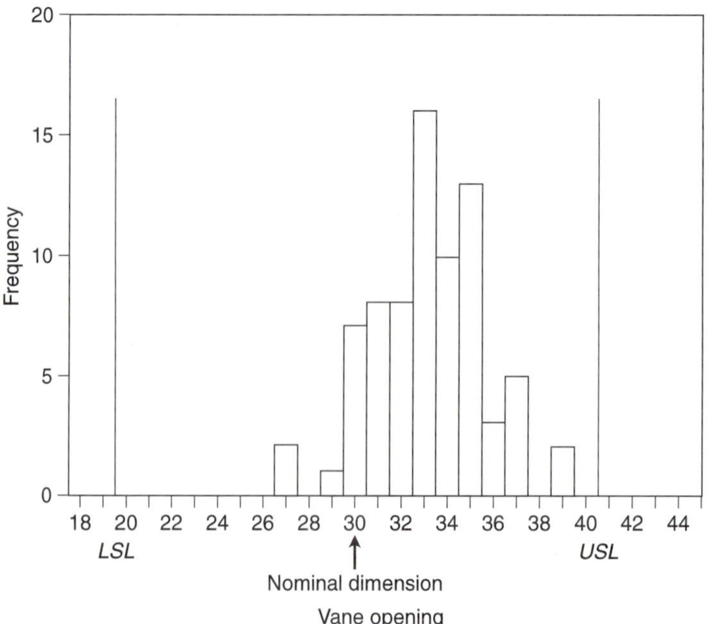

Figure 17-5 Histogram for vane opening.

The *PCR* has a natural interpretation; (1/*PCR*)100 is just the percentage of the tolerance band used by the process. Thus, the vane opening process uses approximately (1/1.62)100 = 61.7% of the tolerance band.

Figure 17-6a shows a process for which the *PCR* exceeds unity. Since the process natural tolerance limits lie inside the specifications, very few defective or nonconforming units will be produced. If *PCR* = 1, as shown in Fig. 17-6b, more nonconforming units result. In fact, for a normally distributed process, if *PCR* = 1, the fraction nonconforming is 0.27%, or 2700 parts per million. Finally, when the *PCR* is less than unity, as in Fig. 17-6c, the process is very yield sensitive and a large number of nonconforming units will be produced.

The definition of the *PCR* given in equation 17-8 implicitly assumes that the process is centered at the nominal dimension. If the process is running off center, its *actual capability* will be less than that indicated by the *PCR*. It is convenient to think of *PCR* as a measure of *potential capability*, that is, capability with a *centered process.* If the process is not centered, then a measure of actual capability is given by

$$PCR_k = \min\left[\frac{USL - \overline{\overline{X}}}{3\sigma}, \frac{\overline{\overline{X}} - LSL}{3\sigma}\right]. \tag{17-9}$$

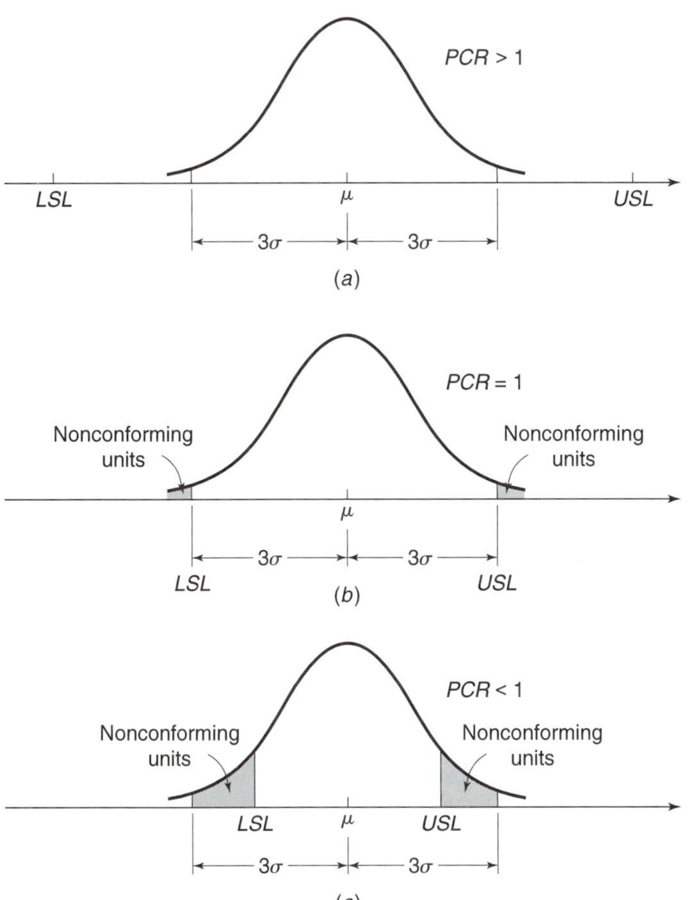

Figure 17-6 Process fallout and the process capability ratio (*PCR*).

In effect, PCR_k is a one-sided process capability ratio that is calculated relative to the specification limit nearest to the process mean. For the vane opening process, we find that

$$
\begin{aligned}
PCR_k &= \min\left[\frac{USL - \overline{\overline{X}}}{3\sigma}, \frac{\overline{\overline{X}} - LSL}{3\sigma}\right] \\
&= \min\left[\frac{40 - 33.19}{3(2.06)} = 1.10, \frac{33.19 - 20}{3(2.06)} = 2.13\right] \\
&= 1.10.
\end{aligned}
$$

Note that if $PCR = PCR_k$ the process is centered at the nominal dimension. Since $PCR_k = 1.10$ for the vane opening process, and $PCR = 1.62$, the process is obviously running off center, as was first noted in Figs. 17-4 and 17-5. This off-center operation was ultimately traced to an oversized wax tool. Changing the tooling resulted in a substantial improvement in the process.

Montgomery (2001) provides guidelines on appropriate values of the PCR and a table relating fallout for a normally distributed process in statistical control as a function of PCR. Many U.S. companies use $PCR = 1.33$ as a minimum acceptable target and $PCR = 1.66$ as a minimum target for strength, safety, or critical characteristics. Also, some U.S. companies, particularly the automobile industry, have adopted the Japanese terminology $C_p = PCR$ and $C_{pk} = PCR_k$. As C_p has another meaning in statistics (in multiple regression; see Chapter 15), we prefer the traditional notation PCR and PCR_k.

17-3.3 Control Charts for Individual Measurements

Many situations exist in which the sample consists of a single observation; that is, $n = 1$. These situations occur when production is very slow or costly and it is impractical to allow the sample size to be greater than one. Other cases include processes where every observation can be measured due to automated inspection, for example. The *Shewhart control chart for individual measurements* is appropriate for this type of situation. We will see later in this chapter that the exponentially weighted moving average control chart and the cumulative sum control chart may be more informative than the individual chart.

The Shewhart control chart uses the moving range, MR, of two successive observations for estimating the process variability. The moving range is defined

$$MR_i = |x_i - x_{i-1}|.$$

For example, for m observations, $m - 1$ moving ranges are calculated as $MR_2 = |x_2 - x_1|$, $MR_3 = |x_3 - x_2|$, …, $MR_m = |x_m - x_{m-1}|$. Simultaneous control charts can be established on the individual observations and on the moving range.

The control limits for the individuals control chart are calculated as

$$
\begin{aligned}
UCL &= \bar{x} + 3\frac{\overline{MR}}{d_2}, \\
\text{Centerline} &= \bar{x}, \\
LCL &= \bar{x} - 3\frac{\overline{MR}}{d_2},
\end{aligned}
\tag{17-10}
$$

where $\overline{MR}$ is the sample mean of the MR_i.

If a moving range of size $n = 2$ is used, then $d_2 = 1.128$ from Table XIII of the Appendix. The control limits for the moving range control chart are

$$\begin{aligned} UCL &= D_4\overline{MR}, \\ \text{Centerline} &= \overline{MR}, \\ LCL &= D_3\overline{MR}. \end{aligned} \tag{17-11}$$

Example 17-2

Batches of a particular chemical product are selected from a process and the purity measured on each. Data for 15 successive batches have been collected and are given in Table 17-2. The moving ranges of size $n = 2$ are also displayed in Table 17-2.

To set up the control chart for individuals, we first need the sample average of the 15 purity measurements. This average is found to be $\bar{x} = 0.757$. The average of the moving ranges of two observations is $\overline{MR} = 0.046$. The control limits for the individuals chart with moving ranges of size 2 using the limits in equation 17-10 are

$$\begin{aligned} UCL &= 0.757 + 3\frac{0.046}{1.128} = 0.879, \\ \text{Centerline} &= 0.757, \\ LCL &= 0.757 - 3\frac{0.046}{1.128} = 0.635. \end{aligned}$$

The control limits for the moving range chart are found using the limits given in Equation 17-11:

$$\begin{aligned} UCL &= 3.267(0.046) = 0.150, \\ \text{Centerline} &= 0.046, \\ LCL &= 0(0.046) = 0. \end{aligned}$$

Table 17-2 Purity of Chemical Product

Batch	Purity, x	Moving Range, MR
1	0.77	
2	0.76	0.01
3	0.77	0.01
4	0.72	0.05
5	0.73	0.01
6	0.73	0.00
7	0.85	0.12
8	0.70	0.15
9	0.75	0.05
10	0.74	0.01
11	0.75	0.01
12	0.84	0.09
13	0.79	0.05
14	0.72	0.07
15	0.74	0.02
	$\bar{x} = 0.757$	$\overline{MR} = 0.046$

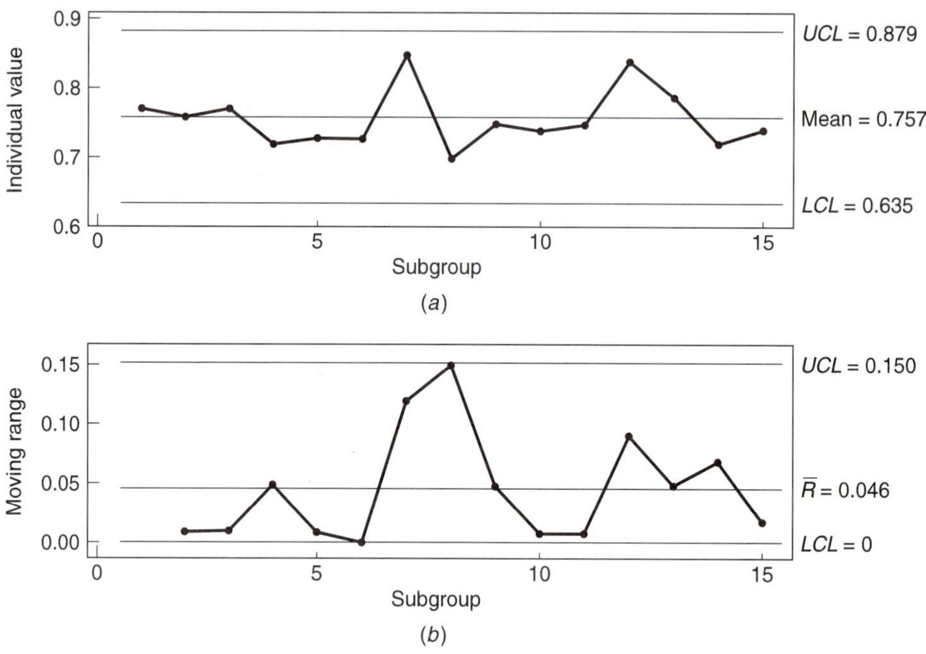

Figure 17-7 Control charts for (a) the individual observations and (b) the moving range on purity.

The control charts for individual observations and for the moving range are provided in Fig. 17-7. Since there are no points beyond the control limits, the process appears to be in statistical control.

The individuals chart can be interpreted much like the $\overline{X}$ control chart. An out-of-control situation would be indicated by either a point (or points) plotting beyond the control limits or a pattern such as a run on one side of the centerline.

The moving range chart cannot be interpreted in the same way. Although a point (or points) plotting beyond the control limits would likely indicate an out-of-control situation, a pattern or run on one side of the centerline is not necessarily an indication that the process is out of control. This is due to the fact that the moving ranges are correlated, and this correlation may naturally cause patterns or trends on the chart.

17-3.4 Control Charts for Attributes

The *p* Chart (Fraction Defective or Nonconforming)

Often it is desirable to classify a product as either defective or nondefective on the basis of comparison with a standard. This is usually done to achieve economy and simplicity in the inspection operation. For example, the diameter of a ball bearing may be checked by determining whether it will pass through a gauge consisting of circular holes cut in a template. This would be much simpler than measuring the diameter with a micrometer. Control charts for attributes are used in these situations. However, attribute control charts require a considerably larger sample size than do their measurements counterparts. We will discuss the fraction-defective chart, or p chart, and two charts for defects, the c and u charts. Note that it is possible for a unit to have many defects, and be either defective or nondefective. In some applications a unit can have several defects, yet be classified as nondefective.

Suppose D is the number of defective units in a random sample of size n. We assume that D is a binomial random variable with unknown parameter p. Now the sample fraction defective is an estimator of p, that is

$$\hat{p} = \frac{D}{n}. \tag{17-12}$$

Furthermore, the variance of the statistic $\hat{p}$ is

$$\sigma_{\hat{p}}^2 = \frac{p(1-p)}{n},$$

so we may estimate $\sigma_{\hat{p}}^2$ as

$$\hat{\sigma}_{\hat{p}}^2 = \frac{\hat{p}(1-\hat{p})}{n}. \tag{17-13}$$

The centerline and control limits for the fraction-defective control chart may now be easily determined. Suppose k preliminary samples are available, each of size n, and D_i is the number of defectives in the ith sample. Then we may take

$$\overline{p} = \frac{\sum_{i=1}^{k} D_i}{kn} \tag{17-14}$$

as the centerline and

$$\begin{aligned} UCL &= \overline{p} + 3\sqrt{\frac{\overline{p}(1-\overline{p})}{n}}, \\ LCL &= \overline{p} - 3\sqrt{\frac{\overline{p}(1-\overline{p})}{n}} \end{aligned} \tag{17-15}$$

as the upper and lower control limits, respectively. These control limits are based on the normal approximation to the binomial distribution. When p is small, the normal approximation may not always be adequate. In such cases, it is best to use control limits obtained directly from a table of binomial probabilities or, perhaps, from the Poisson approximation to the binomial distribution. If p is small, the lower control limit may be a negative number. If this should occur, it is customary to consider zero as the lower control limit.

Example 17-3

Suppose we wish to construct a fraction-defective control chart for a ceramic substrate production line. We have 20 preliminary samples, each of size 100; the number of defectives in each sample are shown in Table 17-3. Assume that the samples are numbered in the sequence of production. Note that $\overline{p} = 800/2000 = 0.40$, and therefore the trial parameters for the control chart are

$$\text{Centerline} = 0.395$$

$$UCL = 0.395 + 3\sqrt{\frac{(0.395)(0.605)}{100}} = 0.5417,$$

$$LCL = 0.395 - 3\sqrt{\frac{(0.395)(0.605)}{100}} = 0.2483.$$

Table 17-3 Number of Defectives in Samples of 100 Ceramic Substrates

Sample	No. of Defectives	Sample	No. of Defectives
1	44	11	36
2	48	12	52
3	32	13	35
4	50	14	41
5	29	15	42
6	31	16	30
7	46	17	46
8	52	18	38
9	44	19	26
10	38	20	30

The control chart is shown in Fig. 17-8. All samples are in control. If they were not, we would search for assignable causes of variation and revise the limits accordingly.

Although this process exhibits statistical control, its capability ($\overline{p} = 0.395$) is very poor. We should take appropriate steps to investigate the process to determine why such a large number of defective units are being produced. Defective units should be analyzed to determine the specific types of defects present. Once the defect types are known, process changes should be investigated to determine their impact on defect levels. Designed experiments may be useful in this regard.

Example 17-4

Attributes Versus Measurements Control Charts

The advantage of measurement control charts relative to the p chart with respect to size of sample may be easily illustrated. Suppose that a normally distributed quality characteristic has a standard deviation of 4 and specification limits of 52 and 68. The process is centered at 60, which results in a fraction defective of 0.0454. Let the process mean shift to 56. Now the fraction defective is 0.1601. If the

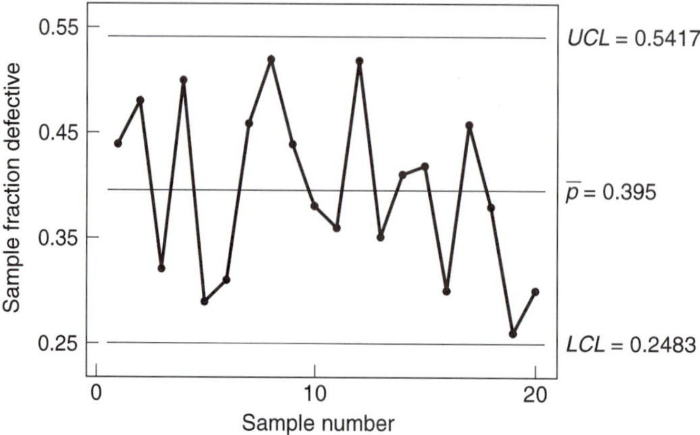

Figure 17-8 The p chart for a ceramic substrate.

probability of detecting the shift on the first sample following the shift is to be 0.50, then the sample size must be such that the lower 3-sigma limit will be at 56. This implies

$$60 - \frac{3(4)}{\sqrt{n}} = 56,$$

whose solution is $n = 9$. For a p chart, using the normal approximation to the binomial, we must have

$$0.0454 + 3\sqrt{\frac{(0.0454)(0.9546)}{n}} = 0.1601,$$

whose solution is $n = 30$. Thus, unless the cost of measurement inspection is more than three times as costly as the attributes inspection, the measurement control chart is cheaper to operate.

The *c* Chart (Defects)

In some situations it may be necessary to control the number of defects in a unit of product rather than the fraction defective. In these situations we may use the control chart for defects, or the c chart. Suppose that in the production of cloth it is necessary to control the number of defects per yard, or that in assembling an aircraft wing the number of missing rivets must be controlled. Many defects-per-unit situations can be modeled by the Poisson distribution.

Let c be the number of defects in a unit, where c is a Poisson random variable with parameter α. Now the mean and variance of this distribution are both α. Therefore, if k units are available and c_i is the number of defects in unit i, the centerline of the control chart is

$$\bar{c} = \frac{1}{k}\sum_{i=1}^{k} c_i, \tag{17-16}$$

and

$$UCL = \bar{c} + 3\sqrt{\bar{c}}, \qquad LCL = \bar{c} - 3\sqrt{\bar{c}} \tag{17-17}$$

are the upper and lower control limits, respectively.

Example 17-5

Printed circuit boards are assembled by a combination of manual assembly and automation. A flow solder machine is used to make the mechanical and electrical connections of the leaded components to the board. The boards are run through the flow solder process almost continuously, and every hour five boards are selected and inspected for process-control purposes. The number of defects in each sample of five boards is noted. Results for 20 samples are shown in Table 17-4. Now $\bar{c} = 160/20 = 8$, and therefore

$$UCL = 8 + 3\sqrt{8} = 16.484,$$
$$LCL = 8 - 3\sqrt{8} < 0, \text{ set to } 0.$$

From the control chart in Fig. 17-9, we see that the process is in control. However, eight defects per group of five printed circuit boards is too many (about 8/5 = 1.6 defects/board), and the process needs improvement. An investigation needs to be made of the specific types of defects found on the printed circuit boards. This will usually suggest potential avenues for process improvement.

Table 17-4 Number of Defects in Samples of Five Printed Circuit Boards

Sample	No. of Defects	Sample	No. of Defects
1	6	11	9
2	4	12	15
3	8	13	8
4	10	14	10
5	9	15	8
6	12	16	2
7	16	17	7
8	2	18	1
9	3	19	7
10	10	20	13

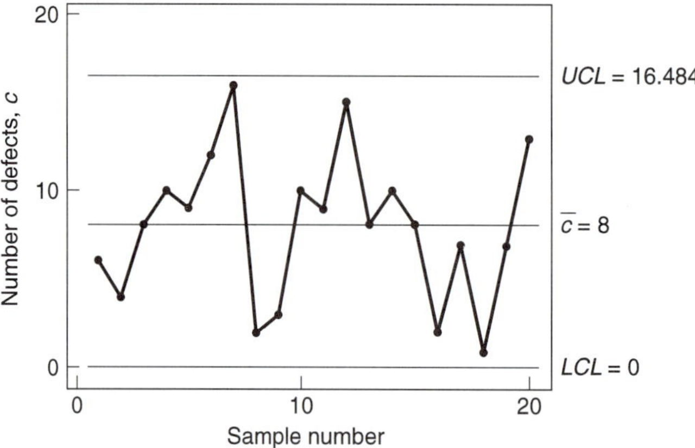

Figure 17-9 The c chart for defects in samples of five printed circuit boards.

The u Chart (Defects per Unit)

In some processes it may be preferable to work with the number of defects per unit rather than the total number of defects. Thus, if the sample consists of n units and there are c total defects in the sample, then

$$u = \frac{c}{n}$$

is the average number of defects per unit. A u chart may be constructed for such data. If there are k preliminary samples, each with $u_1, u_2, \ldots, u_k$ defects per unit, then the center-line on the u chart is

$$\bar{u} = \frac{1}{k}\sum_{i=1}^{k} u_i, \tag{17-18}$$

and the control limits are given by

$$\begin{aligned} UCL &= \bar{u} + 3\sqrt{\frac{\bar{u}}{n}}, \\ LCL &= \bar{u} - 3\sqrt{\frac{\bar{u}}{n}}. \end{aligned} \tag{17-19}$$

Example 17-6

A u chart may be constructed for the printed circuit board defect data in Example 17-5. Since each sample contains $n = 5$ printed circuit boards, the values of u for each sample may be calculated as shown in the following display:

Sample	Sample size, n	Number of defects, c	Defects per unit
1	5	6	1.2
2	5	4	0.8
3	5	8	1.6
4	5	10	2.0
5	5	9	1.8
6	5	12	2.4
7	5	16	3.2
8	5	2	0.4
9	5	3	0.6
10	5	10	2.0
11	5	9	1.8
12	5	15	3.0
13	5	8	1.6
14	5	10	2.0
15	5	8	1.6
16	5	2	0.4
17	5	7	1.4
18	5	1	0.2
19	5	7	1.4
20	5	13	2.6

The centerline for the u chart is

$$\bar{u} = \frac{1}{20}\sum_{i=1}^{20} u_i = \frac{32}{20} = 1.6,$$

and the upper and lower control limits are

$$UCL = \bar{u} + 3\sqrt{\frac{\bar{u}}{n}} = 1.6 + 3\sqrt{\frac{1.6}{5}} = 3.3,$$

$$LCL = \bar{u} - 3\sqrt{\frac{\bar{u}}{n}} = 1.6 - 3\sqrt{\frac{1.6}{5}} < 0, \text{ set to } 0.$$

The control chart is plotted in Fig. 17-10. Notice that the u chart in this example is equivalent to the c chart in Fig. 17-9. In some cases, particularly when the sample size is not constant, the u chart will be preferable to the c chart. For a discussion of variable sample sizes on control charts, see Montgomery (2001).

17-3.5 CUSUM and EWMA Control Charts

Up to this point in Chapter 17 we have presented the most basic of control charts, the Shewhart control charts. A major disadvantage of these control charts is their insensitivity to small shifts in the process (shifts often less than 1.5σ). This disadvantage is due to the fact that the Shewhart charts use information only from the current observation.

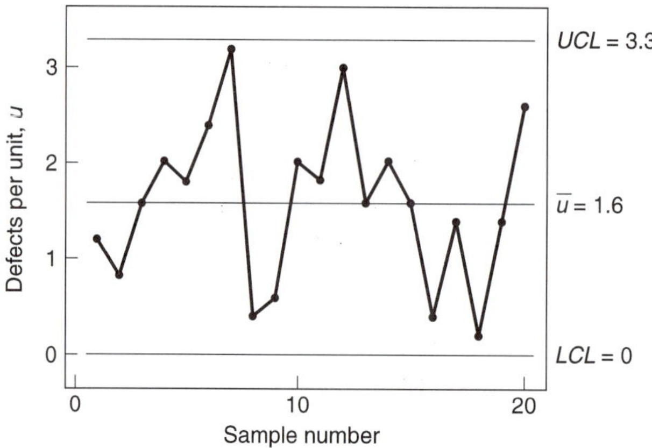

Figure 17-10 The *u* chart of defects per unit on printed circuit boards, Example 17-6.

Alternatives to Shewhart control charts include the *cumulative sum control chart* and the *exponentially weighted moving average control chart.* These control charts are more sensitive to small shifts in the process because they incorporate information from current and *recent past* observations.

Tabular CUSUM Control Charts for the Process Mean

The cumulative sum (CUSUM) control chart was first introduced by Page (1954) and incorporates information from a sequence of sample observations. The chart plots the cumulative sums of deviations of the observations from a target value. To illustrate, let $\bar{x}_j$ represent the jth sample mean, let μ_0 represent the target value for the process mean, and say the sample size is $n \geq 1$. The CUSUM control chart plots the quantity

$$C_i = \sum_{j=1}^{n} \left(\bar{x}_j - \mu_0\right) \tag{17-20}$$

against the sample i. The quantity C_i is the cumulative sum up to and including the ith sample. As long as the process is in control at the target value μ_0, then C_i in equation 17-20 represents a random walk with a mean of zero. On the other hand, if the process shifts away from the target mean, then either an upward or downward drift in C_i will be evident. By incorporating information from a sequence of observations, the CUSUM chart is able to detect a small shift in the process more quickly than a standard Shewhart chart. The CUSUM charts can be easily implemented for both subgroup data and individual observations. We will present the tabular CUSUM for individual observations.

The *tabular* CUSUM involves two statistics, C_i^+ and C_i^-, which are the accumulation of deviations above and below the target mean, respectively. C_i^+ is called the one-sided upper CUSUM and C_i^- is called the one-sided lower CUSUM. The statistics are computed as follows:

$$C_i^+ = \max[0, x_i - (\mu_0 + K) + C_{i-1}^+], \tag{17-21}$$

$$C_i^- = \max[0, (\mu_0 - K) - x_i + C_{i-1}^-], \tag{17-22}$$

with initial values of $C_0^+ = C_0^- = 0$. The constant, K, is referred to as the *reference value* and is often chosen approximately halfway between the target mean, μ_0, and the out-of-control mean that we are interested in detecting, denoted μ_1. In other words, K is half of the magnitude of the shift from μ_0 to μ_1, or

$$K = \frac{|\mu_1 - \mu_0|}{2}.$$

The statistics given in equations 17-21 and 17-22 accumulate the deviations from target that are larger than K and reset to zero when either quantity becomes negative. The CUSUM control chart plots the values of C_i^+ and C_i^- for each sample. If either statistic plots beyond the *decision interval*, H, the process is considered out of control. We will discuss the choice of H later in this chapter, but a good rule of thumb is often $H = 5\sigma$.

Example 17-7

A study presented in *Food Control* (2001, p. 119) gives the results of measuring the dry-matter content in buttercream from a batch process. One goal of the study is to monitor the amount of dry matter from batch to batch. Table 17-5 displays some data that may be typical of this type of process. The reported values, x_i, are percentage of dry-matter content examined after mixing. The target amount of dry-matter content is 45% and assume that $\sigma = 0.84\%$. Let us also assume that we are interested in detecting a shift in the process of mean of at least 1σ; that is, $\mu_1 = \mu_0 + 1\sigma = 45 + 1(0.84) = 45.84\%$. We will use the

Table 17-5 CUSUM Calculations for Example 17-7

Batch, i	x_i	$x_i - 45.42$	C_i^+	$44.58 - x_i$	C_i^-
1	46.21	0.79	0.79	−1.63	0
2	45.73	0.31	1.10	−1.15	0
3	44.37	−1.05	0.05	0.21	0.21
4	44.19	−1.23	0	0.39	0.60
5	43.73	−1.69	0	0.85	1.45
6	45.66	0.24	0.24	−1.08	0.37
7	44.24	−1.18	0	0.34	0.71
8	44.48	−0.94	0	0.10	0.81
9	46.04	0.62	0.62	−1.46	0
10	44.04	−1.38	0	0.54	0.54
11	42.96	−2.46	0	1.62	2.16
12	46.02	0.60	0.60	−1.44	0.72
13	44.82	−0.60	0	−0.24	0.48
14	45.02	−0.40	0	−0.44	0.04
15	45.77	0.35	0.35	−1.19	0
16	47.40	1.98	2.33	−2.82	0
17	47.55	2.13	4.46	−2.97	0
18	46.64	1.22	5.68	−2.06	0
19	46.31	0.89	6.57	−1.73	0
20	44.82	−0.60	5.97	−0.24	0
21	45.39	−0.03	5.94	−0.81	0
22	47.80	2.38	8.32	−3.22	0
23	46.69	1.27	9.59	−2.11	0
24	46.99	1.57	11.16	−2.41	0
25	44.53	−0.89	10.27	0.05	0

recommended decision interval of $H = 5\sigma = 5(0.84) = 4.2$. The reference value, K, is found to be

$$K = \frac{|45.84 - 45|}{2} = 0.42.$$

The values of C_i^+ and C_i^- are given in Table 17-5. To illustrate the calculations, consider the first two sample batches. Recall that $C_0^+ = C_0^- = 0$, and using equations 17-21 and 17-22 with $K = 0.42$ and $\mu_0 = 45$, we have

$$\begin{aligned} C_1^+ &= \max[0, x_1 - (45 + 0.42) + C_0^+] \\ &= \max[0, x_1 - 45.42 + C_0^+] \end{aligned}$$

and

$$\begin{aligned} C_1^- &= \max[0, (45 - 0.42) - x_1 + C_0^-] \\ &= \max[0, 44.58 - x_1 + C_0^-]. \end{aligned}$$

For batch 1, $x_1 = 46.21$,

$$\begin{aligned} C_1^+ &= \max[0, 46.21 - 45.42 + 0] \\ &= \max[0, 0.79] \\ &= 0.79, \end{aligned}$$

and

$$\begin{aligned} C_1^- &= \max[0, 44.58 - 46.21 + 0] \\ &= \max[0, -1.63] \\ &= 0. \end{aligned}$$

For batch 2, $x_2 = 45.73$,

$$\begin{aligned} C_2^+ &= \max[0, 45.73 - 45.42 + 0.79] \\ &= \max[0, 1.10] \\ &= 1.10, \end{aligned}$$

and

$$\begin{aligned} C_2^- &= \max[0, 44.58 - 45.73 + 0] \\ &= \max[0, -1.15] \\ &= 0. \end{aligned}$$

The CUSUM calculations given in Table 17-5 indicate that the upper-sided CUSUM for batch 17 is $C_{17}^+ = 4.46$, which exceeds the decision value of $H = 4.2$. Therefore, the process appears to have shifted out of control. The CUSUM *status* chart created using Minitab® with $H = 4.2$ is given in Fig. 17-11. The out-of-control control situation is also evident on this chart at batch 17.

The CUSUM control chart is a powerful quality tool for detecting a process that has shifted from the target process mean. The correct choices of H and K can greatly improve the sensitivity of the control chart while protecting against the occurrence of *false alarms* (the process is actually in control, but the control chart signals out of control). Design recommendations for the CUSUM will be provided later in this chapter when the concept of average run length is introduced.

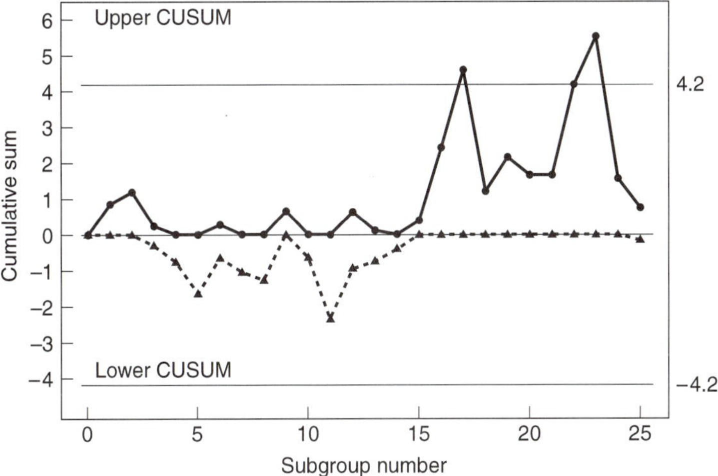

Figure 17-11 CUSUM status chart for Example 17-7.

We have presented the upper and lower CUSUM control charts for situations in which a shift in either direction away from the process target is of interest. There are many instances when we may be interested in a shift in only one direction, either upward or downward. One-sided CUSUM charts can be constructed for these situations. For a thorough development of these charts and more details, see Montgomery (2001).

EWMA Control Charts

The exponentially weighted moving average (EWMA) control chart is also a good alternative to the Shewhart control chart when detecting a small shift in the process mean is of interest. We will present the EWMA for individual measurements although the procedure can also be modified for subgroups of size $n > 1$.

The EWMA control chart was first introduced by Roberts (1959). The EWMA is defined as

$$z_i = \lambda x_i + (1 - \lambda)z_{i-1}, \tag{17-23}$$

where λ is a weight, $0 < \lambda \leq 1$. The procedure to be presented is initialized with $z_0 = \mu_0$, the process target mean. If a target mean is unknown, then the average of preliminary data, $\bar{x}$, is used as the initial value of the EWMA. The definition given in equation 17-23 demonstrates that information from past observations is incorporated into the current value of z_i. The value z_i is a weighted average of all previous sample means. To illustrate, we can replace z_{i-1} on the right-hand side of equation 17-23 to obtain

$$\begin{aligned} z_i &= \lambda x_i + (1 - \lambda)[\lambda x_{i-1} + (1 - \lambda)z_{i-2}] \\ &= \lambda x_i + \lambda(1 - \lambda)x_{i-1} + (1 - \lambda)^2 z_{i-2}. \end{aligned}$$

By recursively replacing z_{i-j}, $j = 1, 2, \ldots, t$, we find

$$z_i = \lambda \sum_{j=0}^{i-1} (1-\lambda)^j x_{i-j} + (1-\lambda)^i z_0.$$

The EWMA can be thought of as a weighted average of all past and current observations. Note that the weights decrease geometrically with the age of the observation, giving less weight to observations that occurred early in the process. The EWMA is often used in forecasting, but the EWMA control chart has been used extensively for monitoring many types of processes.

If the observations are independent random variables with variance σ^2, the variance of the EWMA, z_i, is

$$\sigma_{z_i}^2 = \sigma^2\left(\frac{\lambda}{2-\lambda}\right)\left[1-(1-\lambda)^{2i}\right].$$

Given a target mean, μ_0, and the variance of the EWMA, the upper control limit, centerline, and lower control limit for the EWMA control chart are

$$UCL = \mu_0 + L\sigma\sqrt{\left(\frac{\lambda}{2-\lambda}\right)\left[1-(1-\lambda)^{2i}\right]},$$

$$\text{Centerline} = \mu_0,$$

$$LCL = \mu_0 - L\sigma\sqrt{\left(\frac{\lambda}{2-\lambda}\right)\left[1-(1-\lambda)^{2i}\right]},$$

where L is the width of the control limits. Note that the term $1-(1-\lambda)^{2i}$ approaches 1 as i increases. Therefore, as the process continues running, the control limits for the EWMA approach the steady state values

$$\begin{aligned} UCL &= \mu_0 + L\sigma\sqrt{\frac{\lambda}{2-\lambda}}, \\ LCL &= \mu_0 - L\sigma\sqrt{\frac{\lambda}{2-\lambda}}. \end{aligned} \tag{17-24}$$

Although the control limits given in equation 17-24 provide good approximations, it is recommended that the exact limits be used for small values of i.

Example 17-8

We will now implement the EWMA control chart with $\lambda = 0.2$ and $L = 2.7$ for the dry-matter content data provided in Table 17-5. Recall that the target mean is $\mu_0 = 45\%$ and the process standard deviation is assumed to be $\sigma = 0.84\%$. The EWMA calculations are provided in Table 17-6. To demonstrate some of the calculations, consider the first observation with $x_1 = 46.21$. We find

$$\begin{aligned} z_1 &= \lambda x_1 + (1-\lambda)z_0 \\ &= (0.2)(46.21) + (0.80)(45) \\ &= 45.24. \end{aligned}$$

The second EWMA value is then

$$\begin{aligned} z_2 &= \lambda x_2 + (1-\lambda)z_1 \\ &= (0.2)(45.73) + (0.80)(45.24) \\ &= 45.34. \end{aligned}$$

Table 17-6 EWMA Calculations for Example 17-8

Batch, i	x_i	z_i	UCL	LCL
1	46.21	45.24	45.45	44.55
2	45.73	45.34	45.58	44.42
3	44.37	45.15	45.65	44.35
4	44.19	44.95	45.69	44.31
5	43.73	44.71	45.71	44.29
6	45.66	44.90	45.73	44.27
7	44.24	44.77	45.74	44.26
8	44.48	44.71	45.75	44.25
9	46.04	44.98	45.75	44.25
10	44.04	44.79	45.75	44.25
11	42.96	44.42	45.75	44.25
12	46.02	44.74	45.75	44.25
13	44.82	44.76	45.75	44.25
14	45.02	44.81	45.76	44.24
15	45.77	45.00	45.76	44.24
16	47.40	45.48	45.76	44.24
17	47.55	45.90	45.76	44.24
18	46.64	46.04	45.76	44.24
19	46.31	46.10	45.76	44.24
20	44.82	45.84	45.76	44.24
21	45.39	45.75	45.76	44.24
22	47.80	46.16	45.76	44.24
23	46.69	46.27	45.76	44.24
24	46.99	46.41	45.76	44.24
25	44.53	46.04	45.76	44.24

The EWMA values are plotted on a control chart along with the upper and lower control limits given by

$$UCL = \mu_0 + L\sigma\sqrt{\left(\frac{\lambda}{2-\lambda}\right)\left[1-(1-\lambda)^{2i}\right]}$$

$$= 45 + 2.7(0.84)\sqrt{\left(\frac{0.2}{2-0.2}\right)\left[1-(1-0.2)^{2i}\right]},$$

$$LCL = \mu_0 - L\sigma\sqrt{\left(\frac{\lambda}{2-\lambda}\right)\left[1-(1-\lambda)^{2i}\right]}$$

$$= 45 - 2.7(0.84)\sqrt{\left(\frac{0.2}{2-0.2}\right)\left[1-(1-0.2)^{2i}\right]}.$$

Therefore, for $i = 1$,

$$UCL = 45 + 2.7(0.84)\sqrt{\left(\frac{0.2}{2-0.2}\right)\left[1-(1-0.2)^{2(1)}\right]}$$

$$= 45.45,$$

$$LCL = \mu_0 - L\sigma\sqrt{\left(\frac{\lambda}{2-\lambda}\right)\left[1-(1-\lambda)^{2(1)}\right]}$$

$$= 44.55.$$

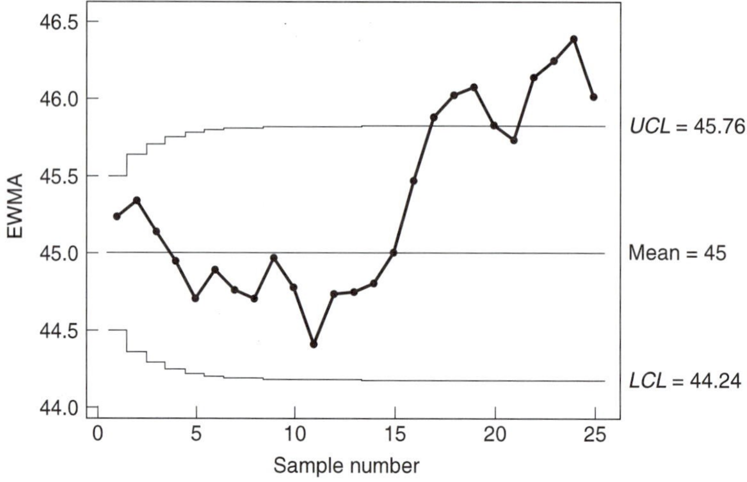

Figure 17-12 EWMA chart for Example 17-8.

The remaining control limits are calculated similarly and plotted on the control chart given in Fig. 17-12. The control limits tend to increase as i increases, but then tend to the steady state values given by equations in 17-24:

$$UCL = \mu_0 + L\sigma\sqrt{\frac{\lambda}{2-\lambda}}$$

$$= 45 + 2.7(0.84)\sqrt{\frac{0.2}{2-0.2}}$$

$$= 45.76,$$

$$LCL = \mu_0 - L\sigma\sqrt{\frac{\lambda}{2-\lambda}}$$

$$= 45 - 2.7(0.84)\sqrt{\frac{0.2}{2-0.2}}$$

$$= 44.24.$$

The EWMA control chart signals at observation 17, indicating that the process is out of control.

The sensitivity of the EWMA control chart for a particular process will depend on the choices of L and λ. Various choices of these parameters will be presented later in this chapter, when the concept of the average run length is introduced. For more details and developments regarding the EWMA, see Crowder (1987), Lucas and Saccucci (1990), and Montgomery (2001).

17-3.6 Average Run Length

In this chapter we have presented control-charting techniques for a variety of situations and made some recommendations about the design of the control charts. In this section, we will present the *average run length* (ARL) of a control chart. The ARL can be used to assess the performance of the control chart or to determine the appropriate values of various parameters for the control charts presented in this chapter.

The ARL is the expected number of samples taken before a control chart signals out of control. In general, the ARL is

$$\text{ARL} = \frac{1}{p},$$

where p is the probability of any point exceeding the control limits. If the process is in control and the control chart signals out of control, then we say that a *false alarm* has occurred. To illustrate, consider the $\overline{X}$ control chart with the standard 3σ limits. For this situation, $p = 0.0027$ is the probability that a single point falls outside the limits when the process is in control. The in-control ARL for the $\overline{X}$ control chart is

$$\text{ARL} = \frac{1}{p} = \frac{1}{0.0027} = 370.$$

In other words, even if the process remains in control we should expect, on the average, an out-of-control signal (or false alarm) every 370 samples. In general, if the process is actually in control, then we desire a large value of the ARL. More formally, we can define the in-control ARL as

$$\text{ARL}_0 = \frac{1}{\alpha},$$

where α is the probability that a sample point plots beyond the control limit.

If on the other hand the process is out of control, then a small ARL value is desirable. A small value of the ARL indicates that the control chart will signal out of control soon after the process has shifted. The out-of-control ARL is

$$\text{ARL}_1 = \frac{1}{1-\beta},$$

where β is the probability of not detecting a shift on the first sample after a shift has occurred. To illustrate, consider the $\overline{X}$ control chart with 3σ limits. Assume the target or in-control mean is μ_0 and that the process has shifted to an out-of-control mean given by $\mu_1 = \mu_0 + k\sigma$. The probability of not detecting this shift is given by

$$\beta = P[LCL \le \overline{X} \le UCL | \mu = \mu_1].$$

That is, β is the probability that the next sample mean plots in control, when in fact the process has shifted out of control. Since $\overline{X} \sim N(\mu, \sigma^2/n)$ and $LCL = \mu_0 - L\sigma/\sqrt{n}$ and $UCL = \mu_0 + L\sigma/\sqrt{n}$ we can rewrite β as

$$\begin{aligned}
\beta &= P\left[\mu_0 - L\sigma/\sqrt{n} \le \overline{X} \le \mu_0 + L\sigma/\sqrt{n} | \mu = \mu_1\right] \\
&= P\left[\frac{\left(\mu_0 - L\sigma/\sqrt{n}\right) - \mu_1}{\sigma/\sqrt{n}} \le \frac{\overline{X} - \mu_1}{\sigma/\sqrt{n}} \le \frac{\left(\mu_0 + L\sigma/\sqrt{n}\right) - \mu_1}{\sigma/\sqrt{n}} | \mu = \mu_1\right] \\
&= P\left[\frac{\left(\mu_0 - L\sigma/\sqrt{n}\right) - (\mu_0 + k\sigma)}{\sigma/\sqrt{n}} \le Z \le \frac{\left(\mu_0 + L\sigma/\sqrt{n}\right) - (\mu_0 + k\sigma)}{\sigma/\sqrt{n}}\right] \\
&= P\left[-L - k\sqrt{n} \le Z \le L - k\sqrt{n}\right],
\end{aligned}$$

where Z is a standard normal random variable. If we let Φ denote the standard normal cumulative distribution function, then

$$\beta = \Phi\left(L - k\sqrt{n}\right) - \Phi\left(-L - k\sqrt{n}\right).$$

From this, $1 - \beta$ is the probability that a shift in the process is detected on the first sample after the shift has occurred. That is, the process has shifted and a point exceeds the control limits—signaling the process is out of control. Therefore, ARL_1 is the *expected* number of samples observed before a shift is detected.

The ARLs have been used to evaluate and design control charts for variables and for attributes. For more discussion on the use of ARLs for these charts, see Montgomery (2001).

ARLs for the CUSUM and EWMA Control Charts

Earlier in this chapter, we presented the CUSUM and EWMA control charts. The ARL can be used to specify some of the parameter values needed to design these control charts.

To implement the tabular CUSUM control chart, values of the decision interval, H, and the reference value, K, must be chosen. Recall that H and K are multiples of the process standard deviation, specifically $H = h\sigma$ and $K = k\sigma$, where $k = 1/2$ is often used as a standard. The proper selection of these values is important. The ARL is one criterion that can be used to determine the values of H and K. As stated previously, a large value of the ARL when the process is in control is desirable. Therefore, we can set ARL_0 to an acceptable level, and determine h and k accordingly. In addition, we would want the control chart to quickly detect a shift in the process mean. This would require values of h and k such that the values of ARL_1 are quite small. To illustrate, Montgomery (2001) provides the ARL for a CUSUM control chart with $h = 5$ and $k = 1/2$. These values are given in Table 17-7. The in-control average run length, ARL_0, is 465. If a small shift, say, 0.50σ, is important to detect, then with $h = 5$ and $k = 1/2$, we would expect to detect this shift within 38 samples (on the average) after the shift has occurred. Hawkins (1993) presents a table of h and k values that will result in an in-control average run length of $ARL_0 = 370$. The values are reproduced in Table 17-8.

Design of the EWMA control chart can also be based on the ARLs. Recall that the design parameters of the EWMA control chart are the multiple of the standard deviation, L, and the value of the weighting factor, λ. The values of these design parameters can be chosen so that the ARL performance of the control charts is satisfactory.

Several authors discuss the ARL performance of the EWMA control chart, including Crowder (1987) and Lucas and Saccucci (1990). Lucas and Saccucci (1990) provide the ARL performance for several combinations of L and λ. The results are reproduced in Table 17-9. Again, it is desirable to have a large value of the in-control ARL and small values of out-of-control ARLs. To illustrate, if $L = 2.8$ and $\lambda = 0.10$ are used, we would expect $ARL_0 \cong 500$ while the ARL for detecting a shift of $0.5\ \sigma$ is $ARL_1 \cong 31.3$. To detect smaller shifts

Table 17-7 Tabular CUSUM Performance with $h = 5$ and $k = 1/2$

Shift in Mean (multiple of σ)	0	0.25	0.50	0.75	1.00	1.50	2.00	2.50	3.00	4.00
ARL	465	139	38.0	17.0	10.4	5.75	4.01	3.11	2.57	2.01

Table 17-8 Values of h and k Resulting in $ARL_0 = 370$ (Hawkins 1993)

k	0.25	0.50	0.75	1.0	1.25	1.5
h	8.01	4.77	3.34	2.52	1.99	1.61

Table 17-9 ARLs for Various EWMA Control Schemes (Lucas and Saccucci 1990)

Shift in Mean (multiple of σ)	$L = 3.054$ $\lambda = 0.40$	$L = 2.998$ $\lambda = 0.25$	$L = 2.962$ $\lambda = 0.20$	$L = 2.814$ $\lambda = 0.10$	$L = 2.615$ $\lambda = 0.05$
0	500	500	500	500	500
0.25	224	170	150	106	84.1
0.50	71.2	48.2	41.8	31.3	28.8
0.75	28.4	20.1	18.2	15.9	16.4
1.00	14.3	11.1	10.5	10.3	11.4
1.50	5.9	5.5	5.5	6.1	7.1
2.00	3.5	3.6	3.7	4.4	5.2
2.50	2.5	2.7	2.9	3.4	4.2
3.00	2.0	2.3	2.4	2.9	3.5
4.00	1.4	1.7	1.9	2.2	2.7

in the process mean, it is found that small values of λ should be used. Note that for $L = 3.0$ and $\lambda = 1.0$, the EWMA reduces to the standard Shewhart control chart with 3-sigma limits.

Cautions in the Use of ARLs

Although the ARL provides valuable information for designing and evaluating control schemes, there are drawbacks to relying on the ARL as a design criterion. It should be noted that run length follows a geometric distribution, since it represents the number of samples before a "success" occurs (a success being a point falling beyond the control limits). One drawback is the standard deviation of the run length is quite large. Second, because the distribution of the run length follows a *geometric* distribution, the mean of the distribution (ARL) may not be a reliable estimate of the true run length.

17-3.7 Other SPC Problem-Solving Tools

While the control chart is a very powerful tool for investigating the causes of variation in a process, it is most effective when used with other SPC problem-solving tools. In this section we illustrate some of these tools, using the printed circuit board defect data in Example 17-5.

Figure 17-9 shows a c chart for the number of defects in samples of five printed circuit boards. The chart exhibits statistical control, but the number of defects must be reduced, as the average number of defects per board is $8/5 = 1.6$, and this level of defects would require extensive rework.

The first step in solving this problem is to construct a *Pareto diagram* of the individual defect types. The Pareto diagram, shown in Fig. 17-13, indicates that insufficient solder and solder balls are the most frequently occurring defects, accounting for $(109/160)100 = 68\%$ of the observed defects. Furthermore, the first five defect categories on the Pareto chart are all solder-related defects. This points to the flow solder process as a potential opportunity for improvement.

To improve the flow solder process, a team consisting of the flow solder operator, the shop supervisor, the manufacturing engineer responsible for the process, and a quality engineer meets to study potential causes of solder defects. They conduct a brainstorming session and produce the *cause-and-effect* diagram shown in Fig. 17-14. The cause-and-effect

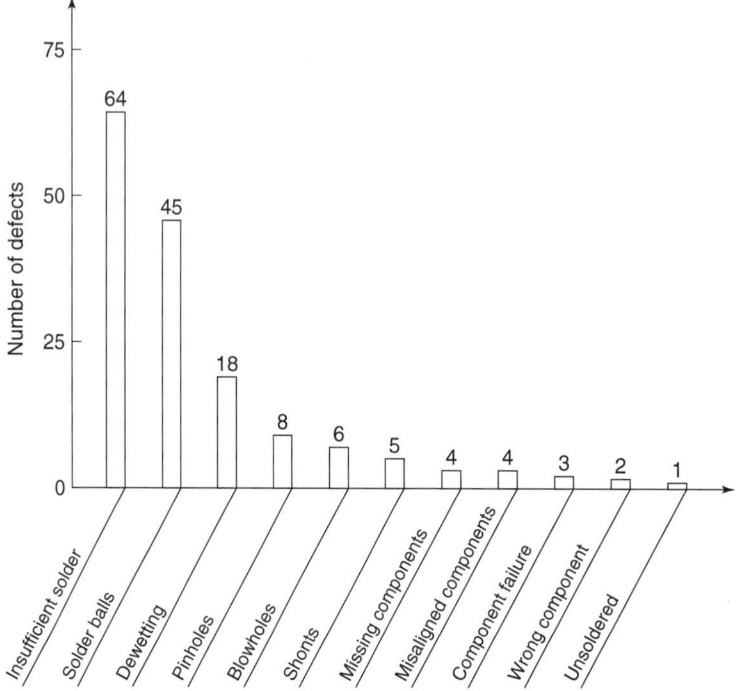

Figure 17-13 Pareto diagram for printed circuit board defects.

diagram is widely used to clearly display the various potential causes of defects in products and their interrelationships. It is useful in summarizing knowledge about the process.

As a result of the brainstorming session, the team tentatively identifies the following variables as potentially influential in creating solder defects:

1. Flux specific gravity
2. Solder temperature
3. Conveyor speed
4. Conveyor angle
5. Solder wave height
6. Preheat temperature
7. Pallet loading method

A statistically *designed experiment* could be used to investigate the effect of these seven variables on solder defects. Also, the team constructed a *defect concentration diagram* for the product. A defect concentration diagram is just a sketch or drawing of the product, with the most frequently occurring defects shown on the part. This diagram is used to determine whether defects occur in the same location on the part. The defect concentration diagram for the printed circuit board is shown in Fig. 17-15. This diagram indicates that most of the insufficient solder defects are near the front edge of the board, where it makes initial contact with the solder wave. Further investigation showed that one of the pallets used to carry the boards across the wave was bent, causing the front edge of the board to make poor contact with the solder wave.

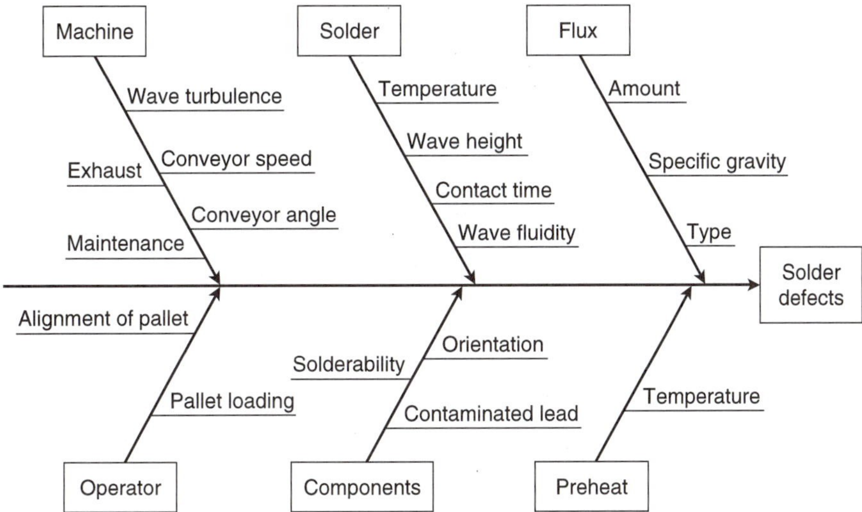

Figure 17-14 Cause-and-effect diagram for the printed circuit board flow solder process.

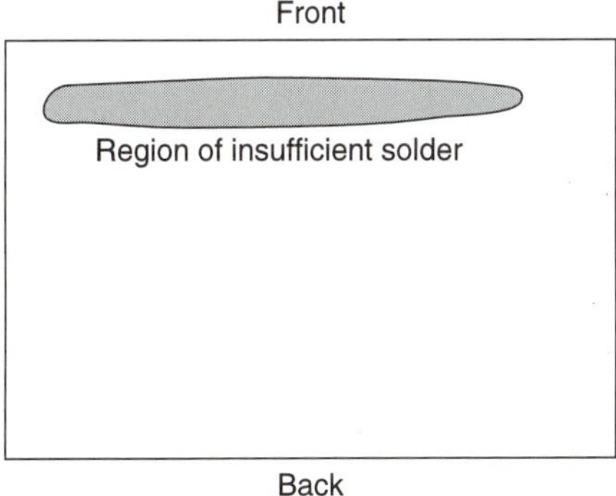

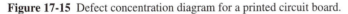

Figure 17-15 Defect concentration diagram for a printed circuit board.

When the defective pallet was replaced, a designed experiment was used to investigate the seven variables discussed earlier. The results of this experiment indicated that several of these factors were influential and could be adjusted to reduce solder defects. After the results of the experiment were implemented, the percentage of solder joints requiring rework was reduced from 1% to under 100 parts per million (0.01%).

17-4 RELIABILITY ENGINEERING

One of the challenging endeavors of the past three decades has been the design and development of large-scale systems for space exploration, new generations of commercial and military aircraft, and complex electromechanical products such as office copiers and computers. The performance of these systems, and the consequences of their failure, is of vital concern. For example, the military community has historically placed strong emphasis on equipment reliability. This emphasis stems largely from increasing ratios of maintenance cost to procurement costs and the strategic and tactical implications of system failure. In the area of consumer product manufacture, high reliability has come to be expected as much as conformance to other important quality characteristics.

Reliability engineering encompasses several activities, one of which is reliability modeling. Essentially, the system survival probability is expressed as a function of a subsystem of component reliabilities (survival probabilities). Usually, these models are time dependent, but there are some situations where this is not the case. A second important activity is that of life testing and reliability estimation.

17-4.1 Basic Reliability Definitions

Let us consider a component that has just been manufactured. It is to be operated at a stated "stress level" or within some range of stress such as temperature, shock, and so on. The random variable T will be defined as time to failure, and the *reliability* of the component (or subsystem or system) at time t is $R(t) = P[T > t]$. R is called the *reliability function*. The failure process is usually complex, consisting of at least three types of failures: initial failures, wear-out failures, and those that fail between these. A hypothetical composite distribution of time to failure is shown in Fig. 17-16. This is a mixed distribution, and

$$p(0) + \int_0^{\infty} g(t)dt = 1. \tag{17-25}$$

Since for many components (or systems) the initial failures or time zero failures are removed during testing, the random variable T is conditioned on the event that $T > 0$, so that the failure density is

$$\begin{aligned} f(t) &= \frac{g(t)}{1 - p(0)}, \qquad t > 0, \\ &= 0 \qquad\qquad \text{otherwise}. \end{aligned} \tag{17-26}$$

Thus, in terms of f, the reliability function, R, is

$$R(t) = 1 - F(t) = \int_t^{\infty} f(x)dx. \tag{17-27}$$

The term *interval failure rate* denotes the rate of failure on a particular interval of time $[t_1, t_2]$ and the terms *failure rate*, *instantaneous failure rate*, and *hazard* will be used synonymously as a limiting form of the interval failure rate as $t_2 \to t_1$. The interval failure rate $FR(t_1, t_2)$ is as follows:

$$FR(t_1, t_2) = \left[\frac{R(t_1) - R(t_2)}{R(t_1)}\right] \cdot \left[\frac{1}{t_2 - t_1}\right]. \tag{17-28}$$

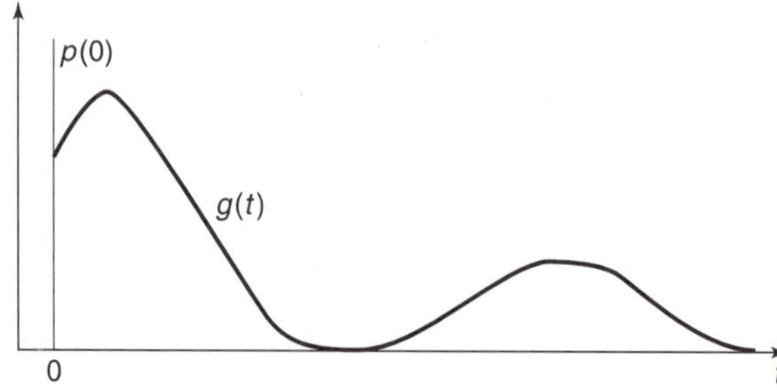

Figure 17-16 A composite failure distribution.

The first bracketed term is simply

$$P\{\text{Failure during } [t_1, t_2] | \text{Survival to time } t_1\}. \tag{17-29}$$

The second term is for the dimensional characteristic, so that we may express the conditional probability of equation 17-29 on a per-unit time basis.

We will develop the instantaneous failure rate (as a function of t). Let $h(t)$ be the hazard function. Then

$$\begin{aligned} h(t) &= \lim_{\Delta t \to 0} \frac{R(t) - R(t+\Delta t)}{R(t)} \frac{1}{\Delta t} \\ &= -\lim_{\Delta t \to 0} \frac{R(t+\Delta t) - R(t)}{\Delta t} \cdot \frac{1}{R(t)}, \end{aligned}$$

or

$$h(t) = \frac{-R'(t)}{R(t)} = \frac{f(t)}{R(t)}, \tag{17-30}$$

since $R(t) = 1 - F(t)$ and $-R'(t) = f(t)$. A typical hazard function is shown in Fig. 17-17. Note that $h(t) \cdot dt$ might be thought of as the instantaneous probability of failure at t, given survival to t.

A useful result is that the reliability function R may be easily expressed in terms of h as

$$R(t) = e^{-\int_0^t h(x)dx} = e^{-H(t)}, \tag{17-31}$$

where

$$H(t) = \int_0^t h(x)dx.$$

Equation 17-31 results from the definition given in equation 17-30

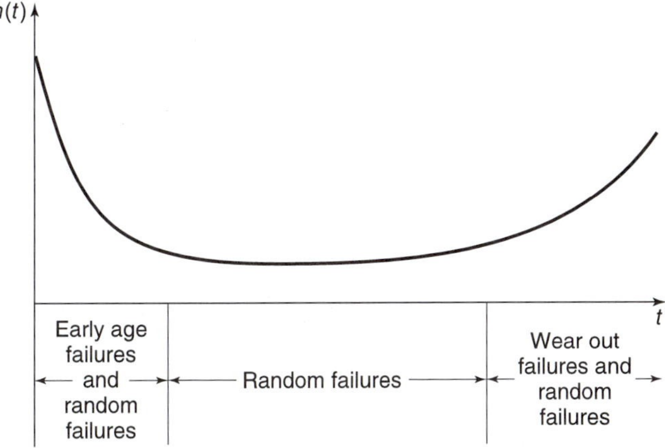

Figure 17-17 A typical hazard function.

and the integration of both sides

$$\int_0^t h(x)dx = -\int_0^t \frac{R'(x)}{R(x)}dx = -\ln R(x)\Big|_0^t,$$

so that

$$\int_0^t h(x)dx = -\ln R(t) + \ln R(0).$$

Since $F(0) = 0$, we see that $\ln R(0) = 0$ and

$$e^{-\int_0^t h(x)dx} = e^{\ln R(t)} = R(t).$$

The mean time to failure (MTTF) is

$$E[T] = \int_0^\infty t f(t)dt.$$

A useful alternate form is

$$E[T] = \int_0^\infty R(t)dt. \tag{17-32}$$

Most complex system modeling assumes that only random component failures need be considered. This is equivalent to stating that the time-to-failure distribution is exponential, that is,

$$\begin{aligned} f(t) &= \lambda e^{-\lambda t}, && t \geq 0, \\ &= 0, && \text{otherwise,} \end{aligned}$$

so that

$$h(t) = \frac{f(t)}{R(t)} = \frac{\lambda e^{-\lambda t}}{e^{-\lambda t}} = \lambda$$

is a constant. When all early-age failures have been removed by *burn in*, and the time to occurrence of wearout failures is very great (as with electronic parts), then this assumption is reasonable.

The normal distribution is most generally used to model wearout failure or stress failure (where the random variable under study is stress level). In situations where most failures are due to wear, the normal distribution may very well be appropriate.

The lognormal distribution has been found to be applicable in describing time to failure for some types of components, and the literature seems to indicate an increased utilization of this density for this purpose.

The Weibull distribution has been extensively used to represent time to failure, and its nature is such that it may be made to approximate closely the observed phenomena. When a system is composed of a number of components and failure is due to the most serious of a large number of defects or possible defects, the Weibull distribution seems to do particularly well as a model.

The gamma distribution frequently results from modeling standby redundancy where components have an exponential time-to-failure distribution. We will investigate standby redundancy in Section 17-4.5.

17-4.2 The Exponential Time-to-Failure Model

In this section we assume that the time-to-failure distribution is exponential; that is, only "random failures" are considered. The density, reliability function, and hazard functions are given in equations 17-33 through 17-35, are shown in Fig. 17-18:

$$\begin{aligned} f(t) &= \lambda e^{-\lambda t}, && t \geq 0, \\ &= 0, && \text{otherwise}, \end{aligned} \tag{17-33}$$

$$\begin{aligned} R(t) = P[T > t] &= e^{-\lambda t}, && t \geq 0, \\ &= 0, && \text{otherwise}, \end{aligned} \tag{17-34}$$

$$\begin{aligned} h(t) = \frac{f(t)}{R(t)} &= \lambda, && t \geq 0, \\ &= 0, && \text{otherwise}. \end{aligned} \tag{17-35}$$

The constant hazard function is interpreted to mean that the failure process has no memory; that is,

$$P\{t \leq T \leq t + \Delta t | T > t\} = \frac{e^{-\lambda t} - e^{-\lambda(t+\Delta t)}}{e^{-\lambda t}} = 1 - e^{-\lambda \Delta t}, \tag{17-36}$$

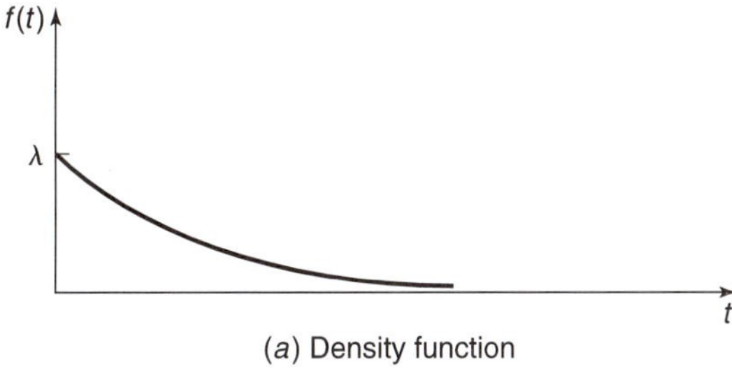

(a) Density function

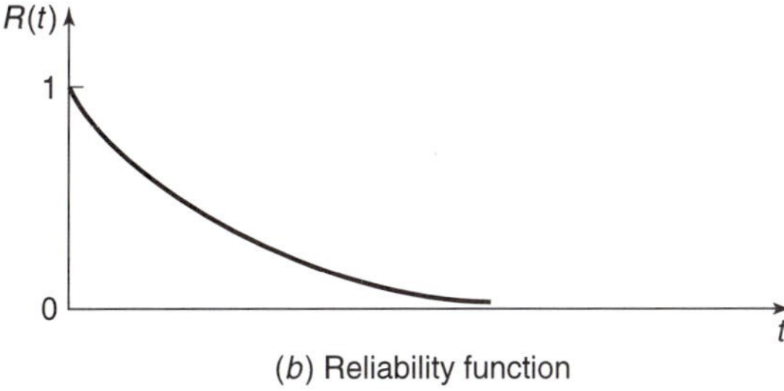

(b) Reliability function

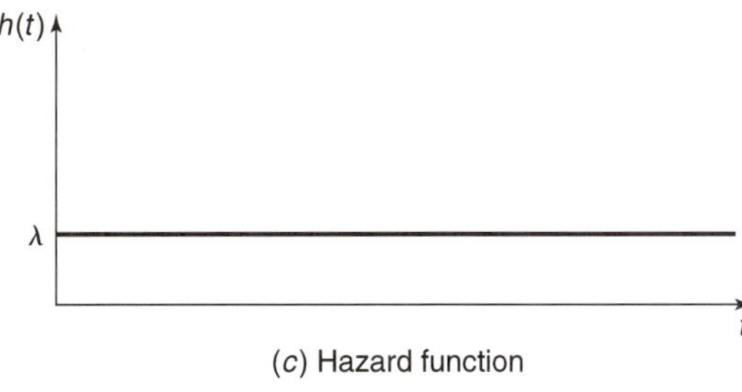

(c) Hazard function

Figure 17-18 Density, reliability function, and hazard function for the exponential failure model.

a quantity that is independent of t. Thus if a component is functioning at time t, it is as good as new. The remaining life has the same density as f.

Example 17-9

A diode used on a printed circuit board has a rated failure rate of 2.3×10^{-8} failures per hour. However, under an increased temperature stress, it is felt that the rate is about 1.5×10^{-5} failures per hour. The time to failure is exponentially distributed, so that we have

$$\begin{aligned} f(t) &= (1.5 \times 10^{-5})e^{-(1.5\times10^{-5})t}, && t \geq 0, \\ &= 0, && \text{otherwise}, \\ R(t) &= e^{-(1.5\times10^{-5})t}, && t \geq 0, \\ &= 0, && \text{otherwise}, \end{aligned}$$

and

$$\begin{aligned} h(t) &= 1.5 \times 10^{-5}, && t \geq 0, \\ &= 0, && \text{otherwise}. \end{aligned}$$

To determine the reliability at $t = 10^4$ and $t = 10^5$, we evaluate $R(10^4) = e^{-0.15} = 0.86$, and $R(10^5) = e^{-1.5} = 0.223$.

17-4.3 Simple Serial Systems

A simple serial system is shown in Fig. 17-19. In order for the system to function, all components must function, and it is assumed that the components function *independently*. We let T_j be the time to failure for component c_j for $j = 1, 2, \ldots, n$ and let T be system time to failure. The reliability model is thus

$$R(t) = P[T > t] = P(T_1 > t) \cdot P(T_2 > t) \cdot \cdots \cdot P(T_n > t),$$

or

$$R(t) = R_1(t) \cdot R_2(t) \cdot \cdots \cdot R_n(t), \tag{17-37}$$

where

$$P[T_j > t] = R_j(t).$$

Example 17-10

Three components must all function for a simple system to function. The random variables T_1, T_2, and T_3 representing time to failure for the components are independent with the following distributions:

$$\begin{aligned} &T_1 \sim N(2\times10^3,\ 4\times10^4), \\ &T_2 \sim \text{Weibull}\left(\gamma = 0,\ \delta = 1,\ \beta = \frac{1}{7}\right), \\ &T_3 \sim \text{lognormal}(\mu = 10,\ \sigma^2 = 4). \end{aligned}$$

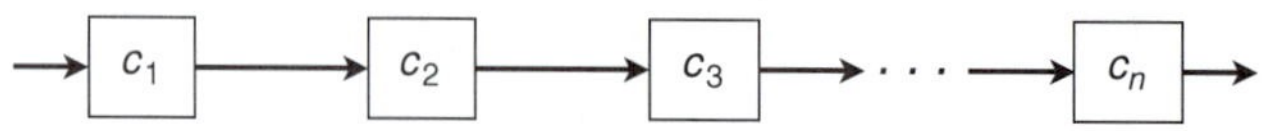

Figure 17-19 A simple serial system.

It follows that

$$R_1(t) = 1 - \Phi\left(\frac{t - 2\times10^3}{200}\right),$$

$$R_2(t) = e^{-t^{(1/7)}},$$

$$R_3(t) = 1 - \Phi\left(\frac{\ln t - 10}{2}\right),$$

so that

$$R(t) = \left[1 - \Phi\left(\frac{t - 2\times10^3}{200}\right)\right]\cdot\left[e^{-t^{(1/7)}}\right]\cdot\left[1 - \Phi\left(\frac{\ln t - 10}{2}\right)\right].$$

For example, if $t = 2187$ hours, then

$$\begin{aligned} R(2187) &= [1 - \Phi(0.935)][e^{-3}][1 - \Phi(-1.154)] \\ &= [0.175][0.0498][0.876] \\ &\simeq 0.0076. \end{aligned}$$

For the simple serial system, system reliability may be calculated using the product of the component reliability functions as demonstrated; however, when all components have an exponential distribution, the calculations are greatly simplified, since

$$R(t) = e^{-\lambda_1 t}\cdot e^{-\lambda_2 t}\cdots e^{-\lambda_n t} = e^{-(\lambda_1 + \lambda_2 + \cdots + \lambda_n)t},$$

or

$$R(t) = e^{-\lambda_s t}, \tag{17-38}$$

where $\lambda_s = \sum_{j=1}^{n} \lambda_j$ represents the *system failure rate.* We also note that the system reliability function is of the same form as the component reliability functions. The system failure rate is simply the sum of the component failure rates, and this makes application very easy.

Example 17-11

Consider an electronic circuit with three integrated circuit devices, 12 silicon diodes, 8 ceramic capacitors, and 15 composition resistors. Suppose under given stress levels of temperature, shock, and so on that each component has failure rates as shown in the following table, and the component failures are independent.

	Failures per Hour
Integrated circuits	1.3×10^{-9}
Diodes	1.7×10^{-7}
Capacitors	1.2×10^{-7}
Resistors	6.1×10^{-8}

Therefore,

$$\begin{aligned} \lambda_s &= 3(0.013\times10^{-7}) + 12(1.7\times10^{-7}) + 8(1.2\times10^{-7}) + 15(0.61\times10^{-7}) \\ &= 3.9189\times10^{-6}, \end{aligned}$$

and

$$R(t) = e^{-(3.9189\times10^{-6})t}.$$

The circuit mean time to failure is

$$\text{MTTF} = E[T] = \frac{1}{\lambda_s} = \frac{1}{3.9189} \times 10^6 = 2.55 \times 10^5 \text{ hours.}$$

If we wish to determine, say, $R(10^4)$, we get $R(10^4) = e^{-0.039189} \simeq 0.96$.

17-4.4 Simple Active Redundancy

An active redundant configuration is shown in Fig. 17-20. The assembly functions if k or more of the components function ($k \leq n$). All components begin operation at time zero, thus the term "active" is used to describe the redundancy. Again, independence is assumed.

A general formulation is not convenient to work with, and in most cases it is unnecessary. When all components have the same reliability function, as is the case when the components are the same type, we let $R_j(t) = r(t)$ for $j = 1, 2, \ldots, n$, so that

$$\begin{aligned} R(t) &= \sum_{x=k}^{n} \binom{n}{x} [r(t)]^x [1 - r(t)]^{n-x} \\ &= 1 - \sum_{x=0}^{k-1} \binom{n}{x} [r(t)]^x [1 - r(t)]^{n-x}. \end{aligned} \tag{17-39}$$

Equation 17-39 is derived from the definition of reliability.

Example 17-12

Three identical components are arranged in active redundancy, operating independently. In order for the assembly to function, at least two of the components must function ($k = 2$). The reliability function for the system is thus

$$\begin{aligned} R(t) &= \sum_{x=2}^{3} \binom{3}{x} [r(t)]^x [1 - r(t)]^{n-x} \\ &= 3[r(t)]^2 [1 - r(t)] + [r(t)]^3 \\ &= [r(t)]^2 [3 - 2r(t)]. \end{aligned}$$

It is noted that R is a function of time, t.

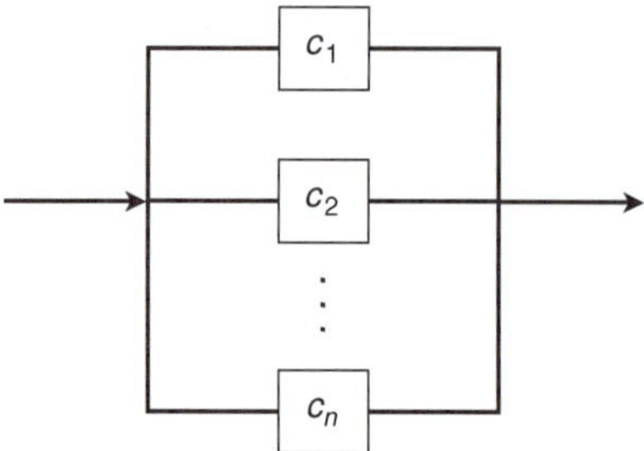

Figure 17-20 An active redundant configuration.

When only one of the n components is required, as is often the case, and the components are not identical, we obtain

$$R(t) = 1 - \prod_{j=1}^{n} \left[1 - R_j(t)\right]. \tag{17-40}$$

The product is the probability that all components fail, and, obviously, if they do not fail the system survives. When the components are identical and only one is required, equation 17-40 reduces to

$$R(t) = 1 - [1 - r(t)]^n, \tag{17-41}$$

where $r(t) = R_j(t)$, $j = 1, 2, \ldots, n$.

When the components have exponential failure laws, we will consider two cases. First, when the components are identical with failure rate λ and at least k components are required for the assembly to operate, equation 17-39 becomes

$$R(t) = \sum_{x=k}^{n} \binom{n}{x} \left[e^{-\lambda t}\right]^x \left[1 - e^{-\lambda t}\right]^{n-x}. \tag{17-42}$$

The second case is considered for the situation where the components have identical exponential failure densities and where only one component must function for the assembly to function. Using equation 17-41, we get

$$R(t) = 1 - [1 - e^{-\lambda t}]^n. \tag{17-43}$$

Example 17-13

In Example 17-12, where three identical components were arranged in an active redundancy, and at least two were required for system operation, we found

$$R(t) = [r(t)]^2[3 - 2r(t)].$$

If the component reliability functions are

$$r(t) = e^{-\lambda t},$$

then

$$R(t) = e^{-2\lambda t}[3 - 2e^{-\lambda t}]$$
$$= 3e^{-2\lambda t} - 2e^{-3\lambda t}.$$

If two components are arranged in an active redundancy as described, and only one must function for the assembly to function, and, furthermore, if the time-to-failure densities are exponential with failure rate λ, then from equation 17-42, we obtain

$$R(t) = 1 - [1 - e^{-\lambda t}]^2 = 2e^{-\lambda t} - e^{-2\lambda t}.$$

17-4.5 Standby Redundancy

A common form of redundancy, called standby redundancy, is shown in Fig. 17-21. The unit labeled DS is a decision switch that we will assume has reliability of 1 for all t. The operating rules are as follows. Component 1 is initially "online," and when this component fails, the decision switch switches in component 2, which remains online until it fails.

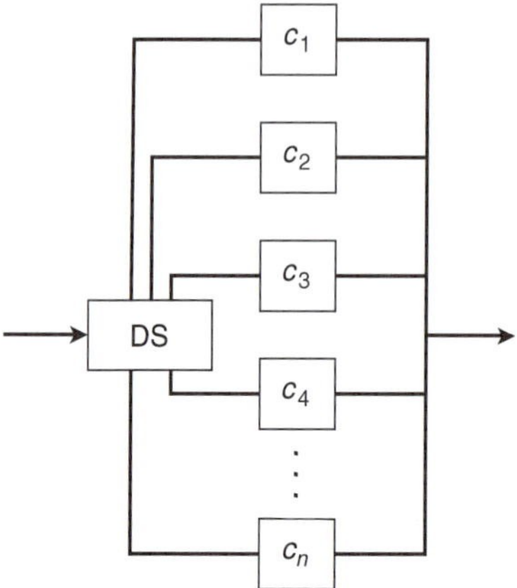

Figure 17-21 Standby redundancy.

Standby units are not subject to failure until activated. The time to failure for the assembly is

$$T = T_1 + T_2 + \cdots + T_n,$$

where T_i is the time to failure for the ith component and $T_1, T_2, \ldots, T_n$ are independent random variables. The most common value for n in practice is two, so the Central Limit Theorem is of little value. However, we know from the property of linear combinations that

$$E[T] = \sum_{i=1}^{n} E(T_i)$$

and

$$V[T] = \sum_{i=1}^{n} V(T_i).$$

We must know the distributions of the random variables T_i in order to find the distribution of T. The most common case occurs when the components are identical and the time-to-failure distributions are assumed to be exponential. In this case, T has a gamma distribution

$$\begin{aligned} f(t) &= \frac{\lambda}{(n-1)!}(\lambda t)^{n-1} e^{-\lambda t}, && t > 0, \\ &= 0 && \text{otherwise,} \end{aligned}$$

so that the reliability function is

$$R(t) = \sum_{k=0}^{n-1} e^{-\lambda t} (\lambda t)^k / k!, \qquad t > 0. \tag{17-44}$$

The parameter λ is the component failure rate; that is, $E(T_i) = 1/\lambda$. The mean time to failure and variance are

$$\text{MTTF} = E[T] = n/\lambda \tag{17-45}$$

and

$$V[T] = n/\lambda^2, \tag{17-46}$$

respectively.

Example 17-14

Two identical components are assembled in a standby redundant configuration with perfect switching. The component lives are identically distributed, independent random variables having an exponential distribution with failure rate 100^{-1}. The mean time to failure is

$$\text{MTTF} = 2/100^{-1} = 200$$

and the variance is

$$V[T] = 2/(100^{-1})^2 = 20{,}000.$$

The reliability function R is

$$R(t) = \sum_{k=0}^{1} e^{-t/100}\left(-t/100\right)^k/k!,$$

or

$$R(t) = e^{-t/100}[1 + t/100].$$

17-4.6 Life Testing

Life tests are conducted for different purposes. Sometimes, n units are placed on test and aged until all or most units have failed; the purpose is to test a hypothesis about the form of the time-to-failure density with certain parameters. Both formal statistical tests and probability plotting are widely used in life testing.

A second objective in life testing is to estimate reliability. Suppose, for example, that a manufacturer is interested in estimating $R(1000)$ for a particular component or system. One approach to this problem would be to place n units on test and count the number of failures, r, occurring before 1000 hours of operation. Failed units are not to be replaced in this example. An estimate of unreliability is $\hat{p} = r/n$, and an estimate of reliability is

$$\hat{R}(1000) = 1 - \frac{r}{n}. \tag{17-47}$$

A $100(1 - \alpha)\%$ lower-confidence limit on $R(1000)$ is given by [1 – upper limit on p], where p is the unreliability. This upper limit on p may be determined using a table of the binomial distribution. In thc case where n is large, an estimate of the upper limit on p is

$$\hat{p} + Z_{1-\alpha}\sqrt{\frac{\hat{p}(1-\hat{p})}{n}}. \tag{17-48}$$

Example 17-15

One hundred units are placed on life test, and the test is run for 1000 hours. There are two failures during test, so $\hat{p} = 0.02$, and $\hat{R}(1000) = 0.98$. Using a table of the binomial distribution, a 95% upper confidence limit on p is 0.06, so that a lower limit on $R(1000)$ is given by 0.94.

In recent years, there has been much work on the analysis of failure-time data, including plotting methods for identification of appropriate failure-time models and parameter estimation. For a good summary of this work, refer to Elsayed (1996).

17-4.7 Reliability Estimation with a Known Time-to-Failure Distribution

In the case where the form of the reliability function is assumed known and there is only one parameter, the maximum likelihood estimator for $R(t)$ is $\hat{R}(t)$, which is formed by substituting $\hat{\theta}$ for the parameter θ in the expression for $R(t)$, where $\hat{\theta}$ is the maximum likelihood estimator of θ. For more details and results for specific time-to-failure distributions, refer to Elsayed (1996).

17-4.8 Estimation with the Exponential Time-to-Failure Distribution

The most common case for the one-parameter situation is where the time-to-failure distribution is exponential, $R(t) = e^{-t/\theta}$. The parameter $\theta = E[T]$ is called the mean time to failure and the estimator for R is $\hat{R}(t)$, where

$$\hat{R}(t) = e^{-t/\hat{\theta}}$$

and $\hat{\theta}$ is the maximum likelihood estimator of θ.

Epstein (1960) developed the maximum likelihood estimators for θ under a number of different conditions and, furthermore, showed that a $100(1 - \alpha)\%$ confidence interval on $R(t)$ is given by

$$[e^{-t/\hat{\theta}_L}, e^{-t/\hat{\theta}_U}], \tag{17-49}$$

for the two-sided case, or

$$[e^{-t/\hat{\theta}_L}, 1] \tag{17-50}$$

for the lower, one-sided interval. In these cases, the values $\hat{\theta}_L$ and $\hat{\theta}_U$ are the lower- and upper-confidence limits on θ.

The following symbols will be used:

n = number of units placed on test at $t = 0$.

Q = total test time in unit hours.

t^* = time at which the test is terminated.

r = number of failures accumulated at time t.

r^* = preassigned number of failures.

$1 - \alpha$ = confidence level.

$\chi^2_{\alpha,k}$ = the upper α percentage point of the chi-square distribution with k degrees of freedom.

There are four situations to consider, according to whether the test is stopped after a preassigned time or after a preassigned number of failures and whether failed items are replaced or not replaced during test.

For the replacement test, the total test time in unit hours is $Q = nt^*$, and for the nonreplacement test

$$Q = \sum_{i=1}^{r} t_i + (n - r)t^*. \tag{17-51}$$

If items are censored (withdrawn items that have not failed), and if failures are replaced while censored items are not replaced, then

$$Q = \sum_{j=1}^{c} t_j + (n - c)t^*, \tag{17-52}$$

where c represents the number of censored items and t_j is the time of the jth censorship. If neither censored items nor failed items are replaced, then

$$Q=\sum_{i=1}^{r} t_i+\sum_{j=1}^{c} t_j+(n-r-c)t^*. \tag{17-53}$$

The development of the maximum likelihood estimators for θ is rather straightforward. In the case where the test is nonreplacement, and the test is discontinued after a fixed number of items have failed, the likelihood function is

$$\begin{aligned} L &= \prod_{i=1}^{r} f(t_i)\cdot\prod_{i=r}^{n} R(t^*) \\ &= \frac{1}{\theta^r}e^{-(1/\theta)\sum_{i=1}^{r} t_i}\cdot e^{-(n-r)t^*/\theta}. \end{aligned} \tag{17-54}$$

Then

$$l=\ln L=-r\ln\theta-\frac{1}{\theta}\sum_{i=1}^{r} t_i-(n-r)t^*/\theta$$

and solving $(\partial l/\partial\theta)=0$ yields the estimator

$$\hat{\theta}=\frac{\sum_{i=1}^{r} t_i+(n-r)t^*}{r}=\frac{Q}{r}. \tag{17-55}$$

It turns out that

$$\hat{\theta}=Q/r \tag{17-56}$$

is the maximum likelihood estimator of θ for all cases considered for the test design and operation.

The quantity $2r\hat{\theta}/\theta$ has a chi-square distribution with $2r$ degrees of freedom in the case where the test is terminated after a fixed number of failures. For fixed termination time t^*, the degrees of freedom becomes $2r+2$.

Since the expression $2r\hat{\theta}/\theta=2Q/\theta$, confidence limits on θ may be expressed as indicated in Table 17-10. The results presented in the table may be used directly with equations 17-49 and 17-50 to establish confidence limits on $R(t)$. It should be noted that this testing procedure does not require that the test be run for the time at which a reliability estimate is required. For example, 100 units may be placed on a nonreplacement test for 200 hours, the parameter θ estimated, and $\hat{R}(1000)$ calculated. In the case of the binomial testing mentioned earlier, it would have been necessary to run the test for 1000 hours.

Table 17-10 Confidence Limits on θ

Nature of Limit	Fixed Number of Failures, r^*	Fixed Termination Time, t^*
Two-sided limits	$\left[\frac{2Q}{\chi^2_{\alpha/2,2r}},\frac{2Q}{\chi^2_{1-\alpha/2,2r}}\right]$	$\left[\frac{2Q}{\chi^2_{\alpha/2,2r+2}},\frac{2Q}{\chi^2_{1-\alpha/2,2r+2}}\right]$
Lower, one-sided limit	$\left[\frac{2Q}{\chi^2_{\alpha,2r}},\infty\right]$	$\left[\frac{2Q}{\chi^2_{\alpha,2r+2}},\infty\right]$

The results are, however, dependent on the assumption that the distribution is exponential.

It is sometimes necessary to estimate the time t_R for which the reliability will be R. For the exponential model, this estimate is

$$\hat{t}_R = \hat{\theta} \cdot \ln \frac{1}{R} \tag{17-57}$$

and confidence limits on t_R are given in Table 17-11.

Example 17-16

Twenty items are placed on a replacement test that is to be operated until 10 failures occur. The tenth failure occurs at 80 hours, and the reliability engineer wishes to estimate the mean time to failure, 95% two-sided limits on θ, $R(100)$, and 95% two-sided limits on $R(100)$. Finally, she wishes to estimate the time for which the reliability will be 0.8 with point and 95% two-sided confidence interval estimates.

According to equation 17-56 and the results presented in Tables 17-10 and 17-11,

$$\hat{\theta} = \frac{nt^*}{r} = \frac{20(80)}{10} = 160 \text{ hours},$$

$$Q = nt^* = 1600 \text{ unit hours},$$

$$\left[\frac{2Q}{\chi^2_{0.025,20}}, \frac{2Q}{\chi^2_{0.975,20}}\right] = \left[\frac{3200}{34.17}, \frac{3200}{9.591}\right]$$

$$= [93.65, 333.65],$$

$$\hat{R}(100) = e^{-100/\hat{\theta}} = e^{-100/160} = 0.535.$$

According to equation 17-49, the confidence interval on $R(100)$ is

$$[e^{-100/93.65}, e^{-100/333.65}] = [0.344, 0.741].$$

Also,

$$\hat{t}_{0.80} = \hat{\theta} \ln \frac{1}{R} = 160 \ln \frac{1}{0.8} = 35.70 \text{ hours}.$$

The two-sided 95% confidence limit is determined from Table 17-11 as

$$\left[\frac{2(1600)(0.223)}{34.17}, \frac{2(1600)(0.223)}{9.591}\right] = [20.9, 74.45].$$

Table 17-11 Confidence Limits on t_R

Nature of Limit	Fixed Number of Failures, r^*	Fixed Termination Time, t^*
Two-sided limits	$\left[\frac{2Q\ln(1/R)}{\chi^2_{\alpha/2,2r}}, \frac{2Q\ln(1/R)}{\chi^2_{1-\alpha/2,2r}}\right]$	$\left[\frac{2Q\ln(1/R)}{\chi^2_{\alpha/2,2r+2}}, \frac{2Q\ln(1/R)}{\chi^2_{1-\alpha/2,2r+2}}\right]$
Lower, one-sided limit	$\left[\frac{2Q\ln L(1/R)}{\chi^2_{\alpha,2r}}, \infty\right]$	$\left[\frac{2Q\ln(1/R)}{\chi^2_{\alpha,2r-2}}, \infty\right]$

17-4.9 Demonstration and Acceptance Testing

It is not uncommon for a purchaser to test incoming products to assure that the vendor is conforming to reliability specifications. These tests are destructive tests and, in the case of attribute measurement, the test design follows that of acceptance sampling discussed earlier in this chapter.

A special set of sampling plans that assumes an exponential time-to-failure distribution has been presented in a Department of Defense handbook (DOD H-108), and these plans are in wide use.

17-5 SUMMARY

This chapter has presented several widely used methods for statistical quality control. Control charts were introduced and their use as process surveillance devices discussed. The $\overline{X}$ and R control charts are used for measurement data. When the quality characteristic is an attribute, either the p chart for fraction defective or the c or u chart for defects may be used.

The use of probability as a modeling technique in reliability analysis was also discussed. The exponential distribution is widely used as the distribution of time to failure, although other plausible models include the normal, lognormal, Weibull, and gamma distributions. System reliability analysis methods were presented for serial systems, as well as for systems having active or standby redundancy. Life testing and reliability estimation were also briefly introduced.

17-6 EXERCISES

17-1. An extrusion die is used to produce aluminum rods. The diameter of the rods is a critical quality characteristic. Below are shown $\overline{X}$ and R values for 20 samples of five rods each. Specifications on the rods are 0.5035 ± 0.0010 inch. The values given are the last three digits of the measurements; that is, 34.2 is read as 0.50342.

Sample	$\overline{X}$	R	Sample	$\overline{X}$	R
1	34.2	3	11	35.4	8
2	31.6	4	12	34.0	6
3	31.8	4	13	36.0	4
4	33.4	5	14	37.2	7
5	35.0	4	15	35.2	3
6	32.1	2	16	33.4	10
7	32.6	7	17	35.0	4
8	33.8	9	18	34.4	7
9	34.8	10	19	33.9	8
10	38.6	4	20	34.0	4

(a) Set up the $\overline{X}$ and R charts, revising the trial control limits if necessary, assuming assignable causes can be found.

(b) Calculate PCR and PCR_k. Interpret these ratios.

(c) What percentage of defectives is being produced by this process?

17-2. Suppose a process is in control, and 3-sigma control limits are in use on the $\overline{X}$ chart. Let the mean shift by 1.5σ. What is the probability that this shift will remain undetected for three consecutive samples? What would this probability be if 2-sigma control limits are used? The sample size is 4.

17-3. Suppose that an $\overline{X}$ chart is used to control a normally distributed process, and that samples of size n are taken every h hours and plotted on the chart, which has k sigma limits.

(a) Find the expected number of samples that will be taken until a false action signal is generated. This is called the in-control average run length (ARL).

(b) Suppose that the process shifts to an out-of-control state. Find the expected number of samples that will be taken until a false action is generated. This is the out-of-control ARL.

(c) Evaluate the in-control ARL for $k = 3$. How does this change if $k = 2$? What do you think about the use of 2-sigma limits in practice?

(d) Evaluate the out-of-control ARL for a shift of one sigma, given that $n = 5$.

17-4. Twenty-five samples of size 5 are drawn from a process at regular intervals, and the following data are obtained:

$$\sum_{i=1}^{25} \overline{X}_i = 362.75, \qquad \sum_{i=1}^{25} R_i = 8.60.$$

(a) Compute the control limits for the $\overline{X}$ and R charts.

(b) Assuming the process is in control and specification limits are 14.50 ± 0.50, what conclusions can you draw about the ability of the process to operate within these limits? Estimate the percentage of defective items that will be produced.

(c) Calculate PCR and PCR_k. Interpret these ratios.

17-5. Suppose an $\overline{X}$ chart for a process is in control with 3-sigma limits. Samples of size 5 are drawn every 15 minutes, on the quarter hour. Now suppose the process mean shifts out of control by 1.5σ 10 minutes after the hour. If D is the expected number of defectives produced per quarter hour in this out-of-control state, find the expected loss (in terms of defective units) that results from this control procedure.

17-6. The overall length of a cigar lighter body used in an automobile application is controlled using $\overline{X}$ and R charts. The following table gives length for 20 samples of size 4 (measurements are coded from 5.00 mm; that is, 15 is 5.15 mm).

	Observation			
Sample	1	2	3	4
1	15	10	8	9
2	14	14	10	6
3	9	10	9	11
4	8	6	9	13
5	14	8	9	12
6	9	10	7	13
7	15	10	12	12
8	14	16	11	10
9	11	7	16	10
10	11	14	11	12
11	13	8	9	5
12	10	15	8	10
13	8	12	14	9
14	15	12	14	6
15	13	16	9	5
16	14	8	8	12
17	8	10	16	9
18	8	14	10	9
19	13	15	10	8
20	9	7	15	8

(a) Set up the $\overline{X}$ and R charts. Is the process in statistical control?

(b) Specifications are 5.10 ± 0.05 mm. What can you say about process capability?

17-7. Montgomery (2001) presents 30 observations of oxide thickness of individual silicon wafers. The data are

Wafer	Oxide Thickness	Wafer	Oxide Thickness
1	45.4	16	58.4
2	48.6	17	51.0
3	49.5	18	41.2
4	44.0	19	47.1
5	50.9	20	45.7
6	55.2	21	60.6
7	45.5	22	51.0
8	52.8	23	53.0
9	45.3	24	56.0
10	46.3	25	47.2
11	53.9	26	48.0
12	49.8	27	55.9
13	46.9	28	50.0
14	49.8	29	47.9
15	45.1	30	53.4

(a) Construct a normal probability plot of the data. Does the normality assumption seem reasonable?

(b) Set up an individuals control chart for oxide thickness. Interpret the chart.

17-8. A machine is used to fill bottles with a particular brand of vegetable oil. A single bottle is randomly selected every half hour and the weight of the bottle recorded. Experience with the process indicates that the variability is quite stable, with $\sigma = 0.07$ oz. The process target is 32 oz. Twenty-four samples have been recorded in a 12-hour time period with the results given below.

Sample Number	x	Sample Number	x
1	32.03	13	31.97
2	31.98	14	32.01
3	32.02	15	31.93
4	31.85	16	32.09
5	31.91	17	31.96
6	32.09	18	31.88
7	31.98	19	31.82
8	32.03	20	31.92
9	31.98	21	31.81
10	31.91	22	31.95
11	32.01	23	31.97
12	32.12	24	31.94

(a) Construct a normal probability plot of the data. Does the normality assumption appear to be satisfied?

(b) Set up an individuals control chart for the weights. Interpret the results.

17-9. The following are the number of defective solder joints found during successive samples of 500 solder joints.

Day	No. of Defectives	Day	No. of Defectives
1	106	11	42
2	116	12	37
3	164	13	25
4	89	14	88
5	99	15	101
6	40	16	64
7	112	17	51
8	36	18	74
9	69	19	71
10	74	20	43
—	—	21	80

Construct a fraction-defective control chart. Is the process in control?

17-10. A process is controlled by a p chart using samples of size 100. The centerline on the chart is 0.05. What is the probability that the control chart detects a shift to 0.08 on the first sample following the shift? What is the probability that the shift is detected by at least the third sample following the shift?

17-11. Suppose a p chart with centerline at $\overline{p}$ with k sigma units is used to control a process. There is a critical fraction defective p_c that must be detected with probability 0.50 on the first sample following the shift to this state. Derive a general formula for the sample size that should be used on this chart.

17-12. A normally distributed process uses 66.7% of the specification band. It is centered at the nominal dimension, located halfway between the upper and lower specification limits.

(a) What is the process capability ratio PCR?

(b) What fallout level (fraction defective) is produced?

(c) Suppose the mean shifts to a distance exactly 3 standard deviations below the upper specification limit. What is the value of PCR_k? How has PCR changed?

(d) What is the actual fallout experienced after the shift in the mean?

17-13. Consider a process where specifications on a quality characteristic are 100 ± 15. We know that the standard deviation of this quality characteristic is 5. Where should we center the process to minimize the fraction defective produced? Now suppose the mean shifts to 105 and we are using a sample size of 4 on an $\overline{X}$ chart. What is the probability that such a shift will be detected on the first sample following the shift? What sample size would be needed on a p chart to obtain a similar degree of protection?

17-14. Suppose the following fraction defective had been found in successive samples of size 100 (read down):

0.09	0.03	0.12
0.10	0.05	0.14
0.13	0.13	0.06
0.08	0.10	0.05
0.14	0.14	0.14
0.09	0.07	0.11
0.10	0.06	0.09
0.15	0.09	0.13
0.13	0.08	0.12
0.06	0.11	0.09

Is the process in control with respect to its fraction defective?

17-15. The following represent the number of solder defects observed on 24 samples of five printed circuit boards: 7, 6, 8, 10, 24, 6, 5, 4, 8, 11, 15, 8, 4, 16, 11, 12, 8, 6, 5, 9, 7, 14, 8, 21. Can we conclude that the process is in control using a c chart? If not, assume assignable causes can be found and revise the control limits.

17-16. The following represent the number of defects per 1000 fcct in rubber-covered wire: 1, 1, 3, 7, 8, 10, 5, 13, 0, 19, 24, 6, 9, 11, 15, 8, 3, 6, 7, 4, 9, 20, 11, 7, 18, 10, 6, 4, 0, 9, 7, 3, 1, 8, 12. Do the data come from a controlled process?

17-17. Suppose the number of defects in a unit is known to be 8. If the number of defects in a unit shifts to 16, what it the probability that it will be detected by the c chart on the first sample following the shift?

17-18. Suppose we are inspecting disk drives for defects per unit, and it is known that there is an average of two defects per unit. We decided to make our inspection unit for the c chart five disk drives, and we control the total number of defects per inspection unit. Describe the new control chart.

17-19. Consider the data in Exercise 17-15. Set up a u chart for this process. Compare it to the c chart in Exercise 17-15.

17-20. Consider the oxide thickness data given in Exercise 17-7. Set up an EWMA control chart with $\lambda = 0.20$ and $L = 2.962$. Interpret the chart.

17-21. Consider the oxide thickness data given in Exercise 17-7. Construct a CUSUM control chart with $k = 0.75$ and $h = 3.34$ if the target thickness is 50. Interpret the chart.

17-22. Consider the weights provided in Exercise 17-8. Set up an EWMA control chart with $\lambda = 0.10$ and $L = 2.7$. Interpret the chart.

17-23. Consider the weights provided in Exercise 17-8. Set up a CUSUM control chart with $k = 0.50$ and $h = 4.0$. Interpret the chart.

17-24. A time-to-failure distribution is given by a uniform distribution:

$$f(t) = \frac{1}{\beta - \alpha}, \qquad \alpha \le t \le \beta,$$
$$= 0 \qquad \text{otherwise.}$$

(a) Determine the reliability function.

(b) Show that

$$\int_0^\infty R(t)dt = \int_0^\infty tf(t)dt.$$

(c) Determine the hazard function.

(d) Show that

$$R(t) = e^{-H(t)},$$

where H is defined as in equation 17-31.

17-25. Three units that operate and fail independently form a series configuration, as shown in the figure at the bottom of this page.

The time-to-failure distribution for each unit is exponential with the failure rates indicated.

(a) Find $R(60)$ for the system.

(b) What is the mean time-to-failure (MTTF) for this system?

17-26. Five identical units are arranged in an active redundancy to form a subsystem. Unit failure is independent, and at least two of the units must survive 1000 hours for the subsystem to perform its mission.

(a) If the units have exponential time-to-failure distributions with failure rate 0.002, what is the subsystem reliability?

(b) What is the reliability if only one unit is required?

17-27. If the units described in the previous exercise are operated in a standby redundancy with a perfect decision switch and only one unit is required for subsystem survival, determine the subsystem reliability.

17-28. One hundred units are placed on test and aged until all units have failed. The following results are obtained, and a mean life of $\bar{t} = 160$ hours is calculated from the serial data.

Time Interval	Number of Failures
0–100	50
100–200	18
200–300	17
300–400	8
400–500	4
After 500 hours	3

Use the chi-square goodness-of-fit test to determine whether you consider the exponential distribution to represent a reasonable time-to-failure model for these data.

17-29. Fifty units are placed on a life test for 1000 hours. Eight units fail during the period. Estimate $R(1000)$ for these units. Determine a lower 95% confidence interval on $R(1000)$.

17-30. In Section 17-4.7 it was noted that for one-parameter reliability functions, $R(t;\theta), \hat{R}(t;\theta) = R(t;\hat{\theta})$, where $\hat{\theta}$ and $\hat{R}$ are the maximum likelihood estimators. Prove this statement for the case

$$R(t;\theta) = e^{-t/\theta}, \qquad t \ge 0,$$
$$= 0, \qquad \text{otherwise.}$$

Hint: Express the density function f in terms of R.

17-31. For a nonreplacement test that is terminated after 200 hours of operation, it is noted that failures occur at the following times: 9, 21, 40, 55, and 85 hours. The units are assumed to have an exponential time-to-failure distribution, and 100 units were on test initially.

(a) Estimate the mean time to failure.

(b) Construct a 95% lower-confidence limit on the mean time to failure.

17-32. Use the statement in Exercise 17-31.

(a) Estimate $R(300)$ and construct a 95% lower-confidence limit on $R(300)$.

(b) Estimate the time for which the reliability will be 0.9, and construct a 95% lower limit on $t_{0.9}$.

→ $\lambda_1 = 3 \times 10^{-2}$ → $\lambda_2 = 6 \times 10^{-3}$ → $\lambda_3 = 4 \times 10^{-2}$ →

Figure for Exercise 17-25.

Chapter 18

Stochastic Processes and Queueing

18-1 INTRODUCTION

The term *stochastic process* is frequently used in connection with observations from a time-oriented, physical process that is controlled by a random mechanism. More precisely, a stochastic process is a sequence of random variables $\{X_t\}$, where $t \in T$ is a time or sequence index. The range space for X_t may be discrete or continuous; however, in this chapter we will consider only the case where at a particular time t the process is in exactly one of $m + 1$ mutually exclusive and exhaustive *states*. The states are labeled 0, 1, 2, 3, ..., m.

The variables $X_1, X_2, \ldots$ might represent the number of customers awaiting service at a ticket booth at times 1 minute, 2 minutes, and so on, after the booth opens. Another example would be daily demands for a certain product on successive days. X_0 represents the initial state of the process.

The chapter will introduce a special type of stochastic process called a *Markov process.* We will also discuss the *Chapman–Kolmogorov equations*, various special properties of *Markov chains*, the *birth–death equations*, and some applications to waiting-line, or *queueing*, and interference problems.

In the study of stochastic processes, certain assumptions are required about the joint probability distribution of the random variables $X_1, X_2, \ldots$. In the case of Bernoulli trials, presented in Chapter 5, recall that these variables were defined to be independent and that the range space (state space) consisted of two values (0, 1). Here we will first consider discrete-time Markov chains, the case where time is discrete and the independence assumption is relaxed to allow for a one-stage dependence.

18-2 DISCRETE-TIME MARKOV CHAINS

A stochastic process exhibits the *Markovian property* if

$$P\{X_{t+1}=j|X_t=i\} = P\{X_{t+1}=j|X_t=i, X_{t-1}=i_1, X_{t-2}=i_2, \ldots, X_0=i_t\} \tag{18-1}$$

for $t = 0, 1, 2, \ldots$, and every sequence $j, i, i_1, \ldots, i_t$. This is equivalent to stating that the probability of an event at time $t + 1$ *given* only the outcome at time t is equal to the probability of the event at time $t + 1$ *given* the entire state history of the system. In other words, the probability of the event at $t + 1$ is not dependent on the state history prior to time t.

The conditional probabilities

$$P\{X_{t+1}=j|X_t=i\} = p_{ij} \tag{18-2}$$

are called one-step transition probabilities, and they are said to be *stationary* if

$$P\{X_{t+1}=j|X_t=i\} = P\{X_1=j|X_0=i\}, \qquad \text{for } t = 0, 1, 2, \ldots, \tag{18-3}$$

so that the transition probabilities remain unchanged through time. These values may be displayed in a matrix $\mathbf{P} = [p_{ij}]$, called the one-step transition matrix. The matrix $\mathbf{P}$ has $m + 1$ rows and $m + 1$ columns, and

$$0 \le p_{ij} \le 1,$$

while

$$\sum_{j=0}^{m} p_{ij} = 1 \qquad \text{for } i = 0, 1, 2 \ldots, m.$$

That is, each element of the $\mathbf{P}$ matrix is a probability, and each row of the matrix sums to one.

The existence of the one-step, stationary transition probabilities implies that

$$p_{ij}^{(n)} = P\{X_{t+n} = j | X_t = i\} = P\{X_n = j | X_0 = i\} \tag{18-4}$$

for all $t = 0, 1, 2, \ldots$. The values $p_{ij}^{(n)}$ are called n-step transition probabilities, and they may be displayed in an n-step transition matrix

$$\mathbf{P}^{(n)} = [p_{ij}^{(n)}],$$

where

$$0 \le p_{ij}^{(n)} \le 1, \qquad n = 0, 1, 2, \ldots, \qquad i = 0, 1, 2, \ldots, m, \qquad j = 0, 1, 2, \ldots, m,$$

and

$$\sum_{j=0}^{m} p_{ij}^{(n)} = 1, \qquad n = 0, 1, 2, \ldots, \qquad i = 0, 1, 2, \ldots, m.$$

The 0-step transition matrix is the identity matrix.

A *finite-state Markov chain* is defined as a stochastic process having a finite number of states, the Markovian property, stationary transition probabilities, and an initial set of probabilities $\mathbf{A} = [a_0^{(0)}, a_1^{(0)}, a_2^{(0)}, \ldots, a_m^{(0)}]$, where $a_i^{(0)} = P\{X_0 = i\}$.

The *Chapman–Kolmogorov* equations are useful in computing n-step transition probabilities. These equations are

$$p_{ij}^{(n)} = \sum_{l=0}^{m} p_{il}^{(v)} \cdot p_{lj}^{(n-v)}, \qquad \begin{array}{l} i = 0, 1, 2, \ldots, m, \\ j = 0, 1, 2, \ldots, m, \\ 0 \le v \le n, \end{array} \tag{18-5}$$

and they indicate that in passing from state i to state j in n steps the process will be in some state, say l, after exactly v steps ($v \le n$). Therefore $p_{il}^{(v)} \cdot p_{lj}^{(n-v)}$ is the conditional probability that given state i as the starting state, the process goes to state l in v steps and from l to j in $(n - v)$ steps. When summed over l, the sum of the products yields $p_{ij}^{(n)}$.

By setting $v = 1$ or $v = n - 1$, we obtain

$$p_{ij}^{(n)} = \sum_{l=0}^{m} p_{il} p_{lj}^{(n-1)} = \sum_{l=0}^{m} p_{il}^{(n-1)} \cdot p_{lj}, \qquad \begin{array}{l} i = 0, 1, 2, \ldots, m, \\ j = 0, 1, 2, \ldots, m, \\ n = 1, 2, \ldots. \end{array}$$

It follows that the n-step transition probabilities, $\mathbf{P}^{(n)}$, may be obtained from the one-step probabilities, and

$$\mathbf{P}^{(n)} = \mathbf{P}^n. \tag{18-6}$$

The unconditional probability of being in state j at time $t = n$ is

$$\mathbf{A}^{(n)} = [a_0^{(n)}, a_1^{(n)}, \ldots, a_m^{(n)}], \tag{18-7}$$

where

$$a_j^{(n)} = P\{X_n = j\} = \sum_{i=0}^{m} a_i^{(0)} \cdot p_{ij}^{(n)}, \qquad \begin{array}{l} j = 0,1,2,\ldots,m, \\ n = 1,2,\ldots. \end{array}$$

Thus, $\mathbf{A}^{(n)} = \mathbf{A} \cdot \mathbf{P}^{(n)}$. Further, we note that the rule for matrix multiplication solves the total probability law of Theorem 1-8 so that $\mathbf{A}^{(n)} = \mathbf{A}^{(n-1)} \cdot \mathbf{P}$.

Example 18-1

In a computing system, the probability of an error on each cycle depends on whether or not it was preceded by an error. We will define 0 as the error state and 1 as the nonerror state. Suppose the probability of an error if preceded by an error is 0.75, the probability of an error if preceded by a nonerror is 0.50, the probability of a nonerror if preceded by an error is 0.25, and the probability of nonerror if preceded by nonerror is 0.50. Thus,

$$\mathbf{P} = \begin{bmatrix} 0.75 & 0.25 \\ 0.50 & 0.50 \end{bmatrix}.$$

Two-step, three-step, …, seven-step transition matrices are shown below:

$$\mathbf{P}^2 = \begin{bmatrix} 0.688 & 0.312 \\ 0.625 & 0.375 \end{bmatrix}, \quad \mathbf{P}^3 = \begin{bmatrix} 0.672 & 0.328 \\ 0.656 & 0.344 \end{bmatrix},$$
$$\mathbf{P}^4 = \begin{bmatrix} 0.668 & 0.332 \\ 0.664 & 0.336 \end{bmatrix}, \quad \mathbf{P}^5 = \begin{bmatrix} 0.667 & 0.333 \\ 0.666 & 0.334 \end{bmatrix},$$
$$\mathbf{P}^6 = \begin{bmatrix} 0.667 & 0.333 \\ 0.667 & 0.333 \end{bmatrix}, \quad \mathbf{P}^7 = \begin{bmatrix} 0.667 & 0.333 \\ 0.667 & 0.333 \end{bmatrix}.$$

If we know that initially the system is in the nonerror state, then $a_1^{(0)} = 1$, $a_2^{(0)} = 0$, and $\mathbf{A}^{(n)} = [a_j^{(n)}] = \mathbf{A} \cdot \mathbf{P}^{(n)}$. Thus, for example, $\mathbf{A}^{(7)} = [0.667, 0.333]$.

18-3 CLASSIFICATION OF STATES AND CHAINS

We will first consider the notion of *first passage times.* The length of time (number of steps in discrete-time systems) for the process to go from state i to state j for the first time is called the first passage time. If $i = j$, then this is the number of steps needed for the process to return to state i for the first time, and this is termed the *first return time* or *recurrence time* for state i.

First passage times under certain conditions are random variables with an associated probability distribution. We let $f_{ij}^{(n)}$ denote the probability that the first passage time from state i to j is equal to n, where it can be shown directly from Theorem 1-5 that

$$\begin{aligned} f_{ij}^{(1)} &= p_{ij}^{(1)} = p_{ij}, \\ f_{ij}^{(2)} &= p_{ij}^{(2)} - f_{ij}^{(1)} \cdot p_{jj}, \\ &\vdots \\ f_{ij}^{(n)} &= p_{ij}^{(n)} - f_{ij}^{(1)} \cdot p_{jj}^{(n-1)} - f_{ij}^{(2)} \cdot p_{jj}^{(n-2)} - \cdots - f_{ij}^{(n-1)} p_{jj}. \end{aligned} \tag{18-8}$$

Thus, recursive computation from the one-step transition probabilities yields the probability that the first passage time is n for given i, j.

Example 18-2

Using the one-step transition probabilities presented in Example 18-1, the distribution of the passage time index n for $i = 0, j = 1$ is determined as

$$f_{01}^{(1)} = p_{01} = 0.250,$$

$$f_{01}^{(2)} = (0.312) - (0.25)(0.5) = 0.187,$$

$$f_{01}^{(3)} = (0.328) - (0.25)(0.375) - (0.187)(0.5) = 0.141,$$

$$f_{01}^{(4)} = (0.332) - (0.25)(0.344) - (0.187)(0.375) - (0.141)(0.5) = 0.105.$$

$$\vdots$$

There are four such distributions corresponding to i, j value: (0, 0), (0, 1), (1, 0), (1, 1).

If i and j are fixed, then $\sum_{n=1}^{\infty} f_{ij}^{(n)} \le 1$. When the sum is equal to one, the values $f_{ij}^{(n)}$, for $n = 1, 2, \ldots$, represent the probability distribution of first passage time for specific i, j. In the case where a process in state i may never reach state j, $\sum_{n=1}^{\infty} f_{ij}^{(n)} < 1$.

Where $i = j$ and $\sum_{n=1}^{\infty} f_{ii}^{(n)} = 1$, the state i is termed a *recurrent state*, since given that the process is in state i it will always eventually return to i.

If $p_{ii} = 1$ for some state i, then that state is called an *absorbing state*, and the process will never leave it after it is entered.

The state i is called a *transient state* if

$$\sum_{n=1}^{\infty} f_{ii}^{(n)} < 1,$$

since there is a positive probability that given the process is in state i, it will never return to this state. It is not always easy to classify a state as transient or recurrent, since it is sometimes difficult to calculate first passage time probabilities $f_{ij}^{(n)}$ for all n, as was the case in Example 18-2. Nevertheless, the expected first passage time is

$$\mu_{ij} = \begin{cases} \infty, & \sum_{n=1}^{\infty} f_{ij}^{(n)} < 1, \\ \sum_{n=1}^{\infty} n \cdot f_{ij}^{(n)}, & \sum_{n=1}^{\infty} f_{ij}^{(n)} = 1, \end{cases} \tag{18-9}$$

and if $\sum_{n=1}^{\infty} f_{ij}^{(n)} = 1$, a simple conditioning argument shows that

$$\mu_{ij} = 1 + \sum_{l \ne j} p_{il} \cdot \mu_{lj}. \tag{18-10}$$

If we take $i = j$, the expected first passage time is called the *expected recurrence time.* If $\mu_{ii} = \infty$ for a recurrent state, it is called *null*; if $\mu_{ii} < \infty$, it is called *nonnull* or *positive recurrent.*

There are no null recurrent states in a finite-state Markov chain. All of the states in such chains are either positive recurrent or transient.

A state is called *periodic* with period $\tau > 1$ if a return is possible only in $\tau, 2\tau, 3\tau, \ldots$, steps; so $p_{ii}^{(n)} = 0$ for all values of n that are not divisible by $\tau > 1$, and τ is the smallest integer having this property.

A state j is termed *accessible* from state i if $p_{ij}^{(n)} > 0$ for some $n = 1, 2, \ldots$. In our example of the computing system, each state, 0 and 1, is accessible from the other, since $p_{ij}^{(n)} > 0$ for all i, j and all n. If state j is accessible from i and state i is accessible from j, then the states are said to *communicate.* This is the case in Example 18-1. We note that any state communicates with itself. If state i communicates with j, j also communicates with i. Also, if i communicates with l and l communicates with j, then i also communicates with j.

If the state space is partitioned into disjoint sets (called equivalence classes) of states, where communicating states belong to the same class, then the Markov chain may consist of one or more classes. If there is only one class so that all states communicate, the Markov chain is said to be *irreducible.* The chain represented by Example 18-1 is thus also irreducible. For finite-state Markov chains, the states of a class are either all positive recurrent or all transient. In many applications, the states will all communicate. This is the case if there is a value of n for which $p_{ij}^{(n)} > 0$ for all values of i and j.

If state i in a class is aperiodic (not periodic), and if the state is also positive recurrent, then the state is said to be *ergodic.* An irreducible Markov chain is ergodic if all of its states are ergodic. In the case of such Markov chains the distribution

$$\mathbf{A}^{(n)} = \mathbf{A} \cdot \mathbf{P}^n$$

converges as $n \to \infty$, and the limiting distribution is independent of the initial probabilities, **A**. In Example 18-1, this was clearly observed to be the case, and after five steps ($n > 5$), $P\{X_n = 0\} = 0.667$ and $P\{X_n = 1\} = 0.333$ when three significant figures are used.

In general, for irreducible, ergodic Markov chains,

$$\lim_{n\to\infty} p_{ij}^{(n)} = \lim_{n\to\infty} a_j^{(n)} = p_j,$$

and, furthermore, these values p_j are independent of i. These "steady state" probabilities, p_j, satisfy the following *state equations*:

$$p_j > 0, \tag{18-11a}$$

$$\sum_{j=0}^{m} p_j = 1, \tag{18-11b}$$

$$p_j = \sum_{i=0}^{m} p_i \cdot p_{ij} \qquad j = 0, 1, 2, \ldots, m. \tag{18-11c}$$

Since there are $m + 2$ equations in 18-11b and 18-11c, and since there are $m + 1$ unknowns, one of the equations is redundant. Therefore, we will use m of the $m + 1$ equations in equation 18-11c with equation 18-11b.

Example 18-3

In the case of the computing system presented in Example 18-1, we have from equations 18-11b and 18-11c,

$$1 = p_0 + p_1,$$

$$p_0 = p_0\,(0.75) + p_1(0.50),$$

or

$$p_0 = 2/3 \qquad \text{and} \qquad p_1 = 1/3,$$

which agrees with the emerging result as $n > 5$ in Example 18-1.

The steady state probabilities and the mean recurrence time for irreducible, ergodic Markov chains have a reciprocal relationship,

$$\mu_{jj} = \frac{1}{p_j}, \qquad j = 0, 1, 2, \ldots, m. \tag{18-12}$$

In Example 18-3 note that $\mu_{00} = 1/p_0 = 1.5$ and $\mu_{11} = 1/p_1 = 3$.

Example 18-4

The mood of a corporate president is observed over a period of time by a psychologist in the operations research department. Being inclined toward mathematical modeling, the psychologist classifies mood into three states as follows:

0: Good (cheerful)

1: Fair (so-so)

2: Poor (glum and depressed)

The psychologist observes that mood changes occur only overnight: thus, the data allow estimation of the transition probabilities

$$\mathbf{P} = \begin{bmatrix} 0.6 & 0.2 & 0.2 \\ 0.3 & 0.4 & 0.3 \\ 0.0 & 0.3 & 0.7 \end{bmatrix}.$$

The equations

$$\begin{aligned} p_0 &= 0.6p_0 + 0.3p_1 + 0p_2, \\ p_1 &= 0.2p_0 + 0.4p_1 + 0.3p_2, \\ 1 &= p_0 + p_1 + p_2 \end{aligned}$$

are solved simultaneously for the steady state probabilities

$$\begin{aligned} p_0 &= 3/13, \\ p_1 &= 4/13, \\ p_2 &= 6/13. \end{aligned}$$

Given that the president is in a bad mood, that is, state 2, the mean time required to return to that state is μ_{22}, where

$$\mu_{22} = \frac{1}{p_2} = \frac{13}{6} \text{ days.}$$

As noted earlier, if $p_{kk} = 1$, state k is called an *absorbing state*, and the process remains in state k once that state is reached. In this case, b_{ik} is called the absorption probability, which is the conditional probability of absorption into state k given state i. Mathematically, we have

$$b_{ik} = \sum_{j=0}^{m} p_{ij} \cdot b_{jk}, \qquad i = 0, 1, 2, \ldots, m, \tag{18-13}$$

where

$$b_{kk} = 1$$

and

$$b_{ik} = 0 \qquad \text{for } i \text{ recurrent, } i \neq k.$$

18-4 CONTINUOUS-TIME MARKOV CHAINS

If the time parameter is continuous rather than a discrete index, as assumed in the previous sections, the Markov chain is called a *continuous-parameter* chain. It is customary to use a slightly different notation for continuous-parameter Markov chains, namely $X(t) = X_t$, where $\{X(t)\}$, $t \geq 0$, will be considered to have states 0, 1, …, m. The discrete nature of the state space [range space for $X(t)$] is thus maintained, and

$$p_{ij}(t) = P\,[X(t+s) = j | X(s) = i], \qquad \begin{array}{l} i = 0, 1, 2, \ldots, m, \\ j = 0, 1, 2, \ldots, m, \\ s \geq 0, t \geq 0, \end{array}$$

is the stationary transition probability function. It is noted that these probabilities are not dependent on s but only on t for a specified i, j pair of states. Furthermore, at time $t = 0$, the function is continuous with

$$\lim_{t \to 0} p_{ij}(t) = \begin{cases} 0 & i \neq j, \\ 1 & i = j. \end{cases}$$

There is a direct correspondence between the discrete-time and continuous-time models. The Chapman–Kolmogorov equations become

$$p_{ij}(t) = \sum_{l=0}^{m} p_{il}(v) \cdot p_{lj}(t-v) \tag{18-14}$$

for $0 \leq v \leq t$, and for the specified state pair i, j and time t. If there are times t_1 and t_2 such that $p_{ij}(t_1) > 0$ and $p_{ji}(t_2) > 0$, then states i and j are said to communicate. Once again states that communicate form an equivalence class, and where the chain is irreducible (all states form a single class)

$$p_{ij}(t) > 0, \quad \text{for } t > 0,$$

for each state pair i, j.

We also have the property that

$$\lim_{t \to \infty} p_{ij}(t) = p_j,$$

where p_j exists and is independent of the initial state probability vector **A**. The values p_j are again called the steady state probabilities and they satisfy

$$p_j > 0, \qquad j = 0, 1, 2, \ldots, m,$$

$$\sum_{j=0}^{m} p_j = 1,$$

$$p_j = \sum_{i=0}^{m} p_i \cdot p_{ij}(t), \qquad j = 0, 1, 2, \ldots, m, \qquad t \geq 0.$$

The *intensity of transition*, given that the state is j, is defined as

$$u_j = \lim_{\Delta t \to 0} \left\{ \frac{1 - p_{jj}(\Delta t)}{\Delta t} \right\} = -\frac{d}{dt} p_{jj}(t)\Big|_{t=0}, \tag{18-15}$$

where the limit exists and is finite. Likewise, the *intensity of passage* from state i to state j, given that the system is in state i, is

$$u_{ij} = \lim_{\Delta t \to 0} \left\{ \frac{p_{ij}(\Delta t)}{\Delta t} \right\} = \frac{d}{dt} p_{ij}(t)\Big|_{t=0}, \tag{18-16}$$

again where the limit exists and is finite. The interpretation of the intensities is that they represent an instantaneous rate of transition from state i to j. For a small Δt, $p_{ij}(\Delta t) = u_{ij}\Delta t + o(\Delta t)$, where $o(\Delta t)/\Delta t \to 0$ as $\Delta t \to 0$, so that u_{ij} is a proportionality constant by which $p_{ij}(\Delta t)$ is proportional to Δt as $\Delta t \to 0$. The transition intensities also satisfy the balance equations

$$p_j \cdot u_j = \sum_{i \neq j} p_i \cdot u_{ij}, \qquad j = 0, 1, 2, \ldots, m. \tag{18-17}$$

These equations indicate that in steady state, the rate of transition out of state j is equal to the rate of transition into j.

Example 18-5

An electronic control mechanism for a chemical process is constructed with two identical modules, operating as a parallel, active redundant pair. The function of at least one module is necessary for the mechanism to operate. The maintenance shop has two identical repair stations for these modules and, furthermore, when a module fails and enters the shop, other work is moved aside and repair work is immediately initiated. The "system" here consists of the mechanism and repair facility and the states are as follows:

0: Both modules operating

1: One unit operating and one unit in repair

2: Two units in repair (mechanism down)

The random variable representing time to failure for a module has an exponential density, say

$$\begin{aligned} f_T(t) &= \lambda e^{-\lambda t}, & t &\geq 0, \\ &= 0, & t &< 0, \end{aligned}$$

and the random variable describing repair time at a repair station also has an exponential density, say

$$\begin{aligned} r_T(t) &= \mu e^{-\mu t}, & t &\geq 0, \\ &= 0, & t &< 0. \end{aligned}$$

Interfailure and interrepair times are independent, and $\{X(t)\}$ can be shown to be a continuous-parameter, irreducible Markov chain with transitions only from a state to its neighbor states: $0 \rightarrow 1$, $1 \rightarrow 0$, $1 \rightarrow 2$, $2 \rightarrow 1$. Of course, there may be no state change.

The transition intensities are

$$\begin{aligned} u_0 &= 2\lambda, & u_1 &= \lambda + \mu, \\ u_{01} &= 2\lambda, & u_{12} &= \lambda, \\ u_{02} &= 0, & u_{20} &= 0, \\ u_{10} &= \mu, & u_{21} &= 2\mu, \\ & & u_2 &= 2\mu. \end{aligned}$$

Using equation 18-17,

$$\begin{aligned} 2\lambda p_0 &= \mu p_1, \\ (\lambda + \mu)\, p_1 &= 2\lambda p_0 + 2\mu p_2, \\ 2\mu p_2 &= \lambda p_1, \end{aligned}$$

and since $p_0 + p_1 + p_2 = 1$, some algebra gives

$$\begin{aligned} p_0 &= \frac{\mu^2}{(\lambda+\mu)^2}, \\ p_1 &= \frac{2\lambda\mu}{(\lambda+\mu)^2}, \\ p_2 &= \frac{\lambda^2}{(\lambda+\mu)^2}. \end{aligned}$$

The system availability (probability that the mechanism is up) in the steady state condition is thus

$$\text{Availability} = 1 - \frac{\lambda^2}{(\lambda+\mu)^2}.$$

The matrix of transition probabilities for time increment Δt may be expressed as

$$\mathbf{P} = \left[p_{ij}(\Delta t)\right]$$
$$= \begin{bmatrix} 1-u_0\Delta t & u_{01}\Delta t & \cdots & u_{0j}\Delta t & \cdots & u_{0m}\Delta t \\ u_{10}\Delta t & 1-u_1\Delta t & \cdots & u_{1j}\Delta t & \cdots & u_{1m}\Delta t \\ \vdots & & & & & \\ u_{i0}\Delta t & u_{i1}\Delta t & \cdots & u_{ij}\Delta t & \cdots & u_{1m}\Delta t \\ \vdots & & & & & \\ u_{m0}\Delta t & u_{m1}\Delta t & \cdots & u_{mj}\Delta t & \cdots & 1-u_m\Delta t \end{bmatrix} \tag{18-18}$$

and

$$p_j(t+\Delta t) = \sum_{i=0}^{m} p_i(t)\cdot p_{ij}(\Delta t), \qquad j = 0,1,2,\ldots,m, \tag{18-19}$$

where

$$p_j(t) = P\,[X(t) = j].$$

From the jth equation in the $m + 1$ equations of equation 18-19,

$$p_j(t+\Delta t) = p_0(t)\cdot u_{0j}\Delta t + \cdots + p_i(t)\cdot u_{ij}\Delta t + \cdots + p_j(t)[1 - u_j\Delta t] + \cdots + p_m(t)\cdot u_{mj}\,\Delta t,$$

which may be rewritten as

$$\frac{d}{dt}p_j(t) = \lim_{\Delta t \to 0}\left[\frac{p_j(t+\Delta t) - p_j(t)}{\Delta t}\right] = -u_j \cdot p_j(t) + \sum_{i \neq j} u_{ij} \cdot p_i(t). \tag{18-20}$$

The resulting system of differential equations is

$$p_j'(t) = -u_j \cdot p_j(t) + \sum_{i \neq j} u_{ij} \cdot p_i(t), \qquad j = 0, 1, 2, \ldots, m, \tag{18-21}$$

which may be solved when m is finite, given initial conditions (probabilities) **A**, and using the result that $\sum_{j=0}^{m} p_j(t) = 1$. The solution

$$[p_0(t), p_1(t), \ldots, p_m(t)] = \mathbf{P}(t) \tag{18-22}$$

presents the state probabilities as a function of time in the same manner that $p_j^{(n)}$ presented state probabilities as a function of the number of transitions, n, given an initial condition vector **A** in the discrete-time model. The solution to equations 18-21 may be somewhat difficult to obtain, and in general practice, transformation techniques are employed.

18-5 THE BIRTH–DEATH PROCESS IN QUEUEING

The major application of the so-called birth–death process that we will study is in *queueing* or *waiting-line* theory. Here birth will refer to an *arrival* and death to a *departure* from a physical system, as shown in Fig. 18-1.

Queueing theory is the mathematical study of queues or waiting lines. These waiting lines occur in a variety of problem environments. There is an input process or "calling population," from which arrivals are drawn, and a queueing *system*, which in Fig. 18-1 consists of the queue and service facility. The calling population may be finite or infinite. Arrivals occur in a probabilistic manner. A common assumption is that the interarrival times are exponentially distributed. The queue is generally classified according to whether its capacity is infinite or finite, and the service discipline refers to the order in which the customers in the queue are served. The service mechanism consists of one or more servers, and the elapsed service time is commonly called the holding time.

The following notation will be employed:

$X(t)$ = Number of customers in system at time t

States = $0, 1, 2, \ldots, j, j+1, \ldots$

s = Number of servers

$p_j(t) = P\{X(t) = j|\mathbf{A}\}$

$p_j = \lim_{t \to \infty} p_j(t)$

λ_n = Arrival rate given that n customers are in the system

μ_n = Service rate given that n customers are in the system

The birth–death process can be used to describe how $X(t)$ changes through time. It will be assumed here that when $X(t) = j$, the probability distribution of the time to the next birth (arrival) is exponential with parameter λ_j, $j = 0, 1, 2, \ldots$. Furthermore, given $X(t) = j$, the remaining time to the next service completion is taken to be exponential with parameter μ_j, $j = 1, 2, \ldots$. Poisson-type postulates are assumed to hold, so that the probability of more than one birth or death at the same instant is zero.

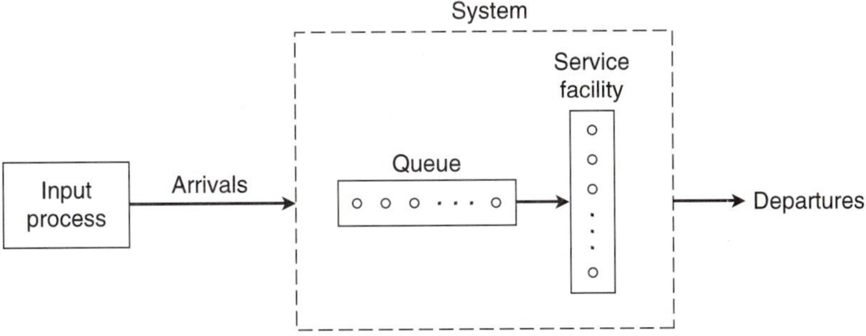

Figure 18-1 A simple queueing system.

A transition diagram is shown in Fig. 18-2. The transition matrix corresponding to equation 18-18 is

$$\mathbf{P} = \begin{bmatrix} 1-\lambda_0\Delta t & \lambda_0\Delta t & 0 & \cdots & 0 & \cdots \\ \mu_1\Delta t & 1-(\lambda_1+\mu_1)\Delta t & \lambda_1\Delta t & \cdots & 0 & \cdots \\ 0 & \mu_2\Delta t & 1-(\lambda_2+\mu_2)\Delta t & \cdots & 0 & \cdots \\ 0 & 0 & \mu_3\Delta t & \cdots & 0 & \cdots \\ \vdots & \vdots & \vdots & & \vdots & \\ 0 & 0 & 0 & \cdots & 0 & \cdots \\ \vdots & \vdots & \vdots & \cdots & \lambda_{j-1}\Delta t & \cdots \\ 0 & 0 & 0 & \cdots & 1-(\lambda_j+\mu_j)\Delta t & \cdots \\ 0 & 0 & 0 & \cdots & \mu_{j+1}\Delta t & \cdots \\ \vdots & \vdots & \vdots & & \vdots & \cdots \\ 0 & 0 & 0 & \cdots & 0 & \cdots \end{bmatrix}.$$

We note that $p_{ij}(\Delta t) = 0$ for $j < i - 1$ or $j > i + 1$. Furthermore, the transition intensities and intensities of passage shown in equation 18-17 are

$$\begin{aligned} u_0 &= \lambda_0, & & \\ u_j &= \lambda_j + \mu_j & & \text{for } j = 1, 2, \ldots, \\ u_{ij} &= \lambda_i & & \text{for } j = i + 1, \\ &= \mu_i & & \text{for } j = i - 1, \\ &= 0 & & \text{for } j < i - 1, j > i + 1. \end{aligned}$$

The fact that the transition intensities and intensities of passage are constant with time is important in the development of this model. The nature of transition can be viewed to be specified by assumption, or it may be considered as a result of the prior assumption about the distribution of time between occurrences (births and deaths).

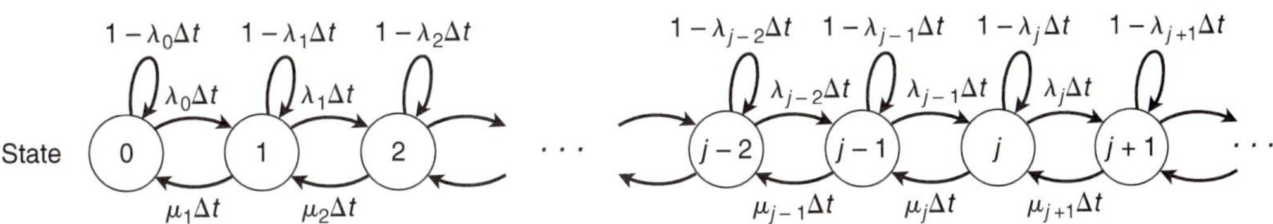

Figure 18-2 Transition diagram for the birth–death process.

The assumptions of independent, exponentially distributed service times and independent, exponentially distributed interarrival times yield transition intensities that are constant in time. This was also observed in the development of the Poisson and exponential distributions in Chapters 5 and 6.

The methods used in equations 18-19 through 18-21 may be used to formulate an infinite set of differential state equations from the transition matrix of equation 18-22. Thus, the time-dependent behavior is described in the following equations:

$$p_0'(t) = -\lambda_0 p_0(t) + \mu_1 p_1(t), \tag{18-23}$$

$$p_j'(t) = -(\lambda_j + \mu_j)\, p_j(t) + \lambda_{j-1} p_{j-1}(t) + \mu_{j+1}\, p_{j+1}(t), \qquad j = 1, 2, \ldots, \tag{18-24}$$

$$\sum_{j=0}^{\infty} p_j(t) = 1, \qquad \text{and} \qquad \mathbf{A} = \left[a_0^{(0)}, a_1^{(0)}, \ldots, a_j^{(0)}, \ldots\right].$$

In the steady state ($t \to \infty$), we have $p_j'(t) = 0$, so the steady state equations are obtained from equations 18-23 and 18-24:

$$\begin{aligned}
\mu_1 p_1 &= \lambda_0 p_0, \\
\lambda_0 p_0 + \mu_2 p_2 &= (\lambda_1 + \mu_1) \cdot p_1, \\
\lambda_1 p_1 + \mu_3 p_3 &= (\lambda_2 + \mu_2) \cdot p_2, \\
&\vdots \\
\lambda_{j-2}\, p_{j-2} + \mu_j p_j &= (\lambda_{j-1} + \mu_{j-1})\, p_{j-1}, \\
\lambda_{j-1} p_{j-1} + \mu_{j+1} p_{j+1} &= (\lambda_j + \mu_j)\, p_j, \\
&\vdots
\end{aligned} \tag{18-25}$$

and $\sum_{j=0}^{\infty} p_j = 1$.

Equations 18-25 could have also been determined by the direct application of equation 18-17, which provides a "rate balance" or "intensity balance." Solving equations 18-25 we obtain

$$\begin{aligned}
p_1 &= \frac{\lambda_0}{\mu_1} \cdot p_0, \\
p_1 &= \frac{\lambda_1}{\mu_2} \cdot p_1 = \frac{\lambda_1 \lambda_0}{\mu_2 \mu_1} \cdot p_0, \\
p_3 &= \frac{\lambda_2}{\mu_3} \cdot p_2 = \frac{\lambda_2 \lambda_1 \lambda_0}{\mu_3 \mu_2 \mu_1} \cdot p_0, \\
&\vdots \\
p_j &= \frac{\lambda_{j-1}}{\mu_j} \cdot p_{j-1} = \frac{\lambda_{j-1} \lambda_{j-2} \cdots \lambda_0}{\mu_j \mu_{j-1} \cdots \mu_1} \cdot p_0, \\
p_{j+1} &= \frac{\lambda_j}{\mu_{j+1}} \cdot p_j = \frac{\lambda_j \lambda_{j-1} \cdots \lambda_0}{\mu_{j+1} \mu_j \cdots \mu_1} \cdot p_0.
\end{aligned}$$

If we let

$$C_j = \frac{\lambda_{j-1}\lambda_{j-2}\cdots\lambda_0}{\mu_j\mu_{j-1}\cdots\mu_1}, \tag{18-26}$$

then

$$p_j = C_j \cdot p_0, \qquad j = 1, 2, 3, \dots,$$

and since

$$\sum_{j=0}^{\infty} p_j = 1 \quad \text{or} \quad p_0 + \sum_{j=1}^{\infty} p_j = 1, \tag{18-27}$$

we obtain

$$p_0 = \frac{1}{1 + \sum_{j=1}^{\infty} C_j}.$$

These steady state results assume that the λ_j, μ_j values are such that a steady state can be reached. This will be true if $\lambda_j = 0$ for $j > k$, so that there are a finite number of states. It is also true if $\rho = \lambda/s\mu < 1$, where λ and μ are constant and s denotes the number of servers. The steady state will not be reached if $\sum_{j=1}^{\infty} C_j = \infty$.

18-6 CONSIDERATIONS IN QUEUEING MODELS

When the arrival rate λ_j is constant for all j, the constant is denoted λ. Similarly, when the service rate per busy server is constant, it will be denoted μ, so that $\mu_j = s\mu$ if $j \geq s$ and $\mu_j = j\,\mu$ if $j < s$. The exponential distributions

$$\begin{aligned} f_T(t) &= \lambda e^{-\lambda t}, && t \geq 0, \\ &= 0, && t < 0, \\ r_T(t) &= \mu e^{-\mu t}, && t \geq 0, \\ &= 0, && t < 0, \end{aligned}$$

for interarrival times and service times in a busy channel produce rates λ and μ, which are constant. The mean interarrival time is $1/\lambda$, and the mean time for a busy channel to complete service is $1/\mu$.

A special set of notation has been widely employed in the steady state analysis of queueing systems. This notation is given in the following list:

$L = \sum_{j=0}^{\infty} j \cdot p_j$ = Expected number of customers in the queueing system
$L_q = \sum_{j=s}^{\infty} (j - s) \cdot p_j$ = Expected queue length
W = Expected time in the system (including service time)
W_q = Expected waiting time in the queue (excluding service time)

If λ is constant for all j, then it has been shown that

$$L = \lambda W \tag{18-28}$$

and

$$L_q = \lambda W_q$$

(These results are special cases of what is known as Little's law.) If the λ_j are not equal, $\bar{\lambda}$ replaces λ, where

$$\bar{\lambda} = \sum_{j=0}^{\infty} \lambda_j \cdot p_j. \tag{18-29}$$

The system utilization coefficient $\rho = \lambda/s\mu$ is the fraction of time that the servers are busy. In the case where the mean service time is $1/\mu$ for all $j \geq 1$,

$$W = W_q + \frac{1}{\mu}. \tag{18-30}$$

The birth–death process rates, $\lambda_0, \lambda_1, \ldots, \lambda_j, \ldots$ and $\mu_1, \mu_2, \ldots, \mu_j, \ldots$, may be assigned any positive values as long as the assignment leads to a steady state solution. This allows considerable flexibility in using the results given in equation 18-27. The specific models subsequently presented will differ in the manner in which λ_j and μ_j vary as a function of j.

18-7 BASIC SINGLE-SERVER MODEL WITH CONSTANT RATES

We will now consider the case where $s = 1$, that is, a single server. We will also assume an unlimited potential queue length with exponential interarrivals having a constant parameter λ, so that $\lambda_0 = \lambda_1 = \cdots = \lambda$. Furthermore, service times will be assumed to be independent and exponentially distributed with $\mu_1 = \mu_2 = \cdots = \mu$. We will assume $\lambda < \mu$. As a result of equation 18-26, we have

$$C_j = \left(\frac{\lambda}{\mu}\right)^j = \rho^j, \qquad j = 1, 2, 3, \ldots, \tag{18-31}$$

and from equation 18-27

$$\begin{aligned} p_j &= \rho^j p_0, \qquad j = 1, 2, 3, \ldots, \\ p_0 &= \frac{1}{1 + \sum_{j=1}^{\infty} \rho^j} = 1 - \rho. \end{aligned} \tag{18-32}$$

Thus, the steady state equations are

$$p_j = (1 - \rho)\rho^j, \qquad j = 0, 1, 2, \ldots . \tag{18-33}$$

Note that the probability that there are j customers in the system p_j is given by a geometric distribution with parameter ρ. The mean number of customers in the system, L, is determined as

$$\begin{aligned} L &= \sum_{j=0}^{\infty} j \cdot (1 - \rho)\rho^j \\ &= (1 - \rho) \cdot \rho \sum_{j=0}^{\infty} \frac{d}{d\rho}\left(\rho^j\right) \\ &= (1 - \rho) \cdot \rho \frac{d}{d\rho} \sum_{j=0}^{\infty} \rho^j \\ &= \frac{\rho}{1 - \rho}. \end{aligned} \tag{18-34}$$

And the expected queue length is

$$\begin{aligned} L_q &= \sum_{j=1}^{\infty} (j-1) \cdot p_j \\ &= L - (1 - p_0) \\ &= \frac{\rho^2}{1-\rho} = \frac{\lambda^2}{\mu(\mu - \lambda)}. \end{aligned} \tag{18-35}$$

Using equations 18-28 and 18-34, we find that the expected waiting time in the system is

$$W = \frac{L}{\lambda} = \frac{\rho}{\lambda(1-\rho)} = \frac{1}{\mu - \lambda}, \tag{18-36}$$

and the expected waiting time in the queue is

$$W_q = \frac{L_q}{\lambda} = \frac{\lambda^2}{\mu(\mu - \lambda) \cdot \lambda} = \frac{\lambda}{\mu(\mu - \lambda)}. \tag{18-37}$$

These results could have been developed directly from the distributions of time in the system and time in the queue, respectively. Since the exponential distribution reflects a memoryless process, an arrival finding j units in the system will wait through $j + 1$ services, including its own, and thus its waiting time T_{j+1} is the sum of $j + 1$ independent, exponentially distributed random variables. This random variable was shown in Chapter 6 to have a gamma distribution. This is a conditional density given that the arrival finds j units in the system. Thus, if S represents time in the system,

$$\begin{aligned} P(S > w) &= \sum_{j=0}^{\infty} p_j \cdot P(T_{j+1} > w) \\ &= \sum_{j=0}^{\infty} (1-\rho)\rho^j \cdot \int_w^{\infty} \frac{\mu^{j+1}}{\Gamma(j+1)} t^j e^{-\mu t} dt \\ &= \int_w^{\infty} (1-\rho)\mu e^{-\mu t} \sum_{j=0}^{\infty} \frac{(\rho \mu t)^j}{j!} dt \\ &= \int_w^{\infty} (1-\rho)\mu e^{-(1-\rho)\mu t} dt \\ &= e^{-\mu(1-\rho)w}, \qquad w \geq 0, \\ &= 0, \qquad w < 0, \end{aligned} \tag{18-38}$$

which is seen to be the complement of the distribution function for an exponential random variable with parameter $\mu(1 - \rho)$. The mean value $W = 1/\mu(1 - \rho) = 1/(\mu - \lambda)$ follows directly.

If we let S_q represent time in the queue, excluding service time, then

$$P(S_q = 0) = p_0 = 1 - \rho.$$

If we take T_j as the sum of j service times, T_j will again have a gamma distribution. Then, as in the previous manipulations,

$$\begin{aligned} P(S_q > w_q) &= \sum_{j=1}^{\infty} p_j \cdot P(T_j > w_q) \\ &= \sum_{j=1}^{\infty} (1-\rho)\rho^j \cdot P(T_j > w_q) \\ &= \rho e^{-\mu(1-\rho)w_q}, \qquad w_q > 0, \\ &= 0, \qquad w_q < 0, \end{aligned} \tag{18-39}$$

and we find the distribution of time in the queue $g(w_q)$, for $w_q > 0$, to be

$$g(w_q) = \frac{d}{dw_q}\left[1 - \rho e^{-\mu(1-\rho)w_q}\right] = \rho(1-\rho)\mu e^{-\mu(1-\rho)w_q}, \qquad w_q > 0.$$

Thus, the probability distribution is

$$\begin{aligned} g(w_q) &= 1 - \rho, \qquad w_q = 0, \\ &= \lambda(1-\rho)e^{-(\mu-\lambda)w_q}, \qquad w_q > 0, \end{aligned} \tag{18-40}$$

which was noted in Section 2-2 as being for a mixed-type random variable (in equation 2-2, $G \neq 0$ and $H \neq 0$). The expected waiting time in the queue W_q could be determined directly from this distribution as

$$\begin{aligned} W_q &= (1-\rho)\cdot 0 + \int_0^{\infty} w_q \cdot \lambda(1-\rho)e^{-(\mu-\lambda)w_q}\, dw_q \\ &= \frac{\lambda}{\mu(\mu-\lambda)}. \end{aligned} \tag{18-41}$$

When $\lambda \geq \mu$, the summation of the terms p_j in equation 18-32 diverges. In this case, there is no steady state solution since the steady state is never reached. That is, the queue would grow without bound.

18-8 SINGLE SERVER WITH LIMITED QUEUE LENGTH

If the queue is limited so that at most N units can be in the system, and if the exponential service times and exponential interarrival times are retained from the prior model, we have

$$\begin{aligned} \lambda_0 &= \lambda_1 = \cdots = \lambda_{N-1} = \lambda, \\ \lambda_j &= 0, \qquad j \geq N, \end{aligned}$$

and

$$\mu_1 = \mu_2 = \cdots = \mu_N = \mu.$$

It follows from equation 18-26 that

$$\begin{aligned} C_j &= \left(\frac{\lambda}{\mu}\right)^j, \qquad j \leq N, \\ &= 0, \qquad j > N. \end{aligned} \tag{18-42}$$

Thus,

$$p_j = \left(\frac{\lambda}{\mu}\right)^j p_0 = \rho^j p_0 \qquad j = 0,1,2,\ldots,N,$$

so that

$$p_0 \sum_{j=0}^{N} \rho^j = 1$$

and

$$p_0 = \frac{1}{1+\sum_{j=1}^{N} \rho^j} = \frac{1-\rho}{1-\rho^{N+1}}. \tag{18-43}$$

As a result, the steady state equations are given by

$$p_j = \rho^j \left[\frac{1-\rho}{1-\rho^{N+1}}\right], \qquad j = 0,1,2,\ldots,N. \tag{18-44}$$

The mean number of customers in the system in this case is

$$\begin{aligned} L &= \sum_{j=0}^{N} j \cdot \rho^j \left[\frac{1-\rho}{1-\rho^{N+1}}\right] \\ &= \rho\left[\frac{1-(N+1)\rho^N + N\rho^{N+1}}{(1-\rho)\left(1-\rho^{N+1}\right)}\right]. \end{aligned} \tag{18-45}$$

The mean number of customers in the queue is

$$\begin{aligned} L_q &= \sum_{j=1}^{N} (j-1) \cdot p_j \\ &= \sum_{j=0}^{N} j p_j - \sum_{j=1}^{N} p_j \\ &= L - (1-p_0). \end{aligned} \tag{18-46}$$

The mean time in the system is found as

$$W = \frac{L}{\lambda}, \tag{18-47}$$

and the mean time in the queue is

$$W_q = \frac{L_q}{\lambda} = \frac{L-1+p_0}{\lambda}, \tag{18-48}$$

where L is given by equation 18-45.

18-9 MULTIPLE SERVERS WITH AN UNLIMITED QUEUE

We now consider the case where there are multiple servers. We also assume that the queue is unlimited and that exponential assumptions hold for interarrival times and service times. In this case, we have

$$\lambda_0 = \lambda_1 = \cdots = \lambda_j = \cdots = \lambda \tag{18-49}$$

and

$$\begin{aligned} \mu_j &= j\mu && \text{for } j \le s, \\ &= s\mu && \text{for } j > s. \end{aligned}$$

Thus, defining $\phi = \lambda/\mu$, we have

$$\begin{aligned} C_j &= \frac{\lambda^j}{j!\mu^j} = \frac{\phi^j}{j!} && j \le s \\ &= \frac{\lambda^j}{s!s^{j-s}\mu^j} = \frac{\phi^j}{s!s^{j-s}}, && j > s. \end{aligned} \tag{18-50}$$

It follows from equation 18-27 that the state equations are developed as

$$\begin{aligned} p_j &= \frac{\phi^j p_0}{j!} && j \le s \\ &= \frac{\phi^j p_0}{s!s^{j-s}} && j > s \\ p_0 &= \frac{1}{1 + \sum_{j=1}^{s} \frac{\phi^j}{j!} + \sum_{j=s+1}^{\infty} \frac{\phi^j}{s!s^{j-s}}} \\ &= \frac{1}{\sum_{j=0}^{s-1} \frac{\phi^j}{j!} + \frac{\phi^s}{s!}\left(\frac{1}{1-\rho}\right)} \end{aligned} \tag{18-51}$$

where $\rho = \lambda/s\mu = \phi/s$ is the utilization coefficient, assuming $\rho < 1$.

The value L_q, representing the mean number of units in the queue, is developed as follows:

$$\begin{aligned} L_q &= \sum_{j=s}^{\infty} (j-s)p_j = \sum_{j=0}^{\infty} j \cdot p_{s+j} = \left[\frac{\phi^s}{s!}\sum_{j=0}^{\infty} j\rho^j\right] \cdot p_0 \\ &= \left[\frac{\phi^s}{s!} \cdot \rho \frac{d}{d\rho}\left(\sum_{j=0}^{\infty} \rho^j\right)\right] \cdot p_0 \\ &= \left[\frac{\phi^s \rho}{s!(1-\rho)^2}\right] \cdot p_0. \end{aligned} \tag{18-52}$$

Then

$$W_q = \frac{L_q}{\lambda} \tag{18-53}$$

and

$$W = W_q + \frac{1}{\mu}, \tag{18-54}$$

so that

$$L = \lambda W = \lambda\left(W_q + \frac{1}{\mu}\right) = L_q + \phi. \tag{18-55}$$

18-10 OTHER QUEUEING MODELS

There are numerous other queueing models that can be developed from the birth–death process. In addition, it is also possible to develop queueing models for situations involving nonexponential distributions. One useful result, given without proof, is for a single-server system having exponential interarrivals and arbitrary service time distribution with mean $1/\mu$ and variance σ^2. If $\rho = \lambda/\mu < 1$, then steady state measures are given by equations 18-56:

$$\begin{aligned} p_0 &= 1-\rho, \\ L_q &= \frac{\lambda^2\sigma^2 + \rho^2}{2(1-\rho)}, \\ L &= \rho + L_q, \\ W_q &= \frac{L_q}{\lambda}, \\ W &= W_q + \frac{1}{\mu}. \end{aligned} \tag{18-56}$$

In the case where service times are constant at $1/\mu$, the foregoing relationships yield the measures of system performance by taking the variance $\sigma^2 = 0$.

18-11 SUMMARY

This chapter introduced the notion of discrete-state space stochastic processes for discrete-time and continuous-time orientations. The Markov process was developed along with the presentation of state properties and characteristics. This was followed by a presentation of the birth–death process and several important applications to queueing models for the description of waiting-time phenomena.

18-12 EXERCISES

18-1. A shoe repair shop in a suburban mall has one shoesmith. Shoes are brought in for repair and arrive according to a Poisson process with a constant arrival rate of two pairs per hour. The repair time distribution is exponential with a mean of 20 minutes, and there is independence between the repair and arrival processes. Consider a pair of shoes to be the unit to be served, and do the following:

(a) In the steady state, find the probability that the number of pairs of shoes in the system exceeds 5.

(b) Find the mean number of pairs in the shop and the mean number of pairs waiting for service.

(c) Find the mean turnaround time for a pair of shoes (time in the shop waiting plus repair, but excluding time waiting to be picked up).

18-2. Weather data are analyzed for a particular locality, and a Markov chain is employed as a model for weather change as follows. The conditional probability of change from rain to clear weather in one day is 0.3. Likewise, the conditional probability of transition from clear to rain in one day is 0.1. The model is to be a discrete-time model, with transitions occurring only between days.

(a) Determine the matrix **P** of one-step transition probabilities.

(b) Find the steady state probabilities.

(c) If today is clear, find the probability that it will be clear exactly 3 days hence.

(d) Find the probability that the first passage from a clear day to a rainy day occurs in exactly 2 days, given a clear day is the initial state.

(e) What is the mean recurrence time for the rainy day state?

18-3. A communication link transmits binary characters, (0, 1). There is a probability p that a transmitted character will be received correctly by a receiver, which then transmits to another link, etc. If X_0 is the initial character and X_1 is the character received after the first transmission, X_2 after the second, etc., then with independence $\{X_n\}$ is a Markov chain. Find the one-step and steady state transition matrices.

18-4. Consider a two-component active redundancy where the components are identical and the time-to-failure distributions are exponential. When both units are operating, each carries load $L/2$ and each has failure rate λ. However, when one unit fails, the load carried by the other component is L, and its failure rate under this load is $(1.5)\lambda$. There is only one repair facility available, and repair time is exponentially distributed with mean $1/\mu$. The system is considered failed when both components are in the failed state. Both components are initially operating. Assume that $\mu > (1.5)\lambda$. Let the states be as follows:

0: No components are failed.
1: One component is failed and is in repair.
2: Two components are failed, one is in repair, one is waiting, and the system is in the failed condition.

(a) Determine the matrix **P** of transition probabilities associated with interval Δt.

(b) Determine the steady state probabilities.

(c) Write the system of differential equations that present the transient or time-dependent relationships for transition.

18-5. A communication satellite is launched via a booster system that has a discrete-time guidance control system. Course correction signals form a sequence $\{X_n\}$ where the state space for X is as follows:

0: No correction required.
1: Minor correction required.
2: Major correction required.
3: Abort and system destruct.

If $\{X_n\}$ can be modeled as a Markov chain with one-step transition matrix as

$$\mathbf{P} = \begin{bmatrix} 1 & 0 & 0 & 0 \\ 2/3 & 1/6 & 1/6 & 0 \\ 0 & 2/3 & 1/6 & 1/6 \\ 0 & 0 & 0 & 1 \end{bmatrix},$$

do the following:

(a) Show that states 0 and 1 are absorbing states.

(b) If the initial state is state 1, compute the steady state probability that the system is in state 0.

(c) If the initial probabilities are (0,1/2,1/2,0), compute the steady state probability p_0.

(d) Repeat (c) with $A = (1/4,1/4,1/4,1/4)$.

18-6. A gambler bets \$1 on each hand of blackjack. The probability of winning on any hand is p, and the probability of losing is $1 - p = q$. The gambler will continue to play until either \$$Y$ has been accumulated, or he has no money left. Let X_t denote the accumulated winnings on hand t. Note that $X_{t+1} = X_t + 1$, with probability p, that $X_{t+1} = X_t - 1$, with probability q, and $X_{t+1} = X_t$ if $X_t = 0$ or $X_t = Y$. The stochastic process X_t is a Markov chain.

(a) Find the one-step transition matrix **P**.

(b) For $Y = 4$ and $p = 0.3$, find the absorption probabilities: b_{10}, b_{14}, b_{30}, and b_{34}.

18-7. An object moves between four points on a circle, which are labeled 1, 2, 3, and 4. The probability of moving one unit to the right is p, and the probability of moving one unit to the left is $1 - p = q$. Assume that the object starts at 1, and let X_n denote the location on the circle after n steps.

(a) Find the one-step transition matrix **P**.

(b) Find an expression for the steady state probabilities p_j.

(c) Evaluate the probabilities p_j for $p = 0.5$ and $p = 0.8$.

18-8. For the single-server queueing model presented in Section 18-7, sketch the graphs of the following quantities as a function of $\rho = \lambda/\mu$, for $0 < \rho < 1$.

(a) Probability of no units in the system.

(b) Mean time in the system.

(c) Mean time in the queue.

18-9. Interarrival times at a telephone booth are exponential, with an average time of 10 minutes. The length of a phone call is assumed to be exponentially distributed with a mean of 3 minutes.

(a) What is the probability that a person arriving at the booth will have to wait?

(b) What is the average queue length?

(c) The telephone company will install a second booth when an arrival would expect to have to wait 3 minutes or more for the phone. By how much must the rate of arrivals be increased in order to justify a second booth?

(d) What is the probability that an arrival will have to wait more than 10 minutes for the phone?

(e) What is the probability that it will take a person more than 10 minutes altogether, for the phone and to complete the call?

(f) Estimate the fraction of a day that the phone will be in use.

18-10. Automobiles arrive at a service station in a random manner at a mean rate of 15 per hour. This station has only one service position, with a mean servicing rate of 27 customers per hour. Service times are exponentially distributed. There is space for only the automobile being served and two waiting. If all three spaces are filled, an arriving automobile will go on to another station.

(a) What is the average number of units in the station?

(b) What fraction of customers will be lost?

(c) Why is $L_q \neq L - 1$?

18-11. An engineering school has three secretaries in its general office. Professors with jobs for the secretaries arrive at random, at an average rate of 20 per 8-hour day. The amount of time that a secretary spends on a job has an exponential distribution with a mean of 40 minutes.

(a) What fraction of the time are the secretaries busy?

(b) How much time does it take, on average, for a professor to get his or her jobs completed?

(c) If an economy drive reduced the secretarial force to two secretaries, what will be the new answers to (a) and (b)?

18-12. The mean frequency of arrivals at an airport is 18 planes per hour, and the mean time that a runway is tied up with an arrival is 2 minutes. How many runways will have to be provided so that the probability of a plane having to wait is 0.20? Ignore finite population effects and make the assumption of exponential interarrival and service times.

18-13. A hotel reservations facility uses inward WATS lines to service customer requests. The mean number of calls that arrive per hour is 50, and the mean service time for a call is 3 minutes. Assume that interarrival and service times are exponentially distributed. Calls that arrive when all lines are busy obtain a busy signal and are lost from the system.

(a) Find the steady state equations for this system.

(b) How many WATS lines must be provided to ensure that the probability of a customer obtaining a busy signal is 0.05?

(c) What fraction of the time are all WATS lines busy?

(d) Suppose that during the evening hours call arrivals occur at a mean rate of 10 per hour. How does this affect the WATS line utilization?

(e) Suppose the estimated mean service time (3 minutes) is in error, and the true mean service time is really 5 minutes. What effect will this have on the probability of a customer finding all lines busy if the number of lines in (b) are used?

Chapter 19

Computer Simulation

One of the most widespread applications of probability and statistics lies in the use of computer simulation methods. A simulation is simply an imitation of the operation of a real-world system for purposes of evaluating that system. Over the past 20 years, computer simulation has enjoyed a great deal of popularity in the manufacturing, production, logistics, service, and financial industries, to name just a few areas of application. Simulations are often used to analyze systems that are too complicated to attack via analytic methods such as queueing theory. We are primarily interested in simulations that are:

1. Dynamic—that is, the system state changes over time.
2. Discrete—that is, the system state changes as the result of discrete events such as customer arrivals or departures.
3. Stochastic (as opposed to deterministic).

The stochastic nature of simulation prompts the ensuing discussion in the text.

This chapter is organized as follows. It begins in Section 19-1 with some simple motivational examples designed to show how one can apply simulation to answer interesting questions about stochastic systems. These examples invariably involve the generation of random variables to drive the simulation, for example customer interarrival times and service times. The subject of Section 19-2 is the development of techniques to generate random variables. Some of these techniques have already been alluded to in previous chapters, but we will give a more complete and self-contained presentation here. After a simulation run is completed, one must conduct a rigorous analysis of the resulting output, a task made difficult because simulation output, for example customer waiting times, is almost never independent or identically distributed. The problem of output analysis is studied in Section 19-3. A particularly attractive feature of computer simulation is its ability to allow the experimenter to analyze and compare certain scenarios quickly and efficiently. Section 19-4 discusses methods for reducing the variance of estimators arising from a single scenario, thus resulting in more-precise statements about system performance, at no additional cost in simulation run time. We also extend this work by mentioning methods for selecting the best of a number of competing scenarios. We point out here that excellent general references for the topic of stochastic simulation are Banks, Carson, Nelson, and Nicol (2001) and Law and Kelton (2000).

19-1 MOTIVATIONAL EXAMPLES

This section illustrates the use of simulation through a series of simple, motivational examples. The goal is to show how one uses random variables within a simulation to answer questions about the underlying stochastic system.

Example 19-1

Coin Flipping

We are interested in simulating independent flips of a fair coin. Of course, this is a trivial sequence of Bernoulli trials with success probability $p = 1/2$, but this example serves to show how one can use simulation to analyze such a system. First of all, we need to generate realizations of heads (H) and tails (T), each with probability 1/2. Assuming that the simulation can somehow produce a sequence of independent uniform (0,1) random numbers, $U_1, U_2, \ldots$, we will arbitrarily designate flip i as H if we observe $U_i < 0.5$, and a flip as T if we observe $U_i \geq 0.5$. How one generates independent uniforms is the subject of Section 19-2. In any case, suppose that the following uniforms are observed:

0.32 0.41 0.06 0.93 0.82 0.49 0.21 0.77 0.71 0.08.

This sequence of uniforms corresponds to the outcomes HHHTTHHTTH. The reader is asked to study this example in various ways in Exercise 19-1. This type of "static" simulation, in which we simply repeat the same type of trials over and over, has come to be known as Monte Carlo simulation, in honor of the European city-state, where gambling is a popular recreational activity.

Example 19-2

Estimate π

In this example, we will estimate π using Monte Carlo simulation in conjunction with a simple geometric relation. Referring to Fig. 19-1, consider a unit square with an inscribed circle, both centered at (1/2,1/2). If one were to throw darts randomly at the square, the probability that a particular dart will land in the circle is $\pi/4$, the ratio of the circle's area to that of the square. How can we use this simple fact to estimate π? We shall use Monte Carlo simulation to throw many darts at the square. Specifically, generate independent pairs of independent uniform (0,1) random variables, (U_{11}, U_{12}), $(U_{21}, U_{22}), \ldots$. These pairs will fall randomly on the square. If, for pair i, it happens that

$$(U_{i1} - 1/2)^2 + (U_{i2} - 1/2)^2 \leq 1/4, \tag{19-1}$$

then that pair will also fall within the circle. Suppose we run the experiment for n pairs (darts). Let $X_i = 1$ if pair i satisfies inequality 19-1, that is, if the ith dart falls in the circle; otherwise, let $X_i = 0$. Now count up the number of darts $X = \sum_{i=1}^{n} X_i$ falling in the circle. Clearly, X has the binomial distribution with parameters n and $p = \pi/4$. Then the proportion $\hat{p} = X/n$ is the maximum likelihood estimate for $p = \pi/4$, and so the maximum likelihood estimator for π is just $\hat{\pi} = 4\hat{p}$. If, for instance,

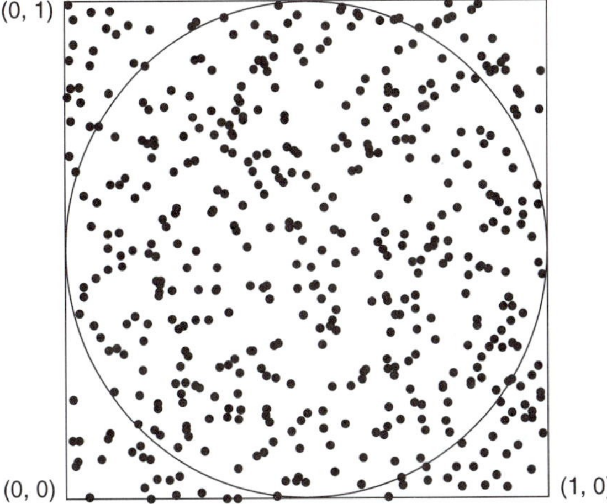

Figure 19-1 Throwing darts to estimate π.

we conducted $n = 1000$ trials and observed $X = 753$ darts in the circle, our estimate would be $\hat{\pi} = 3.12$. We will encounter this estimation technique again in Exercise 19-2.

Example 19-3

Monte Carlo Integration

Another interesting use of computer simulation involves Monte Carlo integration. Usually, the method becomes efficacious only for high-dimensional integrals, but we will fall back to the basic one-dimensional case for ease of exposition. To this end, consider the integral

$$I = \int_a^b f(x)dx = (b-a)\int_0^1 f(a+(b-a)u)du. \tag{19-2}$$

As described in Fig. 19-2, we shall estimate the value of this integral by summing up n rectangles, each of width $1/n$ centered randomly at point U_i on [0,1], and of height $f(a + (b - a)U_i)$. Then an estimate for I is

$$\hat{I}_n = \frac{b-a}{n}\sum_{i=1}^{n} f(a+(b-a)U_i). \tag{19-3}$$

One can show (see Exercise 19-3) that $\hat{I}_n$ is an unbiased estimator for I, that is, $E[\hat{I}_n] = I$ for all n. This makes $\hat{I}_n$ an intuitive and attractive estimator.

To illustrate, suppose that we want to estimate the integral

$$I = \int_0^1 [1+\cos(\pi x)]dx$$

and the following $n = 4$ numbers are a uniform (0,1) sample:

0.419 0.109 0.732 0.893.

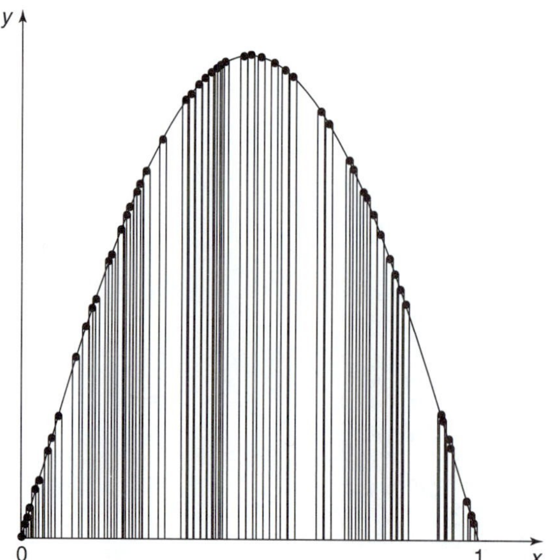

Figure 19-2 Monte Carlo integration.

Plugging into equation 19-3, we obtain

$$\hat{I}_4 = \frac{1-0}{4}\sum_{i=1}^{4}\left[1+\cos\left(\pi\left(0+(1-0)U_i\right)\right)\right] = 0.896,$$

which is close to the actual answer of 1. See Exercise 19-4 for additional Monte Carlo integration examples.

Example 19-4

A Single-Server Queue

Now the goal is to simulate the behavior of a single-server queueing system. Suppose that six customers arrive at a bank at the following times, which have been generated from some appropriate probability distribution:

3 4 6 10 15 20.

Upon arrival, customers queue up in front of a single teller and are processed sequentially, in a first-come-first-served manner. The service times corresponding to the arriving customers are

7 6 4 6 1 2.

For this example, we assume that the bank opens at time 0 and closes its doors at time 20 (just after customer 6 arrives), serving any remaining customers.

Table 19-1 and Fig. 19-3 trace the evolution of the system as time progresses. The table keeps track of the times at which customers arrive, begin service, and leave. Figure 19-3 graphs the status of the queue as a function of time; in particular, it graphs $L(t)$, the number of customers in the system (queue + service) at time t.

Note that customer i can begin service only at time max (A_i, D_{i-1}), that is, the maximum of his arrival time and the previous customer's departure time. The table and figure are quite easy to interpret. For instance, the system is empty until time 3, when customer 1 arrives. At time 4, customer 2 arrives, but must wait in line until customer 1 finishes service at time 10. We see from the figure that between times 20 and 26, customer 4 is in service, while customers 5 and 6 wait in the queue. From the table, the average waiting time for the six customers is $\sum_{i=1}^{6} W_i/6 = 44/6$. Further, the average number of customers in the system is $\int_0^{29} L(t)dt/29 = 70/29$, where we have computed the integral by adding up the rectangles in Fig. 19-3. Exercise 19-5 looks at extensions of the single-server queue. Many simulation software packages provide simple ways to model and analyze more-complicated queueing networks.

Table 19-1 Bank Customers in Single-Server Queueing System

i, customer	A_i, arrival time	B_i, begin service	S_i, service time	D_i, depart time	W_i, wait
1	3	3	7	10	0
2	4	10	6	16	6
3	6	16	4	20	10
4	10	20	6	26	10
5	15	26	1	27	11
6	20	27	2	29	7

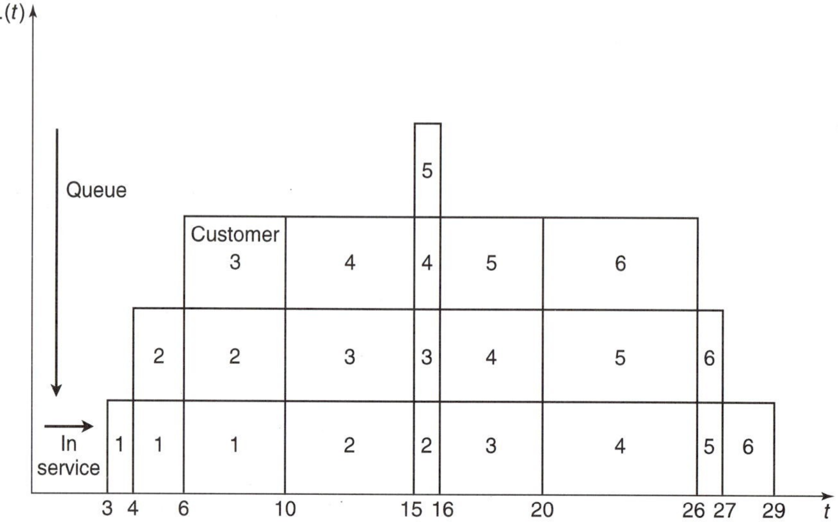

Figure 19-3 Number of customers $L(t)$ in single-server queueing system.

Example 19-5

(s, S) Inventory Policy

Customer orders for a particular good arrive at a store every day. During a certain one-week period, the quantities ordered are

$$10 \quad 6 \quad 11 \quad 3 \quad 20 \quad 6 \quad 8.$$

The store starts the week off with an initial stock of 20. If the stock falls to 5 or below, the owner orders enough from a central warehouse to replenish the stock to 20. Such replenishment orders are placed only at the end of the day and are received before the store opens the next day. There are no customer back orders, so any customer orders that are not filled immediately are lost. This is called an (s, S) inventory system, where the inventory is replenished to $S = 20$ whenever it hits level $s = 5$.

The following is a history for this system:

Day	Initial Stock	Customer Order	End Stock	Reorder?	Lost Orders
1	20	10	10	No	0
2	10	6	4	Yes	0
3	20	11	9	No	0
4	9	3	6	No	0
5	6	20	0	Yes	14
6	20	6	14	No	0
7	14	8	6	No	0

We see that at the end of days 2 and 5, replenishment orders were made. In particular, on day 5, the store ran out of stock and lost 14 orders as a result. See Exercise 19-6.

19-2 GENERATION OF RANDOM VARIABLES

All the examples described in Section 19-1 required random variables to drive the simulation. In Examples 19-1 through 19-3, we needed uniform (0,1) random variables; Examples 19-4 and 19-5 used more-complicated random variables to model customer arrivals, service times, and order quantities. This section discusses methods to generate such random variables automatically. The generation of uniform (0,1) random variables is a good place to start, especially since it turns out that uniform (0,1) generation forms the basis for the generation of all other random variables.

19-2.1 Generating Uniform (0,1) Random Variables

There are a variety of methods for generating uniform (0,1) random variables, among them are the following:

1. Sampling from certain physical devices, such as an atomic clock.
2. Looking up predetermined random numbers from a table.
3. Generating *pseudorandom* numbers (PRNs) from a deterministic algorithm.

The most widely used techniques in practice all employ the latter strategy of generating PRNs from a deterministic algorithm. Although, by definition, PRNs are not truly random, there are many algorithms available that produce PRNs that *appear* to be perfectly random. Further, these algorithms have the advantages of being computationally fast and repeatable—speed is a good property to have for the obvious reasons, while repeatability is desirable for experimenters who want to be able to replicate their simulation results when the runs are conducted under identical conditions.

Perhaps the most popular method for obtaining PRNs is the linear congruential generator (LCG). Here, we start with a nonnegative "seed" integer, X_0, use the seed to generate a sequence of nonnegative integers, $X_1, X_2, \ldots$, and then convert the X_i to PRNs, $U_1, U_2, \ldots$. The algorithm is simple.

1. Specify a nonnegative seed integer, X_0.
2. For $i = 1, 2, \ldots$, let $X_i = (aX_{i-1} + c) \bmod (m)$, where a, c, and m are appropriately chosen integer constants, and where "mod" denotes the modulus function, for example, $17 \bmod (5) = 2$ and $-1 \bmod (5) = 4$.
3. For $i = 1, 2, \ldots$, let $U_i = X_i/m$.

Example 19-6

Consider the "toy" generator $X_i = (5X_{i-1} + 1) \bmod (8)$, with seed $X_0 = 0$. This produces the integer sequence $X_1 = 1, X_2 = 6, X_3 = 7, X_4 = 4, X_5 = 5, X_6 = 2, X_7 = 3, X_8 = 0$, whereupon things start repeating, or "cycling." The PRNs corresponding to the sequence starting with seed $X_0 = 0$ are therefore $U_1 = 1/8, U_2 = 6/8, U_3 = 7/8, U_4 = 4/8, U_5 = 5/8, U_6 = 2/8, U_7 = 3/8, U_8 = 0$. Since any seed eventually produces all integers $0, 1, \ldots, 7$, we say that this is a *full-cycle* (or *full period*) generator.

Example 19-7

Not all generators are full period. Consider another "toy" generator $X_i = (3X_{i-1} + 1) \bmod (7)$, with seed $X_0 = 0$. This produces the integer sequence $X_1 = 1, X_2 = 4, X_3 = 6, X_4 = 5, X_5 = 2, X_6 = 0$, whereupon cycling ensues. Further, notice that for this generator, a seed of $X_0 = 3$ produces the sequence $X_1 = 3 = X_2 = X_3 = \cdots$, not very random looking!

The cycle length of the generator from Example 19-7 obviously depends on the seed chosen, which is a disadvantage. Full-period generators, such as that studied in Example 19-6, obviously avoid this problem. A full-period generator with a long cycle length is given in the following example.

Example 19-8

The generator $X_i = 16807\, X_{i-1} \bmod (2^{31} - 1)$ is full period. Since $c = 0$, this generator is termed a *multiplicative* LCG and must be used with a seed $X_0 \neq 0$. This generator is used in many real-world applications and passes most statistical tests for uniformity and randomness. In order to avoid integer overflow and real-arithmetic round-off problems, Bratley, Fox, and Schrage (1987) offer the following Fortran implementation scheme for this algorithm.

```
FUNCTION UNIF(IX)
K1 = IX/127773
IX = 16807*(IX - K1*127773) - K1*2836
IF (IX.LT.0)IX = IX + 2147483647
UNIF = IX * 4.656612875E-10
RETURN
END
```

In the above program, we input an integer seed `IX` and receive a PRN `UNIF`. The seed `IX` is automatically updated for the next call. Note that in Fortran, integer division results in *truncation*, for example 15/4 = 3; thus `K1` is an integer.

19-2.2 Generating Nonuniform Random Variables

The goal now is to generate random variables from distributions other than the uniform. The methods we will use to do so always start with a PRN and then apply an appropriate transformation to the PRN that gives the desired nonuniform random variable. Such nonuniform random variables are important in simulation for a number of reasons: for example, customer arrivals to a service facility often follow a Poisson process; service times may be normal; and routing decisions are usually characterized by Bernoulli random variables.

Inverse Transform Method for Random Variate Generation

The most basic technique for generating random variables from a uniform PRN relies on the remarkable *Inverse Transform Theorem.*

Theorem 19-1

If X is a random variable with cumulative distribution function (CDF) $F(x)$, then the random variable $Y = F(X)$ has the uniform (0,1) distribution.

Proof

For ease of exposition, suppose that X is a continuous random variable. Then the CDF of Y is

$$\begin{aligned} G(y) &= Pr(Y \leq y) \\ &= Pr(F(X) \leq y) \\ &= Pr(X \leq F^{-1}(y)) \text{ (the inverse exists since } F(x) \text{ is continuous)} \\ &= F(F^{-1}(y)) \\ &= y. \end{aligned}$$

Since $G(y) = y$ is the CDF of the uniform (0,1) distribution, we are done.

With Theorem 19-1 in hand, it is easy to generate certain random variables. All one has to do is the following:

1. Find the CDF of X, say $F(x)$.

2. Set $F(X) = U$, where U is a uniform (0,1) PRN.
3. Solve for $X = F^{-1}(U)$.

We illustrate this technique with a series of examples, for both continuous and discrete distributions.

Example 19-9

Here we generate an exponential random variable with rate λ, following the recipe outlined above.

1. The CDF is $F(x) = 1 - e^{-\lambda x}$.
2. Set $F(X) = 1 - e^{-\lambda X} = U$.
3. Solving for X, we obtain $X = F^{-1}(U) = -[\ln(1 - U)]/\lambda$.

Thus, if one supplies a uniform (0,1) PRN U, we see that $X = -[\ln(1 - U)]/\lambda$ is an exponential random variable with parameter λ.

Example 19-10

Now we try to generate a standard normal random variable, call it Z. Using the special notation $\Phi(\cdot)$ for the standard normal (0,1) CDF, we set $\Phi(Z) = U$, so that $Z = \Phi^{-1}(U)$. Unfortunately, the inverse CDF does not exist in closed form, so one must resort to the use of standard normal tables (or other approximations). For instance, if we have $U = 0.72$, then Table II (Appendix) yields $Z = \Phi^{-1}(0.72) \approx 0.583$.

Example 19-11

We can extend the previous example to generate any normal random variable, that is, one with arbitrary mean and variance. This follows easily, since if Z is standard normal, then $X = \mu + \sigma Z$ is normal with mean μ and variance σ^2. For instance, suppose we are interested in generating a normal variate X with mean $\mu = 3$ and variance $\sigma^2 = 4$. Then if, as in the previous example, $U = 0.72$, we obtain $Z \approx 0.583$, and, as a consequence, $X \approx 3 + 2(0.583) = 4.166$.

Example 19-12

We can also use the ideas from Theorem 19-1 to generate realizations from discrete random variables. Suppose that the discrete random variable X has probability function

$$p(x) = \begin{cases} 0.3 & \text{if } x = -1, \\ 0.6 & \text{if } x = 2.3, \\ 0.1 & \text{if } x = 7, \\ 0 & \text{otherwise.} \end{cases}$$

To generate variates from this distribution, we set up the following table, where $F(x)$ is the associated CDF and U denotes the set of uniform (0,1) PRNs corresponding to each x-value:

x	$p(x)$	$F(x)$	U
–1	0.3	0.3	[0,0.3)
2.3	0.6	0.9	[0.3,0.9)
7	0.1	1.0	[0.9,1.0)

To generate a realization of X, we first generate a PRN U and then read the corresponding x-value from the table. For instance, if $U = 0.43$, then $X = 2.3$.

Other Random Variate Generation Methods

Although the inverse transform method is intuitively pleasing to use, its real-life application may sometimes be difficult to apply in practice. For instance, closed-form expressions for the inverse CDF, $F^{-1}(U)$, might not exist, as is the case for the normal distribution, or application of the method might be unnecessarily tedious. We now present a small potpourri of interesting methods to generate a variety of random variables.

Box–Müller Method

The Box–Müller (1958) method is an exact technique for generating independent and identically distributed (IID) standard normal (0,1) random variables. The appropriate theorem, stated without proof, is

Theorem 19-2

Suppose that U_1 and U_2 are IID uniform (0,1) random variables. Then

$$Z_1 = \sqrt{-2\ln(U_1)}\cos(2\pi U_2)$$

and

$$Z_2 = \sqrt{-2\ln(U_1)}\sin(2\pi U_2)$$

are IID standard normal random variates.

Note that the sine and cosine evaluations must be carried out in *radians*.

Example 19-13

Suppose that $U_1 = 0.35$ and $U_2 = 0.65$ are two IID PRNs. Using the Box–Müller method to generate two normal (0,1) random variates, we obtain

$$Z_1 = \sqrt{-2\ln(0.35)}\cos(2\pi(0.65)) = -0.8517$$

and

$$Z_2 = \sqrt{-2\ln(0.35)}\sin(2\pi(0.65)) = -1.172.$$

Central Limit Theorem

One can also use the Central Limit Theorem (CLT) to generate "quick-and-dirty" random variables that are *approximately* normal. Suppose that $U_1, U_2, \ldots, U_n$ are IID PRNs. Then for large enough n, the CLT says that

$$\begin{aligned}
&\frac{\Sigma_{i=1}^n U_i - \mathrm{E}\left[\Sigma_{i=1}^n U_i\right]}{\sqrt{\mathrm{Var}\left(\Sigma_{i=1}^n U_i\right)}} \\
&= \frac{\Sigma_{i=1}^n U_i - \Sigma_{i=1}^n \mathrm{E}[U_i]}{\sqrt{\Sigma_{i=1}^n \mathrm{Var}(U_i)}} \\
&= \frac{\Sigma_{i=1}^n U_i - (n/2)}{\sqrt{n/12}} \\
&\approx \mathrm{N}\,(0,1).
\end{aligned}$$

In particular, the choice $n = 12$ (which turns out to be "large enough") yields the convenient approximation

$$\sum_{i=1}^{12} U_i - 6 \approx \mathrm{N}(0,1).$$

Example 19-14

Suppose we have the following PRNs:

0.28 0.87 0.44 0.49 0.10 0.76 0.65 0.98 0.24 0.29 0.77 0.90.

Then

$$\sum_{i=1}^{12} U_i - 6 = 0.77$$

is a realization from a distribution that is approximately standard normal.

Convolution

Another popular trick involves the generation of random variables via *convolution*, indicating that some sort of sum is involved.

Example 19-15

Suppose that $X_1, X_2, \ldots, X_n$ are IID exponential random variables with rate λ. Then $Y = \sum_{i=1}^{n} X_i$ is said to have an *Erlang* distribution with parameters n and λ. It turns out that this distribution has probability density function

$$f(y) = \begin{cases} \lambda^n e^{-\lambda y} y^{n-1}/(n-1)! & \text{if } y > 0, \\ 0 & \text{otherwise,} \end{cases} \tag{19-4}$$

which readers may recognize as a special case of the gamma distribution (see Exercise 19-16).

This distribution's CDF is too difficult to invert directly. One way that comes to mind to generate a realization from the Erlang is simply to generate and then add up n IID exponential(λ) random variables. The following scheme is an efficient way to do precisely that. Suppose that $U_1, U_2, \ldots, U_n$ are IID PRNs. From Example 19-9, we know that $X_i = -\frac{1}{\lambda}\ln(1 - U_i)$, $i = 1, 2, \ldots, n$, are IID exponential(λ) random variables. Therefore, we can write

$$\begin{aligned} Y &= \sum_{i=1}^{n} X_i \\ &= \sum_{i=1}^{n} \left[-\frac{1}{\lambda} \ln(1 - U_i) \right] \\ &= -\frac{1}{\lambda} \ln\left(\prod_{i=1}^{n} (1 - U_i) \right). \end{aligned}$$

This implementation is quite efficient, since it requires only one execution of a natural log operation. In fact, we can even do slightly better from an efficiency point of view—simply note that both U_i and $(1 - U_i)$ are uniform (0,1). Then

$$Y = -\frac{1}{\lambda} \ln\left(\prod_{i=1}^{n} U_i \right)$$

is also Erlang.

To illustrate, suppose that we have three IID PRNs at our disposal, $U_1 = 0.23$, $U_2 = 0.97$, and $U_3 = 0.48$. To generate an Erlang realization with parameters $n = 3$ and $\lambda = 2$,we simply take

$$Y = -\frac{1}{\lambda}\ln(U_1U_2U_3) = -\frac{1}{2}\ln((0.23)(0.97)(0.48)) = 1.117.$$

Acceptance–Rejection

One of the most popular classes of random variate generation procedures proceeds by sampling PRNs until some appropriate "acceptance" criterion is met.

Example 19-16

An easy example of the acceptance–rejection technique involves the generation of a geometric random variable with success probability p. To this end, consider a sequence of PRNs $U_1, U_2, \ldots$. Our aim is to generate a geometric realization X, that is, one that has probability function

$$p(x) = \begin{cases} (1-p)^{x-1}p & \text{if } x = 1,2,..., \\ 0, & \text{otherwise.} \end{cases}$$

In words, X represents the number of Bernoulli trials until the first success is observed. This English characterization immediately suggests an elementary acceptance–rejection algorithm.

1. Initialize $i \leftarrow 0$.
2. Let $i \leftarrow i + 1$.
3. Take a Bernoulli(p) observation,

$$Y_i = \begin{cases} 1 & \text{if } U_i < p, \\ 0, & \text{otherwise.} \end{cases}$$

4. If $Y_i = 1$, then we have our first success and we stop, in which case we *accept* $X = i$. Otherwise, if $Y_i = 0$, then we *reject* and go back to step 2.

To illustrate, let us generate a geometric variate having success probability $p = 0.3$. Suppose we have are at our disposal the following PRNs:

$$0.38 \quad 0.67 \quad 0.24 \quad 0.89 \quad 0.10 \quad 0.71.$$

Since $U_1 = 0.38 \geq p$, we have $Y_1 = 0$, and so we reject $X = 1$. Since $U_2 = 0.67 \geq p$, we have $Y_2 = 0$, and so we reject $X = 2$. Since $U_3 = 0.24 < p$, we have $Y_3 = 1$, and so we *accept* $X = 3$.

19-3 OUTPUT ANALYSIS

Simulation output analysis is one of the most important aspects of any proper and complete simulation study. Since the input processes driving a simulation are usually random variables (e.g., interarrival times, service times, and breakdown times), we must also regard the output from the simulation as random. Thus, runs of the simulation only yield *estimates* of measures of system performance (e.g., the mean customer waiting time). These estimators are themselves random variables and are therefore subject to sampling error—and sampling error must be taken into account to make valid inferences concerning system performance.

The problem is that simulations almost never produce convenient raw output that is IID normal data. For example, consecutive customer waiting times from a queueing system

- are not independent—typically, they are serially correlated; if one customer at the post office waits in line a long time, then the next customer is also likely to wait a long time.
- are not identically distributed; customers showing up early in the morning might have a much shorter wait than those who show up just before closing time.
- are not normally distributed—they are usually skewed to the right (and are certainly never less than zero).

The point is that it is difficult to apply "classical" statistical techniques to the analysis of simulation output. Our purpose here is to give methods to perform statistical analysis of output from discrete-event computer simulations. To facilitate the presentation, we identify two types of simulations with respect to output analysis: Terminating and steady state simulations.

1. *Terminating (or transient) simulations.* Here, the nature of the problem explicitly defines the length of the simulation run. For instance, we might be interested in simulating a bank that closes at a specific time each day.
2. *Nonterminating (steady state) simulations.* Here, the long-run behavior of the system is studied. Presumedly this "steady state" behavior is independent of the simulation's initial conditions. An example is that of a continuously running production line for which the experimenter is interested in some long-run performance measure.

Techniques to analyze output from terminating simulations are based on the method of independent replications, discussed in Section 19-3.1. Additional problems arise for steady state simulations. For instance, we must now worry about the problem of starting the simulation—how should it be initialized at time zero, and how long must it be run before data representative of steady state can be collected? Initialization problems are considered in Section 19-3.2. Finally, Section 19-3.3 deals with point and confidence interval estimation for steady state simulation performance parameters.

19-3.1 Terminating Simulation Analysis

Here we are interested in simulating some system of interest over a finite time horizon. For now, assume we obtain *discrete* simulation output $Y_1, Y_2, \ldots, Y_m$, where the number of observations m can be a constant or a random variable. For example, the experimenter can specify the number m of customer waiting times $Y_1, Y_2, \ldots, Y_m$ to be taken from a queueing simulation. Or m could denote the random number of customers observed during a specified time period $[0, T]$.

Alternatively, we might observe *continuous* simulation output $\{Y(t) | 0 \leq t \leq T\}$ over a specified interval $[0, T]$. For instance, if we are interested in estimating the time-averaged number of customers waiting in a queue during $[0, T]$, the quantity $Y(t)$ would be the number of customers in the queue at time t.

The easiest goal is to estimate the expected value of the sample mean of the observations,

$$\theta \equiv \mathrm{E}[\bar{Y}_m],$$

where the sample mean in the discrete case is

$$\bar{Y}_m \equiv \frac{1}{m}\sum_{i=1}^{m} Y_i$$

(with a similar expression for the continuous case). For instance, we might be interested in estimating the expected average waiting time of all customers at a shopping center during the period 10 a.m. to 2 p.m.

Although $\overline{Y}_m$ is an unbiased estimator for θ, a proper statistical analysis requires that we also provide an estimate of $Var(\overline{Y}_m)$. Since the Y_i are not necessarily IID random variables, it may be that $Var(\overline{Y}_m) \neq Var(Y_i)/m$ for any i, a case not covered in elementary statistics courses.

For this reason, the familiar sample variance

$$S^2 = \frac{1}{m-1}\sum_{i=1}^{m}\left(Y_i - \overline{Y}_m\right)^2$$

is likely to be highly *biased* as an estimator of $mVar(\overline{Y}_m)$. Thus, one should *not* use S^2/m to estimate $Var(\overline{Y}_m)$.

The way around the problem is via the method of *independent replications* (IR). IR estimates $Var(\overline{Y}_m)$ by conducting b independent simulation runs (replications) of the system under study, where each replication consists of m observations. It is easy to make the replications independent—simply reinitialize each replication with a different pseudorandom number seed.

To proceed, denote the sample mean from replication i by

$$Z_i \equiv \frac{1}{m}\sum_{j=1}^{m} Y_{i,j},$$

where $Y_{i,j}$ is observation j from replication i, for $i = 1, 2, \ldots, b$ and $j = 1, 2, \ldots, m$.

If each run is started under the same operating conditions (e.g., all queues empty and idle), then the replication sample means $Z_1, Z_2, \ldots, Z_b$ are IID random variables. Then the obvious point estimator for $Var(\overline{Y}_m) = Var(Z_i)$ is

$$\hat{V}_R \equiv \frac{1}{b-1}\sum_{i=1}^{b}\left(Z_i - \overline{Z}_b\right)^2,$$

where the grand mean is defined as

$$\overline{Z}_b \equiv \frac{1}{b}\sum_{i=1}^{b} Z_i.$$

Notice how closely the forms of $\hat{V}_R$ and S^2/m resemble each other. But since the replicate sample means are IID, $\hat{V}_R$ is usually much less biased for $Var(\overline{Y}_m)$ than is S^2/m.

In light of the above discussion, we see that $\hat{V}_R/b$ is a reasonable estimator for $Var(\overline{Z}_b)$. Further, if the number of observations per replication, m, is large enough, the Central Limit Theorem tells us that the replicate sample means are approximately IID *normal.*

Then basic statistics (Chapter 10) yields an approximate $100(1-\alpha)\%$ two-sided confidence interval (CI) for θ,

$$\theta \in \overline{Z}_b \pm t_{\alpha/2,b-1}\sqrt{\hat{V}_R/b}, \tag{19-5}$$

where $t_{\alpha/2,b-1}$ is the $1-\alpha/2$ percentage point of the t distribution with $b-1$ degrees of freedom.

Example 19-17

Suppose we want to estimate the expected average waiting time for the first 5000 customers in a certain queueing system. We will make five independent replications of the system, with each run initialized empty and idle and consisting of 5000 waiting times. The resulting replicate means are

i	1	2	3	4	5
Z_i	3.2	4.3	5.1	4.2	4.6

Then $\overline{Z}_5 = 4.28$ and $\hat{V}_R = 0.487$. For level $\alpha = 0.05$, we have $t_{0.025,4} = 2.78$, and equation 19-5 gives [3.41, 5.15] as a 95% CI for the expected average waiting time for the first 5000 customers.

Independent replications can be used to calculate variance estimates for statistics other than sample means. Then the method can be used to obtain CIs for quantities other than $E[\overline{Y}_m]$, for example quantiles. See any of the standard simulation texts for additional uses of independent replications.

19-3.2 Initialization Problems

Before a simulation can be run, one must provide initial values for all of the simulation's state variables. Since the experimenter may not know what initial values are appropriate for the state variables, these values might be chosen somewhat arbitrarily. For instance, we might decide that it is "most convenient" to initialize a queue as empty and idle. Such a choice of initial conditions can have a significant but unrecognized impact on the simulation run's outcome. Thus, the *initialization bias* problem can lead to errors, particularly in steady state output analysis.

Some examples of problems concerning simulation initialization are as follows.

1. Visual detection of initialization effects is sometimes difficult—especially in the case of stochastic processes having high intrinsic variance, such as queueing systems.
2. How should the simulation be initialized? Suppose that a machine shop closes at a certain time each day, even if there are jobs waiting to be served. One must therefore be careful to start each day with a demand that depends on the number of jobs remaining from the previous day.
3. Initialization bias can lead to point estimators for steady state parameters having high mean squared error, as well as for CIs having poor coverage.

Since initialization bias raises important concerns, how do we detect and deal with it? We first list methods to detect it.

1. *Attempt to detect the bias visually* by scanning a realization of the simulated process. This might not be easy, since visual analysis can miss bias. Further, a visual scan can be tedious. To make the visual analysis more efficient, one might transform the data (e.g., take logs or square roots), smooth it, average it across several independent replications, or construct moving average plots.

2. *Conduct statistical tests for initialization bias.* Kelton and Law (1983) give an intuitively appealing sequential procedure to detect bias. Various other tests check to see whether the initial portion of the simulation output contains more variation than latter portions.

If initialization bias is detected, one may want to do something about it. There are two simple methods for dealing with bias. One is to *truncate the output* by allowing the simulation to "warm up" before data are retained for analysis. The experimenter would then hope that the remaining data are representative of the steady state system. Output truncation is probably the most popular method for dealing with initialization bias, and all of the major simulation languages have built-in truncation functions. But how can one find a good truncation point? If the output is truncated "too early," significant bias might still exist in the remaining data. If it is truncated "too late," then good observations might be wasted. Unfortunately, all simple rules to determine truncation points do not perform well in general. A common practice is to average observations across several replications and then visually choose a truncation point based on the averaged run. See Welch (1983) for a good visual/graphical approach.

The second method is to *make a very long run* to overwhelm the effects of initialization bias. This method of bias control is conceptually simple to carry out and may yield point estimators having lower mean squared errors than the analogous estimators from truncated data (see, e.g., Fishman 1978). However, a problem with this approach is that it can be wasteful with observations; for some systems, an excessive run length might be required before the initialization effects are rendered negligible.

19-3.3 Steady State Simulation Analysis

Now assume that we have on hand stationary (steady state) simulation output, $Y_1, Y_2, \ldots, Y_n$. Our goal is to estimate some parameter of interest, possibly the mean customer waiting time or the expected profit produced by a certain factory configuration. As in the case of terminating simulations, a good statistical analysis must accompany the value of any point estimator with a measure of its variance.

A number of methodologies have been proposed in the literature for conducting steady state output analysis: batch means, independent replications, standardized time series, spectral analysis, regeneration, time series modeling, as well as a host of others. We will examine the two most popular: batch means and independent replications. (Recall: As discussed earlier, confidence intervals for *terminating* simulations usually use independent replications.)

Batch Means

The method of batch means is often used to estimate $Var(\bar{Y}_n)$ or calculate CIs for the steady state process mean μ. The idea is to divide one long simulation run into a number of contiguous *batches*, and then to appeal to a Central Limit Theorem to assume that the resulting batch sample means are approximately IID normal.

In particular, suppose that we partition $Y_1, Y_2, \ldots, Y_n$ into b nonoverlapping, contiguous batches, each consisting of m observations (assume that $n = bm$). Thus, the ith batch, $i = 1, 2, \ldots, b$, consists of the random variables

$$Y_{(i-1)m+1}, Y_{(i-1)m+2}, \ldots, Y_{im}.$$

The ith batch mean is simply the sample mean of the m observations from batch i, $i = 1, 2, \ldots, b$,

$$Z_i = \frac{1}{m}\sum_{j=1}^{m} Y_{(i-1)m+j}.$$

Similar to independent replications, we define the batch means estimator for $Var(Z_i)$ as

$$\hat{V}_B = \frac{1}{b-1}\sum_{i=1}^{b}\left(Z_i - \bar{Z}_b\right)^2,$$

where

$$\bar{Y}_n = \bar{Z}_b = \frac{1}{b}\sum_{i=1}^{b} Z_i$$

is the grand sample mean. If m is large, then the batch means are approximately IID *normal*, and (as for IR) we obtain an approximate $100(1 - \alpha)\%$ CI for μ,

$$\mu \in \bar{Z}_b \pm t_{\alpha/2,b-1}\sqrt{\hat{V}_B/b}.$$

This equation is very similar to equation 19-5. Of course, the difference here is that batch means divides one long run into a number of batches, whereas independent replications uses a number of independent shorter runs. Indeed, consider the old IR example from Section 19-3.1 with the understanding that the Z_i must now be regarded as batch means (instead of replicate means); then the same numbers carry through the example.

The technique of batch means is intuitively appealing and easy to understand. But problems can come up if the Y_j are not stationary (e.g., if significant initialization bias is present), if the batch means are not normal, or if the batch means are not independent. If any of these assumption violations exist, poor confidence interval coverage may result—unbeknownst to the analyst.

To ameliorate the initialization bias problem, the user can truncate some of the data or make a long run, as discussed in Section 19-3.2. In addition, the lack of independence or normality of the batch means can be countered by increasing the batch size m.

Independent Replications

Of the difficulties encountered when using batch means, the possibility of correlation among the batch means might be the most troublesome. This problem is explicitly avoided by the method of IR, described in the context of terminating simulations in Section 19-3.1. In fact, the replicate means are independent by their construction. Unfortunately, since *each* of the b replications has to be started properly, initialization bias presents more trouble when using IR than when using batch means. The usual recommendation, in the context of steady state analysis, is to use batch means over IR because of the possible initialization bias in each of the replications.

19-4 COMPARISON OF SYSTEMS

One of the most important uses of simulation output analysis regards the comparison of competing systems or alternative system configurations. For example, suppose we wish to evaluate two different "restart" strategies that an airline can evoke following a major traffic disruption, such as a snowstorm in the Northeast. Which policy minimizes a certain cost function associated with the restart? Simulation is uniquely equipped to help the experimenter conduct this type of comparison analysis.

There are many techniques available for comparing systems, among them (i) classical statistical CIs, (ii) common random numbers, (iii) antithetic variates, and (iv) ranking and selection procedures.

19-4.1 Classical Confidence Intervals

With our airline example in mind, let $Z_{i,j}$ be the cost from the jth simulation replication of strategy i, $i = 1, 2$, $j = 1, 2, \ldots, b_i$. Assume that $Z_{i,1}, Z_{i,2}, \ldots, Z_{i,b_i}$ are IID normal with unknown mean μ_i and unknown variance, $i = 1, 2$, an assumption that can be justified by arguing that we can do the following:

1. Get independent data by controlling the random numbers between replications.
2. Get identically distributed costs between replications by performing the replications under identical conditions.
3. Get approximately normal data by adding up (or averaging) many subcosts to obtain overall costs for both strategies.

The goal here is to calculate a $100(1 - \alpha)\%$ CI for the difference $\mu_1 - \mu_2$. To this end, suppose that the $Z_{1,j}$ are independent of the $Z_{2,j}$ and define

$$\bar{Z}_{i,b_i} = \frac{1}{b_i}\sum_{j=1}^{b_i} Z_{i,j}, \qquad i = 1, 2,$$

and

$$S_i^2 = \frac{1}{b_i - 1}\sum_{j=1}^{b_i}\left(Z_{i,j} - \bar{Z}_{i,b_i}\right)^2, \qquad i = 1, 2.$$

An approximate $100(1 - \alpha)\%$ CI is

$$\mu_1 - \mu_2 \in \bar{Z}_{1,b_1} - \bar{Z}_{2,b_2} \pm t_{\alpha/2,v}\sqrt{\frac{S_1^2}{b_1} + \frac{S_2^2}{b_2}},$$

where the approximate degrees of freedom v (a function of the sample variances) is given in Chapter 10.

Suppose (as in airline example) that small cost is good. If the interval lies entirely to the left [right] of zero, then system 1 [2] is better; if the interval contains zero, then the two systems must be regarded, in a statistical sense, as about the same.

An alternative classical strategy is to use a CI that is analogous to a paired t-test. Here we take b replications from *both* strategies and set the differences $D_j = Z_{1,j} - Z_{2,j}$ for $j = 1, 2, \ldots, b$. Then we calculate the sample mean and variance of the differences:

$$\bar{D}_b = \frac{1}{b}\sum_{j=1}^{b} D_j \qquad \text{and} \qquad S_D^2 = \frac{1}{b-1}\sum_{j=1}^{b}\left(D_j - \bar{D}_b\right)^2.$$

The resulting $100(1 - \alpha)\%$ CI is

$$\mu_1 - \mu_2 \in \bar{D}_b \pm t_{\alpha/2,b-1}\sqrt{S_D^2/b}.$$

These paired t intervals are very efficient if $Corr(Z_{1,j}, Z_{2,j}) > 0, j = 1, 2, \ldots, b$ (where we still assume that $Z_{1,1}, Z_{1,2}, \ldots, Z_{1,b}$ are IID and $Z_{2,1}, Z_{2,2}, \ldots, Z_{2,b}$ are IID). In that case, it turns out that

$$V\left(\bar{D}_b\right) < \frac{V\left(Z_{1,j}\right) + V\left(Z_{2,j}\right)}{b}.$$

If $Z_{1,j}$ and $Z_{2,j}$ had been simulated *independently*, then we would have *equality* in the above expression. Thus, the trick may result in relatively small S_D^2 and, hence, small CI length. So how do we evoke the trick?

19-4.2 Common Random Numbers

The idea behind the above trick is to use *common random numbers*, that is, to use the same pseudorandom numbers in exactly the same ways for corresponding runs of each of the competing systems. For example, we might use precisely the same customer arrival times when simulating different proposed configurations of a job shop. By subjecting the alternative systems to identical experimental conditions, we hope to make it easy to distinguish which systems are best even though the respective estimators are subject to sampling error.

Consider the case in which we compare two queueing systems, A and B, on the basis of their expected customer transit times, θ_A and θ_B, where the smaller θ-value corresponds to the better system. Suppose we have estimators $\hat{\theta}_A$ and $\hat{\theta}_B$ for θ_A and θ_B, respectively. We will declare A as the better system if $\hat{\theta}_A < \hat{\theta}_B$. If $\hat{\theta}_A$ and $\hat{\theta}_B$ are simulated independently, then the variance of their difference,

$$V(\hat{\theta}_A - \hat{\theta}_B) = V(\hat{\theta}_A) + V(\hat{\theta}_B),$$

could be very large, in which case our declaration might lack conviction. If we could reduce $V(\hat{\theta}_A - \hat{\theta}_B)$, then we could be much more confident about our declaration.

CRN sometimes induces a high positive correlation between the point estimators $\hat{\theta}_A$ and $\hat{\theta}_B$. Then we have

$$\begin{aligned} V(\hat{\theta}_A - \hat{\theta}_B) &= V(\hat{\theta}_A) + V(\hat{\theta}_B) - 2Cov(\hat{\theta}_A, \hat{\theta}_B) \\ &< V(\hat{\theta}_A) + V(\hat{\theta}_B), \end{aligned}$$

and we obtain a savings in variance.

19-4.3 Antithetic Random Numbers

Alternatively, if we can induce *negative* correlation between two unbiased estimators, $\hat{\theta}_1$ and $\hat{\theta}_2$, for some parameter θ, then the unbiased estimator $(\hat{\theta}_1 + \hat{\theta}_2)/2$ might have low variance.

Most simulation texts give advice on how to run the simulations of the competing systems so as to induce positive or negative correlation between them. The consensus is that if conducted properly, common and antithetic random numbers can lead to tremendous variance reductions.

19-4.4 Selecting the Best System

Ranking, selection, and multiple comparisons methods form another class of statistical techniques used to compare alternative systems. Here, the experimenter is interested in selecting the best of a number of competing processes. Typically, one specifies the desired probability of correctly selecting the best process, especially if the best process is significantly better than its competitors. These methods are simple to use, fairly general, and intuitively appealing. See Bechhofer, Santner, and Goldsman (1995) for a synopsis of the most popular procedures.

19-5 SUMMARY

This chapter began with some simple motivational examples illustrating various simulation concepts. After this, the discussion turned to the generation of pseudorandom numbers, that is, numbers that appear to be IID uniform (0,1). PRNs are important because they drive the generation of a number of other important random variables, for example normal, exponential, and Erlang. We also spent a great deal of discussion on simulation output analysis—simulation output is almost never IID, so special care must be taken if we are to make statistically valid conclusions about the simulation's results. We concentrated on output analysis for both terminating and steady state simulations.

19-6 EXERCISES

19-1. Extension of Example 19-1.

(a) Flip a coin 100 times. How many heads to do you observe?

(b) How many times do you observe two heads in a row? Three in a row? Four? Five?

(c) Find 10 friends and repeat (a) and (b) based on a total of 1000 flips.

(d) Now simulate coin flips via a spreadsheet program. Flip the simulated coin 10,000 times and answer (a) and (b).

19-2. Extension of Example 19-2. Throw n darts randomly at a unit square containing an inscribed circle. Use the results of your tosses to estimate π. Let $n = 2^k$ for $k = 1, 2, \ldots, 15$, and graph your estimates as a function of k.

19-3. Extension of Example 19-3. Show that $\hat{I}_n$, defined in equation 19-3, is unbiased for the integral I, defined in equation 19-2.

19-4. Other extensions of Example 19-3.

(a) Use Monte Carlo integration with $n = 10$ observations to estimate $\int_0^2 \frac{1}{2\pi} e^{-x^2/2} dx$. Now use $n = 1000$. Compare to the answer that you can obtain via normal tables.

(b) What would you do if you had to estimate $\int_0^{10} \frac{1}{2\pi} e^{-x^2/2} dx$?

(c) Use Monte Carlo integration with $n = 10$ observations to estimate $\int_0^1 \cos(2\pi x)\, dx$. Now use $n = 1000$. Compare to the actual answer.

19-5. Extension of Example 19-4. Suppose that 10 customers arrive at a post office at the following times:

3 4 6 7 13 14 20 25 28 30

Upon arrival, customers queue up in front of a single clerk and are processed in a first-come-first-served manner. The service times corresponding to the arriving customers are as follows:

6.0 5.5 4.0 1.0 2.5 2.0 2.0 2.5 4.0 2.5

Assume that the post office opens at time 0, and closes its doors at time 30 (just after customer 10 arrives), serving any remaining customers.

(a) When does the last customer finally leave the system?

(b) What is the average waiting time for the 10 customers?

(c) What is the maximum number of customers in the system? When is this maximum achieved?

(d) What is the average number of customers in line during the first 30 minutes?

(e) Now repeat parts (a)–(d) assuming that the services are performed last-in-first-out.

19-6. Repeat Example 19-5, which deals with an (s, S) inventory policy, except now use order level $s = 6$.

19-7. Consider the pseudorandom number generator $X_i = (5X_{i-1} + 1) \bmod (16)$, with seed $X_0 = 0$.

(a) Calculate X_1 and X_2, along with the corresponding PRNs U_1 and U_2.

(b) Is this a full-period generator?

(c) What is X_{150}?

19-8. Consider the "recommended" pseudorandom number generator $X_i = 16807\, X_{i-1} \bmod (2^{31} - 1)$, with seed $X_0 = 1234567$.

(a) Calculate X_1 and X_2, along with the corresponding PRNs U_1 and U_2.

(b) What is $X_{100,000}$?

19-9. Show how to use the inverse transform method to generate an exponential random variable with rate $\lambda = 2$. Demonstrate your technique using the PRN $U = 0.75$.

19-10. Consider the inverse transform method to generate a standard normal (0,1) random variable.

(a) Demonstrate your technique using the PRN $U = 0.25$.

(b) Using your answer in (a), generate an N (1,9) random variable.

19-11. Suppose that X has probability density function $f(x) = |x/4|, -2 < x < 2$.

(a) Develop an inverse transform technique to generate a realization of X.

(b) Demonstrate your technique using $U = 0.6$.

(c) Sketch out $f(x)$ and see if you can come up with another method to generate X.

19-12. Suppose that the discrete random variable X has probability function

$$p(x) = \begin{cases} 0.35 & \text{if } x = -2.5, \\ 0.25 & \text{if } x = 1.0, \\ 0.40 & \text{if } x = 10.5, \\ 0, & \text{otherwise.} \end{cases}$$

As in Example 19-12, set up a table to generate realizations from this distribution. Illustrate your technique with the PRN $U = 0.86$.

19-13. The Weibull (α, β) distribution, popular in reliability theory and other applied statistics disciplines, has CDF

$$F(x) = \begin{cases} 1 - e^{-(x/\alpha)^{\beta}} & \text{if } x > 0, \\ 0 & \text{otherwise.} \end{cases}$$

(a) Show how to use the inverse transform method to generate a realization from the Weibull distribution.

(b) Demonstrate your technique for a Weibull (1.5,2.0) random variable using the PRN $U = 0.66$.

19-14. Suppose that $U_1 = 0.45$ and $U_2 = 0.12$ are two IID PRNs. Use the Box–Müller method to generate two N (0,1) variates.

19-15. Consider the following PRNs:

0.88 0.87 0.33 0.69 0.20 0.79 0.21
0.96 0.11 0.42 0.91 0.70

Use the Central Limit Theorem method to generate a realization that is approximately standard normal.

19-16. Prove equation 19-4 from the text. This shows that the sum of n IID exponential random variables is Erlang. *Hint*: Find the moment-generating function of Y, and compare it to that of the gamma distribution.

19-17. Using two PRNs, $U_1 = 0.73$ and $U_2 = 0.11$, generate a realization from an Erlang distribution with $n = 2$ and $\lambda = 3$.

19-18. Suppose that $U_1, U_2, \ldots, U_n$ are PRNs.

(a) Suggest an easy inverse transform method to generate a sequence of IID Bernoulli random variables, each with success parameter p.

(b) Show how to use your answer to (a) to generate a binomial random variate with parameters n and p.

19-19. Use the acceptance–rejection technique to generate a geometric random variable with success probability 0.25. Use as many of the PRNs from Exercise 19-15 as necessary.

19-20. Suppose that $Z_1 = 3$, $Z_2 = 5$, and $Z_3 = 4$ are three batch means resulting from a long simulation run. Find a 90% two-sided confidence interval for the mean.

19-21. Suppose that $\mu \in [-2.5, 3.5]$ is a 90% confidence interval for the mean cost incurred by a certain inventory policy. Further suppose that this interval was based on five independent replications of the underlying inventory system. Unfortunately, the boss has decided that she wants a 95% confidence interval. Can you supply it?

19-22. The yearly unemployment rates for Andorra during the past 15 years are as follows:

6.9 8.3 8.8 11.4 11.8 12.1 10.6 11.0
9.9 9.2 12.3 13.9 9.2 8.2 8.9

Use the method of batch means on the above data to obtain a two-sided 95% confidence interval for the mean unemployment. Use five batches, each consisting of three years' data.

19-23. Suppose that we are interested in steady state confidence intervals for the mean of simulation output $X_1, X_2, \ldots, X_{10000}$. (You can pretend that these are waiting times.) We have conveniently divided the run up into five batches, each of size 2000; suppose that the resulting batch means are as follows:

100 80 90 110 120

Use the method of batch means on the above data to obtain a two-sided 95% confidence interval for the mean.

19-24. The yearly total snowfall figures for Siberacuse, NY, during the past 15 years are as follows:

100 103 88 72 98 121 106 110 99
162 123 139 92 142 169

(a) Use the method of batch means on the above data to obtain a two-sided 95% confidence interval for the mean yearly snowfall. Use five batches, each consisting of three years' data.

(b) The corresponding yearly total snowfall figures for Buffoonalo, NY (which is down the road from Siberacuse), are as follows:

90 95 72 68 95 110 112 90 75
144 110 123 81 130 145

How does Buffoonalo's snowfall compare to Siberacuse's? Just give an eyeball answer.

(c) Now find a 95% confidence interval for the difference in means between the two cities. *Hint*: Think common random numbers.

19-25. Antithetic variates. Suppose that $X_1, X_2, \ldots, X_n$ are IID with mean μ and variance σ^2. Further suppose that $Y_1, Y_2, \ldots, Y_n$ are also IID with mean μ and variance σ^2. The interesting trick here is that we will also assume that $Cov(X_i, Y_i) < 0$ for all i. So, in other words, the observations *within* one of the two sequences are IID, but they are negatively correlated *between* sequences.

(a) Here is an example showing how can we end up with the above scenario using simulations. Let $X_i = -\ln(U_i)$ and $Y_i = -\ln(1 - U_i)$, where the U_i are the usual IID uniform (0,1) random variables.

i. What is the distribution of X_i? Of Y_i?

ii. What is $Cov(U_i, 1 - U_i)$?

iii. Would you expect that $Cov(X_i, Y_i) < 0$? Answer: Yes.

(b) Let $\overline{X}_n$ and $\overline{Y}_n$ denote the sample means of the X_i and Y_i, respectively, each based on n observations. Without actually calculating $Cov(X_i, Y_i)$, state how $V((\overline{X}_n + \overline{Y}_n)/2)$ compares to $V(\overline{X}_{2n})$. In other words, should we do two negatively correlated runs, each consisting of n observations, or just one run consisting of $2n$ observations?

(c) What if you tried to use this trick when using Monte Carlo simulation to estimate $\int_0^1 \sin(\pi x)\,dx$?

19-26. Another variance reduction technique. Suppose that our goal is to estimate the mean μ of some steady state simulation output process, $X_1, X_2, \ldots, X_n$. Suppose we somehow know the expected value of some other RV Y, and we also know that $Cov(\overline{X}, Y) > 0$, where $\overline{X}$ is the sample mean. Obviously, $\overline{X}$ is the "usual" estimator for μ. Let us look at another estimator for μ, namely, the *control-variate* estimator,

$$C = \overline{X} - k(Y - E[Y]),$$

where k is some constant.

(a) Show that C is unbiased for μ.

(b) Find an expression for $V(C)$. Comments?

(c) Minimize $V(C)$ with respect to k.

19-27. A miscellaneous computer exercise. Make a histogram of $X_i = -\ln(U_i)$, for $i = 1, 2, \ldots, 20{,}000$, where the U_i are IID uniform (0,1). What kind of distribution does it look like?

19-28. Another miscellaneous computer exercise. Let us see if the Central Limit Theorem works. In Exercise 19-27, you generated 20,000 exponential(1) observations. Now form 1000 averages of 20 observations each from the original 20,000. More precisely, let

$$Y_i = \frac{1}{20}\sum_{j=1}^{20} X_{20(i-1)+j}.$$

Make a histogram of the Y_i. Do they look approximately normal?

19-29. Yet another miscellaneous computer exercise. Let us generate some normal observations via the Box–Müller method. To do so, first generate 1000 pairs of IID uniform (0,1) random numbers, $(U_{1,1}, U_{2,1}), (U_{1,2}, U_{2,2}), \ldots, (U_{1,1000}, U_{2,1000})$. Set

$$X_i = \sqrt{-2\ln(U_{1,i})}\cos(2\pi U_{2,i})$$

and

$$Y_i = \sqrt{-2\ln(U_{1,i})}\sin(2\pi U_{2,i})$$

for $i = 1, 2, \ldots, 1000$. Make a histogram of the resulting X_i. [The X_i's are N(0,1).] Now graph X_i vs. Y_i. Any comments?

Appendix

Table I Cumulative Poisson Distribution[a]

x	0.01	0.05	0.10	$c = \lambda t$ 0.20	0.30	0.40	0.50	0.60
0	0.990	0.951	0.904	0.818	0.740	0.670	0.606	0.548
1	0.999	0.998	0.995	0.982	0.963	0.938	0.909	0.878
2		0.999	0.999	0.998	0.996	0.992	0.985	0.976
3				0.999	0.999	0.999	0.998	0.996
4					0.999	0.999	0.999	0.999
5							0.999	0.999

x	0.70	0.80	0.90	$c = \lambda t$ 1.00	1.10	1.20	1.30	1.40
0	0.496	0.449	0.406	0.367	0.332	0.301	0.272	0.246
1	0.844	0.808	0.772	0.735	0.699	0.662	0.626	0.591
2	0.965	0.952	0.937	0.919	0.900	0.879	0.857	0.833
3	0.994	0.990	0.986	0.981	0.974	0.966	0.956	0.946
4	0.999	0.998	0.997	0.996	0.994	0.992	0.989	0.985
5	0.999	0.999	0.999	0.999	0.999	0.998	0.997	0.996
6		0.999	0.999	0.999	0.999	0.999	0.999	0.999
7				0.999	0.999	0.999	0.999	0.999
8							0.999	0.999

x	1.50	1.60	1.70	$c = \lambda t$ 1.80	1.90	2.00	2.10	2.20
0	0.223	0.201	0.182	0.165	0.149	0.135	0.122	0.110
1	0.557	0.524	0.493	0.462	0.433	0.406	0.379	0.354
2	0.808	0.783	0.757	0.730	0.703	0.676	0.649	0.622
3	0.934	0.921	0.906	0.891	0.874	0.857	0.838	0.819
4	0.981	0.976	0.970	0.963	0.955	0.947	0.937	0.927
5	0.995	0.993	0.992	0.989	0.986	0.983	0.979	0.975
6	0.999	0.998	0.998	0.997	0.996	0.995	0.994	0.992
7	0.999	0.999	0.999	0.999	0.999	0.998	0.998	0.998
8	0.999	0.999	0.999	0.999	0.999	0.999	0.999	0.999
9			0.999	0.999	0.999	0.999	0.999	0.999
10							0.999	0.999

Table I Cumulative Poisson Distribution[a] (*continued*)

x	2.30	2.40	2.50	$c = \lambda t$ 2.60	2.70	2.80	2.90	3.00
0	0.100	0.090	0.082	0.074	0.067	0.060	0.055	0.049
1	0.330	0.308	0.287	0.267	0.248	0.231	0.214	0.199
2	0.596	0.569	0.543	0.518	0.493	0.469	0.445	0.423
3	0.799	0.778	0.757	0.736	0.714	0.691	0.669	0.647
4	0.916	0.904	0.891	0.877	0.862	0.847	0.831	0.815
5	0.970	0.964	0.957	0.950	0.943	0.934	0.925	0.916
6	0.990	0.988	0.985	0.982	0.979	0.975	0.971	0.966
7	0.997	0.996	0.995	0.994	0.993	0.991	0.990	0.988
8	0.999	0.999	0.998	0.998	0.998	0.997	0.996	0.996
9	0.999	0.999	0.999	0.999	0.999	0.999	0.999	0.998
10	0.999	0.999	0.999	0.999	0.999	0.999	0.999	0.999
11			0.999	0.999	0.999	0.999	0.999	0.999
12							0.999	0.999

x	3.50	4.00	4.50	$c = \lambda t$ 5.00	5.50	6.00	6.50	7.00
0	0.030	0.018	0.011	0.006	0.004	0.002	0.001	0.000
1	0.135	0.091	0.061	0.040	0.026	0.017	0.011	0.007
2	0.320	0.238	0.173	0.124	0.088	0.061	0.043	0.029
3	0.536	0.433	0.342	0.265	0.201	0.151	0.111	0.081
4	0.725	0.628	0.532	0.440	0.357	0.285	0.223	0.172
5	0.857	0.785	0.702	0.615	0.528	0.445	0.369	0.300
6	0.934	0.889	0.831	0.762	0.686	0.606	0.526	0.449
7	0.973	0.948	0.913	0.866	0.809	0.743	0.672	0.598
8	0.990	0.978	0.959	0.931	0.894	0.847	0.791	0.729
9	0.996	0.991	0.982	0.968	0.946	0.916	0.877	0.830
10	0.998	0.997	0.993	0.986	0.974	0.957	0.933	0.901
11	0.999	0.999	0.997	0.994	0.989	0.979	0.966	0.946
12	0.999	0.999	0.999	0.997	0.995	0.991	0.983	0.973
13	0.999	0.999	0.999	0.999	0.998	0.996	0.992	0.987
14		0.999	0.999	0.999	0.999	0.998	0.997	0.994
15			0.999	0.999	0.999	0.999	0.998	0.997
16				0.999	0.999	0.999	0.999	0.999
17					0.999	0.999	0.999	0.999
18						0.999	0.999	0.999
19							0.999	0.999
20								0.999

(*continues*)

Table I Cumulative Poisson Distribution[a] (*continued*)

x	$c = \lambda t$ 7.50	8.00	8.50	9.00	9.50	10.0	15.0	20.0
0	0.000	0.000	0.000	0.000	0.000	0.000	0.000	0.000
1	0.004	0.003	0.001	0.001	0.000	0.000	0.000	0.000
2	0.020	0.013	0.009	0.006	0.004	0.002	0.000	0.000
3	0.059	0.042	0.030	0.021	0.014	0.010	0.000	0.000
4	0.132	0.099	0.074	0.054	0.040	0.029	0.000	0.000
5	0.241	0.191	0.149	0.115	0.088	0.067	0.002	0.000
6	0.378	0.313	0.256	0.206	0.164	0.130	0.007	0.000
7	0.524	0.452	0.385	0.323	0.268	0.220	0.018	0.000
8	0.661	0.592	0.523	0.455	0.391	0.332	0.037	0.002
9	0.776	0.716	0.652	0.587	0.521	0.457	0.069	0.005
10	0.862	0.815	0.763	0.705	0.645	0.583	0.118	0.010
11	0.920	0.888	0.848	0.803	0.751	0.696	0.184	0.021
12	0.957	0.936	0.909	0.875	0.836	0.791	0.267	0.039
13	0.978	0.965	0.948	0.926	0.898	0.864	0.363	0.066
14	0.989	0.982	0.972	0.958	0.940	0.916	0.465	0.104
15	0.995	0.991	0.986	0.977	0.966	0.951	0.568	0.156
16	0.998	0.996	0.993	0.988	0.982	0.972	0.664	0.221
17	0.999	0.998	0.997	0.994	0.991	0.985	0.748	0.297
18	0.999	0.999	0.998	0.997	0.995	0.992	0.819	0.381
19	0.999	0.999	0.999	0.998	0.998	0.996	0.875	0.470
20	0.999	0.999	0.999	0.999	0.999	0.998	0.917	0.559
21	0.999	0.999	0.999	0.999	0.999	0.999	0.946	0.643
22		0.999	0.999	0.999	0.999	0.999	0.967	0.720
23			0.999	0.999	0.999	0.999	0.980	0.787
24					0.999	0.999	0.988	0.843
25						0.999	0.993	0.887
26							0.996	0.922
27							0.998	0.947
28							0.999	0.965
29							0.999	0.978
30							0.999	0.986
31							0.999	0.991
32							0.999	0.995
33							0.999	0.997
34								0.998

[a]Entries in the table are values of $F(x) = P(X \leq x) = \sum_{i=0}^{x} e^{-c}c^i/i!$. Blank spaces below the last entry in any column may be read as 1.0.

Table II Cumulative Standard Normal Distribution

$$\Phi(z)=\int_{-\infty}^{z}\frac{1}{\sqrt{2\pi}}e^{-u^2/2}\,du$$

z	0.00	0.01	0.02	0.03	0.04	z
0.0	0.500 00	0.503 99	0.507 98	0.511 97	0.515 95	0.0
0.1	0.539 83	0.543 79	0.547 76	0.551.72	0.555 67	0.1
0.2	0.579 26	0.583 17	0.587 06	0.590 95	0.594 83	0.2
0.3	0.617 91	0.621 72	0.625 51	0.629 30	0.633 07	0.3
0.4	0.655 42	0.659 10	0.662 76	0.666 40	0.670 03	0.4
0.5	0.691 46	0.694 97	0.698 47	0.701 94	0.705 40	0.5
0.6	0.725 75	0.729 07	0.732 37	0.735 65	0.738 91	0.6
0.7	0.758 03	0.761 15	0.764 24	0.767 30	0.770 35	0.7
0.8	0.788 14	0.791 03	0.793 89	0.796 73	0.799 54	0.8
0.9	0.815 94	0.818 59	0.821 21	0.823 81	0.826 39	0.9
1.0	0.841 34	0.843 75	0.846 13	0.848 49	0.850 83	1.0
1.1	0.864 33	0.866 50	0.868 64	0.870 76	0.872 85	1.1
1.2	0.884 93	0.886 86	0.888 77	0.890 65	0.892 51	1.2
1.3	0.903 20	0.904 90	0.906 58	0.908 24	0.909 88	1.3
1.4	0.919 24	0.920 73	0.922 19	0.923 64	0.925 06	1.4
1.5	0.933 19	0.934 48	0.935 74	0.936 99	0.938 22	1.5
1.6	0.945 20	0.946 30	0.947 38	0.948 45	0.949 50	1.6
1.7	0.955 43	0.956 37	0.957 28	0.958 18	0.959 07	1.7
1.8	0.964 07	0.964 85	0.965 62	0.966 37	0.967 11	1.8
1.9	0.971 28	0.971 93	0.972 57	0.973 20	0.973 81	1.9
2.0	0.977 25	0.977 78	0.978 31	0.978 82	0.979 32	2.0
2.1	0.982 14	0.982 57	0.983 00	0.983 41	0.983 82	2.1
2.2	0.986 10	0.986 45	0.986 79	0.987 13	0.987 45	2.2
2.3	0.989 28	0.989 56	0.989 83	0.990 10	0.990 36	2.3
2.4	0.991 80	0.992 02	0.992 24	0.992 45	0.992 66	2.4
2.5	0.993 79	0.993 96	0.994 13	0.994 30	0.994 46	2.5
2.6	0.995 34	0.995 47	0.995 60	0.995 73	0.995 85	2.6
2.7	0.996 53	0.996 64	0.996 74	0.996 83	0.996 93	2.7
2.8	0.997 44	0.997 52	0.997 60	0.997 67	0.997 74	2.8
2.9	0.998 13	0.998 19	0.998 25	0.998 31	0.998 36	2.9
3.0	0.998 65	0.998 69	0.998 74	0.998 78	0.998 82	3.0
3.1	0.999 03	0.999 06	0.999 10	0.999 13	0.999 16	3.1
3.2	0.999 31	0.999 34	0.999 36	0.999 38	0.999 40	3.2
3.3	0.999 52	0.999 53	0.999 55	0.999 57	0.999 58	3.3
3.4	0.999 66	0.999 68	0.999 69	0.999 70	0.999 71	3.4
3.5	0.999 77	0.999 78	0.999 78	0.999 79	0.999 80	3.5
3.6	0.999 84	0.999 85	0.999 85	0.999 86	0.999 86	3.6
3.7	0.999 89	0.999 90	0.999 90	0.999 90	0.999 91	3.7
3.8	0.999 93	0.999 93	0.999 93	0.999 94	0.999 94	3.8
3.9	0.999 95	0.999 95	0.999 96	0.999 96	0.999 96	3.9

(*continues*)

Table II Cumulative Standard Normal Distribution (*continued*)

$$\Phi(z)=\int_{-\infty}^{z}\frac{1}{\sqrt{2\pi}}e^{-u^2/2}\,du$$

z	0.05	0.06	0.07	0.08	0.09	z
0.0	0.519 94	0.523 92	0.527 90	0.531 88	0.535 86	0.0
0.1	0.559 62	0.563 56	0.567 49	0.571 42	0.575 34	0.1
0.2	0.598 71	0.602 57	0.606 42	0.610 26	0.614 09	0.2
0.3	0.636 83	0.640 58	0.644 31	0.648 03	0.651 73	0.3
0.4	0.673 64	0.677 24	0.680 82	0.684 38	0.687 93	0.4
0.5	0.708 84	0.712 26	0.715 66	0.719 04	0.722 40	0.5
0.6	0.742 15	0.745 37	0.748 57	0.751 75	0.754 90	0.6
0.7	0.773 37	0.776 37	0.779 35	0.782 30	0.785 23	0.7
0.8	0.802 34	0.805 10	0.807 85	0.810 57	0.813 27	0.8
0.9	0.828 94	0.831 47	0.833 97	0.836 46	0.838 91	0.9
1.0	0.853 14	0.855 43	0.857 69	0.859 93	0.862 14	1.0
1.1	0.874 93	0.876 97	0.879 00	0.881 00	0.882 97	1.1
1.2	0.894 35	0.896 16	0.897 96	0.899 73	0.901 47	1.2
1.3	0.911 49	0.913 08	0.914 65	0.916 21	0.917 73	1.3
1.4	0.926 47	0.927 85	0.929 22	0.930 56	0.931 89	1.4
1.5	0.939 43	0.940 62	0.941 79	0.942 95	0.944 08	1.5
1.6	0.950 53	0.951 54	0.952 54	0.953 52	0.954 48	1.6
1.7	0.959 94	0.960 80	0.961 64	0.962 46	0.963 27	1.7
1.8	0.967 84	0.968 56	0.969 26	0.969 95	0.970 62	1.8
1.9	0.974 41	0.975 00	0.975 58	0.976 15	0.976 70	1.9
2.0	0.979 82	0.980 30	0.980 77	0.981 24	0.981 69	2.0
2.1	0.984 22	0.984 61	0.985 00	0.985 37	0.985 74	2.1
2.2	0.987 78	0.988 09	0.988 40	0.988 70	0.988 99	2.2
2.3	0.990 61	0.990 86	0.991 11	0.991 34	0.991 58	2.3
2.4	0.992 86	0.993 05	0.993 24	0.993 43	0.993 61	2.4
2.5	0.994 61	0.994 77	0.994 92	0.995 06	0.995 20	2.5
2.6	0.995 98	0.996 09	0.996 21	0.996 32	0.996 43	2.6
2.7	0.997 02	0.997 11	0.997 20	0.997 28	0.997 36	2.7
2.8	0.997 81	0.997 88	0.997 95	0.998 01	0.998 07	2.8
2.9	0.998 41	0.998 46	0.998 51	0.998 56	0.998 61	2.9
3.0	0.998 86	0.998 89	0.998 93	0.998 97	0.999 00	3.0
3.1	0.999 18	0.999 21	0.999 24	0.999 26	0.999 29	3.1
3.2	0.999 42	0.999 44	0.999 46	0.999 48	0.999 50	3.2
3.3	0.999 60	0.999 61	0.999 62	0.999 64	0.999 65	3.3
3.4	0.999 72	0.999 73	0.999 74	0.999 75	0.999 76	3.4
3.5	0.999 81	0.999 81	0.999 82	0.999 83	0.999 83	3.5
3.6	0.999 87	0.999 87	0.999 88	0.999 88	0.999 89	3.6
3.7	0.999 91	0.999 92	0.999 92	0.999 92	0.999 92	3.7
3.8	0.999 94	0.999 94	0.999 95	0.999 95	0.999 95	3.8
3.9	0.999 96	0.999 96	0.999 96	0.999 97	0.999 97	3.9

Table III Percentage Points of the χ^2 Distribution[a]

v \ α	0.995	0.990	0.975	0.950	0.900	0.500	0.100	0.050	0.025	0.010	0.005
1	0.00+	0.00+	0.00+	0.00+	0.02	0.45	2.71	3.84	5.02	6.63	7.88
2	0.01	0.02	0.05	0.10	0.21	1.39	4.61	5.99	7.38	9.21	10.60
3	0.07	0.11	0.22	0.35	0.58	2.37	6.25	7.81	9.35	11.34	12.84
4	0.21	0.30	0.48	0.71	1.06	3.36	7.78	9.49	11.14	13.28	14.86
5	0.41	0.55	0.83	1.15	1.61	4.35	9.24	11.07	12.83	15.09	16.75
6	0.68	0.87	1.24	1.64	2.20	5.35	10.65	12.59	14.45	16.81	18.55
7	0.99	1.24	1.69	2.17	2.83	6.35	12.02	14.07	16.01	18.48	20.28
8	1.34	1.65	2.18	2.73	3.49	7.34	13.36	15.51	17.53	20.09	21.96
9	1.73	2.09	2.70	3.33	4.17	8.34	14.68	16.92	19.02	21.67	23.59
10	2.16	2.56	3.25	3.94	4.87	9.34	15.99	18.31	20.48	23.21	25.19
11	2.60	3.05	3.82	4.57	5.58	10.34	17.28	19.68	21.92	24.72	26.76
12	3.07	3.57	4.40	5.23	6.30	11.34	18.55	21.03	23.34	26.22	28.30
13	3.57	4.11	5.01	5.89	7.04	12.34	19.81	22.36	24.74	27.69	29.82
14	4.07	4.66	5.63	6.57	7.79	13.34	21.06	23.68	26.12	29.14	31.32
15	4.60	5.23	6.27	7.26	8.55	14.34	22.31	25.00	27.49	30.58	32.80
16	5.14	5.81	6.91	7.96	9.31	15.34	23.54	26.30	28.85	32.00	34.27
17	5.70	6.41	7.56	8.67	10.09	16.34	24.77	27.59	30.19	33.41	35.72
18	6.26	7.01	8.23	9.39	10.87	17.34	25.99	28.87	31.53	34.81	37.16
19	6.84	7.63	8.91	10.12	11.65	18.34	27.20	30.14	32.85	36.19	38.58
20	7.43	8.26	9.59	10.85	12.44	19.34	28.41	31.41	34.17	37.57	40.00
21	8.03	8.90	10.28	11.59	13.24	20.34	29.62	32.67	35.48	38.93	41.40
22	8.64	9.54	10.98	12.34	14.04	21.34	30.81	33.92	36.78	40.29	42.80
23	9.26	10.20	11.69	13.09	14.85	22.34	32.01	35.17	38.08	41.64	44.18
24	9.89	10.86	12.40	13.85	15.66	23.34	33.20	36.42	39.36	42.98	45.56
25	10.52	11.52	13.12	14.61	16.47	24.34	34.28	37.65	40.65	44.31	46.93
26	11.16	12.20	13.84	15.38	17.29	25.34	35.56	38.89	41.92	45.64	48.29
27	11.81	12.88	14.57	16.15	18.11	26.34	36.74	40.11	43.19	46.96	49.65
28	12.46	13.57	15.31	16.93	18.94	27.34	37.92	41.34	44.46	48.28	50.99
29	13.12	14.26	16.05	17.71	19.77	28.34	39.09	42.56	45.72	49.59	52.34
30	13.79	14.95	16.79	18.49	20.60	29.34	40.26	43.77	46.98	50.89	53.67
40	20.71	22.16	24.43	26.51	29.05	39.34	51.81	55.76	59.34	63.69	66.77
50	27.99	29.71	32.36	34.76	37.69	49.33	63.17	67.50	71.42	76.15	79.49
60	35.53	37.48	40.48	43.19	46.46	59.33	74.40	79.08	83.30	88.38	91.95
70	43.28	45.44	48.76	51.74	55.33	69.33	85.53	90.53	95.02	100.42	104.22
80	51.17	53.54	57.15	60.39	64.28	79.33	96.58	101.88	106.63	112.33	116.32
90	59.20	61.75	65.65	69.13	73.29	89.33	107.57	113.14	118.14	124.12	128.30
100	67.33	70.06	74.22	77.93	82.36	99.33	118.50	124.34	129.56	135.81	140.17

[a] v = degrees of freedom.

Table IV Percentage Points of the t Distribution

v \ α	0.40	0.25	0.10	0.05	0.025	0.01	0.005	0.0025	0.001	0.0005
1	0.325	1.000	3.078	6.314	12.706	31.821	63.657	127.32	318.31	636.62
2	0.289	0.816	1.886	2.920	4.303	6.965	9.925	14.089	23.326	31.598
3	0.277	0.765	1.638	2.353	3.182	4.541	5.841	7.453	10.213	12.924
4	0.271	0.741	1.533	2.132	2.776	3.747	4.604	5.598	7.173	8.610
5	0.267	0.727	1.476	2.015	2.571	3.365	4.032	4.773	5.893	6.869
6	0.265	0.718	1.440	1.943	2.447	3.143	3.707	4.317	5.208	5.959
7	0.263	0.711	1.415	1.895	2.365	2.998	3.499	4.029	4.785	5.408
8	0.262	0.706	1.397	1.860	2.306	2.896	3.355	3.833	4.501	5.041
9	0.261	0.703	1.383	1.833	2.262	2.821	3.250	3.690	4.297	4.781
10	0.260	0.700	1.372	1.812	2.228	2.764	3.169	3.581	4.144	4.587
11	0.260	0.697	1.363	1.796	2.201	2.718	3.106	3.497	4.025	4.437
12	0.259	0.695	1.356	1.782	2.179	2.681	3.055	3.428	3.930	4.318
13	0.259	0.694	1.350	1.771	2.160	2.650	3.012	3.372	3.852	4.221
14	0.258	0.692	1.345	1.761	2.145	2.624	2.977	3.326	3.787	4.140
15	0.258	0.691	1.341	1.753	2.131	2.602	2.947	3.286	3.733	4.073
16	0.258	0.690	1.337	1.746	2.120	2.583	2.921	3.252	3.686	4.015
17	0.257	0.689	1.333	1.740	2.110	2.567	2.898	3.222	3.646	3.965
18	0.257	0.688	1.330	1.734	2.101	2.552	2.878	3.197	3.610	3.922
19	0.257	0.688	1.328	1.729	2.093	2.539	2.861	3.174	3.579	3.883
20	0.257	0.687	1.325	1.725	2.086	2.528	2.845	3.153	3.552	3.850
21	0.257	0.686	1.323	1.721	2.080	2.518	2.831	3.135	3.527	3.819
22	0.256	0.686	1.321	1.717	2.074	2.508	2.819	3.119	3.505	3.792
23	0.256	0.685	1.319	1.714	2.069	2.500	2.807	3.104	3.485	3.767
24	0.256	0.685	1.318	1.711	2.064	2.492	2.797	3.091	3.467	3.745
25	0.256	0.684	1.316	1.708	2.060	2.485	2.787	3.078	3.450	3.725
26	0.256	0.684	1.315	1.706	2.056	2.479	2.779	3.067	3.435	3.707
27	0.256	0.684	1.314	1.703	2.052	2.473	2.771	3.057	3.421	3.690
28	0.256	0.683	1.313	1.701	2.048	2.467	2.763	3.047	3.408	3.674
29	0.256	0.683	1.311	1.699	2.045	2.462	2.756	3.038	3.396	3.659
30	0.256	0.683	1.310	1.697	2.042	2.457	2.750	3.030	3.385	3.646
40	0.255	0.681	1.303	1.684	2.021	2.423	2.704	2.971	3.307	3.551
60	0.254	0.679	1.296	1.671	2.000	2.390	2.660	2.915	3.232	3.460
120	0.254	0.677	1.289	1.658	1.980	2.358	2.617	2.860	3.160	3.373
∞	0.253	0.674	1.282	1.645	1.960	2.326	2.576	2.807	3.090	3.291

Source: This table is adapted from *Biometrika Tables for Statisticians*, Vol. 1, 3rd edition, 1966, by permission of the Biometrika Trustees.

Table V Percentage Points of the F Distribution

$F_{0.25, \nu_1, \nu_2}$

ν_2 \ ν_1 (Degrees of freedom for the denominator (ν_2))	Degrees of freedom for the numerator (ν_1)																		
	1	2	3	4	5	6	7	8	9	10	12	15	20	24	30	40	60	120	∞
1	5.83	7.50	8.20	8.58	8.82	8.98	9.10	9.19	9.26	9.32	9.41	9.49	9.58	9.63	9.67	9.71	9.76	9.80	9.85
2	2.57	3.00	3.15	3.23	3.28	3.31	3.34	3.35	3.37	3.38	3.39	3.41	3.43	3.43	3.44	3.45	3.46	3.47	3.48
3	2.02	2.28	2.36	2.39	2.41	2.42	2.43	2.44	2.44	2.44	2.45	2.46	2.46	2.46	2.47	2.47	2.47	2.47	2.47
4	1.81	2.00	2.05	2.06	2.07	2.08	2.08	2.08	2.08	2.08	2.08	2.08	2.08	2.08	2.08	2.08	2.08	2.08	2.08
5	1.69	1.85	1.88	1.89	1.89	1.89	1.89	1.89	1.89	1.89	1.89	1.89	1.88	1.88	1.88	1.88	1.87	1.87	1.87
6	1.62	1.76	1.78	1.79	1.79	1.78	1.78	1.78	1.77	1.77	1.77	1.76	1.76	1.75	1.75	1.75	1.74	1.74	1.74
7	1.57	1.70	1.72	1.72	1.71	1.71	1.70	1.70	1.70	1.69	1.68	1.68	1.67	1.67	1.66	1.66	1.65	1.65	1.65
8	1.54	1.66	1.67	1.66	1.66	1.65	1.64	1.64	1.63	1.63	1.62	1.62	1.61	1.60	1.60	1.59	1.59	1.58	1.58
9	1.51	1.62	1.63	1.63	1.62	1.61	1.60	1.60	1.59	1.59	1.58	1.57	1.56	1.56	1.55	1.54	1.54	1.53	1.53
10	1.49	1.60	1.60	1.59	1.59	1.58	1.57	1.56	1.56	1.55	1.54	1.53	1.52	1.52	1.51	1.51	1.50	1.49	1.48
11	1.47	1.58	1.58	1.57	1.56	1.55	1.54	1.53	1.53	1.52	1.51	1.50	1.49	1.49	1.48	1.47	1.47	1.46	1.45
12	1.46	1.56	1.56	1.55	1.54	1.53	1.52	1.51	1.51	1.50	1.49	1.48	1.47	1.46	1.45	1.45	1.44	1.43	1.42
13	1.45	1.55	1.55	1.53	1.52	1.51	1.50	1.49	1.49	1.48	1.47	1.46	1.45	1.44	1.43	1.42	1.42	1.41	1.40
14	1.44	1.53	1.53	1.52	1.51	1.50	1.49	1.48	1.47	1.46	1.45	1.44	1.43	1.42	1.41	1.41	1.40	1.39	1.38
15	1.43	1.52	1.52	1.51	1.49	1.48	1.47	1.46	1.46	1.45	1.44	1.43	1.41	1.41	1.40	1.39	1.38	1.37	1.36
16	1.42	1.51	1.51	1.50	1.48	1.47	1.46	1.45	1.44	1.44	1.43	1.41	1.40	1.39	1.38	1.37	1.36	1.35	1.34
17	1.42	1.51	1.50	1.49	1.47	1.46	1.45	1.44	1.43	1.43	1.41	1.40	1.39	1.38	1.37	1.36	1.35	1.34	1.33
18	1.41	1.50	1.49	1.48	1.46	1.45	1.44	1.43	1.42	1.42	1.40	1.39	1.38	1.37	1.36	1.35	1.34	1.33	1.32
19	1.41	1.49	1.49	1.47	1.46	1.44	1.43	1.42	1.41	1.41	1.40	1.38	1.37	1.36	1.35	1.34	1.33	1.32	1.30
20	1.40	1.49	1.48	1.47	1.45	1.44	1.43	1.42	1.41	1.40	1.39	1.37	1.36	1.35	1.34	1.33	1.32	1.31	1.29
21	1.40	1.48	1.48	1.46	1.44	1.43	1.42	1.41	1.40	1.39	1.38	1.37	1.35	1.34	1.33	1.32	1.31	1.30	1.28
22	1.40	1.48	1.47	1.45	1.44	1.42	1.41	1.40	1.39	1.39	1.37	1.36	1.34	1.33	1.32	1.31	1.30	1.29	1.28
23	1.39	1.47	1.47	1.45	1.43	1.42	1.41	1.40	1.39	1.38	1.37	1.35	1.34	1.33	1.32	1.31	1.30	1.28	1.27
24	1.39	1.47	1.46	1.44	1.43	1.41	1.40	1.39	1.38	1.38	1.36	1.35	1.33	1.32	1.31	1.30	1.29	1.28	1.26
25	1.39	1.47	1.46	1.44	1.42	1.41	1.40	1.39	1.38	1.37	1.36	1.34	1.33	1.32	1.31	1.29	1.28	1.27	1.25
26	1.38	1.46	1.45	1.44	1.42	1.41	1.39	1.38	1.37	1.37	1.35	1.34	1.32	1.31	1.30	1.29	1.28	1.26	1.25
27	1.38	1.46	1.45	1.43	1.42	1.40	1.39	1.38	1.37	1.36	1.35	1.33	1.32	1.31	1.30	1.28	1.27	1.26	1.24
28	1.38	1.46	1.45	1.43	1.41	1.40	1.39	1.38	1.37	1.36	1.34	1.33	1.31	1.30	1.29	1.28	1.27	1.25	1.24
29	1.38	1.45	1.45	1.43	1.41	1.40	1.38	1.37	1.36	1.35	1.34	1.32	1.31	1.30	1.29	1.27	1.26	1.25	1.23
30	1.38	1.45	1.44	1.42	1.41	1.39	1.38	1.37	1.36	1.35	1.34	1.32	1.30	1.29	1.28	1.27	1.26	1.24	1.23
40	1.36	1.44	1.42	1.40	1.39	1.37	1.36	1.35	1.34	1.33	1.31	1.30	1.28	1.26	1.25	1.24	1.22	1.21	1.19
60	1.35	1.42	1.41	1.38	1.37	1.35	1.33	1.32	1.31	1.30	1.29	1.27	1.25	1.24	1.22	1.21	1.19	1.17	1.15
120	1.34	1.40	1.39	1.37	1.35	1.33	1.31	1.30	1.29	1.28	1.26	1.24	1.22	1.21	1.19	1.18	1.16	1.13	1.10
∞	1.32	1.39	1.37	1.35	1.33	1.31	1.29	1.28	1.27	1.25	1.24	1.22	1.19	1.18	1.16	1.14	1.12	1.08	1.00

Source: Adapted with permission from *Biometrika Tables for Statisticians*, Vol. 1, 3rd edition, by E. S. Pearson and H. O. Hartley, Cambridge University Press, Cambridge, 1966.

(continues)

Table V Percentage Points of the F Distribution (*continued*)

$F_{0.10, v_1, v_2}$

v_2 \ v_1 (Degrees of freedom for the numerator (v_1))	1	2	3	4	5	6	7	8	9	10	12	15	20	24	30	40	60	120	∞
1	39.86	49.50	53.59	55.83	57.24	58.20	58.91	59.44	59.86	60.19	60.71	61.22	61.74	62.00	62.26	62.53	62.79	63.06	63.33
2	8.53	9.00	9.16	9.24	9.29	9.33	9.35	9.37	9.38	9.39	9.41	9.42	9.44	9.45	9.46	9.47	9.47	9.48	9.49
3	5.54	5.46	5.39	5.34	5.31	5.28	5.27	5.25	5.24	5.23	5.22	5.20	5.18	5.18	5.17	5.16	5.15	5.14	5.13
4	4.54	4.32	4.19	4.11	4.05	4.01	3.98	3.95	3.94	3.92	3.90	3.87	3.84	3.83	3.82	3.80	3.79	3.78	3.76
5	4.06	3.78	3.62	3.52	3.45	3.40	3.37	3.34	3.32	3.30	3.27	3.24	3.21	3.19	3.17	3.16	3.14	3.12	3.10
6	3.78	3.46	3.29	3.18	3.11	3.05	3.01	2.98	2.96	2.94	2.90	2.87	2.84	2.82	2.80	2.78	2.76	2.74	2.72
7	3.59	3.26	3.07	2.96	2.88	2.83	2.78	2.75	2.72	2.70	2.67	2.63	2.59	2.58	2.56	2.54	2.51	2.49	2.47
8	3.46	3.11	2.92	2.81	2.73	2.67	2.62	2.59	2.56	2.54	2.50	2.46	2.42	2.40	2.38	2.36	2.34	2.32	2.29
9	3.36	3.01	2.81	2.69	2.61	2.55	2.51	2.47	2.44	2.42	2.38	2.34	2.30	2.28	2.25	2.23	2.21	2.18	2.16
10	3.29	2.92	2.73	2.61	2.52	2.46	2.41	2.38	2.35	2.32	2.28	2.24	2.20	2.18	2.16	2.13	2.11	2.08	2.06
11	3.23	2.86	2.66	2.54	2.45	2.39	2.34	2.30	2.27	2.25	2.21	2.17	2.12	2.10	2.08	2.05	2.03	2.00	1.97
12	3.18	2.81	2.61	2.48	2.39	2.33	2.28	2.24	2.21	2.19	2.15	2.10	2.06	2.04	2.01	1.99	1.96	1.93	1.90
13	3.14	2.76	2.56	2.43	2.35	2.28	2.23	2.20	2.16	2.14	2.10	2.05	2.01	1.98	1.96	1.93	1.90	1.88	1.85
14	3.10	2.73	2.52	2.39	2.31	2.24	2.19	2.15	2.12	2.10	2.05	2.01	1.96	1.94	1.91	1.89	1.86	1.83	1.80
15	3.07	2.70	2.49	2.36	2.27	2.21	2.16	2.12	2.09	2.06	2.02	1.97	1.92	1.90	1.87	1.85	1.82	1.79	1.76
16	3.05	2.67	2.46	2.33	2.24	2.18	2.13	2.09	2.06	2.03	1.99	1.94	1.89	1.87	1.84	1.81	1.78	1.75	1.72
17	3.03	2.64	2.44	2.31	2.22	2.15	2.10	2.06	2.03	2.00	1.96	1.91	1.86	1.84	1.81	1.78	1.75	1.72	1.69
18	3.01	2.62	2.42	2.29	2.20	2.13	2.08	2.04	2.00	1.98	1.93	1.89	1.84	1.81	1.78	1.75	1.72	1.69	1.66
19	2.99	2.61	2.40	2.27	2.18	2.11	2.06	2.02	1.98	1.96	1.91	1.86	1.81	1.79	1.76	1.73	1.70	1.67	1.63
20	2.97	2.59	2.38	2.25	2.16	2.09	2.04	2.00	1.96	1.94	1.89	1.84	1.79	1.77	1.74	1.71	1.68	1.64	1.61
21	2.96	2.57	2.36	2.23	2.14	2.08	2.02	1.98	1.95	1.92	1.87	1.83	1.78	1.75	1.72	1.69	1.66	1.62	1.59
22	2.95	2.56	2.35	2.22	2.13	2.06	2.01	1.97	1.93	1.90	1.86	1.81	1.76	1.73	1.70	1.67	1.64	1.60	1.57
23	2.94	2.55	2.34	2.21	2.11	2.05	1.99	1.95	1.92	1.89	1.84	1.80	1.74	1.72	1.69	1.66	1.62	1.59	1.55
24	2.93	2.54	2.33	2.19	2.10	2.04	1.98	1.94	1.91	1.88	1.83	1.78	1.73	1.70	1.67	1.64	1.61	1.57	1.53
25	2.92	2.53	2.32	2.18	2.09	2.02	1.97	1.93	1.89	1.87	1.82	1.77	1.72	1.69	1.66	1.63	1.59	1.56	1.52
26	2.91	2.52	2.31	2.17	2.08	2.01	1.96	1.92	1.88	1.86	1.81	1.76	1.71	1.68	1.65	1.61	1.58	1.54	1.50
27	2.90	2.51	2.30	2.17	2.07	2.00	1.95	1.91	1.87	1.85	1.80	1.75	1.70	1.67	1.64	1.60	1.57	1.53	1.49
28	2.89	2.50	2.29	2.16	2.06	2.00	1.94	1.90	1.87	1.84	1.79	1.74	1.69	1.66	1.63	1.59	1.56	1.52	1.48
29	2.89	2.50	2.28	2.15	2.06	1.99	1.93	1.89	1.86	1.83	1.78	1.73	1.68	1.65	1.62	1.58	1.55	1.51	1.47
30	2.88	2.49	2.28	2.14	2.03	1.98	1.93	1.88	1.85	1.82	1.77	1.72	1.67	1.64	1.61	1.57	1.54	1.50	1.46
40	2.84	2.44	2.23	2.09	2.00	1.93	1.87	1.83	1.79	1.76	1.71	1.66	1.61	1.57	1.54	1.51	1.47	1.42	1.38
60	2.79	2.39	2.18	2.04	1.95	1.87	1.82	1.77	1.74	1.71	1.66	1.60	1.54	1.51	1.48	1.44	1.40	1.35	1.29
120	2.75	2.35	2.13	1.99	1.90	1.82	1.77	1.72	1.68	1.65	1.60	1.55	1.48	1.45	1.41	1.37	1.32	1.26	1.19
∞	2.71	2.30	2.08	1.94	1.85	1.77	1.72	1.67	1.63	1.60	1.55	1.49	1.42	1.38	1.34	1.30	1.24	1.17	1.00

Degrees of freedom for the denominator (v_2)

Table V Percentage Points of the F Distribution (*continued*)

$F_{0.05, \nu_1, \nu_2}$

Degrees of freedom for the numerator (ν_1); rows: Degrees of freedom for the denominator (ν_2)

ν_2 \ ν_1	1	2	3	4	5	6	7	8	9	10	12	15	20	24	30	40	60	120	∞
1	161.4	199.5	215.7	224.6	230.2	234.0	236.8	238.9	240.5	241.9	243.9	245.9	248.0	249.1	250.1	251.1	252.2	253.3	254.3
2	18.51	19.00	19.16	19.25	19.30	19.33	19.35	19.37	19.38	19.40	19.41	19.43	19.45	19.45	19.46	19.47	19.48	19.49	19.50
3	10.13	9.55	9.28	9.12	9.01	8.94	8.89	8.85	8.81	8.79	8.74	8.70	8.66	8.64	8.62	8.59	8.57	8.55	8.53
4	7.71	6.94	6.59	6.39	6.26	6.16	6.09	6.04	6.00	5.96	5.91	5.86	5.80	5.77	5.75	5.72	5.69	5.66	5.63
5	6.61	5.79	5.41	5.19	5.05	4.95	4.88	4.82	4.77	4.74	4.68	4.62	4.56	4.53	4.50	4.46	4.43	4.40	4.36
6	5.99	5.14	4.76	4.53	4.39	4.28	4.21	4.15	4.10	4.06	4.00	3.94	3.87	3.84	3.81	3.77	3.74	3.70	3.67
7	5.59	4.74	4.35	4.12	3.97	3.87	3.79	3.73	3.68	3.64	3.57	3.51	3.44	3.41	3.38	3.34	3.30	3.27	3.23
8	5.32	4.46	4.07	3.84	3.69	3.58	3.50	3.44	3.39	3.35	3.28	3.22	3.15	3.12	3.08	3.04	3.01	2.97	2.93
9	5.12	4.26	3.86	3.63	3.48	3.37	3.29	3.23	3.18	3.14	3.07	3.01	2.94	2.90	2.86	2.83	2.79	2.75	2.71
10	4.96	4.10	3.71	3.48	3.33	3.22	3.14	3.07	3.02	2.98	2.91	2.85	2.77	2.74	2.70	2.66	2.62	2.58	2.54
11	4.84	3.98	3.59	3.36	3.20	3.09	3.01	2.95	2.90	2.85	2.79	2.72	2.65	2.61	2.57	2.53	2.49	2.45	2.40
12	4.75	3.89	3.49	3.26	3.11	3.00	2.91	2.85	2.80	2.75	2.69	2.62	2.54	2.51	2.47	2.43	2.38	2.34	2.30
13	4.67	3.81	3.41	3.18	3.03	2.92	2.83	2.77	2.71	2.67	2.60	2.53	2.46	2.42	2.38	2.34	2.30	2.25	2.21
14	4.60	3.74	3.34	3.11	2.96	2.85	2.76	2.70	2.65	2.60	2.53	2.46	2.39	2.35	2.31	2.27	2.22	2.18	2.13
15	4.54	3.68	3.29	3.06	2.90	2.79	2.71	2.64	2.59	2.54	2.48	2.40	2.33	2.29	2.25	2.20	2.16	2.11	2.07
16	4.49	3.63	3.24	3.01	2.85	2.74	2.66	2.59	2.54	2.49	2.42	2.35	2.28	2.24	2.19	2.15	2.11	2.06	2.01
17	4.45	3.59	3.20	2.96	2.81	2.70	2.61	2.55	2.49	2.45	2.38	2.31	2.23	2.19	2.15	2.10	2.06	2.01	1.96
18	4.41	3.55	3.16	2.93	2.77	2.66	2.58	2.51	2.46	2.41	2.34	2.27	2.19	2.15	2.11	2.06	2.02	1.97	1.92
19	4.38	3.52	3.13	2.90	2.74	2.63	2.54	2.48	2.42	2.38	2.31	2.23	2.16	2.11	2.07	2.03	1.98	1.93	1.88
20	4.35	3.49	3.10	2.87	2.71	2.60	2.51	2.45	2.39	2.35	2.28	2.20	2.12	2.08	2.04	1.99	1.95	1.90	1.84
21	4.32	3.47	3.07	2.84	2.68	2.57	2.49	2.42	2.37	2.32	2.25	2.18	2.10	2.05	2.01	1.96	1.92	1.87	1.81
22	4.30	3.44	3.05	2.82	2.66	2.55	2.46	2.40	2.34	2.30	2.23	2.15	2.07	2.03	1.98	1.94	1.89	1.84	1.78
23	4.28	3.42	3.03	2.80	2.64	2.53	2.44	2.37	2.32	2.27	2.20	2.13	2.05	2.01	1.96	1.91	1.86	1.81	1.76
24	4.26	3.40	3.01	2.78	2.62	2.51	2.42	2.36	2.30	2.25	2.18	2.11	2.03	1.98	1.94	1.89	1.84	1.79	1.73
25	4.24	3.39	2.99	2.76	2.60	2.49	2.40	2.34	2.28	2.24	2.16	2.09	2.01	1.96	1.92	1.87	1.82	1.77	1.71
26	4.23	3.37	2.98	2.74	2.59	2.47	2.39	2.32	2.27	2.22	2.15	2.07	1.99	1.95	1.90	1.85	1.80	1.75	1.69
27	4.21	3.35	2.96	2.73	2.57	2.46	2.37	2.31	2.25	2.20	2.13	2.06	1.97	1.93	1.88	1.84	1.79	1.73	1.67
28	4.20	3.34	2.95	2.71	2.56	2.45	2.36	2.29	2.24	2.19	2.12	2.04	1.96	1.91	1.87	1.82	1.77	1.71	1.65
29	4.18	3.33	2.93	2.70	2.55	2.43	2.35	2.28	2.22	2.18	2.10	2.03	1.94	1.90	1.85	1.81	1.75	1.70	1.64
30	4.17	3.32	2.92	2.69	2.53	2.42	2.33	2.27	2.21	2.16	2.09	2.01	1.93	1.89	1.84	1.79	1.74	1.68	1.62
40	4.08	3.23	2.84	2.61	2.45	2.34	2.25	2.18	2.12	2.08	2.00	1.92	1.84	1.79	1.74	1.69	1.64	1.58	1.51
60	4.00	3.15	2.76	2.53	2.37	2.25	2.17	2.10	2.04	1.99	1.92	1.84	1.75	1.70	1.65	1.59	1.53	1.47	1.39
120	3.92	3.07	2.68	2.45	2.29	2.17	2.09	2.02	1.96	1.91	1.83	1.75	1.66	1.61	1.55	1.55	1.43	1.35	1.25
∞	3.84	3.00	2.60	2.37	2.21	2.10	2.01	1.94	1.88	1.83	1.75	1.67	1.57	1.52	1.46	1.39	1.32	1.22	1.00

(*continues*)

Table V Percentage Points of the F Distribution (*continued*)

$F_{0.025, \nu_1, \nu_2}$

Degrees of freedom for the numerator (ν_1); degrees of freedom for the denominator (ν_2)

ν_2 \ ν_1	1	2	3	4	5	6	7	8	9	10	12	15	20	24	30	40	60	120	∞
1	647.8	799.5	864.2	899.6	921.8	937.1	948.2	956.7	963.3	968.6	976.7	984.9	993.1	997.2	1001	1006	1010	1014	1018
2	38.51	39.00	39.17	39.25	39.30	39.33	39.36	39.37	39.39	39.40	39.41	39.43	39.45	39.46	39.46	39.47	39.48	39.49	39.50
3	17.44	16.04	15.44	15.10	14.88	14.73	14.62	14.54	14.47	14.42	14.34	14.25	14.17	14.12	14.08	14.04	13.99	13.95	13.90
4	12.22	10.65	9.98	9.60	9.36	9.20	9.07	8.98	8.90	8.84	8.75	8.66	8.56	8.51	8.46	8.41	8.36	8.31	8.26
5	10.01	8.43	7.76	7.39	7.15	6.98	6.85	6.76	6.68	6.62	6.52	6.43	6.33	6.28	6.23	6.18	6.12	6.07	6.02
6	8.81	7.26	6.60	6.23	5.99	5.82	5.70	5.60	5.52	5.46	5.37	5.27	5.17	5.12	5.07	5.01	4.96	4.90	4.85
7	8.07	6.54	5.89	5.52	5.29	5.12	4.99	4.90	4.82	4.76	4.67	4.57	4.47	4.42	4.36	4.31	4.25	4.20	4.14
8	7.57	6.06	5.42	5.05	4.82	4.65	4.53	4.43	4.36	4.30	4.20	4.10	4.00	3.95	3.89	3.84	3.78	3.73	3.67
9	7.21	5.71	5.08	4.72	4.48	4.32	4.20	4.10	4.03	3.96	3.87	3.77	3.67	3.61	3.56	3.51	3.45	3.39	3.33
10	6.94	5.46	4.83	4.47	4.24	4.07	3.95	3.85	3.78	3.72	3.62	3.52	3.42	3.37	3.31	3.26	3.20	3.14	3.08
11	6.72	5.26	4.63	4.28	4.04	3.88	3.76	3.66	3.59	3.53	3.43	3.33	3.23	3.17	3.12	3.06	3.00	2.94	2.88
12	6.55	5.10	4.47	4.12	3.89	3.73	3.61	3.51	3.44	3.37	3.28	3.18	3.07	3.02	2.96	2.91	2.85	2.79	2.72
13	6.41	4.97	4.35	4.00	3.77	3.60	3.48	3.39	3.31	3.25	3.15	3.05	2.95	2.89	2.84	2.78	2.72	2.66	2.60
14	6.30	4.86	4.24	3.89	3.66	3.50	3.38	3.29	3.21	3.15	3.05	2.95	2.84	2.79	2.73	2.67	2.61	2.55	2.49
15	6.20	4.77	4.15	3.80	3.58	3.41	3.29	3.20	3.12	3.06	2.96	2.86	2.76	2.70	2.64	2.59	2.52	2.46	2.40
16	6.12	4.69	4.08	3.73	3.50	3.34	3.22	3.12	3.05	2.99	2.89	2.79	2.68	2.63	2.57	2.51	2.45	2.38	2.32
17	6.04	4.62	4.01	3.66	3.44	3.28	3.16	3.06	2.98	2.92	2.82	2.72	2.62	2.56	2.50	2.44	2.38	2.32	2.25
18	5.98	4.56	3.95	3.61	3.38	3.22	3.10	3.01	2.93	2.87	2.77	2.67	2.56	2.50	2.44	2.38	2.32	2.26	2.19
19	5.92	4.51	3.90	3.56	3.33	3.17	3.05	2.96	2.88	2.82	2.72	2.62	2.51	2.45	2.39	2.33	2.27	2.20	2.13
20	5.87	4.46	3.86	3.51	3.29	3.13	3.01	2.91	2.84	2.77	2.68	2.57	2.46	2.41	2.35	2.29	2.22	2.16	2.09
21	5.83	4.42	3.82	3.48	3.25	3.09	2.97	2.87	2.80	2.73	2.64	2.53	2.42	2.37	2.31	2.25	2.18	2.11	2.04
22	5.79	4.38	3.78	3.44	3.22	3.05	2.93	2.84	2.76	2.70	2.60	2.50	2.39	2.33	2.27	2.21	2.14	2.08	2.00
23	5.75	4.35	3.75	3.41	3.18	3.02	2.90	2.81	2.73	2.67	2.57	2.47	2.36	2.30	2.24	2.18	2.11	2.04	1.97
24	5.72	4.32	3.72	3.38	3.15	2.99	2.87	2.78	2.70	2.64	2.54	2.44	2.33	2.27	2.21	2.15	2.08	2.01	1.94
25	5.69	4.29	3.69	3.35	3.13	2.97	2.85	2.75	2.68	2.61	2.51	2.41	2.30	2.24	2.18	2.12	2.05	1.98	1.91
26	5.66	4.27	3.67	3.33	3.10	2.94	2.82	2.73	2.65	2.59	2.49	2.39	2.28	2.22	2.16	2.09	2.03	1.95	1.88
27	5.63	4.24	3.65	3.31	3.08	2.92	2.80	2.71	2.63	2.57	2.47	2.36	2.25	2.19	2.13	2.07	2.00	1.93	1.85
28	5.61	4.22	3.63	3.29	3.06	2.90	2.78	2.69	2.61	2.55	2.45	2.34	2.23	2.17	2.11	2.05	1.98	1.91	1.83
29	5.59	4.20	3.61	3.27	3.04	2.88	2.76	2.67	2.59	2.53	2.43	2.32	2.21	2.15	2.09	2.03	1.96	1.89	1.81
30	5.57	4.18	3.59	3.25	3.03	2.87	2.75	2.65	2.57	2.51	2.41	2.31	2.20	2.14	2.07	2.01	1.94	1.87	1.79
40	5.42	4.05	3.46	3.13	2.90	2.74	2.62	2.53	2.45	2.39	2.29	2.18	2.07	2.01	1.94	1.88	1.80	1.72	1.64
60	5.29	3.93	3.34	3.01	2.79	2.63	2.51	2.41	2.33	2.27	2.17	2.06	1.94	1.88	1.82	1.74	1.67	1.58	1.48
120	5.15	3.80	3.23	2.89	2.67	2.52	2.39	2.30	2.22	2.16	2.05	1.94	1.82	1.76	1.69	1.61	1.53	1.43	1.31
∞	5.02	3.69	3.12	2.79	2.57	2.41	2.29	2.19	2.11	2.05	1.94	1.83	1.71	1.64	1.57	1.48	1.39	1.27	1.00

Table V Percentage Points of the F Distribution (*continued*)

$F_{0.01, \nu_1, \nu_2}$

ν_2 \ ν_1	Degrees of freedom for the numerator (ν_1)																		
	1	2	3	4	5	6	7	8	9	10	12	15	20	24	30	40	60	120	∞
1	4052	4999.5	5403	5625	5764	5859	5928	5982	6022	6056	6106	6157	6209	6235	6261	6287	6313	6339	6366
2	98.50	99.00	99.17	99.25	99.30	99.33	99.36	99.37	99.39	99.40	99.42	99.43	99.45	99.46	99.47	99.47	99.48	99.49	99.50
3	34.12	30.82	29.46	28.71	28.24	27.91	27.67	27.49	27.35	27.23	27.05	26.87	26.69	26.00	26.50	26.41	26.32	26.22	26.13
4	21.20	18.00	16.69	15.98	15.52	15.21	14.98	14.80	14.66	14.55	14.37	14.20	14.02	13.93	13.84	13.75	13.65	13.56	13.46
5	16.26	13.27	12.06	11.39	10.97	10.67	10.46	10.29	10.16	10.05	9.89	9.72	9.55	9.47	9.38	9.29	9.20	9.11	9.02
6	13.75	10.92	9.78	9.15	8.75	8.47	8.26	8.10	7.98	7.87	7.72	7.56	7.40	7.31	7.23	7.14	7.06	6.97	6.88
7	12.25	9.55	8.45	7.85	7.46	7.19	6.99	6.84	6.72	6.62	6.47	6.31	6.16	6.07	5.99	5.91	5.82	5.74	5.65
8	11.26	8.65	7.59	7.01	6.63	6.37	6.18	6.03	5.91	5.81	5.67	5.52	5.36	5.28	5.20	5.12	5.03	4.95	4.86
9	10.56	8.02	6.99	6.42	6.06	5.80	5.61	5.47	5.35	5.26	5.11	4.96	4.81	4.73	4.65	4.57	4.48	4.40	4.31
10	10.04	7.56	6.55	5.99	5.64	5.39	5.20	5.06	4.94	4.85	4.71	4.56	4.41	4.33	4.25	4.17	4.08	4.00	3.91
11	9.65	7.21	6.22	5.67	5.32	5.07	4.89	4.74	4.63	4.54	4.40	4.25	4.10	4.02	3.94	3.86	3.78	3.69	3.60
12	9.33	6.93	5.95	5.41	5.06	4.82	4.64	4.50	4.39	4.30	4.16	4.01	3.86	3.78	3.70	3.62	3.54	3.45	3.36
13	9.07	6.70	5.74	5.21	4.86	4.62	4.44	4.30	4.19	4.10	3.96	3.82	3.66	3.59	3.51	3.43	3.34	3.25	3.17
14	8.86	6.51	5.56	5.04	4.69	4.46	4.28	4.14	4.03	3.94	3.80	3.66	3.51	3.43	3.35	3.27	3.18	3.09	3.00
15	8.68	6.36	5.42	4.89	4.36	4.32	4.14	4.00	3.89	3.80	3.67	3.52	3.37	3.29	3.21	3.13	3.05	2.96	2.87
16	8.53	6.23	5.29	4.77	4.44	4.20	4.03	3.89	3.78	3.69	3.55	3.41	3.26	3.18	3.10	3.02	2.93	2.84	2.75
17	8.40	6.11	5.18	4.67	4.34	4.10	3.93	3.79	3.68	3.59	3.46	3.31	3.16	3.08	3.00	2.92	2.83	2.75	2.65
18	8.29	6.01	5.09	4.58	4.25	4.01	3.84	3.71	3.60	3.51	3.37	3.23	3.08	3.00	2.92	2.84	2.75	2.66	2.57
19	8.18	5.93	5.01	4.50	4.17	3.94	3.77	3.63	3.52	3.43	3.30	3.15	3.00	2.92	2.84	2.76	2.67	2.58	2.59
20	8.10	5.85	4.94	4.43	4.10	3.87	3.70	3.56	3.46	3.37	3.23	3.09	2.94	2.86	2.78	2.69	2.61	2.52	2.42
21	8.02	5.78	4.87	4.37	4.04	3.81	3.64	3.51	3.40	3.31	3.17	3.03	2.88	2.80	2.72	2.64	2.55	2.46	2.36
22	7.95	5.72	4.82	4.31	3.99	3.76	3.59	3.45	3.35	3.26	3.12	2.98	2.83	2.75	2.67	2.58	2.50	2.40	2.31
23	7.88	5.66	4.76	4.26	3.94	3.71	3.54	3.41	3.30	3.21	3.07	2.93	2.78	2.70	2.62	2.54	2.45	2.35	2.26
24	7.82	5.61	4.72	4.22	3.90	3.67	3.50	3.36	3.26	3.17	3.03	2.89	2.74	2.66	2.58	2.49	2.40	2.31	2.21
25	7.77	5.57	4.68	4.18	3.85	3.63	3.46	3.32	3.22	3.13	2.99	2.85	2.70	2.62	2.54	2.45	2.36	2.27	2.17
26	7.72	5.53	4.64	4.14	3.82	3.59	3.42	3.29	3.18	3.09	2.96	2.81	2.66	2.58	2.50	2.42	2.33	2.23	2.13
27	7.68	5.49	4.60	4.11	3.78	3.56	3.39	3.26	3.15	3.06	2.93	2.78	2.63	2.55	2.47	2.38	2.29	2.20	2.10
28	7.64	5.45	4.57	4.07	3.75	3.53	3.36	3.23	3.12	3.03	2.90	2.75	2.60	2.52	2.44	2.35	2.26	2.17	2.06
29	7.60	5.42	4.54	4.04	3.73	3.50	3.33	3.20	3.09	3.00	2.87	2.73	2.57	2.49	2.41	2.33	2.23	2.14	2.03
30	7.56	5.39	4.51	4.02	3.70	3.47	3.30	3.17	3.07	2.98	2.84	2.70	2.55	2.47	2.39	2.30	2.21	2.11	2.01
40	7.31	5.18	4.31	3.83	3.51	3.29	3.12	2.99	2.89	2.80	2.66	2.52	2.37	2.29	2.20	2.11	2.02	1.92	1.80
60	7.08	4.98	4.13	3.65	3.34	3.12	2.95	2.82	2.72	2.63	2.50	2.35	2.20	2.12	2.03	1.94	1.84	1.73	1.60
120	6.85	4.79	3.95	3.48	3.17	2.96	2.79	2.66	2.56	2.47	2.34	2.19	2.03	1.95	1.86	1.76	1.66	1.53	1.38
∞	6.63	4.61	3.78	3.32	3.02	2.80	2.64	2.51	2.41	2.32	2.18	2.04	1.88	1.79	1.70	1.59	1.47	1.32	1.00

Degrees of freedom for the denominator (ν_2)

Chart VI Operating Characteristic Curves

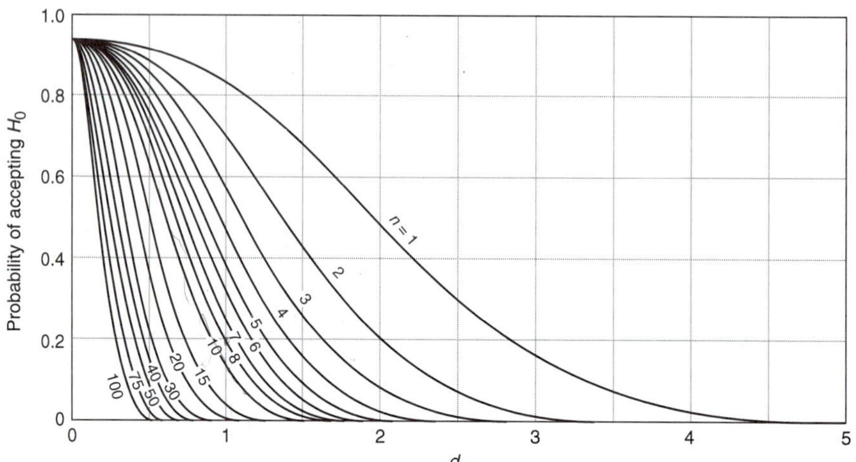

(a) OC curves for different values of n for the two-sided normal test for a level of significance $\alpha = 0.05$.

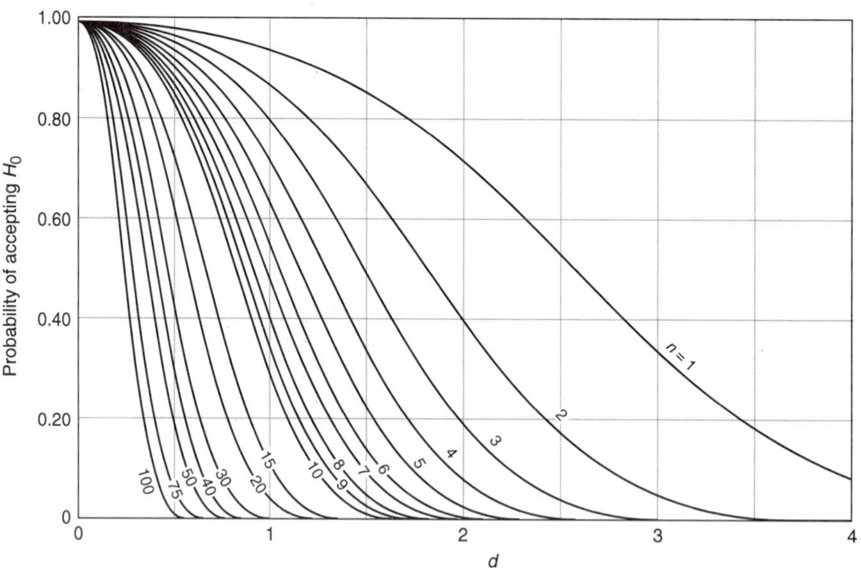

(b) OC curves for different values of n for the two-sided normal test for a level of significance $\alpha = 0.01$.

Source: Charts VI*a*, *e*, *f*, *k*, *m*, and *q* are reproduced with permission from "Operating Characteristics for the Common Statistical Tests of Significance," by C. L. Ferris, F. E. Grubbs, and C. L. Weaver, *Annals of Mathematical Statistics*, June 1946.

Charts VI*b*, *c*, *d*, *g*, *h*, *i*, *j*, *l*, *n*, *o*, *p*, and *r* are reproduced with permission from *Engineering Statistics*, 2nd edition, by A. H. Bowker and G. J. Lieberman, Prentice-Hall, Englewood Cliffs, NJ, 1972.

Chart VI Operating Characteristic Curves (*continued*)

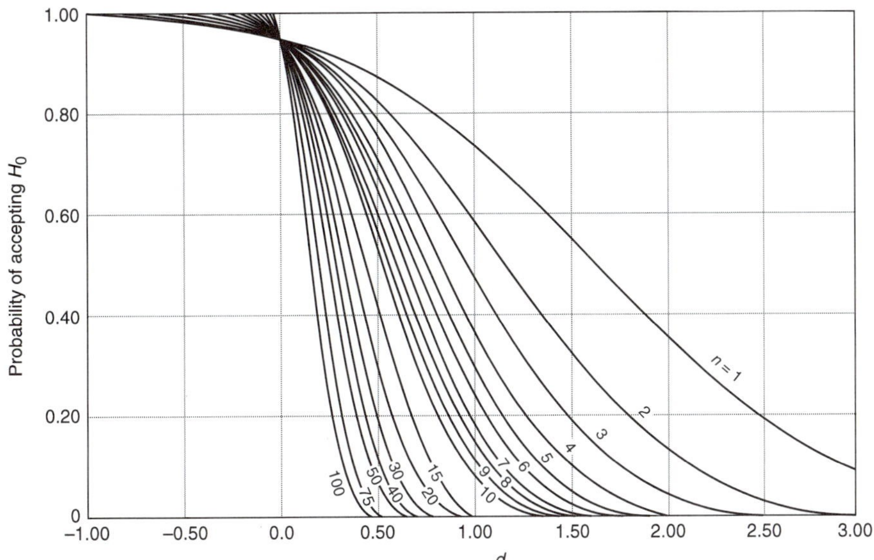

(c) OC curves for different values of n for the one-sided normal test for a level of significance $\alpha = 0.05$.

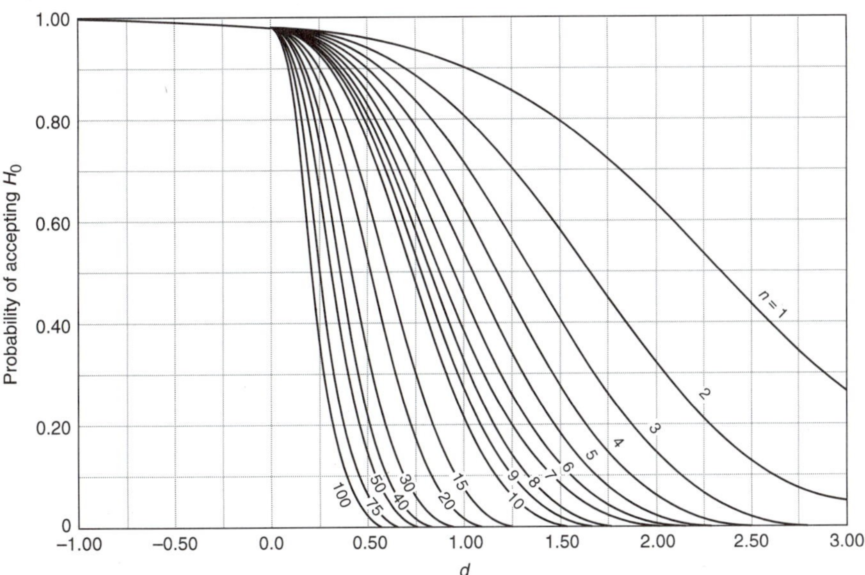

(d) OC curves for different values of n for the one-sided normal test for a level of significance $\alpha = 0.01$.

(*continues*)

Chart VI Operating Characteristic Curves (*continued*)

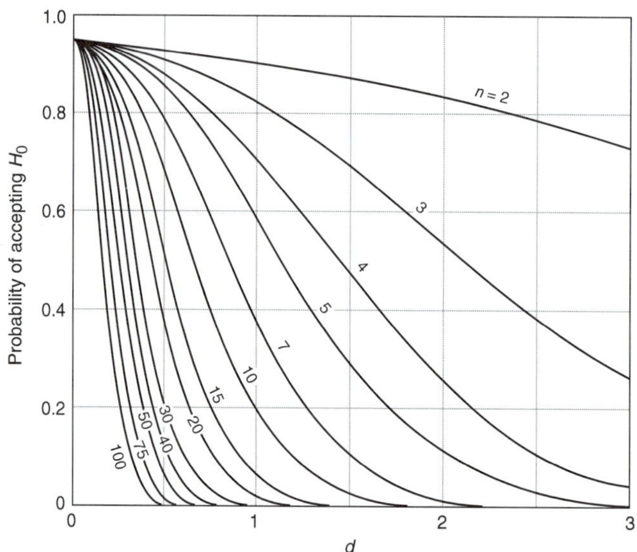

(e) OC curves for different values of n for the two-sided t test for a level of significance $\alpha = 0.05$.

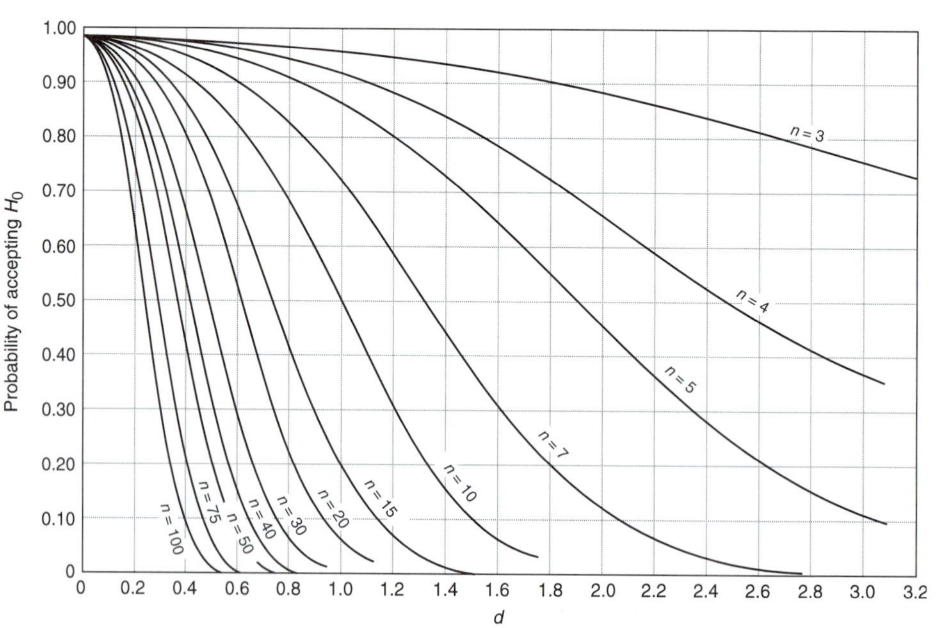

(f) OC curves for different values of n for the two-sided t test for a level of significance $\alpha = 0.01$.

Chart VI Operating Characteristic Curves (*continued*)

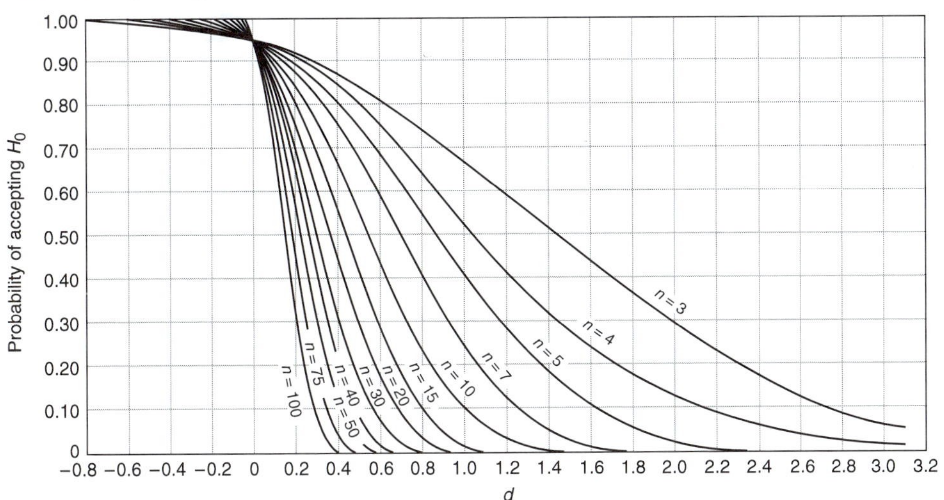

(*g*) OC curves for different values of *n* for the one-sided *t* test for a level of significance $\alpha = 0.05$.

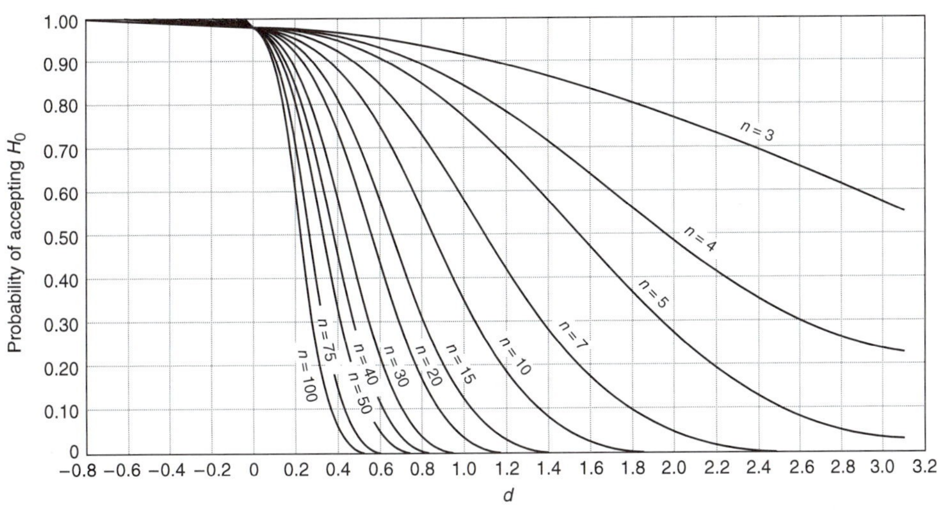

(*h*) OC curves for different values of *n* for the one-sided *t* test for a level of significance $\alpha = 0.01$.

(*continues*)

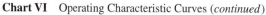

Chart VI Operating Characteristic Curves (*continued*)

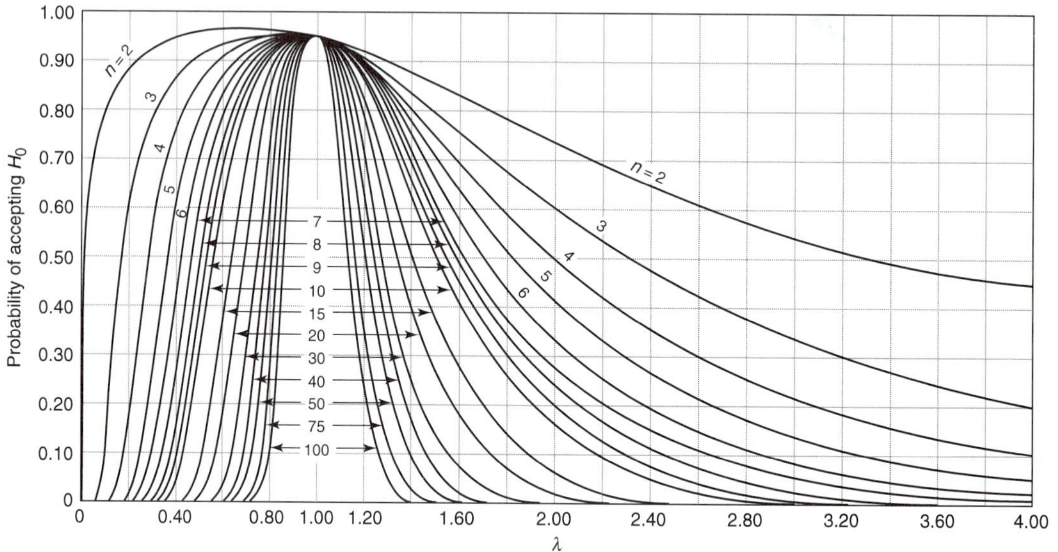

(*i*) OC curves for different values of *n* for the two-sided chi-square test for a level of significance $\alpha = 0.05$.

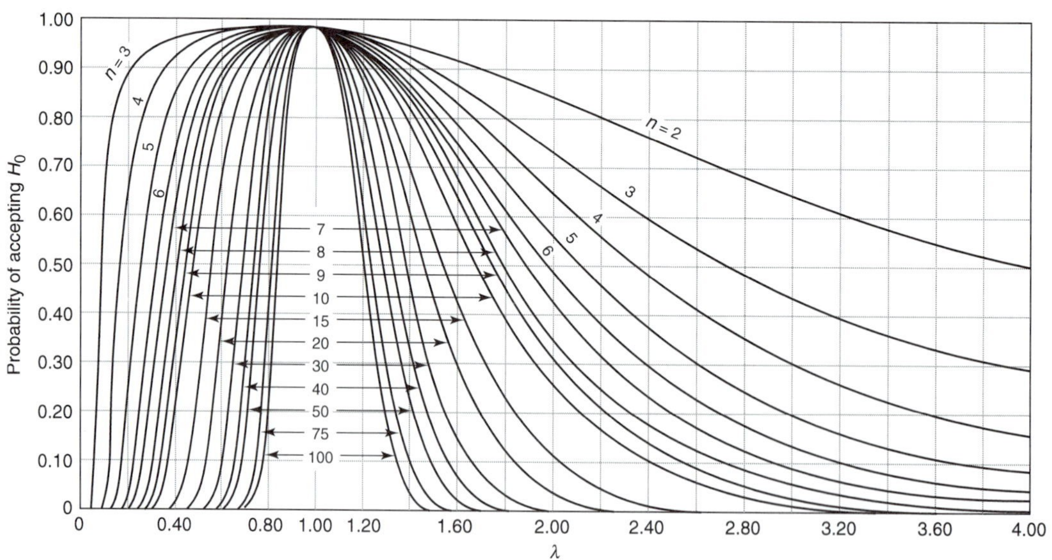

(*j*) OC curves for different values of *n* for the two-sided chi-square test for a level of significance $\alpha = 0.01$.

Chart VI Operating Characteristic Curves (*continued*)

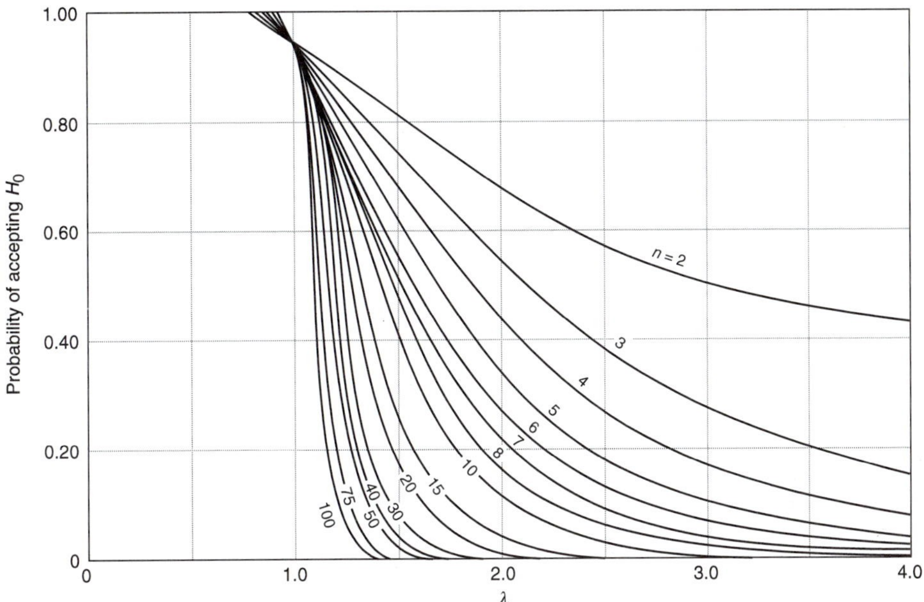

(*k*) OC curves for different values of *n* for the one-sided (upper tail) chi-square test for a level of significance $\alpha = 0.05$.

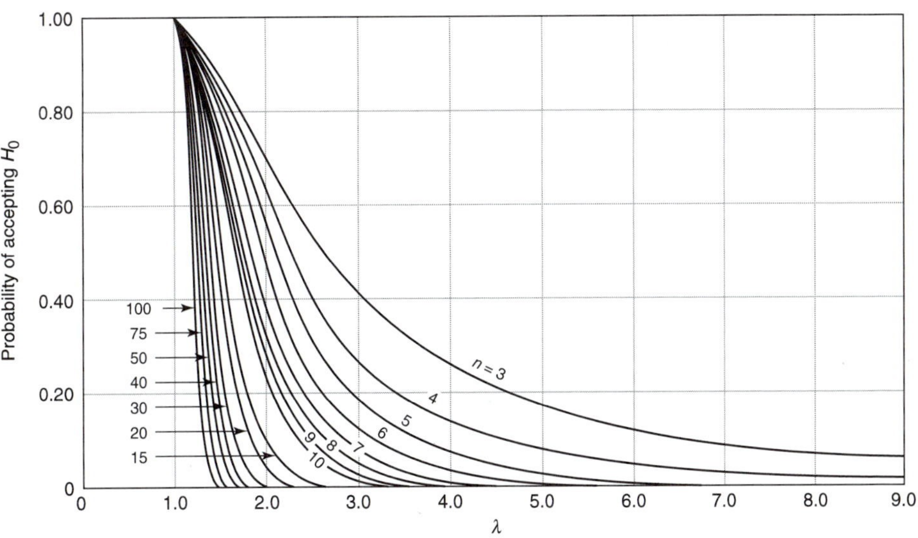

(*l*) OC curves for different values of *n* for the one-sided (upper tail) chi-square test for a level of significance $\alpha = 0.01$.

(*continues*)

Chart VI Operating Characteristic Curves (*continued*)

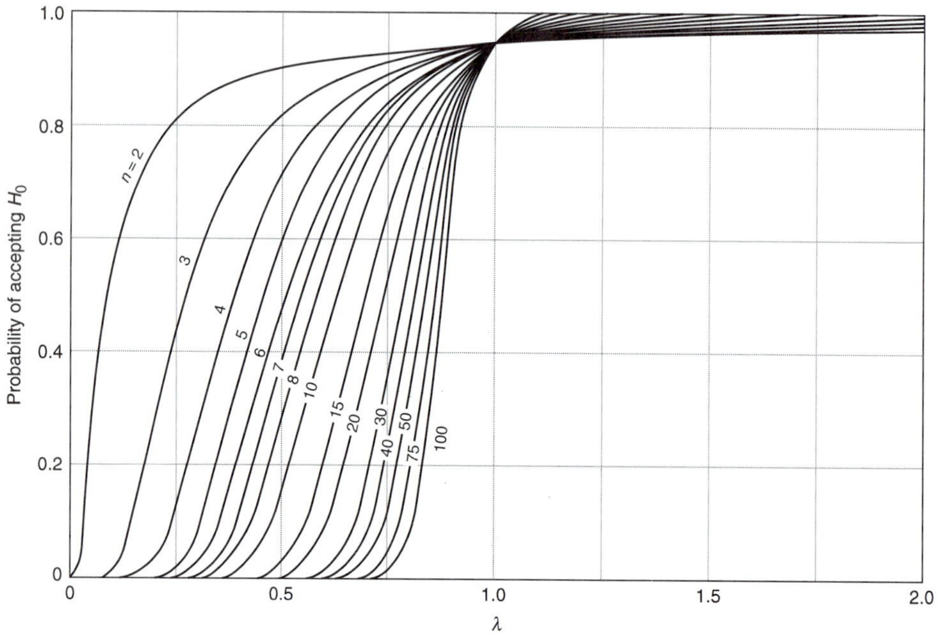

(m) OC curves for different values of n for the one-sided (lower tail) chi-square test for a level of significance $\alpha = 0.05$.

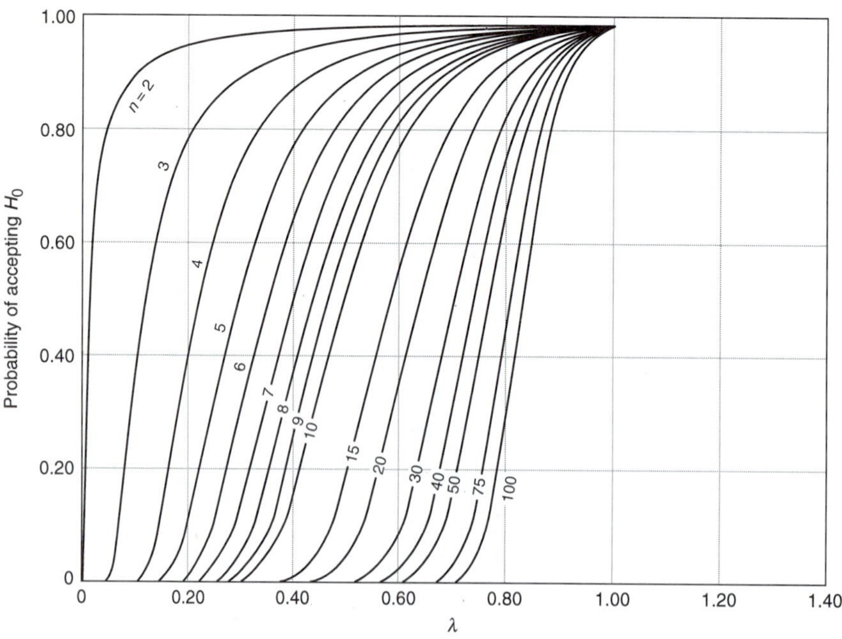

(n) OC curves for different values of n for the one-sided (lower tail) chi-square test for a level of significance $\alpha = 0.01$.

Chart VI Operating Characteristic Curves (*continued*)

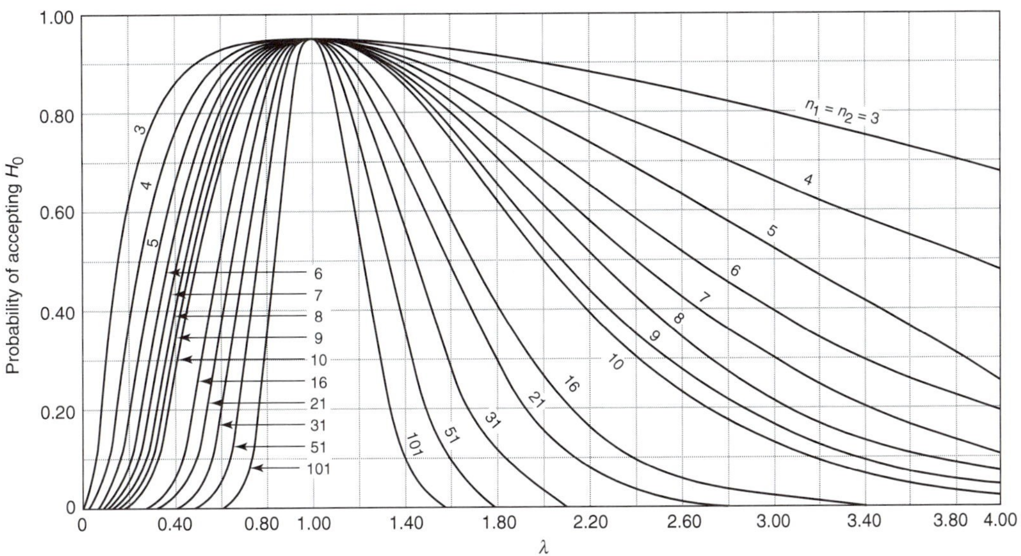

(*o*) OC curves for different values of *n* for the two-sided *F*-test for a level of significance $\alpha = 0.05$.

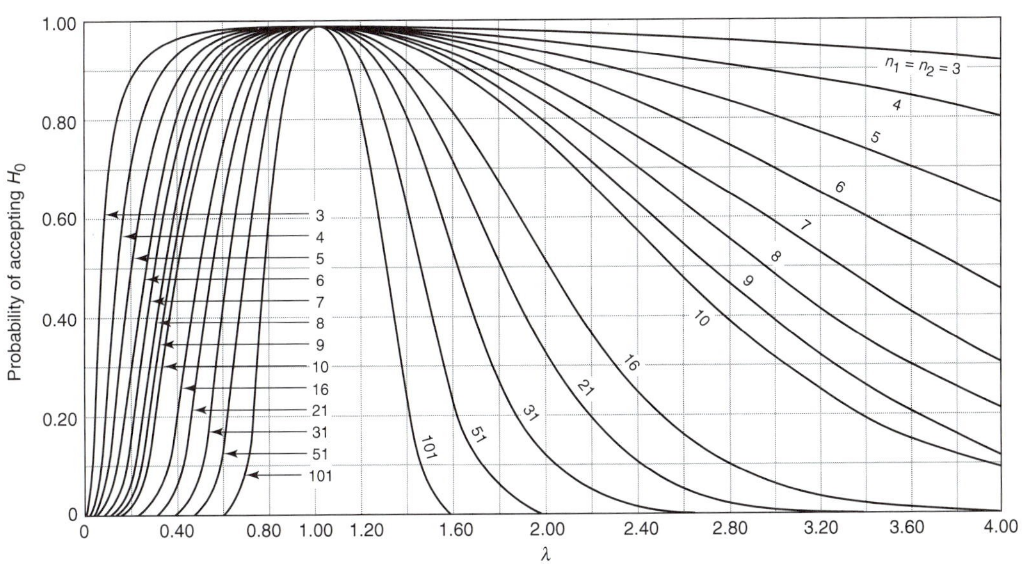

(*p*) OC curves for different values of *n* for the two-sided *F*-test for a level of significance $\alpha = 0.01$

(*continues*)

Chart VI Operating Characteristic Curves (*continued*)

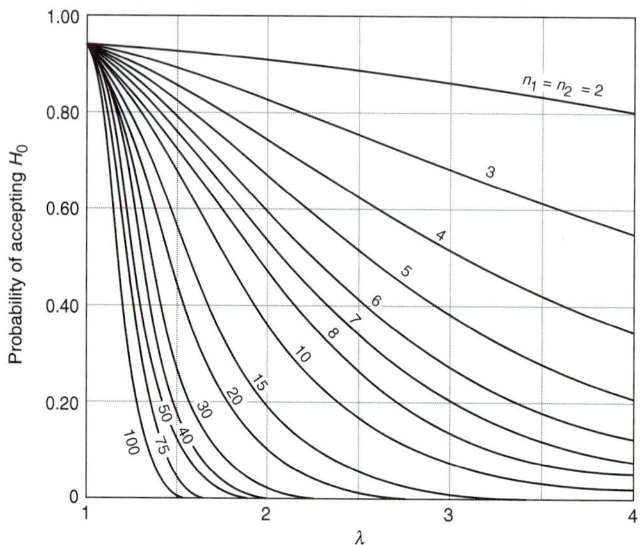

(q) OC curves for different values of n for the one-sided F-test for a level of significance $\alpha = 0.05$.

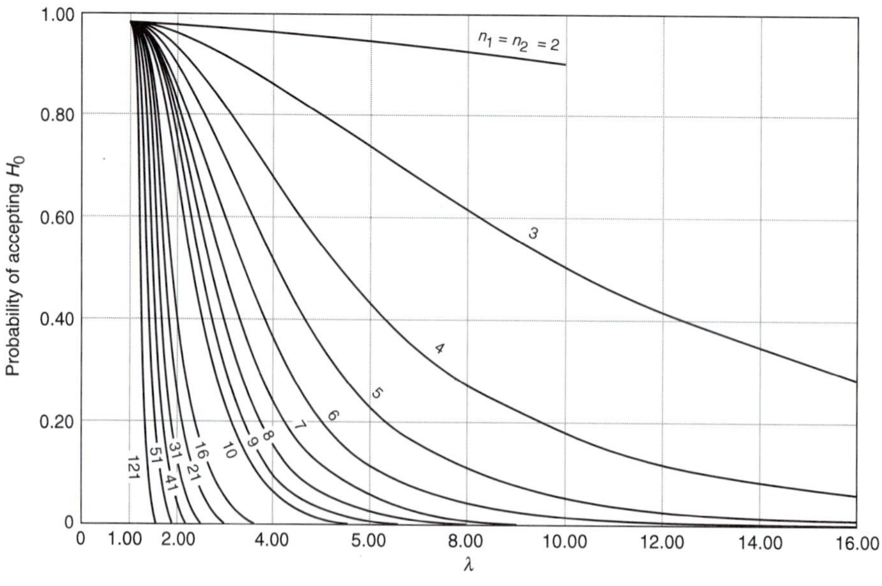

(r) OC curves for different values of n for the one-sided F-test for a level of significance $\alpha = 0.01$.

Chart VII Operating Characteristic Curves for the Fixed-Effects Model Analysis of Variance

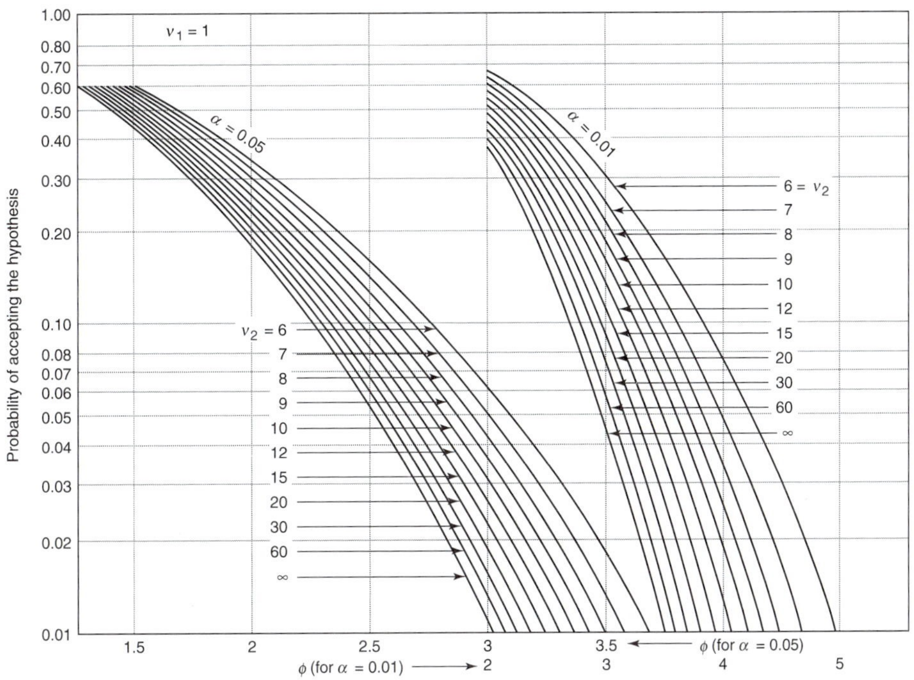

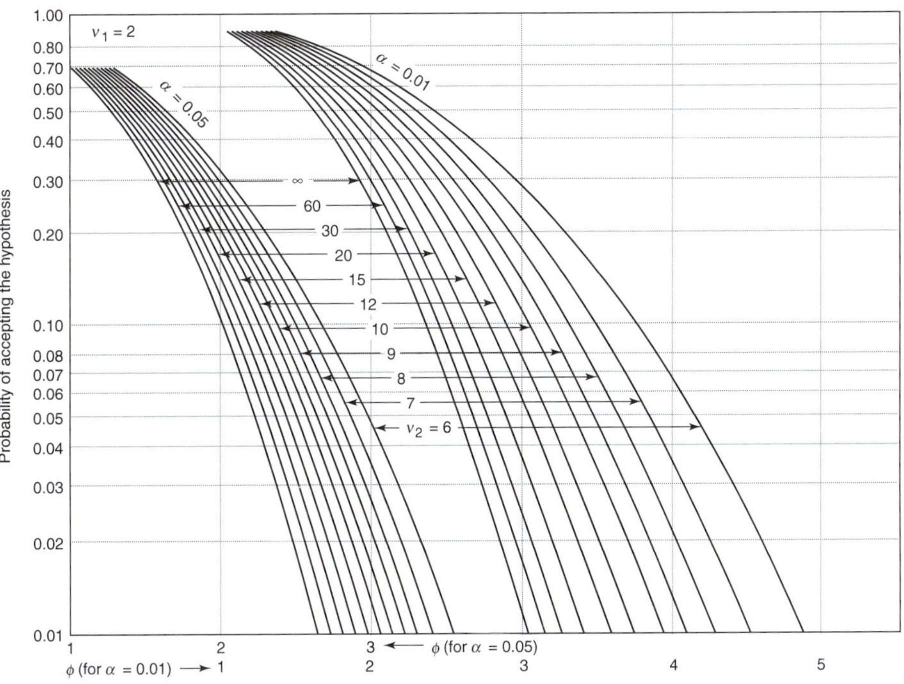

ν_1 = numerator degrees of freedom. ν_2 = denominator degrees of freedom.

Source: Chart VII is adapted with permission from *Biometrika Tables for Statisticians*, Vol. 2, by E. S. Pearson and H. O. Hartley, Cambridge University Press, Cambridge, 1972.

(*continues*)

Chart VII Operating Characteristic Curves for the Fixed-Effects Model Analysis of Variance (*continued*)

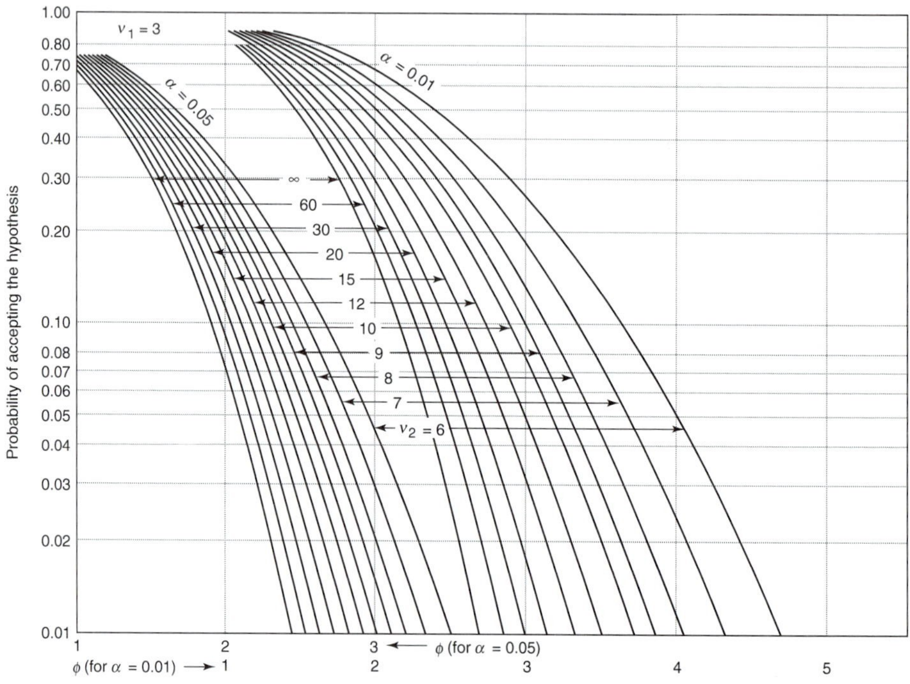

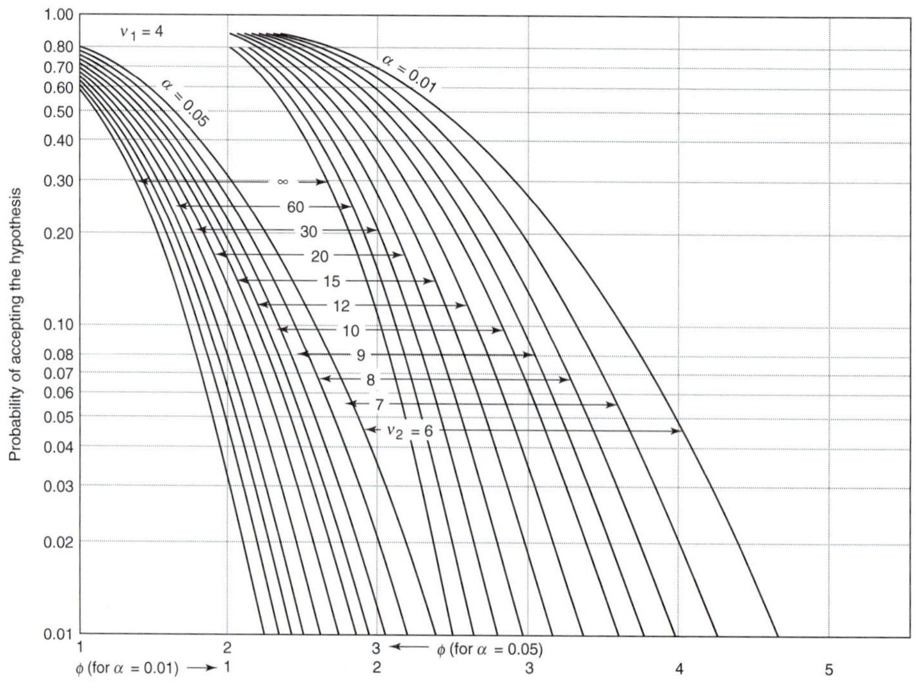

Chart VII Operating Characteristic Curves for the Fixed-Effects Model Analysis of Variance (*continued*)

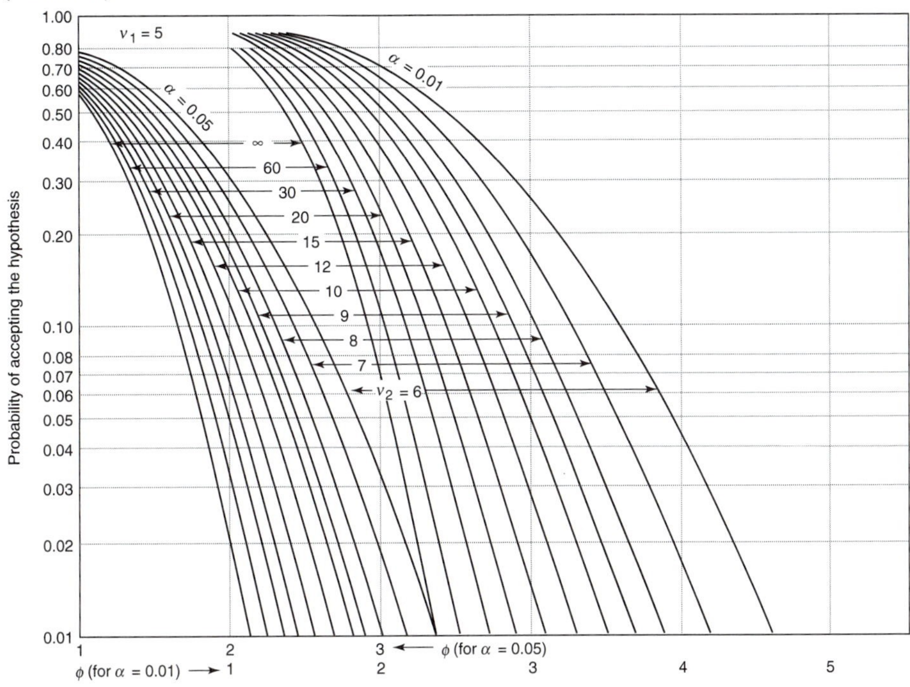

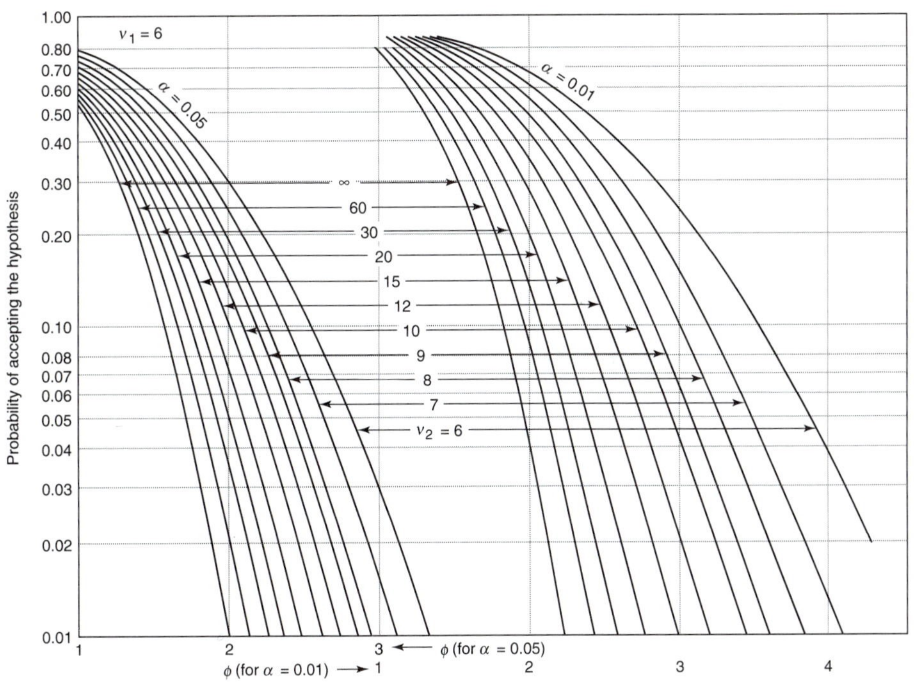

(continues)

Chart VII Operating Characteristic Curves for the Fixed-Effects Model Analysis of Variance (*continued*)

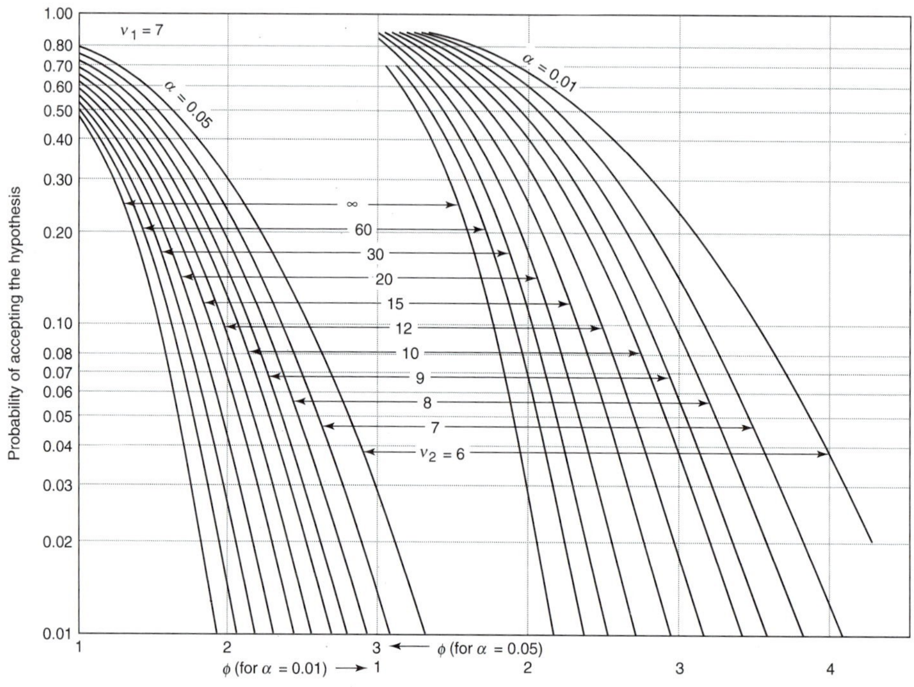

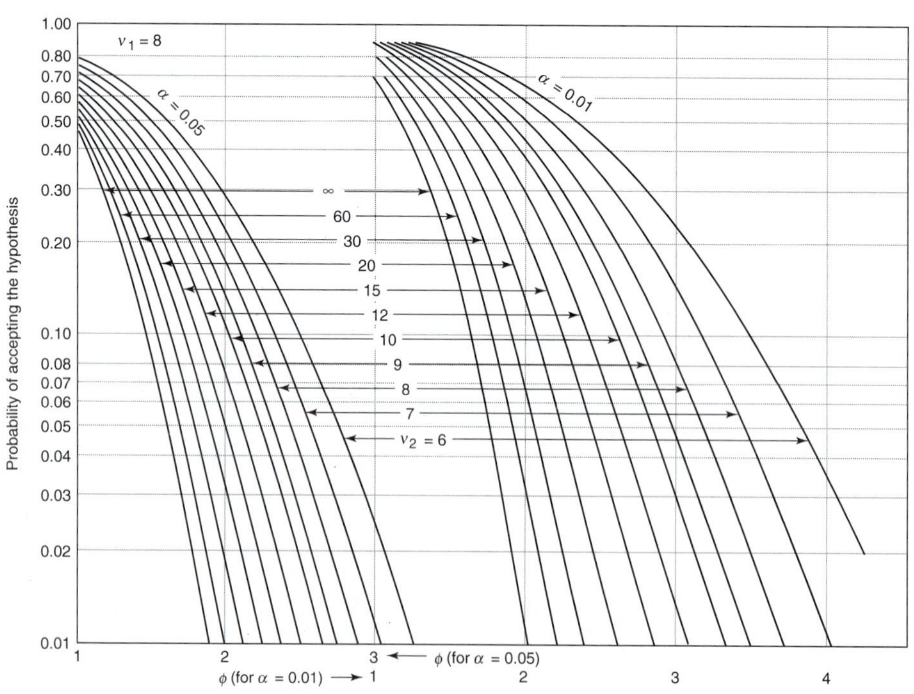

Chart VIII Operating Characteristic Curves for the Random-Effects Model Analysis of Variance

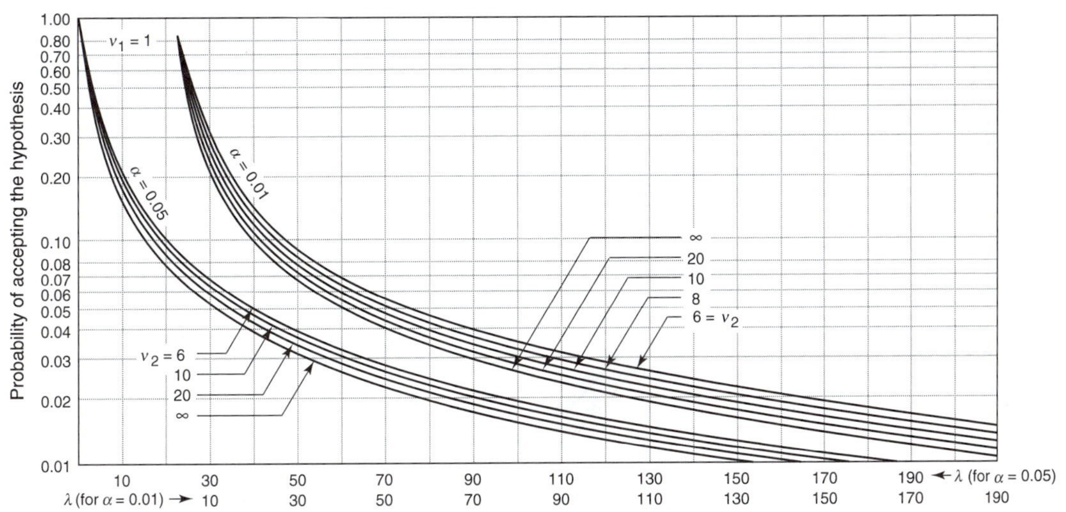

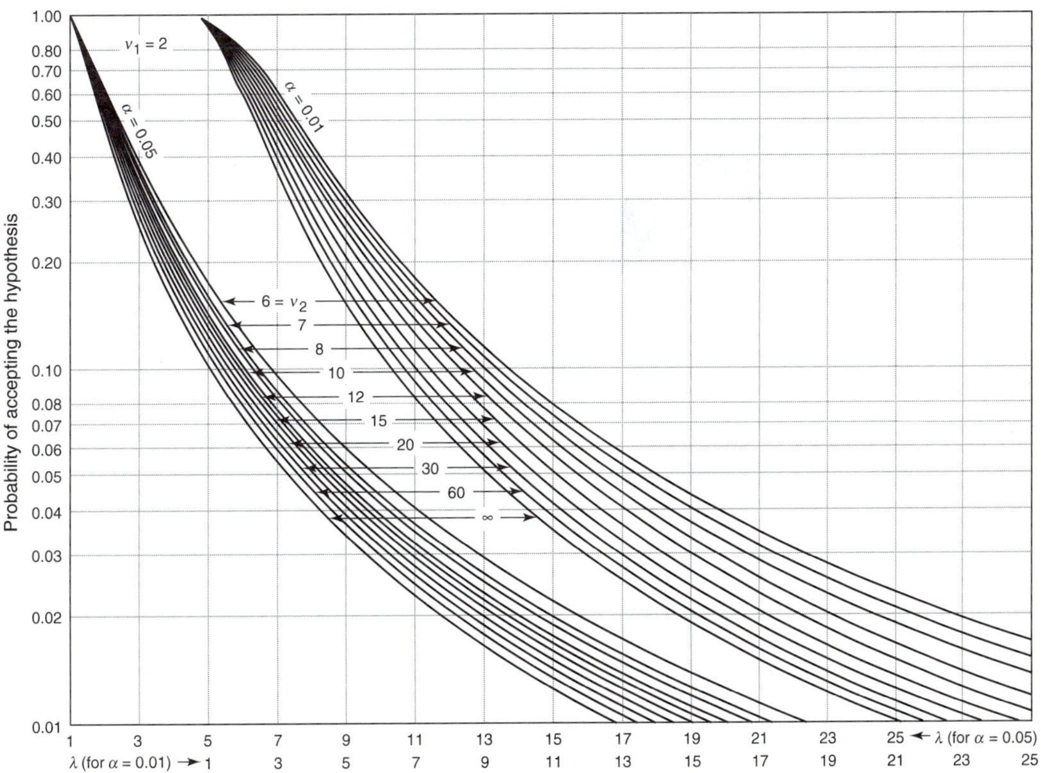

Source: Reproduced with permission from *Engineering Statistics*, 2nd edition, by A. H. Bowker and G. J. Lieberman, Prentice-Hall, Englewood Cliffs, NJ, 1972.

(*continues*)

Chart VIII Operating Characteristic Curves for the Random-Effects Model Analysis of Variance (*continued*)

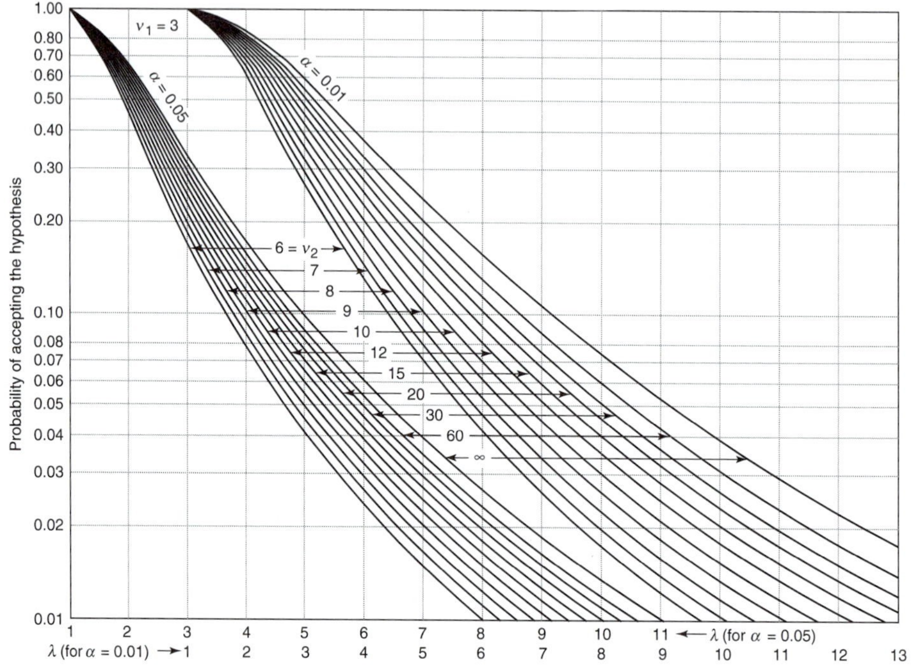

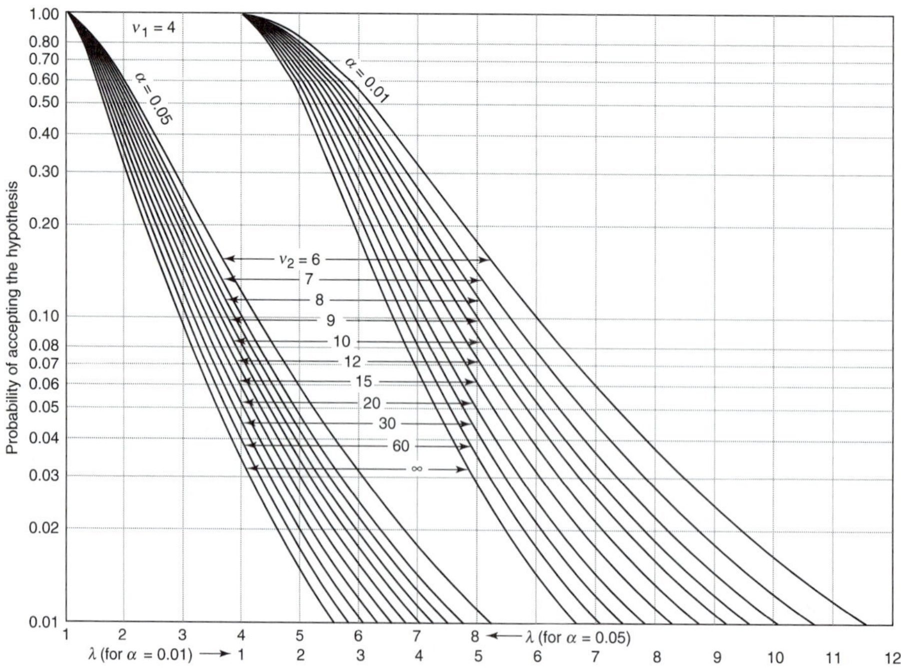

Chart VIII Operating Characteristic Curves for the Random-Effects Model Analysis of Variance (*continued*)

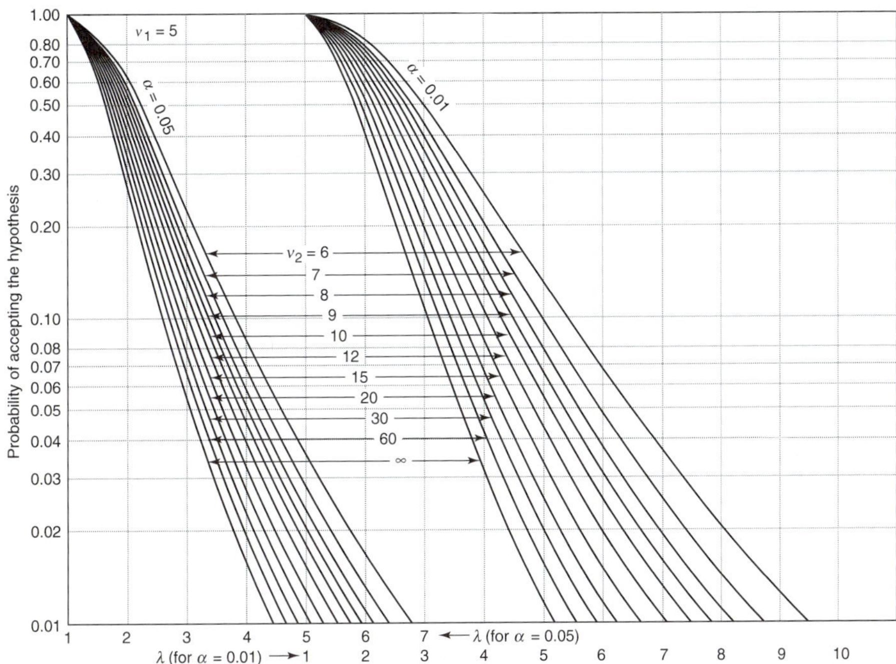

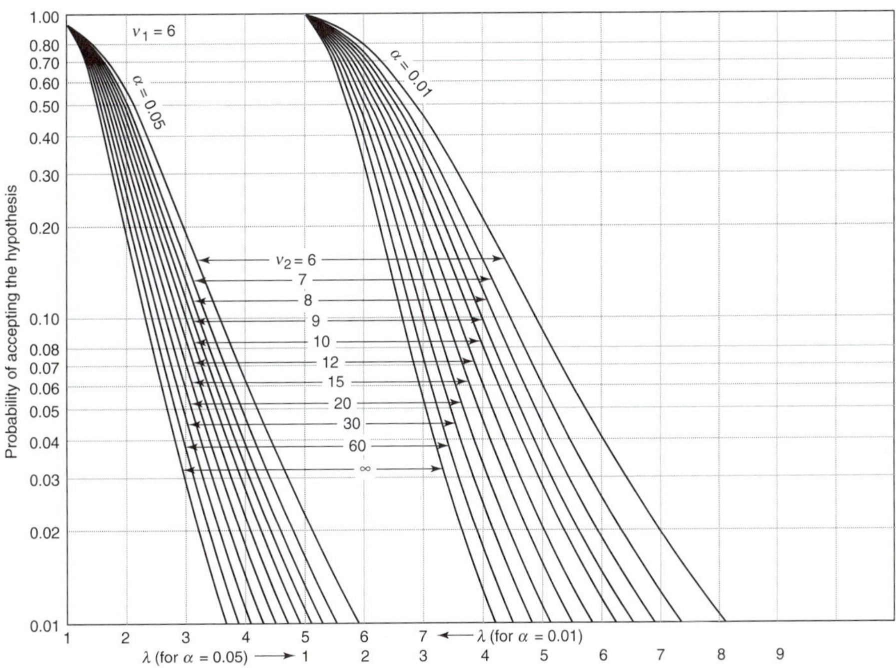

(*continues*)

Chart VIII Operating Characteristic Curves for the Random-Effects Model Analysis of Variance (*continued*)

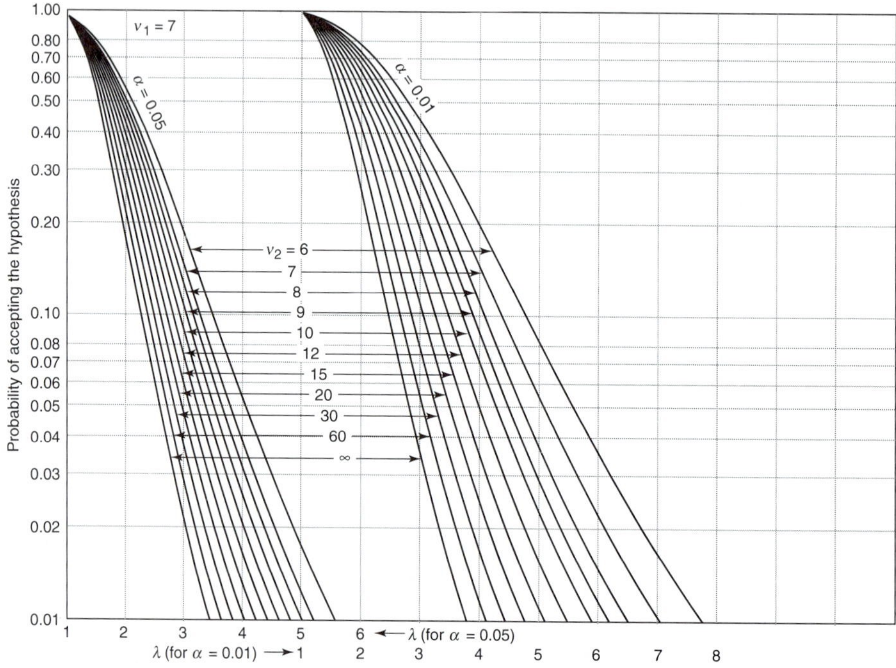

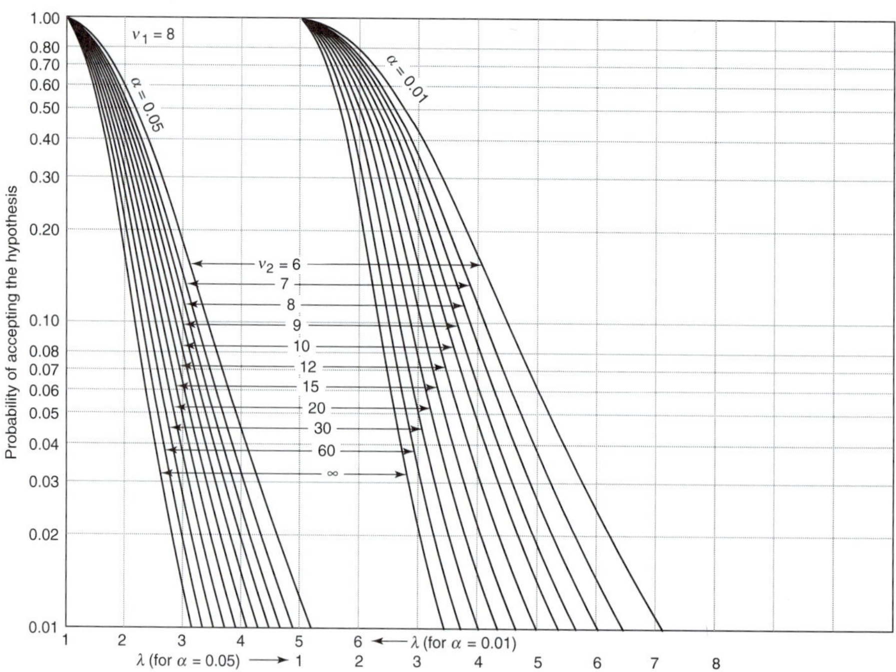

Table IX Critical Values for the Wilcoxon Two-Sample Test[a]

$R^*_{0.05}$

$n_2 \backslash n_1$	2	3	4	5	6	7	8	9	10	11	12	13	14	15
4			10											
5		6	11	17										
6		7	12	18	26									
7		7	13	20	27	36								
8	3	8	14	21	29	38	49							
9	3	8	15	22	31	40	51	63						
10	3	9	15	23	32	42	53	65	78					
11	4	9	16	24	34	44	55	68	81	96				
12	4	10	17	26	35	46	58	71	85	99	115			
13	4	10	18	27	37	48	60	73	88	103	119	137		
14	4	11	19	28	38	50	63	76	91	106	123	141	160	
15	4	11	20	29	40	52	65	79	94	110	127	145	164	185
16	4	12	21	31	42	54	67	82	97	114	131	150	169	
17	5	12	21	32	43	56	70	84	100	117	135	154		
18	5	13	22	33	45	58	72	87	103	121	139			
19	5	13	23	34	46	60	74	90	107	124				
20	5	14	24	35	48	62	77	93	110					
21	6	14	25	37	50	64	79	95						
22	6	15	26	38	51	66	82							
23	6	15	27	39	53	68								
24	6	16	28	40	55									
25	6	16	28	42										
26	7	17	29											
27	7	17												
28	7													

Source: Reproduced with permission from "The Use of Ranks in a Test of Significance for Comparing Two Treatments," by C. White, *Biometrics*, 1952, Vol. 8, p. 37.

[a]For large n_1 and n_2, R is approximately normally distributed, with mean $n_1(n_1 + n_2 + 1)/2$ and variance $n_1n_2(n_1 + n_2 + 1)/12$.

(*continues*)

Table IX Critical Values for the Wilcoxon Two-Sample Test[a] (*continued*)

$R^*_{0.01}$

n_2 \ n_1	2	3	4	5	6	7	8	9	10	11	12	13	14	15
5				15										
6			10	16	23									
7			10	17	24	32								
8			11	17	25	34	43							
9		6	11	18	26	35	45	56						
10		6	12	19	27	37	47	58	71					
11		6	12	20	28	38	49	61	74	87				
12		7	13	21	30	40	51	63	76	90	106			
13		7	14	22	31	41	53	65	79	93	109	125		
14		7	14	22	32	43	54	67	81	96	112	129	147	
15		8	15	23	33	44	56	70	84	99	115	133	151	171
16		8	15	24	34	46	58	72	86	102	119	137	155	
17		8	16	25	36	47	60	74	89	105	122	140		
18		8	16	26	37	49	62	76	92	108	125			
19	3	9	17	27	38	50	64	78	94	111				
20	3	9	18	28	39	52	66	81	97					
21	3	9	18	29	40	53	68	83						
22	3	10	19	29	42	55	70							
23	3	10	19	30	43	57								
24	3	10	20	31	44									
25	3	11	20	32										
26	3	11	21											
27	4	11												
28	4													

Table X Critical Values for the Sign Test[a]

		R^*_α					
n \ α	0.10	0.05	0.01	n \ α	0.10	0.05	0.01
5	0			23	7	6	4
6	0	0		24	7	6	5
7	0	0		25	7	7	5
8	1	0	0	26	8	7	6
9	1	1	0	27	8	7	6
10	1	1	0	28	9	8	6
11	2	1	0	29	9	8	7
12	2	2	1	30	10	9	7
13	3	2	1	31	10	9	7
14	3	2	1	32	10	9	8
15	3	3	2	33	11	10	8
16	4	3	2	34	11	10	9
17	4	4	2	35	12	11	9
18	5	4	3	36	12	11	9
19	5	4	3	37	13	12	10
20	5	5	3	38	13	12	10
21	6	5	4	39	13	12	11
22	6	5	4	40	14	13	11

[a]For $n > 40$, R is approximately normally distributed, with mean $n/2$ and variance $n/4$.

Table XI Critical Values for the Wilcoxon Signed Rank Test[a]

n \ α	0.10	0.05	0.02	0.01	n \ α	0.10	0.05	0.02	0.01
4					28	130	116	101	91
5	0				29	140	126	110	100
6	2	0			30	151	137	120	109
7	3	2	0		31	163	147	130	118
8	5	3	1	0	32	175	159	140	128
9	8	5	3	1	33	187	170	151	138
10	10	8	5	3	34	200	182	162	148
11	13	10	7	5	35	213	195	173	159
12	17	13	9	7	36	227	208	185	171
13	21	17	12	9	37	241	221	198	182
14	25	21	15	12	38	256	235	211	194
15	30	25	19	15	39	271	249	224	207
16	35	29	23	19	40	286	264	238	220
17	41	34	27	23	41	302	279	252	233
18	47	40	32	27	42	319	294	266	247
19	53	46	37	32	43	336	310	281	261
20	60	52	43	37	44	353	327	296	276
21	67	58	49	42	45	371	343	312	291
22	75	65	55	48	46	389	361	328	307
23	83	73	62	54	47	407	378	345	322
24	91	81	69	61	48	426	396	362	339
25	100	89	76	68	49	446	415	379	355
26	110	98	84	75	50	466	434	397	373
27	119	107	92	83					

Source: Adapted with permission from "Extended Tables of the Wilcoxon Matched Pair Signed Rank Statistic" by Robert L. McCornack, *Journal of the American Statistical Association*, Vol. 60, September, 1965.

[a]If $n > 50$, R is approximately normally distributed, with mean $n(n + 1)/4$ and variance $n(n + 1)(2n + 1)/24$.

Table XII Percentage Points of the Studentized Range Statistic[a]

	$q_{0.01}(p, f)$																		
	p																		
f	2	3	4	5	6	7	8	9	10	11	12	13	14	15	16	17	18	19	20
1	90.0	135	164	186	202	216	227	237	246	253	260	266	272	272	282	286	290	294	298
2	14.0	19.0	22.3	24.7	26.6	28.2	29.5	30.7	31.7	32.6	33.4	31.4	34.8	35.4	36.0	36.5	37.0	37.5	37.9
3	8.26	10.6	12.2	13.3	14.2	15.0	15.6	16.2	16.7	17.1	17.5	17.9	18.2	18.5	18.8	19.1	19.3	19.5	19.8
4	6.51	8.12	9.17	9.96	10.6	11.1	11.5	11.9	12.3	12.6	12.8	13.1	13.3	13.5	13.7	13.9	14.1	14.2	14.4
5	5.70	6.97	7.80	8.42	8.91	9.32	9.67	9.97	10.24	10.48	10.70	10.89	11.08	11.24	11.40	11.55	11.68	11.81	11.93
6	5.24	6.33	7.03	7.56	7.97	8.32	8.61	8.87	9.10	9.30	9.49	9.65	9.81	9.95	10.08	10.21	10.32	10.43	10.54
7	4.95	5.92	6.54	7.01	7.37	7.68	7.94	8.17	8.37	8.55	8.71	8.86	9.00	9.12	9.24	9.35	9.46	9.55	9.65
8	4.74	5.63	6.20	6.63	6.96	7.24	7.47	7.68	7.87	8.03	8.18	8.31	8.44	8.55	8.66	8.76	8.85	8.94	9.03
9	4.60	5.43	5.96	6.35	6.66	6.91	7.13	7.32	7.49	7.65	7.78	7.91	8.03	8.13	8.23	8.32	8.41	8.49	8.57
10	4.48	5.27	5.77	6.14	6.43	6.67	6.87	7.05	7.21	7.36	7.48	7.60	7.71	7.81	7.91	7.99	8.07	8.15	8.22
11	4.39	5.14	5.62	5.97	6.25	6.48	6.67	6.84	6.99	7.13	7.25	7.36	7.46	7.56	7.65	7.73	7.81	7.88	7.95
12	4.32	5.04	5.50	5.84	6.10	6.32	6.51	6.67	6.81	6.94	7.06	7.17	7.26	7.36	7.44	7.52	7.59	7.66	7.73
13	4.26	4.96	5.40	5.73	5.98	6.19	6.37	6.53	6.67	6.79	6.90	7.01	7.10	7.19	7.27	7.34	7.42	7.48	7.55
14	4.21	4.89	5.32	5.63	5.88	6.08	6.26	6.41	6.54	6.66	6.77	6.87	6.96	7.05	7.12	7.20	7.27	7.33	7.39
15	4.17	4.83	5.25	5.56	5.80	5.99	6.16	6.31	6.44	6.55	6.66	6.76	6.84	6.93	7.00	7.07	7.14	7.20	7.26
16	4.13	4.78	5.19	5.49	5.72	5.92	6.08	6.22	6.35	6.46	6.56	6.66	6.74	6.82	6.90	6.97	7.03	7.09	7.15
17	4.10	4.74	5.14	5.43	5.66	5.85	6.01	6.15	6.27	6.38	6.48	6.57	6.66	6.73	6.80	6.87	6.94	7.00	7.05
18	4.07	4.70	5.09	5.38	5.60	5.79	5.94	6.08	6.20	6.31	6.41	6.50	6.58	6.65	6.72	6.79	6.85	6.91	6.96
19	4.05	4.67	5.05	5.33	5.55	5.73	5.89	6.02	6.14	6.25	6.34	6.43	6.51	6.58	6.65	6.72	6.78	6.84	6.89
20	4.02	4.64	5.02	5.29	5.51	5.69	5.84	5.97	6.09	6.19	6.29	6.37	6.45	6.52	6.59	6.65	6.71	6.76	6.82
24	3.96	4.54	4.91	5.17	5.37	5.54	5.69	5.81	5.92	6.02	6.11	6.19	6.26	6.33	6.39	6.45	6.51	6.56	6.61
30	3.89	4.45	4.80	5.05	5.24	5.40	5.54	5.65	5.76	5.85	5.93	6.01	6.08	6.14	6.20	6.26	6.31	6.36	6.41
40	3.82	4.37	4.70	4.93	5.11	5.27	5.39	5.50	5.60	5.69	5.77	5.84	5.90	5.96	6.02	6.07	6.12	6.17	6.21
60	3.76	4.28	4.60	4.82	4.99	5.13	5.25	5.36	5.45	5.53	5.60	5.67	5.73	5.79	5.84	5.89	5.93	5.98	6.02
120	3.70	4.20	4.50	4.71	4.87	5.01	5.12	5.21	5.30	5.38	5.44	5.51	5.56	5.61	5.66	5.71	5.75	5.79	5.83
∞	3.64	4.12	4.40	4.60	4.76	4.88	4.99	5.08	5.16	5.23	5.29	5.35	5.40	5.45	5.49	5.54	5.57	5.61	5.65

f = degress of freedom.

[a]From J. M. May, "Extended and Corrected Tables of the Upper Percentage Points of the Studentized Range," *Biometrika*, Vol. 39, pp. 192–193, 1952. Reproduced by permission of the trustees of Biometrika.

(continues)

Table XII Percentage Points of the Studentized Range Statistic[a] (*continued*)

$q_{0.05}(p, f)$

f	p = 2	3	4	5	6	7	8	9	10	11	12	13	14	15	16	17	18	19	20
1	18.1	26.7	32.8	37.2	40.5	43.1	45.4	47.3	49.1	50.6	51.9	53.2	54.3	55.4	56.3	57.2	58.0	58.8	59.6
2	6.09	8.28	9.80	10.89	11.73	12.43	13.03	13.54	13.99	14.39	14.75	15.08	15.38	15.65	15.91	16.14	16.36	16.57	16.77
3	4.50	5.88	6.83	7.51	8.04	8.47	8.85	9.18	9.46	9.72	9.95	10.16	10.35	10.52	10.69	10.84	10.98	11.12	11.24
4	3.93	5.00	5.76	6.31	6.73	7.06	7.35	7.60	7.83	8.03	8.21	8.37	8.52	8.67	8.80	8.92	9.03	9.14	9.24
5	3.64	4.60	5.22	5.67	6.03	6.33	6.58	6.80	6.99	7.17	7.32	7.47	7.60	7.72	7.83	7.93	8.03	8.12	8.21
6	3.46	4.34	4.90	5.31	5.63	5.89	6.12	6.32	6.49	6.65	6.79	6.92	7.04	7.14	7.24	7.34	7.43	7.51	7.59
7	3.34	4.16	4.68	5.06	5.35	5.59	5.80	5.99	6.15	6.29	6.42	6.54	6.65	6.75	6.84	6.93	7.01	7.08	7.16
8	3.26	4.04	4.53	4.89	5.17	5.40	5.60	5.77	5.92	6.05	6.18	6.29	6.39	6.48	6.57	6.65	6.73	6.80	6.87
9	3.20	3.95	4.42	4.76	5.02	5.24	5.43	5.60	5.74	5.87	5.98	6.09	6.19	6.28	6.36	6.44	6.51	6.58	6.65
10	3.15	3.88	4.33	4.66	4.91	5.12	5.30	5.46	5.60	5.72	5.83	5.93	6.03	6.12	6.20	6.27	6.34	6.41	6.47
11	3.11	3.82	4.26	4.58	4.82	5.03	5.20	5.35	5.49	5.61	5.71	5.81	5.90	5.98	6.06	6.14	6.20	6.27	6.33
12	3.08	3.77	4.20	4.51	4.75	4.95	5.12	5.27	5.40	5.51	5.61	5.71	5.80	5.88	5.95	6.02	6.09	6.15	6.21
13	3.06	3.73	4.15	4.46	4.69	4.88	5.05	5.19	5.32	5.43	5.53	5.63	5.71	5.79	5.86	5.93	6.00	6.06	6.11
14	3.03	3.70	4.11	4.41	4.64	4.83	4.99	5.13	5.25	5.36	5.46	5.56	5.64	5.72	5.79	5.86	5.92	5.98	6.03
15	3.01	3.67	4.08	4.37	4.59	4.78	4.94	5.08	5.20	5.31	5.40	5.49	5.57	5.65	5.72	5.79	5.85	5.91	5.96
16	3.00	3.65	4.05	4.34	4.56	4.74	4.90	5.03	5.15	5.26	5.35	5.44	5.52	5.59	5.66	5.73	5.79	5.84	5.90
17	2.98	3.62	4.02	4.31	4.52	4.70	4.86	4.99	5.11	5.21	5.31	5.39	5.47	5.55	5.61	5.68	5.74	5.79	5.84
18	2.97	3.61	4.00	4.28	4.49	4.67	4.83	4.96	5.07	5.17	5.27	5.35	5.43	5.50	5.57	5.63	5.69	5.74	5.79
19	2.96	3.59	3.98	4.26	4.47	4.64	4.79	4.92	5.04	5.14	5.23	5.32	5.39	5.46	5.53	5.59	5.65	5.70	5.75
20	2.95	3.58	3.96	4.24	4.45	4.62	4.77	4.90	5.01	5.11	5.20	5.28	5.36	5.43	5.50	5.56	5.61	5.66	5.71
24	2.92	3.53	3.90	4.17	4.37	4.54	4.68	4.81	4.92	5.01	5.10	5.18	5.25	5.32	5.38	5.44	5.50	5.55	5.59
30	2.89	3.48	3.84	4.11	4.30	4.46	4.60	4.72	4.83	4.92	5.00	5.08	5.15	5.21	5.27	5.33	5.38	5.43	5.48
40	2.86	3.44	3.79	4.04	4.23	4.39	4.52	4.63	4.74	4.82	4.90	4.98	5.05	5.11	5.17	5.22	5.27	5.32	5.36
60	2.83	3.40	3.74	3.98	4.16	4.31	4.44	4.55	4.65	4.73	4.81	4.88	4.94	5.00	5.06	5.11	5.15	5.20	5.24
120	2.80	3.36	3.69	3.92	4.10	4.24	4.36	4.47	4.56	4.64	4.71	4.78	4.84	4.90	4.95	5.00	5.04	5.09	5.13
∞	2.77	3.32	3.63	3.86	4.03	4.17	4.29	4.39	4.47	4.55	4.62	4.68	4.74	4.80	4.84	4.98	4.93	4.97	5.01

Table XIII Factors for Quality-Control Charts

	$\overline{X}$ Chart		R Chart		
	Factors for Control Limits		Factors for Central Line	Factors for Control Limits	
n^a	A_1	A_2	d_2	D_3	D_4
2	3.760	1.880	1.128	0	3.267
3	2.394	1.023	1.693	0	2.575
4	1.880	0.729	2.059	0	2.282
5	1.596	0.577	2.326	0	2.115
6	1.410	0.483	2.534	0	2.004
7	1.277	0.419	2.704	0.076	1.924
8	1.175	0.373	2.847	0.136	1.864
9	1.094	0.337	2.970	0.184	1.816
10	1.028	0.308	3.078	0.223	1.777
11	0.973	0.285	3.173	0.256	1.744
12	0.925	0.266	3.258	0.284	1.716
13	0.884	0.249	3.336	0.308	1.692
14	0.848	0.235	3.407	0.329	1.671
15	0.816	0.223	3.472	0.348	1.652
16	0.788	0.212	3.532	0.364	1.636
17	0.762	0.203	3.588	0.379	1.621
18	0.738	0.194	3.640	0.392	1.608
19	0.717	0.187	3.689	0.404	1.596
20	0.697	0.180	3.735	0.414	1.586
21	0.679	0.173	3.778	0.425	1.575
22	0.662	0.167	3.819	0.434	1.566
23	0.647	0.162	3.858	0.443	1.557
24	0.632	0.157	3.895	0.452	1.548
25	0.619	0.153	3.931	0.459	1.541

[a] $n > 25$; $A_1 = 3/\sqrt{n}$. n = number of observations in sample.

Table XIV *k* Values for One-Sided and Two-Sided Tolerance Intervals

One-Sided Tolerance Intervals									
Confidence Level	0.90			0.95			0.99		
Percent Coverage	0.90	0.95	0.99	0.90	0.95	0.99	0.90	0.95	0.99
2	10.253	13.090	18.500	20.581	26.260	37.094	103.029	131.426	185.617
3	4.258	5.311	7.340	6.155	7.656	10.553	13.995	17.370	23.896
4	3.188	3.957	5.438	4.162	5.144	7.042	7.380	9.083	12.387
5	2.742	3.400	4.666	3.407	4.203	5.741	5.362	6.578	8.939
6	2.494	3.092	4.243	3.006	3.708	5.062	4.411	5.406	7.335
7	2.333	2.894	3.972	2.755	3.399	4.642	3.859	4.728	6.412
8	2.219	2.754	3.783	2.582	3.187	4.354	3.497	4.285	5.812
9	2.133	2.650	3.641	2.454	3.031	4.143	3.240	3.972	5.389
10	2.066	2.568	3.532	2.355	2.911	3.981	3.048	3.738	5.074
11	2.011	2.503	3.443	2.275	2.815	3.852	2.898	3.556	4.829
12	1.966	2.448	3.371	2.210	2.736	3.747	2.777	3.410	4.633
13	1.928	2.402	3.309	2.155	2.671	3.659	2.677	3.290	4.472
14	1.895	2.363	3.257	2.109	2.614	3.585	2.593	3.189	4.337
15	1.867	2.329	3.212	2.068	2.566	3.520	2.521	3.102	4.222
16	1.842	2.299	3.172	2.033	2.524	3.464	2.459	3.028	4.123
17	1.819	2.272	3.137	2.002	2.486	3.414	2.405	2.963	4.037
18	1.800	2.249	3.105	1.974	2.453	3.370	2.357	2.905	3.960
19	1.782	2.227	3.077	1.949	2.423	3.331	2.314	2.854	3.892
20	1.765	2.028	3.052	1.926	2.396	3.295	2.276	2.808	3.832
21	1.750	2.190	3.028	1.905	2.371	3.263	2.241	2.766	3.777
22	1.737	2.174	3.007	1.886	2.349	3.233	2.209	2.729	3.727
23	1.724	2.159	2.987	1.869	2.328	3.206	2.180	2.694	3.681
24	1.712	2.145	2.969	1.853	2.309	3.181	2.154	2.662	3.640
25	1.702	2.132	2.952	1.838	2.292	3.158	2.129	2.633	3.601
30	1.657	2.080	2.884	1.777	2.220	3.064	2.030	2.515	3.447
40	1.598	2.010	2.793	1.697	2.125	2.941	1.902	2.364	3.249
50	1.559	1.965	2.735	1.646	2.065	2.862	1.821	2.269	3.125
60	1.532	1.933	2.694	1.609	2.022	2.807	1.764	2.202	3.038
70	1.511	1.909	2.662	1.581	1.990	2.765	1.722	2.153	2.974
80	1.495	1.890	2.638	1.559	1.964	2.733	1.688	2.114	2.924
90	1.481	1.874	2.618	1.542	1.944	2.706	1.661	2.082	2.883
100	1.470	1.861	2.601	1.527	1.927	2.684	1.639	2.056	2.850

Table XIV *k* Values for One-Sided and Two-Sided Tolerance Intervals (*continued*)

Two-Sided Tolerance Intervals									
Confidence Level	0.90			0.95			0.99		
Percent Coverage	0.90	0.95	0.99	0.90	0.95	0.99	0.90	0.95	0.99
2	15.978	18.800	24.167	32.019	37.674	48.430	160.193	188.491	242.300
3	5.847	6.919	8.974	8.380	9.916	12.861	18.930	22.401	29.055
4	4.166	4.943	6.440	5.369	6.370	8.299	9.398	11.150	14.527
5	3.949	4.152	5.423	4.275	5.079	6.634	6.612	7.855	10.260
6	3.131	3.723	4.870	3.712	4.414	5.775	5.337	6.345	8.301
7	2.902	3.452	4.521	3.369	4.007	5.248	4.613	5.488	7.187
8	2.743	3.264	4.278	3.136	3.732	4.891	4.147	4.936	6.468
9	2.626	3.125	4.098	2.967	3.532	4.631	3.822	4.550	5.966
10	2.535	3.018	3.959	2.839	3.379	4.433	3.582	4.265	5.594
11	2.463	2.933	3.849	2.737	3.259	4.277	3.397	4.045	5.308
12	2.404	2.863	3.758	2.655	3.162	4.150	3.250	3.870	5.079
13	2.355	2.805	3.682	2.587	3.081	4.044	3.130	3.727	4.893
14	2.314	2.756	3.618	2.529	3.012	3.955	3.029	3.608	4.737
15	2.278	2.713	3.562	2.480	2.954	3.878	2.945	3.507	4.605
16	2.246	2.676	3.514	2.437	2.903	3.812	2.872	3.421	4.492
17	2.219	2.643	3.471	2.400	2.858	3.754	2.808	3.345	4.393
18	2.194	2.614	3.433	2.366	2.819	3.702	2.753	3.279	4.307
19	2.172	2.588	3.399	2.337	2.784	3.656	2.703	3.221	4.230
20	2.152	2.564	3.368	2.310	2.752	3.615	2.659	3.168	4.161
21	2.135	2.543	3.340	2.286	2.723	3.577	2.620	3.121	4.100
22	2.118	2.524	3.315	2.264	2.697	3.543	2.584	3.078	4.044
23	2.103	2.506	3.292	2.244	2.673	3.512	2.551	3.040	3.993
24	2.089	2.489	3.270	2.225	2.651	3.483	2.522	3.004	3.947
25	2.077	2.474	3.251	2.208	2.631	3.457	2.494	2.972	3.904
30	2.025	2.413	3.170	2.140	2.529	3.350	2.385	2.841	3.733
40	1.959	2.334	3.066	2.052	2.445	3.213	2.247	2.677	3.518
50	1.916	2.284	3.001	1.996	2.379	3.126	2.162	2.576	3.385
60	1.887	2.248	2.955	1.958	2.333	3.066	2.103	2.506	3.293
70	1.865	2.222	2.920	1.929	2.299	3.021	2.060	2.454	3.225
80	1.848	2.202	2.894	1.907	2.272	2.986	2.026	2.414	3.173
90	1.834	2.185	2.872	1.889	2.251	2.958	1.999	2.382	3.130
100	1.822	2.172	2.854	1.874	2.233	2.934	1.977	2.355	3.096

Table XV Random Numbers

10480	15011	01536	02011	81647	91646	69179	14194	62590
22368	46573	25595	85393	30995	89198	27982	53402	93965
24130	48360	22527	97265	76393	64809	15179	24830	49340
42167	93093	06243	61680	07856	16376	39440	53537	71341
37570	39975	81837	16656	06121	91782	60468	81305	49684
77921	06907	11008	42751	27756	53498	18602	70659	90655
99562	72905	56420	69994	98872	31016	71194	18738	44013
96301	91977	05463	07972	18876	20922	94595	56869	69014
89579	14342	63661	10281	17453	18103	57740	84378	25331
85475	36857	53342	53988	53060	59533	38867	62300	08158
28918	69578	88231	33276	70997	79936	56865	05859	90106
63553	40961	48235	03427	49626	69445	18663	72695	52180
09429	93969	52636	92737	88974	33488	36320	17617	30015
10365	61129	87529	85689	48237	52267	67689	93394	01511
07119	97336	71048	08178	77233	13916	47564	81056	97735
51085	12765	51821	51259	77452	16308	60756	92144	49442
02368	21382	52404	60268	89368	19885	55322	44819	01188
01011	54092	33362	94904	31273	04146	18594	29852	71585
52162	53916	46369	58586	23216	14513	83149	98736	23495
07056	97628	33787	09998	42698	06691	76988	13602	51851
48663	91245	85828	14346	09172	30168	90229	04734	59193
54164	58492	22421	74103	47070	25306	76468	26384	58151
32639	32363	05597	24200	13363	38005	94342	28728	35806
29334	27001	87637	87308	58731	00256	45834	15398	46557
02488	33062	28834	07351	19731	92420	60952	61280	50001
81525	72295	04839	96423	24878	82651	66566	14778	76797
29676	20591	68086	26432	46901	20849	89768	81536	86645
00742	57392	39064	66432	84673	40027	32832	61362	98947
05366	04213	25669	26422	44407	44048	37937	63904	45766
91921	26418	64117	94305	26766	25940	39972	22209	71500
00582	04711	87917	77341	42206	35126	74087	99547	81817
00725	69884	62797	56170	86324	88072	76222	36086	84637
69011	65795	95876	55293	18988	27354	26575	08625	40801
25976	57948	29888	88604	67917	48708	18912	82271	65424
09763	83473	73577	12908	30883	18317	28290	35797	05998
91567	42595	27958	30134	04024	86385	29880	99730	55536
17955	56349	90999	49127	20044	59931	06115	20542	18059
46503	18584	18845	49618	02304	51038	20655	58727	28168
92157	89634	94824	78171	84610	82834	09922	25417	44137
14577	62765	35605	81263	39667	47358	56873	56307	61607
98427	07523	33362	64270	01638	92477	66969	98420	04880
34914	63976	88720	82765	34476	17032	87589	40836	32427
70060	28277	39475	46473	23219	53416	94970	25832	69975
53976	54914	06990	67245	68350	82948	11398	42878	80287
76072	29515	40980	07391	58745	25774	22987	80059	39911
90725	52210	83974	29992	65831	38857	50490	83765	55657
64364	67412	33339	31926	14883	24413	59744	92351	97473
08962	00358	31662	25388	61642	34072	81249	35648	56891
95012	68379	93526	70765	10592	04542	76463	54328	02349
15664	10493	20492	38391	91132	21999	59516	81652	27195

References

Agresti, A., and B. Coull (1998), "Approximate is Better than 'Exact' for Interval Estimation of Binomial Proportions." *The American Statistician*, 52(2).

Anderson, V. L., and R. A. McLean (1974), *Design of Experiments: A Realistic Approach*, Marcel Dekker, New York.

Banks, J., J. S. Carson, B. L. Nelson, and D. M. Nicol (2001), *Discrete-Event System Simulation*, 3rd edition, Prentice-Hall, Upper Saddle River, NJ.

Bartlett, M. S. (1947), "The Use of Transformations," *Biometrics*, Vol. 3, pp. 39–52.

Bechhofer, R. E., T. J. Santner, and D. Goldsman (1995), *Design and Analysis of Experiments for Statistical Selection, Screening and Multiple Comparisons*, John Wiley and Sons, New York.

Belsley, D. A., E. Kuh, and R. E. Welsch (1980), *Regression Diagnostics*, John Wiley & Sons, New York.

Berrettoni, J. M. (1964), "Practical Applications of the Weibull Distribution," *Industrial Quality Control*, Vol. 21, No. 2, pp. 71–79.

Box, G. E. P., and D. R. Cox (1964), "An Analysis of Transformations," *Journal of the Royal Statistical Society*, B, Vol. 26, pp. 211–252.

Box, G. E. P., and M. F. Müller (1958), "A Note on the Generation of Normal Random Deviates," *Annals of Mathematical Statistics*, Vol. 29, pp. 610–611.

Bratley, P., B. L. Fox, and L. E. Schrage (1987), *A Guide to Simulation*, 2nd edition, Springer-Verlag, New York.

Cheng, R. C. (1977), "The Generation of Gamma Variables with Nonintegral Shape Parameters," *Applied Statistics*, Vol. 26, No. 1, pp. 71–75.

Cochran, W. G. (1947), "Some Consequences When the Assumptions for the Analysis of Variance Are Not Satisfied," *Biometrics*, Vol. 3, pp. 22–38.

Cochran, W. G. (1977), *Sampling Techniques*, 3rd edition, John Wiley & Sons, New York.

Cochran, W. G., and G. M. Cox (1957), *Experimental Designs*, John Wiley & Sons, New York.

Cook, R. D. (1979), "Influential Observations in Linear Regression," *Journal of the American Statistical Association*, Vol. 74, pp. 169–174.

Cook, R. D. (1977), "Detection of Influential Observations in Linear Regression," *Technometrics*, Vol. 19, pp. 15–18.

Crowder, S. (1987), "A Simple Method for Studying Run-Length Distributions of Exponentially Weighted Moving Average Charts," *Technometrics*, Vol. 29, pp. 401–407.

Daniel, C., and F. S. Wood (1980), *Fitting Equations to Data*, 2nd edition, John Wiley & Sons, New York.

Davenport, W. B., and W. L. Root (1958), *An Introduction to the Theory of Random Signals and Noise*, McGraw-Hill, New York.

Draper, N. R., and W. G. Hunter (1969), "Transformations: Some Examples Revisited," *Technometrics*, Vol. 11, pp. 23–40.

Draper, N. R., and H. Smith (1998), *Applied Regression Analysis*, 3rd edition, John Wiley & Sons, New York.

Duncan, A. J. (1986), *Quality Control and Industrial Statistics*, 5th edition, Richard D. Irwin, Homewood, IL.

Duncan, D. B. (1955), "Multiple Range and Multiple *F* Tests," *Biometrics*, Vol. 11, pp. 1–42.

Efron, B. and R. Tibshirani (1993), *An Introduction to the Bootstrap*, Chapman and Hall, New York.

Elsayed, E. (1996), *Reliability Engineering*, Addison Wesley Longman, Reading, MA.

Epstein, B. (1960), "Estimation from Life Test Data," *IRE Transactions on Reliability*, Vol. RQC-9.

Feller, W. (1968), *An Introduction to Probability Theory and Its Applications*, 3rd edition, John Wiley & Sons, New York.

Fishman, G. S. (1978), *Principles of Discrete Event Simulation*, John Wiley & Sons, New York.

Furnival, G. M., and R. W. Wilson, Jr. (1974), "Regression by Leaps and Bounds," *Technometrics*, Vol. 16, pp. 499–512.

Hahn, G., and S. Shapiro (1967), *Statistical Models in Engineering*, John Wiley & Sons, New York.

Hald, A. (1952), *Statistical Theory with Engineering Applications*, John Wiley & Sons, New York.

Hawkins, S. (1993), "Cumulative Sum Control Charting: An Underutilized SPC Tool," *Quality Engineering*, Vol. 5, pp. 463–477.

Hocking, R. R. (1976), "The Analysis and Selection of Variables in Linear Regression," *Biometrics*, Vol. 32, pp. 1–49.

Hocking, R. R., F. M. Speed, and M. J. Lynn (1976). "A Class of Biased Estimators in Linear Regression," *Technometrics*, Vol. 18, pp. 425–437.

Hoerl, A. E., and R. W. Kennard (1970a), "Ridge Regression: Biased Estimation for Non-Orthogonal Problems," *Technometrics*, Vol. 12, pp. 55–67.

Hoerl, A. E., and R. W. Kennard (1970b), "Ridge Regression: Application to Non-Orthogonal Problems," *Technometrics*, Vol. 12, pp. 69-82.

Kelton, W. D., and A. M. Law (1983), "A New Approach for Dealing with the Startup Problem in Discrete Event Simulation," *Naval Research Logistics Quarterly*, Vol. 30, pp. 641–658.

Kendall, M. G., and A. Stuart (1963), *The Advanced Theory of Statistics*, Hafner Publishing Company, New York.

Keuls, M. (1952), "The Use of the Studentized Range in Connection with an Analysis of Variance," *Euphytics*, Vol. 1, p. 112.

Law, A. M., and W. D. Kelton (2000), *Simulation Modeling and Analysis*, 3rd edition, McGraw-Hill, New York.

Lloyd, D. K., and M. Lipow (1972), *Reliability: Management, Methods, and Mathematics*, Prentice-Hall, Englewood Cliffs, N.J.

Lucas, J., and M. Saccucci (1990), "Exponentially Weighted Moving Average Control Schemes: Properties and Enhancements," *Technometrics*, Vol. 32, pp. 1–12.

Marquardt, D. W., and R. D. Snee (1975), "Ridge Regression in Practice," *The American Statistician*, Vol. 29, pp. 3–20.

Montgomery, D. C. (2001), *Design and Analysis of Experiments*, 5th edition, John Wiley & Sons, New York.

Montgomery, D. C. (2001), *Introduction to Statistical Quality Control*, 4th edition, John Wiley & Sons, New York.

Montgomery, D. C., E. A. Peck, and G. G. Vining (2001), *Introduction to Linear Regression Analysis*, 3rd edition, John Wiley & Sons, New York.

Montgomery, D.C., and G.C. Runger (2003), *Applied Statistics and Probability for Engineers*, 3rd edtion, John Wiley & Sons, New York.

Mood, A. M., F. A. Graybill, and D. C. Boes (1974), *Introduction to the Theory of Statistics*, 3rd edition, McGraw-Hill, New York.

Neter, J., M. Kutner, C. Nachtsheim, and W. Wasserman (1996), *Applied Linear Statistical Models*, 4th edition, Irwin Press, Homewood, IL.

Newman, D. (1939), "The Distribution of the Range in Samples from a Normal Population Expressed in Terms of an Independent Estimate of Standard Deviation," *Biometrika*, Vol. 31, p. 20.

Odeh, R., and D. Owens (1980), *Tables for Normal Tolerance Limits, Sampling Plans, and Screening*, Marcel Dekker, New York.

Owen, D. B. (1962), *Handbook of Statistical Tables*, Addison-Wesley Publishing Company, Reading, Mass.

Page, E. S. (1954), "Continuous Inspection Schemes." *Biometrika*, Vol. 14, pp. 100–115.

Roberts, S. (1959), "Control Chart Tests Based on Geometric Moving Averages," *Technometrics*, Vol. 1, pp. 239–250.

Scheffé, H. (1953), "A Method for Judging All Contrasts in the Analysis of Variance," *Biometrika*, Vol. 40, pp. 87–104.

Snee, R. D. (1977), "Validation of Regression Models: Methods and Examples," *Technometrics*, Vol. 19, No. 4, pp. 415–428.

Tucker, H. G. (1962), *An Introduction to Probability and Mathematical Statistics*, Academic Press, New York.

Tukey, J. W. (1953), "The Problem of Multiple Comparisons," unpublished notes, Princeton University.

Tukey, J. W. (1977), *Exploratory Data Analysis*, Addison-Wesley, Reading, MA.

United States Department of Defense (1957), *Military Standard Sampling Procedures and Tables for Inspection by Variables for Percent Defective* (MIL-STD-414), Government Printing Office, Washington, DC.

Welch, P. D. (1983), "The Statistical Analysis of Simulation Results," in *The Computer Performance Modeling Handbook* (ed. S. Lavenberg), Academic Press, Orlando, FL.

Answers to Selected Exercises

Chapter 1

1-1. **(a)** 0.75. **(b)** 0.18.

1-3. **(a)** $\overline{A} \cap B = \{5\}$. **(b)** $\overline{A} \cup B = \{1, 3, 4, 5, 6, 7, 8, 9, 10\}$. **(c)** $\overline{\overline{A} \cap \overline{B}} = \{2, 3, 4, 5\}$.
(d) $U = \{1, 2, 3, 4, 5, 6, 7, 8, 9, 10\}$. **(e)** $\overline{A \cap (B \cup C)} = \{1, 2, 5, 6, 7, 8, 9, 10\}$.

1.5 $\mathcal{S} = \{(t_1, t_2), t_1 \geq 0, t_2 \geq 0\}$.
$A = \{(t_1, t_2), t_1 \geq 0, t_2 \geq 0, (t_1 + t_2)/2 \leq 0.15\}$.
$B = \{(t_1, t_2): t_1 \geq 0, t_2 \geq 0, \max(t_1, t_2) \leq 0.15\}$.
$C = \{(t_1, t_2): t_1 \geq 0, t_2 \geq 0, (t_1 - t_2)/2 \leq 0.06\}$.

1-7. $\mathcal{S}$ = {NNNNN, NNNND, NNNDN, NNNDD, NNDNN, NNDND, NNDD, NDNNN, NDNND, NDND, NDD, DNNNN, DNNND, DNND, DND, DD}.

1-9. **(a)** N = not defective, D = defective.
$\mathcal{S}$ = {NNN, NND, NDN, NDD, DNN, DND, DDN, DDD}.
(b) $\mathcal{S}$ = {NNNN, NNND, NNDN, NDNN, DNNN}.

1-11. 30 routes. **1-13.** 560,560 ways.

1-15. $$P(\text{Accept} \mid p') = \sum_{x=0}^{1} \frac{\binom{300p'}{x}\binom{300(1-p')}{10-x}}{\binom{300}{10}}.$$

1-17. 28 comparisons. **1-19.** (40)(39) = 1560 tests.

1-21. **(a)** $\binom{5}{1}\binom{5}{1} = 25$ ways. **(b)** $\binom{5}{2}\binom{5}{2} = 100$ ways.

1-23. $R_s = [1 - (0 - 2)(0.1)(0.1)]\,[1 - (0.2)(0.1)]\,(0.9) = 0.880$.

1-25. S = Siberia, U = Ural, $P(S) = 0.6$, $P(U) = 0.4$, $P(F|S) = P(\overline{F}|S) = 0.5$,
$$P(\overline{F} \mid U) = 0.3, \quad P(S \mid \overline{F}) = \frac{(0.6)(0.5)}{(0.6)(0.5) + (0.4)(0.3)} \doteq 0.714.$$

1-27. $$\frac{1}{m-1}\cdot\frac{1}{m} + \frac{1}{m}\cdot\frac{m-1}{m} = \frac{m^2 - m + 1}{m^2(m-1)}.$$ **1-29.** $P(\text{women}|6') \doteq 0.03226$. **1-31.** 1/4.

1-35. $$P(\overline{B}) = \frac{(365)(364)\cdots(365-n+1)}{365^n}.$$

n	10	20	21	22	23	24	25	30	40	50	60
$P(B)$	0.117	0.411	0.444	0.476	0.507	0.538	0.569	0.706	0.891	0.970	0.994

1-37. 8! = 40320. **1-39.** 0.441.

Chapter 2

2-1. $$P_X(X = x) = \frac{\binom{4}{x}\binom{48}{5-x}}{\binom{52}{5}}, \quad x = 0, 1, 2, 3, 4.$$ **2-3.** $c = 1$, $\mu = 1$, $\sigma^2 = 1$.

2-5. **(a)** Yes. **(b)** No. **(c)** Yes. **2-7.** a, b. **2-9.** $P(X \leq 29) = 0.978$.

2-11. **(a)** $k = \frac{1}{4}$. **(b)** $\mu = 2$, $\sigma^2 = \frac{2}{3}$.

(c)
$$\begin{aligned} F_X(x) &= 0, \quad x < 0, \\ &= x^2/8, \quad 0 \leq x < 2, \\ &= -1 + x - \frac{x^2}{8}, \quad 2 \leq x < 4, \\ &= 1, \quad x \geq 4. \end{aligned}$$

2-13. $k = 2$, $[14 - 2\sqrt{2},\ 14 + 2\sqrt{2}]$.

2-15. **(a)** $k = \frac{8}{7}$. **(b)** $\mu = \frac{11}{7}$, $\sigma^2 = \frac{26}{49}$.

(c)
$$\begin{aligned} F_X(x) &= 0, \ x < 1, \\ &= \tfrac{8}{14}, \ 1 \leq x < 2, \\ &= \tfrac{12}{14}, \ 2 \leq x < 3, \\ &= 1, \ x \geq 3. \end{aligned}$$

2-17. $k = 10$ and $2 + k\sqrt{0.4} \doteq 8.3$ days.

2-21. **(a)**
$$\begin{aligned} F_X(x) &= 0, \ x < 0, \\ &= x^2/9, \ 0 \leq x < 3, \\ &= 1, \ x \geq 3. \end{aligned}$$

(b) $\mu = 2$, $\sigma^2 = \frac{1}{2}$. **(c)** $\mu_3' = \frac{54}{5}$. **(d)** $m = \dfrac{3}{\sqrt{2}}$.

2-23. $F_X(x) = 1 - e^{-x^2/2t^2}$, $x \geq 0$, **2-25.** $k = 1$, $\mu = 1$.

$= 0$, $x < 0$.

Chapter 3

3-1. **(a)**

y	$P_Y(y)$
0	0.6
20	0.3
80	0.1
otherwise	0.0

(b) $E(Y) = 14$, $V(Y) = 564$.

3-3. **(a)** 0.221. **(b)** \$155.80.

3-5. $f_Z(z) = e^{-z}$, $z \geq 0$,

$= 0$, otherwise.

3-7. 93.8 c/gal.

3-9. **(a)** $f_Y(y) = \frac{1}{4}\left(\frac{y}{2}\right)^{-1/2} \cdot e^{-(y/2)^{1/2}}$, $y > 0$,

$= 0$, otherwise.

(b) $f_V(v) = 2ve^{-v^2}$, $v > 0$,

$= 0$, otherwise.

(c) $f_U(u) = e^{-(e^u - u)}$, $u > 0$,

$= 0$, otherwise.

3-11. $s = (5/3) \times 10^6$.

3-13. **(a)** $f_Y(y) = \frac{1}{2}(4 - y)^{-1/2}$, $0 \leq y \leq 3$,

$= 0$, otherwise.

(b) $f_Y(y) = \frac{1}{y}$, $e \leq y \leq e^2$,

$= 0$, otherwise.

3-15. $M_X(t)=\sum_{x=1}^{6}\left(\frac{1}{6}\right)e^{tx}$,

$E(X)=M'_X(0)=\frac{7}{2}$,

$V(X)=M''_X(0)-\left[M'_X(0)\right]^2=\frac{35}{12}$.

3-17. $E(Y) = 1$, $V(Y) = 1$, $E(X) = 6.16$, $V(X) = 0.027$.

3-19. $M_X(t) = (1 - t/2)^{-2}$, $E(X) = M'_X(0) = 1$, $V(X) = M''_X(0) - [M'_X(0)]^2 = \frac{1}{2}$.

3-21. $M_Y^{(t)} = E(e^{ty}) = E(e^{t(aX+b)})$, $= e^{tb}E(e^{(at)X})$, $= e^{tb}M_X^{(at)}$.

3-23. **(a)** $M_X(t) = \frac{1}{2} + \frac{1}{4}e^t + \frac{1}{8}e^{2t} + \frac{1}{8}e^{3t}$, $E(X) = M'_X(0) = \frac{7}{8}$, $V(X) = M''_X(0) - \left(\frac{7}{8}\right)^2 = \frac{71}{64}$.

(b) $F_Y(y) = 0$, $y < 0$,

$= \frac{1}{8}$, $0 \le y < 1$,

$= \frac{1}{2}$, $1 \le y < 4$,

$= 1$, $y > 4$.

Chapter 4

4-1. **(a)**

x	0	1	2	3	4	5
$P_X(x)$	27/50	11/50	6/50	3/50	2/50	1/50

y	0	1	2	3	4
$P_Y(y)$	20/50	15/50	10/50	4/50	1/50

(b)

y	0	1	2	3	4
$P_{Y\mid 0}(y)$	11/27	8/27	4/27	3/27	1/27

(c)

x	0	1	2	3	4	5
$P_{X\mid 0}(x)$	11/20	4/20	2/20	1/20	1/20	1/20

4-3. **(a)** $k = 1$.

(b) $fx_1(x_1) = \frac{1}{100}$, $0 \le x_1 \le 100$,

$= 0$, otherwise,

$fx_2(x_2) = \frac{1}{10}$, $0 \le x_2 \le 10$,

$= 0$, otherwise.

(c) $f_{X_1,X_2}{}^{(x_1,x_2)} = 0$, if x_1 or $x_2 < 0$,

$= \frac{x_1x_2}{1000}$, $0 < x_1 < 100$, $0 < x_2 < 10$,

$= \frac{x_1}{100}$, $0 < x_1 < 100$, $x_2 \ge 10$,

$= \frac{x_2}{10}$, $x_1 \ge 100$, $0 < x_2 < 10$,

$= 1$, $x_1 \ge 100$, $x_2 \ge 10$.

4-5. **(a)** $\frac{1}{9}$. **(b)** $\frac{1}{64}$.

(c) $f_w(w) = 2w$, $0 \le w \le 1$,

$= 0$, otherwise.

4-7. $\frac{33}{17}$ **4-9.** $E(X_1|x_2) = \frac{3}{4}$, $E(X_2|x_1) = \frac{2}{3}$. **4-15.** $E(Y) = 80$, $V(Y) = 36$.

4-19. $\rho = \frac{1}{2}$, $\rho = -0.135$. **4-21.** X and Y are not independent.

4-23. **(a)** $f_X(x)=\frac{2}{\pi}\sqrt{1-x^2}$, $-1<x<1$,

$=0$, otherwise;

$f_Y(y)=\frac{4}{\pi}\sqrt{1-y^2}$, $0<y<1$,

$=0$, otherwise.

4-23. **(b)** $f_{X|y}(x) = \dfrac{1}{2\sqrt{1-y^2}}, \quad -\sqrt{1-y^2} < x < \sqrt{1-y^2},$

$= 0, \quad \text{otherwise};$

$f_{Y|x}(y) = \dfrac{1}{\sqrt{1-x^2}}, \quad 0 < y < \sqrt{1-x^2},$

$= 0, \quad \text{otherwise}.$

(c) $E(X \mid y) = 0,$

$E(Y \mid x) = \frac{1}{2}\sqrt{1-x^2}.$

4-27. **(a)** $E(X \mid y) = \dfrac{2+3y}{3(1+2y)}, \quad 0 < y < 1.$ **(b)** $E(X) = \frac{7}{12}$, **(c)** $E(Y) = \frac{7}{12}$.

4-29. **(a)** $k = (n-1)(n-2)$. **(b)** $F(x, y) = 1 - (1+x)^{2-n} - (1+y)^{2-n} + (1+x+y)^{2-n}$, $x > 0$, $y > 0$.

4-31. **(a)** Independent. **(b)** Not independent. **(c)** Not independent.

4-33. **(a)** $\frac{1}{4}$ **(b)** $\frac{1}{2}$

4-35. **(a)** $F_Z(z) = F_X[(z-a)/b]$. **(b)** $F_Z(z) = 1 - F_X(1/z)$. **(c)** $F_Z(z) = F_X(e^z)$. **(d)** $F_Z(z) = F_X(\ln z)$.

Chapter 5

5-1. $P(X = x) = \binom{4}{x} p^x (1-p)^{4-x}$, $x = 0, 1, 2, 3, 4$.

5-3. Assuming independence, $P(W \geq 4) = 1 - (0.5)^{12} \sum_{w=0}^{3} \binom{12}{w} \doteq 0.927.$

5-5. $p(X > 2) = 1 - \sum_{x=0}^{2} \binom{50}{x} (0.02)^x (0.98)^{50-x} \doteq 0.078.$

5-7. $P(\hat{p} \leq 0.03) \doteq 0.98$. **5-9.** $P(X = 5) \doteq 0.0407$. **5-11.** $p \doteq 0.8$.

5-13. $E(X) = M_X'(0) = \dfrac{1}{p}, E(X^2) = M_X''(0) = \dfrac{1+q}{p^2}, \sigma_x^2 = E(X^2) - (E(X))^2 = \dfrac{q}{p^2}.$

5-15. $P(X = 36) \doteq 0.0083$.

5-17. $P(X = 4) = 0.077$, $P(X < 4) = 0.896$. **5-19.** $E(X) = 6.25$, $V(X) = 1.5625$.

5-21. $p(3, 0, 0) + p(0, 3, 0) + p(0, 0, 3) \doteq 0.118$. **5-23.** $p(4, 1, 3, 2) \doteq 0.005$.

5-25. $P(X \leq 2) \doteq 0.98$. Binomial approx., $P(X \leq 2) \doteq 0.97$.

5-27. $P(X \geq 1) \doteq 0.95$. Binomial approx., $\Rightarrow n = 9$.

5-29. $P(X < 10) = (1.3888 \times 10^{-11}) \sum_{x=0}^{9} \dfrac{(25)^x}{x!}$. **5-31.** $P(X > 5) \doteq 0.215$.

5-33. Poisson model, $c = 30$.

$$P(X \leq 3) = e^{-30} \sum_{x=0}^{3} \frac{30^x}{x!},$$

$$P(X \geq 5) = 1 - e^{-30} \sum_{x=0}^{4} \frac{30^x}{x!}.$$

5-35. Poisson model, $c = 2.5$. $P(X \leq 2) \doteq 0.544$.

5-37. X = number of errors on n pages ~ Bin (5, n/200). $p(x \geq 1) = 0.763$.

(a) $n = 50$. **(b)** $P(X \geq 3) \geq 0.90$, if $n = 151$.

5-39. $P(X \geq 2) = 0.0047$.

Chapter 6

6-1. $\frac{5}{16}, \frac{9}{32}$.

6-3. $f_Y(y) = \frac{1}{4}$, $5 < y < 9$,
$= 0$, otherwise.

6-5. $E(X) = M'_X(0) = (\beta + \alpha)/2$, $V(X) = M''_X(0) - [M'_X(0)]^2 = (\beta - \alpha)^2/12$.

6-7.

y	$F_y(y)$
$y < 1$	0
$1 \leq y < 2$	0.3
$2 \leq y < 3$	0.5
$3 \leq y < 4$	0.9
$y > 4$	1.0

Generate realizations $u_i \sim$ uniform [0, 1] as random numbers, as is described in Section 6-6; use these in the inverse as $y_i = F_Y^{-1}(u_i)$, $i = 1, 2, \ldots$.

6-9. $E(X) = M'_X(0) = 1/\lambda$, $V(X) = M''_X(0) - [M'_X(0)]^2 = 1/\lambda^2$.

6-11. $1 - e^{-1/6} \doteq 0.154$. **6-13.** $1 - e^{-1/3} \doteq 0.283$.

6-15. $C_{\text{I}} = C, x > 15,$ $\quad C_{\text{II}} = 3C, x > 15,$
$= C + Z, x \leq 15;$ $\quad = 3C + Z, x \leq 15;$
$E(C_{\text{I}}) = Ce^{-3/5} + (C + Z)[1 - e^{-3/5}] \doteq C + (0.4512)Z;$
$E(C_{\text{II}}) = 3Ce^{-3/7} + (3C + Z)[1 - e^{-3/7}] \doteq 3C + (0.3486)Z;$
$\Rightarrow$ Process I, if $C > (0.0513)Z$.

6-19. 0.8305. **6-23.** 0.8488.

6-27. For $\lambda = 1$, $r = 2$, $f_X(x) = \dfrac{\Gamma(3)}{\Gamma(1) \cdot \Gamma(2)} \cdot x^0(1-x) = 2(1-x), 0 < x < 1.$
$= 0$, otherwise.

6-31. $1 - e^{-1} \doteq 0.63$. **6-33.** ≈ 0.24. **6-35.** **(a)** ≈ 0.22. **(b)** 4800. **6-37.** ≈ 0.3528.

Chapter 7

7-1. **(a)** 0.4772. **(b)** 0.6827. **(c)** 0.9505. **(d)** 0.9750.
(e) 0.1336. **(f)** 0.9485. **(g)** 0.9147. **(h)** 0.9898.

7-3. **(a)** $c = 1.56$. **(b)** $c = 1.96$. **(c)** $c = 2.57$. **(d)** $c = -1.645$.

7-5. **(a)** 0.9772. **(b)** 0.50. **(c)** 0.6687. **(d)** 0.6915. **(e)** 0.95.

7-7. 30.85%. **7-9.** 2376.63 fc.

7-13. **(a)** 0.0455. **(b)** 0.0730. **(c)** 0.3085. **(d)** 0.3085.

7-15. B, if cost of $A < 0.1368$. **7-17.** $\mu = 7$.

7-19. **(a)** 0.6687. **(b)** 7.84. **(c)** 6.018.

7-23. 0.616. **7-25.** 0.00714.

7-27. **(a)** 0.276. **(b)** At $\mu = 12.0$.

7-29. **(a)** 0.552. **(b)** 0.100. **(c)** 0.758. **(d)** 0.09.

7-30. $n = 139$. **7-36.** 2497.24.

7-37. $E(X) = e^{62.5}$, $V(X) = e^{25}(e^{25}-1)$, MED $= e^{50}$, MODE $= e^{25}$. **7-41.** 0.9788. **7-42.** 0.4681.

Chapter 8

8-1. $\bar{x} = 131.30$, $s^2 = 113.85$, $s = 10.67$. **8-21.** $\bar{x} = 74.002$, $s^2 = 6.875 \times 10^{-6}$, $s = 0.0026$.

8-23. **(a)** The sample average will be reduced by 63.

(b) The sample mean and standard deviation will be 100 units larger. The sample variance will be 10,000 units larger.

8-25. $a = \bar{x}$. **8-29.** **(a)** $\bar{x} = 120.22$, $s^2 = 5.66$, $s = 2.38$. **(b)** $\sim x = 120$, mode = 121.

8-31. For 8-29, $cv = 0.0198$; for 8-30, $cv = 9.72$. **8-33.** $\bar{x} = 22.41$, $s^2 = 208.25$, $\tilde{x} = 22.81$, mode = 23.64.

Chapter 9

9-1. $f(x_1, x_2, \ldots, x_5) = (1/(2\pi\sigma^2))^{5/2} e^{-1/2\sigma^2} \sum_{i=1}^{5} (x_i - \mu)^2$.

9-3. $f(x_1, x_2, x_3, x_4) = 1$. **9-5.** $N(5.00, 0.00125)$.

9-7. Use $S/\sqrt{n}$.

9-9. The standard error of $\bar{X}_1 - \bar{X}_2$ is $\sqrt{\dfrac{\sigma_1^2}{n_1} + \dfrac{\sigma_2^2}{n_2}} = \sqrt{\dfrac{(1.5)^2}{25} + \dfrac{(2.0)^2}{30}} = 0.47$. **9-11.** $N(0,1)$.

9-13. $se(\hat{p}) = \sqrt{p(1-p)/n}$, $\widehat{se}(\hat{p}) = \sqrt{\hat{p}(1-\hat{p})/n}$. **9-15.** $\mu = u$, $\sigma^2 = 2u$.

9-17. For $F_{m,n}$, we have $\mu = n/(n-2)$ for $n > 2$ and $\sigma^2 = \dfrac{2n^2(m+n-2)}{m(n-2)^2(n-4)}$ for $n > 4$.

9-21. $f_{X(1)}(t) = 1 - e^{-n\lambda t}$, $F_{X_{(n)}}(t) = (1 - e^{-\lambda t})^n$.

9-23. **(a)** 2.73. **(b)** 11.34. **(c)** 34.17. **(d)** 20.48.

9-25. **(a)** 1.63. **(b)** 2.85. **(c)** 0.241. **(d)** 0.588.

Chapter 10

10-1. Both estimators are unbiased. Now, $V(\bar{X}_1) = \sigma^2/2n$ whereas $V(\bar{X}_2) = \sigma^2/n$. Since $V(\bar{X}_1) < V(\bar{X}_2)$, $\bar{X}_1$ is a more efficient estimator than $\bar{X}_2$.

10-3. $\hat{\theta}_2$, because it would have a smaller MSE.

10-7. $\hat{\alpha} = \sum_{i=1}^{n} \dfrac{X_i}{n} = \bar{X}$. **10-9.** $(\bar{t})^{-1}$.

10-11. $\hat{\lambda} = \bar{X} \Big/ \left[(1/n)\sum_{i=1}^{n} X_i^2 - \bar{X}^2 \right]$, $\hat{r} = \bar{X}^2 \Big/ \left[(1/n)\sum_{i=1}^{n} X_i^2 - \bar{X}^2 \right]$.

10-13. $1/\bar{X}$. **10-15.** $\bar{X}_N/n$. **10-17.** $\bar{X}/n$.

10-21. $-1 - \bar{n} \Big/ \sum_{i=1}^{n} \ln X_i$. **10-23.** $X_{(1)}$.

10-25. $f(\mu \mid x_1, x_2, \ldots x_n) = C^{1/2} (2\pi)^{-(1/2)} \exp\left\{ -\dfrac{6}{2}\left[\mu - \dfrac{1}{c}\left(\dfrac{n\bar{x}}{\sigma^2} + \dfrac{\mu_0}{\sigma_0^2} \right) \right]^2 \right\}$,

where $C = \dfrac{n}{\sigma^2} + \dfrac{1}{\sigma_0^2}$.

10-27. The posterior density for p is a beta distribution with parameters $a + n$ and $b + \Sigma, x_i - n$.

10-29. The posterior density for λ is gamma with parameters $r = m + \sum x_i + 1$ and $\delta = n + (m+1)/\lambda_0$.

10-31. 0.967. **10-33.** 0.3783.

10-35. **(a)** $f(\theta \mid x_1) = \dfrac{f(x_1, \theta)}{f(x_1)} = \dfrac{2x}{e^2(2-2x)}$. **(b)** $\hat{\theta} = 1/2$. **10-37.** $\alpha_1 = \alpha_2 = \alpha/2$ is shorter.

10-39. **(a)** $74.03533 \leq \mu \leq 74.03666$. **(b)** $74.0356 \leq \mu$.

10-41. **(a)** $3232.11 \leq \mu \leq 3267.89$. **(b)** $1004.80 \leq \mu$. **10-43.** 150 or 151.

10-45. **(a)** $0.0723 \leq \mu_1 - \mu_2 \leq 3267.89$. **(b)** $0.0499 \leq \mu_1 - \mu_2 \leq 0.33$. **(c)** $\mu_1 - \mu_2 \leq 0.3076$.

10-47. $-3.68 \leq \mu_1 - \mu_2 \leq -2.12$. **10-49.** $183.0 \leq \mu \leq 256.6$. **10-51.** 13.

10-53. $94.282 \leq \mu \leq 111.518$. **10-55.** $-0.839 \leq \mu_1 - \mu_2 \leq -0.679$. **10-57.** $0.355 \leq \mu_1 - \mu_2 \leq 0.455$.

10-59. **(a)** $649.60 \leq \sigma^2 \leq 2853.69$. **(b)** $714.56 \leq \sigma^2$. **(c)** $\sigma^2 \leq 2460.62$. **10-61.** $0.0039 \leq \sigma^2 \leq 0.0124$.

10-63. $0.574 \leq \sigma^2 \leq 3.614$. **10-65.** $0.11 \leq \sigma_1^2/\sigma_2^2 \leq 0.86$. **10-67.** $0.088 \leq p \leq 0.152$. **10-69.** 16577.

10-71. $-0.0244 \leq p_1 - p_2 \leq 0.0024$. **10-73.** $-2038 \leq \mu_1 - \mu_2 \leq 3774.8$.

10-75. $-3.1529 \leq \mu_1 - \mu_2 \leq 0.1529$; $-1.9015 \leq \mu_1 - \mu_2 \leq 0.9015$; $-0.1775 \leq \mu_1 - \mu_2 \leq 2.1775$.

Chapter 11

11-1. **(a)** $z_0 = -1.333$, do not reject H_0. **(b)** 0.05. **11-3.** **(a)** $Z_0 = -12.65$, reject H_0. **(b)** 3.

11-5. $Z_0 = 2.50$, reject H_0. **11-7.** **(a)** $Z_0 = 1.349$, do not reject H_0. **(b)** 2. **(c)** 1.

11-9. $Z_0 = 2.656$, reject H_0. **11-11.** $Z_0 = -7.25$, reject H_0. **11-13.** $t_0 = 1.842$, do not reject H_0.

11-15. $t_0 = 1.47$, do not reject H_0 at $\sigma = 0.05$. **11-17.** 3.

11-19. **(a)** $t_0 = 8.49$, reject H_0. **(b)** $t_0 = -2.35$, do not reject H_0. **(c)** 1. **(d)** 5.

11-21. $F_0 = 0.8832$, do not reject H_0. **11-23.** **(a)** $F_0 = 1.07$, do not reject H_0. **(b)** 0.15. **(c)** 75.

11-25. $t_0 = 0.56$, do not reject H_0.

11-27. **(a)** $x_0^2 = 43.75$, reject H_0. **(b)** 0.3078×10^{-4}. **(c)** 0.30. **(d)** 17.

11-29. **(a)** $x_0^2 = 2.28$, reject H_0. **(b)** 0.58. **11-31.** $F_0 = 30.69$, reject H_0; $\beta \cong 0.65$.

11-33. $t_0 = 2.4465$, do not reject H_0. **11-35.** $t_0 = 5.21$, reject H_0. **11-37.** $z_0 = 1.333$, do not reject H_0.

11-41. $Z_0 = -2.023$, do not reject H_0. **11-47.** $\chi_0^2 = 2.915$, do not reject H_0.

11-49. $\chi_0^2 = 4.724$, do not reject H_0. **11-53.** $\chi_0^2 = 0.0331$, do not reject H_0.

11-55. $\chi_0^2 = 2.465$, do not reject H_0. **11-57.** $\chi_0^2 = 34.896$, reject H_0. **11-59.** $\chi_0^2 = 22.06$, reject H_0.

Chapter 12

12-1. **(a)** $F_0 = 3.17$. **12-3.** **(a)** $F_0 = 12.73$. **(b)** Mixing technique 4 is different from 1, 2, and 3.

12-5. **(a)** $F_0 = 2.62$. **(b)** $\hat{\mu} = 21.70$, $\hat{\tau}_1 = 0.023$, $\hat{\tau}_2 = -0.166$, $\hat{\tau}_3 = 0.029$, $\hat{\tau}_4 = 0.059$.

12-7. **(a)** $F_0 = 4.01$. **(b)** Mean 3 differs from 2. **(c)** $SS_{c2} = 246.33$. **(d)** 0.88.

12-9. **(a)** $F_0 = 2.38$. **(b)** None. **12-11.** $n = 3$.

12-15. **(a)** $\hat{\mu} = 20.47$, $\hat{\tau}_1 = 0.33$, $\hat{\tau}_2 = 1.73$, $\hat{\tau}_3 = 2.07$. **(b)** $\hat{\tau}_1 - \hat{\tau}_2 = -1.40$.

Chapter 13

13-1.

Source	DF	SS	MS	F	P
CS	2	0.0317805	0.0158903	15.94	0.000
DC	2	0.0271854	0.0135927	13.64	0.000
CS*DC	4	0.0006873	0.0001718	0.17	0.950
Error	18	0.0179413	0.0009967		
Total	26	0.0775945			

Main effects are significant; interaction is not significant.

13-3. $-23.93 \leq \mu_1 - \mu_2 \leq 5.15$. **13-5.** No change in conclusions.

13-7.

Source	DF	SS	MS	F	P
glass	1	14450.0	14450.0	273.79	0.000
phos	2	933.3	466.7	8.84	0.004
glass*phos	2	133.3	66.7	1.26	0.318
Error	12	633.3	52.8		
Total	17	16150.0			

Significant main effects.

13-9.

Source	DF	SS	MS	F	P
Conc	2	7.7639	3.8819	10.62	0.001
Freeness	2	19.3739	9.6869	26.50	0.000
Time	1	20.2500	20.2500	55.40	0.000
Conc*Freeness	4	6.0911	1.5228	4.17	0.015
Conc*Time	2	2.0817	1.0408	2.85	0.084
Freeness*Time	2	2.1950	1.0975	3.00	0.075
Conc*Freeness*Time	4	1.9733	0.4933	1.35	0.290
Error	18	6.5800	0.3656		
Total	35	66.3089			

Concentration, Time, Freeness, and the interaction Time*Freeness are significant at 0.05.

13-15. Main effects A, B, D, E and the interaction AB are significant.

13-17. Block 1: (1), ab, ac, bc, Block 2: a, b, c, abc.

13-19. Block 1: (1) ab, bcd, acd, Block 2: a, b, cd, abcd, Block 3: c, abc, bd, ad, Block 4: d, abd, bc, ac.

13-21. A and C are significant. **13-25.** **(a)** D = ABC. **(b)** A is significant.

13-27. 2^{3-1} with two replicates. **13-29.** 2^{5-2} design. Estimates for A, B, and AB are large.

Chapter 14

14-1. **(a)** $\hat{y} = 10.4397 - 0.00156x$. **(b)** $F_0 = 2.052$. **(c)** $-0.0038 \le \beta_1 \le 0.00068$. **(d)** 7.316%.

14-3. **(a)** $\hat{y} = 31.656 - 0.041x$. **(b)** $F_0 = 57.639$. **(c)** 81.59%. **(d)** (19.374, 21.388).

14-7. **(a)** $\hat{y} = 93.3399 + 15.6485x$. **(b)** Lack of fit not significant, regression significant.
(c) $7.997 \le \beta_1 \le 23.299$. **(d)** $74.828 \le \beta_0 \le 111.852$. **(e)** (126.012, 138.910).

14-9. **(a)** $\hat{y} = -6.3378 + 9.20836x$. **(b)** Regression is significant. **(c)** $t_0 = -23.41$, reject H_0.
(d) (525.58, 529.91). **(e)** (521.22, 534.28).

14-11. **(a)** $\hat{y} = 77.7895 + 11.8634x$. **(b)** Lack of fit not significant, regression is significant.
(c) 0.3933. **(d)** (4.5661, 19.1607).

14-13. **(a)** $\hat{y} = 3.96 + 0.00169x$. **(b)** Regression is significant.
(c) (0.0015, 0.0019). **(d)** 95.2%.

14-15. **(a)** $\hat{y} = 69.1044 + 0.4194x$. **(b)** 77.35%. **(c)** $t_0 = 5.85$, reject H_0.
(d) $Z_0 = 1.61$. **(e)** (0.5513, 0.8932).

Chapter 15

15-1. **(a)** $\hat{y} = 7.30 + 0.0183x_1 - 0.399x_4$. **(b)** $F_0 = 15.19$.

15-3. (−0.8024, 0.0044). **15-5.** $\hat{y} = -1.808372 + 0.003598x_2 + 0.1939360x_7 - 0.004815x_8$.

15-7. **(a)** $\hat{y} = -102.713 + 0.605x_1 + 8.924x_2 + 1.437x_3 + 0.014x_4$.
(b) $F_0 = 5.106$. **(c)** β_3, $F_0 = 0.361$; β_1, $F_0 = 0.0004$.

15-9. **(a)** $\hat{y} = -13729 \;\; 105.02x - 0.18954x^2$.

15-13. **(a)** $\hat{y} = -4.459 + 1.384x + 1.467x^2$. **(b)** Significant lack of fit. **(c)** $F_0 = 16.68$.

15-15. $t_0 = 1.7898$. **15-21.** $\text{VIF}_1 = \text{VIF}_2 = 1.4$.

Chapter 16

16-1. $R = 2$. **16-5.** $R = 2$. **16-9.** $R = 88.5$. **16-13.** $R_1 = 75$.

16-15. $Z_0 = -2.117$. **16-17.** $K = 4.835$.

Chapter 17

17-1. **(a)** $\bar{\bar{x}} = 34.32$ $\bar{R} = 5.65$. **(b)** $\text{PCR}_k = 1.228$. **(c)** 0.205%.

17-5. $D/2$. **17-7.** LCL = 34.55, CL = 49.85, UCL = 65.14. **17-9.** Process is not in control.

17-13. 0.1587, $n = 6$ or 7. **17-15.** Revised control limits: LCL = 0, UCL = 17.32.

17-17. UCL = 16.485; 0.434. **17-19.** LCL = 0.282, UCL = 4.378.

Chapter 18

18-1. **(a)** ≈ 0.088. **(b)** $L = 2$. $L_q = 1.33$. **(c)** $W = 1$ h.

18-3. $P=\begin{bmatrix} p & 1-p \\ 1-p & p \end{bmatrix}$,

$P^{\infty}=\begin{bmatrix} 1/2 & 1/2 \\ 1/2 & 1/2 \end{bmatrix}$.

18-7. **(a)** $P=\begin{bmatrix} 0 & p & 0 & 1-p \\ 1-p & 0 & p & 0 \\ 0 & 1-p & 0 & p \\ p & 0 & 1-p & 0 \end{bmatrix}$; **(b)** $p_1 = p_2 = p_3 = p_4 = \frac{1}{4}$. **(c)** $p_1 = p_2 = p_3 = p_4 = \frac{1}{4}$.

$A = [1\,0\,0\,0]$.

18-9. **(a)** $\frac{3}{10}$. **(b)** $\frac{9}{10}$. **(c)** 3. **(d)** 0.03. **(e)** 0.10. **(f)** $\frac{3}{10}$.

18-11. **(a)** 0.555. **(b)** 56.378 min. **(c)** 244.18 min.

18-13. **(a)** $p_j = \left[\left(\lambda/\mu^j\right)/j!\right]\cdot p_0,\ j = 0,1,2,\ldots,s,$ **(b)** $s = 6$, $\rho = 0.417$.

$= 0$, otherwise;

$$p_0=\frac{1}{\sum_{j=0}^{s}\frac{(\lambda/\mu)^j}{j!}}.$$

(c) $p_6 = 0.354$. **(d)** From 41.6 to 8.33%. **(e)** $\varnothing = 4.17$, $p_6 = 0.377$.

Chapter 19

19-3.

$$\begin{aligned} E\left(\hat{I}_n\right) &= \frac{b-a}{n}E\left(\sum_{i=1}^{n} f\left(a+(b-a)U_i\right)\right) \\ &= (b-a)E\left(f\left(a+(b-a)U_i\right)\right) \\ &= (b-a)\int_0^1 f\left(a+(b-a)u\right)du \\ &= I. \end{aligned}$$

19-5. **(a)** 35. **(b)** 4.75. **(c)** 5 (at time 14).

19-7. **(a)** $X_1 = 1$, $U_1 = 1/16$, $X_2 = 6$, $U_2 = 6/16$. **(b)** Yes. **(c)** $X_{150} = 2$.

19-9. **(a)** $X = -(1 = \lambda)\ \ell n(1 - U)$. **(b)** 0.693.

19-11. **(a)** $X = -2\sqrt{1-2U}$, if $0 < U < 1/2$, **(b)** $X = 0.894$.

$= 2\sqrt{2U-1}$, if $1/2 < U < 1$.

19-13. **(a)** $X = \sigma[-\ell n(1 - U)]^{1/\beta}$. **(b)** 1.558.

19-15. $\sum_{i=1}^{12} U_i - 6 = 1.07$. **19-17.** $X = -(1 = \lambda)\ \ell\, U_1 U_2 = 0.841$.

19-19. $X = 5$ trials. **19-21.** [-3:41, 4:41].

19-23. [80.4,119.6]. **19-25.** Exponential with parameter 1; $-V(U_l) = -1/12$.

Index

P

Q

R

S